AF826941

Immunomodulators in Aquaculture and Fish Health

This reference book provides updated information about different immunomodulators for managing fish health and sustainable aquaculture. Immunomodulators are dietary additives that enhance innate defense mechanisms and increase resistance against specific pathogens and diseases. The book covers the different types of immunostimulants, their modes of action, and their efficacies. It also reviews safety concerns, ethical regulations, limitations, and outreach to farmers. It discusses the application of herbal immunomodulators, antioxidants, and pre- and pro-biotics in disease management.

Features:

- Reviews the pressing topic of reduction of antibiotic use in aquaculture
- Discusses herbal immunomodulators, nutrients, antioxidants, and pre- and pro-biotics
- Covers the topic of progressive immunomodulation using nanotechnology
- Discusses fish health management in the ever-growing aquaculture industry
- Includes natural and synthetic immunomodulators

The book is meant for researchers and industry experts in aquaculture, fisheries science, and veterinary medicine.

Immunomodulators in Aquaculture and Fish Health

Edited by
Preetham Elumalai
Mehdi Soltani
Sreeja Lakshmi

CRC Press
Taylor & Francis Group
Boca Raton London New York

CRC Press is an imprint of the
Taylor & Francis Group, an informa business

Design cover image: image courtesy Preetham Elumalai, Mehdi Soltani and Sreeja Lakshmi

First edition published 2024
by CRC Press
2385 Executive Center Drive, Suite 320, Boca Raton FL 33431

and by CRC Press
4 Park Square, Milton Park, Abingdon, Oxon, OX14 4RN

CRC Press is an imprint of Taylor & Francis Group, LLC

© 2024 selection and editorial matter, Preetham Elumalai, Mehdi Soltani and Sreeja Lakshmi; individual chapters, the contributors

Reasonable efforts have been made to publish reliable data and information, but the author and publisher cannot assume responsibility for the validity of all materials or the consequences of their use. The authors and publishers have attempted to trace the copyright holders of all material reproduced in this publication and apologize to copyright holders if permission to publish in this form has not been obtained. If any copyright material has not been acknowledged please write and let us know so we may rectify in any future reprint.

Except as permitted under U.S. Copyright Law, no part of this book may be reprinted, reproduced, transmitted, or utilized in any form by any electronic, mechanical, or other means, now known or hereafter invented, including photocopying, microfilming, and recording, or in any information storage or retrieval system, without written permission from the publishers.

For permission to photocopy or use material electronically from this work, access www.copyright.com or contact the Copyright Clearance Center, Inc. (CCC), 222 Rosewood Drive, Danvers, MA 01923, 978-750-8400. For works that are not available on CCC please contact mpkbookspermissions@tandf.co.uk

Trademark notice: Product or corporate names may be trademarks or registered trademarks and are used only for identification and explanation without intent to infringe.

Library of Congress Cataloging-in-Publication Data
Names: Elumalai, Preetham, editor. | Soltani, Mehdi, editor. | Lakshmi, Sreeja, editor.
Title: Immunomodulators in aquaculture and fish health / edited by Preetham Elumalai, Mehdi Soltani, and Sreeja Lakshmi
Description: First edition | Boca Raton, FL : CRC Press, 2024 | Includes bibliographical references and index. | Summary: "This reference book provides updated information about different immunomodulators for managing fish health and sustainable aquaculture. Immunomodulators are dietary additives that enhance innate defense mechanisms and increase resistance against specific pathogens and diseases. The book covers the different types of immunostimulants, their modes of action, and their efficacies. It also reviews safety concerns, ethical regulations, limitations, and outreach to farmers. It discusses the application of herbal immunomodulators, antioxidants, pre- and pro-biotics, in disease management.
Features: Reviews the pressing topic of reduction of antibiotic use in aquaculture Discusses herbal immunomodulators, nutrients, antioxidants and pre- and pro-biotics Covers the topic of progressive immunomodulation using nanotechnology Discusses fish health management in the ever-growing aquaculture industry Includes natural and synthetic immunomodulators The book is meant for researchers and industry experts in aquaculture, fisheries science, and veterinary medicine"-- Provided by publisher.
Identifiers: LCCN 2023015962 (print) | LCCN 2023015963 (ebook) | ISBN 9781032407388 (hardback) | ISBN 9781032420905 (paperback) | ISBN 9781003361183 (ebook)
Subjects: LCSH: Fish culture. | Immunological adjuvants. | Biological response modifiers.
Classification: LCC SH151 .I46 2024 (print) | LCC SH151 (ebook) | DDC 639.3--dc23/eng/20230726
LC record available at https://lccn.loc.gov/2023015962
LC ebook record available at https://lccn.loc.gov/2023015963

ISBN: 978-1-032-40738-8 (hbk)
ISBN: 978-1-032-42090-5 (pbk)
ISBN: 978-1-003-36118-3 (ebk)

DOI: 10.1201/9781003361183

Typeset in Times
by MPS Limited, Dehradun

Contents

Foreword ... vii
Editors ... viii
Contributors ... ix

Section I Immunomodulators: An Overview

1. **Immunomodulators: An Introduction** ... 3
   Falco F, Banaee M, Mauro M, Faggio C, Arathi Kollath, and Preetham Elumalai

2. **Natural and Synthetic Immunomodulators: Inferences for Stress Responses in Aquaculture Fish** ... 18
   Shubhajit Saha, Azubuike V. Chukwuka, Nimai Chandra Saha, Caterina Faggio, and Hamed Mousavi Sabet

3. **Immunomodulators: Mode of Action** ... 29
   Subramaniam Sivakumar, C. Shanmuga Sundaram, Maderi Velayutham Dassprakash, and Rantham Subramaniam Venkatesan

4. **Immunomodulators and Stress Oxidative** ... 43
   Tamilselvan Gokul, Paulraj Balaji, Karthikeyan Venkatachalam, Subramanian Ramya, Ramaraj Jayakumararaj, Chinnathambi Pothiraj, and Kamatchi Ramesh Kumar

Section II Immunomodulators and Sustainable Aquaculture Development

5. **Immunomodulators to Prevent Diseases and Minimize Antimicrobial Use** ... 59
   Akshay Thuruthiyil Rajesh, Sajna Beegum, Neha Omgy, Sreeja Lakshmi, Hethesh Chellapandian, Sivakamavalli Jeyachandran, Einar Ringø, and Preetham Elumalai

6. **Immunomodulation in Aquaculture Health Management: Opportunities and Obstacles** ... 76
   Ramchandran Ishwarya, Baskaralingam Vaseeharan, Rengarajan Jayakumar, Subramaniam Sivakumar, and Preetham Elumalai

7. **Disease Management and Prophylaxis by Immunostimulants** ... 89
   Chinnathambi Pothiraj, Divya Jyoti, Subramanian Ramya, Ramaraj Jayakumararaj, Aseem Grover, Reshma Sinha, Palanichamy Ayyappan, Caterina Faggio, and Paulraj Balaji

8. **Application of Immunostimulants for Aquaculture Health Management** ... 103
   Femi John Fawole, Shamna Nazeemashahul, Thongam Ibemcha Chanu, Arun Sharma, Gbadamosi Oluyemi Kazeem, S. Ferosekhan, and Tejaswini Kinnera

Section III Immunomodulators in Aquaculture Health Management

9. **Herbal Immunomodulators for Aquaculture** ... 119
   Shamna Nazeemashahul, Femi John Fawole, Babitha Rani A.M., Manish Jayant, Neha Qureshi, Hussain Nottanalan, Ashutosh D. Deo, and Parimal Sardar

10. **Prebiotics and Probiotics as Effective Immunomodulators in Aquaculture** ... 136
    Mehdi Soltani, Koushik Ghosh, Dipanjan Dutta, and Einar Ringø

11. Immunomodulation in Fish Through Nutrients, Antioxidants and Hormones 169
Chiranjiv Pradhan, Nikhila Peter, and Sweta Das

12. Cytokines and Fish Health 186
Aifa Fathima, Yaser Arafath, Saqib Hassan, George Seghal Kiran, and Joseph Selvin

13. Progressive Immunomodulation Through Nanotechnology 200
Heba Mahboub, Hiam Elabd, Mian Adnan Kakakhel, Gehad E. Elshopakey, Maram H. Abduljabbar, and Manal E. Alosaimi

Section IV Current Status of Immunomodulators in Aquaculture

14. Efficacy and Limitations of Immunomodulators 213
Arathi Kollath, Lokesh Pawar, Ankeet Bhagat, Sunil Sharma, Owias Iqbal Dar, and Preetham Elumalai

15. Current Status and Recent Advancements with Immunostimulants in Aquaculture 233
Parasuraman Aiya Subramani, S. Kalaivani Priyadarshini, Ramalakshmi Balasubramanian, M. Divya Gnaneswari, Devasree Ganesh Kumar, Priyatharsini Rajendran, Catherine Alexander, and R. Dinakaran Michael

Index 263

Foreword

It is my great pleasure to write this foreword for a very timely book, **"Immunomodulators in Aquaculture and Fish Health,"** edited by my colleagues, Preetham Elumalai, Mehdi Soltani, and Sreeja Lakshmi.

As a believer in identifying opportunities and linking scientific evidence, innovation with improved sustainable aquatic production, together with my professional commitment in the aquaculture industry of over 40 years, I have found this volume of work to be a very extensive review. The wealth of information on the potential therapeutic and preventative roles of immunomodulators, in combating diseases in farmed aquatic species is truly insightful.

The importance of this book defines an era, where the performance of aquatic foods has been greatly recognized and concerted efforts have been initiated to enhance production and to bridge the ever-expanding supply-demand gap, for aquatic blue foods worldwide.

The burden of disease is high in aquatic production and currently estimated as $10 billion USD annually. With the decades of experience and lessons learned in aquatic animal health management, it is convinced that prevention is better than cure for aquatics and investing in prevention is more cost-effective than investing in therapy.

In this regard, I believe we should be aiming for more tools and procedures such as vaccines and vaccination and more research efforts should be supported, both at academic and commercial levels.

I congratulate the editors for this comprehensive volume and hope it will serve the purpose of increasing awareness of immunomodulators towards the implementation within aquatic animal health management.

Dr Rohana Subasinghe
Founder + Director,
FUTUREFISH Co. Ltd.
www.futurefish.org

Aquaculture plays a vital role in global food security and economic development. It is the fastest-growing food production sector in the world and has been for some considerable time, with an average annual growth rate over the last 50 years of 8%. The latest FAO State of the World Fisheries and Aquaculture Report (2022) estimated global aquaculture production of aquatic animals at a record 87.5 million tonnes, with a value of USD 264.8 billion; this equates to 49% of total aquatic animal supply by volume and 65% by value. Approximately 2.5 million people are directly employed in the aquaculture sector around the world, with most of these in Asia, followed by Africa and Latin America. Women comprise 28% of these employment figures, slightly more than the average of 25% for the agriculture sector as a whole.

Infectious diseases represent a major constraint to the continued growth of global aquaculture, with estimated annual losses of at least USD 6 billion. The industry is particularly prone to disease outbreaks because of high stocking densities which increase pathogen transmission rates and reduce water quality, and also because of low genetic diversity in many breeding stocks, which may compromise the immune response to infection of cultured animals. Antimicrobials and antiparasitics are used therapeutically and prophylactically, but these treatments are often expensive and there is concern over the potential adverse effects of their widespread use, particularly the promotion of antibiotic resistance. Vaccination is another option for disease control, but vaccines often have limited efficacy, particularly for juvenile fish which do not have a fully developed immune response. There is therefore increasing interest in alternative approaches to disease control in aquaculture.

Immunomodulators are substances that affect the functioning of the immune system. A range of natural and synthetic products have been used or proposed, with varying degrees of scientific evaluation, to control infectious diseases in aquaculture. This book provides a very comprehensive and timely exploration of their efficacy and potential role in an aquatic animal health management system. The various sections of the book provide an overview of immunomodulators and their mode of action; the potential of immunomodulators to provide a more sustainable approach to disease control; the current use of immunomodulator products in aquaculture; and finally, their efficacy, limitations, and future prospects for the aquaculture industry. The book will be of great benefit to researchers in aquatic animal health, aquatic veterinarians, aquaculture managers, and all of us who wish to promote an economically viable and environmentally sustainable aquaculture industry.

Professor Alan Lymbery
Director, Centre for Sustainable Aquatic Ecosystems
Harry Butler Institute, Murdoch University
Australia

Editors

Preetham Elumalai, PhD, is an associate professor at the Department of Marine Biology, Cochin University of Science and Technology, Kochi, Kerala, India. He earned a master's degree from the University of Madras and a PhD in biochemistry and molecular immunology from the Institute for Immunology, University of Regensburg, Germany. His research practice includes bioassay-guided identification of novel marine compounds, unveiling fish lectins in innate immune defense, aquatic vaccine development, evaluation of cost-effective feed additives and nutrigenomics, and effects of environmental pollutants on marine ecosystems. He has been a partner in numerous EU-, Indian-, and UK-funded projects (e.g., IVVN, BactiVac). He has written more than 70 peer-reviewed articles and has two patents in his name apart from editing five books (Springer, CRC Press) and has presented his work at more than 60 national and international conferences. He has been awarded the prestigious INSA fellowship (2018); MASTS, Fellowship (2019); IVVN award, UK (2020); FRSB award (2021); and BactiVac award, UK (2022).

Mehdi Soltani, PhD, is a distinguished professor at the University of Tehran and an adjunct professor at Murdoch University, Australia. He earned a DVM from the University of Tehran and PhD in aquatic animal health from the University of Tasmania, Australia. Professor Soltani has an international reputation for research on aquatic animal health, with 290 published scientific papers, collaborations with researchers throughout the world, and editorship of scientific journals in fisheries and veterinary science. He chaired government advisory committees in fisheries and aquaculture and has also worked closely with the aquaculture industry. He developed and patented a number of fish vaccines, which are registered throughout the Middle East. He also taught numerous undergraduate courses and supervised many higher degree students. His research interests include vaccine development for fish pathogens; immunopathogenesis of infectious agents in fish/shellfish; and development of alternative therapies such as immunostimulants, probiotics, and phytobiotics for disease control in farmed fish and shellfish.

Sreeja Lakshmi, PhD, is a postdoctoral research scientist in collaboration with Moredun Research Institute (MRI), UK. She graduated from Calicut University and earned a PhD in biochemistry and functional genomics from the Institute for Molecular Biology, University of Regensburg, Germany. She has published research articles in peer-reviewed international journals and authored books and book chapters. She has been awarded prestigious research grants from the Bavarian Research Foundation (Bayerische Forschungsstiftung), Government of Bavaria, Germany; an HRD-Fellowship for Women Scientists from the Department of Health Research, Government of India; and a MASTS (Marine Alliance Science and Technology, Scotland) Award for Postdoctoral and Early Career Research Exchanges (PECRE). She has visited the University of Aberdeen, Scotland, and received an IVVN Fellowship grant from the International Veterinary Vaccinology Network (IVVN), UK.

Contributors

Maram H. Abduljabbar
Department of Pharmacology and Toxicology, College of Pharmacy
Taif University
Taif, Saudi Arabia

Catherine Alexander
Jayaraj Annapackiam College for Women
Thamaraikulam, Periyakulam, Tamil Nadu

Manal E. Alosaimi
Department of Basic Science, College of Medicine
Princess Nourah Bint Abdulrahman University
Riyadh, Saudi Arabia

Babitha Rani A.M.
Aquaculture Division
ICAR – Central Institute of Fisheries Education
Mumbai, Maharashtra, India

Yaser Arafath
Department of Microbiology, School of Life Sciences
Pondicherry University
Puducherry, India

Palanichamy Ayyappan
Department of Botany
Government Arts College
Melur, Madurai, Tamil Nadu, India

Paulraj Balaji
PG and Research Centre in Biotechnology
MGR College
Hosur, Tamil Nadu, India

Ramalakshmi Balasubramanian
Department of Zoology
Lady Doak College
Tallakulam, Tamil Nadu

Sajna Beegum
Department of Marine Biology, Microbiology and Biochemistry, School of Marine Sciences
Cochin University of Science and Technology (CUSAT)
Lakeside Campus, Fine Arts Avenue, Cochin, Kerala, India

Ankeet Bhagat
Aquatic Toxicology Lab, Department of Zoology
Guru Nanak Dev University
Amritsar, Punjab, India

Faggio C
Department of Chemical, Biological, Pharmaceutical and Environmental Sciences
University of Messina
Messina, Italy

Thongam Ibemcha Chanu
Aquaculture Division
ICAR – Central Institute of Fisheries Education
Mumbai, Maharashtra, India

Hethesh Chellapandian
PG and Research Department of Biotechnology and Microbiology
National College (Autonomous)
Tiruchirappalli, Tamil Nadu, India

Azubuike V. Chukwuka
National Environmental Standards and Regulations Enforcement Agency (NESREA)
Nigeria

Owias Iqbal Dar
Aquatic Toxicology Lab, Department of Zoology
Guru Nanak Dev University
Amritsar, Punjab, India
and
Department of Biosciences
Chandigarh University
Gharuan, Punjab, India

Sweta Das
Department of Aquatic Animal Health Management
Kerala University of Fisheries and Ocean Studies
Panangad, Kochi, Kerala

Maderi Velayutham Dassprakash
Department of Biochemistry
Sri Sankara Arts and Science College
Enathur, Kanchipuram, Tamilnadu, India

Ashutosh D. Deo
Fish Nutrition, Biochemistry and Physiology Division
ICAR – Central Institute of Fisheries Education
Mumbai, Maharashtra, India

Dipanjan Dutta
Aquaculture Laboratory, Department of Zoology
The University of Burdwan
Golapbag, Burdwan, West Bengal, India
and
Post Graduate Department of Zoology
Hooghly Mohsin College
Chinsurah, Hooghly, West Bengal, India

Hiam Elabd
Aquatic Animals' Medicine Department, Faculty of Veterinary Medicine
Benha University
Moshtohor, Toukh, Egypt

Gehad E. Elshopakey
Department of Clinical Pathology, Faculty of Veterinary Medicine
Mansoura University
Mansoura, Dakahlia, Egypt

Preetham Elumalai
Department of Marine Biology, Microbiology and Biochemistry, School of Marine Sciences
Cochin University of Science and Technology (CUSAT)
Lakeside Campus Fine Arts Avenue, Cochin, India

Falco F
Institute of Marine Biological Resources and Biotechnologies, National Research Council (CNR)
Mazara Del Vallo, Italy

Caterina Faggio
Department of Chemical, Biological, Pharmaceutical and Environmental Sciences
University of Messina
Messina, Italy

Aifa Fathima
Department of Microbiology, School of Life Sciences
Pondicherry University
Puducherry, India

Femi John Fawole
Fish Nutrition and Biochemistry Unit, Department of Aquaculture and Fisheries
University of Ilorin
Ilorin, Nigeria

S. Ferosekhan
Department of Aquaculture
ICAR – Central Institute of Freshwater Aquaculture
Bhubaneswar, Odisha, India

Koushik Ghosh
Aquaculture Laboratory, Department of Zoology
The University of Burdwan
Golapbag, Burdwan, West Bengal, India

M. Divya Gnaneswari
Department of Zoology
Gargi College
New Delhi, India

Tamilselvan Gokul
PG and Research Department of Zoology
Vivekananda College
Tiruvedakam (West), Madurai, Tamil Nadu, India

Aseem Grover
Department of Chemistry, Biochemistry and Forensic Sciences, Amity School of Applied Sciences
Amity University Haryana
Panchgaon, Haryana, India

Saqib Hassan
Department of Biotechnology, School of Bio and Chemical Engineering
Sathyabhama Institute of Science and Technology
Chennai, Tamilnadu, India
and
FutureLeaders Mentoring Fellow
American Society for Microbiology
Washington, USA

Ramchandran Ishwarya
Mandapam Regional Centre, ICAR – Central
Marine Fisheries Research Institute
Mandapam, Tamil Nadu, India

Rengarajan Jayakumar
ICAR – Central Institute of Brackishwater Aquaculture (CIBA)
MRC Nagar, Chennai, Tamil Nadu, India

Ramaraj Jayakumararaj
Department of Botany
Government Arts College
Melur, Madurai, Tamil Nadu, India

Manish Jayant
Fish Nutrition, Biochemistry and Physiology Division
ICAR – Central Institute of Fisheries Education
Mumbai, Maharashtra, India

Sivakamavalli Jeyachandran
PG and Research Department of Biotechnology and Microbiology
National College (Autonomous)
Tiruchirappalli, Tamil Nadu, India

Divya Jyoti
School of Biological and Environmental Sciences
Shoolini University
Solan, India

Mian Adnan Kakakhel
MOE Key Laboratory of Cell Activities and Stress Adaptations, School of Life Sciences
Lanzhou University
Lanzhou, P.R. China

Gbadamosi Oluyemi Kazeem
Department of Fisheries and Aquaculture Technology
Federal University of Technology
Akure, Ondo State, Nigeria

Tejaswini Kinnera
Fish Nutrition, Biochemistry and Physiology Division
ICAR – Central Institute of Fisheries Education
Mumbai, Maharashtra, India

George Seghal Kiran
Department of Food Science and Technology, School of Life Sciences
Pondicherry University
Puducherry, India

Arathi Kollath
Department of Marine Biology, Microbiology and Biochemistry, School of Marine Sciences
Cochin University of Science and Technology (CUSAT)
Lakeside Campus, Fine Arts Avenue, Cochin, India

Devasree Ganesh Kumar
The Madura College (Autonomous)
Vidya Nagar, Madurai, Tamil Nadu

Kamatchi Ramesh Kumar
PG and Research Department of Zoology
Vivekananda College
Tiruvedakam (West), Madurai, Tamil Nadu, India

Sreeja Lakshmi
Kerala University of Fisheries and Ocean Studies, Panangad
Kochi, India

Banaee M
Aquaculture Department, Faculty of Natural Resources and the Environment
Behbahan Khatam Alanbia University of Technology
Behbahan, Khuzestan Province, Iran

Mauro M
Department of Biological, Chemical and Pharmaceutical Sciences and Technologies (STEBICEF)
University of Palermo
Palermo, Italy

Heba Mahboub
Fish Diseases and Management Department, Faculty of Veterinary Medicine
Zagazig University
Zagazig, Sharkia, Egypt

R. Dinakaran Michael
Centre for Fish Immunology
Vels Institute of Science, Technology, and Advanced Studies (VISTAS)
Pallavaram, Chennai, Tamil Nadu

Shamna Nazeemashahul
Fish Nutrition, Biochemistry and Physiology Division
ICAR – Central Institute of Fisheries Education
Mumbai, Maharashtra, India

Hussain Nottanalan
Fish Nutrition, Biochemistry and Physiology Division
ICAR – Central Institute of Fisheries Education
Mumbai, Maharashtra, India

Neha Omgy
Kerala University of Fisheries and Ocean Studies, Panangad
Kochi, India

Lokesh Pawar
College of Fisheries (Central Agricultural University – Imphal)
Agartala, Tripura, India

Nikhila Peter
Fish Nutrition and Feed Technology Laboratory, Department of Aquaculture
Kerala University of Fisheries and Ocean Studies
Panangad, Kochi, Kerala

Chinnathambi Pothiraj
Department of Botany
Government Arts College
Melur, Madurai, Tamil Nadu, India

Chiranjiv Pradhan
Fish Nutrition and Feed Technology Laboratory, Department of Aquaculture
Kerala University of Fisheries and Ocean Studies
Panangad, Kochi, Kerala

S. Kalaivani Priyadarshini
Department of Biotechnology
Lady Doak College
Hakim Ajmal Khan Rd, Tallakulam, Tamil Nadu

Neha Qureshi
Fisheries Economics, Extension and Statistics Division
ICAR-CIFE, Mumbai, Maharashtra, India

Priyatharsini Rajendran
Department of Zoology
Lady Doak College
Tallakulam, Tamil Nadu, India

Akshay Thuruthiyil Rajesh
Department of Marine Biology, Microbiology and Biochemistry, School of Marine Sciences
Cochin University of Science and Technology (CUSAT)
Lakeside Campus, Cochin, Kerala, India

Subramanian Ramya
Department of Zoology
Yadava College (Men)
Madurai, Tamil Nadu, India

Einar Ringø
Norwegian College of Fishery Science, Faculty of Biosciences, Fisheries and Economics
UiT, The Arctic University of Norway
Tromsø, Norway

Nimai Chandra Saha
University of Burdwan
Purba Barddhaman, India

Shubhajit Saha
Fisheries and Ecotoxicology Research Laboratory
Department of Zoology
The University of Burdwan, Burdwan
West Bengal, India

Parimal Sardar
Fish Nutrition, Biochemistry and Physiology Division
ICAR – Central Institute of Fisheries Education
Mumbai, Maharashtra, India

Joseph Selvin
Department of Microbiology, School of Life Sciences
Pondicherry University
Puducherry, India

Arun Sharma
Aquatic Animal Health Management Division
ICAR – Central Institute of Fisheries Education
Mumbai, Maharashtra, India

Sunil Sharma
Aquatic Toxicology Lab, Department of Zoology
Guru Nanak Dev University
Amritsar, Punjab, India

Reshma Sinha
Department of Animal Sciences
Central University of Himachal Pradesh
Kangra, India

Subramaniam Sivakumar
Department of Biochemistry
Sri Sankara Arts and Science College
Enathur, Kanchipuram, Tamilnadu, India

Hamed Mousavi Sabet
Department of Fisheries, Faculty of Natural Resources
University of Guilan
Sowmeh Sara, Iran

Mehdi Soltani
Centre for Sustainable Aquatic Ecosystems, Harry Butler Institute
Murdoch University
Perth, Australia
and
Department of Aquatic Animal Health Faculty of Veterinary Medicine
University of Tehran
Tehran, Iran

Parasuraman Aiya Subramani
Centre for Fish Immunology
Vels Institute of Science, Technology, and Advanced Studies (VISTAS)
Pallavaram, Chennai, Tamil Nadu

C. Shanmuga Sundaram
Department of Microbiology
Hindustan College of Arts and Science
Kelambakkam, Chennai, Tamilnadu, India

Baskaralingam Vaseeharan
Biomaterials and Biotechnology in Animal Health Lab, Department of Animal Health and Management
Alagappa University
Karaikudi, Tamil Nadu, India

Karthikeyan Venkatachalam
Faculty of Innovative Agriculture and Fishery Establishment Project, Prince of Songkla
University Surat Thani Campus
Makham Tia, Mueang, Surat Thani, Thailand

Rantham Subramaniam Venkatesan
Lakshmi Bangaru Arts and Science College
Melmaruvathur, Chengalpet, Tamilnadu, India

Section I

Immunomodulators: An overview

1

Immunomodulators: An Introduction

Falco F[1], Banaee M[2], Mauro M[3], Faggio C[4], Arathi Kollath[5], and Preetham Elumalai[5]
[1]*Institute of Marine Biological Resources and Biotechnologies, National Research Council (CNR), Mazara Del Vallo, Italy*
[2]*Aquaculture Department, Faculty of Natural Resources and the Environment, Behbahan Khatam Alanbia University of Technology, Behbahan, Khuzestan Province, Iran*
[3]*Department of Biological, Chemical and Pharmaceutical Sciences and Technologies (STEBICEF), University of Palermo, Palermo, Italy*
[4]*Department of Chemical, Biological, Pharmaceutical and Environmental Sciences, University of Messina, Messina, Italy*
[5]*Department of Marine Biology, Microbiology and Biochemistry, School of Marine Sciences, Cochin University of Science and Technology (CUSAT), Lakeside Campus, Fine Arts Avenue, Cochin, India*

1.1 Introduction

Before going into detail about immunomodulators, it's important to understand how immune systems work and how different variables can stimulate the immune system in different ways. Many substances are known to exist in our environment, and they are always capable of affecting the immune systems of living beings. For example, it has been shown that the immune system plays an important role in aquatic organisms subjected to various physical stress conditions (Mauro et al. 2021), noise pollution (Mauro et al. 2020), drug pollution (Mauro et al. 2022), or bacterial activity (Vizzini et al. 2021). Several biomarkers are also used to evaluate the health status of important animals in aquaculture (Mauro et al. 2022). In addition, aquatic organisms are also an excellent source of molecules with antimicrobial and antitumor activity (Mauro et al. 2022). In this chapter, we briefly introduce the function and role of immunomodulators, especially in the fish immune system. Fish have an immune system that is similar to that of higher vertebrates. As a result, every living entity must preserve its integrity and health status when challenged and must be able to recognise and distinguish between "self" (its molecules, cells, and tissues) and "non-self" (all other organisms or substances). The purpose of the immune system is to recognise the millions of non-self organisms that are potentially harmful to the self and to eliminate them or reduce their impact so that no damage occurs to the self (Takx-Köhlen 1992). This chapter does not explain self-immunity or self-tolerance phenomena that lead to suppression of the immune system and the spread of many autoimmune diseases in fish. We prefer to divide the immune system into two broad categories based on function, namely the innate immune system (non-specific immune system) as the first line of defence against pathogens (Carbone & Faggio 2016) and the adaptive immune system (specific or acquired immune system) (Marshall et al. 2018).

1.2 How Do the Immunostimulants Work in Fish?

Since their embryonic life stage, fish have relied on their innate immune system, and their survival depends on it. The skin is the principal non-specific defence in fish and plays a key role in protecting and preventing the entry of pathogens into the epithelium through the secretion of a mucus layer involved in the immunity system (Salinas et al. 2011). The cells and mediators involved will differ depending on the time, the trigger, the anatomical location (inflammation can affect any tissue), and the severity of the inflammation (Calder et al. 2013). Teleosts have a cellular defence system that includes macrophage-like phagocytic cells, neutrophils, and natural killer (NK) cells, as well as T and B lymphocytes, as well as various humoral defence components like complement (classical and alternative pathways), lysozyme, natural hemolysin, transferring factor, and C-reactive protein (Watts et al. 2001). Furthermore, teleosts and elasmobranchs are the most primitive groups that possess the major histocompatibility complex (MHC) and T-cell receptors, which are the primary components of the immune response against

DOI: 10.1201/9781003361183-2

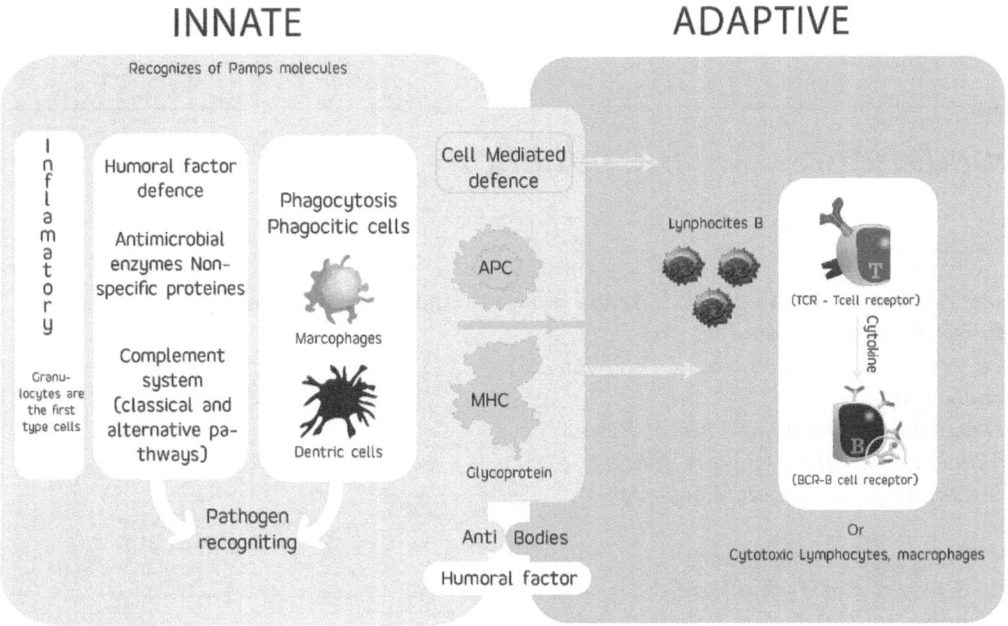

FIGURE 1.1 The immune system assumption in teleostfish.

The above figure represents types of immune system in fish. Innate immunity provides primary defence against pathogens by activating phagocytosis and antimicrobial, complement activation. On the other hand, adaptive immune system functions by stimulating specific lymphocytes.

pathogenic organisms (Zou & Secombes 2016). Sakai et al. (2021) investigated the role and function of cytokines in fish (interferon, interleukin 2, and macrophage activating factors), whereas among the lymphoid organs found in fish to mediate the responses have been the thymus, spleen, and kidney (Zapata 1996). The collaboration of fish innate and adaptive (memory) immune systems to eliminate intruders or activate defensive mechanisms is now well recognised; these two systems are classified into cell-mediated defence and humoral components (soluble substances). Figure 1.1 describes their main functions.

The innate system is made up of three parts: the tegument (skin and mucus), cellular components (granulocytes, monocytes, macrophages, and natural killer cells), and humoral components (granulocytes, monocytes, macrophages, and natural killer cells) (the complement system, antimicrobial enzyme system, and non-specific mediators such as interferon and interleukin). Furthermore, inflammation is thought to be an innate immune response driven by complex interactions between cellular and humoral components. Granulocytes are the first cells to arrive at the site of inflammation and are responsible for killing infections. Innate and adaptive immune systems normally work in concert, with innate responses serving as the host's first line of defence and enabling adaptive responses by antigen-specific T and B cells to produce antibodies in the presence of specific humoral components called histocompatibility molecules and glycoprotein receptors encoded by genes in the major histocompatibility complex (MHC).

1.3 Immunomodulators

The immunomodulators are all antigens (chemical or physical) that manage to vary the immune system's activities. When an agent depresses the immune system, also known as an immunosuppressant, they can cause a negative response. Immunosuppression is described by Dohms and Saif (1984) as "a state of transient or persistent dysfunction of the immune response resulting from insults to the immune system and increasing susceptibility to disease," and the qualifier "and frequently a suboptimal antibody response." Otherwise, if an immunomodulator can increase or promote activity, it is called an immunostimulant. Finally, another category is immune adjuvants, which hold the promise of being the actual modulators (De Paula Barbosa 2014) of the immune response, especially to enhance the vaccine's efficacy. An example in this regard is Freund's complete adjuvant, which is being used to enhance the potency of poor immunogenic substances (Tengjaroenkul & Yowarach 2011). Very often, the same agent can have both immunostimulant and immunosuppressive effects on an organism. Studies showed that the physiological response of fish to different doses of immune system stimulants could be different (Petit 2019). In other words, in some cases, excessive use of several immunostimulants simultaneously may cause immunosuppression in fish (Raa 1996).

1.4 Immune Suppressors

Immunosuppressive compounds are chemical, biochemical, and physical agents that suppress, decrease, or disrupt the immune system functions in fish. Studies show that all immunosuppressant compounds carry a severe risk of infection. Various types of immunosuppressive materials may change the immune response in fish when challenged by pathogens (Hidasi 2017). Most immunosuppressants may weaken the immune system by altering the gene expression of immune parameters. However, some immunosuppressive materials may decrease the absorption of vitamins and dietary supplements. Furthermore, immunosuppressive compounds reduce the ability of fish's innate and specific immune systems to respond to foreign objects. The interaction of immune suppressants with drugs may also reduce their effectiveness. The following part explains some immune suppression effects on the fish immune system.

1.4.1 Environmental Stressors

The functions of the immune system of fish depend on nutrition, environmental conditions, health, gender, and ontogeny. Temperature, dissolved oxygen, salinity, pH, and hardness ranges may also differ between fish species. Therefore, environmental fluctuations could alter the fish's physiological status and immunological response. For example, hyperosmotic stress could decrease the immune system response of *Scatophagusargus* to bacterial infection (Lu et al. 2022). In hostile biological conditions, fish may not be able to feed properly. As a result, they will not have enough energy to allocate to the immune system (Estensoro et al. 2012). Moreover, breeder malnutrition may lead to epigenetic changes in the offspring's immune and metabolic genes. Improper fish nutrition under stressful conditions can also lead to impaired immune priming by dendritic cells and monocytes and impair the function of effective memory T-cells. Acute and chronic stress are critical agents in suppressing the immune functions of fish (Guo et al. 2021).

Environmental stress can suppress the immune response in fish. As a result, fish that live in stressful environments may have a weaker immune system than fish that live in normal conditions. Guo et al. (2022) discovered that exposing Wuchang Bream (Megalobramaamblycephala) to ammonia nitrogen reduced immunoglobulin M (IgM), interleukin 1 (IL-1), and tumour necrosis factor (TNF-) levels while decreasing TLR mRNA expression. Following exposure to ammonia, there was a considerable drop in IgM and component C3 levels and lysozyme activity in the spleen and head kidney of Pelteobagrusvachellii (Qi et al. 2017). Moreover, those latter authors showed that high ammonia concentrations in the environment could disrupt the expression of immune-related genes in crucian carp (*Carassius auratus*) (Mazini et al. 2022). It was observed that the immune system's response would be decreased when fish were exposed to the stress of transporting between different farms. They discovered that an increase in corticosteroids was linked to immunological suppression in fish. Corticosteroids may affect the effectiveness of the immune system by altering mRNA expressions implicated in immunological parameters. Therefore, to better understand the impact of environmental stressors, each of the stressors has been discussed separately.

1.4.1.1 Agrochemicals

Agrochemicals include fertilisers, phytohormones, and pesticides (e.g., insecticides, pesticides, herbicides, and fungicides) used in the agriculture industry. Agrochemicals also include medications, disinfectants, hormones, and growth stimulants used in cattle, poultry, and aquaculture. Suppression of the immune system in fish exposed to agrochemicals can increase the susceptibility and vulnerability of fish to various pathogens (Banaee et al. 2019; Farag et al. 2021). Changes in intrinsic and specific immunological indices are perhaps the most important reason for suppressing the immune system of fish exposed to agrochemicals (Hassan et al. 2022). This section explains the reasons for the decay of the fish immune system after exposure to agrochemicals.

Due to their lipophilic nature, most pesticides easily cross biological barriers and enter the aquatic body. Pesticides can suppress the immune system by interacting with immune agents or causing oxidative damage in tissues involved in the immune system (Farag et al. 2021). Previous studies have shown that fish exposure to pesticides can cause changes in haematological parameters, including a decrease in leucocytes and an alteration in the differential count of white blood cells (Banaee et al. 2008). In fish treated with pesticides, decreased total immunoglobulin, C3, and C4 complement activities have also been reported (Hatami et al. 2019). A significant decrease was reported in lysozyme activity, respiratory burst activity, and total immunoglobulin levels in Nile tilapia (Oreochromis *niloticus*) exposed to cypermethrin (Abdel-Tawwab et al. 2020). Exposure to pesticides can also lead to changes in the gene expression of inflammatory cytokines such as TNF-, IL-1, and IL-6 (Acar et al. 2021; Wang et al. 2020).One study found that glyphosate exposure altered the levels of interferon- (IFN-) and IL-1 in the hematopoietic tissues of common carp (*Cyprinus carpio*).

1.4.1.2 Heavy Metals

Exposure to heavy metals at levels higher than the accepted dosages could suppress the immune response in fish. Also, the transmission of heavy metal contamination through the food chain can affect fish immune systems (Mohiseni et al. 2017). A significant decrease in immune functions of Vardar chub (*Squalius vardarensis,* Karaman) may be due to histopathological damage to the kidney and spleen (Jordanova et al. 2017). Bernier et al. (1995) demonstrated that bioaccumulation of heavy metals could cause suppression of the immune system, autoimmune diseases, increased susceptibility to pathogens, and inflammation reactions in fish. Banaee et al. (2019) demonstrated that alterations in humoral immune parameters in fish exposed to heavy metals indicated quenching of innate immune responses. Manganese reduced lysozyme activity and IgM levels while increasing the expression of tlr3, tnf-, il-1, and il-6 in juvenile yunlong groupers (*Epinephelus maureus* and *E. lanceolatus*) (Wang et al. 2022). Heavy metals were exposed to *Centrarcus lal.rax.* changes in bactericidal activity in the skin mucosa of gilthead seabream (*Sparus aurata*) after heavy metal exposure have also been reported (Guardiola et al. 2015). A significant decrease in HK-B cell proliferation, IgM level, and serum bactericide potential was observed in catfish exposed to arsenic (Ghosh et al. 2007). Heavy metal exposure in fish may affect leucocyte counts. Furthermore, changes in the granulocyte/granulocyte ratio suggest that heavy metals trigger the fish immune system. A significant increase in monocyte and neutrophil numbers and a significant decrease in lymphocyte numbers were reported in goldfish (*C. auratus*) after exposure to manganese (Aliko et al. 2018).

Genotoxicity may be linked to oxidative stress, DNA damage, and changes in mRNA expression. Indeed, Ghazy et al. (2017) found that the gene expression in the immune system changed in Nile tilapia that lived in waters contaminated with heavy metals. Heavy metals altered the expression of IL1, TNF-, IFN, Mx, Lyz, C3B, and CXCL-Clc in zebrafish (*D. rerio*) embryos (Cobbina et al. 2015). Moreover, significant changes in the mRNA transcription of immune-related genes were observed in the leucocytes of European sea bass exposed to cadmium, lead, and mercury (Morcillo et al. 2015).

1.4.2 Other Xenobiotics

Segner et al. (2021) showed that xenobiotics could bind to the aryl hydrocarbon receptor (AhR) as a vital transcription factor. Then, complex AHR and xenobiotics were set on sequence response elements and changed immune-related gene expression. Suppression of the immune system in aquatic animals exposed to xenobiotics may be due to energy shifts for detoxification. Decreased energy allocated to the immune system may reduce its performance.

1.4.3 Natural Toxins

Qiao et al. (2013) investigated the immunotoxicity of cyanobacteria in Crucian carp. The immunological fish response might be boosted by blood cyanobacteria, according to authors Rymuszka and Adaszek (2013), who assessed the effects of microcystin derived from cyanobacteria on carp leucocyte proliferation under *in vitro* conditions. They showed microcystin could increase apoptosis rates in white blood cells. Rymuszka and Sieroslawska (2018) showed that carp leucocytes' half-life decreased after exposure to nodularin, a cyanobacteria toxin. Contamination of food with aflatoxin toxins can reduce the potency of non-specific immune systems in fish (Bitsayah et al. 2018). In common carp fed with aflatoxins-contaminated feed, Bitsayah et al. (2018) found a substantial shift in complement C3, C4, and CH50, lysozyme activity, and total immunoglobulin content.

1.4.4 Anti-nutritional Agents

Some antinutritional compounds in the diet can suppress the immune system in aquatic animals. However, the immunodepressive effects of anti-nutritional compounds on the immune system depend on their bioavailability and their dosages in feed. By generating cytotoxicity, anti-nutritional substances may have an impact on immune function. Furthermore, by producing inflammation or modifying the quantity of inflammatory markers in the blood, these substances may impair immunological function. Inhibiting the absorption of micronutrients in the intestine by anti-nutritional compounds can also play a role in suppressing the aquatic immune system (Abdel-Tawwab et al. 2018). The immune system may be harmed by phytotoxins such as cyanogenic glycosides (Cho et al. 2013). Glycinin is a dietary allergen in soy, which can harm an animal's immune system (Sun et al. 2008). Gossypol, an anti-nutritional compound found in cottonseed, can cause apoptosis by activating caspase-3 (Sadahira et al. 2014).

1.4.5 Diseases

Increased inflammatory cytokines in chronic diseases may be a reason to suppress the immune system. Furthermore, toxins and enzymes secreted by the primary pathogen can reduce the immunity of fish against secondary pathogens (Ilgová et al. 2021). Simultaneous fish infection with two or more pathogens can weaken the immune system (Shameena et al. 2021). Ilgová et al. (2021) found that chronic pathogen infections could significantly delay the immune system response of fish. They showed that infection with monogeneans could increase the susceptibility of fish to secondary pathogens. Ilgová et al. (2017, 2020) demonstrated that infection with *Eudiplozoon nipponicum* reduced TNF-gene expression in common carp macrophages in vitro.

1.4.6 Sex Hormones

Hormone manipulation and changes in the steroid hormones may affect gene expression in a fish's immune response. Therefore, a decrease in the immune system capacity of adult fish may be related to changes in sex hormones. Szwejser et al. (2017) found that fish leucocytes have receptors and cytochrome P450 aromatase. As a result, oestrogen levels in the blood might affect the fish's immune system. Cabas et al. (2018) found a link between oestrogen levels and autoimmune illness and chronic inflammation in fish and discovered that fluctuations in sex steroids affected the immune systems of spotted snakeheads (*Channa punctatus*). They found that increased sex steroids could mitigate innate and cellular immune responses. Dietrich et al. (2021) showed that hormone therapy in common carp could change mRNA expression in hematopoietic tissues.

1.4.7 Drugs and Antibiotics

Antibiotics can suppress the fish immune system by disrupting the regulation of NF-B signalling and immunotoxic pathways (Qiu et al. 2020). Yang et al. 2020 indicated that antibiotics are toxic to fish at high dosages. They also showed that treatment with a high dose of antibiotics would suppress the fish's immune system. Increased mortality and an inflammatory response were observed in zebrafish treated with 260 ng of L-1 sulfamethoxazole (Zhou et al. 2016). Common carp exposed to sulfamethoxazole has reportedly seen a similar outcome (Iftikhar et al. 2022). Liu et al. (2020) found that exposure of zebrafish larvae to high doses of sulfamethoxazole could change the mRNA expression of cytokines such as IL-1, IFN-, IL-11, and TNF-. The genotoxicity effects of antibiotics on leucocytes could mitigate the efficiency of cellular immunity in fish (Grondel et al. 1985).

1.5 Immunostimulants

Immunostimulants have a crucial role in activating the non-specific defence mechanism in fish protection against pathogens, and they are valuable for controlling fish diseases. Previous studies have shown that numerous immunostimulants may be useful to fish cultures in aquafeed. Particularly for enhancing and improving the immunity and disease resistance in fish, increasing non-specific defence mechanisms immune stimulants may also be an effective strategy to increase fish performance. Although the use of immunostimulants in the aquaculture industry has been successful in some cases, research in this field is still ongoing (Jadhav et al. 2006).

Chemical agents, bacterial components such as probiotics (Abdel-Latif et al. 2022), polysaccharides (e.g., from plants) (Faggio et al. 2016, 2015), animal or vegetable extracts (plant-based) (Rashidian et al. 2021), feed additives and herbal extract (Elumalai, as pointed out above, immunostimulants) might mainly facilitate the function of phagocytic cells and increase their bactericidal activities (Abarike et al. 2019). Moreover, natural killer cells, complement, lysozyme, and other antibody responses may be stimulated by different types of immunostimulants. Immunological function and activation are associated with improved protection against infectious diseases in aquaculture due to their ability to serve as an alternative and supplement to vaccination. Moreover, they also have additional effects on growth performance and the survival rates of the fish under stress (Heo et al. 2001) (Figure 1.2).

1.5.1 Chemical Agents

Since immune system stimulants contain many chemical compounds that may affect the function of the aquatic immune system, they can be administered alone or in combination with vaccination. Some immunostimulants have been shown in studies to improve vaccination efficiency. Shahbazi and Bolhassani (2016) found that some biochemical compounds such as vitamin C and E, lactoferrin, interferon, growth hormone, prolactin, and recombinant cytokines can act as immunostimulants. The chemical agents could disrupt immunological structures and functions, making animals more susceptible to both infections and non-infectious agents. A wide range of substances target the immune system and prolonged exposure to these compounds can result in immunological dysfunction (Koller 2001). For example, levamisole is the most useful chemical agent used and is a

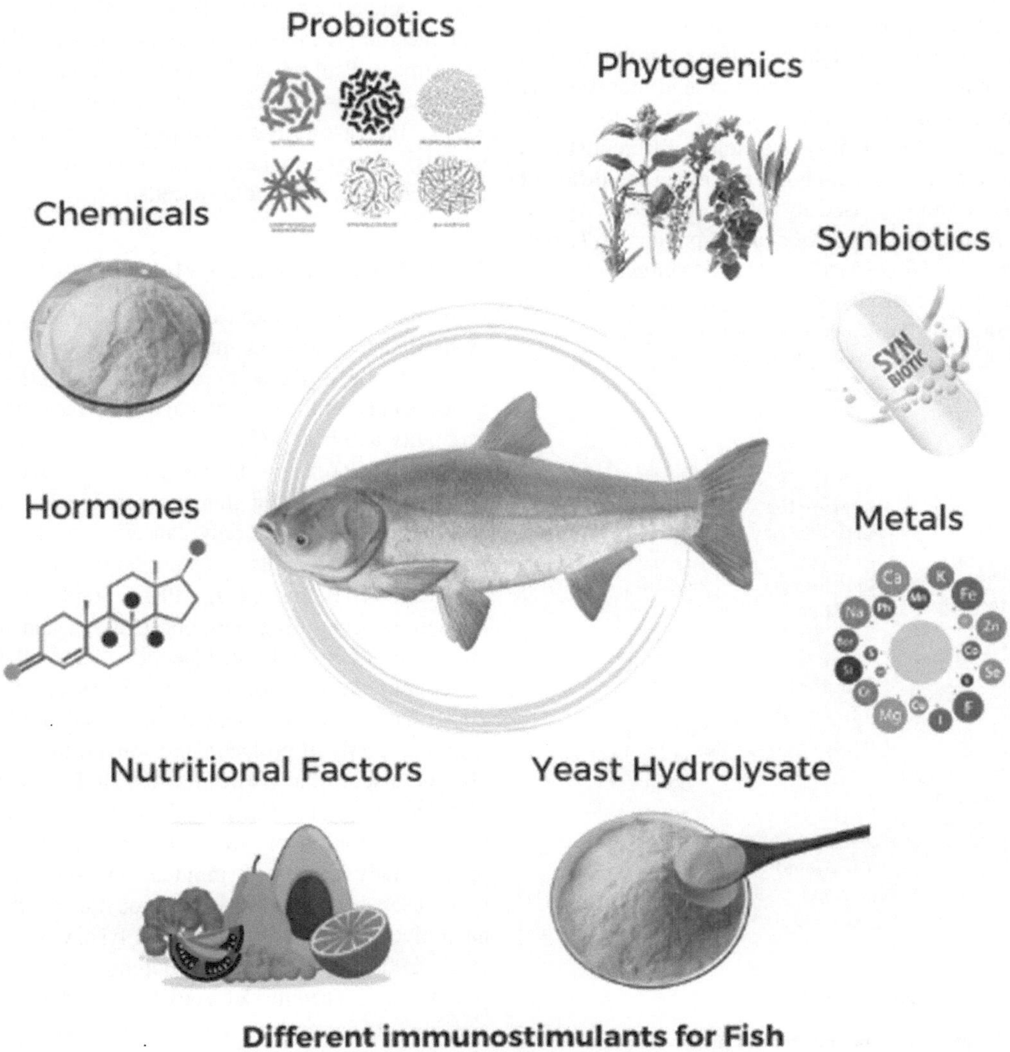

FIGURE 1.2 Different categories of immunostimulants for fish.

Various natural and synthetic agents that enhance immunological activity in fish are depicted in the figure. Bacterial agents, specifically prebiotics and probiotics are of high potential, along with immensely used phytochemicals, yeast derivatives, chemical agents and metals.

levo-isomer of tetramisole (Findlay & Munday 2000). It is a polysaccharide that can increase macrophage activity and provide resistance to specific harmful microorganisms. On the other hand, chitosan is a de-N-acetylated version of chitin. Both chitin and chitosan have the potential to be significant components in aquaculture. Chitosan treatment by injection or immersion was found to enhance brook trout (*Salvelinus fontinalis*) resistance to an *A. salmonicida* infection. Chitosan treatments had a substantial influence on the non-specific immunity and immunological response of both healthy and cortisol-treated *Labeo rohita* (Barman et al. 2013).

1.5.2 Biological Substances and Bacterial Derivates

The most common biological substances are bacterial derivatives, also known as killed pathogens, and their products. The use of immunostimulants such as probiotics and prebiotics has always been considered (Bachère 2003). Probiotics are a collection of non-pathogenic microorganisms often found in aquatic animal digestive systems. Oral administration of probiotics can change bacterial flora, boost the immune system, and stimulate growth performance (Villamil et al. 2002). Previous studies show that the administration of probiotics can significantly increase antibody production and non-specific immune parameters in fish (Abareethan & Amsath 2015). Therefore, it is essential to identify and isolate different strains of microorganisms to produce probiotics. Among the different probiotic strains commonly used in fish are *Bacillus,* e.g., *B. subtilis* and *B. licheniforms*; *Lactobacillus* sp. such as *L. delbrueckii subsp., L. bulgaricus, and L. acidophilus*; and *Bifidobacterium* sp. Numerous reports on the use of probiotics in aquatic environments have been published (Van Doan et al. 2020; Jahangiri & Esteban 2018; Chauhan & Singh 2019), and for more details, readers are referred to Chapter 10.

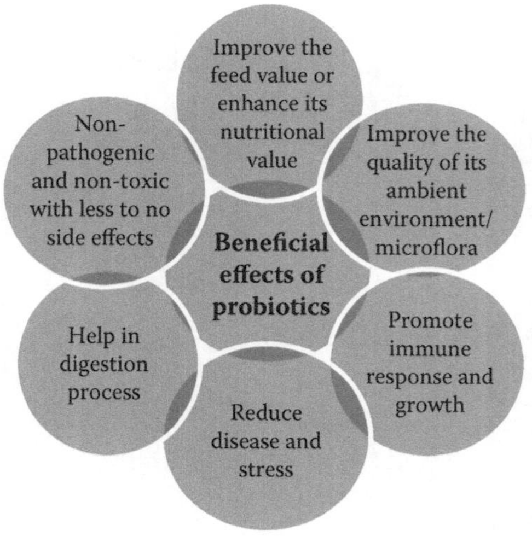

FIGURE 1.3 Benefits of probiotics.

Figure 1.3 illustrates the functions of probiotics as an immunostimulant in aquaculture. Probiotics promote immunity, growth, digestion, reduce stress and improve the feed value and quality of microflora in fish.

1.5.3 Lipopolysaccharide

Lipopolysaccharide (LPS) is the main component of the outer membrane of gram-negative bacteria such as *Salmonella typhimurium* and *Escherichia coli* (Miura & Mizushima 1968), and its preparations include O-antigens and endotoxins. The biological activity of LPS is a consequence of both hydrophobic domains known as lipid A (or endotoxin), a "core" oligosaccharide, and a distal polysaccharide (or O-antigen) (Neidhardt 1996). Moreover, LPS has been used as a potential immunostimulant. Toll-like receptor (TLR)-4 is mainly involved in the activation of the immune system by LPS through the specific recognition of its endotoxin (lipid A) moiety. LPS studies were made on fish both in vitro and in vivo, and they reported that LPS influences the growth and health status of fish. (Guttvik et al. 2002) showed that Atlantic salmon fry fed with LPS-coated feed (0.1% LPS) for 63 days had a reduced survival rate when challenged with a virulent strain of *A. salmonicida.* Furthermore, Paulsen et al. (2003) discovered that LPS stimulates plasma lysozyme activity originating from macrophages in various organs (e.g., blood polymorphonuclear and cells isolated from the head, kidney, and intestine) in their experiment on *Salmo salar*. In an in vitro experiment, Paulsen et al. (2001) in *Salmo salar* found that in head kidney macrophages grown in the presence of LPS, there was an increase in lysozyme production in the culture supernatants, which coincided with an accumulation of lysozyme gene transcript in stimulated cells.

1.5.4 Hormones and Cytokines

Hormones and cytokines are part of the neuroendocrine system. Their role as immunomodulators in the immune system has been studied in recent years. Acute stress may often be associated with fish life stages and have an impact on fish immunity and health (Figure 1.4). An example could be stress (resulting in potential advantages), thus involving short-term challenges resulting in immune activation or enhancing processes. Hormones generally can directly affect macrophages, lymphocytes, NK cells, and mitotic activity. Cortisol, growth hormone (GH), prolactin (PRL), reproductive hormones, melanin-concentrating hormone (MCH), and pro-opioid melanocortin (POMC)-derived peptides have all been shown to affect immune function in many fish species (Harris & Bird 2000). The growth hormone (GH, or somatropin) is a hormone from the family of prolactin and somatostatin; the main role of GH and insulin-like growth factor-I (IGF-I) is in the regulation of body size in growing animals. Previous studies showed that the administration of exogenous growth hormone (GH) improves many aspects of immune function, for example, cytotoxic (Sakai et al. 1996), phagocytic (Yada et al. 2006), haemolytic (Kim et al. 2013), and lysozyme activities (Harris et al. 2000) as non-specific defences, and immunoglobulin production as specific defences Yada (2007) observed that the activation of immune function during seawater adaptation was closely associated with increased plasma GH levels in some euryhaline fishes. Moreover, a previous study (Yada et al. 2006) showed that ghrelin (an important regulator of GH

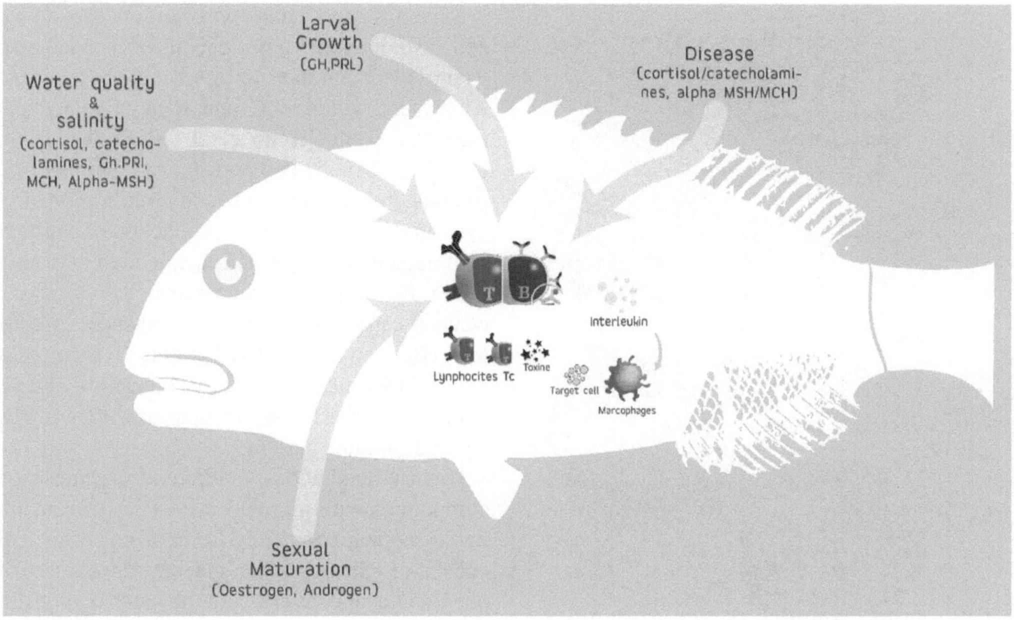

FIGURE 1.4 Interconnections between the general immune system and the endocrine system about environmental processes and fish life stages.

secretion) stimulates superoxide production associated with phagocytosis in trout leucocytes. While GH increased the mRNA levels of superoxide dismutase, which catalyses the dismutation of superoxide into oxygen and hydrogen peroxide, Also, Pontigo & Vargas-Chacoff (2021) found that GH may modulate the immune response in the SHK-1 cell line and leucocyte cultures of the head kidney in Atlantic salmon. Therefore, their work points out the independent action of GH on the immune system and the GH/IGF axis.

Cytokines are low-molecular-weight glycoproteins involved in regulating the immune system. These molecules are mainly secreted by cells of the innate and adaptive immune systems, but they play an important role in the innate immune response in fish. Kono et al. (1996) reported that IL-1 plays an essential role in fish immunity by activating lymphocytes and phagocytic cells and increasing resistance to *A. hydrophila* infection. Type I IFNs (homologs to human IFN- and IFN-) also have antiviral activity. Secombes & Belmonte (2016) discovered that type II IFN (IFN-) had bactericidal activity against intracellular parasitic bacteria. Studies to elucidate cytokines' functions in fish have recently begun. Still, more work is needed to select their appropriate functions, such as immunostimulants and vaccine adjuvants, to prevent infection in farmed fish. Sakai et al. (2021) developed a multiplex reverse transcription-polymerase chain reaction assay to investigate the immune response of fish when activated by an immunostimulant. Moreover, cytokines are also defined as biological response modifiers because of their ability to enable communication between different cell populations, in agreement with what was reported by Wilson et al. (2002).

Conforming to changes in environmental factors such as water quality, salinity and diseased conditions, the T-cell proliferation and cytokine expression also range, which is mediated through hormonal regulations. Hormones expressed during sexual maturation and larval growth enhance the expression of these immunological mediators.

1.5.5 Phytogenic Immunostimulants

(Plant extracts, herbals, garlic, ginger, triterpenic acid, polyphenols, olive oil, seaweeds)

Phytogenics are plant-derived natural products characterised by their richness in biologically active compounds that are mainly incorporated into the feed to enhance the innate immunity, health status, and growth performance of the animals. In farmed fish, phytogenics have been reported to contribute as antimicrobials, antioxidants, anti-inflammatory agents, immunostimulants, and sedatives. They work as promoters of growth and appetite stimulators, and they could influence the bile secretion and several enzymes associated with digestion (Chakraborty et al. 2011, Firmino et al. 2021, and Caipang et al. 2021). Botanicals, including herbs and spices, contain aromatic compounds and essential oils (extracted from parts of plants such as leaves, flowers, roots, and fruits), and many other medicinal plants come under phytogenics (Caipang et al. 2021). Plant extracts are active substances with desirable properties that are extracted from plant tissue for specific purposes, such as immunostimulant use. They have been known to have increased lysozyme activity, complement activity, phagocytic activity, an antibody response, elevated respiratory burst activity, and higher plasma protein (albumin and globulin) (Reverter et al. 2014; Harikrishnan R et al. 2011a). Saponin compounds, herbs, ginger, triterpenes from fungi and plants and seaweeds, etc. are some of the compounds and products which have an immunostimulatory effects on fish health and welfare and are discussed in detail in this book.

1.5.6 Nutritional and Dietary Factors

(Dietary amino acids, vitamin C, vitamin E, dietary nucleotides, organic acids, polysaccharides, probiotics, and food waste)

Nutrition plays a crucial role in maintaining the body's functioning and health. It provides all the essentials required by the

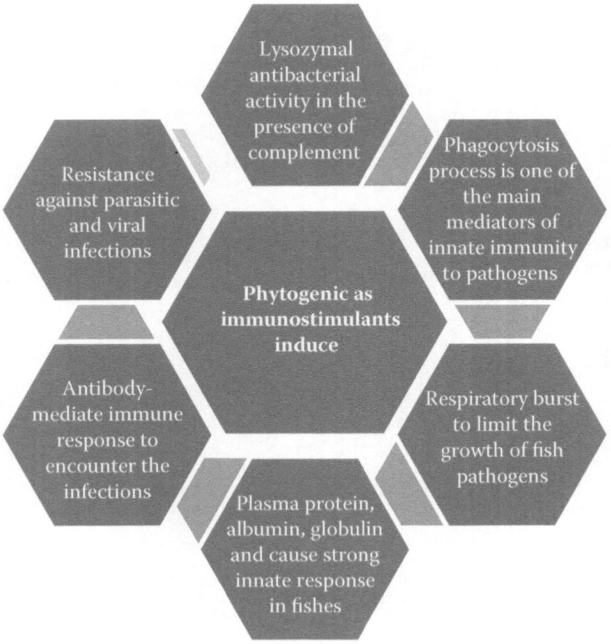

FIGURE 1.5 Activities of plant extract as an immunostimulant.

Figure 1.5 represents the role of plant extracts in enhancing immunity in fishes. Plant extract stimulate immunity by different cellular mechanism like phagocytosis, respiratory burst, antibody mediated responses, lysosymal antibacterial activity, and stimulate immunogenic plasma proteins.

body to maintain life, like metabolic energy and elements and compounds that act as co-factors for various physiological processes. When we talk about immunostimulants, dietary nutritional components include dietary amino acids (AA), vitamin C, vitamin E, dietary nucleotides, organic acids, polysaccharides, probiotics, food waste, etc. Dietary amino acids are essential acids that can improve the haemocyte count, phagocytic activity, respiratory burst in haemolymph, and lysozyme activity in cell-free haemolymph (CFH). Moreover, they can remarkably downregulate the malondialdehyde content (Luo et al. 2021). In addition to these, AAs are necessary for endogenous synthesis of protein and act as important energy substrates. These can also modulate the necessary metabolic pathways (Dawood et al. 2021) (Figure 1.6).

Vitamins are organic compounds necessary for animal growth and development. They are required in small quantities and must be provided with food, as they cannot be synthesised in the body. At present, vitamins C and E have gained popularity as immunostimulants (Kono et al. 1996). Vit E (tocopherols) are bioactive phenolic compounds, and a proper dose of this can (i) promote the differentiation and proliferation of lymphocytes and cytokines, (ii) enhance the production of antibodies and enhance complement activity when encountered with an antigen, and (iii) it can also improve phagocytosis activity and cytotoxicity. In channel catfish and turbot, VE enhances macrophage phagocytosis (Barman et al. 2013). Vitamin C (ascorbic acid) is involved as a cofactor in many bioactive processes, like neuromodulation, collagen synthesis, and cellular activities related to hormones, and the immune system.

Dietary nucleotides, chitin, and organic waste do also have compounds with immunostimulatory effects and are discussed in subsequent chapters. Bacterial and yeast cell walls mostly consist of glucans. When glucan was given to feed, it activated phagocytic cells in fish, and they also demonstrated an increase in the development rate of *Litopenaeus vannamei* juveniles, enhancing phagocytosis and the capacity of the cells to eliminate harmful pathogens. Additionally, they increase complement and lysozyme activity (Kono et al. 1996). Additionally, glucans improve the non-specific defence mechanisms of fish and shellfish and offer defence against bacterial infections (Barman et al. 2013). Prebiotic compounds called fungal polysaccharides are commonly regarded as a dietary component for controlling growth and health issues. Higher fungi are excellent providers of a variety of crucial natural compounds (Mohan et al. 2019). More details are discussed in subsequent chapters (Figure 1.7).

1.5.7 Trace Elements and Metals

Minerals that are less abundantly present in living tissues are referred to as trace elements (or trace metals). They are considered to be nutritionally essential, although if consumed at sufficiently high levels, they may prove toxic. Copper, chromium, iron, iodine, fluoride, manganese, molybdenum, selenium, and zinc are considered as essential trace elements. General functions of minerals include structural constituents of tissues, formation of the exoskeleton, osmotic pressure balance, muscle contractions, and nerve impulse transmission. They are

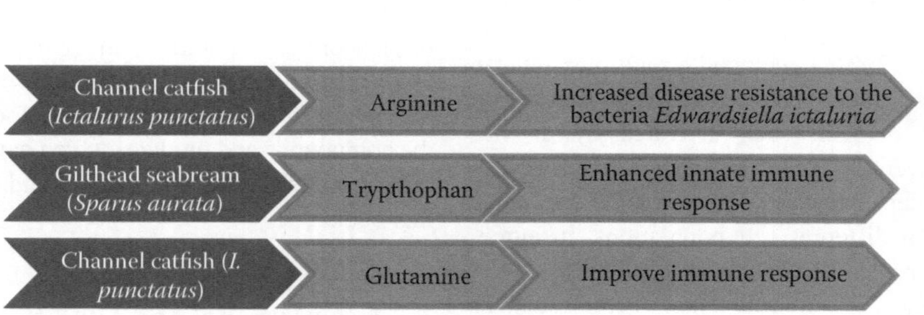

FIGURE 1.6 Effects of different amino acids on fish health.

Essential amino acids such as arginine and tryptophan and non essential aminoacid like glutamine have increased effect on fish immunity. Arginine, in channel catfish improve the resistance in fish against *Edwardsiella ictaluria*. Tryptophan in Gilthead seabream has a transient immune enhancement activity in the fish. Similarly, glutamine improve the immune response in channel catfish.

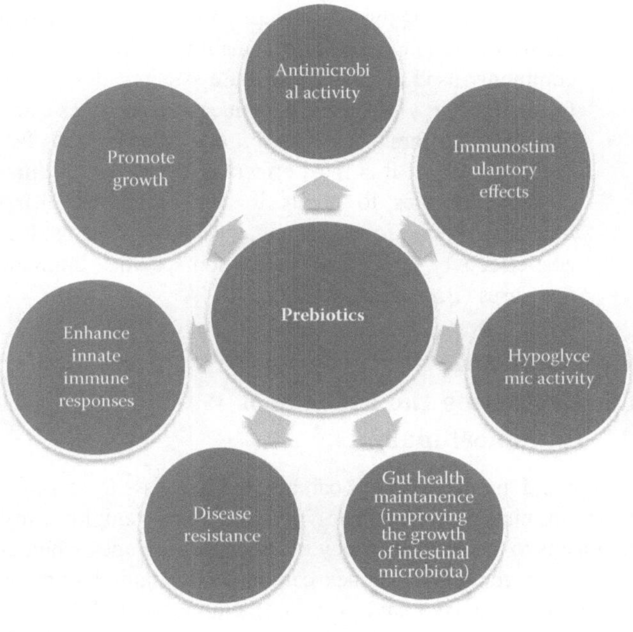

FIGURE 1.7 Functional properties of prebiotics.

Prebiotics have an enhancing effect on fish growth, antimicrobial activity, and disease resistance. They improve favourable intestinal microbiota and stimulate immune system in teleost.

also prime components for co-factors in metabolism, catalysts, enzymes, enzyme activators, hormones, pigments, and vitamins.

1.5.8 Synbiotics

Synbiotics are a combination of both prebiotics and probiotics and work as growth and immunity promoters. These have been used in aquaculture for over a decade, but the functional mechanism is still not very clear. Prebiotics, which are parts of synbiotics when hydrolysed to simpler mono- or disaccharides, show an exceptional increase in biomass and colonisation of probiotic bacteria on the surface of intestinal epithelial cells in the host. By releasing extracellular bacterial enzymes and bioactive substances from their metabolic activities, they also contribute to the growth of aquatic animals. These enzymes also improve the nutrient absorption capacity, which in turn help in effective utilisation of feed. Synbiotics stimulate the immune system's synthesis of nitric oxide, phagocytosis, and respiratory burst activity in fish.

1.5.9 Yeast Hydrolysate

Yeast is known for keeping the digestive system of an animal's body healthy and in balance; hence, it is widely used as a supplement in aquatic feeds. Yeast has a high protein and energy content, as well as high micronutrient content. Aside from being high in amino acids and proteins, yeast products also have immunomodulatory compounds like mannan oligosaccharides (MOS), chitin, glucans, and nucleic acids. Mannan oligosaccharides are well known to enhance the growth performance of rainbow trout (*Oncorhynchus mykiss*), *Aespeciallyes cerevisiae* is a well-known yeast from which glucan is extracted and purified, and an intraperitoneal injection of this improves both specific and non-specific immune responses in carp *(Cyprinus carpio) to* the bacterial challenge posed by *A. hydrophila.* Yeast hydrolysate is a hydrolysate of yeast cells obtained through various methods like acids, enzymes, or other hydrolysis. The extract may be obtained through autolysis, where the enzymes found in yeast itself are used to break down the protein, or through hydrolysis, where enzymes are added from external sources as depicted in Figure 1.8 (Gong et al. 2019).

1.6 How to Administer the Immunomodulators to Fish?

Overall, the immunostimulants could be administered through different routes. Although injection methods are the best strategy to enhance non-specific immune system responses, this method is costly and time-consuming. Furthermore, injection methods are only performed by experts. Therefore, it is applied in experiments where the fish are intended as brook stock in genetic studies. Another method is immersion, but its efficiency is less than that of injection; however, it requires crowding and an increase in the handling of fish stocks. Immune system stimulants may be prescribed as oral supplements. However, the immunostimulants' concentration in the diet depends on the size, ontogeny stages, and initial weight of the fish. This method consists of oral ingestion, produces a suitable non-specific immune response, and can be the most cost-effective method of administration. Top dressing can help you achieve these. The surface of the food is treated with the pure immunostimulant in this case. This is comparable to employing a layer of fish oil to top-dress antibiotic granules. This technique produces variable results depending on how well the immunostimulant adheres to the feed. At last, one method that is much more advanced is bio-encapsulation.

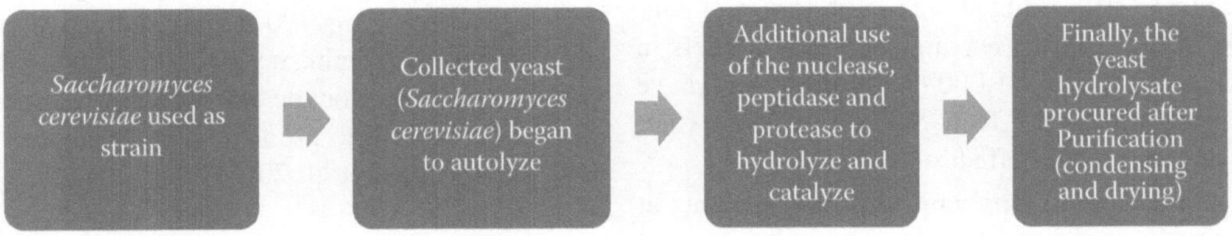

FIGURE 1.8 Extraction of hydrolysate.

Demonstrate the pathway for the extraction of hydrolysate from yeast to use as an immunostimulant for fishes. The yeast (*Saccharomyces cerevisiae*) are collected, autolyzed yeast is subjected to nuclease, peptidase, and protease to get the hydrolysed product, which is subsequently purified using condensing and drying.

1.7 Limitations of Immunostimulants

- Though showing extraordinary growth in the field of aquaculture, one of the important disadvantages of some immunostimulants is their high cost.
- The administration of the drug is important to consider as it depends on its efficacy. Immunostimulants show limited efficiency upon parental administration.
- Immunostimulants are not completely effective against all diseases.
- Overdoses of immunostimulants in feeds may cause immunosuppression.
- In some cases, aquatic animals may also fail to provide enhanced protection or an increase in immunity.
- Immunostimulants are successfully used in aquaculture against various infections and pathogens; however, the ability to improve innate resistance against many diseases (e.g., columnaris disease) has not been studied.

1.8 Factors Affecting the Efficiency of Immunostimulants

The effectiveness of an immunostimulant can depend on various factors, as follows:

- Solubility: Laminaran is an algal extract that can boost respiratory burst activity in leucocytes of the anterior kidney and activate macrophages in Atlantic salmon because it is more soluble than the fungal and yeast glucans. It has also been demonstrated to be a potential chemical for diet usage due to its greater solubility, so the solubility is considered an important aspect for immunostimulants.
- Duration of dose: Salmon that received M-glucan injections only took 2 days to produce their peak leucocyte responses. After 4–7 days of therapy with yeast beta-glucan, the respiratory burst activity increased. This demonstrated how immunostimulants might improve non-specific immunity with very brief dosages.
- Dosage: A high dose or overdose certainly does not seem to have an enhancing effect and can, in turn, inhibit the immune responses. At concentrations of 0.1–1 g/mL, the respiratory burst activity of glucans-treated macrophages increased. Whereas glucan had no effect at a concentration of 10 g/mL, it was inhibitory at a concentration of 50 g/mL. Very high vitamin E levels in feed, such as those of a >1,000–5,000 IE/kg diet, have an immunosuppressive effect. So, the dose per unit weight has a significant effect on efficacy.
- Time of administration: applying immunostimulants at the right time is very important in aquaculture. Mostly, application is needed before the outbreak of disease so that losses due to disease can be reduced.
- Method of Administration: The administration of immunostimulants through injection has been reported to be most effective against a range of pathogens. Vaccination seems to be impractical for small fish. Immersion is commonly used in intensive culture systems, despite the fact that it is less expensive than injection, produces a less non-specific immune response, and stresses the fish during handling. It is most effective during the acclimation of juveniles to ponds in field conditions. Oral ingestion is good for extensive aquaculture systems. It is cost-effective and enhances non-specific immune responses (Barman et al. 2013).

1.9 Evaluating the Efficacy of Immunostimulants

In vivo and in vitro methods can be used to assess the efficacy of an immunostimulant. The in-vivo method employs fish pathogens to assess the efficacy of immunostimulants, whereas the in-vitro method examines cellular and humoral immune mechanisms. In-vivo and in-vitro methods should be performed together to check the basic mechanisms for providing protection. In preliminary studies, in vitro methods are preferred. In vitro evaluation is based on lymphocyte proliferation, complement activation, total erythrocyte and leucocyte counts, chemokinesis, chemotaxis, lysozyme activity, and RBA phagocytosis. Other parameters include monitoring natural cytotoxic activity, macrophage-activating factor (MAF) levels, and C-reactive protein levels. However, these tests are too expensive to be conducted to check the efficacy of immunostimulants. A deep level of research is required to check the efficacy of various compounds for aquaculture species and their pathogens and to ultimately decrease the cost of the immunostimulants.

1.10 Timing of Administration

It is very important to use immunostimulants at the correct time and in the right concentration to boost the immune system. Anderson proposed in 1992 that it is best to use immunostimulants prior to the possibility of disease outbreaks in order to minimise disease-related loss. Furthermore, the effective dose and timing of exposure have a significant impact and are complicated by the culture system and feeding scenario. Studies in Atlantic salmon showed that the maximum non-specific disease resistance is attained only after the third week of injecting glucan at 10 mg/100 g, whereas the effects of low dosing at 1 mg/100 g last only for 1 week. Similarly, in African catfish, administration of glucan led to a maximum increase in the phagocytic cells at 7 days but not after 14 days. So, it is preferable to use them well in advance and at regular intervals (Barman et al. 2013).

1.11 Detection of Immunostimulants

Detection of immunostimulants in fish body is detected by using 'omic' technologies.

Methods of detection of immunostimulation are as follows and it is discussed in further chapters in detail:

- In vitro measurement
- In vivo measurement
- Phagocytic activity

1.12 Attributes of Immunostimulants

The most important attribute of an immunostimulant is that it directly influences the animal's health. It is biodegradable and biocompatible, and therefore safe for the environment. It enhances the immune system of animals, promoting good health, and is non-toxic to both fish and shellfish with no side effects observed. In aquaculture, it provides disease resistance to animals against a broad spectrum of pathogens and reduces mortality caused by opportunistic pathogens. It can also keep the host safe by providing enhanced immune stimulation to fight viral diseases. It increases the effectiveness of many antimicrobial substances, vaccines, and antibiotics. Moreover, it is cheap, easily available, and most importantly, an eco-friendly method for immune stimulation (Barman et al. 2013).

1.13 Vaccine

A vaccination is a biological treatment that increases immunity to a specific disease. Vaccine often comprises an agent that resembles a microorganism that causes a disease and is frequently created from weaker or dead versions of the pathogen. In order for the immune system to more quickly identify and eliminate any further interactions with this disease-causing microorganism, The agent prompts the body's immune system to identify the agent as foreign, eliminate it, and "remember" it so that the immune system will be better able to identify and eliminate any of these microorganisms that it comes into contact with in the future. "Prevention is better than cure" is the core tenet of a vaccination. The name "vaccine" originated from Edward Jenner's usage of the phrase "cow pox" in 1796 (Latin "variolvaccin," which was borrowed from the Latin "vaccn-us," from "vacca" cow). He was a pioneer in the use of cowpox vaccinations to stop the spread of smallpox.

FIGURE 1.9 Vaccination via injection.

Figure 1.9 represents vaccine administration through injection via intramuscular or intraperitoneal method to increase resistance against pathogens by stimulating immune system.

Fish immunisation started in 1942 when David C. B. Duff successfully immunised trout orally against the bacteria *Aeromonas salmonicida* (the first fish vaccine). He is known "Father of Fish Vaccination". The first commercially approved fish vaccination was a dead *Yersinia ruckeri* vaccine against enteric redmouth disease that was administered by immersion in 1976.

1.14 Immunostimulants vs Vaccine

Table 1.1 is the comparison of immunostimulants and vaccines (Dawood et al. 2021).

1.15 New Paradigm

1.15.1 Nutrigenomics

Nutrigenomics is a branch of science that integrates bioinformatics, nutrition, genomics, molecular biology, and epidemiology. It links the relationship between nutrients and cellular processes and shows how the dietary components alter the genetic makeup. Although the relationship between nutrition and the immune system is generally known, it is still unclear

TABLE 1.1

The Comparison of Immunostimulant with Vaccines

Immunostimulants (IS)	Vaccine
• More treatments are required as the prophylactic effect is short term • Efficacy of immunostimulants is good • IS possess wide spectrum of activity • Nontoxic with less side effects • No toxic residue accumulation • Positive/no environmental impact • Mainly enhance non-specific immune system before specific immune system matures. • Can be used at any stage of life cycle • Easy to supply to larvae of fish and shrimp • Cost-effective	• One or two treatments are enough as prophylactic effect is long • Efficacy of vaccine is excellent • Vaccine possess limited spectrum of activity • Nontoxic with less side effects • No toxic residues accumulation • No environmental impact • Enhance specific and nonspecific immune response • Cannot be used at any stage of life cycle • Difficult to supply to larvae of fish and shrimp • Costly

how nutrition, animal energy status, and immune function are linked together. The effects of diet on the immune system are becoming more transparent because of emerging omics technologies like transcriptomics (microarray and RNA-seq) and proteomics. Modules of genes can reveal changes in both local (intestinal) and systemic immune function by applying molecular pathway enrichment analysis. Using the omics, researchers can now investigate the effects of dietary manipulations such as fasting, feed additives, and protein replacement on gene expression, protein synthesis, and immune functions. It is a relatively new approach in aquaculture, but the scope it provides to understand the mechanism behind gene alteration through nutrition may lead to more intense research and the development of aquaculture (Samuel and Martin 2017).

1.15.2 Trained Innate Immunity

It is a concept that argues that not only can adaptive immunity provide immunity by memorising the pathogen, but an innate immune response can also recognise the pathogen and adapt to provide an immune response after exposure. It stimulates defence and increases nonspecific resistance to infection. One such example is the prophylactic effects of glucan injection in fish against *Vibrio salmonicida*. Though technological advancements and research have revealed the mechanism responsible for such immune responses to be effective programming of cells like monocytes, NK cells, macrophages, etc. through pattern recognition (MAP kinase dependent), we are still far from knowing the actual effectiveness, mechanism, and potential side effects it may cause. The approach is opening doors of application in various aquaculture fields like brood stocking, larval rearing, and first-feeding fish; however, assessments of this approach using modern tools like transcriptomics, epigenetics, proteomics, and metabolomics are needed (Zhangzuobing et al. 2019).

REFERENCES

Abareethan, M. & Amsath, A. Characterization and evaluation of probiotic fish feed. *International Journal of Pure and Applied Zoology*, 3, 148–153 (2015).

Abarike, E. D. et al. Influences of immunostimulants on phagocytes in cultured fish: A mini review. *Reviews in Aquaculture*, 11, 1219–1227 (2019).

Abdel-Latif, H. M. et al. Health benefits and potential applications of fucoidan (FCD) extracted from brown seaweeds in aquaculture: An updated review. *Fish & Shellfish Immunology*, 122, 115–130 (2022).

Abdel-Tawwab, M. & Hamed, H. S. Antagonistic effects of dietary guava (*Psidium guajava*) leaves extract on growth, hemato-biochemical, and immunity response of cypermethrin-intoxicated Nile tilapia, *Oreochromis niloticus*, fingerlings. *Aquaculture*, 529, 735668 (2020), doi: 735610.731016/j.aquaculture.732020.735668.

Abdel-Tawwab, M., Adeshina, I., Jenyo-Oni, A., Ajani, E. K. & Emikpe, B. O. Growth, physiological, antioxidants, and immune response of African catfish, *Clarias gariepinus* (B.), to dietary clove basil, *Ocimum gratissimum*, leaf extract and its susceptibility to Listeria monocytogenes infection. *Fish & Shellfish Immunology*, 78, 346–354 (2018).

Acar, Ü., İnanan, B. E., Navruz, F. Z. & Yılmaz, S. Alterations in blood parameters, DNA damage, oxidative stress and antioxidant enzymes and immune-related genes expression in Nile tilapia (*Oreochromis niloticus*) exposed to glyphosate-based herbicide. *Comparative Biochemistry and Physiology Part C: Toxicology & Pharmacology*, 249, 109147 (2021), doi: 109110.101016/j.cbpc.102021.109147.

Aliko, V., Qirjo, M., Sula, E., Morina, V. & Faggio, C. Antioxidant defense system, immune response and erythron profile modulation in gold fish, *Carassius auratus*, after acute manganese treatment. *Fish & Shellfish Immunology*, 76, 101–109 (2018).

Bachère, E. Anti-infectious immune effectors in marine invertebrates: Potential tools for disease control in larviculture. *Aquaculture*, 227, 427–438 (2003).

Banaee, M., Mirvagefei, A. R., Rafei, G. R. & Majazi Amiri, B. Effect of sub-lethal diazinon concentrations on blood plasma biochemistry. *International Journal of Environmental Research*, 2(2), 189–198 (2008).

Banaee, M., Tahery, S., Nematdoost Haghi, B., Shahafve, S. & Vaziriyan, M. Blood biochemical changes in common carp (Cyprinus carpio) upon co-exposure to titanium dioxide nanoparticles and paraquat. *Iranian Journal of Fisheries Sciences*, 18(12), 242–255 (2019), doi: 210.22092/ijfs.22019.118174.

Barman, D., Nen, P., Mandal, S. C. & Kumar, V. Immunostimulants for aquaculture health management. *Journal of Marine Science Research and Development*, 3(3), 1–11 (2013).

Bernier, J., Brousseau, P., Krzystyniak, K., Tryphonas, H. & Fournier, M. Immunotoxicity of heavy metals in relation to Great Lakes. *Environmental Health Perspectives*, 103 (Suppl 109), 123–134 (1995), doi: 110.1289/ehp.95103s95923.

Bitsayah, A., Banaee, M. & Nematdoost Haghi, B. Effects of aflatoxin-contaminated feed on immunological parameters of common carp (*Cyprinus carpio*). *Iranian Journal of Toxicology* 12 (11), 7–12 (2018), doi: 10.29252/arakmu.12.1.7.

Cabas, I., Chaves-Pozo, E., Mulero, V. & García-Ayala, A. Role of estrogens in fish immunity with special emphasis on GPER1. *Developmental and Computational Immunology*, 89, 102–110 (2018), doi: 110.1016/j.dci.2018.1008.1001.

Caipang, C. M. A., Suharman, I., Avillanosa, A. L. & Gonzales-Plasus, M. M. Influence of phytogenic feed additives on the health status in the gut and disease resistance of cultured fish. *IOP Conference Series: Earth and Environmental Science*, 695(1), 012024 (2021).

Calder, P. C. et al. A consideration of biomarkers to be used for evaluation of inflammation in human nutritional studies. *British Journal of Nutrition*, 09, S1–S34 (2013).

Carbone, D. & Faggio, C. Importance of prebiotics in aquaculture as immunostimulants. Effects on immune system of *Sparus aurata* and *Dicentrarchus labrax*. *Fish & Shellfish Immunology*, 54, 172–178 (2016).

Chakraborty, S., Shukla, D., Vuddanda, P. R., Mishra, B., & Singh, S. Effective in-vivo utilization of lipid-based nanoparticles as drug carrier for carvedilol phosphate. *Journal of Pharmacy and Pharmacology*, 63(6), 774–779 (2011).

Chauhan, A. & Singh, R. Probiotics in aquaculture: a promising emerging alternative approach. *Symbiosis*, 77, 99–113 (2019).

Cho, H. J. et al. Determination of cyanogenic compounds in edible plants by ion chromatography. *Toxicology Research*, 29, 143–147 (2013), doi:10.5487/tr.2013.29.2.143.

Cobbina, S. J. et al. A multivariate assessment of innate immune-related gene expressions due to exposure to low concentration individual and mixtures of four kinds of heavy metals on zebrafish (*Danio rerio*) embryos. *Fish & Shellfish Immunology*, 47 (42), 1032–1042 (2015), doi: 1010.1016/j.fsi.2015.1011.1003.

Dawood, M. A. Nutritional immunity of fish intestines: Important insights for sustainable aquaculture. *Reviews in Aquaculture*, 13(1), 642–663 (2021).

De Paula Barbosa, A. Anti-inflammatory properties and immunoadjuvant activitiy of Samanea saman extract. *Emirates Journal of Food and Agriculture*, 818–821 (2014).

Dietrich, M. A. et al. 2D-DIGE proteomic analysis of blood plasma reveals changes in immune- and stress-associated proteins following hormonal stimulation of carp males. *Fish & Shellfish Immunology*, 118, 354–368 (2021), doi: 110.1016/j.fsi.2021.1009.1018.

Dohms, J. E. & Saif, Y. M. Guest editorial: Criteria for evaluating immunosuppression. *Avian Diseases*, 28(2), 305–310 (1984).

Estensoro, I., Calduch-Giner, J. A., Kaushik, S., Pérez-Sánchez, J. & Sitjà-Bobadilla, A. Modulation of the IgM gene expression and IgM immunoreactive cell distribution by the nutritional background in gilthead sea bream (*Sparus aurata*) challenged with Enteromyxum leei (Myxozoa). *Fish & Shellfish Immunology*, 33 (32), 401–410 (2012), doi: 410.1016/j.fsi.2012.1005.1029.

Faggio, C. et al. Potential use of polysaccharides from the brown alga*Undaria pinnatifida* as anticoagulants. *Brazilian Archives of Biology and Technology*, 58, 798–804 (2015).

Faggio, C., Pagano, M., Dottore, A., Genovese, G. & Morabito, M. Evaluation of anticoagulant activity of two algal polysaccharides. *Natural Product Research*, 30, 1934–1937 (2016).

Farag, M. R. et al. Immune response and susceptibility of Nile tilapia fish to Aeromonas hydrophila infection following the exposure to Bifenthrin and/or supplementation with Petroselinum crispum essential oil. *Ecotoxicology and Environmental Safety*, 216, 112205 (2021), doi: 112210.111016/j.ecoenv.112021.112205.

Findlay, V. L. & Munday, B. L. The immunomodulatory effects of levamisole on the nonspecific immune system of Atlantic salmon, Salmo salar L. *Journal of Fish Diseases*, 23(6), 369–378 (2000).

Firmino, J. P., Galindo-Villegas, J., Reyes-López, F. E. & Gisbert, E.. Phytogenic bioactive compounds shape fish mucosal immunity. *Frontiers in Immunology*, 12, 695973 (2021).

Ghazy, H. A., Abdel-Razek, M. A. S., El Nahas, A. F. & Mahmoud, S. Assessment of complex water pollution with heavy metals and Pyrethroid pesticides on transcript levels of metallothionein and immune related genes. *Fish & Shellfish Immunology*, 68, 318–326 (2017), doi: 310.1016/j.fsi.2017.1007.1034.

Ghosh, D., Datta, S., Bhattacharya, S. & Mazumder, S. Long-term exposure to arsenic affects head kidney and impairs humoral immune responses of *Clarias batrachus*. *Aquatic Toxicology*, 81 (81), 79–89 (2007), doi: 10.1016/j.aquatox.2006.1011.1004.

Gong, Y., Yang, F., Hu, J., Liu, C., Liu, H., Han, D., ... & Xie, S. Effects of dietary yeast hydrolysate on the growth, antioxidant response, immune response and disease resistance of largemouth bass (Micropterus salmoides). *Fish & Shellfish Immunology*, 94, 548–557 (2019).

Grondel, J. L., Angenent, G. C. & Egberts, E. The influence of antibiotics on the immune system. III. *Investigations on the cellular functions of chicken leukocytes in vitro. Veterinary Immunology and Immunopathology*, 10(14), 307–316 (1985), doi: 310.1016/0165-2427(1085)90020-90020.

Guardiola, F. A. et al. Evaluation of waterborne exposure to heavy metals in innate immune defences present on skin mucus of gilthead seabream (*Sparus aurata*). *Fish Shellfish Immunology*, 45(41), 112–123 (2015), doi: 110.1016/j.fsi.2015.1002.1010.

Guo, H. & Dixon, B. Understanding acute stress-mediated immunity in teleost fish. *Fish and Shellfish Immunology Reports*, 2, 100010 (2021), doi: 100010.101016/j.fsirep.102021.100010.

Guo, H. et al. Evaluation of ammonia nitrogen exposure in immune defenses present on spleen and head-kidney of wuchang bream (*Megalobrama amblycephala*). *International Journal of Molecular Sciences*, 23 (26), 3129 (2022), doi: 3110.3390/ijms23063129.

Guttvik, A. et al. Oral administration of lipopolysaccharide to Atlantic salmon (*Salmo salar L.*) fry. Uptake, distribution, influence on growth and immune stimulation. *Aquaculture*, 214, 35–53 (2002).

Harikrishnan, R., Balasundaram, C. & Heo, M. Impact of plant products on innate and adaptive immune system of cultured finfish. *Fish and Shellfish Immunolgy*, 29, 668–673 (2011).

Harris, J. & Bird, D. J. Modulation of the fish immune system by hormones. *Veterinary Immunology and Immunopathology*, 77, 163–176 (2000).

Hassan, A., Gulzar, S., Javid, H. & Nawchoo, I. A. in *Bacterial Fish Diseases* 87–101 (Academic Press, 2022), doi: 110.1016/B1978-1010-1323-85624-85629.00020-85628.

Hatami, M., Banaee, M. & Nematdoost Haghi, B. Sub-lethal toxicity of chlorpyrifos alone and in combination with polyethylene glycol to common carp (*Cyprinus carpio*). *Chemosphere*, 219, 981–988 (2019), doi: 210.1016/j.chemosphere.2018.1012.1077.

Heo, G., Kim, J., Jeon, B., Park, K. & Ra, J. Effects of FST-Chitosan mixture on cultured rockfish (*Sebastes schlegeli*) and olive flounder (*Paralichthys olivaceus*). *Korean Journal of Veterinary Public Health*, 25, 141–150 (2001).

Hidasi, A. O., Groh, K. J., Suter, M. J.-F. & Schirmer, K. Clobetasol propionate causes immunosuppression in zebrafish (*Danio rerio*) at environmentally relevant concentrations. *Ecotoxicology and Environmental Safety*, 138, 16–24 (2017).

Iftikhar, N., Zafar, R. & Hashmi, I. Multi-biomarkers approach to determine the toxicological impacts of sulfamethoxazole antibiotic on freshwater fish *Cyprinus carpio*. *Ecotoxicology and Environmental Safety*, 233, 113331 (2022), doi: 113310.111016/j.ecoenv.112022.113331.

Ilgová, J., Jedličková, L., Dvořáková, H., Benovics, M., Mikeš, L., Janda, L., ... Kašný, M. A novel type I cystatin of parasite origin with atypical legumain-binding domain. *Scientific Reports*, 7(1), 1–12 (2017).

Ilgová, J., Kavanová, L., Matiašková, K., Salát, J. & Kašný, M. Effect of cysteine peptidase inhibitor of *Eudiplozoon nipponicum* (Monogenea) on cytokine expression of macrophages in vitro. *Molecular and Biochemical Parasitology*, 235, 111248 (2020), doi: 111210.111016/j.molbiopara.112019.111248.

Ilgová, J., Salát, J. & Kašný, M. Molecular communication between the monogenea and fish immune system. *Fish & Shellfish Immunology*, 112, 179–190 (2021), doi: 110.1016/j.fsi.2020.1008.1023.

Jadhav, V. S., Khan, S. I., Girkar, M. M. & Gitte, M. J. The role of immunostimulants in fish and shrimp aquaculture. *Aquaculture Asia*, 11, 24 (2006).

Jahangiri, L. & Esteban, M. Á. Administration of probiotics in the water in finfish aquaculture systems: A review. *Fishes3*, 33 (2018).

Jordanova, M. et al. Effects of heavy metal pollution on pigmented macrophages in kidney of Vardar chub (*Squalius vardarensis* Karaman). *Microscopic Research and Technique*, 80(88), 930–935 (2017), doi: 910.1002/jemt.22884.

Kim, J.-H., Balfry, S. & Devlin, R. H. Disease resistance and health parameters of growth-hormone transgenic and wild-type coho salmon, *Oncorhynchus kisutch*. *Fish & Shellfish Immunology*, 34, 1553–1559 (2013), doi: 10.1016/j.fsi.2013.03.365.

Koller, L. D. A perspective on the progression of immunotoxicology. *Toxicology*, 160, 105–110 (2001).

Kono, K., Salazar-Onfray, F., Petersson, M., Hansson, J., Masucci, G., Wasserman, K., ... & Kiessling, R. Hydrogen peroxide secreted by tumor-derived macrophages down-modulates signal-transducing zeta molecules and inhibits tumor-specific T cell-and natural killer cell-mediated cytotoxicity. *European Journal of Immunology*, 26(6), 1308–1313 (1996).

Liu, J. et al. Early exposure to environmental levels of sulfamethoxazole triggers immune and inflammatory response of healthy zebrafish larvae. *Science Total Environment*, 703, 134724 (2020). doi: 134710.131016/j.scitotenv.132019.134724.

Lu, M., Su, M., Liu, N. & Zhang, J. Effects of environmental salinity on the immune response of the coastal fish Scatophagus argus during bacterial infection. *Fish & Shellfish Immunology*, 124, 401–410 (2022), doi: 110.1016/j.fsi.2022.1004.1029.

Luo, K., Li, X., Wang, L., Rao, W., Wu, Y., Liu, Y., Pan, M., Huang, D., Zhang, W. & Mai, K. Ascorbic acid regulates the immunity, anti-oxidation and apoptosis in abalone Haliotis discus hannaiIno. *Antioxidants*, 10(9), 1449 (2021).

Marshall, J. S., Warrington, R., Watson, W. & Kim, H. L. An introduction to immunology and immunopathology. *Allergy, Asthma & Clinical Immunology*, 14, 49, doi:10.1186/s13223-018-0278-1 (2018).

Mauro, M. et al. The effect of low frequency noise on the behaviour of juvenile *Sparus aurata*. *The Journal of the Acoustical Society of America*, 147, 3795–3807 (2020).

Mauro, M. et al. Humoral responses during wound healing in *Holothuria tubulosa* (Gmelin, 1788). *Comparative Biochemistry and Physiology Part B: Biochemistry and Molecular Biology*, 253, 110550 (2021).

Mauro, M. et al. Haemolymphatic parameters in two aquaculture crustacean species cherax destructor (Clark, 1836) and *Cherax quadricarinatus* (Von Martens, 1868). *Animals* 12, 543 (2022).

Mauro, M. et al. Effects of diclofenac on the gametes and embryonic development of *Arbacia lixula*. *The European Zoological Journal* 89, 535–545 (2022).

Mazini, B. S. M., Martins, G. P., de Castro Menezes, L. L. & Guimarães, I. G. Nutritional feed additives reduce the adverse effects of transport stress in the immune system of Tambaqui (*Colossoma macropomum*). *Fish and Shellfish Immunology Reports*, 3, 100051 (2022), doi: 100010.101016/j.fsirep.102022.100051.

Miura, T. & Mizushima, S. Separation by density gradient centrifugation of two types of membranes from spheroplast membrane of *Escherichia coli* K12. *Biochimica et Biophysica Acta (BBA) – Biomembranes*, 150, 159–161 (1968), doi: 10.1016/0005-2736(68)90020-5.

Mohan, K., Ravichandran, S., Muralisankar, T., Uthayakumar, V., Chandirasekar, R., Seedevi, P. & Rajan, D. K. Potential uses of fungal polysaccharides as immunostimulants in fish and shrimp aquaculture: A review. *Aquaculture*, 500, 250–263 (2019).

Mohiseni, M., Sepidnameh, M., Bagheri, D., Banaee, M. & Nematdust Haghi, B. Comparative effects of shirazi thyme and vitamin E on some growth and plasma biochemical changes in common carp (*Cyprinus carpio*) during cadmium exposure. *Aquaculture Research*, 48(49), 4811–4821 (2017), doi: 4810.1111/are.13301.

Morcillo, P., Cordero, H., Meseguer, J., Esteban, M. & Cuesta, A. In vitro immunotoxicological effects of heavy metals on European sea bass (Dicentrarchus labrax L.) head-kidney leucocytes. *Fish Shellfish Immunol*, 47, 245–254 (2015), doi:10.1016/j.fsi.2015.09.011.

Neidhardt, F. C. *Escherichia coli* and *Salmonella*: Cellular and molecular biology. (1996).

Paulsen, S. M., Engstad, R. E. & Robertsen, B. Enhanced lysozyme production in Atlantic salmon (*Salmo salar L.*) macrophages treated with yeast β-glucan and bacterial lipopolysaccharide. *Fish & Shellfish Immunology*, 11, 23–37 (2001).

Paulsen, S. M., Lunde, H., Engstad, R. E. & Robertsen, B. In vivo effects of β-glucan and LPS on regulation of lysozyme activity and mRNA expression in Atlantic salmon (*Salmo salar L.*). *Fish & Shellfish Immunology*, 14, 39–54 (2003).

Petit, J., Embregts, C. W., Forlenza, M. & Wiegertjes, G. F. Evidence of trained immunity in a fish: Conserved features in carp macrophages. *The Journal of Immunology*, 203, 216–224 (2019).

Pontigo, J. P. & Vargas-Chacoff, L. Growth hormone (GH) and growth hormone release factor (GRF) modulate the immune response in the SHK-1 cell line and leukocyte cultures of head kidney in Atlantic salmon. *General and Comparative Endocrinology*, 300, 113631 (2021), doi: 10.1016/j.ygcen.2020.113631.

Qi, X. Z. et al. Ammonia exposure alters the expression of immune-related and antioxidant enzymes-related genes and the gut microbial community of crucian carp (*Carassius auratus*). *Fish & Shellfish Immunology*, 70, 485–492 (2017), doi: 410.1016/j.fsi.2017.1009.1043.

Qiao, Q., Liang, H. & Zhang, X. Effect of cyanobacteria on immune function of crucian carp (*Carassius auratus*) via chronic exposure in diet. *Chemosphere*, 90 (93), 1167–1176 (2013), doi: 1110.1016/j.chemosphere.2012.1109.1025.

Qiu, W. et al. Evidence linking exposure of fish primary macrophages to antibiotics activates the NF-kB pathway. *Environmental International*, 138, 105624 (2020), doi: 105610.101016/j.envint.102020.105624.

Raa, J. The use of immunostimulatory substances in fish and shellfish farming. *Reviews in Fisheries Science*, 4, 229–288 (1996).

Rashidian, G., Boldaji, J. T., Rainis, S., Prokić, M. D. & Faggio, C. Oregano (*Origanum vulgare*) extract enhances zebrafish (*Danio rerio*) growth performance, serum and mucus innate immune responses and resistance against Aeromonas hydrophila challenge. *Animals*, 11, 299 (2021).

Reverter, M., Bontemps, N., Lecchini, D., Banaigs, B. & Sasal, P. Use of plant extracts in fish aquaculture as an alternative to chemotherapy: Current status and future perspectives. *Aquaculture*, 433, 50–61 (2014).

Rymuszka, A. & Adaszek, Ł. Cytotoxic effects and changes in cytokine gene expression induced by microcystin-containing extract in fish immune cells – An in vitro and in vivo study. *Fish & Shellfish Immunology*, 34, 1524–1532 (2013).

Rymuszka, A. & Sieroslawska, A. Comparative studies on the cytotoxic effects induced by nodularin in primary carp leukocytes and the cells of the fish CLC line. *Toxicon*, 148, 147-115 (2018), doi: 110.1016/j.toxicon.2018.1004.1001.

Sadahira, K. et al. Gossypol induces apoptosis in multiple myeloma cells by inhibition of interleukin-6 signaling and Bcl-2/Mcl-1 pathway. *International Journal of Oncology*, 45(46), 2278–2286 (2014), doi: 2210.3892/ijo.2014.2652.

Sakai, M., Hikima, J.-i. & Kono, T. Fish cytokines: Current research and applications. *Fisheries Science*, 87, 1–9 (2021).

Sakai, M., Okajita, Y., Kobayashi, M. & Kawauchi, H. Increase in haemolytic activity of serum from rainbow trout Oncorhynchus mykissinjected with exogenous growth hormone. *Fish & Shellfish Immunology*, 6(8), 615–617 (1996).

Salinas, I., Zhang, Y.-A. & Sunyer, J. O. Mucosal immunoglobulins and B cells of teleost fish. *Developmental & Comparative Immunology*, 35, 1346–1365 (2011).

Samuel, A. M., Martin, ElżbietaKról, Nutrigenomics and immune function in fish: New insights from omics technologies. *Developmental & Comparative Immunology*, 75, 86–98 (2017), ISSN 0145-305X,10.1016/j.dci.2017.02.024m.

Secombes, C. J. & Belmonte, R. Overview of the fish adaptive immune system. In A. Adams (Ed.), Birkhauser Advances in Infectious Diseases (pp. 35–52). Springer Basel (2016). https://doi.org/10.1007/978-3-0348-0980-1_2.

Segner, H., Bailey, C., Tafalla, C. & Bo, J. Immunotoxicity of xenobiotics in fish: A role for the aryl hydrocarbon receptor (AhR)? *International Journal of Molecular Sciences*, 22(17), 9460 (2021).

Shahbazi, S. & Bolhassani, A. Immunostimulants: Types and functions. *Journal of Medical Microbiology and Infectious Diseases*, 4, 45–51 (2016).

Shameena, S. S. et al. Dose-dependent co-infection of *Argulus* sp. and *Aeromonas* hydrophila in goldfish (*Carassius auratus*) modulates innate immune response and antioxidative stress enzymes. *Fish & Shellfish Immunology*, 114, 199–206 (2021), doi: 110.1016/j.fsi.2021.1004.1026.

Sun, P., Li, D., Dong, B., Qiao, S. & Ma, X. Effects of soybean glycinin on performance and immune function in early weaned pigs. *Archives of Animal Nutrition*, 62, 313–321 (2008).

Szwejser, E. et al. Stress differentially affects the systemic and leukocyte estrogen network in common carp. *Fish & Shellfish Immunology*, 68, 190–201 (2017), doi: 110.1016/j.fsi.2017.1007.1011.

Takx-Köhlen, B. C. M. J. Immunomodulators. *Pharmaceutisch Weekblad*, 14, 245–252 (1992), doi:10.1007/BF01962546.

Tengjaroenkul, B. & Yowarach, S. Efficacy of vaccine combined Freud's complete adjuvant to prevent streptococcosis in Nile tilapia. *KKU Veterinary Journal*, 19, 188–196 (2011).

Van Doan, H. et al. Host-associated probiotics: A key factor in sustainable aquaculture. *Reviews in Fisheries Science & Aquaculture*, 28, 16–42 (2020), doi: 10.1080/23308249.2019.1643288.

Villamil, L. A., Tafalla, C., Figueras, A. & Novoa, B. Evaluation of immunomodulatory effects of lactic acid bacteria in turbot (*Scophthalmus maximus*). *Clinical and Vaccine Immunology*, 9(6), 1318–1323 (2002).

Vizzini, A. et al. Transcriptomic analyses reveal 2 and 4 family members of cytochromes P450 (CYP) Involved in LPS inflammatory response in pharynx of *Ciona robusta*. *International Journal of Molecular Sciences*, 22, 11141 (2021).

Wang, W. et al. Glyphosate induces lymphocyte cell dysfunction and apoptosis via regulation of miR-203 targeting of PIK3R1 in common carp (*Cyprinus carpio* L.). *Fish & Shellfish Immunology*, 101, 51–57 (2020).

Wang, X. et al. Bioaccumulation of manganese and its effects on oxidative stress and immune response in juvenile groupers (*Epinephelus moara* ♀ × *E. lanceolatus* ♂). *Chemosphere*, 297, 134235 (2022), doi: 134210.131016/j.chemosphere.132022.134235.

Watts, M., Munday, B. & Burke, C. Immune responses of teleost fish. *Australian Veterinary Journal*, 79, 570–574 (2001).

Wilson, C. J., Finch, C. E. & Cohen, H. J. Cytokines and cognition—The case for a head-to-toe inflammatory paradigm. *Journal of the American Geriatrics Society*, 50, 2041–2056 (2002).

Yada, T. Growth hormone and fish immune system. *General and Comparative Endocrinology*, 152, 353–358 (2007).

Yada, T. et al. Ghrelin stimulates phagocytosis and superoxide production in fish leukocytes. *Journal of Endocrinology*, 189, 57–65 (2006).

Yang, C., Song, G. & Lim, W. A review of the toxicity in fish exposed to antibiotics. *Comparative Biochemistry and Physiology C: Toxicology & Pharmacology*, 237, 108840 (2020), doi: 108810.101016/j.cbpc.102020.108840.

Zapata, A. G. Cells and tissues of the immune system of fish. The fish immune system. *Organism, Pathogen, and Environment*. Academic Press (1996).

Zhang, Z., Chi, H. & Dalmo, R. A. Trained innate immunity of fish is a viable approach in larval aquaculture. *Journal Frontiers in Immunology*, 10 (2019), doi: 10.3389/fimmu.2019.00042. https://www.frontiersin.org/articles/10.3389/fimmu.2019.00042

Zhou, L. J., Wu, Q. L., Zhang, B. B., Zhao, Y. G. & Zhao, B. Y. Occurrence, spatiotemporal distribution, mass balance and ecological risks of antibiotics in subtropical shallow lake Taihu, China. *Environmental Science: Processes and Impacts*, 18(14), 500–513 (2016), doi: 510.1039/c1036em00062b.

Zou, J. & Secombes, C. J. The function of fish cytokines. *Biology*, 5, 23 (2016).

2

Natural and Synthetic Immunomodulators: Inferences for Stress Responses in Aquaculture Fish

Shubhajit Saha[1], Azubuike V. Chukwuka[2], Nimai Chandra Saha[3], Caterina Faggio[4], and Hamed Mousavi Sabet[5]

[1]*Fisheries and Ecotoxicology Research Laboratory, Department of Zoology, The University of Burdwan, Burdwan, West Bengal, India*

[2]*National Environmental Standards and Regulations Enforcement Agency (NESREA), Nigeria*

[3]*University of Burdwan, Purba Barddhaman, India*

[4]*Department of Chemical, Biological, Pharmaceutical and Environmental Sciences, University of Messina, Messina, Italy*

[5]*Department of Fisheries, Faculty of Natural Resources, University of Guilan, Sowmeh Sara, Iran*

2.1 Introduction

Fish are frequently stressed because they are exposed to several pathogenic agents, including bacteria, fungi, viruses, and parasites transferred on their farms [1]. Disease-related issues have been one of the industry's biggest roadblocks over the past 20 years. The rising prevalence of sickness has caused the spread of infectious organisms between nations. As a result, the shrimp business in various climes has experienced enormous financial losses. Immunostimulants have emerged as the preferred strategy for controlling fish and shrimp infections because there are no effective treatments for these viral diseases [2].

Fish that successfully combat pathogen infection either live or die depending on how well they can stop the spread of the infection. The potency of immune responses of an organism to attenuate the onset and proliferation of pathogens determines outcomes of eventual survival or mortality. Innate immunity and acquired immunity (specific) are two categories of the immune system of fish (non-specific). Both rely on cellular, humoral, and physical defenses against pathogenic microorganisms. The particular immune systems offer a defense against a particular infection by identifying particular antigens on that organism. Innate immune systems offer a collection of defense responses that are on standby to put a wide range of infections under check [3].

The adaptive immune system is highly dependent on the innate immune system as its first line of defense against antigenic assault. Macrophages are thought to be a crucial component of the cellular immune system in fish and work to defend the host by phagocytizing foreign substances, including pathogens. Additionally, macrophages actively participate in the preparation and presentation of phagocytosed antigens, which have an impact on cell-mediated reactions.

Antibiotics are typically used in feed or water to manage bacterial diseases, boost production, and reduce mortalities in fish farms since immune systems may not have the full potential to effectively prevent and eliminate such infections [4]. However, the emergence of antibiotic resistance makes its haphazard use potentially harmful to the ecosystem [5]. As a result, different aquaculture systems have explored and adopted other approaches to enhance or modify the immunological activity to improve fish responses to xenobiotics and pathogenic agents [6].

Immunostimulants are substances that directly interact and activate immune system cells to improve innate or non-specific immune responses when administered to an organism. They are primarily divided into two categories: synthetic (chemical) agents and natural agents, such as those derived from cytokines, polysaccharides, animal or plant extracts, or bacterial preparations [7]. Immunostimulants have been used to successfully control a variety of pathogens in fish and shrimp, including bacteria like *Aeromonas hydrophila*, *A. salmonicida*, *Edwardsiella tarda*, *E. ictaluri*, *Vibrio anguillarum*, *V. vulnificus*, *V. salmonicida*, *Yersinia ruckeri*, and *Streptococcus* spp. Viral pathogens that have been overcome by immunostimulation of fisheries include acute hematopoietic necrosis infections, yellow head disease, Egtved disease, and the Ich parasitic ciliate.

An immunostimulant is a substance that enhances either the immune response or the non-specific defensive mechanism [8]. Aquaculture uses immunostimulants to combat the immunosuppressive effects of stressors [9] as a preventative measure against anticipated seasonal breakouts of known endemic illnesses [10] or as a suppressive measure against latent or sub-

 DOI: 10.1201/9781003361183-3

TABLE 2.1

Immunostimulants Evaluated in Fish and Shrimp (Adapted from Barman, Nen [2])

Immunostimulants	Effects of Immunostimulants	References
Synthetic chemicals	Levamisole, fK.565 (Lactoylretrapeptide from *Streptomyces olvaccogriseus*). Quaternary ammonium compounds (QACs)	
Natural substances		
1. Bacterial derivatives	MOP (Muramyl dipeptide from *Mycobocterlum species*). Lipopolysaccharides (LPS), Freund's complete adjuvant (FCA), EF203 (fermented egg product), Peptidoglucan (from *Brevibacrerium lactofermentum* and *Vibrio sp.*), *Clostridium butyricum* cells, *Achromobacter stenohalis* cells, *Vibrio anguillarum* cells (Vibrio vaccine)	[11]
2. Yeast derivatives	b-1, 3 glucan, b- 1, 6 glucan	
3. Nutritional factors	Vitamin C and E, n-3 fatty acid	
4. Hormones	Growth hormone, Prolactin, triiodothyronine	
5. Cytokines	Interferon, interleukins	
6. Polysaccharides	Chitosan (brook trout), Chitin, Lentinan, Schizophyllan, Oligosaccharide	[12]
7. animal extracts	Fermented products of chicken egg (EF203) on rainbow trout [11]	[11]
8. plant extracts	*Ocimum sanctum* (Tulsi), *Phyllanthus emblica* (Amla), *Azadirachta indica* (Neem), *Solanum trilobatum* (Purple Fruited Pea Eggplant), *Eclipta alba* (Bhringraj), *Zingiber officinale* (Ginger), *Echinacea* (Purple coneflowers), *Allium sativum* (Garlic), *Camellia sinensis* (Green tea), *Cynodon dactylon* (Bermuda Grass), *Achyranthes aspera* (Prickly Chaff Flower, *Nyctanthes arbortristis* (Night-flowering Jasmine)	[12]
9. Fermented vegetable product (FVP)	The phagocytic activities and superoxide generation of peritoneal induced leukocytes were significantly higher in fish Japanese flounder (*Paralichthys olivaceus*) fed the FVP supplemented diet	[13]
10. *Mucor circinelloides* (Fungi)	Gilthead seabream growth was slightly higher in specimens fed *M. circinelloides*-supplemented diets	[14]
11. *Spirulina platensis*	*Spirulina* enhanced responses of bactericidal, phagocytic activity and lysozyme activity in tilapia, *Oreochromis niloticus*	[15]
12. Others	Lactoferrin, Quil A, *Achyranthes aspera* (Herb)	

lethal infections. Additionally, immunostimulants can aid in the recovery from stress-induced immunosuppression states [7]. Table 2.1 provides a summary of the various immunostimulants tested on fish and shrimp.

Chemicals called immunostimulants cause leukocytes to become active [16]. One of the first immunostimulants to be used in animals to increase a particular immune response was adjuvant (FCA), which recorded desirable results in combined administration with fish bacterins injection [8]. Glucose molecule variants i.e., the glucans polymer extracted from plant cell walls, and microbes (fungi, and bacteria), have been documented with a greater level of potency particularly through oral administration for aquaculture fish and shrimp [17]. It is practical to use these various immunostimulants to improve the immunological capacity and disease resistance of fish and shellfish. Numerous drugs are now being used in aquaculture, and research on immune stimulants is ongoing.

The two broad classes of immunomodulators, their peculiar dimensions of contributing to fish health and aspects of their limitations are discussed in the following section.

2.2 Natural Immunomodulators

As already mentioned, fish rely on innate immunity, but immunostimulants significantly improve innate immune responses. To control disease in aquaculture, chemotherapeutics, vaccinations, and antibiotics have been employed. However, the indiscriminate use of antibiotics has led to the emergence of bacterial strains that are resistant to them. In both treatments for parasites, trichlorfon or praziquantel, for instance, have drawbacks such as the emergence of resistance, potential risks to animal health, and environmental pollution [5]. Additionally, a single vaccination only protects against a specific kind of infection, and immunizing young fish is time-consuming and expensive. Sequel to the above-mentioned challenges, the use of natural immunostimulants, i.e., substances of natural origin have been a major alternative approach to ensure sustainable aquaculture (Table 2.1).

Particular plant extracts have been found to activate the immune system in fish cultures, increase growth, increase appetite, and have anti-parasite and antibacterial characteristics. Popular examples of herbal immunostimulants include saponins, phenolics, glycosides, and essential oils. Others include alkaloids, flavonoids, terpenoids, tannins, and steroids [6]. The antistress, antipathogen, appetite-stimulating, growth-promoting, tonicity-enhancing, and immunostimulatory actions of plants and their extracts have been demonstrated in aquacultures [18–20].

Over the past ten years, various aquaculture systems have used medicinal plants and their derivatives extensively as immunostimulants [21]. In both freshwater and marine fish species, herbal medicine has been shown to enhance the immune response and hematological parameters as well as reduce losses from bacteria, viruses, and parasites [22].

The health and physiological status of the fish can be monitored using hematological and non-specific immunological markers [23]. Pourmoghim and Haghighi [24] documented plant extracts that enhanced the plasma levels of total protein, globulin, and albumin compared with the control. These serum proteins function as humoral components and detected increases suggested an improvement in the fishes' non-specific immune reactions. In many nations, immune-boosting herbs are utilized to improve health, boost the body's defenses against infection, and treat and prevent a wide range of diseases [25]. The innate immune system can be modulated by herbal immunostimulants, which are less expensive or less effective than the medications, including synthetic antibiotic drugs currently administered to diseased fish and shellfish [26] (Table 2.2).

TABLE 2.2

Effects of Various Compounds on Immune System of Fish

S. No	Name of Compounds	Immunological Responses	References
1	Polyphenols	• Proliferation of the immunocompetent cells. Including T helper 1 (Th1), natural killer cells, dendritic cells (DCs), and macrophages in secondary lymphoid organs. • Increasing splenic T and 8 cells populations.	[27]
	Flavanols	Suppress on of inducible nitric oxide synthase (INOS), cyclooxygenase-2 (COX-2) as well as cytokines TNF-a, IL-1ß, and IL-6 from macrophage.	[28]
a	Epigalocatechin-3-gallate	• Foxp3 and IL-I0 gene expression increases in Jurkat T cells. • Increases regulatory T cells proliferation in spleen, pancreatic lymph nodes, and mesenteric lymph nodes. Decreases the Th1 and Th 17 populations and lower the expression of T-bet and ROR-yt in animal models of experimental autoimmune encephalomyelitis. • Decreases the Th9 cells proliferation along with decreased expression of the PU-I. • Immunosuppressive and decreases IFN-y production in vitro.	[29]
b	Baicalin	Under TGF- ß stimulus increase the Foxp3 gene expression and also induction of functional Treg from splenic CD4+CD25-T cells in HEK 293 T cells.	[30]
c	Procyanidin CI	Decreases the secretion of TNF-a, IL-I-ß, and IL-6.	[31]
d	Fisecin	Inhibits the expression of IL-6 and TNF-a inTHP-1 monocytes cells exposed to high osmolarity.	[32]
e	Theaflavin	Decreases the expression of IL-6, Monocyte chemoattractant protein-I (MCP- 1) and Intracellular adhesion molecule-I (ICAM-1).	[33]
f	Curcumin analog 3,5-Bls (2-pyridinylmethy lidene)-4-piperidone (EF31)	Inhibits secretion and express on of TNF-a, L-1-, and IL-6.	[34]
g	Propolis	It can act as vaccine adjuvant.	[35]
h	Resveratrol	• Reducing the levels of pro-inflammatory cytokines TNF-a, L-I-. and IL-6 and inflammation. • Increases proliferation CDl9+ cells and immunoglobulin synthesis in 8lymphocytes.	[36]
i	Resveratrol analog RVSA-40	Increases anti-inflammatory cytokine IL-I-ß together with down- regulation of TNF-a and IL-6.	[37]
j	7-0-Methylnaringenin	Down regulate TNF-a. L-6 and IL-1-f expression in a dose-dependent manner.	[38]
2	Pentachlorophenols	• Enhances B and T-lymphocyte blastogenesis. • Decreases antibody titers and delayed type hypersensitivity response and increases induced peritoneal macrophage count.	[39]
3	Hydroquinone	• Inhibits IFN-y production and increases Th2 response mediated allergic diseases. Enhances IL-4 production In CD4+ T cells and also increases age levels in antigen-primed mice. • Blocks the IL-12 production via suppression of NF-Kß activity. • Inhibition of the production of IL-1-ß, IL-2, and NO. • Suppresses the production of TNF-a, IL-1-ß, and IL-6. • Inhibits production of NO and ROS phagocytic uptake of FITC- labeled dextran. • Induces production of costimulatory molecules.	[40]
1	**Organophosphate**		
1.a	Chloropyrifos	• Increases CD26 expression on cytotoxic T cells. • Reduces mitogenesis immune cells in response to concanavillin and phytohemagglutinin. • Amplifies autoantibodies production.	[41]
1.b	Diisopropyl methyl phosphonate	• Inhibits activity of human and murine NK cells and murine cytotoxic T cells.	[42]

TABLE 2.2 (Continued)

Effects of Various Compounds on Immune System of Fish

S. No	Name of Compounds	Immunological Responses	References
1.c	Malathion	• Suppresses of NO production and LPS-induced TNF-alpha generation. • Increases antibody production following immunization with a T-lymphocyte dependent antigen and macrophage function and led to mast cell degranulation.	[43]
2	Carbamate		
2.a.	Aldicarb	• Increases TB lymphocytes population and decreases T4: TB cell ratio. • Increases lymphocytes proliferation and alters macrophage functions by decreasing with IL-I production.	[44]
2.b.	Dimethoate	Decreases total serum lgG and lgM and T cells in the thymus, forming autologous rosettes.	[45]
2.c.	Sodium methyl dithiocarbamate	Inhibits expression of IL- Iα, IL- Iβ, IL- 1B, IL- 12, IFN-y, p35, p40 m-RNA level and macrophage migration inhibitory factor (MIF), whereas increases IL- I 0 m-RNA level.	[46]
2.d.	Mancozeb	• NO production decreases with the in vitro exposure. • Suppresses TNF-y secretion in vitro whereas enhance release detected in ex vivo experiment.	[47]
2.e.	Carbendazim	Decreases B lymphocytes proliferation and serum lgG, lgM, and lgA levels.	[48]
3	Organochlorine		
3.a.	Endosulfan	Decreases lgA and lgG production.	[49]
3.b.	Methoxychlor	Decreases in lgM splenic plaque-forming cell responses, splenic T-cell (CD3+) populations and germinal center (GC) B-cell (CDl9+PNA+) populations.	[50]
3.c.	Hexachlorocyclohexane	Increases population of CD3 (+) CD4 (+) T-lymphocytes and expression of CD45RO (+) on CD4 (+) and COB (+) T-lymphocytes. Decreases CD4(+) CD25(+) T-lymphocytes and level of I L-2 and IFN -y in SLE patients.	[51]
3.d.	o, p‘-Dichlorodiphenyl-trichloroethane (DDT)	Increases the percentages of CD3(+) CD4(+) T-lymphocytes and IL-10 level.	[51]
S. No.	**Metals**	**Immunological Responses**	**References**
1	Cadmium	• Increases production of NO. • Inhibits synthesis of lgE by B cells or PBMCs upon IL-4/aCD40 stimulation and decreases proliferation of B cells or PBMCs. • Increase production of chemoattractant Leukotriene B4 from neutrophils and monocytes.	[52]
2	Lead (Pb)	• Decreases NO production in cytokine-induced cell lines. • Induces production of TNF, IL-6, and IL-12 and decreases IL- I 0 production.	[53]
3	Mercury(Inorganic)	• Increases IL-12, IL- 17, IFN-y, and TNF-a production. • Overexpression of CD86 and HLA-DR and production of TNF and IL-8 in vitro.	[54]
4	Arsenic	• Decreases IL-2, I L-4, IL-5, I L-10, TNF-a, and IFN-y secretion from T cells. • Increases lgG, lgA, and lgE levels.	[55]
S. No.	**PAHs**	**Immunological Responses**	**References**
1	2,3,7,8-Tetrachloro-dibenzo-p-dioxin (TCDD)	• Increases the expression of genes involve in antigen processing and presentation. • It also increases expression of genes of dendritic cell maturation pathways.	[56]
2	Polychlorinated Biphenyls 2,2’,4,4‘,5,5’-hexachlorobiphenyl	• Immunosuppressive. • Deceases CCL22 gene expression.	[57]
3	Polybrominated diphenyl ethers	• Decreases peripheral blood monocytes, NK cell activity and splenic CD4 (+) CDS (+) proliferation. • Increases T and B cell proliferation on mitogen stimulation. • Decreases proinflammatory cytokine production (TNF-y and IL-6).	[58]

The use of natural immunostimulants to boost innate immunity may be able to reduce disease outbreaks in commercial fisheries [59], however, this method is currently disregarded by the scientific community because it is not standardized [60]. Herbal remedies are a more affordable source of medicines, are more accurate than chemotherapeutic agents, and provide a workable solution to every issue that aquaculture is currently facing. Numerous plant compounds with antibacterial, anti-stress, immunostimulant, and growth-promoting properties have a big impact on fish or shrimp larval culture [61]. Proteoglycans, flavonoids, polysaccharides, and phenolics from plants all have a significant impact on the prevention or management of pathogenic microorganisms [62]. Herbal substances have the power to scavenge free radicals and prevent the formation of oxygen anions. Shrimps have successfully employed *Picrorhiza kurroa* as an anti-stress chemical [62] and *Ocimum sanctum* significantly improved *Oreochromis mossambicus*' immune stimulatory effects, and antibody response, and disease resistance against *Aeromonas hydrophila* infection [63]. The serum bactericidal activity and phagocytosis in rohu *Labeo rohita* were likewise boosted by an enriched meal that included *Curcuma longa* and *allium sativum.* In fingerling rohu, *L. rohita* fed with diets enriched with *Achyranthus* at various doses, superoxide anions were produced more frequently, serum lysozyme levels were higher, and serum bactericidal levels were also higher [64]. The phagocytic activity in certain fish is greatly increased when immunostimulants, such as plant extracts, are applied [65].

2.3 Synthetic Immunomodulators

With varying degrees of efficacy, substances of non-biological origin (chemically developed antibiotics) have been adopted as interventions of choice in situations of disease and infections in aquaculture fish and shrimp [66]. Depending on the condition, vaccines against various infections have been produced recently with varying degrees of efficacy. The administration of different substances to enhance or potentiate the innate immune response of farmed fish and shrimp has been a different strategy. These substances, also known as immunostimulants, are thought to be an alluring and effective tool for preventing illnesses in fish and shellfish. Immunostimulants are now widely used for disease management in aquaculture techniques due to their proven positive benefits in a variety of living systems [67].

Levamisole is an anthelmintic drug used to treat nematode infections in both humans and animals (Table 2.1). In vitro, it can activate the immune system. Levamisole increased phagocytic activity, the NBT reaction, and the number of cells that produce antibodies. Levamisole increased lysozyme activity in serum, leukocyte count, accelerated NBT decrease, and phagocytic index of phagocytic cells when given orally [68]. Although Levamisole was used in rainbow trout, changes in hematocrit, leucocrit, or immunoglobulin values were not significant [69]. Findlay et al. [70] have recommended the application of levamisole as an immunostimulant in fish. Rainbow trout treated with a bath treatment containing 5, 10, or 25 g/ml of levamisole over two hours showed resistance to *Y. ruckeri* [71].

Levamisole has been approved by the U.S. Food and Drug Administration to treat helminth infections in ruminants, which has sparked interest in its immunostimulatory potential in fish in the USA and abroad. Levamisole, which is the active laevo-isomer of tetramisole has also been discovered to have the potential as a regulator of carp and rainbow trout immune responses [72]. Both species were observed with up-regulated innate immune responses with demonstrated tolerance to pathogenic microorganisms under experimental conditions [73]. Greater phagocyte activity, macrophage cytokine presence, lymphocyte infiltration following the introduction of mitogens, and generation of antibodies may all be associated with the enhanced protection.

2.4 Immunostimulant Administration Methods

Immune stimulants boost the host's natural capacity for immune responses, enhancing its ability to fight off infections. The natural killer cells, complement, lysozyme, and antibody responses of fish are also stimulated by several immunostimulants [74]. Immunostimulants can be administered primarily in three ways: via injection, submersion, and oral absorption. Immunostimulant injections can result in potent non-specific responses, but they are expensive, labor-consuming, and only suited to large fish with a body weight of more than 10–15 g in an intense aquaculture environment. According to reports, injections offer broad protection against a variety of infections, as demonstrated by the intra-peritoneal injection of glucan into channel catfish, which reduced fish mortality when exposed to *Edwardsiella ictaluri* [75]. It is impossible to vaccinate little fish. Although less expensive than the injection, immersion causes more stress to the fish when handled, making it useful in systems of intensive culture. When introducing sub-adult life-staged fish into field ponds, the prior dipping method is particularly effective. Levamisole-immersed fish, showed increased leukocytes circulation, phagocyte activity, and resistance against *P. damselae* sub sp. piscicida [76].

Oral administration has also been shown to be a cost-effective enhanced innate immune capacity. It is suitable for large-scale aquacultural setup and running. Powdered feed is often fortified with immunostimulant substances are mixed with and coated with fish oil. Bioencapsulation is now being utilized to immunize fish larvae with live-feeding animals during their early larval stages.

2.5 Timing of Administration

Immunostimulants must be used properly and at the appropriate time. [8] suggested that to prevent illness-related losses, immunostimulants should be used before disease outbreaks. Based on various culture systems with varied feeding regimens, effective dose, and exposure duration will become even more complicated [77]. Injections of high doses of glucans (100 mg/kg) in Atlantic salmon resulted in a week without protection, although the full advantages are only felt after 3–4 weeks. Additionally, protecting for only one week at modest injection doses of 2–10 mg/kg [59]. Similar to this, it has been observed that over 30 days but not over 45 days, *Clarias spp.* fed with glucan or oligosaccharides exhibit an increase in the number of NBT-positive cells [78].

2.6 Mode of Action

Immunostimulants work by stimulating the immune systems of living things to boost defenses against pathogen invasion. The method is quite varied or may not be well known, and it is also affected by the type or variety of immune stimulants, the dosage administered, the route of uptake, and exposure duration.

Several of the mechanisms of action are as follows:

- T-lymphocyte stimulators: Levamisole, Freund's Complete Adjuvant (FCA), Glucans, Muramyl dipeptide, FK-565 (Lactoyl tetrapeptide from *Streptomyces olivaceogriseus*).
- Bacterial endotoxins and lipopolysaccharides stimulate B-cells.
- Glucans, Chitin, and Chitosan are macrophage activators.
- Inflammatory agents, such as chemo toxins
- Detergents and sodium dodecyl sulfate, quaternary ammonium compounds (QAC), and saponins are cell membrane modifiers.
- Nutritional considerations include vitamins C and E, as well as omega-3 fatty acids.
- Cytokines include leukotriene and interferon.
- Metal – e.g., Cd
- Mitogens extracted from non-plant organisms

Immune stimulants often increase the capacity of macrophages, complements, lymphocytes, and non-specific cytotoxic cells to phagocytose and kill germs, providing resistance and protection against a variety of illnesses and invasive microbes.

2.7 Environmental Immunomodulators

Environmental immunomodulators, also known as xenobiotics, are manmade or natural substances that frequently have no metabolic function and behave as foreign molecules to an organism [79]. The impact of pesticides, synthetic chemicals, metals, polycyclic aromatic hydrocarbons, mycotoxins, and other substances on the immune system has been demonstrated by epidemiological and animal investigations [80]. These immunomodulators fall into an intriguing category because their existence or activity may unintentionally create physiological changes that undermine ongoing immunostimulant treatments or interventions in aquaculture species.

In addition to immune modulators purposely added to fish to increase disease resistance, contaminated fish food can also unintentionally introduce immune modulators to fish. For instance, polyphenols, which are micronutrients found in typical diets and divided into non-flavonoids (phenolic amine) and flavonoids (phenolic acids). Due to their antioxidant qualities and ability to prevent diseases associated with oxidative stress, such as neurodegenerative diseases [81] polyphenols are frequently utilized in the food sector for applications (color and flavor enrichment). By changing epigenetic processes including DNA methylation, histone protein modification, and post-translational modifications, polyphenols influence the immunological response (activation and differentiation of immune cells) [82].

Fish meal and oil, which are feed items made from smaller, wild pelagic fish such as herring and sardines, are POPs and mercury sources for farmed fish. Both POPs and mercury enter the ocean as a result of air emissions, mostly from the combustion of fossil fuels for POPs and agricultural pesticide that is used for mercury [83]. Farmed salmon tended to have lower levels of POPs and mercury when fish-based feed is obtained from regions of the world. These small pelagic fish when sourced from South America contain fewer contaminants, compared to smaller pelagic fish from Northern Europe, where higher levels of contaminants have been recorded [84]. Some fish processors use active carbon filters to eliminate POPs from fish meal and oil to reduce pollution in farmed salmon [85]. Reducing the use of fish meal and oil in aquaculture is another method for lowering POPs in farmed fish. In addition to nutritional sources, exposure to alterations in ambient environmental elements inside aquaculture installations may cause immunological modulation in fish. In addition to intensifying farming practices, global warming reduces the amount of oxygen on the water's surface [23]. Limited quantities of dissolved oxygen in the culture water were revealed to be the cause of immune suppression in tilapia at higher temperatures [22]. According to these research results, hypoxia in fish modulates the immune system.

According to several studies, exposure to metals alters several immune system characteristics in the host, making them more susceptible to allergies and infections [86]. Additionally, Arunkumar and Rajasekaran [87] hypothesized that exposure to metals triggers immunological responses in fish, either directly by binding the tertiary structures of physiologically active molecules or inadvertently by acting as stressors that change corticosteroid concentrations. An essential immunological reaction is a phagocytosis, in which harmful particles are taken up by intracellular vacuoles and eliminated. Pillet and Lesage [88] noted significant increases in phagocytic activity in Zn-exposed harbor seals (*Phoca vitulina*) and gray seals (*Halichoerus grypus*). It has been established that exposure to aquatic toxins can increase the production of lysozyme, a crucial element of the innate immune response in fish [89]. Observable increases in the serum lysozyme were also noted in rainbow trout, *Oncorhynchus mykiss*, subjected to Zn, cadmium, mercury, and metal combinations, according to Sanchez-Dardon, Voccia [90]. When exposed to Zn, Mozambique tilapia, *Oreochromis mossambicus*, showed an increase in both phagocytic and lysozyme activity, according to Çelik, Kaya [91]. Given that lysozyme is a trustworthy indicator for assessing how changes in the environment affect fish's innate immunity [89]. The ectothermic characteristic of fish, which directly affects their physiology and immune function, has been attributed to this interaction between environmental events and fish immunomodulation [92].

The blue mussel, *Mytilus edulis*, showed a considerable increase in phagocytosis at a higher temperature, according to Parry and Pipe [93]. Furthermore, nickel-exposed Pacific abalone, *Haliotis discus hannai*, showed enhanced lysozyme activity in high water temperatures (Figure 2.1).

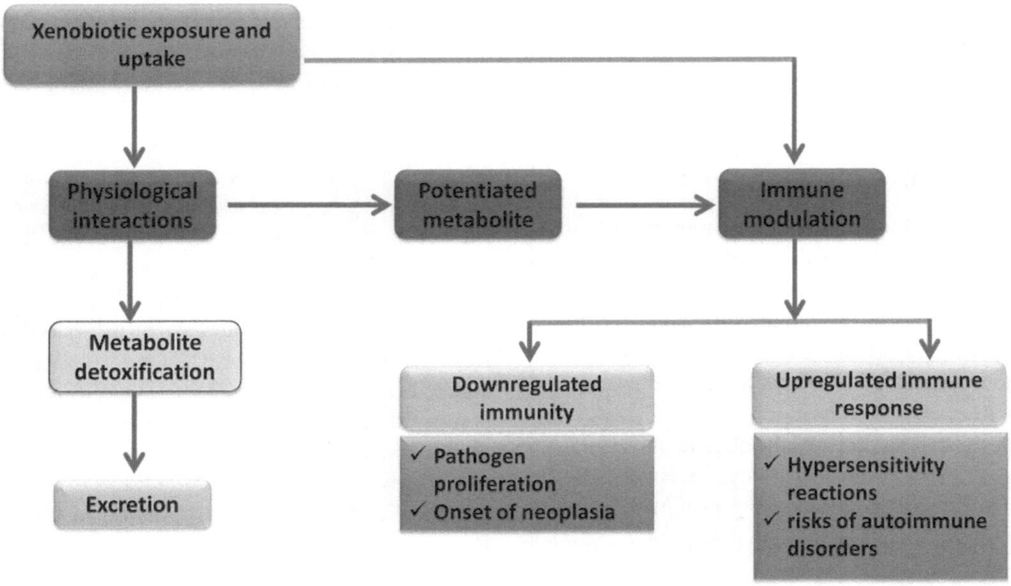

FIGURE 2.1 Pathways of xenobiotic-induced immunomodulation (adapted from Sajid and Cameotra [79]).

2.8 Inferences for Stress Responses in Aquaculture Fish

Regarding innate immunity, fish are a particularly significant animal group because fish take longer to build an adaptive or acquired immune response than mammals, which may make fish suffer more severely from innate immunity impairment. In fish, the idea of species-specific immunological competence has been proven. This suggests that fish can adjust their immunocompetence to the seasonal variations in environmental conditions or can respond favorably to specific immunomodulatory substances like other fish species, but responses are often relative [94]. Numerous innate immunity genes have also been identified in teleost fish, and it has been found that while some of their signaling pathways are preserved from fish to mammals, others have unique characteristics and exhibit significantly more species variation [95]. Monitoring fish health can also identify the state of the aquatic system or medium in which fish are reared [96]. As a consequence, fish innate immunity can serve as a warning system for potential dangers to fish populations [97]. Furthermore, it is economically significant to comprehend how toxicants affect fish's innate immunity. The well-being of other fish species, such as forage species, which may be crucial to the survival of a fishery, is as important as the target species' health is important. Financial success in aquaculture may be hampered by a weakened innate immune system in farm populations. Even though the immune system is a defense mechanism against pathogens and cancer, xenobiotic metabolism produces activated metabolites that cause severe pathophysiological conditions such as immune disruptions, neurological disruptions, cardiovascular, nephrological, and ultimately disturbances in the body's homeostasis. Xenobiotics can disrupt the immune system's normal function, and these immunocompromised conditions increase the likelihood of pathogen encounter and tumor formation. When the immune system is suppressed after being exposed to xenobiotics, it makes people more susceptible to bacterial, viral, and parasite illnesses as well as more likely to develop cancer. On the other side, the stimulation of immune responses might start autoimmune processes or cause overall tissue damage (Figure 2.2).

Different molecular event(s) in immune cells are triggered as a result of the diversity in physio-chemical characteristics of xenobiotics. The main mechanism behind changes in immunological response is the activation of oxidative stress by the creation of ROS. Adduct formation and ROS generation are two aspects of the polycyclic aromatic hydrocarbons' carcinogenic process. These pathways also impair immune cells' abilities and cause the breakdown of their genetic integrity. Dietary polyphenols (secondary metabolites) function as antioxidants and reduce the generation of ROS in immune cells, but they also influence epigenetic pathways that control immunological response. Pesticides used on a global scale have a variety of immunomodulatory effects. ROS production, caspase-3 activation, mitochondrial malfunction, and changes in the transcription factor activity all play a role in the mechanism of pesticide-induced immunotoxicity. Aryl hydrocarbon receptor-bound xenobiotics like dioxins, biphenyls, and dibenzofurans influence immunological response (AhR). Th17 cells produce more xenobiotic metabolizing enzymes and IL22 when they are bound to the AhR receptor. Protein-reactive xenobiotic compounds are produced by the action of xenobiotic metabolizing enzymes. Some of these xenobiotic compound's function as neoantigens and encourage T-cell activation. Increased IL-22 production and T-cell activation worsen autoimmune disease symptoms. Additionally, exposure to heavy metals in the environment and at work increases the likelihood of inflammation and pathogen contact by significantly disrupting immunological responses.

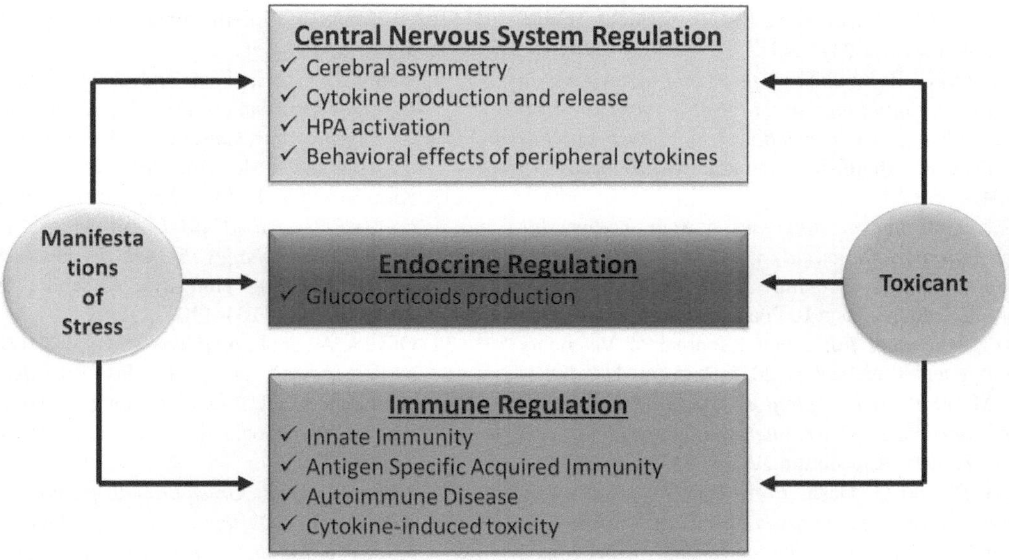

FIGURE 2.2 Toxicants and stress via immunomodulation.

2.9 Conclusion

For the development of novel intervention techniques that increase aquaculture production, it is crucial to sufficiently grasp the dynamics of fish immune systems. This includes understanding the immune system's plasticity and how it helps to better regulate and optimize rearing settings for improved fish welfare and performance. Furthermore, managers must exercise comprehensive management control and comprehend how environmental immunomodulator sources can confound efforts to maintain healthy aquaculture fisheries.

Acknowledgment

The authors express immense gratitude to Dr. Tarun Mandal, Principal of Sundarban Hazi Desarat College, South 24 Parganas, for providing permission to write this chapter on immunomodulators.

Disclosure statement: The authors declare that they have no conflicts of interest.

REFERENCES

1. Walker, P.J. and J.R. Winton, *Emerging viral diseases of fish and shrimp*. Veterinary Research, 2010. **41**(6): p. 51.
2. Barman, D., et al., *Immunostimulants for aquaculture health management*. Journal of Marine Science and Research Development, 2013. **3**(3): p. 1–11.
3. Balfry, S.K. and D.A. Higgs, *Influence of dietary lipid composition on the immune system and disease resistance of finfish*. Nutrition and Fish Health, 2001: p. 213–234.
4. Finlay, B.B. and G. McFadden, *Anti-immunology: evasion of the host immune system by bacterial and viral pathogens*. Cell, 2006. **124**(4): p. 767–782.
5. Cabello, F.C., *Heavy use of prophylactic antibiotics in aquaculture: a growing problem for human and animal health and for the environment*. Environmental Microbiology, 2006. **8**(7): p. 1137–1144.
6. Chakraborty, S.B. and C. Hancz, *Application of phytochemicals as immunostimulant, antipathogenic and antistress agents in finfish culture*. Reviews in Aquaculture, 2011. **3**(3): p. 103–119.
7. Sakai, M., *Current research status of fish immunostimulants*. Aquaculture, 1999. **172**(1-2): p. 63–92.
8. Anderson, D.P., *Immunostimulants, adjuvants, and vaccine carriers in fish: applications to aquaculture*. Annual Review of Fish Diseases, 1992. **2**: p. 281–307.
9. Jeney, G. and D.P. Anderson, *Glucan injection or bath exposure given alone or in combination with a bacterin enhance the non-specific defence mechanisms in rainbow trout (Oncorhynchus mykiss)*. Aquaculture, 1993. **116**(4): p. 315–329.
10. Nikl, L., T.T. Evelyn and L. Albright, *Trials with an orally and immersion-administered β-1, 3 glucan as an immunoprophylactic against Aeromonas salmonicida in juvenile chinook salmon Oncorhynchus tshawytscha*. Diseases of Aquatic Organisms, 1993. **17**(3): p. 191–196.
11. Wang, S.-H. and J.-C. Chen, *The protective effect of chitin and chitosan against Vibrio alginolyticus in white shrimp Litopenaeus vannamei*. Fish & Shellfish Immunology, 2005. **19**(3): p. 191–204.
12. Bairwa, M.K., et al., *Animal and plant originated immunostimulants used in aquaculture*. Journal of Natural Product and Plant Resources, 2012. **2**(3): p. 397–400.
13. Nya, E.J. and B. Austin, *Use of dietary ginger, Zingiber officinale Roscoe, as an immunostimulant to control Aeromonas hydrophila infections in rainbow trout, Oncorhynchus mykiss (Walbaum)*. Journal of fish diseases, 2009. **32**(11): p. 971–977.
14. Ragap, H.M., R.H. Khalil and H.H. Mutawie, *Immunostimulant effects of dietary Spirulina platensis on tilapia Oreochromis niloticus*. Journal of Applied Pharmaceutical Science, 2012. **2**(2): p. 26–31.
15. Rodríguez, A., et al., *The effect of dietary administration of the fungus Mucor circinelloides on non-specific immune*

responses of gilthead seabream. Fish & shellfish immunology, 2004. **16**(2): p. 241–249.
16. Lundén, T. and G. Bylund, *The influence of in vitro and in vivo exposure to antibiotics on mitogen-induced proliferation of lymphoid cells in rainbow trout (Oncorhynchus mykiss)*. Fish & Shellfish Immunology, 2000. **10**(5): p. 395–404.
17. Philip, R., et al. *Immunostimulants – Source, diversity, commercial preparations and mode of application* in *Nature Workshop on Aquaculture Medicine*. 2001.
18. Valladão, G., S. Gallani, and F. Pilarski, *Phytotherapy as an alternative for treating fish disease*. Journal of Veterinary Pharmacology and Therapeutics, 2015. **38**(5): p. 417–428.
19. Reverter, M., et al., *Use of plant extracts in fish aquaculture as an alternative to chemotherapy: current status and future perspectives*. Aquaculture, 2014. **433**: p. 50–61.
20. Ramudu, K.R. and G. Dash, *Prevalence of monogenean parasites on Indian major carps in bheries of West Bengal*. International Journal of Chemical and Biochemical Science, 2013. **4**: p. 13–21.
21. Vaseeharan, B. and R. Thaya, *Medicinal plant derivatives as immunostimulants: an alternative to chemotherapeutics and antibiotics in aquaculture*. Aquaculture International, 2014. **22**(3): p. 1079–1091.
22. Harikrishnan, R., et al., *Immunomodulatory effect of Withania somnifera supplementation diet in the giant freshwater prawn Macrobrachium rosenbergii (de Man) against Aeromonas hydrophila*. Fish & Shellfish Immunology, 2012. **32**(1): p. 94–100.
23. Adeogun, A., et al., *Haematological and serum biochemical profile of the blue crab, Callinectes amnicola from two tropical lagoon ecosystems*. African Journal of Biomedical Research, 2015. **18**(3): p. 233–247.
24. Pourmoghim, H., M. Haghighi, and M.S. Rohani, *Effect of dietary inclusion of Origanum vulgare extract on non-specific immune responses and hematological parameters of rainbow trout (Oncorhynchus mykiss)*. Bulletin of Environment, Pharmacology and Life Sciences, 2015. **4**(3): p. 33–39.
25. Devasagayam, T. and K. Sainis, *Immune system and antioxidants, especially those derived from Indian medicinal plants*. Indian Journal of Experimental Biology, 2002. **40**(6): p. 639–655.
26. Yeh, S.-T. and J.-C. Chen, *White shrimp Litopenaeus vannamei that received the hot-water extract of Gracilaria tenuistipitata showed earlier recovery in immunity after a Vibrio alginolyticus injection*. Fish & Shellfish Immunology, 2009. **26**(5): p. 724–730.
27. John, C.M., et al., *Immunomodulatory activity of polyphenols derived from Cassia auriculata flowers in aged rats*. Cellular Immunology, 2011. **271**(2): p. 474–479.
28. González, R., et al., *Effects of flavonoids and other polyphenols on inflammation*. Critical Reviews in Food Science and Nutrition, 2011. **51**(4): p. 331–362.
29. Wong, C.P., et al., *Induction of regulatory T cells by green tea polyphenol EGCG*. Immunology Letters, 2011. **139**(1-2): p. 7–13.
30. Yang, J., et al., *Identification of Baicalin as an immunoregulatory compound by controlling TH17 cell differentiation*. PLoS One, 2011. **6**(2): p. e17164.
31. Byun, E.-B., et al., *The procyanidin trimer C1 inhibits LPS-induced MAPK and NF-κB signaling through TLR4 in macrophages*. International Immunopharmacology, 2013. **15**(2): p. 450–456.
32. Kim, H.J., S.H. Kim and J.-M. Yun, *Fisetin inhibits hyperglycemia-induced proinflammatory cytokine production by epigenetic mechanisms*. Evidence-Based Complementary and Alternative Medicine, 2012. **2012**.
33. Kim, S. and Y.-E. Joo, *Theaflavin inhibits LPS-induced IL-6, MCP-1, and ICAM-1 expression in bone marrow-derived macrophages through the blockade of NF-κB and MAPK signaling pathways*. Chonnam Medical Journal, 2011. **47**(2): p. 104–110.
34. Olivera, A., et al., *Inhibition of the NF-κB signaling pathway by the curcumin analog, 3, 5-Bis (2-pyridinylmethylidene)-4-piperidone (EF31): anti-inflammatory and anti-cancer properties*. International immunopharmacology, 2012. **12**(2): p. 368–377.
35. Fischer, G., et al., *Green propolis phenolic compounds act as vaccine adjuvants, improving humoral and cellular responses in mice inoculated with inactivated vaccines*. Memórias do Instituto Oswaldo Cruz, 2010. **105**: p. 908–913.
36. Tao, K., et al., *Effects of resveratrol on the treatment of inflammatory response induced by severe burn*. Inflammation, 2015. **38**(3): p. 1273–1280.
37. Capiralla, H., et al., *Identification of potent small-molecule inhibitors of STAT 3 with anti-inflammatory properties in RAW 264.7 macrophages*. The FEBS Journal, 2012. **279**(20): p. 3791–3799.
38. Soromou, L.W., et al., *Regulation of inflammatory cytokines in lipopolysaccharide-stimulated RAW 264.7 murine macrophage by 7-O-methyl-naringenin*. Molecules, 2012. **17**(3): p. 3574–3585.
39. Blakley, B., et al., *Effect of pentachlorophenol on immune function*. Toxicology, 1998. **125**(2-3): p. 141–148.
40. Kim, A.R., et al., *Hydroquinone modulates reactivity of peroxynitrite and nitric oxide production*. Journal of Pharmacy and Pharmacology, 2005. **57**(4): p. 475–481.
41. Thrasher, J.D., G. Heuser and A. Broughton, *Immunological abnormalities in humans chronically exposed to chlorpyrifos*. Archives of Environmental Health: An International Journal, 2002. **57**(3): p. 181–187.
42. Li, Q., et al., *The by-products generated during sarin synthesis in the Tokyo sarin disaster induced inhibition of natural killer and cytotoxic T lymphocyte activity*. Toxicology, 2000. **146**(2-3): p. 209–220.
43. Ayub, S., J. Verma and N. Das, *Effect of endosulfan and malathion on lipid peroxidation, nitrite and TNF-α release by rat peritoneal macrophages*. International Immunopharmacology, 2003. **3**(13-14): p. 1819–1828.
44. Dean, T.N., et al., *Immunosuppression by aldicarb of T cell responses to antigen-specific and polyclonal stimuli results from defective IL-1 production by the macrophages*. Toxicology and Applied Pharmacology, 1990. **106**(3): p. 408–417.
45. Aly, N. and K. El-Gendy, *Effect of dimethoate on the immune system of female mice*. Journal of Environmental Science and Health, Part B, 2000. **35**(1): p. 77–86.
46. Pruett, S.B., et al., *Sodium methyldithiocarbamate inhibits MAP kinase activation through toll-like receptor 4, alters cytokine production by mouse peritoneal macrophages, and suppresses innate immunity*. Toxicological Sciences, 2005. **87**(1): p. 75–85.

47. Chung, A.-H. and M.-Y. Pyo, *Effects of mancozeb on the activities of murine peritoneal macrophagesin vitro andex vivo*. Archives of Pharmacal Research, 2005. **28**(1): p. 100–105.
48. Singhal, L.K., et al., *Down regulation of humoral immunity in chickens due to carbendazim*. Toxicology In Vitro, 2003. **17**(5-6): p. 687–692.
49. Nishimoto, S., et al., *Abnormal response induced by pesticides on mammalian immune system*. Interdisciplinary Studies on Environmental Chemistry-Environmental Research in Asia, 2009. **211**: p. 217.
50. Hayashi, K., et al., *Immunotoxicity of the organochlorine pesticide methoxychlor in female ICR, BALB/c, and C3H/He mice*. Journal of Immunotoxicology, 2013. **10**(2): p. 119–124.
51. Dar, S.A., et al., *Alterations in T-lymphocyte sub-set profiles and cytokine secretion by PBMC of systemic lupus erythematosus patients upon in vitro exposure to organochlorine pesticides*. Journal of Immunotoxicology, 2012. **9**(1): p. 85–95.
52. Ramirez, D.C., et al., *Biphasic effect of cadmium in noncytotoxic conditions on the secretion of nitric oxide from peritoneal macrophages*. Toxicology, 1999. **139**(1-2): p. 167–177.
53. Valentino, M., et al., *Effect of lead on the levels of some immunoregulatory cytokines in occupationally exposed workers*. Human and Experimental Toxicology, 2007. **26**(7): p. 551–556.
54. Migdal, C., et al., *Responsiveness of human monocyte-derived dendritic cells to thimerosal and mercury derivatives*. Toxicology and Applied Pharmacology, 2010. **246**(1-2): p. 66–73.
55. Biswas, R., et al., *Analysis of T-cell proliferation and cytokine secretion in the individuals exposed to arsenic*. Human and Experimental Toxicology, 2008. **27**(5): p. 381–386.
56. Vorderstrasse, B.A. and N.I. Kerkvliet, *2, 3, 7, 8-Tetrachlorodibenzo-p-dioxin affects the number and function of murine splenic dendritic cells and their expression of accessory molecules*. Toxicology and Applied Pharmacology, 2001. **171**(2): p. 117–125.
57. Harper, N., et al., *Immunosuppressive activity of Polychiorinated biphenyl mixtures and congeners: non-additive (antagonistic) interactions*. Toxicological Sciences, 1995. **27**(1): p. 131–139.
58. Hennigar, S.R., J.L. Myers and A.R. Tagliaferro, *Exposure of alveolar macrophages to polybrominated diphenyl ethers suppresses the release of pro-inflammatory products in vitro*. Experimental Biology and Medicine, 2012. **237**(4): p. 429–434.
59. Robertsen, B., et al., *Enhancement of non-specific disease resistance in Atlantic salmon, Salmo salar L., by a glucan from Saccharomyces cerevisiae cell walls*. Journal of Fish Diseases, 1990. **13**(5): p. 391–400.
60. Ponni, A., *Ethanotherapeutics. An approach to study the antimicrobial evaluation of some siddha and ayurvedic medicines for the treatment of infectious disease*. 2002, M. Sc. Dissertation, MS University, Tamil Nadu, India.
61. Citarasu, T., et al., *Influence of the antibacterial herbs, Solanum trilobatum, Andrographis paniculata and Psoralea corylifolia on the survival, growth and bacterial load of Penaeus monodon post larvae*. Aquaculture International, 2003. **11**(6): p. 581–595.
62. Citarasu, T., *Herbal biomedicines: a new opportunity for aquaculture industry*. Aquaculture International, 2010. **18**(3): p. 403–414.
63. Logambal, S., S. Venkatalakshmi and R. Dinakaran Michael, *Immunostimulatory effect of leaf extract of Ocimum sanctum Linn. in Oreochromis mossambicus (Peters)*. Hydrobiologia, 2000. **430**(1): p. 113–120.
64. Rao, Y.V., et al., *Effect of Achyranthes aspera on the immunity and survival of Labeo rohita infected with Aeromonas hydrophila*. Fish & Shellfish Immunology, 2006. **20**(3): p. 263–273.
65. Rao, Y.V. and R. Chakrabarti, *Dietary incorporatino of Achyranthes aspera seed influences the immunity of common carp Cyprinus carpio*. Indian Journal of Animal Sciences, 2005. **75**(9): p. 1097.
66. Labh, S.N. and S.R. Shakya, *Application of immunostimulants as an alternative to vaccines for health management in aquaculture*. International Journal of Fisheries and Aquatic Studies, 2014. **2**(1): p. 153–156.
67. Jadhav, V.S., et al., *The role of immunostimulants in fish and shrimp aquaculture*. Aquaculture Asia, 2006. **11**(3): p. 24.
68. Siwicki, A.K., *Immunostimulating influence of levamisole on nonspecific immunity in carp (Cyprinus carpio)*. Developmental & Comparative Immunology, 1989. **13**(1): p. 87–91.
69. İSPİR, Ü. and M. Dörücü, *A study on the effects of levamisole on the immune system of rainbow trout (Oncorhynchus mykiss, Walbaum)*. Turkish Journal of Veterinary & Animal Sciences, 2005. **29**(5): p. 1169–1176.
70. Findlay, V., D. Zilberg and B. Munday, *Evaluation of levamisole as a treatment for amoebic gill disease of Atlantic salmon, Salmo salar L*. Journal of Fish Diseases, 2000. **23**(3): p. 193–198.
71. Ispir, U., *Prophylactic effect of levamisole on rainbow trout (Oncorhynchus mykiss) against Yersinia ruckeri*. Pesquisa Veterinária Brasileira, 2009. **29**: p. 700–702.
72. Kajita, Y., et al., *The immunomodulatory effects of levamisole on rainbow trout, Oncorhynchus mykiss*. Fish Pathology, 1990. **25**(2): p. 93–98.
73. Gannam, A.L. and R.M. Schrock, *Immunostimulants in fish diets*. Journal of Applied Aquaculture, 1999. **9**(4): p. 53–89.
74. Johansson, M. and K. Soderhall, *Cellular immunity in crustaceans and the proPO system*. Parasitology Today, 1989. **5**(6): p. 171–176.
75. Smith, V.J., K. Söderhäll and M. Hamilton, *β 1, 3-Glucan induced cellular defence reactions in the shore crab, Carcinus maenas*. Comparative Biochemistry and Physiology Part A: Physiology, 1984. **77**(4): p. 635–639.
76. Chen, D. and A. Ainsworth, *Glucan administration potentiates immune defence mechanisms of channel catfish, Ictalurus punctatus Rafinesque*. Journal of Fish Diseases, 1992. **15**(4): p. 295–304.
77. Galeotti, M., D. Volpatti and G. Jeney. *The nature of non-specific immune response of Sea bass (Dicentrarchus labrax) to Pasteurella piscicida following bath exposure to levamisole*. in *Seventh International Conference Palma de Mallorca, Spain: European Assoc. Fish Pathologists*. 1995.

78. GH, K. and S. YL, *Vibriosis resistance induced by glucan treatment in tiger shrimp (Penaeus monodon).* Fish Pathology, 1994. **29**(1): p. 11–17.
79. Sajid, M. and S.S. Cameotra, *Immunomodulatory effect of xenobiotics*. Journal of Environmental of Immunology and Toxicology, 2016. **3**(1): p. 1–8.
80. Lehmann, I., U. Sack and J. Lehmann, *8 Metal Ions Affecting the Immune System.* Metal Ions in Toxicology: Effects, Interactions, Interdependencies, 2015: p. 157–186.
81. Renaud, J., et al., *Epigallocatechin-3-gallate, a promising molecule for Parkinson's disease?* Rejuvenation Research, 2015. **18**(3): p. 257–269.
82. Cuevas, A., et al., *Modulation of immune function by polyphenols: possible contribution of epigenetic factors.* Nutrients, 2013. **5**(7): p. 2314–2332.
83. Lohmann, R., et al., *Global fate of POPs: current and future research directions.* Environmental Pollution, 2007. **150**(1): p. 150–165.
84. Ikonomou, M., et al., *Flesh quality of market-size farmed and wild British Columbia salmon.* Environmental Science & Technology, 2007. **41**(2): p. 437–443.
85. Hardy, R.W. and C. Lee, *Aquaculture feed and seafood quality.* Bulletin of Fisheries Research and Development Agency, 2010. **31**: p. 43–50.
86. Bernier, J., et al., *Immunotoxicity of heavy metals in relation to Great Lakes.* Environmental Health Perspectives, 1995. **103**(suppl 9): p. 23–34.
87. Arunkumar, R.I., P. Rajasekaran and R.D. Michael, *Differential effect of chromium compounds on the immune response of the African mouth breeder Oreochromis mossambicus (Peters).* Fish & Shellfish Immunology, 2000. **10**(8): p. 667–676.
88. Pillet, S., et al., *In vitro exposure of seal peripheral blood leukocytes to different metals reveal a sex-dependent effect of zinc on phagocytic activity.* Marine Pollution Bulletin, 2000. **40**(11): p. 921–927.
89. Bols, N.C., et al., *Ecotoxicology and innate immunity in fish.* Developmental & Comparative Immunology, 2001. **25**(8-9): p. 853–873.
90. Sanchez-Dardon, J., et al., *Immunomodulation by heavy metals tested individually or in mixtures in rainbow trout (Oncorhynchus mykiss) exposed in vivo.* Environmental Toxicology and Chemistry: An International Journal, 1999. **18**(7): p. 1492–1497.
91. Çelik, E.Ş., et al., *Effects of zinc exposure on the accumulation, haematology and immunology of Mozambique tilapia, Oreochromis mossambicus.* African Journal of Biotechnology, 2013. **12**(7).
92. Le Morvan, C., D. Troutaud and P. Deschaux, *Differential effects of temperature on specific and nonspecific immune defences in fish.* The Journal of Experimental Biology, 1998. **201**(2): p. 165–168.
93. Parry, H. and R. Pipe, *Interactive effects of temperature and copper on immunocompetence and disease susceptibility in mussels (Mytilus edulis).* Aquatic Toxicology, 2004. **69**(4): p. 311–325.
94. Mandiki, S., et al., *Immune Status and Immunomodulation in Percid Fish*, in *Biology and Culture of Percid Fishes.* 2015, Springer. p. 761–797.
95. Zhou, Q., et al., *Effect of dietary vitamin C on the growth performance and innate immunity of juvenile cobia (Rachycentron canadum).* Fish & Shellfish Immunology, 2012. **32**(6): p. 969–975.
96. Zelikoff, J., et al., *Biomarkers of immunotoxicity in fish: from the lab to the ocean.* Toxicology Letters, 2000. **112**: p. 325–331.
97. Zelikoff, J.T., *Biomarkers of immunotoxicity in fish and other non-mammalian sentinel species: predictive value for mammals?* Toxicology, 1998. **129**(1): p. 63–71.

3

Immunomodulators: Mode of Action

Subramaniam Sivakumar[1], C. Shanmuga Sundaram[2], Maderi Velayutham Dassprakash[1], and Rantham Subramaniam Venkatesan[3]
[1]*Department of Biochemistry, Sri Sankara Arts and Science College, Enathur, Kanchipuram, Tamilnadu, India*
[2]*Department of Microbiology, Hindustan College of Arts and Science, Kelambakkam, Chennai, Tamilnadu, India*
[3]*Lakshmi Bangaru Arts and Science College, Melmaruvathur, Chengalpet, Tamilnadu, India*

3.1 Introduction

Aquaculture is blooming as one of the most promising and viable industries for keeping pace with the indispensable requirement for animal protein, providing nutritional and healthy food to humans, particularly those living in regions where livestock availability is relatively scarce. With every move toward intensification of increasing production in aquaculture methods, there is a rise of the stress level in the environment as well as the fish. Hence, disease outbreak is being increasingly identified as one of the most important risks to the aquaculture production in most of the countries. The disease control in aquaculture is mostly achieved by the use of chemical compounds and antibiotics. Due to side effects and resistance development, the development of non-antibiotic and environment friendly agents is the need of the present scenario in the field of aquaculture.(1) At this juncture, the current studies are directed towards another promising area i.e., the applications of immunomodulators to increase the immune response and protection against the diseases affecting aquaculture industry.(2)

Immunomodulators are those intrinsic or extrinsic substances which regulate or alter the competency, duration, type or scope of the immune response.(3) The initial immune response to antigenic stimulation induces specific antibody synthesis and production of active effector cells. The nature, magnitude, and specific form is determined by an array of modulatory processes which orchestrate this response. Multiple pathways of regulation are used by the competent immune system, depending on the appropriate response to the organism's need.(4)

Biological response modifiers (BRMs) modify the animal responses by aiding or retrieving effector mechanisms or agents of host defense, reducing the dangerous components of the host's reaction, exchange the defective effector or mediator mechanisms with natural compounds or synthetic derivatives, aiding the host response to vaccines and chemotherapeutic agents, reducing transformation and increasing differentiation of cancerous cells, or by promoting the ability of the host to tolerate damage by cytotoxic pathways.(5)

The following agents are considered as biological response modifiers (BRM)(3)

- Immunoaugmenting agents
- Immunosuppressive agents
- Interferons and interferon inducers
- Lymphokines and monokines
- Growth factors
- Thymic factors
- Tumor antigens
- Antitumor antibodies
- Antitumor effector cells
- Maturation factors
- Recombinant genes rendering normal cells less susceptible to cytotoxic effects of chemotherapeutic agents
- T cell immunoregulators

Probably due to the restricted knowledge of the fish's immune response, non-specific immunomodulators have been used in limited numbers and the method of their application in aquaculture. Along with the rapid growth of aquaculture sector over the last 30 years, an associated increase in disease problems also occurs as result of rapid expansion and amongst other factors like high stocking densities. Immunomodulators have been used as dietary additives to improve feed efficiency, weight gain, and disease resistance in order to maintain fish health and to improve fish growth performance of immunomodulators in aquaculture industry.(6)

3.2 Lymphoid Organs

All the major primary and secondary lymphoid organs present in fish as in all vertebrates, except lymphatic nodules and bone

DOI: 10.1201/9781003361183-4

marrow.[7] The head, kidney, or pronephros has hematopoietic functions, and unlike in higher vertebrates, it is the immune organ involved in phagocytosis, antigen processing, production of IgM, and immune memory through melanomacrophage centers. The thymus, situated near the opercular cavity in teleosts, produces T cells involved in allograft rejection, stimulation of phagocytosis, and antibody production by B cells.[8] The spleen, another lymphoid organ, plays a vital role in trapping and processing of antigens.[9] Melanomacrophage centers are present for clearance of ingested material and can be surrounded by immunoglobulin-positive cells, especially after immunization. Proliferation of granular cells has also been observed in association with ellipsoids and melanomacrophage centers after the immunization of fish.[10–12]

3.3 Immune Response

The immune system is essential for the living organisms to protect them from diseases. Immunity can be divided into two types, namely the innate immune response and the adaptive immune response. The innate immune response is the non-specific type of defense against infection, which are provided by physical barriers and cellular response. The adaptive immune response is the specific type of defense against pathogens.

3.3.1 Innate Immune Response

Through specific receptors like toll receptors (TLRs) of the innate immune system, it identifies conserved molecular patterns common to pathogens such as DNA, RNA, lipopolysaccharides, polysaccharides, and peptidoglycans. This type of recognition may lead to the removal of pathogens, for example by phagocytosis, or may induce adaptive immune responses.[13] The innate immune system possesses different types cells with a diverse array of functions. The phagocytic cells remove pathogens by engulfing and degrading them. Other cells can stimulate the migration of immune cells to the site of infection by producing and secreting the cytokines and chemokines and which further directs the immune response to clear the pathogens.[14]

3.3.2 Adaptive Immune Response

The adaptive immune system recognizes foreign bodies by means of two receptors of B and T lymphocytes, the B cell receptor (BCR), and the T cell receptor (TCR), respectively. Adaptive immunity is attained by variety of mechanisms. It acts by the antigen exposure and elicit immunological memory. The immunological memory property is the basis of vaccine development.[15,16] The effector phase of adaptive response generally starts three to five days after infection and is capable of recognizing specific antigens, which increases the immune response level in both speed and magnitude for the future exposures.[17] The vital players of the adaptive immune response are the B and T lymphocytes. Once the B-cells recognize antigens through BCR, they are activated. After activation, they are capable of differentiating into plasma cells and memory cells. Plasma cells (effector B-cells) can secrete antibodies whereas memory B-cells will play important role for the successive infection by the same pathogen. Once the T-cells recognize antigens through TCR by the antigen presenting cells (APCs), they are activated. Activated T cells differentiate into either helper T (TH) lymphocytes or cytotoxic T (TC) lymphocytes. Helper T cells are capable of activating other immune cells such as B cells and macrophages through cytokines, while cytotoxic T cells upon activation are able to lyse the infected or self-altered cells.[12,17]

3.4 Mechanisms of Immunomodulators

There has been a growing interest in the last decade in the use of plant, herb, and algae extracts as immunomodulators in dietary supplements for fish.[18,19] Several studies in mammals have shown that plants extract modulate TLRs as evidenced by the use of natural extracts for centuries by many civilizations.[20] Some algae and plants derivatives can down-regulate gene expression [21,22] and induction [23,24] of TLRs, regulating the TLR/NF-kB signalling pathway.[25] Some other extracts promote the proinflammatory response[26] and enhance TLR9 and IRF7 activation and in turn inducing Interferon-β (IFN-β) protein expression.[27] They also elicit the M1 and M2 macrophage responses through the TLR/NF-kB signaling pathway.[28]

To date, lipopolysaccharides (LPS), peptidoglycan (PGN), and beta-glucans are the most commonly used PAMPs in fish immunomodulator diets. In mammals, TLR4 recognized LPS, a component of the outer membrane of gram-negative bacteria, and forms a complex with co-receptors lymphocyte antigen 96 (MD2), and CD14 on the cell surface that leads to the activation of multiple signaling components and the subsequent production of proinflammatory cytokines.[29]

The mechanism of action of many immunomodulators depends upon their interaction with TLRs, a class of proteins that plays a key role in the innate immune system as shown in Figure 3.1.[30] Their mode of action is a very specialized pathway that can identify endogenous microorganisms and microbial products as well as induce the specific immune response for each antigen.[31] In fish, at least 11 types of TLRs have been discovered, each with different functions and ligands.[18] The TLR-2, TLR-5, and TLR-9 receptors were highly expressed in zebrafish (*Danio rerio*) in addition to normal TLRs.[32] The eleven types of TLR homologs were also identified in Japanese flounder (*Paralichthys olivaceus*).[33] From the growing use of innovative technologies for knowledge and evaluation of the immune system, such as proteomics analyses and transcriptomics to evaluate the modulation of the immune system, studies on TLRs have opened a new research area with high potential in the field of fish immunology.[34]

TLRs are localized on the cell surface and in the endosome for ligand recognition. A member of the TLR1 family, TLR14, dimerizes with other TLR1 family molecule to recognize lipoproteins. Tiger puffer TLR3 and TLR22 recognize dsRNA. Teleost TLR4 genes have only been reported in Cypriniformes. However, the homologs of the TLR4 co-receptors, CD14 and MD2, are lacking in teleosts. Soluble TLR5 (TLRS5) evolved in the teleost lineage after it separated from the mammalian lineage. Rainbow trout TLRS5 acts synergistically with membrane-bound TLR5 to activate NF-κB activation upon recognition of bacterial flagellin. Teleost TLR7 and TLR8

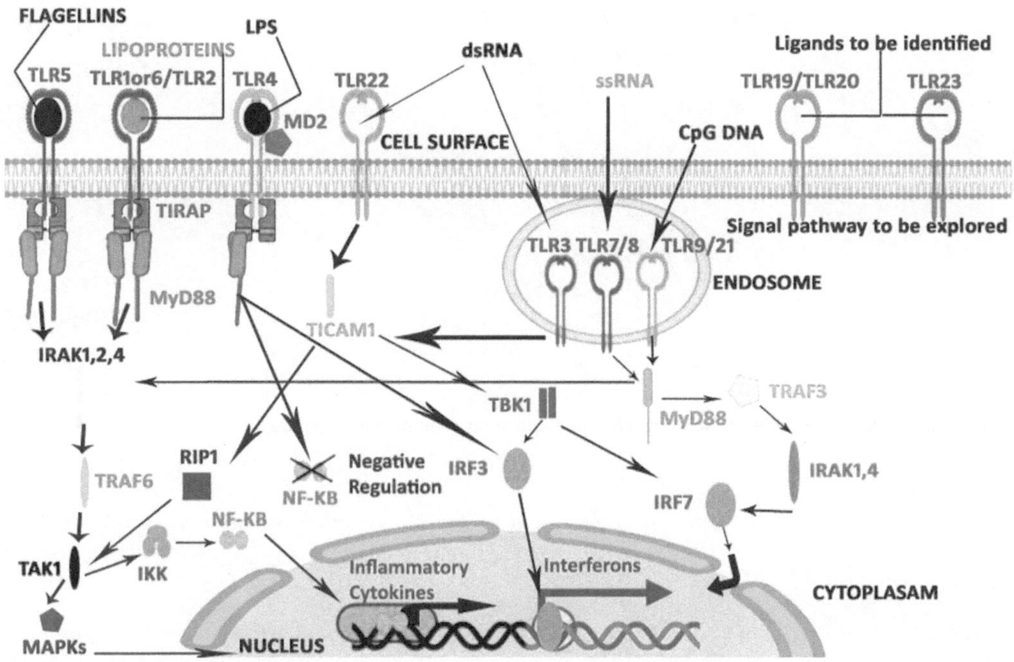

FIGURE 3.1 A putative model of the fish TLR system[18,30,35].

Abbreviations used are IKK - IκB kinase, IL – Interleukin, IRAK - IL-1R-associated kinase, IRF - Interferon regulatory factor, MAPKs - Mitogen-activated protein kinases, MD-2 - Myeloid differentiation factor 2, MyD88 - Myeloid differentiation primary response protein 88, NF-κB - Nuclear factor kappa B, RIP1 - Receptor-interacting protein1, TAK1 - TGF- β-activated kinase 1, TANK - TRAF family-member-associated NF-κB activator, TBK1 TANK-binding kinase 1, TGF – Tumour growth factor, TIR – Toll-interleukin-1 receptor domain, TRIF, TICAM1 – TIR - domain-containing adaptor inducing IFN- γ,TIRAP or Mal TIR domain-containing adaptor protein, TNFR – Tumor necrosis factor receptor, TRAF – TNFR-associated factor.

recognizes sRNA and its analogues, such as the imidazoquinolines. Japanese flounder TLR9 recognizes the CpG DNA analogue, CpG ODN. Chicken TLR21 was also reported to recognize CpG DNA, suggesting a functional similarity with teleost TLR21 as a CpG DNA receptor. Among the teleost-specific TLRs, both TLR19 and TLR20 have been found in zebrafish, whereas only TLR20 was found in channel catfish. TLR19 and TLR20 have a shorter leucine-rich repeats (LRR) containing domain than the other TLR family members. TLR23 is unique to the puffer fish lineage. The function of these teleost-specific TLRs needs to be assessed further to identify novel TLR functions. Regarding the TIR domain-containing adaptors, the universal adaptor, MyD88, is evident in many teleost species. TICAM-1 seems to be essential for the signalling of teleost TLR3 and TLR22. Zebrafish TICAM-1 associates with RIP1 to activate the NFkB cascade. The direct interaction of zebrafish TICAM-1 and TBK1 is reported, however, IRF3/IRF7 activation by the TICAM-1-dependent pathway is not clear in teleosts. TIRAP association with TLR2 and TLR4 seen in mammals is also seen in teleosts. However, TICAM-2, which is involved in the MyD88- independent pathway in mammals, has also not been found in any teleost species. In addition, zebrafish TLR4 negatively regulates NF-κB, unlike mammalian TLR4. This suggests an impairment of LPS recognition by TLR4 in teleosts. IFN production induced via stimulation of salmonid TLR7 and TLR8 has been reported. It may indicate the conservation of the MyD88- dependent pathway among teleost TLR7 family members. Among the IRAK homologs identified in teleosts, IRAK2 is absent from the genome of tiger puffer and zebrafish. However, other IRAK homologs (IRAK1 and IRAK4) are conserved in the teleost lineage. These IRAK homologs may play a pivotal role in signaling via the MyD88-dependent pathway. Immune kinases, including TAK1, IKK, and MAPKs have been identified in teleosts. These kinases activate transcription factors to produce inflammatory cytokines.[18]

3.4.1 Immune Response Followed by TLR-PAMPs Pathway

Once the pathogenic microbes and their products are recognized by the APCs through the TLR-PAMPs recognition process, the pathogenic agents are removed or neutralized by any one of the following immune responses, namely eliminating intracellular pathogens, targeting extracellular parasitic worms, and intoxicating extracellular microbes. The mechanisms were explained through pictorial representations in Figures 3.2, 3.3, and 3.4.[36,37]

3.4.2 Role of Fish Cytokines and Other Factors in Elimination of Intracellular Pathogens

Interleukin-12 is secreted in large amounts by antigen presenting cells (APCs) after their activation by PAMPs and damage/danger-associated molecular patterns (DAMPs). IL-12 induces IFN-γ and IFN-γrel expression by NK cells and by developing T-cells.[38,39] When TH naive cells receive strong TCR signals in the presence of Interleukin-12 and IFN-γ, T-box expressed in T cells (T-bet) or T-box transcription factor (TBX21) expression

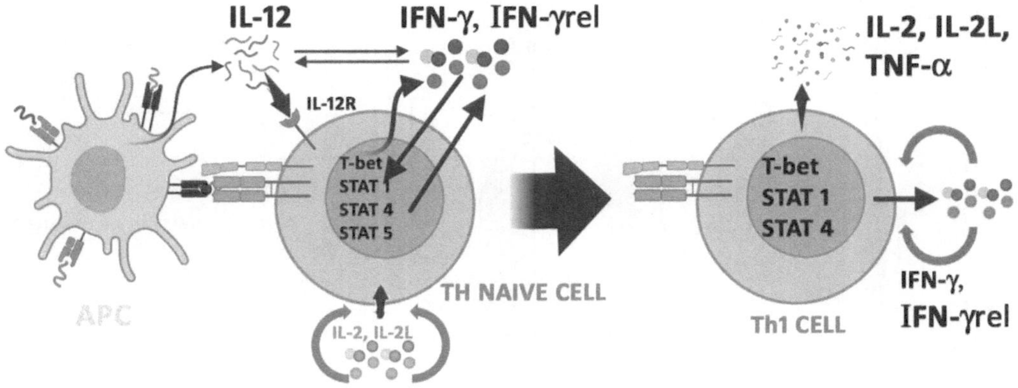

FIGURE 3.2 Potential fish cytokines and other factors involved in elimination of intracellular pathogens.

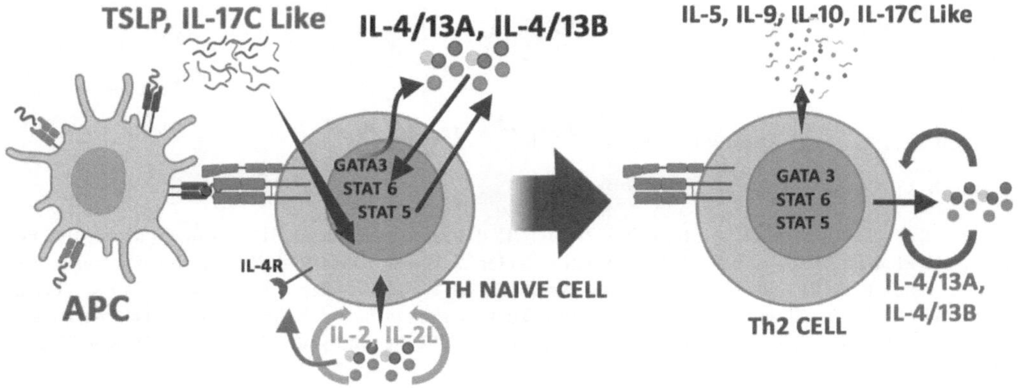

FIGURE 3.3 Potential fish cytokines and other factors involved in the removal of extracellular parasitic worms.

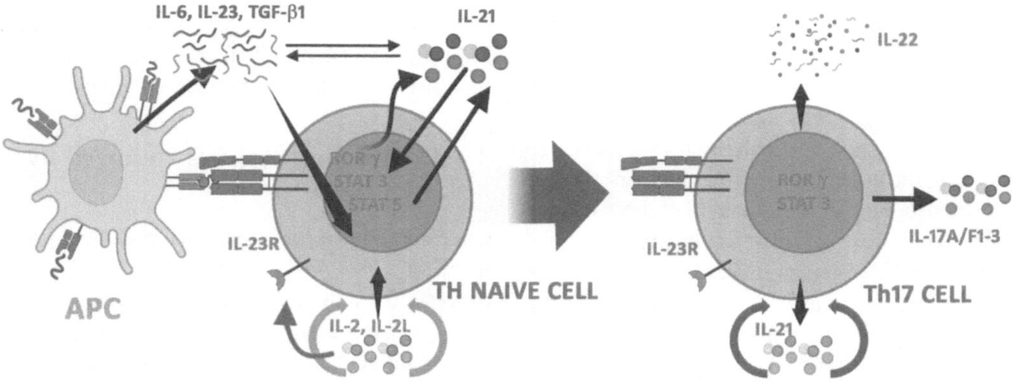

FIGURE 3.4 Potential fish cytokines and other factors involved in the removal of extracellular microbes.

is induced by IFN-γ/signal transducer and activator of transcription (STAT)1 and IL-12/STAT4 signaling.[40] T-bet induces IFN-γ and IFN-γrel production, thereby upregulating T-bet expression and fortify the Th1 phenotype in a positive feedback loop, and increase the expression of IL-12R.[41] In addition, IL-2 and IL2L, produced by T cells after TCR activation, regulates Th1 differentiation via STAT-5-dependent upregulation of IL-12R expression.[42] By secretion of IFN-γ, IFN-γrel, tumour necrosis factor-α(TNF-α), IL2L, and IL-2, Th1 cells help eliminate intracellular pathogens.[37,43]

3.4.3 Role of Fish Cytokines and Other Factors in Removal of Extracellular Parasitic Worms

Cytokines such as thymic stromal lymphopoietein (TSLP) and IL-17C-like cytokine from non-lymphoid cellular sources are essential for Th2 cell differentiation.[44,45] IL-2 and IL2L produced by TCR signaling activate STAT5 and induce IL-4/13 A, IL-4/13B and IL-4R expression. Th2 cells require an IL-4/13 A or IL-4/13B/STAT6 signal for development.[46,47] The initial sources of IL- 4/13 A and IL-4/13B can be from basophils, mast cells, NK cells, follicular Th cells and TH naive cells. IL-4/13 A or IL-4/13B binds to IL-4R and activates STAT6 to control the expression of IL-4 target genes including the Th2 master regulator GATA-binding protein-3 (GATA-3), which is responsible for the expression of Th2 effector cytokines.[40] Th2 cells, by releasing the cytokines IL-4/13 A, IL-4/13B, IL-5, IL-9, IL-10, and IL-17C, are specialized to elicit immune reactions against extracellular parasitic worms.[37,44,48–50]

3.4.4 Role of Fish Cytokines and Other Factors in Removal of Extracellular Microbes

IL-6, IL-21, IL-23, and TGF-β1 are the major signaling cytokines involved in Th 17 cell differentiation, and RORγ is the master regulator.[51–54] TGF-β1 and IL-6 in combination with a strong TCR signal cause naive TH cells to express the IL-23R and to induce retinoid- related orphan receptors γ (RORγ), as well as to produce IL-17A/F1-3 and IL-21.[55,56] IL-6, IL-21 and IL-23 activate STAT3 and induce the expression of RORγ and effector cytokines.[57] IL-2 or IL2L activate STAT5, which compete with STAT3, down-regulate IL-23R expression and block differentiation to the Th17 lineage.[58,59] Th17 cells, by the production of IL-17A/F1- 3, IL-21, and IL-22, induce the expression of numerous chemokines and antimicrobial peptides in epithelial cells and fibroblasts, which are important for the neutrophil-mediated immune reactions against extracellular microbes.[37,54,56,60]

3.5 Actions of Different Immunomodulators

The best-known immunomodulators are cytokines, glucans, hormones, vitamins, minerals, lipopolysaccharides (LPS), peptidoglycans, synthetic compounds (levamisole; FK-565), polysaccharides (chitin, chitosan), and plant extracts that may increase the immune responses of fish.[35] Only few of them are discussed in this chapter.

3.5.1 Cytokines

Immunomodulators can stimulate immune response by increasing lymphocyte proliferation, antibody production and regulating cytokine levels.[61] Cytokines play a vital role in both the innate and adaptive immune responses. All major cytokine families have been shown to existing fish and appear to have similar roles to those observed in mammals. However, multiple forms exist in several fish species which may result in variations in the function that cytokines play compared to their counterparts in other species.[62] The type II cytokines of fish listed along with their immune functions in Table 3.1.[63] The innate immune system stimulated by immunomodulators derived from rice, human milk, and soybean tryptic hydrolysates. Chalamaiah et al. (2018) have reviewed about immunomodulators from food protein.[64] Subhan et al. (2017) proved that peptides could downregulate the expression of cytokines from fish scale collagen *in vitro*.[65] Thus, peptides isolated from fish scale collagen had a useful effect as an anti-inflammatory agent.[66] Figure 3.5 depicts the mechanism of action of cytokines as immunomodulators.[67] Details of fish cytokines are given in Chapter 10.

3.5.1.1 Chemokines

Chemokines are a superfamily of approximately 40 different cytokines which promote the recruitment of immune cells to the infection region. Chemokines exhibit their activity by binding to the G-protein-linked receptors. The chemokine superfamily consists of four distinct subgroups and are designated as CXC (or a), CC (or b), CX3C (or d), and C (or g), which are defined by the arrangement of the first two cysteine amino acid residues within their peptide sequences.[68]

3.5.1.2 Transforming Growth Factor Beta (TGF-beta)

In various leukocyte lineages including lymphocytes, granulocytes, macrophages, NK cells, and dendritic cells; TGF-beta regulates cell development, proliferation, differentiation, migration, and survival since it is a pleiotropic cytokine. TGF-beta1 is to suppress autoimmunity and maintain immune tolerance in the mammalian immune system. TGF-beta1 suppresses Th1 and Th2 cell proliferation, whereas it promotes T regulatory cell generation by inducing Foxp3 expression, thus

TABLE 3.1
Biological Activity of Fish Type II Cytokines[63]

Sl. No.	Cytokine	Human Homologue	Receptors	Immune Functions
1.	IL-10	IL-10	IL-10R1/CRFB12, IL10R2/CRFB4	Suppress immune responses, inhibit inflammation, promote T cell proliferation, memory B-cells and IgM production
2.	IL-22	IL-22	Not characterized	Activate antimicrobial peptide genes and antibacterial immunity
3.	Group I IFN (IFN-a, d, e)	Type I IFN	IFNAR1/CRFB5, IFNAR2/CRFB1	Induce expression of antiviral effectors; promote apoptosis, regulate inflammation
4.	Group I IFN (IFN-b, c, f)	Type I IFN	IFNAR1/CRFB5, IFNAR2/CRFB2	Induce expression of antiviral effectors; promote apoptosis
5.	IFN-γ	Type II IFN	IFN-γR1/CRFB13, IFN-γR2/CRFB6	Activate macrophages, enhance antigen presentation, promote Th1 cytokine expression
6.	IFN-γrel	Type II IFN	CRFB17, IFN-γR2/CRFB6	Regulate antibacterial and antiviral immunity

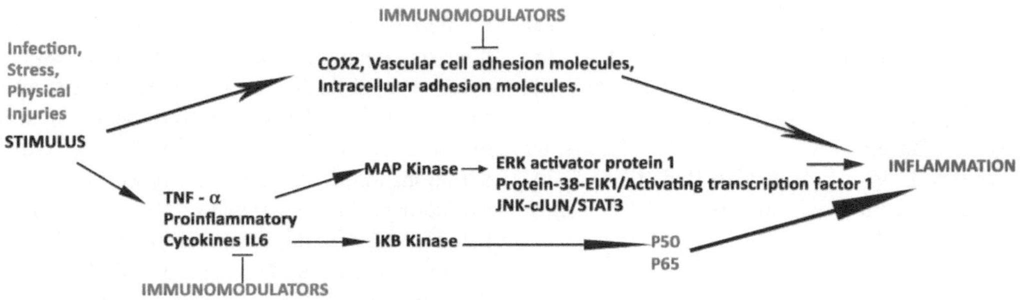

FIGURE 3.5 Mechanism of action of cytokines as immunomodulators[67].

Abbreviations used are COX2 – Cyclooxygenase2, ERK – Extracellular signal-regulated kinase, EIK – Ets-like molecule 1, ETS – E26 transformation-specific protein, JNK – c-Jun N-terminal kinases, cJUN – Proto-oncogene of jun (ju-nana, the Japanese word for 17) gene AP-1 Transcription Factor Subunit.

TGF-beta1 as a potent immunosuppressive agent. On the other hand, TGF-beta induces immune responses by promoting the Th17 cells generation. Therefore, in mammals, the regulatory roles of TGF-beta as a positive or negative control agent in immunity are widely recognized. In teleost, some research works have revealed that TGF-beta1 also exerts powerful immune-suppressing effects on activated leukocytes. For example, TGF-beta1 significantly blocks TNF alpha-induced activation of macrophage in common carp and goldfish, but promotes the proliferation of the fibroblast cell line CCL71 of goldfish. In grass carp, TGF-beta1 suppresses LPS/PHA-stimulated proliferation of peripheral blood lymphocyte in contrast with the stimulatory effect of TGF-beta1 alone in the same cells.[69] Similar phenomenon was also detected during leukocyte migration under TGF-beta1 treatment, with or without LPS challenges in red sea bream.[70] These studies not only define TGF-beta1 as an immune modulator in teleost, but also proved that during the evolution of vertebrates, TGF-beta1 may have retained similar functions in animal immunity.[12,53]

3.5.1.3 Tumor Necrosis Factor Alpha

Tumor necrosis factor alpha (TNF alpha) plays an important role in various host responses, including cell proliferation, differentiation, apoptosis, and necrosis. TNF alpha can induce either apoptosis or NF-kB-mediated survival, depending on the cellular status. TNF alpha induces powerful anti-microbial responses, including killing infected cells, inducing apoptosis, up-regulating diverse host response genes and inhibiting intracellular pathogen replication. In several bony fish, including carp, gilthead seabream, Japanese flounder, turbot, catfish, tilapia, rainbow trout, and goldfish, TNF alpha has been identified, cloned, and characterized. The occurrence of multiple isoforms of TNF alpha in some teleost species has been identified. The high constitutive expression of TNF alpha gene in different tissues of healthy fish and its relatively poor up-regulation by immune challenge *in vitro* and *in vivo are* also reported. The *in vitro effects* of TNF alpha on phagocyte activation in rainbow trout, gilthead seabream, turbot, and goldfish were also demonstrated. The intraperitoneal injection of recombinant TNF alpha in gilthead seabream promotes the recruitment of phagocytes to the injection site and further increases the respiratory burst phenomenon. The endothelial cells are the target cells of fish TNF alpha and involved in the recruitment of immune cells to the inflammatory region rather than their activation. The differential expression of fish TNF alpha has been observed in rainbow trout leucocytes and have shown an increased response to various pro-inflammatory stimuli, such as human recombinant TNF alpha, zymosan, lipopolysaccharides and muramyl dipeptide as a peptidoglycan component of bacteria. However, infectious pancreatic necrosis virus (IPNV)-mediated up-regulation of TNF alpha controls both the RIP1 (receptor-interacting protein-1)/ROS-mediated secondary necrosis pathway and the Bad/Bid-mediated apoptotic pathway.[12,51]

3.5.1.4 Interferons (IFNs)

Interferons are involved in modulating innate and adaptive immune systems and are also responsible for cellular resistance against viral pathogens. The IFNs are mainly classified into two groups namely type I and type II.[71] Type I IFN includes the classical IFN alpha/beta, which is induced by most of the animal cells, whereas type II IFN is only composed of a single gene called IFN gamma and is produced by natural killer (NK) cells and T lymphocytes in response to interleukin-12 (IL-12), IL-18, mitogens, and antigens. Structurally, both interferons types belong to the class II alpha-helical cytokine family, but have different three-dimensional structures and bind to different receptors.[12]

From Atlantic salmon, two IFNs (IFN alpha 1 and IFN alpha 2) have been cloned and characterized with respect to promoter, gene structure, sequence, antiviral activity and induction of ISGs. Salmon IFN alpha 1, similar to mammalian IFN alpha/beta and IFN lambda, induces both Mx and ISG15 proteins in TO cells. Furthermore, salmon IFN alpha1 induces potent antiviral activity against the IPNV under in vitro conditions, but this effect has not been observed in the *in vivo* studies, i.e., Atlantic salmon challenged intraperitoneally with IPNV.[12]

Three types of type I IFNs have been discovered in rainbow trout namely rtIFN1, rtIFN2, and rtIFN3. The three rtIFNs can show a differential expression in cells and tissues. The IFN1 (rtIFN1) and IFN2 (rtIFN2) display high sequence similarity to Atlantic salmon IFN alpha 1 and IFN alpha2, which contain two cysteine residues. However, rtIFN3 contains four cysteine residues, which further confirms the relationship between fish

IFNs and mammalian IFN alpha. Both recombinant rtIFN1 and rtIFN2 have been shown to increase the expression of Mx and inhibit VHSV replication in RTG-2 cells, whereas recombinant rtIFN3 was a poor inducer of Mx and antiviral activity.(72) The three trout IFNs have different functions in fish immune system, which paved the way for further research.(73) The IFN gamma has been identified in several fish species, including Atlantic salmon and rainbow trout. Fish and mammalian IFN gamma are similar in exon-intron structure and display gene synteny. However, a second IFN gamma subtype named IFN gamma rel was identified in some fish species, which is different from the classical IFN gamma. Carp and rainbow trout IFN gamma have several common functions with mammalian IFN gamma, including phagocytosis of bacteria in macrophages, nitric oxide production and the ability to enhance respiratory burst activity. However, fish IFN gamma induces antiviral activity against both IPNV and the salmon alpha virus (SAV) in salmon cell lines.(12)

3.5.2 Beta Glucans

Glucans are potent immunomodulators in both mammals and fish. In fish, they can activate macrophages, increasing their capacity to kill pathogens. They have also been shown to reinforce other innate immune factors such as lysozyme and complement activities. When administered orally, glucans have shown a protective effect in many fish species and against different bacterial pathogens including *Vibrio salmonicida, Aeromonas salmonicida*, and *Pasteurella piscicida*.(74) The administration of glucans can increase a protection against *Aeromonas salmonicida* in trout; *A. salmonicida* in Chinook salmon; *Aeromonas hydrophila* in blue gourami; *Edwardsiella ictalurid* in catfish; *E. tarda* in common carp; *Flexibacter columnaris* in rainbow trout; and *Vibrio anguillarum*, *Vibrio salmonicida*, and *Yersinia ruckeri* in Atlantic salmon.(2)

Beta glucans act in two different manners to improve fish health: first, stimulating the beneficial gastrointestinal bacteria as a typical prebiotic, and second, directly stimulating the immune system through absorption in the small intestine.(35) Consequently, the recognition of beta glucans as a fungus triggers a cascade of events resulting in immunostimulation, as recently discussed by Oliveira et al. (2019).(75) Hisano et al. (2018) (76) fed pacu with beta glucan of varying concentrations in combination with mannanoligosaccharides (MOS) for 30 days and found that the inclusion of 0.2% of beta glucans and mannanoligosaccharides promoted the best feed efficiency, growth response, and intestinal morphology. The same product and different concentrations were treated in pacu by Soares et al. (2018),(77) who reported that 0.2 percent and 0.4 percent diets were sufficient to increase the number of neutrophils, thrombocytes, and monocytes in the blood, the respiratory burst of leukocytes, and lysozyme activity after a stressful handling and bacterial challenge.

3.5.3 Cortisol

Cortisol is a primary corticosteroid, produced from the interrenal gland of teleost and there is small evidence for the significant formation of cortisone or corticosterone in fish. In mammals, the immunomodulatory effects of corticosteroids and stress are well reported and similar effects have been observed in fish also.(78) For example, a number of studies have shown that stress or cortisol administration can decrease fish resistance to bacterial and fungal pathogens.(79) Most studies in fish have focused on the effects of cortisol administration on the immune system, both in vitro and in vivo conditions.(80)

Cortisol administration can reduce the number of circulating T- and B-like lymphocytes, while the number of circulating phagocytes (neutrophils and macrophages) may increase. The fate of lymphocytes following stress is unclear, although retrafficking of cells to lymphoid tissues may be involved. Thus, acute stress or cortisol treatment in juvenile coho salmon (*Oncorhynchus kisutch*) lowers the number of circulating leukocytes, but increases the number of these cells in the thymus and head kidney. Cortisol, however, has also shown to induce an apoptosis in B cells and this could be responsible for their clearance from the blood following stress conditions. Cortisol has depressive effects on a number of fish immune responses, including phagocytosis and lymphocyte mitogenesis. Cortisol also decreases the activity of B cells and circulating titers of IgM.(79) The effects of cortisol are mediated via the glucocorticoid receptor (GR), which has been identified in leukocytes of coho salmon and common carp (*Cyprinus carpio*).(81)

3.5.4 Growth Hormone and Prolactin

Vertebrate polypeptides like growth hormone (GH) and prolactin (PRL) are produced in the adenohypohysis of the pituitary gland and showed many structural and functional similarities. In all vertebrates, GH is involved in the regulation of post-natal somatic growth, but in fish it also influences osmoregulation and stimulates gonadal steroidogenesis. Prolactin is a versatile peptide with diverse functions that may be summarised under three broad categories of growth and development, osmoregulation, and reproduction. Both hormones modulate the immune responses under *in vitro* and *in vivo* conditions. Indeed, GH is structurally similar to a number of cytokines, including interleukin (IL)-2, IL-4, IL-5, granulocyte-colony stimulating factor (G-CSF), granulocyte-macrophage colony-stimulating factor (GM-CSF), and the interferons.(81)

In fish, GH can stimulate lymphopoiesis and phagocytosis in gilt head sea bream (*Sparus aurata*) and silver sea bream (*Sparus sarba*). It also promotes leukocyte mitogenesis in chum salmon (*Oncorhynchus keta*), natural killer cell activity, phagocytosis, antibody production, serum hemolytic activity in rainbow trout (*Oncorhynchus mykiss*), and respiratory burst activity of leukocytes in rainbow trout and Mediterranean sea bass (*Dicentrarchus labrax*).(81)

Other studies have shown that the GH enhances the resistance of rainbow trout to the bacterial pathogen Vibrio anguillarum infection. In addition, transferring brown trout (*Salmo trutta*) from freshwater to seawater increases plasma titres of GH while reducing titres of thyroid hormones and this has been correlated with increased phagocytic activity of head kidney leukocytes and elevated plasma lysozyme concentrations. Prolactin has similar effects on fish immune responses including stimulation of leukocyte mitogenesis, respiratory burst activity, phagocytosis, and an increase of plasma IgM titer.(81)

3.5.5 Vitamins

3.5.5.1 Vitamin C

The ability of vitamin C to induce the immune response has been established in fish. Elevated doses of dietary vitamin C have been shown to enhance several innate immune parameters such as macrophage activities, lysozyme activity, lymphocyte proliferation, complement activity, and natural cytotoxicity. Immunized fish fed high levels of vitamin C, elevated specific antibody level.[74] High levels of dietary vitamin C are reported to increase resistance to *E. tarda* and *E. ictaluri* infections in channel catfish to *V. anguillarum*, IPN, *Ichthyophthirius multifiliis*, *Y. ruckeri* and VHS in rainbow trout, and to *A. salmonicida* and *V. salmonicida* in Atlantic salmon. The enhanced adaptive immune response has been demonstrated in other fish species such as *Labeo rohita* through after fish being fed vitamin C in their diets.[2] Xu et al., (2016) studied the effect of dietary vitamin C deficiency in carp and reported that dietary vitamin C deficiency affects the growth, spleen and head kidney immunity, and structural integrity by regulating NF-kB signalling in young grass carp (*Ctenopharyngodon idella*). The studies on vitamin C dietary supplementation in *Labeo rohita* displayed improved survival rate.[82] Vitamin C with its redox catalyst property reduced the production of free radicals and increased the enzymatic activity of iNOS by chemical stabilization of (6 R)-5,6,7,8-tetrahydrobiopterin or BH4.[83] Gupta et al. (2021) explained the effect of vitamin C dietary supplementation and its stress amelioration efficacy against various stressors.

3.5.5.2 Vitamin E

Vitamin E also enhances both cellular and humoral defenses in animals.[84] For instance, Atlantic salmon fed vitamin E depleted diets had significantly increased mortality rates following *A. salmonicida* infection compared with fish on a commercial diet. Vitamin E deficiency also affects the specific immune response of fish, as demonstrated by the enhanced antibody titres to *E. ictaluri* in channel catfish fed high levels of vitamin E for 180 days.[2] The detoxifying ability of vitamin E mediated by methemoglobin reductase production. Moreover, the use of dietary vitamin E also attenuated oxidative damage resulting from the production of reactive nitrogen species and modulated the expression and activation of signal transduction pathways in *L. rohita.*[85] Belo et al. (2014) fed pacu with vitamin E (12, 58, and 310 mg/kg) during 18 days and evaluated the inflammatory responses to a foreign body (glass coverslip) on connective tissue over 15 days.[86] Fish fed with the lower vitamin E dose had a lower number of lymphocytes and LG-PAS cells and an elevated plasma cortisol level. Vitamin E (α-tocopherols) arrested the chain process of lipid peroxidation by quenching lipid peroxyl radicals in Rohu (*Labeo rohita*) has been reported.[87]

3.5.6 Lysozyme and Lactoferrin

Lysozyme (or muramidase or N-acetylmuramic acid hydrolase E.C. 3.2.1.17) is a protein that exerts its enzymatic activity through the hydrolysis of the β-1,4-glycosidic bonds between N-acetylmuramic acid (NAM) and N-acetylglucosamide (NAG) in the polysaccharide backbone of the peptidoglycans of the Gram-positive bacterial cell wall.[88] Lysozyme is present in lymphoid tissue, mucus and serum of most of the fish species, but not in few species (such as cod and wolfish). Lysozyme activity is an index of innate immunity of fish. Lysozyme activity vary depending on the age and size, sex, water temperature, season, toxicants, pH, infections, and degree of stressors. Lysozyme, which is a strong bacteriolytic agent, has been demonstrated in a variety of fish tissues and also in fish mucus. It is also present at high concentrations in human milk but at lower concentrations in bovine milk. Lysozyme has the ability, as an enzyme, to bring about the breakdown of bacterial cell wall peptidoglycan polysaccharide and destroy susceptible bacterial cells due to osmotic shock.[89]

Lactoferrin is an iron-binding protein, that displays both bacteriostatic and bacteriocidal effects. In terms of the bacteriocidal effects, lactoferrin has the ability to kill species of bacteria, which action is not associated with its ability to compete with the bacteria for iron.[89] In the African cichlid (*Sciaenochromis fryeri*), dietary bovine lactoferrin (LF) effects on the haemato-biochemical parameters, growth, stress responses, and non-specific serum immunity were investigated. The highest lysozyme and bactericidal activities were observed in the fish fed the diets with 200 and 800 mg/kg of LF, respectively. Their study results showed that adding 400 mg/kg LF could be useful as an iron discriminator, for liver function as well as an immune and stress resistance promoter in the African cichlid.[90] Several factors such as nutritional status, seasonal variation, sex and sexual maturation, salinity, pH, water temperature, stresses, and infections could be considered as possible parameters involved in the variations in lysozyme activity in vertebrates.[91] It has been reported that oral administration of LF increases mucus secretion in the Japanese flounder and Siberian sturgeon. Moreover, dietary LF did not reportedly affect the stress indicators (serum glucose and cortisol levels) in the Siberian sturgeon (*Acipenser baeri*) during the rest period.[92]

Esmaeili et al. (2019) studied the effect of the dietary supplementation of bovine LF on hematological and immunological parameters, growth performance, antioxidant enzymes activity, and disease resistance against *Vibrio harveyi* in yellowfin sea bream (*Acanthopagrus latus*) fingerling. The study results showed that diet supplementation in *A. latus* especially with 1,200 mg LF per kg improves fish immune parameters and growth performance, as well as survival rate against *Vibrio harveyi.*[93]

3.5.7 Minerals

3.5.7.1 Selenium

Selenium (Se) is a vital micronutrient for antioxidant defenses, and one of its roles is to prevent the immunosuppression caused by oxidative stress. Selenium deficiency is associated with diminished activity of selenium-dependent glutathione peroxidase in neutrophils, and negatively affects neutrophil bacterial activity.[89] The oxidative stress impairs the immune response and indicates the occurrence of a physiological trade-off between immune and antioxidant systems in fish.[35] The immune response and antioxidant defenses of fish were improved by supplementing the diet of pacu with 0.3, 0.6, 0.9, and 1.8 mg Se-yeast/kg for 65 days.[94] Selenium is generally available as trace element to the fish mainly due to their differential bioavailability.[95] The zinc oxide (ZnO) and selenium (Se) nanoparticles dietary supplementation stimulates immunity and enhances resistance to bacterial infection in *L. rohita*.[96]

3.5.7.2 Chromium

Trivalent chromium is known to reduce plasma cortisol levels in stressful situations. In the swim bladder, Castro et al. (2014) investigated the effect of the inclusion of chromium carbochelate in the diet of pacu on acute inflammation challenged by injection of inactivated *A. hydrophila*. Gatta et al. (2001) reported that chromium is able to modulate the immune response of rainbow trout, and this effect appears to be both dose and time dependent. In the study of Castro et al. (2014), fish-fed diets containing 18 and 36 mg of Cr per kg had plasma cortisol and glucose reduced and, thus, an enhancement of the immune response. In addition, pacu fed with chromium carbochelate and/or in combination with *Saccharomyces cerevisiae* for 105 days by Bortoluzzi et al. (2017). They reported that fish supplemented with chromium and *S. cerevisiae* had a greater degree of organization of collagen fibers, higher neovascularization, and faster reepithelialization, when subjected to skin injury.[35,97] Dietary supplementation with betaine (Bet) and chromium picolinate (Cr-Pic) combination is improving fish immune response and antioxidative status and aquaculture sustainability in Nile tilapia (*Oreochromis niloticus*).[98]

3.5.8 Bacterial Lipopolysaccharides (LPS)

Microbial gut flora is known to modulate fish immunity in different ways and bacterial lipopolysaccharides (LPS) in particular may have a highlighting effect on fish immune system.[89] Bacterial LPS, called endotoxin, are among the most potent immunomodulators among the known ones. Despite the less sensitivity of aquatic animals to LPS, mammalians are very sensitive to LPS thus, it has no application as immunostimulant for practical use in mammals because of its toxic effect which causes inflammation reactions resulting in fever, reduced feed intake, and impaired performance of animals.[99] Although LPS has traditionally been regarded as a B cell mitogen and macrophage activator, it is now apparent that an interaction between T cells and LPS occurs.[99] Bacterial endotoxin LPS potentiates antigen-specific proliferation of T helper cells. This effect is independent of interleukin I action, the T cell activation factor released by macrophages after LPS stimulation. Thus, through LPS, both specific and non-specific protective immune responses could be improved.[89]

3.5.9 Levamisole

Levamisole protects fish from *A. hydrophila* in carp, A. *salmonicida infections* in coho salmon, *V. anguillarum* in rainbow trout, and *V. anguillarum* in gilthead seabream. Levamisole administration to fish also reported to increase the antibody titres values.[2] In mammals, levamisole stimulates the phagocytic activity of neutrophils and the number of phagocytic cells and leukocytes increased and the activity of lysozyme in serum of mammals also increased. The similar effects have also been reported in fish by Siwicki (1987). Olivier et al. (1985) also proved that levamisole provided the protection against *Aeromonas salmonicida*. Protection against infectious fish diseases was achieved with the help of phagocytic cells. Levamisole activated the fish phagocytic cells and NK cell activities. Levamisole activated the alternative complement pathway of rainbow trout, which might play as opsonic complement effects in the nonspecific defense mechanism.[100] To stimulate the immune function of *Cirrhinus mrigala*, a dose of 250 mg of levamisole per kg diet is optimum.[101] The antioxidant action of levamisole is by the enhancing effect on the major cellular redox systems like glutathione, enzymes like superoxide dismutase, catalase, and also possible effects on glutathione-related enzymes. On the other hand, the anti-inflammatory action is by the inhibition of TNF-alpha (tumor necrosis factor) and interleukin 6. Anti-neoplastic action of levamisole is by two ways that is, through its apoptotic action causing cell cycle inhibition and increased endothelial cell adhesion and the second action is through its anti-angiogenic property.[102] The oral administration of levamisole for 15 days modulated circulating cortisol levels during the stress response and improved the innate immune response against *A. hydrophila* infection in pacu (*Piaractus mesopotamicus*).[103] Levamisole showed an increase in circulating leukocytes, phagocytic rate, and increase protection against *P. damselae subs P. piscicida* in European seabass.[104]

3.5.10 Kaempferol 3-a-L-(4-O-acetyl) Rhamnopyranoside-7-a-Lrhamnopyranoside (SA)

The compound isolated from Crown Wood-Fern (*Dryopteris crassirhizoma*) is kaempferol 3-a-L- (4-O-acetyl) rhamnopyranoside-7-a-Lrhamnopyranoside (SA).[105] It is also present in *Zingiber officinale*. This compound has the ability to inhibit Ribosomal S6 kinase (RSK) activity. It can be used to reduce the proliferation of neoplastic cells or for inhibiting the establishment or maintenance of an intracellular pathogenic infection by pathogens whose pathogenicity derives in part from the pathogen's ability to impede endosomal or phagosomal maturation in the host cell [106] (Patent No: US9040673B2) This compound increased the immune responses like lysozyme activity, phagocytic activity, complement C3, IgM level and SOD activity of fish under *in vivo* condition. It also induced expression of immune response-related genes like TNF-α, MyD88, Mx1and IL-1β in fish of *in vitro* studies.[107]

3.5.11 Herbal Immunomodulators

Herbal immunomodulators are promising among the other immunomodulators used in aquaculture because of its propensities to improve growth performance and fish immunity. They are good alternatives to chemical treatments and antibiotics. Medicinal herbs and plant-based immunomodulators gained attention in the prevention and treatment of fish diseases due to being ecofriendly and cost effective and have minimal side effects.[108] Details of herb/plant immunostimulatory effects in fish are given in Chapter 7.

3.6 Conclusion

Immunomodulators can be defined as natural or chemical components that induce both the specific and nonspecific defense mechanisms. They are suitable for aquaculture applications to improve feed efficiency, weight gain, and disease resistance.[35] Examples of immunomodulators are beta-glucans,[109] lactoferrin, and levamisole.[110] In fish and shellfish, immunomodulators promote immunocompetency and disease resistance by augmenting specific and non-specific immune responses.[111] Immunomodulators exhibit broad-spectrum antibiotic activity and promote the fish resistance.[112–114] The use of immunomodulators has been considered as a promising method to prevent diseases by strengthening the immune system of fish, and also made them more effective than vaccines, since vaccines are usually directed toward one microorganism.[115] The immunomodulators are normally administered in fish as dietary supplement or through injection. The effect of immunomodulators on innate immune cells improve the function of phagocytic cells.[112]

In order to develop sustainable and environment friendly aquaculture, the benefits of the immunomodulators are essential. The efficiency relies upon their route of administration, dose, and duration time, and studies to define the best protocol to optimize the benefits and avoid immunosuppression are required.[35] However, one of the current problems faced by scientists in better understanding the fish biology is the lack of transcriptomic and whole-genome sequencing studies. The recent studies by Mareco et al. (2015) for pacu, Pimentel et al. (2016), and Prado-Lima and Val (2016) for tambaqui used the recently developed valuable screening approaches such as transcriptome and proteome analysis, could bring a significant advance in the field of aquaculture.[35] Immunomodulators are already in use as adjuvants in fish vaccines and as additives in aquafeeds, and there is a growing interest in the aquaculture industry to make use of immunomodulators as prophylactic agents prior to situations known to represent an elevated risk of disease outbreak. The efficacy of immunomodulators is highly dependent on dose and mode of administration. Thus, future valuable research areas will be the immunomodulators-based studies to achieve profitable, sustainable, and environment-friendly aquaculture.

Acknowledgment

The authors express immense gratitude to Dr. K.R. Venkatesan, principal and the management of Sri Sankara Arts and Science College, Enathur, for providing permission to write this chapter on vaccines.

Disclosure statement: The authors declare that they have no conflicts of interest.

REFERENCES

1. B. C. De, D. K. Meena, B. K. Behera, P. Das, P. K. Das Mohapatra, and A. P. Sharma. Probiotics in fish and shellfish culture: immunomodulatory and ecophysiological responses. 40(3):921–971. ISSN 1573-5168.
2. P. K. Sahoo and S. C. Mukherjee. The effect of dietary immunomodulation upon *Edwardsiella tarda* vaccination in healthy and immunocompromised Indian major carp (labeo rohita). 12(1):1–16. ISSN 1050-4648.
3. I. J. Lebish and R. M. Moraski. Mechanisms of immunomodulation by drugs. 15(3): 338–345. ISSN 0192-6233.
4. J. Laurence. The immune system in AIDS. 253(6):84–93. ISSN 0036-8733.
5. J. E. Talmadge, R. K. Oldham, and I. J. Fidler. Practical considerations for the establishment of a screening procedure for the assessment of biological response modifiers. 3(1):88–109. ISSN 0732-6580.
6. E. Vallejos-Vidal, F. Reyes-L´opez, M. Teles, and S. MacKenzie. The response of fish to immunostimulant diets. 56:34–69. ISSN 10504648. URL: https://journals.scholarsportal.info/details/10504648/v56icomplete/troftid.xml. Publisher: Elsevier.
7. S. Reyes-Cerpa, F. E. Reyes-Lopez, D. Toro-Ascuy, J. Ibanez, K. Maisey, A. M. Sandino, and M. Imarai. IPNV modulation of pro and anti-inflammatory cytokine expression in Atlantic salmon might help the establishment of infection and persistence. 32(2):291–300. ISSN 1095-9947.
8. T. J. Bowden, P. Cook, and J. H. W. M. Rombout. Development and function of the thymus in teleosts. 19(5):413–427. ISSN 1050-4648.
9. A. Espenes, C. M. Press, B. H. Dannevig, and T. Landsverk. Immune-complex trapping in the splenic ellipsoids of rainbow trout (*Oncorhynchus mykiss*). 282(1): 41–48. ISSN 0302-766X.
10. C. Agius and R. J. Roberts. Melano-macrophage centres and their role in fish pathology. 26(9):499–509. ISSN 0140-7775.
11. A. K. Abbas, A. H. Lichtman, and S. Pillai. *Cellular and Molecular Immunology*. Saunders. ISBN 978-81-312-1034-5.
12. S. Reyes-Cerpa, K. Maisey, F. Reyes-Lopez, D. Toro- Ascuy, A. M. Sandino, and M. Imarai. *Fish Cytokines and Immune Response*. IntechOpen. ISBN 978-953-51-0909-9. URL: https://www.intechopen.com/chapters. PublicationTitle: New Advances and Contributions to Fish Biology.
13. W. E. Paul. *Fundamental Immunology*. Lippincott Williams & Wilkins. ISBN 978-0-7817-6519-0. Google-Books-ID: oPSG1PGmZUkC.
14. C. A. Janeway. How the immune system protects the host from infection. 3(13): 1167–1171. ISSN 1286-4579.
15. L. J. McHeyzer-Williams and M. G. McHeyzer-Williams. Antigen-specific memory b cell development. 23(1):487–513. URL: 10.1146/annurev.immunol.23.021704.115732.
16. F. Sallusto and A. Lanzavecchia. Heterogeneity of CD4+ memory t cells: functional modules for tailored immunity. 39(8):2076–2082. ISSN 1521-4141.

17. B. Dixon and Rj Stet. The relationship between major histocompatibility receptors and innate immunity in teleost fish. 25(8). ISSN 0145-305X. URL: https://pubmed.ncbi.nlm.nih.gov/11602190/. Publisher: Dev Comp Immunol.
18. T. Takano, S. D. Hwang, H. Kondo, I. Hirono, T. Aoki, and M. Sano. Evidence of molecular toll-like receptor mechanisms in teleosts. 45(1):1–16.
19. M. Reverter, N. Bontemps, D. Lecchini, B. Banaigs, and P. Sasal. Use of plant extracts in fish aquaculture as an alternative to chemotherapy: current status and future perspectives. 433:50–61. ISSN 0044-8486. URL: https://www.sciencedirect.com/science/article/pii/S0044848614002798.
20. D. S. Chahal, R. K. Sivamani, R. Rivkah Isseroff, and Mohan R. Dasu. Plant-based modulation of toll-like receptors: an emerging therapeutic model. 27(10):1423–1438. ISSN 1099-1573.
21. R. C. Robertson, F. Guiheneuf, B. Bahar, M. Schmid, D. B. Stengel, G. F. Fitzgerald, R. Paul Ross, and C. Stanton. The anti-inflammatory effect of algae-derived lipid extracts on lipopolysaccharide (LPS)- stimulated human THP-1 macrophages. 13(8):5402–5424. ISSN 1660-3397.
22. N. Arora, K. Shah, and S. Pandey-Rai. Inhibition of imiquimod-induced psoriasis-like dermatitis in mice by herbal extracts from some Indian medicinal plants. 253(2):503–515. ISSN 1615-6102.
23. S. Granica, J. P. Piwowarski, and A. K. Kiss. Ellagitannins modulate the inflammatory response of human neutrophils ex vivo. 22(14):1215–1222. ISSN 1618-095X.
24. C. T. Nguyen, T. T. Luong, S. Y. Lee, G. L. Kim, H. Kwon, H.-G. Lee, C.-K. Park, and D.-K. Rhee. Panax ginseng aqueous extract prevents pneumococcal sepsis in vivo by potentiating cell survival and diminishing inflammation. 22(11):1055–1061. ISSN 1618-095X.
25. S.-J. Kim, S.-J. Yoon, Y.-M. Kim, S.-W. Hong, S. H. Yeon, K.-I. Choe, and S.-M. Lee. HS-23, lonicera japonica extract, attenuates septic injury by suppressing toll-like receptor 4 signaling. 155(1):256–266. ISSN 1872-7573.
26. A. Ghochikyan, A. Pichugin, A. Bagaev, A. Davtyan, A. Hovakimyan, A. Tukhvatulin, H. Davtyan, D. Shcheblyakov, D. Logunov, M. Chulkina, A. Savilova, D. Trofimov, E. L. Nelson, M. G. Agadjanyan, and R. I. Ataullakhanov. Targeting TLR-4 with a novel pharmaceutical grade plant derived agonist, immunomax®, as a therapeutic strategy for metastatic breast cancer. 12(1):322. ISSN 1479-5876. URL: 10.1186/s12967-014-0322-y.
27. C.-J. Lin, H.-J. Lin, T.-H. Chen, Y.-A. Hsu, C.-S. Liu, G.-Y. Hwang, and L. Wan. *Polygonum cuspidatum* and its active components inhibit replication of the influenza virus through toll-like receptor 9-induced interferon beta expression. 10(2):e0117602. ISSN 1932-6203.
28. S. Ryu, C. Hs, Y. Ky, L. Oh, K. Kj, and L. By. Oleuropein suppresses LPS-induced inflammatory responses in RAW 264.7 cell and zebrafish. 63(7). ISSN 1520-5118. URL: https://pubmed.ncbi.nlm.nih.gov/25613688/. Publisher: J Agric Food Chem.
29. B. Park and J. Lee. Recognition of lipopolysaccharide pattern by TLR4 complexes. 45(12). ISSN 2092-6413. URL: https://pubmed.ncbi.nlm.nih.gov/24310172/. Publisher: Exp Mol Med.
30. K. Takeda, T. Kaisho, and S. Akira. Toll-like receptors. 21:335–376. ISSN 0732-0582.
31. H. Heine and E. Lien. Toll-like receptors and their function in innate and adaptive immunity. 130(3):180–192. ISSN 1018-2438.
32. A. H. Meijer, S. F. Gabby Krens, I. A. Medina Rodriguez, S. He, W. Bitter, B. E. Snaar-Jagalska, and H. P. Spaink. Expression analysis of the toll-like receptor and TIR domain adaptor families of zebrafish. 40(11):773–783. ISSN 0161-5890.
33. S. D. Hwang, K. Fuji, T. Takano, T. Sakamoto, H. Kondo, I. Hirono, and T. Aoki. Linkage mapping of toll-like receptors (TLRs) in Japanese flounder, *Paralichthys olivaceus*. 13(6):1086–1091. ISSN 1436-2236.
34. H. Ye, Q. Lin, and H. Luo. Applications of transcriptomics and proteomics in understanding fish immunity. 77:319–327. ISSN 1095-9947.
35. E. C. Urbinati, F´abio Sabbadin Zanuzzo, and J. D. Biller. Chapter 5 - Stress and immune system in fish. In B. Baldisserotto, E. C. Urbinati, and J. E. P. Cyrino, editors, *Biology and Physiology of Freshwater Neotropical Fish*, pages 93–114. Academic Press. ISBN 978-0-12-815872-2. URL: https://www.sciencedirect.com/science/article/pii/B9780128158722000051.
36. J. W. Hadden and A. Szentivanyi. *Immunopharmacology Reviews Volume 2*. Springer Science & Business Media. ISBN 978-1-4613-0349-7. Google-Books-ID: ZyPnBwAAQBAJ.
37. T. Wang and C. J. Secombes. The cytokine networks of adaptive immunity in fish. 35(6):1703–1718. ISSN 1095-9947.
38. D. S. Nascimento, A. do Vale, A. M. Tom´as, J. Zou, C. J. Secombes, and N. M. S. dos Santos. Cloning, promoter analysis and expression in response to bacterial exposure of sea bass (*Dicentrarchus labrax* l.) interleukin-12 p40 and p35 subunits. 44(9):2277–2291. ISSN 0161-5890.
39. C. Furnes, M. Seppola, and B. Robertsen. Molecular characterisation and expression analysis of interferon gamma in Atlantic cod (*Gadus morhua*). 26(2): 285–292. ISSN 1095-9947.
40. L. Wang, N. Shang, H. Feng, Q. Guo, and H. Dai. Molecular cloning of grass carp (*Ctenopharyngodon idellus*) t-bet and GATA-3, and their expression profiles with IFN-gamma in response to grass carp reovirus (GCRV) infection. 39(4):793–805. ISSN 1573-5168.
41. W. Q. Chen, Q. Q. Xu, M. X. Chang, J. Zou, C. J. Secombes, K. M. Peng, and P. Nie. Molecular characterization and expression analysis of the IFN-gamma related gene (IFN-gammarel) in grass carp ctenopharyngodon idella. 134(3):199–207. ISSN 1873-2534.
42. C. J. Secombes, T. Wang, and S. Bird. The interleukins of fish. 35(12):1336–1345. ISSN 1879-0089.
43. T. Kadowaki, H. Harada, Y. Sawada, C. Kohchi, G.-I. Soma, Y. Takahashi, and H. Inagawa. Two types of tumor necrosis factor – alpha in bluefin tuna (*Thunnus orientalis*) genes: molecular cloning and expression profile in response to several immunological stimulants. 27(5): 585–594. ISSN 1095- 9947.
44. A. G. Stanbery, S. Smita, J. von Moltke, E. D. Tait Wojno, and S. F. Ziegler. TSLP, IL-33, and IL-25: not just for allergy and helminth infection. ISSN 0091-6749. URL: https://www.sciencedirect.com/science/article/pii/S0091674922009435.

45. J. Rivers-Auty, M. J. D. Daniels, I. Colliver, D. L. Robertson, and D. Brough. Redefining the ancestral origins of the interleukin-1 superfamily. 9(1):1156. ISSN 2041-1723. URL: https://www.nature.com/articles/s41467-018-03362-1. Number: 1 Publisher: Nature Publishing Group.
46. S. Mitra, A. Alnabulsi, C. J. Secombes, and S. Bird. Identification and characterization of the transcription factors involved in t-cell development, t-bet, stat6 and foxp3, within the zebrafish, *Danio rerio*. 277(1):128–147. ISSN 1742-4658.
47. F. Takizawa, E. O. Koppang, M. Ohtani, T. Nakanishi, K. Hashimoto, U. Fischer, and J. M. Dijkstra. Constitutive high expression of interleukin-4/13a and GATA-3 in gill and skin of salmonid fishes suggests that these tissues form th2-skewed immune environments. 48(12):1360–1368. ISSN 1872-9142.
48. M. Seppola, A. N. Larsen, K. Steiro, B. Robertsen, and I. Jensen. Characterisation and expression analysis of the interleukin genes, IL-1beta, IL-8 and IL-10, in atlantic cod (*Gadus morhua l.*). 45(4):887–897. ISSN 0161-5890.
49. T. Wang, S. A. M. Martin, and C. J. Secombes. Two interleukin- 17c-like genes exist in rainbow trout oncorhynchus mykiss that are differentially expressed and modulated. 34(5):491–500. ISSN 1879-0089.
50. K. Buchmann. Fish immune responses against endoparasitic nematodes – experimental models. 35(9):623–635. ISSN 1365-2761. URL: https://onlinelibrary.wiley.com/doi/abs/10.1111/j.1365-2761.2012.01385.x. eprint: https://onlinelibrary.wiley.com/doi/pdf/10.1111/j.1365-2761.2012.01385.x.
51. T. Wang, P. Diaz-Rosales, M. M. Costa, S. Campbell, M. Snow, B. Collet, S. A. M. Martin, and C. J. Secombes. Functional characterization of a nonmammalian IL-21: rainbow trout oncorhynchus mykiss IL-21 upregulates the expression of the th cell signature cytokines IFN-gamma, IL-10, and IL-22. 186(2):708–721. ISSN 1550-6606.
52. H.-H. Chen, H.-T. Lin, Y.-F. Foung, and J. H.-Y. Lin. The bioactivity of teleost IL-6: IL-6 protein in orange-spotted grouper (*Epinephelus coioides*) induces th2 cell differentiation pathway and antibody production. 38(2):285–294. ISSN 1879-0089.
53. M. Yang, X. Wang, D. Chen, Y. Wang, A. Zhang, and H. Zhou. TGF-beta1 exerts opposing effects on grass carp leukocytes: implication in teleost immunity, receptor signaling and potential self-regulatory mechanisms. 7(4):e35011. ISSN 1932-6203.
54. L. Yin, J. Ren, D. Wang, S. Feng, X. Qiu, M. Lv, X. Wang, and H. Zhou. Functional characterization of three fish-specific interleukin-23 isoforms as regulators of th17 signature cytokine expression in grass carp head kidney leukocytes. 92:315–321. ISSN 1095-9947.
55. M. M. Costa, P. Pereiro, T. Wang, C. J. Secombes, A. Figueras, and B. Novoa. Characterization and gene expression analysis of the two main th17 cytokines (IL- 17a/f and IL-22) in turbot, *Scophthalmus maximus*. 38(4):505–516. ISSN 1879-0089.
56. H. Ashfaq, H. Soliman, M. Saleh, and M. El-Matbouli. CD4: a vital player in the teleost fish immune system. 50(1):1. ISSN 1297-9716. URL: 10.1186/s13567-018-0620-0.
57. L. Du, X. Yang, L. Yang, X. Wang, A. Zhang, and H. Zhou. Molecular evidence for the involvement of RORalpha and RORgamma in immune response in teleost. 33(2):418–426. ISSN 1095-9947.
58. T. Kono, S. Bird, K. Sonoda, R. Savan, C. J. Secombes, and M. Sakai. Characterization and expression analysis of an interleukin-7 homologue in the Japanese pufferfish, *Takifugu rubripes*. 275(6):1213–1226. ISSN 1742-464X.
59. M. Gorissen, E. de Vrieze, Gert Flik, and Mark O. Huising. STAT genes display differential evolutionary rates that correlate with their roles in the endocrine and immune system. 209(2):175–184. ISSN 1479-6805.
60. M. M. Monte, T. Wang, J. W. Holland, J. Zou, and C. J. Secombes. Cloning and characterization of rainbow trout interleukin-17a/f2 (IL-17a/f2) and IL-17 receptor a: Expression during infection and bioactivity of recombinant IL-17a/f2. 81(1):340–353. ISSN 0019-9567. URL: https://www.ncbi.nlm.nih.gov/pmc/articles/PMC3536140/.
61. A. Reyes-Diaz, A. F. Gonzalez-Cordova, A. Hernandez-Mendoza, R. Reyes-Diaz, and B. Vallejo-Cordoba. Immunomodulation by hydrolysates and peptides derived from milk proteins. 71(1):1–9. ISSN 1471-0307. URL: https://onlinelibrary.wiley.com/doi/abs/10.1111/1471-0307.12421. eprint: https://onlinelibrary.wiley.com/doi/pdf/10.1111/1471-0307.12421.
62. L. Whitehouse and B. Dixon. Cytokines in Teleost Fish, Reference Module in Life Sciences, Elsevier, 2019, ISBN 9780128096338, https://doi.org/10.1016/B978-0-12-809633-8.20910-2. Pages 1–7.
63. J. Zou and C. J. Secombes. The function of fish cytokines. 5(2):23. ISSN 2079-7737. URL: https://www.ncbi.nlm.nih.gov/pmc/articles/PMC4929537/.
64. M. Chalamaiah, W. Yu, and J. Wu. Immunomodulatory and anticancer protein hydrolysates (peptides) from food proteins: A review. 245:205–222. ISSN 1873- 7072.
65. F. Subhan, H. Y. Kang, Y. Lim, M. Ikram, S.-Y. Baek, S. Jin, Y. H. Jeong, J. Y. Kwak, and S. Yoon. Fish scale collagen peptides protect against CoCl2/TNF-alpha-induced cytotoxicity and inflammation via inhibition of ROS, MAPK, and NF-kappab pathways in HaCaT cells. 2017:9703609. ISSN 1942-0994.
66. A. V. Le Gouic, P. A. Harnedy, and R. J. FitzGerald. Bioactive peptides from fish protein by-products. In J.-M. M´erillon and K. G. Ramawat, editors, *Bioactive Molecules in Food*, pages 1–35. Springer International Publishing. ISBN 978-3-319-54528-8. URL: 10.1007/978-3-319-54528-8.
67. A. F. G. Cicero, F. Fogacci, and A. Colletti. Potential role of bioactive peptides in prevention and treatment of chronic diseases: a narrative review. 174(11):1378–1394. ISSN 1476-5381.
68. C. J. Secombes, T. Wang, S. Hong, S. Peddie, M. Crampe, K. J. Laing, C. Cunningham, and J. Zou. Cytokines and innate immunity of fish. 25(8):713–723. ISSN 0145-305X. URL: https://www.sciencedirect.com/science/article/pii/S0145305X01000325.
69. M. Yang, Y. Wang, X. Wang, C. Chen, and H. Zhou. Characterization of grass carp (*Ctenopharyngodon idellus*) foxp1a/1b/2: evidence for their involvement in the activation of peripheral blood lymphocyte subpopulations. 28(2):289–295. ISSN 1095-9947.

70. Z. Cai, C. Gao, L. Li, and K. Xing. Bipolar properties of red seabream (pagrus major) transforming growth factor-beta in induction of the leucocytes migration. 28(4):695–700. ISSN 1095-9947.
71. R. Savan and M. Sakai. Genomics of fish cytokines. 1(1):89–101. ISSN 1878-0407.
72. J. Zou, C. Tafalla, J. Truckle, and C. J. Secombes. Identification of a second group of type I IFNs in fish sheds light on IFN evolution in vertebrates. 179(6):3859–3871. ISSN 0022-1767.
73. B. Robertsen. Expression of interferon and interferon-induced genes in salmonids in response to virus infection, interferon-inducing compounds and vaccination. 25(4): 351–357. ISSN 1050-4648.
74. V. Verlhac, A. Obach, J. Gabaudan, W. Schu¨ep, and R. Hole. Immunomodulation by dietary vitamin c and glucan in rainbow trout (*Oncorhynchus mykiss*). 8(6):409–424. ISSN 1050-4648. URL: https://www.sciencedirect.com/science/article/pii/S1050464898901486.
75. C. A. F. de Oliveira, V. Vetvicka, and F´abio S. Zanuzzo. *β*-glucan successfully stimulated the immune system in different jawed vertebrate species. 62:1–6. ISSN 1878-1667.
76. H. Hisano, M. P. Soares, F. G. Luiggi, and A. C. Arena. Dietary *β*-glucans and mannanoligosaccharides improve growth performance and intestinal morphology of juvenile pacu piaractus mesopotamicus (Holmberg, 1887). 26(1): 213–223. ISSN 1573-143X. URL: https://doi.org/10.1007/s10499-017-0210-6.
77. M. P. Soares, F. C. Oliveira, I. L. Cardoso, E. C. Urbinati, C. M. de Campos, and H. Hisano. Glucan-MOS improved growth and innate immunity in pacu stressed and experimentally infected with *Aeromonas hydrophila*. 73:133–140. ISSN 1095-9947.
78. C. B. Schreck. *The Fish Immune System: Organism, Pathogen, and Environment, Volume 15 – 1st Edition*. URL: https://www.elsevier.com/books/978-0-12-350439-5.
79. A. G. Maule, R. A. Tripp, S. L. Kaattari, and C. B. Schreck. Stress alters immune function and disease resistance in chinook salmon (*Oncorhynchus tshawytscha*). 120 (1): 135–142. ISSN 0022-0795.
80. Y. K. Narnaware and B. I. Baker. Evidence that cortisol may protect against the immediate effects of stress on circulating leukocytes in the trout. 103(3):359–366. ISSN 0016-6480.
81. J. Harris and D. J. Bird. Modulation of the fish immune system by hormones. page 14.
82. M. Ahmed, Z. Ferdous, M. Paul, M. H. Rashid, and M. S. Hossain. Effect of dietary vitamin c on the growth and survival rate of rohu (*Labeo rohita*). 16(2):1375–1385. ISSN 23127945. URL: https://www.journalbinet.com/jbar-160218-170.html.
83. N. A. Khan, JaiGopal Sharma, and R. Chakrabarti. The study of ameliorative effect of dietary supplementation of vitamin c, vitamin e, and tryptophan on *Labeo rohita* (cyprinidae) fry exposed to intense light. 45(3):1153–1165. ISSN 1573-5168 .
84. E. D. Lewis, S. N. Meydani, and D. Wu. Regulatory role of vitamin E in the immune system and inflammation. 71(4):487–494. ISSN 1521-6543. URL: https://www.ncbi.nlm.nih.gov/pmc/articles/PMC7011499/.
85. A. Ciji, N. P. Sahu, A. K. Pal, and M. S. Akhtar. Physiological changes in labeo rohita during nitrite exposure: Detoxification through dietary vitamin E. 158(2):122–129. ISSN 1532-0456. URL: https://www.sciencedirect.com/science/article/pii/S1532045613000719.
86. Marco Antonio de Andrade Belo, Flávio Ruas de Moraes, Luciana Yoshida, Ed Johnny da Rosa Prado, Julieta Rodini Engrácia de Moraes, Vando Edésio Soares, and Marta Gomes da Silva. Deleterious effects of low level of vitamin E and high stocking density on the hematology response of pacus, during chronic inflammatory reaction. 422-423:124–128. ISSN 0044-8486. URL: https://www.sciencedirect.com/science/article/pii/S0044848613006522.
87. M. Latif, M. Faheem, Asmatullah, S. H. Hoseinifar, and H. Van Doan. Dietary black seed effects on growth performance, proximate composition, antioxidant and histo-biochemical parameters of a culturable fish, rohu (*Labeo rohita*). 11(1):48. ISSN 2076-2615. URL: https://www.ncbi.nlm.nih.gov/pmc/articles/PMC7824491/.
88. P. Ferraboschi, S. Ciceri, and P. Grisenti. Applications of lysozyme, an innate immune defense factor, as an alternative antibiotic. 10(12):1534. ISSN 2079-6382. URL: https://www.ncbi.nlm.nih.gov/pmc/articles/PMC8698798/.
89. K. S. Sohn, M. K. Kim, J. D. Kim, and In K. Han. The role of immunostimulants in monogastric animal and fish - review. 13(8):1178–1187. ISSN 1011-2367. URL: https://koreascience.kr/article/JAKO200023443159397. Publisher: Asian Australasian Association of Animal Production Societies.
90. A. M. Moradian, S. Dorafshan, F. Paykan Heyrati, and E. Ebrahimi. Effects of dietary bovine lactoferrin on growth, haemato-biochemical parameters, immune functions and tolerance to air exposure stress in the African cichlid *Sciaenochromis fryeri*. 24(1):392–399. ISSN 13535773. URL: https://onlinelibrary.wiley.com/doi/10.1111/anu.12570.
91. A. Alhazmi, J. Stevenson, S. Amartey, and W. Qin. Discovery, modification and production of t4 lysozyme for industrial and medical uses. 6(4):45. ISSN 1916-9671. URL: https://ccsenet.org/journal/index.php/ijb/article/view/38645. Number: 4.
92. K. Eslamloo, B. Falahatkar, and S. Yokoyama. Effects of dietary bovine lactoferrin on growth, physiological performance, iron metabolism and non-specific immune responses of Siberian sturgeon *Acipenser baeri*. 32(6):976–985. ISSN 1095-9947.
93. A. Esmaeili, E. Sotoudeh, V. Morshedi, D. Bagheri, and S. Dorafshan. Effects of dietary supplementation of bovine lactoferrin on antioxidant status, immune response and disease resistance of yellowfin sea bream (*Acanthopagrus latus*) against *Vibrio harveyi*. 93:917–923. ISSN 1050-4648. URL: https://www.sciencedirect.com/science/article/pii/S1050464819308459.
94. L. S. Takahashi, J. D. Biller-Takahashi, C. F. M. Mansano, E. C. Urbinati, R. Y. Gimbo, and M. V. Saita. Long-term organic selenium supplementation overcomes the trade-off between immune and antioxidant systems in pacu (*Piaractus mesopotamicus*). 60:311–317. ISSN 1095-9947.
95. K. U. Khan, A. Zuberi, J. B. K. Fernandes, I. Ullah, and H. Sarwar. An overview of the ongoing insights in selenium research and its role in fish nutrition and fish health. 43(6):1689–1705. ISSN 1573-5168.

96. P. Swain, R. Das, A. Das, S. K. Padhi, K. C. Das, and S. S. Mishra. Effects of dietary zinc oxide and selenium nanoparticles on growth performance, immune responses and enzyme activity in rohu, *Labeo rohita* (Hamilton). 25(2):486–494. ISSN 1365-2095. URL: https://onlinelibrary.wiley.com/doi/abs/10.1111/anu.12874. eprint: https://onlinelibrary.wiley.com/doi/pdf/10.1111/anu.12874.
97. N. L. Bortoluzzi, Ma. P. de Castro, Gustavo da Silva Claudiano, J. Yunis-Aguinaga, V. A. Cueva-Quiroz, Julieta Rodini Engr´acia de Moraes, and Fl´avio Ruas de Moraes. Wound healing in *Piaractus mesopotamicus* supplemented with chromium carbochelate and saccharomyces cerevisiae. URL: https://doaj.org/article/04f40c74648842fb877b1862174d7424.
98. E. Risha, F. Ahmed, A. A. Khaled, F. M. A. Hossain, N. Akhtar, and E. Zahran. Interactive effects of dietary betaine and chromium picolinate on the immunomodulation, antioxidative response and disease resistance of nile tilapia (*Oreochromis niloticus*). 53(9):3464–3477. ISSN 1365-2109. URL: https://onlinelibrary.wiley.com/doi/abs/10.1111/are.15853. eprint: https://onlinelibrary.wiley.com/doi/pdf/10.1111/are.15853.
99. I. Motta, D. Portnoi, and P. Truffa-Bachi. Effect(s) of lipopolysaccharide on lectin-induced t-cell activation. 97(2):267–275. ISSN 0008-8749.
100. Y. Kajita, M. Sakai, S. Atsuta, and M. Kobayashi. The immunomodulatory effects of levamisole on rainbow trout, *Oncorhynchus mykiss*. 25 (2):93–98.
101. A. Bhatnagar. Immunostimulating and growth promoting activity of dietary levamisole on *Cirrhinus mrigala* fingerlings. Volume 4. ISSN 2378-3184. URL: https://medcraveonline.com/JAMB/JAMB-04-00102.pdf. Publisher: MedCrave Publishing.
102. M. L. Chandy, C. Soman, S. P. Kumar, S. Kurup, and R. Jose. Understanding molecular mechanisms in multivariant actions of levamisole as an anti-helminthic, anti-inflammatory, antioxidant, anti-neoplastic and immunomodulatory drug. 28(4):354–357. ISSN 2212-5558. URL: https://www.sciencedirect.com/science/article/pii/S2212555816300023.
103. E. Pahor-Filho, A. S. C. Castillo, N. L. Pereira, F. Pilarski, and E. C. Urbinati. Levamisole enhances the innate immune response and prevents increased cortisol levels in stressed pacu (*Piaractus mesopotamicus*). 65:96–102. ISSN 1095-9947.
104. F. S. Farooqi and W. U. H. Qureshi. Immunostimulants for aquaculture health management. Journal of Pharmacognosy and Phytochemistry 2018; 7(6): 1441-1447, ISSN 2349-8234. URL: https://www.phytojournal.com/archives/2018/vol7issue6/PartY/7-5-684-797.pdf page 7.
105. H. Cao, T.-T. Chai, X. Wang, M. F. B. Morais-Braga, J.-H. Yang, F.-C. Wong, R. Wang, H. Yao, J. Cao, L. Cornara, B. Burlando, Y. Wang, J. Xiao, and H. D. M. Coutinho. Phytochemicals from fern species: potential for medicine applications. 16(3):379–440. ISSN 1572-980X. URL: https://europepmc.org/articles/PMC7089528.
106. S. M. Hecht, D. A. Lannigan-Macara, J. A. Smith, G. A. O'doherty, and M. K. Hilinski. Synthesis and identification of novel RSK-specific inhibitors. URL: https://patents.google.com/patent/US9040673B2/en.
107. C. Cheng, S. S. Giri, J. W. Jun, H. J. Kim, S. Yun, S. G. Kim, and S. C. Park. Immunomodulatory effects of a bioactive compound isolated from *Dryopteris crassirhizoma* on the grass carp *Ctenopharyngodon idella*. 2016:e3068913. ISSN 2314-8861. URL: https://www.hindawi.com/journals/jir/2016/3068913/. Publisher: Hindawi.
108. P. Elumalai, A. Kurian, S. Lakshmi, C. Faggio, M. A. Esteban, and E. Ringø. Herbal immunomodulators in aquaculture. 29(1):33– 57. ISSN 2330-8249. URL: 10.1080/23308249.2020.1779651. Publisher: Taylor & Francis eprint: 10.1080/23308249.2020.1779651.
109. J. K. Chettri, P. W. Kania, and K. Buchmann. Immunomodulation of rainbow trout (*Oncorhynchus mykiss*) fry by bath exposure to a β-glucan from *Euglena gracilis*. 44(9):1407–1415. ISSN 1355-557X. URL: https://journals.scholarsportal.info. Publisher: Wiley.
110. B. T. Hang, N. T. Phuong, and P. Kestemont. Can immunostimulants efficiently replace antibiotic in striped catfish (*Pangasianodon hypophthalmus*) against bacterial infection by *Edwardsiella ictaluri*? 40(2):556–562. ISSN 1095-9947.
111. E. Ringoe, R. Olsen, J. G. Vecino, and S. Wadsworth. Use of immunostimulants and nucleotides in aquaculture: A review. 02.
112. M. Sakai. Current research status of fish immunostimulants. 172(1):63–92. ISSN 0044-8486. URL: https://www.sciencedirect.com/science/article/pii/S0044848698004360.
113. Musthafa M. Saiyad, Asgari Sm, Kurian A., Elumalai P., Jawahar Ali Ar, Paray Ba, and Al-Sadoon Mk. Protective efficacy of mucuna pruriens (l.) seed meal enriched diet on growth performance, innate immunity, and disease resistance in *Oreochromis mossambicus* against *Aeromonas hydrophila*. 75. ISSN 1095-9947. URL: https://pubmed.ncbi.nlm.nih.gov/29458093/. Publisher: Fish Shellfish Immunol.
114. D. Linga Prabu, S. Chandrasekar, K. Ambasankar, J. S. Dayal, S. Ebeneezar, K. Ramachandran, M. Kavitha, and P. Vijayagopal. Effect of dietary syzygium cumini leaf powder on growth and non-specific immunity of *Litopenaeus vannamei* (boone 1931) and defense against virulent strain of vibrio parahaemolyticus. 489:9–20. URL: https://www.sciencedirect.com/science/article/pii/S0044848617311912. Publisher: Elsevier.
115. S. N. Labh and S. R. Shakya. Application of immunostimulants as an alternative to vaccines for health management in aquaculture. International Journal of Fisheries and Aquatic Studies 2014; 2(1): 153-156, ISSN: 2347-5129. URL: https://www.fisheriesjournal.com/vol2issue1/52.1.html page 4.

4

Immunomodulators and Stress Oxidative

Tamilselvan Gokul[1], Paulraj Balaji[2], Karthikeyan Venkatachalam[3], Subramanian Ramya[4], Ramaraj Jayakumararaj[5], Chinnathambi Pothiraj[5], and Kamatchi Ramesh Kumar[1]

[1]PG and Research Department of Zoology, Vivekananda College, Tiruvedakam (West), Madurai, Tamil Nadu, India

[2]PG and Research Centre in Biotechnology, MGR College, Hosur, Tamil Nadu, India

[3]Faculty of Innovative Agriculture and Fishery Establishment Project, Prince of Songkla, University Surat Thani Campus, Makham Tia, Mueang, Surat Thani, Thailand

[4]Department of Zoology, Yadava College (Men), Madurai, Tamil Nadu, India

[5]Department of Botany, Government Arts College, Melur, Madurai, Tamil Nadu, India

4.1 Introduction

Research is increasingly focusing on the development of new immunomodulators as a means to either prevent or treat illness (Safriani et al., 2022). Compounds that may influence the immune response in either a positive or negative manner can be called immunomodulators. These compounds can be obtained from natural or synthetic sources and can modulate the immunological response via specific or non-specific routes (Nair et al., 2019). Fish experience stress when confronted with a particular scenario that threatens their survival (Schreck et al., 2016). Several factors may cause stress, but oxidative stress is the most significant factor affecting fish on a molecular level. These factors can come from many different places. One kind of stress that is known to disrupt the equilibrium between oxidants and antioxidants is referred to as oxidative stress (Dontha et al., 2016). Oxidative stress indicates cellular and molecular stress (Sopinka et al., 2016). Fish erythrocytes are easily damaged by oxidative stress, so they are used as models to study how oxidative stress causes cell death (Li et al., 2013). It can change how a fish's immune system works. This shows that there is a physiological trade-off between the immune and antioxidant systems (Urbinati et al., 2020). Higher energy demand in fish has been linked to poor conditions and elevated oxidative stress, mainly when the immune system is activated to act against a pathogenic infection (Kurtz et al., 2006). During the stress condition, there is an increase in the production of reactive oxygen species (ROS) and the development of an imbalance between the formation of ROS and antioxidant capacity (Pohl et al., 2018; Lee et al., 2017; Aniya et al., 2018). A decrease in antioxidant enzymes and subsequent increased oxidative stress may explain the reduced levels of antioxidant enzymes in various fish tissues (Ghazanfar et al., 2018; Latif et al., 2020). The ecosystem of aquatic life is threatened by the formation of free radicals and ROS (Lushchak et al., 2011). When a cell is exposed to a stressor, it makes ROS, leading to oxidative stress (Sopinka et al., 2016). Environmental stressors can also cause oxidative stress, which leads to cell ageing (Monaghan et al., 2009), and an increase in the level of protein damage (Iwama et al., 1998). Several studies have looked into the possibility that the production of free radicals and oxidative stress is a major cause of DNA damage in various organismal tissues (Sharma et al., 2019; Ali et al., 2020). In addition, many other substances present in microorganisms can cause oxidative stress; for example, the lipopolysaccharide found on the outer surface of Gram-negative bacteria is a well-known oxidative stress inducer (Swain et al., 2008; Giri et al., 2020). Also, other chemicals like polycyclic aromatic hydrocarbons (Madison et al., 2015), potassium dichromate, melanosomes, cortisol, melatonin (Gozdowska et al., 2022), microplastics (Solomando et al., 2022), and other chemical substances can cause oxidative stress in fish (Figure 4.1). There are various signs in fish to understand the oxidative stress level; fish with higher levels of oxidative stress have higher levels of reactive oxygen metabolites (d-ROM) (Hossain et al., 2017). Various antioxidants have been shown in numerous studies to improve stress tolerance and mitigate oxidative stress's negative effects (Choi et al., 2017). Fish must have a healthy equilibrium between oxidative stress and antioxidant defence mechanisms to adjust physiologically to their aquatic system (Wu et al., 2017). There are numerous developments and discoveries to mitigate oxidative stress, the body's antioxidant defences consist of enzymatic and non-enzymatic components that eliminate oxidative stress-inducing

DOI: 10.1201/9781003361183-5

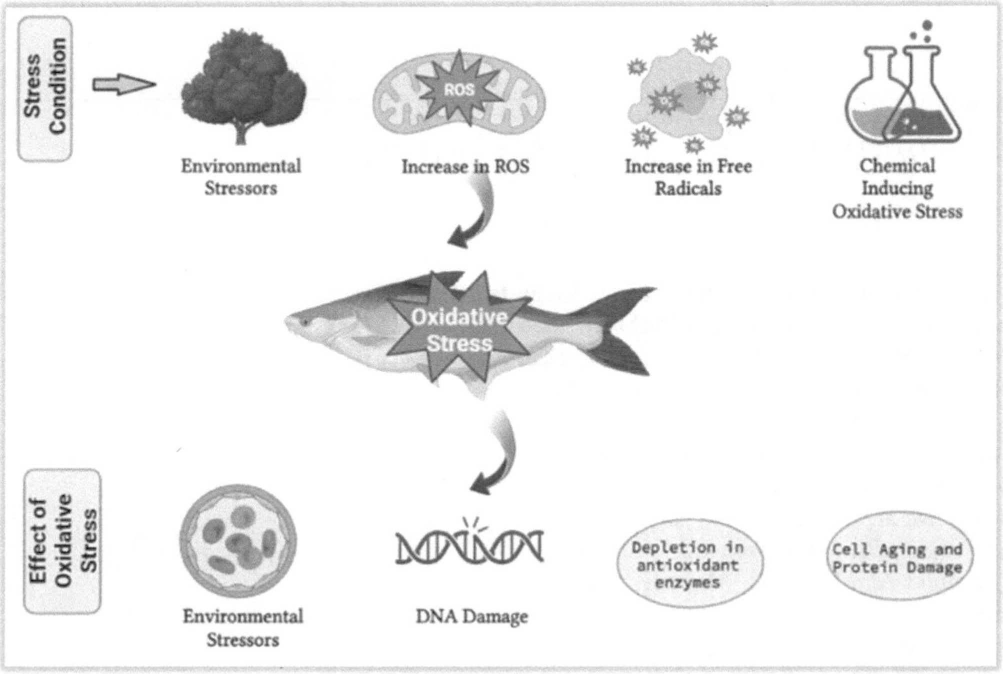

FIGURE 4.1 Different stress conditions responsible for causing oxidative stress and their effects.

free radicals while keeping ROS at a normal physiological level (Ondreicka et al., 1998). In general, feed additives act as immunomodulators, and many natural plant-based phytochemicals, especially phenolic compounds, act as feed additives for fish and are antioxidative (Beltrán, et al., 2022) For several aquatic animals, β-glucans have shown usefulness as an immunostimulant (Dawood et al., 2015), it has been found that glutamate and glutamine protect against oxidative damage (Dawood et al., 2018), antioxidant supplements like vitamins C and E help fish cope better with oxidative stress (Gao et al., 2013). Farmers are willing to employ these immunostimulants and other treatments and vaccines. Immunostimulants appear to have superior efficacy and safety compared to chemotherapy and vaccinations (Table 4.1), making them appropriate for use in aquaculture. As knowledge grew, understanding of immunomodulatory mechanisms increased, and more novel strategies for regulating oxidative stress in fish remain to be discovered. This chapter discusses the immunomodulators, stress indicators, oxidative stress, and responses of fish in the aquatic system.

4.1.1 Concept of Immunostimulants

Many pathogens infect fish causing a systemic infection that leads to disease and death. Several strategies have been used in aquaculture, including using various chemicals to strengthen or enhance the body's immune system of cultured fish and shrimp. Vaccination against certain infections has recently been created with different levels of success depending on the disease, it is an excellent preventive treatment for infections in fish culture, although it can be costly and traumatic for the fish. Due to the complex antigenic nature, a single vaccination is ineffective against many pathogens and only protects against a specific disease (Ardo et al., 2008). Fish and shrimp have been

TABLE 4.1

Comparison of Vaccines and Immunostimulants in Fish Health Management

S. No.	Vaccine	Immunostimulants
1.	Prevention of disease will be for long duration with only one or two treatments	Prevention of disease will be for short duration, require more treatments
2.	Highly effective	Effective
3.	Reduced range of activity	Wide range of activity
4.	No toxic effects/no accumulation of toxic residues	No toxic effects/no accumulation of toxic residues
5.	Negligible effects on the environment	Negligible effects on the environment
6.	Increase the body's immune response, both innate and adaptive	Mainly enhance adaptive immune system of larvae before innate immune system develops
7.	Difficult to vaccinate	Easy to vaccinate
8.	Expensive	Economically viable

treated and prevented with chemical additives and drugs in addition to vaccination.

Nonetheless, the rise of multidrug-resistant infections poses an increasing global danger to veterinary medicine. Consequently, innovative antimicrobial medications are always required to limit the likelihood of increased antimicrobial resistance and undesirable properties on the native microbiota of juvenile or mature fish (Misra et al., 2006). Antibiotics have been used for many years to treat bacterial infections in fish production. Due to the inappropriate use of antibiotics, a worldwide problem has emerged in aquaculture: the emergence of microorganisms resistant to antibiotics associated with fish diseases.

An immunostimulant is a substance, medication, or activity that boosts the non-specific defence mechanism or the specific immunological response among the several disease control strategies. It has been demonstrated to promote phagocytic cell activity and enhance their bactericidal and fungicidal effects and they are substances that stimulate leukocytes and aid in recovery from stress-induced immunosuppression. When used in aquaculture, it can be used to reverse the immunomodulatory properties of stimuli, avert periodic resurgence of recognized chronic illnesses, or inhibit latent or sub-lethal pathogens. These are believed to be an attractive and potential disease prevention treatment for fish and crustacean. The prophylactic injection of immunostimulants is seen as a feasible alternative to chemotherapy and vaccinations in aquaculture because of their wide effectiveness, relatively inexpensive, and environmentally friendly properties. The use of natural immunostimulants for the prevention and/or control of fish diseases in aquaculture has been endorsed by both the WHO and the FAO of the United Nations. This is intended to strengthen and stimulate fish's immune systems while averting future environmental, animal, and human health problems (Reverter et al., 2014; Cunha et al., 2018; Magnadottir et al., 2010). For the further expansion of the global aquaculture industry, immunostimulants that improve the innate defences of aquatic animals and increase their resistance to infection are a feasible alternative to antibiotic treatments (Fuchs et al., 2015). As a result of their ability to increase system efficiency, stimulate output, lessen chemical use, and make aquaculture products more palatable and secure, immunostimulants are likely to become increasingly widely used as feed additives in the aquaculture industry (Figure 4.2).

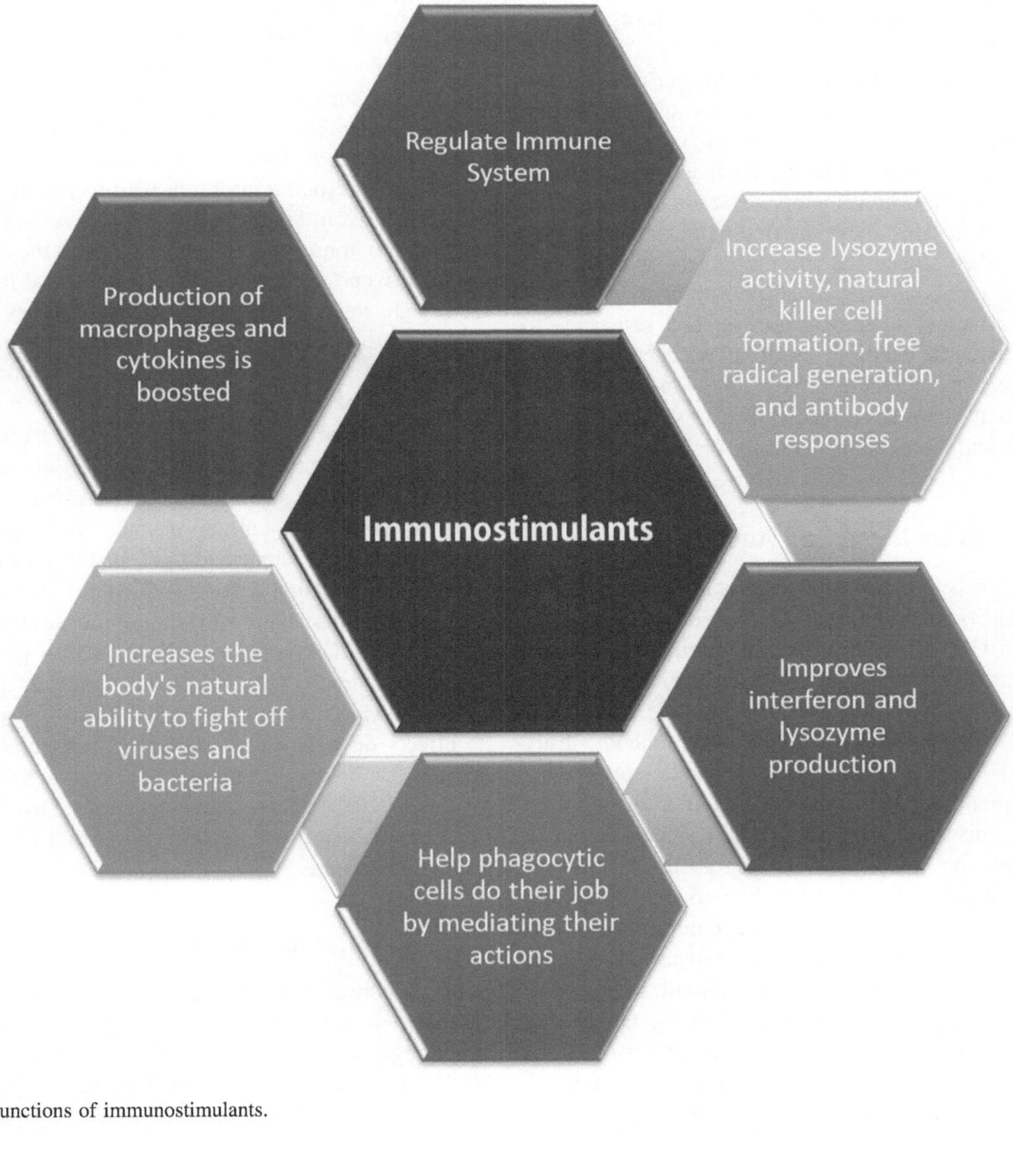

FIGURE 4.2 Functions of immunostimulants.

4.2 Stress and Disease in Fish

Fish and other animals experience stress as a general physiological response to harmful occurrences. When a stressor is seen, a stress response is generated very immediately. Even under low-stress levels situations, adaptive reactions can be elicited, but they can also lead to maladaptive or unfavourable results (distress). Both corticosteroids (mainly cortisol) and catecholamines are involved in the stress response (adrenaline, non-adrenaline, and its pioneer dopamine). As a whole, these components regulate the body's secondary stress response, which in turn affects the flow of vital nutrients and oxygen throughout the body, putting the immune system at danger. A stressor's response to a fish differs widely among taxa, stocks, and species because of the genetic diversity among these groups. The way a fish reacts to stress depends on a number of factors, including its past experiences, its current environment, and its current physiological state. Because unstressed fish can exist even in the absence of clinical stress markers, breakthroughs in fish physiology have made it possible to swiftly identify stressed fish. Stress-related clinical indications may have a tendency to falsely indicate a bad outcome. A stressor's severity cannot be determined with any precision or accuracy from clinical data.

Adding fresh, oxygenated drinkable freshwater can increase solubility, but it can also introduce pollutants and deprive the plants and animals being farmed of oxygen (in this example, fish). Physical and chemical factors, such as those already mentioned, are common contributors to stress. For decades, doctors have been at odds about the origins of the stress syndrome, but now they agree. The specific effects of stress on a human being's body and mind are that it can slow down or even stop the body's natural functions, growth, and resistance to disease, and even cause death. It's well knowledge that immunosuppression can have severe effects, but viruses are more prone to flourish in water. It's no secret that stress contributes to health problems and has been the subject of a lot of research.

4.2.1 Stress and Its Impact on Aquaculture

The concept of stress is both all-encompassing and universally known, although its scientific definition has been heavily disputed. Selye (1936, 1973) invented the term "stress" and described by way of "the non-specific response of the organism to any need for change"; however, he likewise stated that "everyone understands what stress is, but nobody truly knows." Selye (1936) comprised ideas in his attempt to characterise the response of the stress, calling it a "generalised effort of the organism to adjust itself to new situations, called the general adaptation syndrome," and contrasting it with other forms of generic self-defence. According to one definition, stress is "a state in which the dynamic equilibrium of the species, known as homeostasis, is threatened or altered as a result of an action of intrinsic or external stimuli that serve as stressors." The "disruption of homeostasis" by the threatening stimuli is a critical factor shared by most definitions of stress.

These definitions suggest that the stress response is an adaptation that helps an organism survive by restoring a state of homeostasis following a stressful event. Nevertheless, it is known that an animal's adaptive reaction might be insufficient, excessive, or prolonged, which can result in disease. In reaction to a stressor, the general adaptation syndrome develops in three stages. Hormones are released rapidly during the "alarm stage," the first step of the "fight or flight" response. The fight-or-flight reaction has been resolved and the normal condition has been restored during the resistance stage, which takes place within 48 hours of contact with the stressor. It is during this period of tiredness that animals lose their ability to withstand stress and surrender owing to the return of alarm-stage traits to their environment. The so-called chronic stress repercussions are now included in later definitions of stress as a result of the overstimulation of physiological and psychological reactions.

The animal's survival, the onset of a diseased state, or even death are all possible outcomes of a stressor. When it comes to the stress response, the amplitude and duration of physiological reactions are largely dictated by the type of stressor, whether acute or chronic (Davis, 2006). When faced with an acute stressor, the body's fight-or-flight reaction kicks in quickly, and the stress response is characterised by its extreme severity, brief duration, and rapid onset. Because of their longevity, chronic stressors induce more discomfort and maladaptation than short-term ones. This puts people's lives in jeopardy because of the greater energy costs (Ellis et al., 2012; Martnez-Porchas et al., 2009). Chronic strain in farmed fish must be taken into account because acute stress frequently leads in a positive stress response (Davis, 2006). The impacts of the stress response are often seen as negative since farmed fish cannot avoid the difficulties of aquaculture (Davis, 2006).

Many environmental and physical stressors, as well as social (competitiveness, crowding) and behavioural (aggressiveness) factors are involved in the production process of farmed fish compared to wild fish that are not subjected to these stressors (Braithwaite and Ebbesson, 2014). Disruption in the fish's welfare can be difficult to identify because it might be mistaken for normal and adaptive reactions to stress (Huntingford et al., 2006; Segner et al., 2012; Ellis et al., 2012). Reduced appetite, irregular swimming, and repetitive behaviours are some of the behavioural and physiological signs of persistent stress, along with stunted growth, inability to reproduce, and poor nutrition (Huntingford and Kadri, 2014). Aquaculture aims to maximise output by improving growth, resistance to disease, reproductive success, and quality in farmed fish by controlling stress. With this fundamental difficulty in aquaculture, a wide range of experiments on stress physiology and its effects on fish well-being have been published in recent decades, with several reviews of these studies (Pavlidis and Mylonas, 2011). Stress markers have also contributed to improvements in fish husbandry as a result of the rising focus on fish well-being.

4.2.2 Cellular and Molecular Indicators of Stress

Stress responses at the cellular and molecular level function in tandem with those that occur at higher levels of biological structure to provide a temporary resistance to stressors (Kultz, 2005). Heat shock factors (HSFs) interact with cortisol, which raises the production of metallothionein, ubiquitin, and HSPs as part of stress reactions Kassahn et al., 2009). The c-Jun element of transcription factor activation protein-1 (AP-1) and glucocorticoid receptors are also affected by cortisone. Cortisol has different transcriptional effects depending on tissue type, HSP 90 expression levels, and cortisol binding to glucocorticoid receptors. Damaged DNA and proteins are repaired, the cell cycle is stopped, waste products are eliminated, and the body shifts from an anabolic to a catabolic metabolic state as a response to various stresses. These cellular and molecular processes are caused by minimum-stress proteins (Faught et al., 2016; Kultz, 2005).

When a cell is subjected to oxidative stress, it releases more reactive oxygen species (ROSs). Oxidative stress contributes to protein damage. The degree of protein damage can be gauged by measuring the amount of ubiquitin-labelled protein in the cell. Because of the defective and ubiquitinated proteins, the cell's heat shock response focuses on protein repair. When a cell is exposed to a stressor that is both large and long-lasting, they are more likely to suffer from heat shock. When the hypothalamic–pituitary–interrenal (HPI) axis is activated, HSF1 regulates HSP expression by dissociating HSPs from HSF1. After making its way to the nucleus, HSF1 begins transcription of HSP (Kassahn et al., 2009). It's possible that ROS can speed up the telomere shortening that occurs as a result of oxidative stress. Telomeres, or the protective caps on the ends of chromosomes, are critical for genome stability and can hasten cell ageing if damaged too quickly. Growing data suggest that telomeres may be an indicator of cell and, possibly, organismal ageing (Monaghan et al., 2009). Stress, such as emotional straining or increased generative exertion, may impact telomere length (Kotrschal et al., 2007). Oxidative stress, which can lead to cell senescence and eventually organism senescence, has been linked to environmental stress (Monaghan et al., 2009).

Gene expression (quantitatively and qualitatively, as well as in terms of response coefficients) has been connected to a variety of stresses (Krasnov et al., 2005). It is currently common practise to employ genotyping and expression levels analysis to learn more about fish responses to environmental stresses. These well-known stressors include hypoxia (temperature) and hypoxic stress management (toxicants), which have been examined in several research (Jeffries, 2012). Clusters of differentially expressed genes associated to a wide range of metabolic pathways can be identified utilising cDNA microarray technology, which examines thousands of genes simultaneously. Compared to other stress indicators, such as primary, secondary, and tertiary, acute stressor exposure has been researched in a very small number of studies (Caipang et al., 2008). To understand better how environmental influences regulate expression of genes is rather one of the primary goals of functional genetic analysis (Buckley, 2007). Migrating salmon have an unfavourable profile that includes signs of bad health, such as decreased expression of blood clotting proteins and genetic factor linked to aerobic respiration and immunological function (Miller et al., 2011). Several genes and gene families are difficult to understand, and the ramifications of their responses to stimuli are not always evident. A wide variety of fish genes have been studied as a possible signal of various stresses (Jefferies et al., 2012).

Stress factors such as low water and air exposure raise the level of JUNB in rainbow trout (Momoda et al., 2007). In rainbow trout, stress-induced increases in the transcription factor NUPR1, which regulates cell development and death, were seen in a similar way (Momoda et al., 2007). As a result of stress, rainbow trout (*O. mykiss*) livers have already revealed gene expression alterations connected to wound, amino acid degradation, and the host–pathogen response (Momoda et al., 2007). According to the findings of Krasnov et al. (2005) a netting stressor alters gene levels of genes involved to immunological responses, cell growth, cell death, and synthesis of protein in rainbow trout, as well as gene expression in the kidney. Gene expression in the liver that is involved in anaerobic glycolysis, glycolysis, and glucose metabolism may be an indicator of post-stress recuperation (Momoda et al., 2007). Both male and female fish display unique genetic responses to an exercise stressor, according to Donaldson et al (2014). Stress has been tough to study because of this, and it's been difficult to figure out how different species, methods, and fish (Krasnov et al., 2005) all influence the activation of specific genes in the fish (Kassahn et al., 2009).

4.2.3 Primary and Secondary Indicators of Physiological Function

The whole animal's physiology is impacted by stress indicators, which include cellular and molecular reactions and systemic effects. The HPI axis is only one of the many physiological responses triggered by stress (Schreck and Tort, 2016). Researchers may utilise a variety of basic indicators (for example, glucose levels or ion balances) and secondary reactions (such as changes in acid–base balance and immunological processes) to measure stress in fish (for example, catecholamines or stress hormones). The pace of subsequent adjustments slows. Many stress indicators have advantages and disadvantages depending on the stressor that is being studied and how much contextual evidence is obtainable about the population being researched. Short-term field disturbances can have a major impact on stress indicators, including main or secondary stress markers, depending on the aquaculture and handling practises used. Catecholamines are the first stress hormones to be released, but because of their fast release, it is difficult to detect them. As a result, it is extremely difficult to measure catecholamine levels in the field. Also known as the stress response marker cortisol, this hormone is one of the most often used in stress testing.

In contrast to catecholamines, cortisol is stable and can be found in either labs or the field, provided animals are analysed within minutes of capture. Secondary stress indicators include glucose catabolism and an increase in anaerobic metabolism after hypoxia or exercise stressors, in addition to glucose and lactate increases. Three factors can be affected by an increase in heart rate and the permeability of the gills: (1) osmolality, (2) specific ions, and (3) leukocytes. As well as indicating stress responses, these factors can also be used to predict outcomes such as a person's survival or performance. It is essential that these markers be assessed in plasma in order to acquire an accurate picture of their concentrations in the bloodstream. It is possible to collect plasma (or red blood cells for measurements like hematocrit) without injuring the fish, which is a benefit (Beecham et al., 2006). Not all signs make sense when seen in plasma, even if it is the most prevalent. Plasma-signal detection has this as its primary drawback. Glycogen buildup in the liver as a result of chronic stressor exposure cannot be detected in a plasma sample alone. Muscle tissue is preferred for measuring PCr and ATP since these markers are acute exercise stress indicators. Thus, the choice of tissue is determined by the nature of the question being investigated and the primary or secondary physiological signal that is to be analysed. Environmental obstacles cause stress in both people and animals, and these creatures' stress levels can be measured using secondary and main stress markers (Breuner et al., 2008).

4.2.4 Tertiary/Whole-Organism Indicators of Physiological Function

Just a few of the various whole-organism (or tertiary) reactions to stress can be detected, such as changes in development, health, and health-related behaviours. Environmentally relevant stress indicators are often referred to as this species. Bioenergetics and population models are affected by growth rates, fitness, and survival, although fitness and survival are also relevant to demographic processes. Everything from food intake to predator avoidance to habitat selection has a physiological basis. There are several condition-related metrics that rely on simple measures of mass or length (or both). Their fish condition measurements have been criticised for being too simplistic and lacking in detail. Despite their relative complexity, organosomatic indices (such as the ratio of organ mass to body mass) are commonly utilised as a stress and condition assessment.

Researchers can tailor their HAI selections to their individual needs and the various environments in which their studies are conducted. Custom-built swim tunnels and flumes are the most precise means for assessing a swimmer's swimming ability, but other methods are available. Despite its present lack of significance, critical swimming speed is still important for evaluating the swimming ability of fish subjected to various stressor treatments. In 1978, tail-pinning data collected in a round tank and compared to more formal studies of essential swimming speed were shown to be statistically significant. Resilience to stress is measured by the "recovery ratio," which has been applied to a wide range of organisms. Measures of cardiac activity (such as vital signs or myocardial contractility), metabolic activity (oxygen intake), and effective swimming activities are all necessary in studies examining the effects of stress on ecologically relevant metrics of organism performance.

An accurate technique for predicting mortality in the field can be established by scaling metabolic, cardiovascular, and swimming capacity from laboratory data. In comparison to primary or secondary stress reactions, tertiary stress indicators have access to a broader range of data. Tertiary stress response data may be collected and analysed by field ecologist, ethologist or physiologist. In contrast, the physiological basis of tertiary stress markers is clearer. However, physiologists also investigate other parts of behaviour, such as swimming, as well as the heart and metabolism as a whole (both the performance of swimming and general locomotion). Metabolic and locomotor activity have a significant impact on energy consumption. Energy budgets dictate somatic growth rates and circumstances, and behaviour. The metabolic and cognitive functions of fish are more closely linked to their survival than previously thought (Davis, 2010).

Markers of homeostasis as a whole (such as neuroendocrine activity) make it far more challenging to infer a stressed condition than tests that concentrate on secondary or primary determinants of homeostasis (such as temperature or blood pressure). A fish's death isn't always an indication that it was stressed; ageing is an inevitable part of life, and fish can be preyed on even if they weren't stressed. Fish's previous feeding histories and metabolic demands, as well as the availability of food at different times of the year and heredity, all influence their feeding habits. In the case of stress signals like cortisol and osmolality, reference ranges and thresholds are well defined. If a single fish's cortisol levels are greater than the standard, but there are no clear goals at the organism level, it's feasible. This may or may not endanger the fish, one cannot be sure. It's vital to compare tertiary outcomes to acceptable controls while stressing out a system. However, despite these difficulties, collecting and analysing tertiary indicators is simpler than other laboratory-based physiological stress assessments. Survival and response indicators (such as the presence of a given reaction) are two simple, obvious, critical measurements of total stress in an organism (e.g., did an organism live or die).

In the field, researchers frequently use the presence or absence of certain reflex dysfunctions as an indicator of an animal's health and a possible death forecast (Raby et al., 2012; Davis, 2010). This reflex dysfunction has yet to be linked to a specific molecular aetiology. In free-swimming fish, secondary stressors can be measured. Examples of this type of data include locomotor activity and heart activity at small scales (Clark et al., 2010). Either wirelessly (using biotelemetry/biosensors) or on the animal's body for future retrieval, electronic tags can be used to assess blood glucose levels in animals. Biologging is a term used in the field of biometrics to describe this process (Raby et al., 2012).

4.2.5 Considerations for Measuring and Interpreting Stress

4.2.5.1 Interspecific Differences

To be expected, different fish species react differently to the same pressures in the environment. Given the evolutionary, ecological, environmental, and life cycle variables that contribute to species divergence, it is anticipated that HPI activity and other stress markers will vary among species. It is difficult to compare the HPI stress response amplitude across species because of the large diversity of stressor types, severity, and duration in numerous investigations. Cortisol and glucose levels, gene expression, immune function, habitat preference, and avoidance behaviour are all affected simultaneously by stress (Jefries et al., 2014), as well as other factors such as gene expression, immune function, and habitat preference (Jacobsen et al., 2014). Stress responses might be similar amongst various animals in some circumstances. Sablefish juveniles *Anoplopoma fimbria* are as vulnerable to capture mortality as walleye pollock juveniles *Theragra chalcogramma.*

4.2.5.2 Intraspecific Differences

Stress analysis is typically inspired by the emergence of diverse populations. Populations from different parts of the world are likely to have different stress responses due to a variety of environmental and biological factors. Indeed, there was a longitudinal gradient in the stress-induced plasma cortisol responses of killifish. If the fish were caught in an area with low or high levels of predators, the outcomes would be different. After stressor exposure, fitness outcomes can differ between groups even when there are no differences in physiological stress markers (e.g., plasma cortisol and lactate) (Donaldson et al., 2012). From an evolutionary vantage point, it is important to account for population variability in stress response before drawing species-wide generalisations about the sensitivity of an organism to a given stressor.

Gender, size, social rank, and domestication are among the other intraspecific characteristics that influence stress indicator variance. When it comes to some animals, it is easy to see the variations between sexes. High stress-induced plasma cortisol levels and an increased risk of death were seen in adult female Pacific salmon (*Oncorhynchus* spp.) exposed to high temperatures (Donaldson et al., 2014). Size matters when it comes to European sea bass's maximal stressor-induced plasma cortisol response (Fatira et al., 2014). HPI activity varies between dominant and subordinate rainbow trout (Jeffrey et al., 2014). Fish in captivity or put back into the wild may not do as well if they are stressed out to the same degree as hatchery or wild trout or salmon. At long last, there is growing recognition of the heritability of stress-responsiveness and the differences between individuals and families, as well as the connection between physiological and behavioural stress indicators within an individual (Kittilsen and Pottinger, 2009; Hori and Pottinger, 2012; Castanheira et al., 2017). To develop genetically distinct strains of rainbow trout (*O. mykiss*) that respond differently to a three-hour confinement stressor, aquaculture has been using this variation for more than a decade. This variation is now related to a variety of behavioural stress markers that are regularly found in circulation. Tolerance to stress sampling methodologies may be affected by these individual factors. Stress indicators can be influenced by a variety of intraspecific variable components, and it is vital to account for and explain this variation.

4.2.5.3 Context-Specific Differences

There may be differences in HPI axis activity and supplementary molecular physiological and whole-animal responses depending on the specific ecological and environmental conditions to which an individual is subjected. The HPI axis should always be initiated in a consistent, reliable manner. All fitness losses must be accepted and overcome through HPI-mediated physiological and behavioural changes (Schreck and Tort, 2016). An individual can have physiological and behavioural changes if they are confronted with a predator, competition for resources or a lack of food. Stress assessments vary depending on the environment in which they are taken at all organisational levels. Stressor type influenced the concentrations of brain monoamines (e.g., norepinephrine and serotonin) in three-spined sticklebacks (Bell et al., 2010). Plasma responses may not be affected by minor changes in stressor type (Ros et al., 2014). Identical reactions can be triggered by two stressors that couldn't be more dissimilar in terms of environmental significance and severity. The three-spined stickleback's plasma cortisol levels rise in reaction to an implanted elastomer tag, which is visible to the naked eye, much as they would after experiencing a real predator attack (Furtbauer et al., 2015). The stressor's circumstances also have an effect on secondary, physiological processes. *Galaxias maculatus* showed varying degrees of metabolic changes when exposed to visual or olfactory predator cues (Milano et al., 2010). Stressors affect whole-animal breathing rates differently. Stressor intensity is most likely linked to environmental and ecological context when determining which indicator to employ, predicting how the indicator will respond, and interpreting the results.

If stress is to be judged scientifically, it is best to expose animals to a variety of stressors. Experimentation with various stressors can be utilised to determine the magnifying power of stress markers (Jacobsen et al., 2014; Marcogliese et al., 2005). Fish, like humans, are subjected to repeated stressors, and much like humans, fish may exhibit different stress responses depending on how much experience they've had with a certain stressor and their potential for learning (Barcellos et al., 2010). Before exposure to stressors (e.g., isolation versus groups), housing environments (e.g., olfactory indications of conspecific stress) may also alter stress indicators (Giacomini et al., 2015; Barcellos et al., 2014). Fishermen may prefer to test individual fish rather than groups because of the species' schooling and social behaviour. If biological significance of stress markers is the research priority, such intricacies of ecological and environmental context can be investigated. In other cases, a wide range of acceptable indicators can be used. Gene expression, gamete quality, and a number of physiological processes and behaviours are all affected by toxicological stresses (Jeffries et al., 2015; Sloman and McNeil, 2012). Responses can vary even within a single class of symptoms (for example, rises in lactate concentration suggesting insufficient oxygen delivery to muscles but no change in HSP expression indicating no protein degradation). As a result, when determining which stress indicators are better to monitor, it is necessary to take the stressor or stressors' type and the interplay of signs into account.

4.2.5.4 Stressor Severity

Acute or single exposure is characterised by physiological adaptations and brief exposure durations. The consequences of an acute stressor may be reflected in variations in the levels of stress indicators over time (Gesto et al., 2015, 2013). Depending on the anticipated level of impairment (e.g., a two-minute confinement versus a five-minute net pursuit), acute stressors may have varying intensities. It is typical to see relationships between the concentration of pollutant stressors and the rate of indicator reaction or mortality (i.e., LC50). Similar reactions can be provoked by stressors that are thought to be different from one another in terms of severity, but which are actually the same save for their duration (Fatira et al., 2014). Repetitive, consecutive exposure to an acute stressor (such as routine handling over several weeks) or protracted, uninterrupted exposure to a stressor both qualify as examples of what is known as "chronic stress". On the reaction of an indicator to various exposures, short-term (e.g., avoidance behaviour) and long-term (e.g., growth) fitness repercussions may have different effects. Additionally, several animals are exposed to chronic stressors. Adaptive responses may occur in response to indicators of a chronically stressful condition (such as allostatic overload) (Boonstra, 2013). Thus, while determining whether "disrupted negative feedback" or allostatic overload is to blame for stress marker modifications, habituation must be taken into account (Schreck and Tort, 2016).

Hormonal stress indices, distinguish between habituation, desensitisation, and fatigue. Behavioural/performance indicators may not be as useful in detecting habituation as physiological markers (such as HPI axis activity). Repeated acute stressor exposure may be tracked halfway through by monitoring the HPI axis and stress indicator in certain animals (ie, immediately following daily handling stressor delivered over multiple weeks). It is conceivable to expose a subgroup of animals to an acute stressor in the midst of a longer stressor exposure to guarantee the HPI axis and the target stress indicator are responsive throughout the exposure (i.e., single, acute chase stressor during continuous exposure to increased water temperature over numerous weeks). Many chronically stressed species have a lower endocrine response to stress (Barcellos et al., 2006; Jentoft et al., 2005).

According to Wingfield et al. (2011), stress becomes maladaptive after a certain threshold. When thresholds are crossed, a stressor's resistance potential rises due to attenuated stress response (Wingfield et al., 2011). It's feasible to determine thresholds at which stress indicators no longer respond or revert to pre-stressor values by taking multiple samples during chronic exposure (exhaustion). It's vital to incorporate phenotypic plasticity when analysing stress in wild populations because it's a chronic stressor in environments with changing conditions (Crozier and Hutchings, 2014). Due to coping mechanisms, lab-based stress markers may not be reliable in wild animals. In the lab, wild populations may show indications of chronic stress, but these may not affect the population as a whole. The round goby population is steady in contaminated regions but dropping in reference areas (McCallum et al., 2014). These new stable states in organisms can be viewed as stress while not influencing population dynamics.

4.2.5.5 Field versus Laboratory

Both in-lab and on-site stress testing have advantages and disadvantages. Because of the complexity of the instruments used to gather samples, certain stress indicators can only be analysed in a lab (for example, serial blood sampling via cannulations). Since the behaviour cannot be reproduced in a controlled laboratory setting, additional indicators will require measurements to be taken in the wild. It is easy to regulate variables that can skew stress indicator measurements in controlled laboratory investigations, but this is not the case in the field.

For studies examining the mechanistic links between stressors and indicators, this degree of control is essential. Reproducing stresses that fish might experience in the field in a laboratory environment is insufficient to capture the full response to stress in an animal under ecologically suitable settings. Stress indicators, for instance, don't reflect the long-term effects of stress (e.g., post-release predation after fisheries capture, a stressor that upsets equilibrium) because they're tested in a controlled environment (Danylchuk et al., 2007). It's possible that doing stress tests in the wild would be better to a controlled laboratory setting, or vice versa, depending on the study's hypotheses and goals. With the advent of remote sensing tools and the validation of point-of-care devices in fishes, it is now possible to monitor markers that were previously only achievable in the lab (Stoot et al., 2014).

Advances in biologging and biotelemetry now make it possible to track fish before, during, and after they are exposed to stressors. Free-swimming fish experience shifts in their activity, vigour, foraging, and heart rate (Clark et al., 2008; Donaldson et al., 2010; Raby et al., 2015). There are a number of issues with mechanical stress detection in the field, including expensive costs, surgical procedures for tag installation, inefficient detection, and challenging tag retrieval. Studying wild fish that have been subjected to unpredictable, variable pressures (such as cyclones or floods) can be instructive, but it is also quite risky (Wingfield, 2013). Multidisciplinary approaches that combine field research with laboratory investigations to isolate pathways provide the fullest picture of the stress response possible.

4.2.5.6 Temporal Aspects

There are numerous levels of timing to consider when looking at animal stress. First, what age group is the stress indicator appropriate for? Several species have well-developed hyporesponsive periods in the early stages of development. Hyperresponsive phases can be seen in smolting salmon as well. Stress-induced plasma cortisol peaks were shown to be different in concentration and onset time in fingerling, juvenile, and adult jundia (*Rhamdia quelen*). Cortisol, a stress hormone, has been the primary focus of research into the impacts of ageing up until recently (Schreck and Tort, 2016). Changing stress signals during life may make it difficult to accurately measure stress. Some physiological performance and behavioural symptoms of stress might change over time. The fact that some species may be able to change their circadian rhythms should be kept in mind, though.

There will be a variation in induction and recovery timeframes based on the diagnosis (Gesto et al., 2015). Cortisol takes longer to climb above prestressed levels than catecholamines (and other sympathetic nervous system activity like pulse rate and respiration). An interplay between stress and variations in mRNA abundance is hypothesised (i.e., transcription). If the stressor is mild or severe, the induction and recovery periods may also be different. However, when treatment-level repercussions (for example, stressor kind or severity) are important (e.g., the stress response's onset time), temporal influences should also be taken into account.

4.3 Modulation of the Immune System by Immunostimulants

We now have a better understanding of fish's innate immune system because of immunostimulant research, and we also have an alternative technique for preventing common illnesses in fish farming (Galina et al., 2009). The global impact of disease resistance, toxicity, and environmental impact have all contributed to antibiotic resistance; however, dietary variables (vitamin C and selenium) may enhance the effects of immunostimulants, but no nutritional element can be termed an immunostimulant on its own. Specific or non-specific pathways may be activated by Immunostimulants, which modify the immune system and improve the fish's immunocompetence. Several immunostimulants, including glucan, have been demonstrated to establish early immunity, reduce mortality, and increase life span after their administration (Rojo-Cebreros et al., 2018; Petit and Wiegertjes, 2016). The term "trained immunity" which is also known as "prolonged innate defence" refers to the immune system's ability to recognise a pathogen and then mount a more comprehensive defence against the same infectious agent upon a subsequent infection. This can establish cross protection, which may be aided by epigenetic mechanisms (Rojo-Cebreros et al., 2018; Netea et al., 2016). The LPS, peptidoglycans, and the glucans have also been demonstrated to promote non-specific immune responses in fish, as have synthetic chemicals, polysaccharides (chitin), vitamins, and hormones. There have been studies showing that these immunostimulants (Chettri et al., 2013; Maqsood et al., 2009) can improve immune responses and disease resistance in aquaculture.

Anticancer and immunomodulatory properties were discovered in a study using honey bee propolis, a bee honey product. The study looked at how propolis affects antibody and immune cell production, including innate and acquired responses. Inhibition of lymphoproliferation may be a mechanism via which it has anti-inflammatory properties. Fish have also been treated with propolis (Cuesta et al., 2005). Using medicinal herbs in aquaculture, on the other hand, has been shown to be an alternative strategy (Galina et al., 2009). Numerous studies have demonstrated the effectiveness of various plants in boosting fish immunity and shielding them against bacterial infections (Zanuzzo et al., 2017; Zanuzzo et al., 2012). Along with its extracts, garlic (*Allium sativum*) has been utilised for thousands of years as a natural immunostimulant. Many fish diseases, as well as their latent use in aquaculture, have lately been examined by researchers using garlic (Lee and Gao, 2012). *Aeromonas hydrophila* infections are protected by garlic in fish, and the innate immune system is stimulated by increased neutrophil oxidative radical formation, lymphocyte proliferation, and the phagocytic activity of cephalic kidney macrophages (Nya and Austin, 2009).

4.4 Future Perspectives and Conclusion

There are numerous studies concerning the impact of nutritional additives in fish from seaweed, herbal products, and plant phytochemical in a non-specific form based on tradition and folklore transmitted from one generation, and their physiological features are generally analysed in vitro or, at times, in animal experiments. An efficient and ecologically acceptable farming depends on the advantages of immunostimulants. In order to maximise the advantages and minimise immunosuppression, studies to determine the optimum protocol are needed to determine the dose/concentration, duration, and route of administration of these components. The process of receptor identification, extraction structure, and stimulation of fish immune function are yet fragmented, despite certain research showing a clear and dose/concentration dependent stimulatory action of medicinal plants. As a result, reaching an agreement on this issue is a top priority. As a final step, more studies employing high-throughput screening technologies are required to better understand the genomic and proteomic responsiveness of fish to immunostimulants in order to fully understand how the dietary supplementation of immunostimulants affects fish.

REFERENCES

Ali, D., F. A. Falodah, B. Almutairi, S. Alkahtani, and S. Alarifi. "Assessment of DNA damage and oxidative stress in juvenile *Channa punctatus* (Bloch) after exposure to multi-walled carbon nanotubes." *Environmental Toxicology* 35, no. 3 (2020): 359–367.

Aniya, Y. "Development of bioresources in Okinawa: understanding the multiple targeted actions of antioxidant phytochemicals." *Journal of Toxicology and Pathology* 31 (2018) 241–253, 10.1293/tox.2018-0041.

Ardó, L., G. Yin, P. Xu, L. Váradi, G. Szigeti, Z. Jeney, and G. Jeney. "Chinese herbs (*Astragalus membranaceus* and *Lonicera japonica*) and boron enhance the non-specific immune response of Nile tilapia (*Oreochromis niloticus*) and resistance against *Aeromonas hydrophila.*" *Aquaculture* 275, no. 1-4 (2008): 26–33.

Barcellos, L. J. G., F. Ritter, L. C. Kreutz, and L. Cericato. "Can zebrafish *Danio rerio* learn about predation risk? The effect of a previous experience on the cortisol response in subsequent encounters with a predator." *Journal of Fish Biology* 76, no. 4 (2010): 1032–1038.

Barcellos, L. J. G., G. Koakoski, J. G. S. da Rosa, D. Ferreira, R. E. Barreto, P. C. Giaquinto, and G. L. Volpato. "Chemical communication of predation risk in zebrafish does not depend on cortisol increase." *Scientific Reports* 4, no. 1 (2014): 1–7.

Barcellos, L. J. G., L. C. Kreutz, and R. M. Quevedo. "Previous chronic stress does not alter the cortisol response to an additional acute stressor in jundiá (Rhamdia quelen, Quoy and Gaimard) fingerlings." *Aquaculture* 253, no. 1-4 (2006): 317–321.

Beecham, R. V., B. C. Small, and C. D. Minchew. "Using portable lactate and glucose meters for catfish research: acceptable alternatives to established laboratory methods?." *North American Journal of Aquaculture* 68, no. 4 (2006): 291–295.

Bell, A. M., L. Henderson, and F. A. Huntingford. "Behavioral and respiratory responses to stressors in multiple populations of three-spined sticklebacks that differ in predation pressure." *Journal of Comparative Physiology B* 180, no. 2 (2010): 211–220.

Beltrán, J. M. G., and M. Á. Esteban. "Nature-identical compounds as feed additives in aquaculture." *Fish & Shellfish Immunology* 123 (2022): 409–416.

Boonstra, R. "Reality as the leading cause of stress: rethinking the impact of chronic stress in nature." *Functional Ecology* 27, no. 1 (2013): 11–23.

Braithwaite, V. A., and L. O. E. Ebbesson. "Pain and stress responses in farmed fish." *Reviews in Science and Technology* 33, no. 1 (2014): 245–253.

Breuner, C. W., S. H. Patterson, and T. P. Hahn. "In search of relationships between the acute adrenocortical response and fitness." *General and Comparative Endocrinology* 157, no. 3 (2008): 288–295.

Buckley, B. A. "Comparative environmental genomics in non-model species: using heterologous hybridization to DNA-based microarrays." *Journal of Experimental Biology* 210, no. 9 (2007): 1602–1606.

Caipang, C. M. A., M. F. Brinchmann, I. Berg, M. Iversen, R. Eliassen, and V. Kiron. "Changes in selected stress and immune-related genes in Atlantic cod, *Gadus morhua*, following overcrowding." *Aquaculture Research* 39, no. 14 (2008): 1533–1540.

Castanheira, M. F., L. E. C. Conceição, S. Millot, S. Rey, M.-L. Bégout, B. DamsgAard, T. Kristiansen, E. Höglund, Ø. Øverli, and C. I. M. Martins. "Coping styles in farmed fish: consequences for aquaculture." *Reviews in Aquaculture* 9, no. 1 (2017): 23–41.

Chettri, J. K., P. W. Kania, and K. Buchmann. "Immunomodulation of rainbow trout (*Oncorhynchus mykiss*) fry by bath exposure to a β-glucan from *Euglena gracilis*." *Aquaculture Research* 44, no. 9 (2013): 1407–1415.

Choi, H.-I., H.-W. Lee, T.-M. Eom, S.-A. Lim, H.-Y. Ha, I.-C. Seol, Y.-S. Kim, D.-S. Oh, and H.-R. Yoo. "A traditional Korean multiple herbal formulae (Yuk-Mi-Jihwang-Tang) attenuates acute restraint stress-induced brain tissue oxidation." *Drug and Chemical Toxicology* 40, no. 2 (2017): 125–133.

Clark, T. D., B. D. Taylor, R. S. Seymour, D. Ellis, J. Buchanan, Q. P. Fitzgibbon, and P. B. Frappell. "Moving with the beat: heart rate and visceral temperature of free-swimming and feeding bluefin tuna." *Proceedings of the Royal Society B: Biological Sciences* 275, no. 1653 (2008): 2841–2850.

Clark, T. D., E. Sandblom, S. G. Hinch, D. A. Patterson, P. B. Frappell, and A. P. Farrell. "Simultaneous biologging of heart rate and acceleration, and their relationships with energy expenditure in free-swimming sockeye salmon (*Oncorhynchus nerka*)." *Journal of Comparative Physiology B* 180, no. 5 (2010): 673–684.

Crozier, L. G., and J. A. Hutchings. "Plastic and evolutionary responses to climate change in fish." *Evolutionary Applications* 7, no. 1 (2014): 68–87.

Cuesta, A., A. Rodrı́guez, M. A. Esteban, and J. Meseguer. "In vivo effects of propolis, a honeybee product, on gilthead seabream innate immune responses." *Fish & Shellfish Immunology* 18, no. 1 (2005): 71–80.

Cunha, J. A., B. M. Heinzmann, and B. Baldisserotto. "The effects of essential oils and their major compounds on fish bacterial pathogens – a review." *Journal of Applied Microbiology* 125, no. 2 (2018): 328–344.

Danylchuk, S. E., A. J. Danylchuk, S. J. Cooke, T. L. Goldberg, J. Koppelman, and D. P. Philipp. "Effects of recreational angling on the post-release behavior and predation of bonefish (Albula vulpes): the role of equilibrium status at the time of release." *Journal of Experimental Marine Biology and Ecology* 346, no. 1-2 (2007): 127–133.

Davis, K. B. "Management of physiological stress in finfish aquaculture." *North American Journal of Aquaculture* 68, no. 2 (2006): 116–121.

Davis, M. W. "Fish stress and mortality can be predicted using reflex impairment." *Fish and Fisheries* 11, no. 1 (2010): 1–11.

Dawood, M. A. O., S. Koshio, M. Ishikawa, and S. Yokoyama. "Interaction effects of dietary supplementation of heat-killed *Lactobacillus plantarum* and β-glucan on growth performance, digestibility and immune response of juvenile red sea bream, Pagrus major." *Fish & Shellfish Immunology* 45, no. 1 (2015): 33–42.

Dawood, M. A. O., S. Koshio, and M. Á. Esteban. "Beneficial roles of feed additives as immunostimulants in aquaculture: a review." *Reviews in Aquaculture* 10, no. 4 (2018): 950–974.

Donaldson, M. R., S. G. Hinch, G. D. Raby, D. A. Patterson, A. P. Farrell, and S. J. Cooke. "Population-specific consequences of fisheries-related stressors on adult sockeye salmon." *Physiological and Biochemical Zoology* 85, no. 6 (2012): 729–739.

Donaldson, M. R., T. D. Clark, S. G. Hinch, S. J. Cooke, D. A. Patterson, M. K. Gale, P. B. Frappell, and A. P. Farrell. "Physiological responses of free-swimming adult coho salmon to simulated predator and fisheries encounters." *Physiological and Biochemical Zoology* 83, no. 6 (2010): 973–983.

Donaldson, M. R., S. G. Hinch, K. M. Jeffries, D. A. Patterson, S. J. Cooke, A. P. Farrell, and K. M. Miller. "Species-and sex-specific responses and recovery of wild, mature pacific salmon to an exhaustive exercise and air exposure stressor." *Comparative Biochemistry and Physiology Part A: Molecular & Integrative Physiology* 173 (2014): 7–16.

Dontha, S. "A review on antioxidant methods." *Asian Journal of Pharmaceutical and Clinical Research* 9, no. 2 (2016): 14–32.

Ellis, T., H. Y. Yildiz, J. López-Olmeda, M. T. Spedicato, L. Tort, Ø. Øverli, and C. I. M. Martins. "Cortisol and finfish welfare." *Fish Physiology and Biochemistry* 38, no. 1 (2012): 163–188.

Fatira, E., N. Papandroulakis, and M. Pavlidis. "Diel changes in plasma cortisol and effects of size and stress duration on the cortisol response in European sea bass (Dicentrarchus labrax)." *Fish Physiology and Biochemistry* 40, no. 3 (2014): 911–919.

Faught, E., N. Aluru, and M. M. Vijayan. "The molecular stress response." In *Fish physiology*, vol. 35, pp. 113–166. Academic Press, 2016.

Fuchs, V. I., J. Schmidt, M. J. Slater, J. Zentek, B. H. Buck, and D. Steinhagen. "The effect of supplementation with polysaccharides, nucleotides, acidifiers and Bacillus strains in fish meal and soy bean based diets on growth performance in juvenile turbot (Scophthalmus maximus)." *Aquaculture* 437 (2015): 243–251.

Fürtbauer, I., A. J. King, and M. Heistermann. "Visible implant elastomer (VIE) tagging and simulated predation risk elicit similar physiological stress responses in three-spined stickleback *Gasterosteus aculeatus*." *Journal of Fish Biology* 86, no. 5 (2015): 1644–1649.

Gao, J., S. Koshio, M. Ishikawa, S. Yokoyama, B. T. Nguyen, and R. E. Mamauag. "Effect of dietary oxidized fish oil and vitamin C supplementation on growth performance and reduction of oxidative stress in Red Sea Bream *Pagrus major*." *Aquaculture Nutrition* 19, no. 1 (2013): 35–44.

Gesto, M., M. A. López-Patiño, J. Hernández, J. L. Soengas, and J. M. Míguez. "Gradation of the stress response in rainbow trout exposed to stressors of different severity: the role of brain serotonergic and dopaminergic systems." *Journal of Neuroendocrinology* 27, no. 2 (2015): 131–141.

Gesto, M., M. A. López-Patiño, J. Hernández, J. L. Soengas, and J. M. Míguez. "The response of brain serotonergic and dopaminergic systems to an acute stressor in rainbow trout: a time course study." *Journal of Experimental Biology* 216, no. 23 (2013): 4435–4442.

Ghazanfar, M., S. Shahid, and I. Z. Qureshi. "Vitamin C attenuates biochemical and genotoxic damage in common carp (*Cyprinus carpio*) upon joint exposure to combined toxic doses of fipronil and buprofezin insecticides." *Aquatic Toxicology* 196 (2018): 43–52.

Giacomini, A. C. V. V., M. S. de Abreu, G. Koakoski, R. Idalêncio, F. Kalichak, T. A. Oliveira, J. G. S. da Rosa, D. Gusso, A. L. Piato, and L. J. G. Barcellos. "My stress, our stress: blunted cortisol response to stress in isolated housed zebrafish." *Physiology & Behavior* 139 (2015): 182–187.

Giri, S. S., H. J. Kim, S. G. Kim, S. W. Kim, J. Kwon, S. B. Lee, V. Sukumaran, and S. C. Park. "Effectiveness of the guava leaf extracts against lipopolysaccharide-induced oxidative stress and immune responses in Cyprinus carpio." *Fish & Shellfish Immunology* 105 (2020): 164–176.

Gozdowska, M., E. Sokołowska, K. Pomianowski, and E. Kulczykowska. "Melatonin and cortisol as components of the cutaneous stress response system in fish: Response to oxidative stress." *Comparative Biochemistry and Physiology Part A: Molecular & Integrative Physiology* 268 (2022): 111207.

Hossain, M. S., S. Koshio, M. Ishikawa, S. Yokoyama, N. M. Sony, M. A. Kader, M. Maekawa, and T. Fujieda. "Effects of dietary administration of inosine on growth, immune response, oxidative stress and gut morphology of juvenile amberjack, Seriola dumerili." *Aquaculture* 468 (2017): 534–544.

Hori, T. S., A. K. Gamperl, C. E. Hastings, G. E. V. Voort, J. Andrew, B. Robinson, S. C. Johnson, and L. O. B. Afonso. "Inter-individual and-family differences in the cortisol responsiveness of Atlantic cod (Gadus morhua)." *Aquaculture* 324 (2012): 165–173.

Huntingford, F. A., and S. Kadri. "Defining, assessing and promoting the welfare of farmed fish." *Revue scientifique et technique (International Office of Epizootics)* 33, no. 1 (2014): 233–244.

Huntingford, F. A., Adams, C., Braithwaite, V. A., Kadri, S., Pottinger, T. G., Sandoe, P., & Turnbull, J. F. Current issues in fish welfare. *Journal of Fish Biology*, 68, (2006): 332–372. 10.1111/j.0022-1112.2006.001046.x

Iwama, G. K., P. T. Thomas, R. B. Forsyth, and M. M. Vijayan. "Heat shock protein expression in fish." *Reviews in Fish Biology and Fisheries* 8, no. 1 (1998): 35–56.

Jacobsen, L., H. Baktoft, N. Jepsen, K. Aarestrup, S. Berg, and C. Skov. "Effect of boat noise and angling on lake fish behaviour." *Journal of Fish Biology* 84, no. 6 (2014): 1768–1780.

Jeffrey, J. D., M. J. Gollock, and K. M. Gilmour. "Social stress modulates the cortisol response to an acute stressor in rainbow trout (Oncorhynchus mykiss)." *General and Comparative Endocrinology* 196 (2014): 8–16.

Jeffries, K. M., S. G. Hinch, T. Sierocinski, P. Pavlidis, and K. M. Miller. "Transcriptomic responses to high water temperature in two species of P acific salmon." *Evolutionary Applications* 7, no. 2 (2014): 286–300.

Jeffries, K. M., S. G. Hinch, T. Sierocinski, T. D. Clark, E. J. Eliason, M. R. Donaldson, S. Li, P. Pavlidis, and K. M. Miller. "Consequences of high temperatures and premature mortality on the transcriptome and blood physiology of wild adult sockeye salmon (Oncorhynchus nerka)." *Ecology and Evolution* 2, no. 7 (2012): 1747–1764.

Jeffries, K. M., Brander, S. M., Britton, M. T., Fangue, N. A., & Connon, R. E. Chronic exposures to low and high concentrations of ibuprofen elicit different gene response patterns in a euryhaline fish. *Environmental Science and Pollution Research*, 22, (2015): 17397–17413. 10.1007/s11356-015-4227-y

Jentoft, S., A. H. Aastveit, P. A. Torjesen, and Ø. Andersen. "Effects of stress on growth, cortisol and glucose levels in non-domesticated Eurasian perch (Perca fluviatilis) and domesticated rainbow trout (Oncorhynchus mykiss)." *Comparative Biochemistry and Physiology Part A: Molecular & Integrative Physiology* 141, no. 3 (2005): 353–358.

Kassahn, K. S., R. H. Crozier, H. O. Pörtner, and M. J. Caley. "Animal performance and stress: responses and tolerance limits at different levels of biological organisation." *Biological Reviews* 84, no. 2 (2009): 277–292.

Kittilsen, S., T. Ellis, J. Schjolden, B. O. Braastad, and Ø. Øverli. "Determining stress-responsiveness in family groups of Atlantic salmon (Salmo salar) using non-invasive measures." *Aquaculture* 298, no. 1-2 (2009): 146–152.

Kotrschal, A., P. Ilmonen, and D. J. Penn. "Stress impacts telomere dynamics." *Biology Letters* 3, no. 2 (2007): 128–130.

Krasnov, A., H. Koskinen, P. Pehkonen, C. E. Rexroad, S. Afanasyev, and H. Mölsä. "Gene expression in the brain and kidney of rainbow trout in response to handling stress." *BMC Genomics* 6, no. 1 (2005): 1–11.

Kultz, D. "Molecular and evolutionary basis of the cellular stress response." *Annual Review of Physiology* 67, no. 1 (2005): 225–257.

Kurtz, J., K. M. Wegner, M. Kalbe, T. B. H. Reusch, H. Schaschl, D. Hasselquist, and M. Milinski. "MHC genes and oxidative stress in sticklebacks: an immuno-ecological approach." *Proceedings of the Royal Society B: Biological Sciences* 273, no. 1592 (2006): 1407–1414.

Latif, M., and M. Faheem. "Study of Oxidative Stress and Histo-Biochemical Biomarkers of Diethyl Phthalate Induced Toxicity in a Cultureable Fish, *Labeo rohita*." *Pakistan veterinary journal* 40, no. 2 (2020).

Lee, J.-Y., and Y. Gao. "Review of the application of garlic, Allium sativum, in aquaculture." *Journal of the World Aquaculture Society* 43, no. 4 (2012): 447–458.

Lee, M. T., W. C. Lin, B. Yu, and T. T. Lee. "Antioxidant capacity of phytochemicals and their potential effects on oxidative status in animals — a review, Asian-Australas." *Journal of Animal Science* 30 (2017): 299–308, 10.5713/ajas.16.0438.

Li, H.-T., L. Feng, W.-D. Jiang, Y. Liu, J. Jiang, S.-H. Li, and X.-Q. Zhou. "Oxidative stress parameters and anti-apoptotic response to hydroxyl radicals in fish erythrocytes: protective effects of glutamine, alanine, citrulline and proline." *Aquatic Toxicology* 126 (2013): 169–179.

Lushchak, V. I. "Environmentally induced oxidative stress in aquatic animals." *Aquatic toxicology* 101, no. 1 (2011): 13–30.

Madison, B. N., P. V. Hodson, and V. S. Langlois. "Diluted bitumen causes deformities and molecular responses indicative of oxidative stress in Japanese medaka embryos." *Aquatic Toxicology* 165 (2015): 222–230.

Magnadottir, B. "Immunological control of fish diseases." *Marine Biotechnology* 12, no. 4 (2010): 361–379.

Maqsood, S., M. H. Samoon, and P. Singh. "Immunomodulatory and growth promoting effect of dietary levamisole in Cyprinus carpio fingerlings against the challenge of Aeromonas hydrophila." *Turkish Journal of Fisheries and Aquatic Sciences* 9, no. 1 (2009).

Martínez-Porchas M., Luis Rafael M.-C. & Rogelio, R.-E. Cortisol and glucose: reliable indicators of fish stress? *Pan-American Journal of Aquatic Sciences* (2009): 158–178.

Marcogliese, D. J., L. G. Brambilla, F. Gagné, and A. D. Gendron. "Joint effects of parasitism and pollution on oxidative stress biomarkers in yellow perch Perca flavescens." *Diseases of Aquatic Organisms* 63, no. 1 (2005): 77–84.

McCallum, E. S., R. E. Charney, J. R. Marenette, J. A. M. Young, M. A. Koops, D. J. D. Earn, B. M. Bolker, and S. Balshine. "Persistence of an invasive fish (Neogobius melanostomus) in a contaminated ecosystem." *Biological Invasions* 16, no. 11 (2014): 2449–2461.

Milano, D., M. Lozada, and H. E. Zagarese. "Predator-induced reaction patterns of landlocked Galaxias maculatus to visual and chemical cues." *Aquatic Ecology* 44, no. 4 (2010): 741–748.

Miller, K. M., S. Li, K. H. Kaukinen, N. Ginther, E. Hammill, J. M. R. Curtis, D. A. Patterson et al. "Genomic signatures predict migration and spawning failure in wild Canadian salmon." *Science* 331, no. 6014 (2011): 214–217.

Misra, C. K., B. K. Das, S. C. Mukherjee, and P. Pattnaik. "Effect of long term administration of dietary β-glucan on immunity, growth and survival of Labeo rohita fingerlings." *Aquaculture* 255, no. 1-4 (2006): 82–94.

Momoda, T. S., A. R. Schwindt, G. W. Feist, L. Gerwick, C. J. Bayne, and C. B. Schreck. "Gene expression in the liver of rainbow trout, Oncorhynchus mykiss, during the stress response." *Comparative Biochemistry and Physiology Part D: Genomics and Proteomics* 2, no. 4 (2007): 303–315.

Monaghan, P., N. B. Metcalfe, and R. Torres. "Oxidative stress as a mediator of life history trade-offs: mechanisms, measurements and interpretation." *Ecology Letters* 12, no. 1 (2009): 75–92.

Nair, A., D. Chattopadhyay, and B. Saha. "Plant-derived immunomodulators." In *New look to phytomedicine*, pp. 435–499. Academic Press, 2019.

Netea, M. G., L. A. B. Joosten, E. Latz, K. H. G. Mills, G. Natoli, H. G. Stunnenberg, L. A. J. O'Neill, and R. J. Xavier. "Trained immunity: a program of innate immune memory in health and disease." *Science* 352, no. 6284 (2016): aaf1098.

Nya, E. J., and B. Austin. "Use of garlic, Allium sativum, to control Aeromonas hydrophila infection in rainbow trout, Oncorhynchus mykiss (Walbaum)." *Journal of Fish Diseases* 32, no. 11 (2009): 963–970.

Ondreicka, R., I. Beno, O. Cerna, E. Grancicová, M. Staruchová, K. Volkovová, P. Bobek, and M. Tatara. "Relation between levels of vitamins C, E, A and beta-carotene and activity of antioxidant enzymes in the blood." *Bratislavske Lekarske Listy* 99, no. 5 (1998): 250–254.

Pavlidis, M. A., and C. C. Mylonas, eds. *Sparidae: Biology and aquaculture of gilthead sea bream and other species*. John Wiley & Sons, 2011.

Petit, J., and G. F. Wiegertjes. "Long-lived effects of administering β-glucans: Indications for trained immunity in fish." *Developmental & Comparative Immunology* 64 (2016): 93–102.

Pohl, F., and P. K. T. Lin. "The potential use of plant natural products and plant extracts with antioxidant properties for the prevention/treatment of neurodegenerative diseases: In vitro, in vivo and clinical trials." *Molecules* 23 (2018) 3283, 10.3390/molecules23123283.

Raby, G. D., M. R. Donaldson, S. G. Hinch, D. A. Patterson, A. G. Lotto, D. Robichaud, K. K. English et al. "Validation of reflex indicators for measuring vitality and predicting the delayed mortality of wild coho salmon bycatch released from fishing gears." *Journal of Applied Ecology* 49, no. 1 (2012): 90–98.

Raby, G. D., T. D. Clark, A. P. Farrell, D. A. Patterson, N. N. Bett, S. M. Wilson, W. G. Willmore, C. D. Suski, S. G. Hinch, and S. J. Cooke. "Facing the river gauntlet: understanding the effects of fisheries capture and water temperature on the physiology of coho salmon." *PLoS One* 10, no. 4 (2015): e0124023.

Reverter, M., N. Bontemps, D. Lecchini, B. Banaigs, and P. Sasal. "Use of plant extracts in fish aquaculture as an alternative to chemotherapy: current status and future perspectives." *Aquaculture* 433 (2014): 50–61.

Rojo-Cebreros, A. H., L. Ibarra-Castro, and J. M. Martínez-Brown. "Immunostimulation and trained immunity in marine fish larvae." *Fish & Shellfish Immunology* 80 (2018): 15–21.

Ros, A. F. H., P. Vullioud, R. Bruintjes, A. Vallat, and R. Bshary. "Intra-and interspecific challenges modulate cortisol but not androgen levels in a year-round territorial damselfish." *Journal of Experimental Biology* 217, no. 10 (2014): 1768–1774.

Safriani, N., F. Z. Rungkat, E. Prangdimurti, R. Verpoorte, and N. D. Yuliana. "Using metabolomics to discover the immunomodulator activity of food plants." *Heliyon* (2022): e09507.

Schreck, C. B., and L. Tort. "The concept of stress in fish." In *Fish Physiology* (eds. C. B. Schreck, L. Tort, A. P. Farrell, and C. J. Brauner) vol. 35, pp. 1–34. San Diego, CA: Academic Press, 2016.

Segner, H., H. Sundh, K. Buchmann, J. Douxfils, K. S. Sundell, C. Mathieu, N. Ruane, F. Jutfelt, H. Toften, and L. Vaughan. "Health of farmed fish: its relation to fish welfare and its utility as welfare indicator." *Fish Physiology and Biochemistry* 38, no. 1 (2012): 85–105.

Sharma, P., P. Chadha, and H. S. Saini. "Tetrabromobisphenol A induced oxidative stress and genotoxicity in fish *Channa punctatus*." *Drug and Chemical Toxicology* 42, no. 6 (2019): 559–564.

Sloman, K. A., and P. L. McNeil. "Using physiology and behaviour to understand the responses of fish early life stages to toxicants." *Journal of Fish Biology* 81, no. 7 (2012): 2175–2198.

Sopinka, N. M., M. R. Donaldson, C. M. O'Connor, C. D. Suski, and S. J. Cooke. "Stress indicators in fish." *In Fish Physiology*, vol. 35, pp. 405–462. Academic Press, 2016.

Solomando, A., A. Cohen-Sánchez, A. Box, I. Montero, S. Pinya, and A. Sureda. "Microplastic presence in the pelagic fish, Seriola dumerili, from Balearic Islands (Western Mediterranean), and assessment of oxidative stress and detoxification biomarkers in liver." *Environmental Research* 212 (2022): 113369.

Stoot, L. J., N. A. Cairns, F. Cull, J. J. Taylor, J. D. Jeffrey, F. Morin, J. W. Mandelman, T. D. Clark, and S. J. Cooke. "Use of portable blood physiology point-of-care devices for basic and applied research on vertebrates: a review." *Conservation Physiology* 2, no. 1 (2014).

Swain, P., S. K. Nayak, P. K. Nanda, and S. Dash. "Biological effects of bacterial lipopolysaccharide (endotoxin) in fish: a review." *Fish & Shellfish Immunology* 25, no. 3 (2008): 191–201.

Selye, H. A syndrome produced by Diverse Nocuous Agents. Nature, 138, (1936): 32–32. 10.1038/138032a0

Urbinati, E. C., F. S. Zanuzzo, and J. D. Biller. "Stress and immune system in fish." *In Biology and Physiology of Freshwater Neotropical Fish*, pp. 93–114. Academic Press, 2020.

Wingfield, J. C. "Ecological processes and the ecology of stress: the impacts of abiotic environmental factors." *Functional Ecology* 27, no. 1 (2013): 37–44.

Wingfield, J.C., Kelley, J.P., & Angelier, F. What are extreme environmental conditions and how do organisms cope with them? *Current Zoology* 57(2011): 363–374.

Wu, P., Y. Liu, W.-D. Jiang, J. Jiang, J. Zhao, Y.-A. Zhang, X.-Q. Zhou, and L. Feng. "A comparative study on antioxidant system in fish hepatopancreas and intestine affected by choline deficiency: different change patterns of varied antioxidant enzyme genes and Nrf2 signaling factors." *PLoS One* 12, no. 1 (2017): e0169888.

Zanuzzo, F. S., R. E. Sabioni, L. N. F. Montoya, G. Favero, and E. C. Urbinati. "Aloe vera enhances the innate immune response of pacu (Piaractus mesopotamicus) after transport stress and combined heat killed Aeromonas hydrophila infection." *Fish & Shellfish Immunology* 65 (2017): 198–205.

Zanuzzo, F. S., J. D. Biller-Takahashi, and E. C. Urbinati. "Effect of Aloe vera extract on the improvement of the respiratory activity of leukocytes of matrinxa during the transport stress." *Revista Brasileira de Zootecnia* 41 (2012): 2299–2302.

Section II

Immunomodulators and sustainable aquaculture development

5

Immunomodulators to Prevent Diseases and Minimize Antimicrobial Use

Akshay Thuruthiyil Rajesh[1], Sajna Beegum[1], Neha Omgy[2], Sreeja Lakshmi[2], Hethesh Chellapandian[3], Sivakamavalli Jeyachandran[3], Einar Ringø[4], and Preetham Elumalai[1]
[1]*Department of Marine Biology, Microbiology and Biochemistry, School of Marine Sciences, Cochin University of Science and Technology (CUSAT), Lakeside Campus, Cochin, Kerala, India*
[2]*Kerala University of Fisheries and Ocean Studies, Panangad, Kochi, India*
[3]*PG and Research Department of Biotechnology and Microbiology, National College (Autonomous), Tiruchirappalli, Tamil Nadu, India*
[4]*Norwegian College of Fishery Science, Faculty of Biosciences, Fisheries and Economics, UiT, The Arctic University of Norway, Tromsø, Norway*

5.1 Introduction

Through the diagnosis and extermination of pathogens, ageing cells, and tumor cells, the immune response is essential in sustaining human health. Numerous factors, including the presence of pathogens, tissue damage, and myocardial infarction, can impact functional efficiency (Villani et al., 2018). "Immunomodulation" deals with the immune system's capacity to regulate a slew of potentially severe disease, including the human immunodeficiency virus, cancer (Naidoo et al., 2014), aging (Routy et al., 2016) and multiple sclerosis (Havla et al., 2016). Macrophages are mononuclear leukocytes that serve as the host's main defense mechanism and have the ability to identify and phagocytose cancer cells and microbial infections. These cells are also involved in the control of the immune system by releasing cytokines such as interferon (IFN), tumor necrosis factors (TNFs), and interleukin (IL-1β, IL-6, IL-10, and IL-12) (Figure 5.1). Immunoadjuvants, immunosuppressants, and immunostimulants are the clinical categories for immunomodulators. By modifying the immune system of the host, immunotherapy treats diseases. Several drugs, including imiquimod, levamisole, pidotimod, cyclophosphamide, tilorone, prostaglandin, thiocarbamate, cyclosporine A, penicillamine, and niridazole, are now used to suppress immunity of humans. Synthetic compounds that alter immunity show adverse effects and toxicity, which restricts their application to conditions like those of chronic disorders (Wang et al., 2010). In contrast, most naturally occurring proteins or peptides have less complications and are less expensive, indicating the possibility of using them in therapies of immune system. Although the compounds from marine source is full of biologically active molecules, there are to our knowledge less studies available looking at how these resources can modify the immune system, only approximately 190,000 of the estimated 2,210,000 species that are expected in the ocean have been documented (Grienke et al., 2014). Immunostimulants can be produced using proteins, peptides, and protein hydrolysates that are isolated from marine organisms (algae, fish, shellfish, etc.). Current research on the probable immunomodulatory consequence of marine resources is only emerging.

5.2 Immune System – An Overview

The immune system, a cornerstone of the host's defensive mechanism, eliminates pathogens, or potentially harmful molecules. Innate and adaptive immune systems are the two subsystems that make up the immune system. Natural or native immunity, also referred to as innate(non-specific) immunity and acts as the body's initial line defense employing the mucosa, bone marrow, skin, mast cells, neutrophils, monocytes, basophils, dendritic cells, eosinophils, macrophages, as well as other inflammatory components such as defensins, interferons, and cytokines. Important first-line phagocytosis responders include neutrophils and macrophages.

Specific or adaptive immunity (acquired immunity), uses T cells and B lymphocytes and is extremely specific to deadly pathogens. Either cells or humoral mechanisms can mediate the adaptive immune response. B lymphocytes discharge antibodies that are particular for the pathogenic microorganisms, mediate humoral immunity. The pathogen or infected cells are bound by T lymphocytes during cell-mediated reaction, which results in

DOI: 10.1201/9781003361183-7

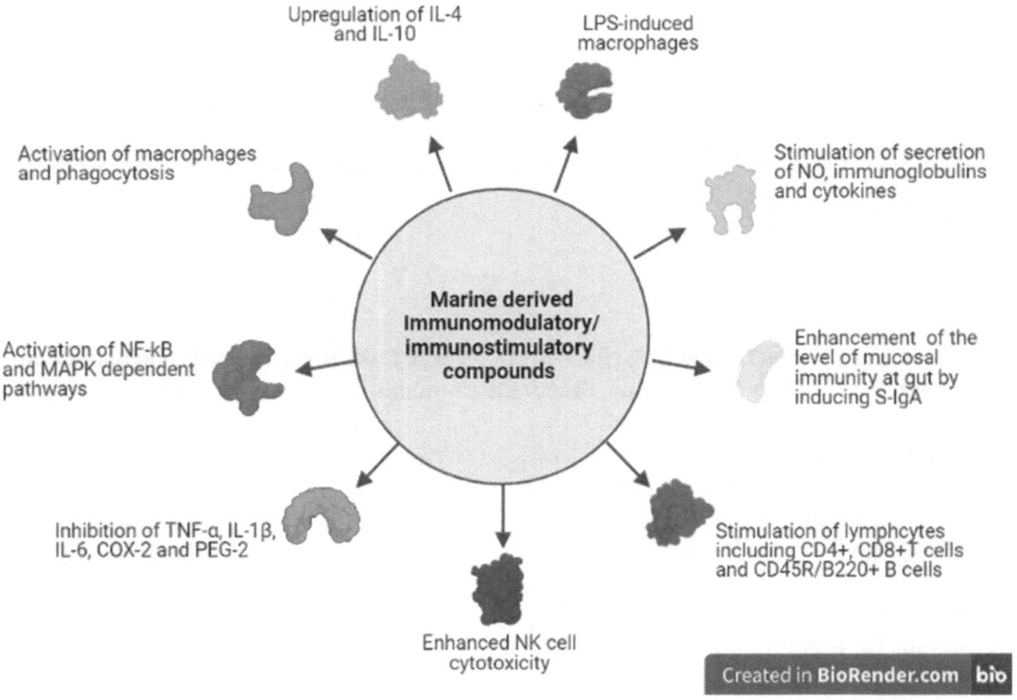

FIGURE 5.1 Immunomodulatory effects of marine-derived bioactive compounds (created by www.biorender.com).

the lysis of these cells and the production of immune regulating molecules, cytokines. Cytotoxic T cells, Helper T (T_H) cells, cytotoxic T cells, and regulatory T (suppressor T)cells are the distinct subgroups of T lymphocytes. Cluster of differentiation (CD)8+, a surface receptor on cytotoxic T lymphocytes, recognizes endogenous antigens linked to major histocompatibility complex class I and eliminates the infected cells. The major histocompatibility complex class II-complexed foreign antigens are recognized by T_H cells that have CD4+ as their surface receptors. Cytokines secreted by T_H cells aid in the activation of B and T lymphocytes and other immune cells, by the prevention of harmful immune activation the immune response is controlled with the help of regulatory T cells and also helps to maintain the self-tolerance to prevent autoimmune diseases.

5.2.1 Fish Immune System

The changes in the cultural methods in fish farming accelerated the outburst of many infections. A more practical way to decrease or keep away from chemicals and antibiotics is to increase the use of compounds like β-glucans, chitin, prolactin, etc. which boosts immune system, but this urges the need to understand the immune system's composition and operation in fish.

The immune system is an intricate web of organs, tissues, and cells. The compounds they create aid the body in its defense against poisons, cancerous cells, and diseases. Both endogenous and exogenous factors stimulate immune system. Innate (nonspecific) and acquired (specific) immune systems are the two types of immune system. These two systems jointly work to eliminate intruders, thus activating defense mechanism. Fish have both innate and adaptive (memory) immune systems, which are further separated into humoral and cell-mediated defense mechanisms (soluble substances). The innate system works as the body's initial line of defense and reacts more quickly than the specialized (adaptive) system but cannot retain memory of past responses. If the innate immune defenses fail to eliminate the pathogen, adaptive immune system is recruited. Innate immune system consists of all elements that were there in the body before the pathogenic agent appeared. Physical barriers include scales, mucous surfaces of the skin, gills, and epidermis. Additionally, the complement system, antimicrobial enzymes, interleukins, interferon, and organic defense cells such granulocytes, monocytes, macrophages, and natural killer cells are present. The complicated interplay between cellular and humoral components that cause inflammation are also considered to be an innate immunological reaction. When tissue becomes infected, mediation factors are released to stretch and make blood capillaries more permeable, allowing defense cells to travel across them. The first cell type to arrive at the site of inflammation are granulocytes, which are responsible for eliminating pathogens. Macrophages, on the other hand, phagocytose the remaining potentially dangerous cells and cellular waste (Magnadottir et al., 2006).

The nonspecific immune system, the simplest mechanism, most likely developed in single-celled creatures early in evolution. The innate immune system is capable of detecting areas in molecules known as PAMPS (Pathogen Associated Molecular Patterns) from pathogens or members of the normal microbiota, such as lipopolysaccharide (LPS), peptidoglycan, bacterial DNA, viral RNA, or other molecules known as "non-self" compounds, which are not typically reported on the surface of multicellular organisms. The PAMPS are well-preserved sections of a species evolutionary history and exists in most pathogens. The specific system originally arose 450 million years ago and are located in all animals, with the exception of Agnatha-class fish. The specific system receptors, also known as membrane immunoglobulin or free antibodies, are found on the cell membranes of T lymphocytes (TCR- T cell receptor) and B lymphocytes (BCR- B cell receptor), which are responsible for detecting harmful substances. The presence of an antigen is necessary for the specific system of defense. A molecule or cell known as an antigen can trigger reactions, boost the circulation of particular antibodies, and promote immunological memory. Antigen-presenting cells (APCs) such as macrophages, dendritic cells, and B lymphocytes identify and process foreign antigens to break down microorganisms into their component molecules and initially elicit an immune response of proliferation and immunological response of memory.

T lymphocytes (the cells of the specific system) will be presented with the APC-processed antigen. T cells can only detect the antigen,when a certain humoral component known as the major histocompatibility complex (MHC) is present. The major histocompatibility complex (MHC) is a group of genes that codes for glycoprotein receptors. Once the invader has been identified, the T cell releases cytokines, which are proteins that encourage the activation of other cells such as B lymphocytes (cells that produce antibodies), cytotoxic lymphocytes, macrophages, and other cells. Phagocytosis (component of innate immunity) is activated as a result of the antibody's recognition and connection to specific microorganisms. This clearly demonstrates the interaction between the nonspecific and specific immunity. Antibodies can aid in agent neutralization or opsonization in addition to binding the complement system and potentially binding to foreign antigens. When the antigen is within the cell, cytotoxic T cells are in charge of providing the defense.

5.2.2 Organs of Immune System

Lymphoid immune system is seen in bony fish. Since they lake bone marrow and lymph nodes, myeloid classification is absent in bony fish. The lymphoid organs include the kidneys (the biggest lymphoid organ), the thymus, the spleen, and the gut-associated lymphoid tissues (GALT). It is crucial for the formation of immune system components such as lymphocytes, monocytes, granulocytes, thrombocytes, mast cells, NK cells, cytotoxic cells, and dendritic cells that lymphoid tissues and organs be arranged in a reticular cell network.

In thymus, development and maturation of T lymphocytes take place. The kidney is crucial for fish immunity and hemopoiesis. Red and white blood cells are produced and matured in the kidney. For the production of defense cells, the anterior kidney region is particularly crucial. Additionally, the anterior portion of the kidney is where leukocytes including granulocytes, macrophages, monocytes, and lymphocytes differentiate and mature. The fish spleen is composed of red and white pulp and is not as highly organized as the spleen in mammals. Red pulp aids in the phagocytosis of damaged or aged cells, while white pulp promotes hematopoiesis and the development of defense cells. Antigens are held on macrophages grouped in centers when blood passes through the spleen so they can be processed and given to T lymphocytes.

Gut-associated lymphoid tissues (GALTs), comprised of gills, skin, and mucosa, secrete mucus that act as first barrier against pathogens. Mucus consists of defense components such as mast cells, granulocytes, macrophages, and lymphocytes. Antigens which are to be processed are captured by these cells, thus promoting immune memory. The liver produces humoral substances such as complement system proteins and acute phase proteins of inflammatory reaction.

5.2.3 Humoral- and Cellular-Mediated Immunity

Fish are susceptible to infections with agents like virus, fungus, bacteria, and parasites. Because of specialized and intrinsic components with coordinated humoral and cellular processes, they can, to some extent, defend against pathogens. The humoral-specific system only uses antibodies, whereas the humoral innate mechanism functions through a variety of soluble components in bodily fluids. The elements of the humoral innate system include antiproteases, lysozyme, transferrin, C-reactive protein, and complement system. Phagocytes, neutrophils, and macrophages are the cells of humoral innate immunity (Takahashi et., 2014).

Antiproteases cause the proteolysis of blood proteins. Amino acids obtained from proteolysis form the source of energy. Monocytes and neutrophils are the principal cells that secrete lysozymes, which are prevalent in mucus, eggs, blood, and tissues, Lysozymes cause lysis of peptidoglycan on pathogens cell wall. The pentraxin family member C-reactive protein has been reported in blood, eggs, and mucus. It can identify and attach to the phosphoryl colin on the pathogen's cell wall. Along with mannose-binding lectin they assist in opsonization, complement activation, and phagocytosis. The inflammation-acute proteins known as C-reactive proteins and mannose-binding lectin are also known as receptors of soluble microbial components. Acute phase proteins are produced mostly by the liver, brain, and leukocytes and are concentrated more when there are pathogens present and high temperatures. For illness diagnosis, the most often utilized tests include the C-reactive protein, serum amyloid A, transferrin, C3 complement, lysozyme, and lectins. The primary mechanism in fish is the complement system, which secretes inflammatory mediators, has lytic action, and encourages chemotaxis, pro-inflammation, and opsonization. Sequential pathway activates the complement system, which is made up of membrane and soluble proteins that are inactive or present in low concentrations. The three mechanisms that activate the complement system are the lectin pathway, which is activated by pathogen's surface carbohydrates, the alternative pathway, which is activated by PAMPS from bacteria, and the classical pathway, which is activated by an antibody-dependent activation. The alternative pathway, which is thought to be crucial, is quickly activated by gram-negative bacteria and LPS and results in cell lysis. The classical route mediates the interaction of the innate immune system with the specialized immune system. Two complement systems C3b and C5a are essential in inflammation and phagocytosis. For neutrophils and macrophages, C5a and C3b are chemotactic proteins that initiate opsonization, phagocytosis, and leukocyte chemotaxis. The complement system is frequently utilized to determine immunological state.

Lymphoid tissues such as the kidney, thymus, spleen, and GALT produce defense cells. Hematopoiesis, or the generation and differentiation of erythrocytes, granulocytes, monocytes, lymphocytes, mast cells, and thrombocytes, is the process of producing lymphoid cells. Cytokines that operate as pluripotent cell receptors affect the survival, proliferation, differentiation,

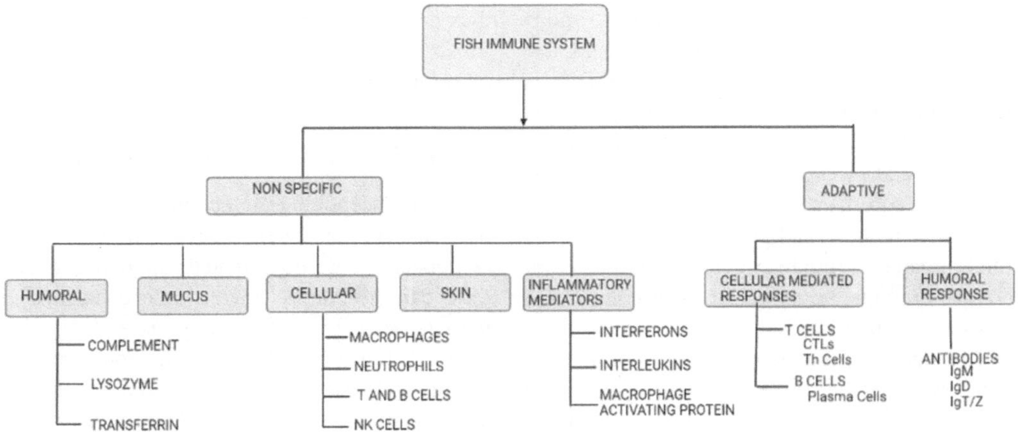

FIGURE 5.2 Immune system of fish.

maturation, and function of hematopoiesis. The evaluation of hematological parameters may serve as a marker for fish disease outbreaks and physical health.

The cells of the adaptive immune system are T and B lymphocytes, that differ from natural killer cells or T cytotoxic cells and destroys injured somatic cells. Dendritic cells and macrophages correlate the function of the innate immune system with that of the acquired immune system (antigen-presenting cells). Major histocompatibility complex (MHC) class 2 receptors assist antigen-presenting cells digest the pathogen and deliver the processed molecule to T lymphocytes, generating cell-mediated acquired immune reaction. MHC molecules are glycoprotein receptors that lack specificity and can recognize both internal and external antigens. They are encoded by a gene complex. As opposed to this, antibodies (Immunoglobulins Ig) are glycoproteins that are expressed on the B lymphocyte membrane (BCR) or are free in body fluids and released by plasma cells. Immunoglobulins IgM, IgD, IgZ, and IgT are detected in fish. The well-known fish protein immunoglobulin IgM is a tetrameric protein with four sites for antigen recognition. Immunoglobulins are reported in bodily fluids, mucous, eggs, and gastrointestinal mucosa. Among the roles that antibodies can play are the anti-adhesin function, anti-toxins functions, and anti-invasive functions. Anti-adhesin function happens in digestive systems epithelial surface, gills, and skin and prevents bacterial adherence. Anti-toxins antibodies make bacterial toxins inert, and anti-invasions antibodies prevent bacterial invasion of unprotected cells.

Immunoglobulin receptors on B lymphocytes that recognize an antigen promote the cellular uptake and multiplication of B memory cells, which in turn activates T lymphocytes. This triggers the production of cytokines and activates macrophages and B cells. Species, age, sexual maturity, physiological events, artificial immunization, and persistent infection can all affect the serum antibody concentration (Figure 5.2).

5.3 Immunomodulatory Compounds from Marine Organisms

Immunomodulatory compounds from more than 16,000 marine species have been isolated from the seas and oceans during the last few decades and extensively studied. Amino acids, peptides, proteins, glycoproteins, lipids, and polysaccharides produced from marine sources possess the potential to modulate the immunity (Kiewiet et al., 2018). This chapter gives an overview of immunomodulatory compounds from marine sources and from different organisms as an alternative for antibiotics (Table 5.1).

5.3.1 Immunomodulatory Peptides, Amino Acids, and Proteins

The described marine biomaterials such as proteins, biopolymers, fatty acids, enzymes, pigments, and minerals are obtained from natural marine sources. Large numbers of different proteins with a range of bioactivities and functions are present in marine biomaterials (10–47% (w/w)). Many proteins in hemolymphs and hemocytes serve important roles in the innate immune system by protecting diverse immunological components like glycoproteins, antimicrobial peptides, metalloproteins, amino sulfonic acid, coagulation factors, and protease inhibitors.

Lectin, a glycoprotein that regulates immune activity, has been implicated in innate immune responses that result in pathogen identification and endocytosis as well as adaptive immunological responses that include B- and T-cell activation and apoptosis (Sharon, 2007). Many different plants and animals have been found to contain lectins. Lectins from clam are similar, and influence the innate immune response(Medzhitov et al., 2000). C-type that are Ca^{2+}-dependent, R-type, and metal-independent (Galectins) have all been classified as lectins. C-type lectins serve a critical function in carbohydrate recognition during the immune response by recognizing a wide range of endogenous ligands and pathogens that control intercellular connections (Brown et al., 2018). These proteins have a significant role in autoimmune disorders as well. Tachylectin is a pathogen-recognition protein that was identified from the hemocytes of the mangrove horseshoe crab (*Carcinoscorpiusrotundicauda*). Phagocytosis or the lectin pathway in the complement system is used to recognize microbes (Ma et al., 2004). Phagocytosis or the lectin pathway in the complement system are used to recognize microbes. In human peripheral blood cells, the lectin from the mussel *Mytilus trossulus* increases the production of proinflammatory cytokines

TABLE 5.1

Immunomodulatory Compounds and its Mechanism of Action

Organism Obtained from	Bioactive Compound	Concentration	Mechanism/Activity	References
Filamentous marine fungi	PolatuzumabVedotin	*In vitro* administration with 125 μg/mL	Induction of human mesenchymal stem cell (hMSC) differentiation	Sehn et al., 2020
Marine fungus (*Phomaherbarum* YS4108)	D-glucan	*In vivo*intraperitoneal injection (40 mg/kg)	Regulation of T lymphocytes and dendritic cells	Newman et al., 2019
Microalgae	β-carotene Phycocyanobilin 9-cis- β-carotene	In silico study	Anti-inflammatory mechanism	Costa et al., 2021
Microalga (*Phaeodactylumtricornutum*)	β-glucans	β-glucans 0.6 g/kg of feed	Promotes the growth of commensal microbiota Antioxidant, anti-inflammatory and immunomodulatory reactions in fishes	Speranza et al., 2012
Microalga (*Porphyridiumcruentum*)	Exopolysaccharides (EPS)	Treated with increasing conc. of EPS	Antibacterial and immunostimulatory effects with elevation in total hemocyte value, phagocytotic activity	Wasana et al., 2021
Alga (*Ulva ohnoi*)	Ulvan	*In vitro*: 100 μg/mL	Immunomodulatory activity	Hafting et al., 2015
Alga (*Gracilarialemaneiformis*)	Sulfated oligosaccharides (GLSO)	–	Immunomodulatory activity	Cumashi et al., 2007
Alga (*Caulerpa cupressoides*)	Sulfated pyruvates (SP1 and SP2) and galactans, sulfated polysaccharides	*In vitro*: 100 μg/mL	Immunostimulatory activity	Lauritano et al., 2018
Algae (*Ecklonia cava, Fucus vesiculosus, Macrocystis pyrifera and Underia pinnatifida*)	Fucoidan	*In vitro*: 100 μg/mL	Pro-proliferative activity Immunostimulatory activity	Yang et al., 2008
Coral (*Pseudopterogorgia Americana*)	PPA polysaccharide	*In vitro*: 10 μg/mL	Immunomodulatory activity	Xiang et al., 2022
Mollusk (*Megathuracrenulata, Concholepasconcholepas, Fissurellalatimarginata*)	Haemocyanin		Immunostimulatory activity against several cancer types with no side effects	Del Campo et al., 2011
Mollusk (*Scapharcabroughtonii*)	Protein HPCG2	*In vitro*: 250 μg/mL and 500 μg/mL	Immunomodulatory activity and promotes phosphorylation	Paolino & Penninger 2010
Mollusk (*Mytilus edulis* and *M. galloprovincialis*)	Mytilin A, B, C, D, G1, Defensin, Myticin, Mytimycin	–	Antimicrobial activity and prevents the breach of microbes to circulatory system	Mitta et al., 2000
Mollusk (*Mytilus coruscus*)	Polysaccharides	*In vitro*: 300 μg/mL and 600 μg/mL	Immunomodulatory activity Anti-inflammatory, promotes abundance of probiotics in the colon	Gunathilake et al., 2020
Mollusk (*Arca subcrenata* Lischke)	ASPG-2 polysaccharide	*In vitro*: 250 μg/mL and 500 μg/mL for 24 h	Immunomodulatory activity	De Almeida Leone et al., 2000
Sponge (*Astrosclerawilleyana*)	Ageliferins extracts	10-50 μM	Immunomodulatory activity by inhibiting ubiquitin ligase activity	Altmann et al., 2001

(Continued)

TABLE 5.1 (Continued)
Immunomodulatory Compounds and its Mechanism of Action

Organism Obtained from	Bioactive Compound	Concentration	Mechanism/Activity	References
Sponge (*Haliclona sp.*)	Crude extract	*In vivo*: 5 mg/kg; 10 mg/kg; 15 mg/kg	Immunomodulatory activity and increases neutrophil: lymphocyte ratio	Talley et al., 1967
Fish (*Danio rerio*)	Phosvitin-derivedpeptide Pt5	–	Extension of the survival rate of zebrafish infected by*Aeromonashydrophila*by elevating Antimicrobial activity and immunomodulatory activity	Ding et al., 2012
Fish (*Salmo salar*)	Salmo salar natural killer(NK)-lysin	*In vivo*: 5 mg/kg; 10 mg/kg; 15 mg/kg	Antimicrobial activity	Acosta et al., 2019
Fish (*Oreochromis niloticus*)	Tilapia piscidin 3 (TP3)and tilapia piscidin4 (TP4)	*In vitro*: 250 µg/mL and 500 µg/mL	Antimicrobial, wound-healing, and antitumor activity	Lin et al., 2016
Fish (*Beleophthalmus pectinirostris*)	BpNK-Lysine	*In vivo*: injection of 1.0 µg/g. *In vitro*: 1.0 µg/mL	Immunomodulatory activity, antibacterial activity	Acosta et al., 2019

including TNF-α and IFN-γ but inhibits the anaphylaxis of the anti-inflammatory cytokine IL-10 (Chikalovets et al., 2013). Lectins, which are generated from the hemolymph of crustaceans, are regarded as active antibodies‘ progenitors since they are crucial for both immune recognition and antimicrobial activity.

In mollusks, hemocyanins, the metalloproteins are present that use copper-binding sites to bind and transport oxygen. Similar to hemoglobin, hemocyanins are multi-component molecules with an oxygen-binding functional subunit. In a rat colon carcinoma model, hemocyanins derived from a variety of marine gastropods exhibit strong immunostimulatory activity against some malignancies with manageable side effects (Del Campo et al., 2011). Hemocyanins have been tested in clinical trials for treating superficial bladder cancer, whereas hemocyanins derived from *Concholepasconcholepas* have been employed as innate immunostimulators to decrease the development of superficial bladder cancer (Lammers et al., 2012). An immunostimulator called keyhole limpet hemocyanin is made from the sea mollusk's circulating glycoproteins (*Megathuracrenulata*). Keyhole limpet hemocyanin enhances the host immune response by interacting with T cell monocytes, macrophages, and polymorphonuclear lymphocytes (Mora Roman et al., 2019). Hemocyaninfrom *Fissurellalatimarginata* (gastropod), is a novel immunostimulator with stronger immunostimulatory activity than other isolated gastropod hemocyanins, like keyhole limpet hemocyanin and *Concholepasconcholepas*.

An amino acid called taurine, also known as 2-amino ethane sulfonic acid, is found in large quantities in animal tissues, including those of saltwater clams. Taurine is produced by numerous different creatures in addition to being a rich source in marine organisms. Lymphocytes, monocytes, and neutrophils are among the immune cells that are enriched in taurine, which has cytoprotective and immunomodulatory properties (Marcinkiewicz & Kontny et al., 2014). This implies that phagocytes that are mediating inflammatory lesions have high taurine levels, which may be related to innate immunity. Taurine builds up in phagocytes; upon coming into touch with a pathogen, activated cells (neutral and macrophages) release harmful oxidants and a variety of antibacterial chemicals via the peroxidase system, and act at inflammatory areas to kill germs. By eradicating pathogens, toxic oxidants—part of the innate immune system—protect the host from infections. However, they may also cause tissue damage to the host. Additionally, taurine functions as a scavenger to flush out undesired or dangerous compounds from cells and safeguard them from oxidative stress.By activating NF-κB, a strong signal transducer for inflammatory cytokines, taurine affects the immune system. Additionally, taurine in the liver activates the peroxisome proliferator-activated receptor-γ (PPAR-γ) to regulate adiposity and glucose metabolism. Therefore, taurine's stimulation of PPAR-γ prevents diabetic retinopathy's retinal neuronal damage (Song et al., 2011).

5.3.2 Antimicrobial and Immunomodulatory Compounds

Active research is being done on the peptides that regulate immune responses. Although it is known that immunomodulatory peptides boost immunological responses like lymphocyte proliferation, NK cell activity, and cytokine modulation, the precise mechanisms behind these actions are yet unknown. For instance, little is known about how immunomodulatory peptides affect the human neurological, digestive, cardiovascular, and immunological systems. Therefore, in order to create unique immunomodulatory peptides, it is essential to comprehend how peptides interact with these systems. Antimicrobial peptides (AMPs) from marine sources are defensive chemicals that may improve innate immunity. Marine AMPs' promise as therapeutic candidates was reinforced by the finding that they target particular cells. Therefore, AMPs with immune regulating qualities originating from marine sources are crucial.

Specificity of marine-derived AMPs to target cells makes them a potential drug candidate. The rationale for the targeting specificity of AMP is the variation in cellular membrane makeup among pathogens and host cells (Zhang et al., 2021). By selectively reacting with the negative charges on the membrane through electrostatic interaction, cationic AMPs exert antibacterial activity. This increases porosity, causes cell membrane rupture and substance leakage and ultimately results in cell death. In order to interact with molecules inside cells, AMPs can potentially penetrate the membrane and enter the cytoplasm. The mechanism of action of anionic AMPs is not clearly known. In addition to penetration and pore formation AMPs have other mechanisms to kill bacteria. AMPs show their action on nuclei acids, protein synthesis, on enzyme activity, on cell wall synthesis. Bacterial resistance to AMPs is prevented by the quick death and intracellular effects, which may not focus on any particular molecules or pathways. Hence, AMPs are good candidates in the management of antibacterial resistance (Figure 5.3).

In spite of antibacterial activity AMPs also show antiviral, antiparasitic, tumor-modulatory activity, and immunomodulatory activity. AMPs, also known as host defence peptides, work by modulating the immune system and preventing pathogen growth. AMPs activate innate immunity by acting directly on dendritic cells, neutrophils, T-lymphocytes, monocytes, and mast cells. B-lymphocytes are indirectly activated by AMPs, thus switching on adaptive immunity. AMPs immunomodulatory activity incorporates effects on the recruitment and activation of other cells

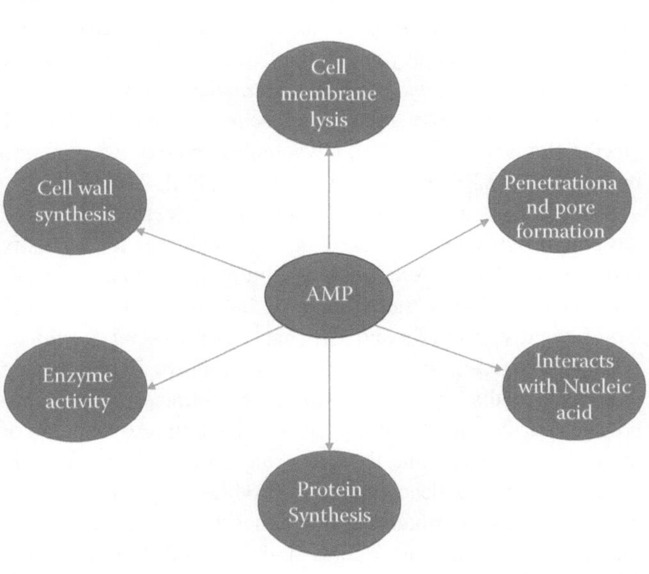

FIGURE 5.3 Mechanism of antibacterial activity of AMP on different molecules.

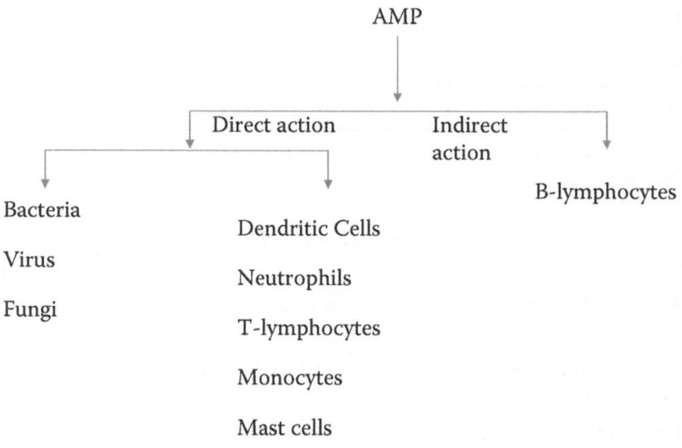

FIGURE 5.4 Immunomodulatory activity of AMP on different cell types.

in the microenvironment, completing the inflammatory process (Figure 5.4).

Cathelicidins, the AMPs in neutrophils and mucosal epithelia, exhibit a broad spectrum of antimicrobial activity. Defensins are AMPs that function as peptides for host defense. Insects, plants, and animals all have defensin. Humans have two different forms of defensins, known as α defensins and β defensins, depending upon their sequence homology and cysteine residues. Defensins destroy microbial membrane and activate immunological response. Defensins adjuvant activity facilitated the concept of "defensin vaccine" as a novel vaccine development.

The AMP, phosvitin-derived peptide Pt5 from zebrafish has equivalent in vitro antibacterial and immunomodulatory action to phosvitin. Pt5 significantly improved the survival rate of infected zebrafish by reducing the amount of *Aeromonas hydrophila* in the spleen, blood, liver, kidney, and muscles. Pt5 also decreases the expression of proinflammatory cytokine genes (IL-1, IL-6, TNF-, and IFN-), while boosting the expression of anti-inflammatory cytokine genes (IL-10 and IL-14) in the head kidney and spleen of zebrafish infected with *A. hydrophila* (Ding et al., 2012). Because they boosted the expression of pro-inflammatory cytokines (IL-1 and IL-8) in the head renal leukocytes, NK-lysin fish from Atlantic salmon (*Salmo salar*) strengthened immune responses (Acosta et al., 2019).

AMP research is of at most significance in studying and analyzing the evolution of immunity of different organisms. Numerous peptides with antimicrobial function can be used to develop as potent drug in the current situation of developing antibiotic resistance in microorganisms.

5.3.3 Immunomodulatory Protein Hydrolysates

Protein hydrolysates made from various proteins have been shown to exhibit a wide range of biological activities, notably anti-inflammatory, anticancer, antioxidant, antibacterial, antihypertensive, and immunomodulatory actions (Kang et al., 2019). Bioactive peptides of various lengths are produced because of peptide bond cleavage during protein hydrolysis. Numerous immunomodulatory protein hydrolysates were produced using a number of proteolytic enzymes. Pancreatin, Kojizyme™, trypsin, Alcalase®, Flavourzyme®, Protamex™, α-chymotrypsin, pepsin, Neutrase®, and thermolysin are some of the enzymes in this group (Ahmad et al., 2019).

TABLE 5.2
Immunomodulatory Marine Protein Hydrolysates

Hydrolysate	Source	Mechanism of Action	Ref.
Cyclina sinensis protein hydrolysate	Venusclam: *Cyclina sinensis*	Boosting NO, TNF-α, IL-6, and IL-1β, and increased protein levels of iNOS, NF-κB, and NLRP3 in RAW 264.7 cells; activation of NF-κB signalling pathway.	Li et al., 2019
Shellfish *Mytilus* Protein hydrolysate	Shellfish: *Mytilus coruscus*	Inhibited lipopolysaccharide (LPS) - NO production is induced in macrophages of RAW264.7 cells	Kim et al., 2013
Alaska Pollock protein hydrolysate	Alaska pollock: *Theragra chalcogramma*	Increased the action of specific, and non-specific immune system in mice with immunosuppression	Hou et al., 2016
Salmon fish protein hydrolysate	Fish: Atlantic salmon fish	IgM, IgG, and IgA and CD4/CD8 ratios changes were observed in Indian children with malnutrition	Nesse et al., 2011
Tuna fish protein hydrolysate	Fish: Tuna:*Thunnasalbacares*	Elevation in the levels of NO, TNF-α, IL-6, and IL-1β, and overexpression protein levels of TLR2 and TLR4, in RAW264.7 cells; activation of the NF-κBsignalling pathway	Cai B et al., 2022

Antimicrobial peptides play a crucial role in the initial line of the human defense system against harmful bacteria which are ingested by the host through the environment. The immune system is stimulated by protein hydrolysates on both the humoral and cellular levels. Oyster and salmon protein hydrolysates increase NK cell activity in the spleen. Protein hydrolysates from algae modulate cytokine action by inhibiting TNF-α and IFN-γ (pro-inflammatory cytokines), boosting IL-10 in rat splenocytes. Table 5.2 summarizes the mechanism of action of marine protein hydrolysates.

5.4 Polysaccharides

A large part of algal biomass is made up of marine polysaccharides. It primarily functions as a structural storage compound. Algal polysaccharides as immunomodulators have gained more attention because of their nontoxicity and variability in structure. Sulfated polysaccharides are produced by the association of sulfur and carbon by photosynthetic organisms. There are three major forms of sulfated polysaccharides, fucoidans, carrageenans, and ulvan based on their sources. *Rhodophyceae*, *Phaeophyceae*, and *Chlorophyceae* are three major classes of seaweeds containing photosynthetic pigments: red, brown, and green, respectively (Figure 5.5).

Some echinoderms(sea urchin and sea cucumber) and marine brown algae (*Phaeophyceae*) are used to make fucoidan. Fucoidan is a complex sulfated polysaccharide, containing fucose residues and other monosaccharides. It is a polymer of α-l-fucose units possesses strong antibacterial activity and is an immunostimulant. Fucoidan shows bacteriostatic effects in *Escherichiacoli, S. epidermidis, S. aureus, and B. licheniformis. E. coli* is the most delicate (Ayrapetyan et al., 2021). Interleukin-2 (IL-2) and tumour necrosis factor-α (TNF-α) were secreted more readily as a result of the immunostimulatory activity of two polysaccharides, SCP1 and SCP2, which were isolated from *Sargassum carpophyllum.* Ulvan from sea lettuce, or *Ulvasp*, is the polymer of glucuronic acid and sulfated rhamnose or iduronic acid and sulfated rhamnosehas antibacterial, antiviral, and immunomodulatory activity. Ulvan has shown inhibition against *Staphylococcus* sp., *Streptococcus* sp., and *Bordetella sp.*

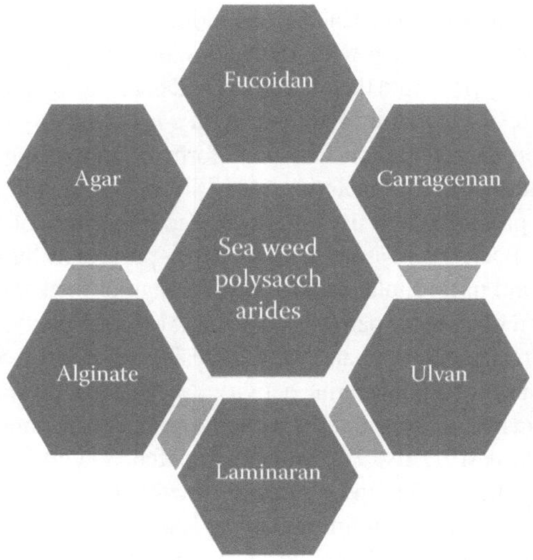

FIGURE 5.5 Seaweed polysaccharides.

Nitric oxide (NO), tumour necrosis factor-α (TNF-α), and interleukin-1 (IL-1) production, as well as macrophage cell proliferation and phagocytic activity, are all stimulated by *U. intestinalis. U. ohnoi* exhibited immune system functional changeson Senegalese sole (*Solea senegalensis*) (Sasikala & Geetha Ramani et al., 2017). Ulvan enhanced the inflammation brought on by LPS and raised IL-1, IL-6, and IL-10 levels.

Carrageenan, a polymer of repeating galactose units and 3,6-anhydrogalactose is used as an additive to emulsify and preserve foods. According to Zhu et al. (2017), oxidized-carrageenan inhibits the growth of both Gram-positive and Gram-negative bacteria and kills bacteria including *S. aureus, L. monocytogenes, E. coli,* and *P. aeruginosa* by destroying their cell walls and cytoplasmic membranes. An amino-polysaccharide called chitin is derived from the exoskeleton of marine creatures, primarily crabs and shrimp. The naturally occurring chitin is deacetylated to produce chitosan. Chitosan directly causes the inflammasome to activate and the release of the pro-inflammatory molecules IL-1 and PGE2 by activating NLRP3. COS causes cytokines and NO to be released from dormant macrophages. A significant amount of NO produced

TABLE 5.3

Polysaccharides from Marine Sources with Immunostimulant Activity

Name of the Compound	Name of the Source	Mode of Action	References
Fucoidans (Sulphated polysaccharides)	Phaeophyceae	Improved CD3+ T cell expression, increased GM-CSF, TNF-α, interleukin (IL)-4, and IL-10 secretion	Ayrapetyan et al., 2021 Peng et al., 2019
Ulvan (Sulphated plysaccahrides)	Chlorophceae	Pro-inflammatory cytokines, including nitric oxide (NO), tumor necrosis factor-α (TNF-α), and interleukin-1β (IL-1β) are produced	Sasikala, & Geetha Ramani et al., 2017
Carrageenan (Sulphated polysaccharides)	Rhodoph-yceae	Damages bacterial cell wall and cytoplasmic membranes. Inhibits bacterial growth. IFN-γ, IL-12, and IL-1β are secreted	Zhu et al., 2017
Chitosan	Exo-skeleton of Crabs and shrimps	Stimulatesdorment macrophages to secrete NO and cytokines. Secretes IL-1β and PGE2	Laurienzo, 2010; Wu et al., 2019
Laminaran	Pheoph-yceae	Influences the up-regulation of immune-related genes and proteins.	Selvaraj et al., 2005

has the ability to kill tumor cells, parasites, and microbes due to its cytotoxic properties. The immune-regulatory mechanisms of COS entail the control of a number of vital pathways, such as the activation of AMP-activated protein kinase (AMPK) and the inhibition of nuclear factor kappa B (NF-κB) and mitogen-activated protein kinases (MAPK).

Laminaran is a 1,3-glucan found in the cell walls of brown seaweed (*Phaeophyceae*). The seaweed species *Durvillaeaantarctica* has a high concentration of -1,3/1,6-glucan, which activates both specific and non-specific immunity and affects the production of immune-related genes and proteins. Glucan administration increased survival in carp, perhaps by promoting both generalized and specialized immune responses. By interacting with particular cell receptors, B-glucans taken from the microalga *Phaeodactylumtricornutum* can function as prebiotics, fostering the development of the commensal microbiota and directly triggering the innate immune system (Table 5.3).

Sulfated polysaccharides primarily regulate macrophage activity, the activity of natural killer cells, T and B lymphocytes, and complement system activation in order to have immunological activity (Cunha et al., 2016). The RAW264.7 macrophage cell line's pinocytic activity has been reported to be promoted by the sulfated polysaccharide UPP-2, which was obtained from the brown alga *Undaria pinnatifida*. Nitric oxide, tumor necrosis factor-α (TNF-α), and interleukin-6(IL-6) secretions have also been shown to be increased. Crude polysaccharides from *Gracilaria dura* and *Chloroidiumellipsoideum* showed similar impacts by stimulating macrophage and RAW264.7 cells and generating significant amounts of different cytokines and nitric oxide. Nuclear factor-kappa B (NF-κB) and mitogen-activated protein kinase (MAPK) pathways are activated as part of the procedure (Kim et al., 2017).

5.5 Alkaloids

Alkaloids are made up of heterocyclic nitrogen substances. Alkaloids in marine algae are rather uncommon when compared to those in terrestrial plants. Halogenated indole alkaloids obtained from red algae has antibacterial properties. When *Channa striata* was exposed to *A. hydrophila*, the N-Oxide-Quaternary Alkaloid Fraction (NOQAF) of the chlorophycean macroalga *Caulerpa scalpelliformis* significantly improved the non-specific immune responses by increasing serum lysozyme, peroxidase, antiprotease, and alternate complement activities as well as the production of reactive oxygen and nitrogen species and peroxidase activity by peripheral blood leucocyte. IL-2 production by Jurkat T cells is inhibited by the bis-heterocyclic quinolone-imidazole alkaloids (+) and (–) spiroreticulatine from the South China sea sponge *Fascaplysinopsis reticulata* (Wang et al., 2015).

5.6 Polyphenols

Potent compounds primarily belonging to polyphenols and sulfated polysaccharides are found in marine plants. Polyphenols have antibacterial and free-radical scavenging properties. Green seaweeds' bromophenols and flavonoids have antioxidant properties. It has been demonstrated through testing that some species, including *Ulva clathrata, Ulva compressa, Ulva intestinalis,* and *Chaetomorphamoniligera,* exhibit high radical scavenging properties. 5′-hydroxyisoavrainvilleol, a bromophenol extracted from the *Avrainvillea nigricans* is toxic to KB cells and has shown some encouraging anti-microbial action. Red seaweeds are the only source of the mycosporine-like amino acid class of phenolic substances, which possesses immunomodulatory effects.

The most typical multidrug-resistant cause of nosocomial infection in Europe is methicillin-resistant *S. aureus* (MRSA). Such hazardous conditions as sepsis, pneumonia, endocarditis, erysipelas, osteomyelitis, and foodborne illnesses are caused by this bacterium. Penicillin-binding protein 2a (PBP2a), whose existence determines a decreased affinity for binding β-lactam antibiotics, is encoded by the mecA gene, which causes MRSA to be resistant to β-lactam antibiotics. β-lactam antibiotic resistance is provided by this protein. According to reports, the algae *E. bicyclis* phlorotannin has potent anti-MRSA properties. The affected MEC operon mediates this characteristic of substances.

5.6.1 Sterols/Steroids

Steroids come from a variety of sources, including those that are synthetic, marine, and terrestrial and are lipophilic molecules generated from cholesterol. Steroidal substances extracted from sponges modify NF-κBactivity, that affect the pregnane X receptors (PXRs). Solomonsterol A, a selective PXR agonist from the marine sponge *Theonellaswinhoei*, decreased immune system abnormalities and systemic inflammation in a rat model of RA.Solomonsterol A prevents transgenic mice carrying a humanized PXR2 from developing arthritis brought on by anti-collagen antibodies. Solomonsterol A decreases the expression of the chemokines MIP1-α and RANTES and the inflammatory markers TNF-α, IFN-γ, and IL-17, which lessens the inflammatory response (Mencarelli et al., 2014). Steroids extracted from the starfish *Astropecten polyacanthus* exhibited immunomodulatory function by reducing the production of IL-6, IL-12 p40, and TNF-α, when dendritic cells derived from bone marrow are activated by LPS (Thao et al., 2014).

5.7 Marine Microorganisms with Immunomodulatory Compounds from Bacteria and Fungi

It is widely recognized that fungi produce anti-inflammatory substances (Xu et al., 2019). Some fungi have been regarded great candidates for immunotherapy because they have been shown to be able to boost the immune system, have anticancer action, and restore the immune system that has been damaged by chemotherapy and radiotherapy (Ayeka 2018). There are numerous substances that have been isolated from various marine fungi species that have been shown to have an impact on the immune system. Examples include semi-vioxanthin and azonazine, which have been tested on RAW264.7 cells for their immunomodulatory and anti-inflammatory activities, respectively (Wu et al., 2010). According to certain research, fungi are also a source of polysaccharides, which are increasingly valued for their potential probiotic effects and use in the treatment of immunological disorders (Goldring & Pattenden, 2004). The mushroom (*Phomaherbarum* YS4108) whose mycelium was used to make the α-D-glucan YCP was purified, and it demonstrated significant immunomodulatory functions by controlling T lymphocytes and dendritic cells in vitro, activating macrophages in vitro, and boosting phagocytotic activity in vitro and in vivo (Liu et al., 2020). In a recent study by Wei Liu et al., (2020), mice with acute colitis brought on by dextran sodium sulphate (DSS) were given the α-D-glucan YCP. After seven days, the YCP administration was successful in alleviating the colitis-related clinical symptoms in rats, reestablishing intestinal immunological homeostasis, and reversing mucosal damage. The overexpression of the pro-inflammatory cytokines IL-1, IL-6, and TNF-αinduced by DSS in the colon was also inhibited by the YCP. The tissues also had much higher amounts of IL-10 and IL-22, which led to the regeneration of injured tissues. Firmicutes, Bacteroidetes, Proteobacteria, Clostridiales, and Lachnospiraceae, which are strongly associated to immune modulation and mucus repair, were significantly altered by YCP in the gut microbiota (Kamada et al., 2013). In order to promote human mesenchymal stem cell (hMSC) development, Marchese et al. (2021) used a high-throughput drug-screening technique to bioprospect a sizable library of Irish deep-sea creature extracts. Additionally, filamentous fungi were present in the library, and they had a 6.8% success rate in lowering the inflammation caused by activated macrophages. In fact, the extracts have demonstrated a considerable decrease in the production of pro-inflammatory cytokines like TNF and IL-1, making them promising candidates for the development of anti-inflammatory medications. (Marchese et al., 2021). Excessive use of antibiotics, antiseptics, and chemotherapeutic drugs has negative effects on the environment and has contributed to the emergence of bacteria that are resistant to a variety of antibiotics. In Hong Kong in 1955, *Shigella* was shown to be resistant to antibiotics like sulphonamides, streptomycin, chloramphenicol, and tetracycline. The natural balance of flora and fauna is being threatened by the buildup of chemical and antibiotic residues in the environment. The World Health Organization defines probiotics as living bacteria that, when taken insufficiently, promote digestion and food absorption while also improving immunity, disease prevention, stress tolerance, and disease control. *Pseudoalteromonasxiamenensis*, which has demonstrated potential as a probiotic candidate to stimulate the host's immune system to fight disease and to improve environmental suitability, has shown to have antimicrobial capabilities against a variety of pathogens. This phenomenon was assessed by using murine macrophages (Wasana et al., 2021). According to the definition of the World Health Organization, probiotics are living microorganisms which, when administered in adequate quantities, confer health benefits such as improvement of immunity, disease control, tolerance to stress, improved digestion, and improved absorption of food.

5.8 Cyanobacteria

Cyanobacteria are biologically important group of microorganisms showing an abundant resource of biologically active chemicals, including toxins with a variety of therapeutic uses (Singh et al., 2011). Bioactive chemicals made by cyanobacteria have a wide range of effects, including antibacterial, antiviral, protease inhibition, anticancer, and immunomodulatory activity, LPS produced by marine cyanobacteria may have TLR4 antagonistic properties (Durai et al., 2015). Cyanobacterial LPS (CyP), from *Oscillatoria planktothrix* FP1, demonstrated protective effects in inflammatory circumstances and inhibits TLR4 activation. In both human and mouse dendritic cells, CyP inhibits the MyD88-dependent and MyD88-independent TLR4 signaling pathways, potentially inhibiting the inflammatory response brought on by LPS (Macagno et al., 2006). A group of substances called malyngamides, which are produced from the marine cyanobacterium *Lyngbyamajuscula*, have strong anti-inflammatory properties. Malyngamide F acetate inhibits nitric oxide (NO) and other inflammatory indicators production in RAW 264.7 cells. Because LPS stimulation decreases IL-1 and increases TNF-α production in MyD88 mutant animals via a MyD88-

independent mechanism, it specifically suppresses the MyD88-dependent pathway (Villa et al., 2010).

5.9 Microalgae

Among the most potential natural source of substances that can be employed as alternatives to popular pharmaceutical medications is algae, specifically microalgae. They have long evolved and adapted to a variety of habitats and harsh environments, which makes them particularly interesting candidates for use as immunomodulating agents (Sayed et al., 2020). They are very simple to sample and cultivate in comparison to macroorganisms. Numerous microalgae have also demonstrated immunomodulatory effects in human and animal models. Polysaccharides that are derived from marine microalgae have potent immune-boosting effects. Due to their high bioactivity and relative lack of toxicity, polysaccharides have drawn a lot of interest (Yu et. al., 2018). The microalgae *Thraustochytriidae* species' exopolysaccharide (EPS) demonstrated anti-cancer efficacy for changes in gene expression and cell proliferation. primarily concerned with the gene expression occurring in the BG-1 ovarian cancer cell line, MCF-7 breast cancer cell line, and SW-620 colon cancer cell line. The impact of EPS is determined by looking at T-cell cytokine production as well as the acceleration of T and B lymphocyte development. Microalgal EPS decreased the production of IL-6 while having no discernible impact on TNF-α (Geon-Tae Park et al., 2017). Microalgae have been found to have phycobilin, carotenoids, and chlorophylls as their main pigment classes (Yen et al., 2015). A 2022 in silico experiment utilizing molecular docking revealed that pro-inflammatory proteins like TNF-α, IL-6, and NF-κB-inducing kinase (NIK) docked with the pigments β -carotene, phycocyanobilin, astaxantin, and 9-cis-carotene generated by microalgae (NIK). The van der Waals interaction is primarily responsible for the binding of the pigments to protein. These connections, therefore, imply that the protein and its receptor interact. Better ligand-protein binding results from reduced binding energies. Studies have revealed the lowest binding energy to IL-6 is possessed by β-carotene, with a binding energy of 7.9 Kcal/mol. Autoimmunity and chronic inflammation were fostered by the dysregulation of IL-6 (Tanaka, & Kishimoto et al., 2014). Therefore, research and development of a substance that may bind to and suppress IL-6 may be helpful in treating autoimmune illness and stopping the cascade of cytokines. The 9-cis-carotene, with a binding energy of 7.9 Kcal/mol, has the greatest affinity for TNF- α. The NF-kB pathway and the mitogen-activated pathway kinase (MAPK) cascade are both activated by TNF-signalling. TNF-α may bind to either TNF receptor type 1 or type 2 to initiate TNF signaling. The pro-inflammatory factor TNFR1 also encourages apoptosis. As a result, it may be effective in reducing the cytokine storm associated with autoimmune illness by preventing TNF-'s attachment to TNFR1 (Lam et al., 2017). With a binding energy of -9.9 Kcal/mol, phycocyanobilin has the highest affinity for NF-κB-inducing kinase (NIK). NF-kB pathway regulation and TNF-signaling cascade support are accomplished by NIK. Discovering natural substances that can block NIK can result in the NF-kB pathway being blocked and the immune system being modulated (Magnani et al., 2000). In this way, microalgae pigment may help to modulate the immune system and prevent and lessen chronic inflammatory diseases (Widyaningrum et al., 2022). The majority of the microalgae-derived substances (alginate, fucoidan, ascophyllan, and porphyrin) function as adjuvants in vaccines by boosting the immune response and activating APCs (Montuori et al., 2022). Mice were used to test the *Chlorella stigmatophora* polysaccharide extract both in vitro and in vivo. Studies showed that the polysaccharide extract from *Chlorella stigmatophora* enhanced the phagocytic nature of macrophages from the peritoneal cavity. The immunostimulatory activity was boosted by the use of microalgae as diet. In mice, the spray-dried version of *D. salina* boosted NK and macrophage activation. *Tetraselmischuii* was administered orally, and the outcome was a high number of immune system-related genes such as major histocompatibility complex genes, IgM, and T-cell receptor beta being expressed. Further research is necessary since the variety of microalgal species that produce immunomodulating chemicals has not yet been sufficiently defined (Riccio et al., 2019).

5.10 Macroalgae

Numerous chemicals found in macroalgae have been shown to have positive effects on human health (Biris-Dorhoi, et al., 2020). These substances have favorable nutritional values due to the high content of carotenoids, proteins, dietary fibers, essential fatty acids, vitamins (vitamins C, D, E, K, and B complex), minerals, and about 60 different elements, including calcium, phosphorus, sodium, magnesium, iron, copper, manganese, potassium, and iodine (Takeiti et al., 2009). Algae are a staple of the diet in many countries across the world, particularly in Asian nations and seaweeds have long been utilized in Southeast Asia as a food source and a component of traditional medical preparations (Wells et al., 2017). Numerous algal extracts have been demonstrated to have immune system-stimulating properties. Ulvan, a gelling sulfated polysaccharide produced from *Ulva ohnoi*, was found to have immunomodulatory activity among the substances obtained from algae. In terms of the immunomodulatory activity, which was evaluated by examining the expression of inflammatory mediators, it was found that the fractions with the highest molecular weights (U21 and U209) demonstrated an immune response at 100 g/mL, improving LPS-induced inflammation and decreasing prostaglandin E2 (a potent inflammatory mediator produced by cyclooxygenase 2 conversion of arachidonic acid). Rhamnose and glucuronic acid were identified as the compound's primary sugar ingredients through chemical characterization. Ulvan is a promising substance to create medications for ectopic usage because it is known that cutaneous fibroblasts and keratinocytes are capable of directly recognizing these two sugars (Montuori, E et al., 2022). Brown algae (phaeophyceae) is a source of biologically active substances like fucoidans, which have anti-inflammatory, antitumor, antiviral, anti-allergic, and anticoagulant properties as well as immunostimulant effects by activating various immune cells (Yende et al., 2014). L-fucose and sulfate ester groups are the main components of the polysaccharides known as fucoidans. The immunomodulatory capabilities of fucoidans derived from brown seaweed have been studied in numerous in vivo and in vitro investigations (Saeed et al., 2021). In T cells, the fucoidans isolated from *Ecklonia cava* displayed stronger proliferative activity and increased INF-γ production (Synytsya et al., 2015). Galactose, xylose, arabinose, mannose, rhamnose, glucuronic acid, and/or glucose make up the majority of the components of green algae SPs (Ngo et al., 2013). Two galactans, sulfates and pyruvates (SP1 and SP2), were isolated from an SP-rich fraction from *Caulerpa cupressoides*in the study by Barbosa and co-authors. Sulphated polysaccharides (SPs) from *Caulerpa cupressoides* have shown potential as new drug candidates with immunomodulatory properties of biomedical interest, with potential uses in the management of immunodeficiency or hypoimmunity diseases. Galactans have an immunostimulant effect, as shown by the increased production of NO, endoplasmic reticulum oxidoreductin (ERO), and the cytokines IL-6 and TNF-α in murine macrophages RAW 264.7. The primary red seaweed cell wall polysaccharides, which possessed broad-spectrum medicinal properties, are the hydrocolloids galactans, carrageenans, and agars (Ismail et al., 2020). By increasing the number of tumor-infiltrating M1 macrophages, dendritic cell antigen "DCs," and additional stimulated CD4 + CD8 + T lymphocytes in the spleen λ-carrageenan injection could suppress malignant development in B16-F10 and 4T1 bearing mice and upgrade tumor resistant reactions (Luo et al., 2015).

5.11 Sponges

Sponges contain an abundance of substances with medicinally significant immunomodulatory, anti-inflammatory, and anticancer properties. Colon cancer cells were sensitive to Avarol (IC50<7 mM), a hydroquinone sesquiterpenoid that was isolated from *Dysideaavara* (Gordaliza et al., 2010). Halaven®, a chemotherapeutic medicine from Eisai Research Institute that is a condensed analog of Halichondrin-B, is offered on the market for the treatment of metastatic breast cancer(Kingston et al., 2009). A study on *Dysidea sp.* in 2000 revealed that the polyoxygenated sterols isolated, have potent, selective immunosuppressive properties because they prevent the interaction of IL-8 with its receptor (Sipkema et al., 2005). By giving albino Wistar mice the crude extract over the course of 14 days at different concentrations of 15, 10, and 5 mg/kg, a recent study evaluated the crude extract of a novel Sri Lankan sponge *Haliclona (Soestella) sp.* in vivo and ex vivo (Gunathilake et al., 2020). In the trials, mice given the greatest quantity of the crude extract of *Haliclona (Soestella) sp.* displayed immunomodulatory action (HSCE). The extract demonstrated a decrease in immune cells (white blood cells, lymphocytes, platelets, mesenchymal cells from bone marrow, and splenocytes) and a rise in the neutrophil: lymphocyte ratio when compared to the reference. The rise in plasma TNF-α levels in mice given 15 mg/kg of HSCE, which was used to treat the mice, was evidence of the immunomodulatory activity.

5.12 Mollusks

A vast range of bioactive metabolites that can be created by mollusks themselves, gathered from food sources, or produced by symbionts have been isolated as a result of chemical research on mollusks (Benkendorff et al., 2010). More than 1145 chemicals, including peptides, sterols, terpenes, polypropylene compounds, macrolides, fatty acid derivatives, nitrogen compounds, and alkaloids, were recovered from marine mollusks. The mollusk *Arca subcrenata Lischke*, which is frequently used as a meal with great nutritional value, has recently yielded an ASPG-2 polysaccharide (Shi et al., 2021), a water-soluble glucan. ASPG-2 boosted NO secretion and improved phagocytosis in murine RAW 264.7 macrophages. It also showed strong immunomodulatory effects and no cellular damage. TLR4-MAPK/Akt-NF-B signaling pathway activation is correlated with its effect. On LPS-stimulated murine RAW 264.7 cells, the polysaccharides from the bivalve mollusk *Mytilus coruscus* (MP) demonstrated an anti-inflammatory and immunomodulatory activity. The intestinal barrier's integrity and the Firmicutes/Bacteroidetes ratio were both improved in mice with dextran sodium sulfate-induced ulcerative colitis (DSS) who received MP treatment. Additionally, the colon saw an increase in the number of various potential probiotics, including *Anaerotruncus, Lactobacillus, Desulfovibrio, Alistipe, Odoribacter,* and *Enterorhabdus.* Consequently, MP might be a good choice for treating ulcerative colitis and boosting immunity. HPCG2, a protein with immunomodulatory function, was isolated from the bivalve (*Scapharcabroughtonii*). The results demonstrated that Akt, ERK, and JNK contributed in the polarization effects generated by HPCG2 on macrophages. HPCG2 enhanced the phosphorylation of Akt, ERK, and JNK. In RAW 264.7 cells, it was connected to the control of the TLR4/JNK/ERK and STAT3 signalling pathways. Thus, HPCG2 could be a possible immunomodulatory agent (Liu et al., 2020). Due to their great nutritional content, oysters are commonly consumed as food in coastal regions (Venugopal & Gopakumar, 2017). In a study, oyster peptides (OPs) were assessed for their ability to regulate the immunity of the mucosa and intestinal microbiota of immunosuppressed mice that had received the chemotherapy drug cyclophosphamide (Cy). The first natural defense against potential environmental harm is the intestinal mucosa's immune system, which is made up of lymphocytes, macrophages, and plasma cells (Brandtzaeg et al., 2010). Intestinal mucosa is damaged by cyclophosphamide (Huang et al., 2021) and the microbiota is diminished, but most importantly, the sIgA content is decreased, which is crucial for intestinal homeostasis and the proper functioning of the immune system. Total short-chain fatty acids SCFA (acetic acid, propionic acid, butyric acid, and acid valeric)are produced as a result of an increase in sIgA levels and Firmicutes/Bacteroidetes, the most prevalent bacteria in the intestinal microbiota (Xiang et al., 2021). SFCAs are metabolites that might keep the immune system functioning normally. Furthermore, it was discovered with precision that mice treated with Cy and given OP had an increase in SCFA in their faeces. This indicates an increase in intestinal bacterial activity and, consequently, an improvement in the bacterial community (Mirzaei et al., 2021).

5.13 Corals

The principal secretory product of hard (*Scleractinia*) and soft (*Alcyonacea*) corals, coral mucus, which contains different ratios of proteins, lipids, and polysaccharides, was the subject of multiple studies on corals in the 1990s (Wessels et al., 2016). Different coral species have different mucus glycoprotein chemical constitutions (Meikle et al., 1988). The polysaccharide PPA, which has immunomodulatory characteristics, was recently isolated from the coral *Pseudopterogorgiaamericana*. It has been demonstrated that it increases the expression of pro-inflammatory mediators in macrophages via the ROS, MAPK, PKC-α/δ-and NF-κB-dependent pathways, as well as the production of TNF-α and IL-6 in macrophages infected with *Shigella sonnei*or *E. coli* and the phagocytosis activity of bacteria-infected macrophages (Chernikov et al., 2021.

Corals such as *Antillogorgiaturgida, A. americana, Rhipidogorgia flabellum, Briareumasbestinum, Plexaurahomomalla, Plexaurelladichotoma, and Plexauropsiscrassa* have shown antibacterial property against marine bacteria. The mRNA expression of IL-1, IL-6, and TNF-α was increased in soft coral *Lobophytumsarcophytoides*, which demonstrated that LCPs-1-A had good immunoenhancing activity (Xueqin Cao et al., 2022).

A study on Vietnamese octocorals by (Kapustina et al., 2014), reported that an extract from *Junceella* sp. exhibited antibacterial activity against *E. coli*. This might have been due to the difference in cell wall composition. *Alcyonacean* soft corals from the Red Sea showed a notable amount of antibacterial action against marine microorganisms. *Alcyonacean* soft corals produce antibiotic chemicals as a form of chemical defense against microbial assault.

5.14 Fishes

In a recent study, it was found that collagen and chitosan solutions made from fishing waste had an immunostimulating effect on human PBMC cells, although neither collagen nor chitosan had cytotoxic effects. They increased the levels of cytosolic calcium and mitochondrial membrane potential, as well as the generation of the cytokines IL-6, IL-10, and TNF-; CD8+ and CD4+ T cell differentiation and activation (Portela et al., 2022). The immunomodulatory properties of host defense peptide (HPD), cationic HPD, such as cathelicidin, defensin, and NK-lysine (NKL), have been the subject of various investigations in recent decades (Montuori et al., 2022). HDPs are antimicrobial peptides (AMPs) with immunomodulatory effects in addition to their ability to destroy bacteria (Haney et al., 2013). NKL displays antibacterial activities against *E. coli* and *Staphylococcus aureus* and also immunomodulatory activity (Baien et al., 2020). When commercial antibiotics (kanamycin) were compared to BpNKL (a homolog of NKL), its actions against the bacteria *E. tarda, Vibrio parahaemolyticus,* and *Vibrio alginolyticus*were shown to be more effective (Ding et al., 2019). Additionally, fish secrete a variety of host defense peptides, including defensine, cathelicidins, hepcidins, piscidins, pleurocidins, and histone-derived peptides (Masso-Silva et al., 2014). Fish-defensin has demonstrated immunomodulatory, chemotactic, and anti-inflammatory effects. A new defense protein is known as On-Def was recently discovered in the gill mRNA of red-toothed triggerfish (*Odonusniger*) (Neelima et al., 2021).

Metabolites produced from omega-3 and omega-6 have a significant role in immunological modulation. Cyclooxygenase, lipoxygenase, and cytochrome P450 enzymes control their synthesis. The ability to phagocytose, the generation and release of cytokines and chemokines, and the polarisation of macrophages into either classically activated or alternatively activated macrophages are all enhanced by omega-3 fatty acids. PUFA is believed to play a significant role in macrophage gene regulatory changes. Docosahexaenoic acid (DHA) and eicosapentaenoic acid (EPA) treatment cause significant alterations in the gene expression in macrophages generated from THP-1 and activated by LPSWhen DHA is incubated with Raw macrophage cell lines, the expression of miRNAs that control gene expression, signal transduction, immunological defense, growth and differentiation, and genes linked with transport is altered. The phagocytic capability of macrophages is improved by omega-3 fatty acids. Neutrophils can synthesize prostaglandins, leukotrienes, thromboxanes, maresins, protectins, and resolvins by incorporating omega-3 fatty acids into the phospholipids in their cell membranes.

The production of reactive oxygen species and cytokines, neutrophil migration, phagocytic capability, and other aspects of neutrophil function are all modulated by omega-3 fatty acids. increase the neutrophils' ability to phagocytose. Omega-3 fatty acids can affect neutrophil ROS production. ROS are produced more frequently when EPA and DHA are combined. DHA and EPA supplementation for two months in humans increased the amount of ROS produced, when phorbol myristate acetate activated neutrophils (Saray Gutiérrez et al., 2019). Proinflammatory cytokines (IL-6, IL-10, and TNF-αProinflammatory cytokines (IL-6, IL-10, and TNF-) and T CD8+ and CD4+ lymphocyte differentiation and activation are boosted by fish collagen and shrimp chitosan. Through the creation of cytosolic and mitochondrial ROS, they support non-cytotoxic peripheral blood mononuclear cells and T CD8+ and CD4+ lymphocyte differentiation and activation are boosted by fish collagen and shrimp chitosan. Through the creation of cytosolic and mitochondrial ROS, they support non-cytotoxic peripheral blood mononuclear cells (Pinho Nascimento Portela et al., 2022).

5.15 Conclusions

Numerous studies are conducted on bioactive peptide, proteins, or protein hydrolysates from marine-derived biomaterials, especially their immunomodulatory effects. Approaches to use various immunomodulatory drugs to control these diseases have been spurred by a recent rise in the occurrence of diverse diseases. Most of the research has concentrated on evaluating the immunological responses of marine-derived bioactive compounds in animal studies or cell culture, but subsequent research in humans is less available. The biocompatibility, safety, and immunological effects of immunomodulatory agents made of amino acids, peptides, proteins hydrolysates, and proteins derived from marine biomaterials in the human body therefore require extensive clinical studies. The investigation is needed into the impact of proteins or peptides obtained from sources on the immune response.

REFERENCES

Acosta, J., Roa, F., Gonzalez-Chavarria, I., Astuya, A., Maura, R., Montesino, R., Munoz, C., Camacho, F., Saavedra, P., Valenzuela, A., et al. (2019). In vitro immunomodulatory activities of peptides derived from *Salmo salar* NK-lysin and cathelicidin in fish cells. *Fish Shellfish Immunol*, *88*, 587–594

Ahmad, B., Shah, M., & Choi, S. (2019). Oceans as a source of immunotherapy. *Marine Drugs*, *17*, 282.

Altmann, K. H. (2001). Microtubule-stabilizing agents: a growing class of important anticancer drugs. *Current Opinion in Chemical Biology*, *5*(4), 424–431.

Arancibia, S., Espinoza, C., Salazar, F., Del Campo, M., Tampe, R., Zhong, T. Y.,... & Becker, M. I. (2014). A novel immunomodulatory hemocyanin from the limpet Fissurella latimarginata promotes potent anti-tumor activity in melanoma. *PloS One*, *9*(1), e87240.

Ayeka, P. A. (2018). Potential of mushroom compounds as immunomodulators in cancer immunotherapy: a review. *Evidence-Based Complementary and Alternative Medicine*, 2018.

Ayrapetyan, O. N., Obluchinskaya, E. D., Zhurishkina, E. V., Skorik, Y. A., Lebedev, D. V., Kulminskaya, A. A., & Lapina, I. M. (2021). Antibacterial properties of fucoidans from the brown algae Fucus vesiculosus L. of the Barents Sea. *Biology*, *10*(1), 67.

Baien, S. H., Seele, J., Henneck, T., Freibrodt, C., Szura, G., Moubasher, H.,... ... & de Buhr, N. (2020). Antimicrobial and immunomodulatory effect of gum arabic on human and bovine granulocytes against *Staphylococcus aureus* and *Escherichia coli*. *Frontiers in Immunology*, *10*, 3119.

Benkendorff, K. (2010). Molluscan biological and chemical diversity: Secondary metabolites and medicinal resources produced by marine molluscs. *Biological Reviews*, *85*(4), 757–775.

Biris-Dorhoi, E. S., Michiu, D., Pop, C. R., Rotar, A. M., Tofana, M., Pop, O. L.,... ... & Farcas, A. C. (2020). Macroalgae—A sustainable source of chemical compounds with biological activities. *Nutrients*, *12*(10), 3085.

Brandtzaeg, P. (2010). The mucosal immune system and its integration with the mammary glands. *The Journal of Pediatrics*, *156*(2), S8–S15.

Brown, G. D., Willment, J. A., & Whitehead, L. (2018). C-type lectins in immunity and homeostasis. *Nature Reviews Immunology*, *18*(6), 374–389.

Cai, B., Chen, H., Wan, P., Luo, L., Ye, Z., Huang, J.,... & Pan, J. (2022). Isolation and identification of immunomodulatory peptides from the protein hydrolysate of tuna trimmings (Thunnas albacares). LWT, 164, 113614.

Cao, X., Zhang, Q., Zhu, Y., Li, S., Cai, Y., Li, P., Liu, D., Leng, Y., Ye, S., Xu, Z. and Li, H. (2022). Structural characterization and immunoenhancing effects of a polysaccharide from the soft coral lobophytumsarcophytoides. *Marine Biotechnology*, *24*(1), 203–215.

Chernikov, O. V., Chiu, H. W., Li, L. H., Kokoulin, M. S., Molchanova, V. I., Hsu, H. T.,... ... & Hua, K. F. (2021). Immunomodulatory properties of polysaccharides from the coral pseudopterogorgiaamericana in macrophages. *Cells*, *10*(12), 3531.

Chikalovets, I. V., Kovalchuk, S. N., Litovchenko, A. P., Molchanova, V. I., Pivkin, M. V., & Chernikov, O. V. (2016). A new Gal/GalNAc-specific lectin from the mussel Mytilus trossulus: structure, tissue specificity, antimicrobial and antifungal activity. *Fish & Shellfish Immunology*, *50*, 27–33.

Costa, J. A. V., Lucas, B. F., Alvarenga, A. G. P., Moreira, J. B., & de Morais, M. G. (2021). Microalgae polysaccharides: an overview of production, characterization, and potential applications. *Polysaccharides*, *2*(4), 759–772.

Cumashi, A., Ushakova, N. A., Preobrazhenskaya, M. E., D'Incecco, A., Piccoli, A., Totani, L.,... & Nifantiev, N. E. (2007). A comparative study of the anti-inflammatory, anticoagulant, antiangiogenic, and antiadhesive activities of nine different fucoidans from brown seaweeds. *Glycobiology*, *17*(5), 541–552.

Cunha, L. & Grenha, A. (2016). Sulfated seaweed polysaccharides as multifunctional materials in drug delivery applications. *Marine Drugs*, *14*(3), 42.

Ding, F. F., Li, C. H., & Chen, J. (2019). Molecular characterization of the NK-lysin in a teleost fish, Boleophthalmuspectinirostris: Antimicrobial activity and immunomodulatory activity on monocytes/macrophages. *Fish & Shellfish Immunology*, *92*, 256–264.

Ding, Y., Liu, X., Bu, L., Li, H., & Zhang, S. (2012). Antimicrobial-immunomodulatory activities of zebrafish phosvitin-derived peptide Pt5. *Peptides*, *37*, 309–313

Durai, P., Batool, M., & Choi, S. (2015). Structure and effects of cyanobacterial lipopolysaccharides. *Marine Drugs*, *13*(7), 4217–4230.

Goldring, W. P., & Pattenden, G. (2004). Total synthesis of (±)-phomactin G, a platelet activating factor antagonist from the marine fungus Phoma sp. *Organic & Biomolecular Chemistry*, *2*(4), 466–473.

Gordaliza, M. (2010). Cytotoxic terpene quinones from marine sponges. *Marine Drugs*, *8*(12), 2849–2870.

Grienke, U., Silke, J., & Tasdemir, D. (2014). Bioactive compounds from marine mussels and their effects on human health. *Food Chemistry*, *142*, 48–60.

Gunathilake, V., Bertolino, M., Bavestrello, G., & Udagama, P. (2020). Immunomodulatory activity of the marine sponge, Haliclona (Soestella) sp.(Haplosclerida: Chalinidae), from Sri Lanka in Wistar albino rats: immunosuppression and Th1-skewed cytokine response. *Journal of Immunology Research*, *2020*, 1–11.

Gutiérrez, S., Svahn, S. L., & Johansson, M. E. (2019). Effects of omega-3 fatty acids on immune cells. *International Journal of Molecular Sciences*, *20*(20), 5028.

Hafting, J. T., Craigie, J. S., Stengel, D. B., Loureiro, R. R., Buschmann, A. H., Yarish, C.,... & Critchley, A. T. (2015). Prospects and challenges for industrial production of seaweed bioactives. *Journal of Phycology*, *51*(5), 821–837.

Haney, E. F., & Hancock, R. E. (2013). Peptide design for antimicrobial and immunomodulatory applications. *Peptide Science*, *100*(6), 572–583.

Havla, J., Warnke, C., Derfuss, T., Kappos, L., Hartung, H.P. & Hohlfeld, R. (2016). Interdisciplinary risk management in the treatment of multiple sclerosis. *Deutsches Ärzteblatt International*, *113* (51-52), 879- 886.

Hou, H., Fan, Y., Wang, S. K., Si, L. L., & Li, B. F. (2016). Immunomodulatory activity of Alaska pollock hydrolysates obtained by glutamic acid biosensor—Artificial neural network and the identification of its active central fragment. *J. Funct. Foods*, *24*, 37–47.

Huang, J., Huang, J., Li, Y., Wang, Y., Wang, F., Qiu, X.,… … & Li, H. (2021). Sodium alginate modulates immunity, intestinal mucosal barrier function, and gut microbiota in cyclophosphamide-induced immunosuppressed BALB/c mice. *Journal of Agricultural and Food Chemistry*, *69*(25), 7064–7073.

Ismail, M. M., Alotaibi, B. S., & El-Sheekh, M. M. (2020). Therapeutic uses of red macroalgae. *Molecules*, *25*(19), 4411.

Kamada, N., Seo, S. U., Chen, G. Y., & Núñez, G. (2013). Role of the gut microbiota in immunity and inflammatory disease. *Nature Reviews Immunology*, *13*(5), 321–335.

Kang, H. K., Lee, H. H., Seo, C. H., & Park, Y. (2019). Antimicrobial and immunomodulatory properties and applications of marine-derived proteins and peptides. *Marine Drugs*, *17*(6), 350.

Kapustina, I. I., Kalinovskii, A. I., Dmitrenok, P. S., Kuz'mich, A. S., Nedashkovskaya, O. I. and Grebnev, B. B. (2014). Diterpenoids and other metabolites from the Vietnamese gorgonians Lophogorgia sp. and Junceella sp. *Chemistry of Natural Compounds*, *50*(6), 1140–1142.

Kiewiet, M. B., Faas, M. M., & De Vos, P. (2018). Immunomodulatory protein hydrolysates and their application. *Nutrients*, *10*(7), 904.

Kim, J., Yung, B. C., Kim, W. J., & Chen, X. (2017). Combination of nitric oxide and drug delivery systems: tools for overcoming drug resistance in chemotherapy. *Journal of Controlled Release*, *263*, 223–230.

Kim, E. K., Kim, Y. S., Hwang, J. W., Kang, S. H., Choi, D. K., Lee, K. H.,... & Park, P. J. (2013). Purification of a novel nitric oxide inhibitory peptide derived from enzymatic hydrolysates of Mytilus coruscus. *Fish & Shellfish Immunology*, 34(6), 1416–1420.

Kingston, D. G. (2009). Tubulin-interactive natural products as anticancer agents. *Journal of Natural Products*, *72*(3), 507–515.

Lam, A., Prabhu, R., Gross, C. M., Riesenberg, L. A., Singh, V., & Aggarwal, S. (2017). Role of apoptosis and autophagy in tuberculosis. *American Journal of Physiology-Lung Cellular and Molecular Physiology*, *313*(2), L218–L229.

Lammers, T., Kiessling, F., Hennink, W. E., & Storm, G. (2012). Drug targeting to tumors: principles, pitfalls and (pre-) clinical progress. *Journal of Controlled Release*, *161*(2), 175–187.

Laurienzo, P. (2010). Marine polysaccharides in pharmaceutical applications: an overview. *Marine drugs*, *8*(9), 2435–2465.

Lauritano, C., Martín, J., de la Cruz, M., Reyes, F., Romano, G., & Ianora, A. (2018). First identification of marine diatoms with anti-tuberculosis activity. *Scientific Reports*, *8*(1), 2284.

Li, W., Ye, S., Zhang, Z., Tang, J., Jin, H., Huang, F.,... & Yu, F. (2019). Purification and characterization of a novel pentadecapeptide from protein hydrolysates of Cyclina sinensis and its immunomodulatory effects on RAW264. 7 cells. *Marine Drugs*, *17*(1), 30.

Lin, W. C., Chang, H. Y., & Chen, J. Y. (2016). Electrotransfer of the tilapia piscidin 3 and tilapia piscidin 4 genes into skeletal muscle enhances the antibacterial and immunomodulatory functions of *Oreochromis niloticus*. *Fish & Shellfish Immunology*, *50*, 200–209.

Liu, W., Bi, S., Li, C., Zheng, H., Guo, Z., Luo, Y.,… … & Yu, R. (2020). Purification and characterization of a new CRISP-related protein from Scapharcabroughtonii and its immunomodulatory activity. *Marine Drugs*, *18*(6), 299.

Luo, M., Shao, B., Nie, W., Wei, X. W., Li, Y. L., Wang, B. L.,... & Wei, Y. Q. (2015). Antitumor and adjuvant activity of λ-carrageenan by stimulating immune response in cancer immunotherapy. *Scientific Reports*, *5*(1), 11062.

Ma, Y. G., Cho, M. Y., Zhao, M., Park, J. W., Matsushita, M., Fujita, T., & Lee, B. L. (2004). Human mannose-binding lectin and L-ficolin function as specific pattern recognition proteins in the lectin activation pathway of complement. *Journal of Biological Chemistry*, *279*(24), 25307–25312.

Macagno, A., Molteni, M., Rinaldi, A., Bertoni, F., Lanzavecchia, A., Rossetti, C., Sallusto, F. (2006). A cyanobacterial LPS antagonist prevents endotoxin shock and blocks sustained TLR4 stimulation required for cytokine expression. *JEM*, *203*(6), 1481–1492

Magnadottir, B. (2006). Innate immunity of fish (overview). *Fish & Shellfish Immunology*, *20*, 137–151

Magnani, M., Crinelli, R., Bianchi, M., & Antonelli, A. (2000). The ubiquitin-dependent proteolytic system and other potential targets for the modulation of nuclear factor-kB (NF-kB). *Current Drug Targets*, *1*(4), 387–399.

Marchese, P., Young, R., O'Connell, E., Afoullouss, S., Baker, B. J., Allcock, A. L.,... & Murphy, J. M. (2021). Deep-sea coral garden invertebrates and their associated fungi are genetic resources for chronic disease drug discovery. *Marine Drugs*, *19*(7), 390.

Marcinkiewicz, J., & Kontny, E. (2014). Taurine and inflammatory diseases. *Amino Acids*, *46*, 7–20.

Masso-Silva, J. A., & Diamond, G. (2014). Antimicrobial peptides from fish. *Pharmaceuticals*, *7*(3), 265–310.

Medzhitov, R., & Janeway Jr, C. (2000). Innate immunity. *New England Journal of Medicine*, *343*(5), 338–344.

Meikle, P., Richards, G. N., & Yellowlees, D. (1988). Structural investigations on the mucus from six species of coral. *Marine Biology*, *99*, 187–193.

Mencarelli, A., Amore, C. D., Renga, B., Cipriani, S., Carino, A., Sepe, V., Perissutti, E., Auria, M. V. D., Zampella, A., Distrutti, E., Fiorucci, S.. (2014). Solomonsterol A, a Marine Pregnane-X-Receptor Agonist, Attenuates Inflammation and Immune Dysfunction in a Mouse Model of Arthritis. *Marine Drugs*, *12*, 36–53.

Mirzaei, R., Bouzari, B., Hosseini-Fard, S. R., Mazaheri, M., Ahmadyousefi, Y., Abdi, M.,… … & Karampoor, S. (2021). Role of microbiota-derived short-chain fatty acids in nervous system disorders. *Biomedicine & Pharmacotherapy*, *139*, 111661.

Mitta, G., Vandenbulcke, F., Hubert, F., Salzet, M., & Roch, P. (2000). Involvement of mytilins in mussel antimicrobial defense. *Journal of Biological Chemistry*, *275*(17), 12954–12962.

Montuori, E., de Pascale, D., & Lauritano, C. (2022). Recent discoveries on marine organism immunomodulatory activities. *Marine Drugs*, *20*(7), 422.

Mora Román, J. J., Del Campo, M., Villar, J., Paolini, F., Curzio, G., Venuti, A.,... ... & Becker, M. I. (2019). Immunotherapeutic potential of mollusk hemocyanins in combination with human vaccine adjuvants in murine models of oral cancer. *Journal of Immunology Research, 2019*.

Naidoo, J., Page, D. B., & Wolchok, J. D. (2014). Immune modulation for cancer therapy. *British Journal of Cancer*, *111*(12), 2214–2219

Neelima, S., Archana, K., Athira, P. P., Anju, M. V., Anooja, V. V., Bright Singh, I. S., & Philip, R. (2021). Molecular characterization of a novel β-defensin isoform from the red-toothed trigger fish, Odonusniger (Ruppel, 1836). *Journal of Genetic Engineering and Biotechnology*, *19*(1), 71.

Nesse, K. O., Nagalakshmi, A. P., Marimuthu, P., and Singh, M. (2011). Efficacy of a fish protein hydrolysate in malnourished children. *Ind. J. Clin. Biochem.*, *26*, 360–365.

Newman, D. J., & Cragg, G. M. (2020). Natural products as sources of new drugs over the nearly four decades from 01/1981 to 09/2019. *Journal of Natural Products*, 83(3), 770–803.

Ngo, D. H., & Kim, S. K. (2013). Sulfated polysaccharides as bioactive agents from marine algae. *International Journal of Biological Macromolecules*, *62*, 70–75.

Nijkamp, J., Kusters, M., Beets-Tan, R. G., Martijn, H., Beets, G. L., van de Velde, C. J., & Marijnen, C. A. (2011). Three-dimensional analysis of recurrence patterns in rectal cancer: the cranial border in hypofractionated preoperative radiotherapy can be lowered. *International Journal of Radiation Oncology, Biology, Physics*, *80*(1), 103–110.

Paolino, M., & Penninger, J. M. (2010, June). Cbl-b in T-cell activation. In *Seminars in immunopathology* (Vol. 32, pp. 137–148). Springer-Verlag.

Park, G. T., Go, R. E., Lee, H. M., Lee, G. A., Kim, C. W., Seo, J. W.,... & Hwang, K. A. (2017). Potential anti-proliferative and immunomodulatory effects of marine microalgal exopolysaccharide on various human cancer cells and lymphocytes in vitro. *Marine Biotechnology*, *19*, 136–146.

Peng, Y., Song, Y., Wang, Q., Hu, Y., He, Y., Ren, D.,... & Zhou, H. (2019). In vitro and in vivo immunomodulatory effects of fucoidan compound agents. *International Journal of Biological Macromolecules*, *127*, 48–56.

Portela, L. C. P. N., Cahú, T. B., Bezerra, T. S., do Nascimento Santos, D. K. D., Sousa, G. F., Portela, R. W. S.,... ... & de Souza Bezerra, R. (2022). Biocompatibility and immunostimulatory properties of fish collagen and shrimp chitosan towards peripheral blood mononuclear cells (PBMCs). *International Journal of Biological Macromolecules*, *210*, 282–291.

Riccio, G., & Lauritano, C. (2019). Microalgae with immunomodulatory activities. *Marine Drugs*, *18*(1), 2.

Routy, J.P., Routy, B., Graziani, G.M. & Mehraj, V. (2016). The kynurenine pathway is a double-edged sword in immune-privileged sites and in cancer: implications for immunotherapy. *International Journal of Tryptophan Research*, *9*, IJTR-S38355.

Saeed, M., Arain, M. A., Ali Fazlani, S., Marghazani, I. B., Umar, M., Soomro, J.,... & Alagawany, M. (2021). A comprehensive review on the health benefits and nutritional significance of fucoidan polysaccharide derived from brown seaweeds in human, animals and aquatic organisms. *Aquaculture Nutrition*, *27*(3), 633–654.

Sasikala, C., & Geetha Ramani, D. (2017). Comparative study on antimicrobial activity of seaweeds. *Asian J. Pharm. Clin. Res.*, *10*, 384–386.

Sayed, A. M., Hassan, M. H., Alhadrami, H. A., Hassan, H. M., Goodfellow, M., & Rateb, M. E. (2020). Extreme environments: Microbiology leading to specialized metabolites. *Journal of Applied Microbiology*, *128*(3), 630–657.

Sehn, L. H., Herrera, A. F., Flowers, C. R., Kamdar, M. K., McMillan, A., Hertzberg, M.,... & Matasar, M. J. (2020). Polatuzumab vedotin in relapsed or refractory diffuse large B-cell lymphoma. *Journal of Clinical Oncology*, 38(2), 155.

Selvaraj, V., Sampath, K., & Sekar, V. (2005). Administration of yeast glucan enhances survival and some non-specific and specific immune parameters in carp (Cyprinus carpio) infected with Aeromonas hydrophila. *Fish & Shellfish Immunology*, *19*(4), 293–306.

Sharon, N. (2007). Lectins: Carbohydrate-specific reagents and biological recognition molecules. *Journal of Biological Chemistry*, *282*(5), 2753–2764.

Shi, H., Bi, S., Li, H., Li, J., Li, C., Yu, R.,... ... & Zhu, J. (2021). Purification and characterization of a novel mixed-linkage α, β-d-glucan from Arca subcrenata and its immunoregulatory activity. *International Journal of Biological Macromolecules*, *182*, 207–216.

Singh, R. K., Tiwari, S. P., Rai, A. K., & Mohapatra, T. M. (2011). Cyanobacteria: an emerging source for drug discovery. *The Journal of Antibiotics*, *64*(6), 401–412.

Sipkema, D., Franssen, M. C., Osinga, R., Tramper, J., &Wijffels, R. H. (2005). Marine sponges as pharmacy. *Marine Biotechnology*, *7*, 142–162.

Song, M. K., Salam, N. K., Roufogalis, B. D., & Huang, T. H. W. (2011). Lyciumbarbarum (Goji Berry) extracts and its taurine component inhibit PPAR-γ-dependent gene transcription in human retinal pigment epithelial cells: possible implications for diabetic retinopathy treatment. *Biochemical Pharmacology*, *82*(9), 1209–1218.

Synytsya, A., Čopíková, J., Kim, W. J., & Park, Y. I. (2015). Cell wall polysaccharides of marine algae. *Springer Handbook of Marine Biotechnology*, 543–590.

Takahashi, B. J. D., & Urbinati, E. C. Fish Immunology. (2014). The modification and manipulation of the innate immune system: Brazilian studies. *An Acad Bras Cienc.*, *86*, 3.

Takeiti, C. Y., Antonio, G. C., Motta, E. M., Collares-Queiroz, F. P., & Park, K. J. (2009). Nutritive evaluation of a non-conventional leafy vegetable (Pereskia aculeata Miller). *International Journal of Food Sciences and Nutrition*, *60*(sup1), 148–160.

Talley, R. W., O'Bryan, R. M., Tucker, W. G., & Loo, R. V. (1967). Clinical pharmacology and human antitumor activity of cytosine arabinoside. *Cancer*, *20*(5), 809–816.

Tanaka, T., & Kishimoto, T. (2014). The biology and medical implications of interleukin-6. *Cancer Immunology Research*, *2*(4), 288–294.

Thao, N. P., Luyen, B. T. T., Kim, E. J., Kang, J. I., Kang, H. K., Cuong, N. X., Nam, N. H., Kiem, P. V., Minh, C. V., & Kim, Y. H. (2014). Steroidal constituents from the edible sea Urchin Diademasavignyi Michelin induce apoptosis in human cancer cells. *J. Med. Food*, *18*, 45–53.

Venugopal, V., & Gopakumar, K. (2017). Shellfish: Nutritive value, health benefits, and consumer safety. *Comprehensive Reviews in Food Science and Food Safety*, *16*(6), 1219–1242.

Villa, F. A., Lieske, K., & Gerwick, L. (2010). Selective MyD88-dependent pathway inhibition by the cyanobacterial natural product Malyngamide F acetate. *Eur. J. Pharmacol.*, *629*, 140–146.

Villani, D., Cognetta, C., Repetto, C., Serino, S., Toniolo, D., Scanzi, F., & Riva, G. (2018). Promoting emotional well-being in older breast cancer patients: Results From an eHealth Intervention. *Frontiers in Psychology*, 9, 2279.

Wang, Y. K., He, H. L., Wang, G. F., Wu, H., Zhou, B. C., Chen, X. L., & Zhang, Y. Z. (2010). Oyster (*Crassostrea gigas*) Hydrolysates produced on a plant scale have antitumor activity and immunostimulating effects in BALB/c mice. *Marine Drugs*, *8*, 255–268.

Wasana, W. P., Senevirathne, A., Nikapitiya, C., Eom, T. Y., Lee, Y., Lee, J. S.,... & De Zoysa, M. (2021). A novel Pseudoalteromonas xiamenensis marine isolate as a potential probiotic: Anti-inflammatory and innate immune modulatory effects against thermal and pathogenic stresses. *Marine Drugs*, *19*(12), 707.

Wells, M. L., Potin, P., Craigie, J. S., Raven, J. A., Merchant, S. S., Helliwell, K. E.,... & Brawley, S. H. (2017). Algae as nutritional and functional food sources: Revisiting our understanding. *Journal of Applied Phycology*, *29*, 949–982.

Wessels, W. (2016). *Molecular bases of soft coral reproduction* (Doctoral dissertation, James Cook University).

Widyaningrum, D., Oktafika, R. A., & Cecilia, D. (2022, February). Microalgae pigments as a promising immunomodulating food ingredient: In silico study. In *IOP Conference Series: Earth and Environmental Science* (*Vol. 998*, No. 1, p. 012056). IOP Publishing.

Wu, M. H., Pan, T. M., Wu, Y. J., Chang, S. J., Chang, M. S., & Hu, C. Y. (2010). Exopolysaccharide activities from probiotic bifidobacterium: Immunomodulatory effects (on J774A. 1 macrophages) and antimicrobial properties. *International Journal of Food Microbiology*, *144*(1), 104–110.

Wu, H., Lei, Y., Lu, J., Zhu, R., Xiao, D., Jiao, C.,... & Li, M. (2019). Effect of citric acid induced crosslinking on the structure and properties of potato starch/chitosan composite films. *Food Hydrocolloids*, *97*, 105208.

Xiang, N., Rädecker, N., Pogoreutz, C., Cárdenas, A., Meibom, A., Wild, C.,... & Voolstra, C. R. (2022). Presence of algal symbionts affects denitrifying bacterial communities in the sea anemone Aiptasia coral model. *ISME Communications*, *2*(1), 105.

Xu, J., Yi, M., Ding, L., & He, S. (2019). A review of anti-inflammatory compounds from marine fungi, 2000–2018. *Marine Drugs*, *17*(11), 636.

Yang, M., Ma, C., Sun, J., Shao, Q., Gao, W., Zhang, Y.,... & Qu, X. (2008). Fucoidan stimulation induces a functional maturation of human monocyte-derived dendritic cells. *International Immunopharmacology*, *8*(13-14), 1754–1760.

Yen, H. W., Yang, S. C., Chen, C. H., & Chang, J. S. (2015). Supercritical fluid extraction of valuable compounds from microalgal biomass. *Bioresource Technology*, *184*, 291–296.

Yen, H. W., Hu, I. C., Chen, C. Y., Ho, S. H., Lee, D. J., & Chang, J. S. (2013). Microalgae-based biorefinery–from biofuels to natural products. *Bioresource Technology*, *135*, 166–174.

Yende, S. R., Harle, U. N., & Chaugule, B. B. (2014). Therapeutic potential and health benefits of Sargassum species. *Pharmacognosy Reviews*, *8*(15), 1.

Yu, Y., Shen, M., Song, Q., & Xie, J. (2018). Biological activities and pharmaceutical applications of polysaccharide from natural resources: a review. *Carbohydrate polymers*, *183*, 91–101.

Zhang, Q. Y., Yan, Z. B., Meng, Y. M., Hong, X. Y., Shao, G., Ma, J. J.,... & Fu, C. Y. (2021). Antimicrobial peptides: mechanism of action, activity and clinical potential. *Military Medical Research*, *8*, 1–25.

Zhu, M., Ge, L., Lyu, Y., Zi, Y., Li, X., Li, D., & Mu, C. (2017). Preparation, characterization and antibacterial activity of oxidized κ-carrageenan. *Carbohydrate Polymers*, *174*, 1051–1058.

6

Immunomodulation in Aquaculture Health Management: Opportunities and Obstacles

Ramchandran Ishwarya[1], Baskaralingam Vaseeharan[2], Rengarajan Jayakumar[3], Subramaniam Sivakumar[4], and Preetham Elumalai[5]

[1]*Mandapam Regional Centre, ICAR – Central Marine Fisheries Research Institute, Mandapam, Tamil Nadu, India*

[2]*Biomaterials and Biotechnology in Animal Health Lab, Department of Animal Health and Management, Alagappa University, Karaikudi, Tamil Nadu, India*

[3]*ICAR - Central Institute of Brackishwater Aquaculture (CIBA), Chennai, Tamil Nadu, India*

[4]*Department of Biochemistry, Sri Sankara Arts and Science College, Enathur, Kanchipuram, Tamilnadu, India*

[5]*Department of Marine Biology, Microbiology and Biochemistry, School of Marine Sciences, Cochin University of Science and Technology (CUSAT), Lakeside Campus. Fine Arts Avenue, Cochin, Kerala, India*

6.1 Introduction

One of the most realistic and promising industries for meeting the growing need for animal protein and ensuring human nutrition and food security, particularly for people living in areas with a dearth of animals, is aquaculture. Recently, aquaculture is becoming more popular around the world. This sector has developed to become the greatest growing food producing sector in the world in the past three decades. Aquaculture industry has grown by 52.7% during the past three decades, producing 82 million tonnes with a first sale value estimated at $250 billion USD (FAO, 2022). Aquaculture residue as an essential food producing industry in the world, excluding is disputed through some discouraging problems on sustainability from biological, environmental, and socio-economic points of observation. In recent years, attention has been focused on addressing the global challenges of hunger, malnutrition, and impaired health in our population. Since 2014, aquaculture has provided more fish for human consumption than capture fisheries, and by 2030 it is expected to contribute 60% of the total fish and shrimp available for human consumption. Global aquatic products aquatic plants, fish, crustaceans, mollusks, and other species represent 16.4% of the world's supply of animal protein and are the third-largest source of food protein for human consumption after cereals and milk. Expanding the role of aquaculture towards ending hunger and all forms of malnutrition is as important as the focus on increasing production. Aquaculture is uniquely positioned to be one of the driving forces needed to address the nutrition challenges that we face in the world today. More than 40% of the aquaculture sector is devoted to the production of fish, which is the most widely farmed and harvested aquatic product species. Compared to other food animal production industries, aquaculture is currently growing more quickly; it now contributes more to the world's supplies of fish, crustaceans and mollusks each year (Boyd et al., 2022). The increased demand for fish and other aquatic foods has altered the landscape of fisheries and aquaculture, particularly accelerated the development of aquaculture and increasing the proportion of aquaculture production to approximately 7.7% by 2020. By that year, total fisheries and aquaculture production reached a record high of 214 million tonnes, including 178 million tonnes of aquatic, and the proportion is predicted to increase by another 15% by 2030 (FAO 2022). Ever-increasing fish density and unscientific fish culture practices results in deterioration of water quality and stress often foremost to eruption of deadly diseases of microbial etiology leading to production and economic losses. This has led to an increasing interest in fish health. Numerous commercial fish species have been raised intensively in cramped or enclosed areas like ponds, cages, or tanks at high densities, which has had a negative impact on their health by creating a potentially stressful environment and infectious diseases. The outbreaks of

DOI: 10.1201/9781003361183-8

contagious diseases have become a barrier to aquaculture's growth. The incidents have spread due to the unrestricted movement of aquatic creatures, which has caused the spread of infectious organisms between the nations (Bashir et al., 2020). Better preventative, diagnostic, and therapeutic methods are required throughout fish farming operations to ensure optimal fish production (Hussan et al., 2016). Unfortunately in some aquaculture regions with a development in aquaculture sector use of antibiotics and other chemicals is gradually rising that can cause several problems including development of antibiotic resistance, chemical residual in the carcasses and destruction of aquatic ecosystems (Araújo et al., 2021). Therefore, application of novel biotechnological tools are necessary to maintain the sustaninablity of aquaculture development worldwide. Application of immunostimulants for disease control measures in aquaculture is receiving more attention than the use of chemotherapy drugs, and immunomodulation has been suggested as a potential strategy to protect cultured fish from infectious pathogens by enhancing their innate immune system. The argument for including the use of immunostimulants in the overall aquaculture health management strategy has recently been validated by many research works. And the results are promising in increasing the economy of aquaculture production and food safety chains. This chapter addressed the opportunities and obstacles encountered in the application of immunomodulation in the aquaculture industry.

6.1.1 Immunomodulation

Immunomodulation in aquaculture has been suggested as a viable strategy for enhancing larval endurance through raising the developing animal's innate responses awaiting its adaptive immune response is strong enough to establish an effective defence against the pathogen. Immunomodulations are alterations to the immune system that occur following contact with a chemical or product that either stimulates or suppresses the immunological response. They encompass all therapeutic involvements expected at changing the immune response. The only option left to combat the illness in that situation is immunomodulation, which aims to boost the body's immune capacity. In general, when the immune response is pathologically diminished (i.e., disease states characterised by insufficient immunological function) and when the "normal" immune response proves insufficient to control a specific infection or pathological condition, a more active immune response may be desirable. Immunodeficiency is the name given to the first condition, which comprises numerous inherited and acquired diseases. The introduction of numerous immunomodulators over time led to the discovery that herbal immunomodulators based on the traditional Indian ayurvedic medical system were the most efficient and side-effect-free. Animal immunomodulators have the capacity to influence an immune response to improve the effectiveness of protection in food-producing animals, and these substances are known as immunomodulators. As a result, they are often referred to as immunomodulators, immunological regenerating agents, immunoaugmenting agents, or biological response modifiers. Immunomodulators refer to all therapeutic interventions that are intended to modulate the immune system, including the manipulation of the immune system. Immunostimulation or immunoptentiation, the argumentation of immunological response, can be used to avoid infection via vaccination through the activation of humoral immunity to fight against an infection that has already taken hold by modifying the immune response. The use of immunostimulants to enhance the particular immune response and protection against infections is another interesting area of current study (Table 6.1). The recovery from immunosuppressive conditions brought on by any type of stress can be aided by immunostimulants. Additionally, these immunostimulants can be categorised as natural or synthetic based on their place of origin, chemical makeup and method of operation. In contrast to synthetic immunostimulants, which are produced chemically, natural immunostimulants are macromolecular substances originating from plants, microorganisms or algae. When combined, they have been shown to have broad-spectrum activities that are independent of the causing factor. Additionally, these greatly lessen the negative effects of stressors on the general wellbeing of the impacted species. In order to control and prevent fish disease, immunostimulants were first introduced to the aquaculture sector decades ago.

Immunostimulants, also known as chemicals, medicines, stressors or behaviours, work by directly interacting with systemic cells to improve the immune response and to further activate the host's defences against pathogenic organisms. Numerous substances that have the power to activate or enhance the immune system are known as immunostimulants. As a result, using immunostimulants in aquaculture as a substitute for antibiotics and vaccinations is growing in popularity. To protect fish from infections, it is crucial to choose an immunostimulant that is both appropriate and effective. This method is viewed as a promising substitute for chemotherapy and vaccines due to immunostimulants' broad spectrum activity, affordability and environmentally friendly disease control nature. Fish and shellfish have been successfully treated with a variety of immunostimulants, including chemical substances, bacterial parts, polysaccharides, and animal or plant extracts. Polysaccharides, minerals, oligosaccharides, herbs, antimicrobial peptides, and microbes are the principal immunostimulants used in aquaculture. By improving both specific and non-specific defensive mechanisms in fish, shellfish, and other animals, immunostimulants are efficient ways to improve immune competence and disease resistance. In this regard, using immunostimulants appears to be a practical strategy with higher potential for boosting the host's immune and so increasing fish health (Wang et al., 2017). The study of immunostimulants is one of the most active fields of applied medical research because of its importance to health care (Figure 6.1). Polysaccharides, hormones, vitamins, various bacterium components, physiologically active substances, Chinese medicines, and synthetic pharmaceuticals are examples of immunostimulants that can encourage and generate a potent immune response in the host. By stimulating an efficient sustained response to pathogens, the host's particular and non-specific defensive mechanisms are stimulated, which typically results in the early relief of infections or stress. The protection so obtained also becomes crucial in situations where it may not be possible to immunise with a specific vaccine due to the unknown nature of the disease. In the end, this strategy helps to counteract stress's immunosuppressive effects. Immunostimulants are used in aquaculture to stimulate aquatic animals' immune systems and increase their ability for disease resistance. Fish have unique immune systems that are made up of cellular immunity and humoral immunity. Cellular immunity in fish consists of lymphocytes, natural killer cells, neutrophils, and phagocytic cells, while humoral immunity contains lysozyme, haemolysin, immunoglobulins, and complement molecules. Immunostimulants in fish coordinate cellular and humoral immunity, increase the phagocytic capability of neutrophils and lymphocytes, promote the release of cytokines from lymphocytes, and elicit antibody and complement responses. Immunostimulants are added to aquaculture systems, which greatly increases growth rates, survival rates, and disease resistance. These effects might be connected to the various immunostimulants‘ diverse structural and functional characteristics. It is anticipated that the in vivo expression of immunostimulant genes in breeding species will increase disease resistance, stop the spread of disease, and breed disease-resistant types for use in aquaculture. The usage of new immunostimulants might be a safer option than using antibiotics. We still know very little about the immune systems of aquatic organisms and have a long way to go before we can identify the specific immunological processes that provide defence against the majority of local infections. Immunostimulants for marine species will probably be industrialised through genetic engineering as more sources become available, eventually taking the place of antibiotics as feed additives. Additionally, this technology has the potential to increase the dependability and safety. The widespread usage of immunostimulants nowadays is a result of their safety, affordability, and little environmental impact. Polysaccharides, microbial derivatives, and plant extracts are used in aquaculture as immunostimulants to raise fish weight, increase feeding effectiveness, and boost disease resistance.

6.1.2 Prospects of Immunomodulation

Immunostimulants are used more frequently in aquaculture, and there are numerous instances of how they have successfully been used to enhance fish welfare, health, and productivity. The use of glucans in fish diets was one of the first uses of immunostimulants in aquaculture. They were also thought to be beneficial in managing illness eruptions following traumatic dealings like rating. When immune function is impaired, natural aspects like immunological molecules from aquatic animals cannot efficiently aim the pathogen surface proteins because impaired immune function prevents pathogen detection, leaving the animals vulnerable to a range of diseases. Consequently, natural substances that serve as immunostimulants can strengthen immune systems and help people stay healthy. Moreover, the immunostimulants have some of the features like health- and environment-friendly, biocompatible, and degradable; encourage a healthy bodily status by activating the host's immune system; not harmful to shellfish or fish; have no documented negative impacts; improve disease resistance to a variety of microbial diseases; decrease opportunistic pathogen-related mortality and virulent disease prevention; increase the effectiveness of antibacterial agents; increase the effectiveness of medicines and vaccinations; and are affordable, green, and readily accessible. Immunostimulants' biological effects are mostly based on the target cells‘ receptors identifying them as potentially dangerous chemicals and activating a variety of defence mechanisms (Figure 6.2). Therefore, it's critical to gain a better understanding of receptor specificity and the inflammatory reactions that various receptors cause when they connect to antigens. Many human receptors, such as NLR (NOD-like receptors), which have been found to bind immunostimulants, have not yet been discovered in fish. Nevertheless, one can predict the biological effects of immunostimulants in fish if one assumes that fish and mammalian cells have many comparable receptors. Some of the common features of immunomodulations are as follows:

- Health and environment-friendly, biocompatible, and degradable
- Encourage a healthy bodily status by activating the host's immune system
- Not harmful to shellfish or fish, and have no documented negative impacts
- Enhance disease resistance against a wide range of pathogens
- Decrease mortality from opportunistic infections
- Prevent viral diseases; improve antimicrobial substance efficacy
- Improve the efficacy of vaccines and medicines
- Make treatment affordable, environmentally friendly, and accessible.
- They are used often and with success to increase fish welfare, health, and output
- It improves phagocytic cell function, boosts their bactericidal capabilities, and stimulates the complement system, natural killer cells, lysozyme activity, and antibody response in fish and shellfish, which improves their resistance to infectious illnesses
- It is thought to boost non-specific immune responses while also improving the quality of the aquatic environment.
- Enhances both specialised and generalised defence mechanisms to control fish illnesses.
- Some immunostimulants increased the myeloperoxidase and phagocytic activity of neutrophils.

6.1.2.1 In Management of Disease Stressors

In aquatic animals, stress affects host immunity and often exacerbates illness incidence in animals exposed to stressful settings. One of the most common reasons for employing immunomodulators in food-producing animals is to reduce stress-related immunosuppression and illness (Campos et al., 1993). This logic has proven to be efficient in the treatment and prevention of cattle respiratory disease and mastitis, as well as in aquaculture (Jorgensen et al., 1993; Raa, 1992; Robertson et al., 1990). Immunomodulators can help stressed animals become more disease resistant. Immune stimulant research in aquaculture is ongoing, and a lot of compounds are presently in use in the sector. Aquaculture employs immunostimulants to boost disease resistance and immunological function in aquatic animals. The cellular and humoral immune systems of fish are comprised. In fish, phagocytic cells, neutrophils, natural killer cells, and lymphocytes make up cellular immunity, whereas lysozyme, haemolysin, immunoglobulins, and complement molecules make up humoral immunity. Furthermore, cytokines have been discovered in fish (Aly, Mohamed & John 2008; Vasta et al., 2011).

6.1.2.2 To Augment Development of Neonatal Immunity

In neonates, the scope and intensity of an immune response are often smaller than in mature adults. As a result, many aquatic animal immunomodulation studies have focused on improving neonatal animals' immune responses in order to reduce illness incidence and risk during this vital period. Finding novel approaches to reduce the risk of infection in young animals will become increasingly crucial as management strategies to improve and shorten the production cycle of food-producing animals. Immunomodulation will remain a key tool for achieving this goal. Immunostimulants have been given to newly hatched larvae in fish either indirectly through feed pellets or directly through bath treatment (Zhang et al., 2019; Magnadottir et al., 2006; Zhang et al., 2009; Dalmo et al., 2000; Bricknell et al., 2005). The transfer of innate immune components from oocytes has been demonstrated, as well as the direct transmission of immunostimulants from mother to child (Mulero et al., 2007; Wang et al., 2011; Hara et al., 2016). Another advantage of innate regulation in fish is the evidence of transgenerational epigenetic alterations to innate immune phenotypes and genes. (Beemelmanns & Roth 2016; 2017). In some areas, fish producers have an evolutionary advantage over mammalian species when it comes to using intrinsic remembrance to avert sickness and together businesses have taken steps to research and use this evolutionary compensation. Researchers discovered increased oxidative burst, cytokine production, and lysozyme activity, as well as reduced death rate (Rodriguez et al., 2009; Zhang et al., 2019; Kawakami et al., 1998). Up to 30 days after the β-glucan feeding stopped, there was a reduction in mortality (Chang et al., 2013). Dietary-glucan therapy causes epigenetic modifications in progenitor cells in fish, allowing for long-term changes in innate immune response.

6.1.2.3 To Influence the Metabolic Cost of Immunity

The immune system is given first priority when it comes to nutrient utilization and the integrity of the host defence is preserved at the expense of practically all other tissues save cerebral function. Vaccination has a metabolic cost connected to immune system activation, which includes the production and development of acute-phase proteins and cytokines. Generation efficiency should improve if the immune response could be modulated to restrict the production of proinflammatory, catabolic cytokines without jeopardising host defensive mechanisms. This hypothesis has been demonstrated in food animals through studies using two immunomodulators, conjugated linoleic acid and β-glucan. For a more thorough examination of innate training in several fish species, readers are urged to consult review publications summarising the proof for innate immunity in fish (Zhang et al., 2019). Since the 1990s, β-glucan has been examined or given to industrial fisheries for its immunomodulating properties. Conversely, a thorough study reveals that showed advantages are depending on the fish species, and the age of the fish (Dalmo et al., 2008). Innate receptors in fish are more diverse and varied than those in mammals and they are both similar and distinct. Petit et al. 2019 discovered numerous potential β-glucan receptor candidates in *Cyprinus carpio*.

Kato et al. (2011, 2010, 2012) found that BCG injection improves immune responses in different fish species and encourages defence against *N. seriolae* defy in Japanese flounder. *S. quinqueradiata* sheltered against *P. piscicada* disease after being subjected to one of various immunostimulants. The most protective treatment was pre-exposure to Freund's complete adjuvant (CFA), which have inactivated *Mycobacterium bovis* (Kawakami et al., 1998). Indeed, whereas glucan pre-treatment improved survival and immunological markers, the impact was significantly amplified by CFA therapy. Cross protection was demonstrated in these recent investigations, which is indicative of innate training and significant proof for innate immunity in fish. It's uncertain if the defence was related to innate immunity or simply nonspecific defence from plasmid DNA. Regardless, there was modest immunomodulation to improve disease resistance. The mechanism of modern learned immunity is still unknown. As the request for fish grows, techniques to improve disease resistance in farmed fish lacking the use antibiotics are sought.

6.1.2.4 Antimicrobial Agent

Immunostimulants are foods that boost the body's natural defence mechanisms and resistance against certain diseases. The immune response only lasts a few minutes, and no memory component is produced. Immunostimulants are substances that cause leukocyte activation. Freund's Complete Adjuvant (FCA) is the earliest immunostimulants used in animals to boost a particular immune response, which has also been used efficiently in conjunction with fish bacterin injection (Mohapatra et al., 2013; Buchmann, 2014; Shrestha et al., 2015). Immunostimulants are important for triggering the immunological responses that can provide complete protection against infections. Pathogens include *A. hydrophila, A. salmonicida, E. tarda, Edwardsiella ictaluri, V. anguillarum, V. vulnificus, V. salmonicida, Yersinia ruckeri,* and *Streptococcus spp.;* viruses like infectious hematopoietic necrosis, yellow head virus, and viral hemorrhagic septicemia; and parasites. According to Merchie et al., vitamin C plays a role as an immunostimulant, as evidenced by the ability of *P. monodon* postlarvae and juveniles to avoid baculovirus and to resist disease caused by *V. harveyi* and saline shock (Merchie et al., 1998). Application of these distinct forms of immunostimulants can help fish and shellfish to improve their immunological competency and disease resistance. Antibiotics may be replaced by new immunostimulants, which may be a safer option. However, many features of aquatic organisms' immune systems remain unclear and we are currently learning which immune pathways provide protection against various resident diseases. Immunostimulants for aquatic animals will probably be developed through genetic engineering as more sources of immunostimulants become available, eventually returning antibiotics as feed additives. This expertise has the probable to increase aquaculture product protection and dependability. Different immunostimulants have effectiveness for different life stages based on solubility and give a different degree of protection against the pathogens (Galeotti et al., 1995). Applications of β-1,3 glucan enhance the nonspecific cellular defence mechanisms of animals by increasing the number of phagocytes and the bacterial killing activity of macrophages in rainbow trout, Atlantic salmon, catfish, and carp (Rao et al., 2006) and through production of superoxide anions by macrophages (Jorgensen et al., 1993). In Indian major carps, *Labeorohita*, yeast glucan have been observed to enhance the phagocytic activity of leucocytes and stimulate generation of reactive oxygen species (ROS) in phagocytes (Ali et al., 1996).

To optimize immunogenicity for a specific immune reaction in the production of efficient immunostimulants, biological factors from animal sources should be integrated with the application of genetic engineering. These immunostimulants may cause certain immunological development to occur lacking causing a widespread reaction with negative side effects. The advantages of immunostimulants inside an organisms are that the aspects are (i) constant and well-organized immune attractive molecules; (ii) they have a particular antibacterial range; (iii) nontoxic to the host. Immunostimulants have been extensively studied in fish and shellfish both at whole animal and on a cellular level. It has been used as prophylactics to control infectious diseases in animals and also plays the role of alarm molecules that activate the immune system (Lopez et al., 2003). (iv) Medication resistance is improbable. Upcoming investigations can give novel immunostimulant source and concepts for their utilisation, potentially replacing existing antibiotics. As a result, gene-engineered immunostimulants may become a valuable new commodity in aquaculture as a contender for novel purposes. The investigation for the true molecular associates of immunostimulant protection must be undertaken with vigour. The immunostimulatory potential of a given chemical should be firmly proven in future immunostimulant research, pursued through investigations on its strength and efficiency in the situation of immunostimulants.

6.1.2.5 Immunomodulation in Aquatic Health Management

Fish and shellfish have been the subject of substantial whole-animal and cellular immunostimulant research. It has been utilised as preventative medicine to manage animal illnesses and diseases, as well as acting as alarm molecules that stimulate the immune system. Immunostimulants are essential to aquatic animals' health management techniques since fish and shrimp rely more on generalised defence mechanisms than do mammals (Figure 6.3). There are at least 20 distinct substances that are utilised as immunostimulants, adjuvants, and vaccine carriers in fish, including levamisole, lipopolysaccharides, glucan, vitamins C and E, etc. Among these substances, glucan is one of the most intriguing immunostimulants in aquatic species and one of the most potential stimulants for nonspecific defence mechanisms. Nucleotides and their metabolites have drawn more interest recently as possible immunomodulators (Li et al., 2004). They are crucial for several crucial physiologic processes, including as signal transduction, controlling energy metabolism, and encoding genetic information (Agett et al., 2002). Oral nucleotide delivery has been shown to have positive effects on immune system performance, vaccine effectiveness, and disease resistance in a variety of fish species, including salmon, rainbow trout, common carp, tilapia, and striped bass. Additionally, dietary nucleotides have been seen to alter gene expression (Sanchez pozo & Gil, 2002), a phenomenon that Low et al. 2003 also proved to occur in fish. It is still unclear, nevertheless, how dietary nucleotides alter gene expression as adaptive immunity develops.

Fish and shellfish health management strategies have paid a lot of attention to the immunostimulants capabilities of entire microorganisms, chitin particles, lactoferrin, sodium alginates, vitamins C and E, and dietary carbs. The impact of chitin on the immune system of fish in vitro is little understood. Studies on the in vivo administration of chitin have been conducted, and it has been discovered that chitin injections or dietary administration may improve the innate immune system of many fish species (Cuesta et al., 2003). The addition of specific vitamins to the diet may help fish become more immune-competent and illness-resistant (Sahoo & Mukherjee, 2002). It has been demonstrated that higher vitamin C doses improve immunological responses in fish, including macrophage activity, cell proliferation, natural killer cell activity, complement, and lysozyme levels (Verlhac et al., 1996). Merchie et al., 1998 claim that vitamin C functions as immunostimulants, which is supported by *P. monodon* postlarvae and juveniles' capacity to evade baculovirus infection and to fend off sickness brought on by *V. harveyi* and saline shock. Although the vitamin C mode of action as an immunostimulant is unclear, its antioxidant properties and subsequent cell protection may serve as a mechanism to conserve haemoglobin, strengthening the shrimp's overall immune system. The findings imply that even though shrimp have a non-specific immune system, virus resistance could be improved in shrimp by judicious and routine administration of immunostimulants. Use of immunostimulants in cultured fish results in macrophage activation, increased phagocytosis by neutrophils and monocytes, increased lymphocyte numbers, increased serum immunoglobulins, and increased lysozyme (Sakai, 1999). Absorption of laminaran from water by yolk-sac larvae of Atlantic halibut, *Hippoglossushippoglosus*, through the skin and intestinal epithelium suggests that it may have the potential to enhance immunity in early life stages before the development of acquired immunity (Strand and Dalmo, 1997).

6.1.3 Cons of Immunomodulation in Aquaculture

Naturally, there is some concern that by using immunostimulants in aquaculture could result in needs and developing, the tolerance is the most evident serious complication, as it would be in fish larvae in the aquaculture industry. Antigen opposition has been seen in Atlantic salmon, and acceptance has been described in several fish species, indicating that the processes for tolerance induction might live but are not well implicit. It is believed that an essential investigation into the processes involved in fish immunological tolerance induction should rank highly among research priorities. Less well understood are other issues that could arise from the use of immunostimulants in aquaculture. Some of the disadvantages of immunomodulation in aquaculture are given below:

- They are costly.
- Limited effectiveness when administered by parents.
- They are not always successful in treating ailments.
- When taken in excess, they suppress the immune system.

Trouble with palatability and hierarchical issues where dominant fish receive far more immunostimulants than subordinate animals, and immunostimulant delivery timing are a few examples of some of the cons of immunomodulation in aquaculture.

6.1.3.1 Timing of Immunomodulant Administration

Immunomodulation can be used alone or with immunisation. Fish could be treated ahead of time to prepare for sickness that is expected, such as seasonal pathogen exposure. It is easiest to provide immunomodulants and vaccinations at the same time, when the fish is exposed to a specific antigen. The two substances are injected into the body or combined with diet. Immunostimulants are more effective when given before to vaccination because they prepare the fish for antigen disclosure. Conversely, each substance's dosage and administration duration must be evaluated. Significant increases ($P<0.05$) in blood lysozyme, gill, and liver HSP concentrations were seen one month after the ending of treatment in the same experimental groups, but a significant increase ($P<0.05$) in complement activity was only observed in fish fed an Ergosan diet. Endurance, growth performance, and adaptation indices in treated and control fish were not substantially different over time in innate or specific immunological measures. Throughout the winter season, innate and acquired immunological markers declined markedly in all groups, followed by a partial rebound as water temperatures rose. Reduced complement and lysozyme activity were substantially associated with changes in water temperature ($P<0.01$) (Bagni et al., 2005; Li, Li & Wang 2010).

Most immunostimulation techniques work by up-regulating the immune response by pulse-feeding the immunostimulant for a brief period of time, typically 4–6 weeks. A similar amount of time is subsequently spent without the immunostimulant, allowing the level of immunostimulation to return to its resting state before another dose is administered. As a result, the host immune system frequently oscillates between a resting state and an increased response before returning to rest. Being able to feed the immunostimulant during times when there is a higher danger of disease means that this method of fish farming obviously gives a great deal of flexibility. Risky times of the year include the spring and fall changes in ambient temperature, the time right before the breeding season for fish that batch-spawn, such cod or halibut, or the time right before sea transfer for smolts. The effect of immune stimulants on nonspecific immune mechanism is normally of short duration (Galeotti, 1998). It has been shown in salmon gave maximum leucocyte responses just two days after injection with beta-glucan (Yoshida et al., 1995), and that yeast β-glucan gave an increase in respiratory burst activity 4–7 days after treatment (Jorgensen et al., 1993). The protection in trout using glucans and chitosan was greatly reduced after 14 days of immune stimulation (Anderson et al., 1995).

6.1.3.2 Dosage of Immunostimulant

The dosage of immunostimulants has a significant impact on their effects. There are only two effective tactics that will work in fish that have undergone metamorphosis and have a fully developed range of immune responses. These include pulsed immunostimulant administration or continuous immunostimulant administration. The third tactic, bathing the fish in an immunostimulant solution, does not appear to have been researched for adult fish. This might be because recirculatory systems experience bio-degradation or have trouble maintaining a steady dose in flow-to-waste systems. Immunostimulant continuous feeding has typically been replaced by pulse feeding in adult fish. When an immunostimulant is fed continuously, there are two probable results. According to studies, the optimal dose of orally delivered VE in gilthead seabream is 1,200 mg kg 1, with lower (1,800 mg kg 1) VE concentrations eliciting unique immunostimulation patterns (Ortuno et al., 2000). Another study discovered that fish-fed antibiotic-supplemented diets outperform fish on standard diets in terms of growth rate and feed efficiency. Fish on 1 g levamisole per kg diets, on the other hand, have shown signs of chronic toxicity, including reduced feed intake and efficiency (Alvarez-Pellitero et al., 2006). The use of several immunostimulants for prolonged periods does not appear to provide additional advantages with respect to a single dose (Galeottiet al., 1995). Tseng et al. (Tseng & Chen, 2004) observed that after 29 days, the protection induced in blue gourami by injection of 20 mg/kg laminaran for 22 days, was not significantly different from the effect of a single injection of 20 mg/kg laminaran. Sakai (1999) reported that the increase in respiratory burst activity of glucans-treated macrophages was maximal at glucan concentrations of 0.1–1 µg/ml, whereas at 10 µg/ml, no effect was seen and at 50 µg/ml glucan, was inhibitory (Sakai, 1999). Vitamin E is immunostimulatory in concentration of 50–300 IE/kg feed, but very high vitamin E levels (>1,000–5,000 IE/kg diet) fed for a prolonged time have an immunosuppressive effect (Galeotti, 1998). The effectiveness and potency of immunostimulatory medicines can be impacted by dosage change. The immune system is first up-regulated by the immunostimulant to enhanced levels, and this state is maintained until the immunostimulant is stopped. This is an extremely uncommon occurrence, though. Furthermore, repeated exposure to an immunostimulant may result in tolerance development. The immunostimulant response is lost as a result of the host immune system being desensitised to the immunostimulant. When an immunostimulant is exposed repeatedly under harsh conditions, the immune response becomes repressed, resulting in a decreased level of innate defences while exposure to that particular immunostimulant is maintained.

6.1.3.3 Mode of Action

Immunostimulants work by stimulating the immune system, to increase the amount of immunity to invading pathogens. The strategy is really varied or could be poor understood, as well as the type, dose, and timing of immunostimulants administration method, duration, and length of exposure. Some of the mechanisms of actions are stimulators of T-lymphocytes; macrophage activators; and inflammatory substances such as chemotoxins, cell membrane modifiers, nutritional components, toxic metals, and extracts of fish and animals. Immunostimulants often providing resistance and protection against a variety of illnesses and invasive microbes. Fish injected intraperitoneally with chitin demonstrated increased humoral and cellular immune responses, whereas fish injected intravenously with chitin showed no effect on immunological parameters. Bathing fish with immunostimulant solution is the third method of administration to them. According to Baba et al. 1993, carp submerged for 24 hours in a 10 µg/ml levamisole solution illustrate accelerated phagocytosis and chemotaxis, the generation of active oxygen in head kidney phagocytes, and improved protection against *Aeromonas hydrophila* for 14 days. According to Anderson et al., 1995 rainbow trout immersed in glucan or chitosan solutions for three days are more resistant to *Aeromonas salmonicida*. More studies are needed to discover the types of immunostimulants and dosages that are appropriate for use in aquaculture.

6.1.3.4 Evaluating the Efficacy of Immunostimulants

The immunostimulant can save from numerous diseases like bacterial species *Vibrio and Streptococcu,* and viral infections like IHN and YHV disease, and parasitic infections like white spot disease and sea lice and lower mortality rates of fishes. Furthermore, immunostimulants do not amplify efficient against some bacterial species like *Renibacterium and Pseudomonas* infections because of their resistances to phagocytosis and ability to stay alive inside macrophages.

Immunostimulant's efficacy can be tested through in vivo, like a confront test with fish pathogens and in vitro for assessing the efficiency of cellular and humoral immune processes. Limited understanding of fish immune systems and much less knowledge on how most immunostimulatory substances work offers enormous avenues for future research. It is recommended to analyse immunostimulants utilising in vitro protocols that test the substance's effects on the immune system in basic research. In vitro research has to be complemented with in vivo research wherever probable to clarify the fundamental mechanisms behind any defence offered. C-reactive protein (CRP) levels, natural cytotoxic activity, and macrophage-activating factor (MAF) levels are all important parameters to monitor.

6.1.3.5 Immunostimulant Effects in Fish Larvae

It is largely agreed upon that feeding immunostimulants to fish that have reached immunological maturity improves their defences against bacterial and, to a lesser extent, viral illnesses. Additionally, it has been demonstrated that adding an immunostimulant to experimental vaccination formulations can lengthen the duration of immunity and boost the intensity of the immunological response, particularly after intraperitoneal injection of the vaccine preparation (Austin, 1997, Mulero et al., 1998). On the other hand, a number of findings indicate that using immunostimulants had no positive benefits. Indeed, after administering the immunostimulant orally or intravenously to the animals, some of these formulations show hazardous effects (Bullock et al., 2000; Salinas et al., 2004). It has been suggested that immunomodulating larval fish could increase innate responses in the developing organism until its adaptive immune response is well developed enough to build a successful defence against the virus. To this purpose, it has been suggested that providing immunostimulants to larval fish as a dietary supplement is quite beneficial, enhancing the animal's natural defences with little harm to the maturing animal. A different school of thinking, however, expresses concern over immunomodulating neotanous animals before their immune systems are fully developed as this could negatively impact the emergence of a typical immune response. As a result, there are clearly insufficient studies in the field of immunostimulation in larval fish, and little is known about the effects the immunostimulant has on the immune system's development.

6.1.3.6 Efficiencies and Considered Use of Immunostimulation Methods

Different farm conditions have resulted in a variety of immunostimulant administration techniques. The three fundamental techniques used are injection, immersion, and oral. The fish must be handled or at the very least confined in a small space during the processes for the injection and immersion methods, which are only appropriate for intensive aquaculture. Numerous writers claimed that administering immunostimulants via injection improved leucocyte activity and pathogen defence. However, this procedure is time-consuming, labor-intensive, and impracticable

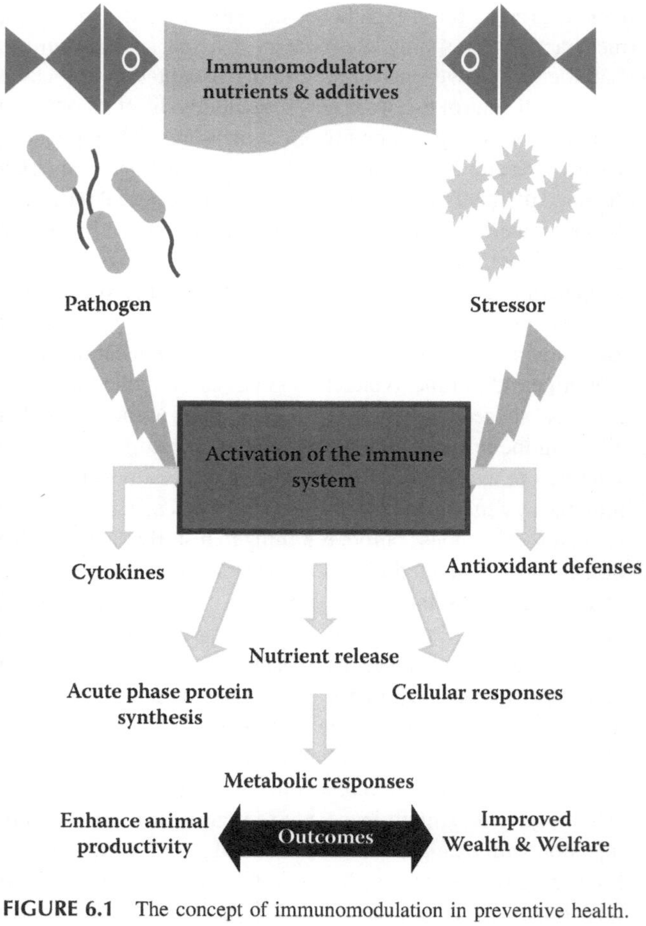

FIGURE 6.1 The concept of immunomodulation in preventive health.

for fish that weigh less than 15 g. Several writers have shown the effectiveness of immersion, however since dilution, exposure period, and levels of efficacy are not clearly defined, care must be exercised when using this technique. Oral administration is the only technique that is financially feasible for large-scale aquaculture, is stress-free, permits bulk administration independent of fish size, but is obviously only possible in artificial diets be given solely as part of an artificial diet. The suggested dose and exposure times, as well as drawbacks, are presented after reviewing certain research trials outcomes in which various compounds were tried as immunostimulants.

- Only extensive aquaculture makes use of the injection technique. Work hard (Anesthetic; handling) with stress and fish must have weigh between 10–15 g.
- Intensive aquaculture only using the immersion method. Handling stress dips and rises. Not as potent as an injectable approach.
- Oral approach, ineffective potency high doses of immunostimulants are necessary to establish protection. Only appropriate for fish fed an artificial feed.

6.2 Conclusion

Aquaculture now has a new chapter and a very promising field thanks to the application of immunostimulants. Immunostimulants may be utilised to stop disease-related losses. Immunomodulators have emerged to be the most capable and constructive strategies for treating farmed fish and shrimp prophylactically. It is less toxic than chemotherapy and has a wider spectrum of efficacy than immunisation. The most effective and practical instruments for the prophylactic treatment of farmed fish and shrimp seem to be immunostimulants. It is less dangerous than chemotherapy, and it has a larger spectrum of effectiveness than immunisation. These substances won't, however, take the place of immunizations, healthy eating, or effective management methods. The strength of these substances appears to lay in their capacity to improve larval culture prior to the development of the particular immune system, the ability to immunise the animals, and their improvement of nonspecific immune function against a wide range of diseases. These substances, on the other hand, will not be able to replace immunizations, correct diet, or superior

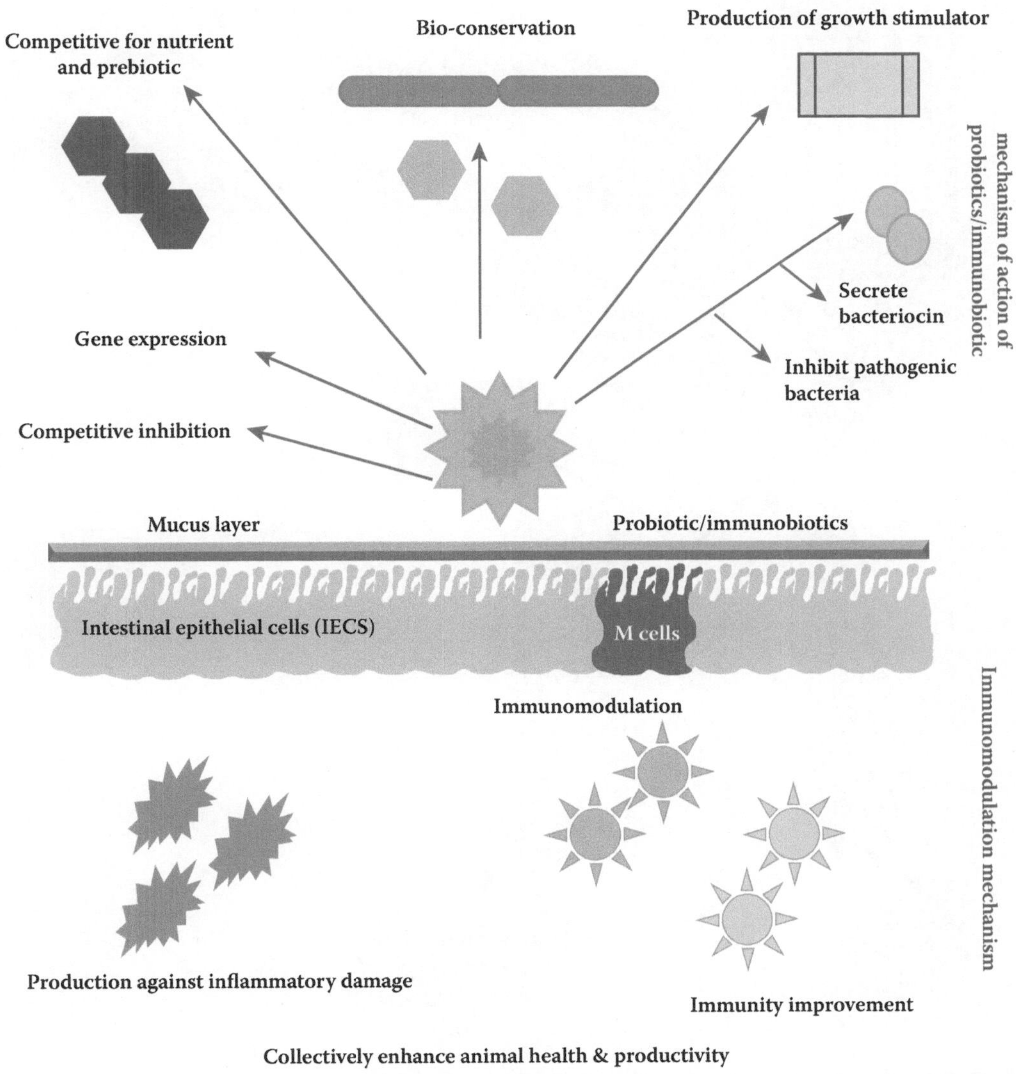

FIGURE 6.2 Immunomodulation mechanism in fishes.

organization approaches. As a result, immunostimulants created through gene engineering could emerge as a significant new product in aquaculture and be a candidate for usage in novel applications. It is important to pursue the study for the true molecular correlates of the defence offered by immunostimulants. There isn't a single commercial vaccination that can protect against all infectious diseases, and vaccines that are imported may not work as well because the disease-causing strains of various infectious diseases can vary. Immunostimulants have a wider spectrum of efficacy than vaccinations and are safer than chemotherapy. As a result, immunostimulants might be an effective approach for managing infectious infections. Future studies on immunostimulants should first clearly demonstrate a substance's immunostimulatory capabilities before examining its potency and effectiveness when used in conjunction with immunostimulants. Thus, the use of immunomodulation to aquatic health management has enormous promise, but understanding the method of action requires a great deal of scientific investigation. Immunostimulants are predicted to have a larger role in aquaculture sustainability in the coming years. Galeotti et al. concluded that genetics of the species, life history stage, and the culture environment all interact with the type and dosage of immunostimulants to contribute to the efficacy of the immunostimulatory substance (Galeotti et al., 1995). It is safer than chemotherapeutics and their range of efficacy is wider than vaccination. However, these compounds will not replace vaccines proper nutrition or good management techniques (Barman et al., 2013). Thus, application of immunostimulants to aquatic health management has immense potential, but in order to capitalize on this issue, a lot of scientific research is necessary to understand the mode of action. It is expected that during future years immunostimulants will find more applications to make aquaculture sustainable. Therefore, immunostimulants may be an effective tool for controlling infectious diseases in aquaculture.

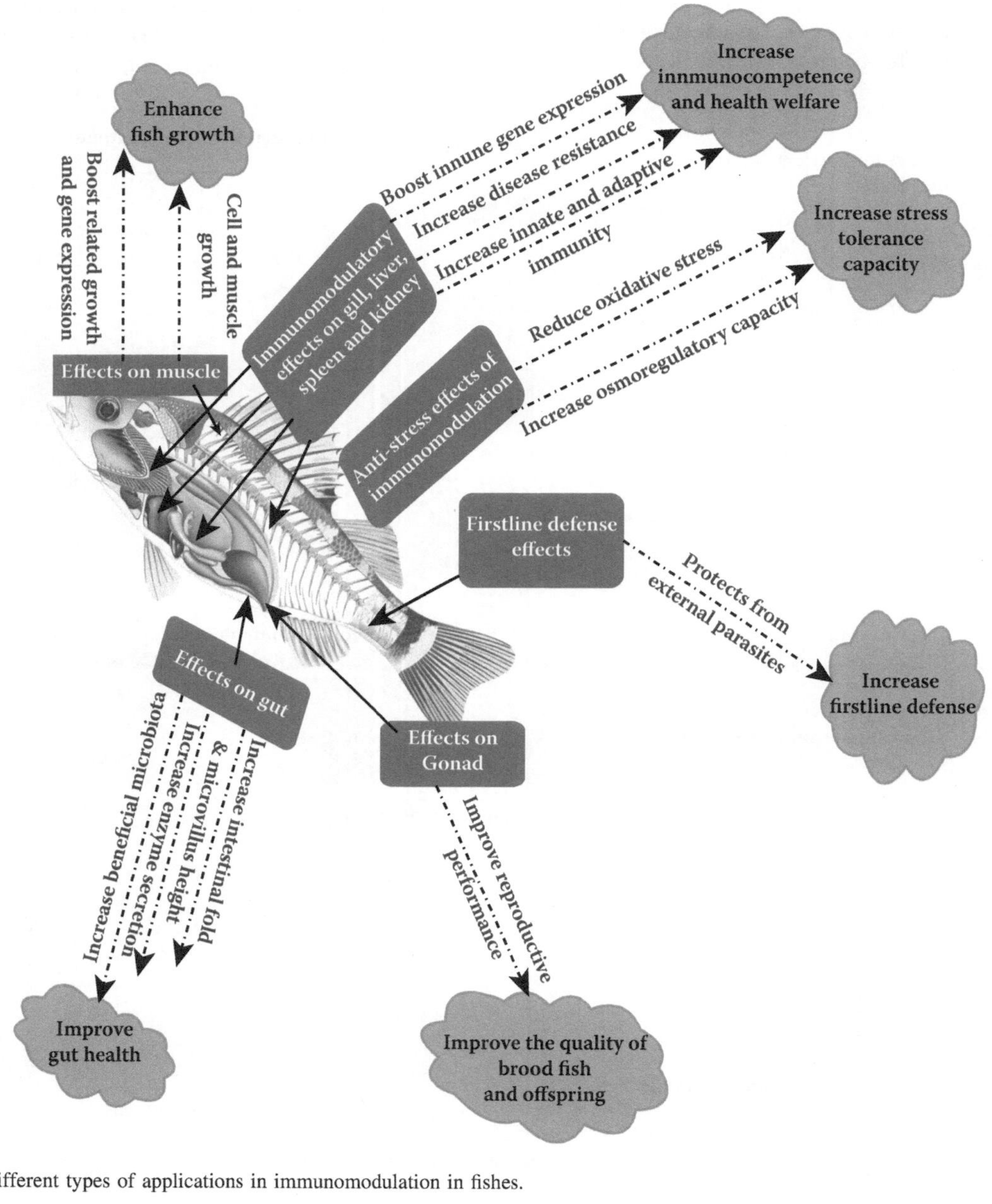

FIGURE 6.3 Different types of applications in immunomodulation in fishes.

TABLE 6.1

List of Some Immunomodulatory Agents and Their Applications in Fishes

S. No	Fish	Immunomodulatory Agent	Application	References
1	Grass carp *(Ctenopharyngodon idella)*	Kaempferol 3-a-L-(4-O-acetyl) rhamnopyranoside-7-a-L-rhamnopyranoside (SA) from *Dryopteris crassirhizoma*	Increased the immune responses (e.g., lysozyme activity, complement C3, phagocytic activity, SOD activity, IgM level) Induced expression of immune-related genes Enhanced the resistance of fish against pathogen infection.	Chi et al., 2016
2	Atlantic salmon *Salmo salar L.*	*Laminaria digitata* from seaweeds	More efficient immune response. Increased the expression of genes involved in *response to* lipopolysaccharide and inflammatory response.	Palstra et al., 2018
3		LPS, laminaran and sulphated laminaran β(1,3)-D-glucan	Higher production of superoxide anion Higher activity of acid phosphatase by macrophages	Dalmo et al., 1995
4		Laminaran and derivatives	Uptake of immunomodulatory laminaran from the intestine into the circulation in Atlantic salmon after peranal administration. Native or slightly modified laminaran may be an interesting candidate as a feed additive in fish farming	Dalmo et al., 1994
5		Levamisole	Enhances the nonspecific immune system as measured by phagocytic ability, Superoxide anion production and lytic lysozyme activity	Findlay and Munday 2000
6		Laminaran glucan	Enhance the nonspecific defence against infectious diseases	Dalmo et al., 1996
7		Aminated β-1,3- polyglucose	Rich immunocompetent tissue Highly perspective of disease prevention in farmed fish	Ingebrigtsen et al., 1993
8	Common carp *(Cyprinus carpio)*	Leaf extract from *Aegle marmelos*	Stimulates the immunity, highly resistant to *Aeromonas hydrophila.*	Pradeepa et al., 2010
9		Levamisole	Enhances the non-specific immunity, increases their resistance to infection, reduces the mortality and enhances the growth of fish.	Maqsood et al., 2009
10		Chitin, chitosan and levamisole	Enhances the innate immune system and survivability of common carp protection against *A. hydrophila* infection	Gopalakannan and Arul 2006
11		Levamisole	Enhanced immune response	Siwicki, 1987
12	Silver catfish *(Rhamdia quelen)*	β-glucan	Improve innate immunity resistance to specific pathogen	Domenico et al., 2017
13	Indian major carp *(Labeo rohita)*	Gelatinized and non-gelatinized starch	Promoting growth, protecting immunity	Kumar et al., 2007
14		Tuftsin	Induced the T-helper cell type I immune response in systemic immunity. Immune system of brood fish can be enhanced	Misra et al., 2006
15		*Edwardsiella tarda*	Increase their resistance to infection reducing mortality rates	Sahoo et al., 2002
16	Rainbow Trout, (*Oncorhynchus mykiss*)	Levamisole	levamisole may be a possible immunomodulatory substance for the immune enhancement to fish disease	Kajita et al., 1990
17		β-1,3-glucan from *Euglena gracilis*	Potent plasma lysozyme activity response	Skov et al., 2012
18		vitamin C and glucan	Enhanced immune response	Verlhac et al., 1998
19	Turbot (*Scophthalmus maximus L.*)	Nisin	Antibacterial activity	Villamil et al., 2003
20	Gilthead seabream (*Sparus aurata*)	Chitin	Enhances immune activity through the non-specific modulation of haemolytic complement activity, leucocyte respiratory burst activity and cytotoxicity	Esteban et al., 2001

Acknowledgement

The author is appreciative of the financial support received from the Science and Engineering Research Board, India, New Delhi, India [Ref: PDF/2020/001027].

There is no conflict of interest.

REFERENCES

Aggett R, Leach JL, Rueda R, MacLean WC (2002) Innovation in infant formula development: a reassessment of ribonucleotides in 2002. *Nutrition* 19; 375–384.

Alvarez-Pellitero P, Sitja-Bobadilla A, Bermudez R, Quiroga MI (2006) Levamisole activates several innate immune factors in *Scophthalmus maximus* (L.) Teleostei. *Int J Immunopathol Pharmacol* 19; 727–738.

Aly SM, Mohamed MF, John G (2008) Effect of probiotics on the survival, growth and challenge infection in Tilapia nilotica (*Oreochromis niloticus*). *Aquaculture Res* 39; 647–656.

Ali A, Karunasagar I, Pais R, Tauro P (1996) Effect of yeast glucans on the immune response of Indian major carp Labeorohita. Abst37thAnn ConfAssocMicrobolIndia.

Anderson DP, Siwicki AK, Rumsey GL (1995) Injection or immersion delivery of selected immunostimulants to trout demonstrate enhancement of nonspecific defense mechanisms and protective immunity. In: *Diseases in Asian Aquaculture*, Vol. 11 (edited by M Shariff, RP Subasighe & JR Arthur) Fish Health Section, Asian Fisheries Society, Manila, Philippines, pp. 413–426.

Austin B (1997). Progress in understanding the fish pathogen Aeromonas salmonicida. *Trends Biotechnol* 15; 131–134.

Baba T, Watase Y, Yoshinaga Y (1993). Activation of mononuclear phagocyte function by levamisole immersion in carp. *Nippon Suisan Gakkaishi*, 59 (2); 301–307.

Bagni M, Romano N, Finoia MG, Abelli L, Scapigliati G, Tiscar P (2005) Short- and long-term effects of a dietary yeast beta-glucan (Macrogard) and alginic acid (Ergosan) preparation on immune response in sea bass (*Dicentrarchus labrax*). *Fish Shellfish Immunol* 18; 311–325.

Bashir I, Lone FA, Bhat RA, Mir SA, Dar ZA, Dar SA (2020) Concerns and threats of contamination on aquatic ecosystems. *InBioremediation and Biotechnology*, Springer, Cham; 1–26.

Beemelmanns A, Roth O (2017) Grandparental immune priming in the pipefish *Syngnathus typhle*. *BMC Evol Biol* 17; 44.

Beemelmanns A, Roth O (2016) Biparental immune priming in the pipefish *Syngnathus typhle*. *Zoology* 119; 262–272.

Boyd CE, McNevin AA, Davis RP (2022) The contribution of fisheries and aquaculture to the global protein supply. *Food Security* 20; 1–23.

Bricknell I, Dalmo R (2005) The use of immunostimulants in fish larval aquaculture. *Fish. Shellfish Immunol.* 19; 457–472.

Buchmann K (2014). Evolution of innate immunity: clues from invertebrates via fish to mammals. *Front. Immunol.* 23; 5:459.

Bullock G, Blazer V, Tsukuda S, Summerfelt S (2000) Toxicity of acidified chitosan for cultured rainbow trout (*Oncorhynchus mykiss*). *Aquaculture* 185; 273–280.

Barman D, Nen P, Mandal SC, Kumar V (2013) Immunostimulants for Aquaculture Health Management. *J Marine Sci Res Dev* 3; 134. doi: 10.4172/2155-9910.1000134.

Campos M, Godson D, Hughes H, Babiuk L, Sordillo L (1993) The role of biological response modifiers in disease control. *J Dairy Sci* 76; 2407–2417.

Caroline Alexandre de Araújo, L, Maria da Silva, S, Artur de Queiroz Cavalcanti de Sá, R, Vitoria Araujo Lima, A, Virginia Barbosa, A, dos Santos Silva, J, Massari Leite, K, Jose do Nascimento Júnior, W, da Mota Silveira-Filho, V, Lucena Mendes-Marques, C, Henrique Da Silva, F, & Betânia Melo de Oliveira, M (2021). Effects of Antibiotics on Impacted Aquatic Environment Microorganisms, *Emerging Contaminants*. IntechOpen. 10.5772/intechopen.93910.

Chang CS, Huang SL, Chen S, Chen SN (2013) Innate immune responses and efficacy of using mushroom beta-glucan mixture (MBG) on orange-spotted grouper, *Epinephelus coioides*, aquaculture. *Fish. Shellfish. Immunol* 35; 115–125.

Chi C, Giri SS, Jun JW, Kim HJ, Yun S, Kim SG, Park SC (2016) Immunomodulatory effects of a bioactive compound isolated from *Dryopteris crassirhizoma* on the grass carp *Ctenopharyngodon idella*. *J Immunol Res* 2016.

Cuesta A, Esteban MA, Meseguer J (2003) In vitro effect of chitin particles as the innate cellular immune system of gilthead seabream (Sparusaurata L.). *Fish Shellfish Immunol* 15; 1–11.

Dalmo RA, Bogwald J (2008). Beta-glucans as conductors of immune symphonies. *Fish Shellfish Immunol* 25; 384–396.

Dalmo RA, Bogwald J, Ingebrigtsen K, Seljelid R (1996) The immunomodulatory effect of laminaran [β (1, 3)-D-glucan] on Atlantic salmon, *Salmo salar* L., anterior kidney leucocytes after intraperitoneal, peroral and peranal administration. *J Fish Dis* 19; 449–457.

Dalmo RA, Ingebrigtsen K, Horsberg TE, Seljelid R (1994) Intestinal absorption of immunomodulatory laminaran and derivatives in Atlantic salmon, *Salmo salar* L. *J Fish Dis* 17; 579–589.

Dalmo RA, Kjerstad AA, Arnesen SM, Tobias PS, Bogwald J (2000) Bath exposure of atlantic halibut (*Hippoglossus hippoglossus* L.) yolk sac larvae to bacterial lipopolysaccharide (LPS): absorption and distribution of the LPS and effect on fish survival. *Fish Shellfish Immunol* 10; 107–128.

Dalmo RA, Seljelid R (1995) The immunomodulatory effect of LPS, laminaran and sulphated laminaran [β (l, 3)-D-glucan] on Atlantic salmon, *Salmo salar* L., macrophages in vitro. *J Fish Dis* 18; 175–185.

Domenico JD, Canova R, Soveral LD, Nied CO, Costa MM, Frandoloso R, Kreutz LC (2017) Immunomodulatory effects of dietary β-glucan in silver catfish (*Rhamdia quelen*). *Pesqui Vet Bras* 37; 73–78.

Esteban MA, Cuesta A, Ortuno J, Meseguer J (2001) Immunomodulatory effects of dietary intake of chitin on gilthead seabream (*Sparus aurata* L.) innate immune system. *Fish Shellfish Immunol* 11; 303–315.

Findlay VL, Munday BL (2000) The immunomodulatory effects of levamisole on the nonspecific immune system of Atlantic salmon, *Salmo salar* L. *J Fish Dis* 23; 369–378.

Galeotti M, Volpatti D, Jeney G (1995) The nature of non-specific immune response of sea bass (Dicentrarchuslabrax) to Pasteurellapiscicida following bath exposure to levamisole. European Assoc Fish Pathologist Seventh IntConfPalma de Mallorca, Spain.

Galeotti M (1998) Some aspects of the application of immunostimulants and a critical review of methods for their evaluation. *J ApplIchthyol* 14; 189–199.

Gopalakannan A, Arul V (2006) Immunomodulatory effects of dietary intake of chitin, chitosan and levamisole on the immune system of *Cyprinus carpio* and control of *Aeromonas hydrophila* infection in ponds. *Aquaculture* 255; 179–187.

Halwart M. Aquaculture in SOFIA (2022) FAO Aquaculture Newsletter. 2022 Dec 1(66); 7–8.

Hara A, Hiramatsu N, Fujita T (2016) Vitellogenesis and choriogenesis in fishes. *Fish Sci.* 82; 187–202.

Hussan A, Gon T (2016) Common problems in aquaculture and their preventive measures. *Aquac Times J.* 2; 6–9.

Ingebrigtsen K, Horsberg TE, Dalmo R, Seljelid R (1993) Tissue distribution of the immunomodulator aminated β1–3 polyglucose in Atlantic salmon (*Salmo salar*) after intravenous, intraperitoneal and peroral administration. *Aquaculture* 117; 29–35.

Jorgensen JB, Sharp GJE, Secombes CJ, Robertsen B (1993) Effect of a yeast-cell-wall glucan on the bactericidal activity of rainbow trout macrophages. *Fish Shellfish Immunol* 3; 267–277.

Jorgensen JB, Lunde H, Robertsen B (1993). Peritoneal and head kidney cell response to intraperitoneally injected yeast glucan in Atlantic salmon, Salmo salar L. *Journal of Fish Diseases*, 16; 313–325.

Kajita Y, Sakai M, Atsuta S, Kobayashi M (1990) The immunomodulatory effects of levamisole on rainbow trout, *Oncorhynchus mykiss*. *Fish Pathol* 25; 93–98.

Kato G, Kato K, Saito K, Pe Y, Kondo H, Aoki T (2011) Vaccine efficacy of *Mycobacterium bovis* BCG against Mycobacterium sp. Infection in amberjack *Seriola dumerili. Fish Shellfish Immunol* 30; 467–472.

Kato G, Kondo H, Aoki T, Hirono I (2010) BCG vaccine confers adaptive immunity against Mycobacterium sp. infection in fish. *Dev Comp Immunol.* 34; 133–140.

Kato G, Kondo H, Aoki T, Hirono I (2012) *Mycobacterium bovis* BCG vaccine induces non-specific immune responses in Japanese flounder against *Nocardia seriolae. Fish Shellfish Immunol* 33; 243–250.

Kawakami H, Shinohara N, Sakai M (1998) The non-specific immunostimulation and adjuvant effect of *Vibrio anguillarum* bacterin, M-glucan, chitin, Freund's complete adjuvant against *Pasteurella piscicida* n in yellowtail. *Fish Pathol* 33; 267–292.

Kumar V, Sahu NP, Pal AK, Kumar S (2007) Immunomodulation of *Labeo rohita* juveniles due to dietary gelatinized and non-gelatinized starch. *Fish Shellfish Immunol* 23; 341–353.

Li P, Lewis DH, Galtin DM (2004) Dietary oligonucleotides from yeast RNA influences immune responses and resistance of hybrid striped bass (Moronechrysops x Moronesaxatilis) to Streptococcus iniae infection. *Fish Shellfish Immunol* 16; 561–569. 73.

Li Y, Li J & Wang Q (2010) Effects of size grading on growth and non-specific immunity factors of the shrimp *Litopenaeus vannamei* Boone. *Agri Sci China* 9; 416–422.

Lopez N, Cuzon G, Gaxiola G, Taboada G, Valenzuela M, et al. (2003) Physiological, nutritional and immunological role of dietary ß 1-3 glucan and ascorbic acid 2-monophosphate in Litopenaeusvannameijuveniles. *Aquaculture* 224; 223–243.

Low C, Wadsworth S, Burrells C, Secombes CJ (2003) Expression of immune genes in turbot (Scophthalmusmaximus) fed a nucleotide-supplemented diet. *Aquaculture*, 221(1-4); 23–40.

Magnadottir B, Gudmundsdottir BK, Lange S, Steinarsson A, Oddgeirsson M, Bowden T (2006) Immunostimulation of larvae and juveniles of cod, *Gadus morhua* L. *J Fish Dis (2006)* 29; 147–155.

Maqsood S, Samoon MH, Singh P (2009) Immunomodulatory and growth promoting effect of dietary levamisole in *Cyprinus carpio* fingerlings against the challenge of *Aeromonas hydrophila. Turk J Fish Aquatic Sci* 9.

Maqsood S, Singh P, Samoon MH & Munir K (2011) Emerging role of immunostimulants in combating the disease outbreak in aquaculture. *International Aquatic Res* 3; 147–163.

Merchie G, Kontara EKM, Lavens P, Robles R, Kurmaly K (1998) Effect of vitamin C and astaxanthin on stress and disease resistance of postlarval tiger shrimp *Penaeus monodon* (Fabricius). *Aquac Res* 29: 579–585.

Misra CK, Das BK, Mukherjee SC, Meher PK (2006) The immunomodulatory effects of tuftsin on the non-specific immune system of Indian Major carp, *Labeo rohita. Fish Shellfish Immunol* 20; 728–738.

Mohapatra S, Chakraborty T, Kumar V, DeBoeck G, Mohanta KN (2013) Aquaculture and stress management: a review of probiotic intervention. *J Anim Physiol Anim Nutr* 97; 405–430.

Mulero I, García-Ayala A, Meseguer J, Mulero V (2007) Maternal transfer of immunity and ontogeny of autologous immuno competence of fish: a mini review. *Aquaculture.* 268; 244–250.

Mulero V, Esteban MA, Munoz J, Meseguer J (1998). Dietary intake of levamisole enhances the immune response and disease resistance of the marine teleost gilthead seabream (Sparus aurata L.). *Fish Shellfish Immunol* 8; 49–62.

Ortuno J, Esteban MA, Meseguer J (2000) High dietary intake of a-tocopherol acetate enhances the nonspecific immune response of gilthead seabream (*Sparus aurata* L.). *Fish Shellfish Immunol* 10; 293–370.

Palstra AP, Kals J, Blanco Garcia A, Dirks RP, Poelman M (2018) Immunomodulatory effects of dietary seaweeds in LPS challenged Atlantic salmon *Salmo salar* as determined by deep RNA sequencing of the head kidney transcriptome. *Front physiol* 9; 625.

Petit J, Bailey EC, Wheeler RT, de Oliveira CAF, Forlenza M, Wiegertjes GF (2019) Studies into beta-glucan recognition in fish suggests a key role for the C-type lectin pathway. *Front Immunol* 10; 280.

Pratheepa V, Ramesh S, Sukumaran N (2010) Immunomodulatory effect of *Aegle marmelos* leaf extract on freshwater fish *Cyprinus carpio* infected by bacterial pathogen *Aeromonas hydrophila. Pharmac Biol* 48 (11); 1224–1239.

Raa J (1992) The use of immunostimulants to increase resistance of aquatic organisms to microbial infections. *In* Shariff M, SubaSinghe RP, Arthur JR (eds): *Diseases in Asian Aquaculture*. Manila, Asian Fisheries Society, Fish Health Section, 39–50.

Rao VY, Das BK, Jyotyrmayee P, Chakrabarti R (2006). Effect of Achyranthes aspera on the immunity and survival of Labeo rohita infected with Aeromonas hydrophila. *Fish & shellfish immunology* 20(3); 263–273.

Robertson B, Relrstad G, Engstad R (1990). Enhancement of non-specific disease resistance in Atlantic salmon, Salmo salar L., by a glucan from Saccharomyces cerevisiae cell walls. *Journal of fish diseases* 13(5); 391–400.

Rodriguez I, Chamorro R, Novoa B, Figueras A (2009) beta-Glucan administration enhances disease resistance and some innate immune responses in zebrafish (*Danio rerio*). *Fish Shellfish Immunol* 27; 369–373.

Sahoo PK, Mukherjee SC (2002) The effect of dietary immunomodulation upon Edwardsiella tarda vaccination in healthy and immunocompromised Indiamajor carp (Labeorohita). *Fish Shellfish Immunol* 12; 1–16.

Salinas I, Lockhart K, Bowden TJ, Collet B, Secombes C, Ellis AE (2004). Assessment of immunostimulants as Mx inducers in Atlantic salmon (*Salmo salar* L.) parr and the effect of temperature on the kinetics of Mx responses. *Fish Shellfish Immunol* 17; 159–170.

Sakai M (1999) Current research status of fish immunostimulants. *Aquaculture* 172; 63–92

Sánchez-Pozo A, Gil A (2002) Nucleotides as semiessential nutritional components. Br J Nutr87: 135-137. 79. Low C, Wadsworth S, Burrells C, Secombes CJ (2003) Expression of immune genes in turbot (*Scophthalmusmaximus*) fed a nucleotide-supplemented diet. *Aquaculture* 221; 23–40.

Shrestha G, St Clair LL & O' Neill KL (2015) The immunostimulating role of lichen polysaccharides: a review. *Phytother Res* 29; 317–322.

Siwicki A (1987) Immunomodulating activity of levamisole in carp spawners, Cyprinus carpio *L. J Fish Boil* 31; 245–246.

Skov J, Kania PW, Holten-Andersen L, Fouz B, Buchmann K (2012) Immunomodulatory effects of dietary β-1, 3-glucan from *Euglena gracilis* in rainbow trout (*Oncorhynchus mykiss*) immersion vaccinated against *Yersinia ruckeri Fish Shellfish Immunol* 33; 111–120.

Strand HK, Dalmo RA (1997) Absorption of immunomodulating beta -glucan in yolk sac larvae of Atlantic halibut. *Hippoglossus hippoglossus Journal of Fish Diseases*, 20 (1); 41–49.

Tseng IT, Chen JC (2004) The immune response of white shrimp Litopenaeusvannamei and its susceptibility to Vibrio alginolyticus under nitrite stress. *Fish Shellfish Immuno* l17; 325–333.

Vasta GR, Nita-Lazar M, Giomarelli B, Ahmed H, Du S, Cammarata M, Parrinello N, Bianchet MA, Amzel LM (2011) Structural and functional diversity of the lectin repertoire in teleost fish: relevance to innate and adaptive immunity. *Dev Comp Immunol* 33; 1388–1399.

Verlhac V, Gabaudan J, Obach A, Schuep W, Hole R (1996) Influence of dietary glucan and vitamin C on non-specific and specific immune responses of rainbow trout (Oncorhynchusmykiss). *Aquaculture* 143; 123–133.

Verlhac V, Obach A, Gabaudan J, Schüep W, Hole R (1998) Immunomodulation by dietary vitamin C and glucan in rainbow trout (*Oncorhynchus mykiss*). *Fish Shellfish Immunol* 8; 409–424.

Villamil L, Figueras A, Novoa B (2003) Immunomodulatory effects of nisin in turbot (*Scophthalmus maximus* L.). *Fish Shellfish Immunol* 14; 157–169.

Wang S, Wang Y, Ma J, Ding Y, Zhang S (2011) Phosvitin plays a critical role in the immunity of zebrafish embryos via acting as a pattern recognition receptor and an antimicrobial effector. *J Biol Chem* 286; 22653–22664.

Wang W, Sun J, Liu C, Xue Z (2017) Application of immunostimulants in aquaculture: current knowledge and future perspectives. *Aquaculture Res* 48; 1–23.

Yoshida T, Kruger R, Inglis V (1995) Augmentation of non-specific protection in African catfish, Clariasgariepinus (Burchell), by the long term oral administration of immunostimulants. *J Fish Dis* 18; 195–198.

Zhang Z, Chi H, Dalmo RA (2019) Trained innate immunity of fish is a viable approach in larval aquaculture. *Front Immunol* 10; 42.

Zhang Z, Chi H, Dalmo RA (2019) Trained innate immunity of fish is a viable approach in larval aquaculture. *Front Immunol* (2019) 10; 42.

Zhang Z, Swain T, Bøgwald J,Dalmo RA, Kumari J (2009) Bath immunostimulation of rainbow trout (*Oncorhynchus mykiss*) fry induces enhancement of inflammatory cytokine transcripts, while repeated bath induce no changes. *Fish Shellfish Immunol* 26; 677–684.

ABBREVIATIONS

A salmonicida	Aeromonas salmonicida
A hydrophila	Aeromonas hydrophila
µg	microgram
BCG	bacille Calmette-Guerin
CRP	C-reactive protein
DNA	Deoxy ribonucleic acid
E. tarda	Edwardsiella tarda
E. ictaluri	Edwardsiella ictaluri
FAO	Food and Agricultural organization
FCA	Freund's complete adjuvant
HSP	Henoch-Schönlein Purpura
IHN	Infectious Hematopoetic Necrosis
IgM	Immunoglobulin M
Kg	Kilogram
MAF	Macrophage-activating factor
M. bovis	Mycobacterium bovis
NLR	NOD-like receptors
N. seriolae	Nocardia seriolae
P. piscicada	Pasteurella piscicada
SOD	Superoxide dismutase
S. quinqueradiata	Seriola quinqueradiata
USD	United States Dollar
VE	Vitamin E
V. anguillarum	Vibrio anguillarum
V. salmonicida	Vibrio salmonicida
V. vulnificus	Vibrio vulnificus
Y. ruckeri	Yersinia ruckeri
YHV	yellow head virus
β-1,3-glucan	Beta 1,3 glucan

7

Disease Management and Prophylaxis by Immunostimulants

Chinnathambi Pothiraj[1], Divya Jyoti[2], Subramanian Ramya[3], Ramaraj Jayakumararaj[1], Aseem Grover[4], Reshma Sinha[5], Palanichamy Ayyappan[1], Caterina Faggio[6], and Paulraj Balaji[7]

[1]*Department of Botany, Government Arts College, Melur, Madurai, Tamil Nadu, India*
[2]*School of Biological and Environmental Sciences, Shoolini University, Solan, India*
[3]*Department of Zoology, Yadava College (Men), Madurai, Tamil Nadu, India*
[4]*Department of Chemistry, Biochemistry and Forensic Sciences, Amity School of Applied Sciences, Amity University Haryana, Panchgaon, Haryana, India*
[5]*Department of Animal Sciences, Central University of Himachal Pradesh, Kangra, India*
[6]*Department of Chemical, Biological, Pharmaceutical and Environmental Sciences, University of Messina, Italy*
[7]*PG and Research Centre in Biotechnology, MGR College, Hosur, Tamil Nadu, India*

7.1 Introduction

Aquaculture is a purposeful activity for breeding, culturing, rearing, raising, and harvesting fish along with other water species such as shrimps, oysters and crabs. Aquaculture industry is a rapidly growing industry which is expected to rise to about 37% by 2030 with a total production of 109 million metric tons. This tremendous increase in aquaculture industry may be accredited to the increased demand for food and technological advances. This industry requires low input but serves as a profitable source of income, food, nutrition, oils, ornaments, and many more, thus consisting of both food and non-food use.

Looking at the profit, aqua culturist is preferably practising intensive or semi-intensive piscine culture techniques whereby with human inputs, great output in terms of productivity can be achieved from a small area. Open net pens and recirculating systems are among the commonly used intensive pisciculture systems for providing high-quality sea food. This sort of farming follows limiting the resources but getting maximum products. At times, higher productivity (food and other resources) can be devastated by significant loss of livestock due to stress, pollution, environmental disbalances, and toxicants.

Besides these, diseases are another major contributing factor to the loss. In aquaculture, stocks may suffer from varied fungal, protozoal, helminth, and bacterial diseases. These pose significant mortality and of serious concern for the hatchery. In an aquatic media, spread of vectors and infection becomes quite easy and is a frequent phenomenon. Moreover, a stock inhabiting ~~a~~ stressed conditions (e.g., overcrowding, overhandling, over-transportation) can be easily impacted by any infection, which may be either due to the lowered immunity after surviving in the cultured conditions or by the increased susceptibility to pathogens. This prevalence of disease can also be accredited to the development of antimicrobial resistance (AMR) due to the abuse of antibiotics and chemicals. In aquaculture, water bodies may suffer overdose of the anti-parasitic drugs, antiseptics, disinfectants, sanitizers, antibiotics, bioremediation products, feed supplements, and growth promoters that target the overproduction of stock. Besides these, some of the anthropogenic wastes may find their way as well.

Persistence to infectious diseases has been seen in aquaculture, which can be effectively circumvented by boosting individual defence system mediated by immunostimulants as prophylactic treatment (Hoseinifar et al., 2020). Organic chemicals, polymers, microbiological formulations, plant or animal extracts, dietary requirements, and cytokines are all

DOI: 10.1201/9781003361183-9

included in the category of immunostimulants that can boost and stimulate the animal immune system.

7.2 Current Research Status of Fish Immunostimulants

Fish farming is seen as an essential agricultural activity that can help reduce global malnutrition and poverty (Kaleem and Sabi, 2021). When it comes to freshwater fish, India's output is second only to China's. India ranks third overall among fish producers and second among freshwater fish producers worldwide. Freshwater aquaculture accounts for most of the expansion in fish farming. India consumes about 12.8% of its animal protein intake from freshwater fish. Estimates for global fish output in 2020 ranged from 15 to 20 million metric tonnes, with 9 million tonnes expected to come from freshwater-farmed species, principally carps, due to their high dietary value.

The Indian major carps *Catla catla* (Catla) and *Cirrhinus mrigala* (Mrigal) are two of the most popular species farmed for their meat and fish. These are the most widely consumed inland water species because of their rapid reproduction, delicious flavour, and widespread appeal (Bais, 2018). The aquaculture and fishing industries face numerous difficulties. Fish in the early stages of development are more vulnerable to sickness because they lack the ability to fight off infection (Mondal et al., 2022). Consequently, disease outbreaks in the aquaculture business hamper aquaculture productivity around the world. Several pathogens, including viruses, bacteria, and fungi, have been suspected as a major limiting factor in fish production in both the wild and aquacultured sectors, with the most prevalent and frequent disease in India's commercial carp fishery being *Aeromonas hydrophil, Edwardsiella tard, Vibrio anguillarum,* and *Aphanomyces invadans*. Drugs, chemicals, and antibiotics are frequently used to cure illnesses brought on by pathogenic germs, although prophylactic use of these substances is no longer recommended and often ineffective. The accumulation of these contaminants and antibiotics in the aquatic system is detrimental to both the ecosystem and fish populations (Kumar et al., 2022; Assefa and Abunna, 2018).

7.3 Immune System of Fish

Fish have well-developed defence systems comprising both innate (non-specific) and adaptive (specific) response system. The fish immune system shows some similarity to the mammalian system. Whenever any antigen enters, initially it is handled by the first line of defence formed by the innate system, while upon persistence of the antigen, it is being taken care by the antibody-mediated specific response. The immunity is mediated by the presence of immune structures like the thymus, kidney, spleen, blood, liver, and intestinal tract.

Thymus: Except agnatha, fish possess the thymus gland like higher vertebrates. The tonsils also develop from the pharyngeal pouches and are the first lymphoid organ to be invaded by lymphocytes throughout development. It is generally found in the upper part of the gill as superficial tissue in the gill chamber and being superficial in location it is easily exposed to pathogens.

Histologically, these may differ in species. It is a paired gland, comprised of white tissue that is lined by an epithelial capsule. It is comprised of the cortex and medulla, where immature T lymphocytes are observed in the medullary region, while the cortical region has smaller lymphocytes. Reticular-epithelial cells provide a scaffold for lymphocytes within the gland. Fish have been shown to have immunological reactions that depend on thymus-derived cells. Normally, the thymic microenvironment is where T lymphocyte specificities are learned. In similarity to vertebrates, the size of the thymus is related to the age, season (spawning), and physiological stress.

Kidney: The kidneys, which are a dark red colour, are located on the ventral side of the body along the spinal column. Both the anterior (in the head) and the posterior (in the trunk) kidneys exist. The anterior kidney has a major role to play in both the process of haematopoiesis and immune action. The kidney's anterior region contains both mitotic and big immature cells. Important immune functions such as phagocytosis and antigen processing are also performed here. Humoral immunity and immunological memory are activated by the presence of B-lymphocytes. Furthermore, the reticulo-endothelial system of circulating macrophages assists in filtering out damaged blood cells and debris.

Liver: Liver sinusoids have been evidenced to have phagocytic mononuclear cells similar to Kupffer cell (mammals) that act in the antigen presentation. Production of the C-reactive protein, CRP, as an inflammation indicator has been observed in rainbow trout (*Oncorhynchus mykiss*).

Spleen: It is a dark-coloured organ localised in the posterior abdominal cavity. Histologically, it exhibits connective tissue and a pulp matrix enclosed in the capsule. The organ collects macrophages and lymphocytes, where macrophages are organized in centres and remain responsible for phagocytosis. The pulp matrix has further differentiation into white pulp (lymphocyte generation) and red pulp for haematopoiesis. In addition, to serve for filtering the blood-borne antigens, it has a major role in the defence mechanism. Although fish lack lymph nodes, the functions of the hematopoiesis and other functions are accomplished by the spleen.

Intestine: Additionally, antigen can be taken up and processed up by the intestinal cells and thus are allocated to the defence action as well. The mucosal defences, in particular, is thought to involve the stratum granulosum's layer of mucosal cells, which is mediated by the gut-associated lymphoid tissue (GALT). GALT provides a local immune response environment for protection against pathogens.

Blood: In fish, large basophilic nuclei (13–16 μm long) are reported in the erythrocytes. Lymphocytes account for a significant portion of the leukocyte population (> 90%), with neutrophilic granulocytes, thrombocytes, monocytes, and natural cytotoxic cells rounding out the rest. Natural cytotoxic cells are tiny lymphocytes that are agranular in structure. Their function is analogous to that of the natural killer (NK) cells found in mammalian immune systems (Barrow et al., 2019).

B and T lymphocytes can be distinguished on basis of Ig (sIgM) present on the membrane, where T cells lack Ig (sIgM) compared to B cells. Although basophilic and eosinophilic granulocytes are uncommon in teleosts, neutrophilic granulocytes develop multi-lobed nuclei. Having phagocytic and bactericidal properties, monocytes and neutrophilic granulocytes are also chemotactic. Melanoma macrophages have been seen in the spleen, liver, kidney, and even the gonads, all of which are lymphoid organs. It is well established that when collagen is exposed by a wound, thrombocytes release clotting factors.

7.3.1 Non-specific Immunity/Innate Immune System

The innate immune system forms the first line of defence by acting against the antigen, but it does not develop any memory. Fish non-specific immune system includes physical barrier (skin), cellular process (phagocytosis), and release of humoral components (soluble proteins) interleukins, interferon, antimicrobial enzymes, and organic defense cells *i.e.,* monocytes, granulocytes, macrophages, and (NK) natural killers cells. The skin is embedded with many goblet cells, secreting mucous containing antibodies and lysozyme. Additionally, it is covered over by scales lining the exposed body in the teleost and trout. In fish exposed directly to pathogens or toxicants, the gill surface has been diagnosed with a large number of macrophages, forming the first line of defence against parasite infection. A strong immune response has been observed in trout exposed to *Yersinia ruckeri* bacteria.

The nonspecific immune response in trout is stimulated by opsonisation and complement. This is mediated by initiation of mononuclear phagocytes (macrophages) and polymorphonuclear granulocytes (PMN). At the site of infection, chemotactic movement of neutrophils, eosinophils, basophils (not in all species), macrophages, and lymphocytes develops an inflammatory set of reactions followed by little increases in temperature. Chemotaxis of immune cells to the site of infection is mediated by soluble substances such as cytokines, eicosanoids, and complement components.

Interleukin (IL)-1, transforming growth factor (TGF)-β, and tumour necrosis factor α (TNF α) show biological activity in fish, indicating the presence of receptors for these cytokines. Molecular cloning and sequencing have furthermore recommended the presence of an IL-2-like, IFN-like, IL-1β, and TGF-β genes in fish.

7.3.2 Specific Immunity

Antigens can be taken in orally, where they trigger MHC-mediated, antigen-specific immunity. Similar to humans, both MHC I and II are implicated in activation of fish lymphocytes (B and T), followed by cytokines synthesis which results in antigen-specific antibody generation from the plasma cells. Immunization in fish generates antibodies with properties such as agglutination, precipitation, complement fixation, opsonisation, and skin sensitization (Wieczorek et al., 2017). In aquaculture, temperature has an important role in immunisation as optimal temperature for antibody production is required by each fish species e.g., 14-16°C for rainbow trout.

However, no information about fish species' ability to swap isotypes or develop affinity for or maturity in specific antigens has been found (Pastoret et al., 1998). The fish antibodies, such as IgM, are tetrameric molecules, not pentameric like those found in mammals. Several fish species, including Atlantic salmon (*Salmo salar*), zebrafish (*Danio rerio*), carp (*Cyprinus carpio*), and rainbow trout, have had full-length MHC class I α-chain cDNA and MHC class II β-chain validated using molecular cloning (Wieczorek et al., 2017). Both carp and Atlantic salmon, albeit to varying degrees, express MHC class II -chain in their thymus, head kidney, spleen, and peripheral blood (Wieczorek et al., 2017).

7.4 Development of Disease

Disease is a state of discomfort which can be mediated either by any physical or living organism. It can develop structural (tissue or organ damage), functional abnormality or slowed growth, or death indicated through specific or non-specific symptoms. In farmed aquatic animals, diseases may be caused by infectious organisms, poor management practices, pollutants, toxicants, and environmental issues, which may be classified as physical, biological, and chemical (Figure 7.1). Persistent disease outbreaks could lead to the failure of aquaculture ventures and jeopardize the industry's long-term viability.

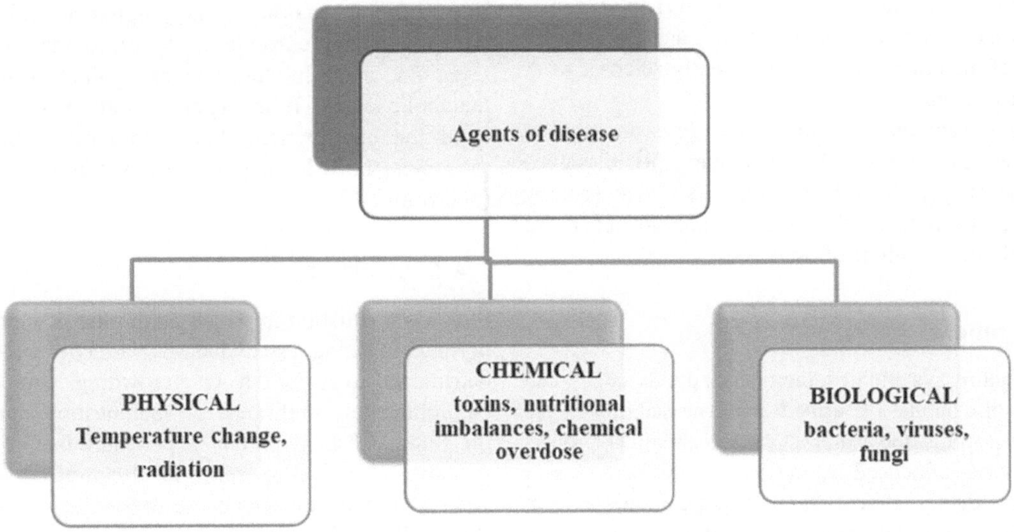

FIGURE 7.1 Disease-causing agents.

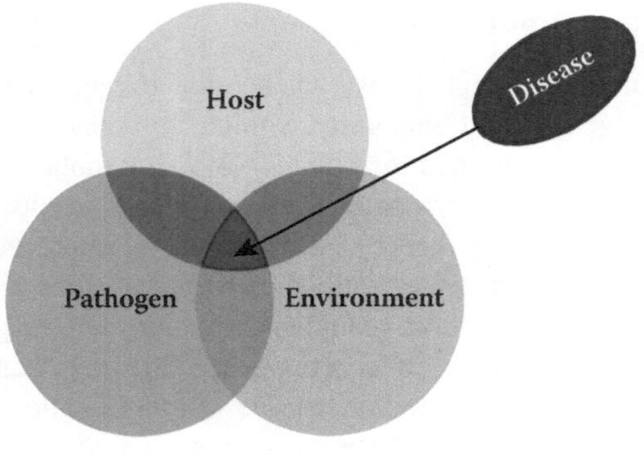

FIGURE 7.2 Interaction of factors for the occurrence of disease in an organism.

Disease development in an aquaculture system is a complex interaction between three major factors of host (farmed fish), pathogens (disease-causing organisms), and environment (surrounding). For a disease situation to exist, there must be a possible pathogen, a host response, and ecological factors that can promote pathogen virulence or diminish host resistance (Figure 7.2).

7.4.1 Host

Host innate disease resistance, physiological, and immunological status all play a significant role when exposed to a pathogen, in determining whether an infection will be established or not. Age, size, species, defence systems used, and fish health all play a role in determining whether a given host is susceptible or resistant to a given pathogen.

7.4.2 Pathogen

Pathogens are defined as the organism which can cause a disease in the host. The virulence, pathogen's capacity to live and/or reproduce in the absence of a host, the contact time required to commence infection, and the mode of transmission all influence the pathogen's ability to cause a disease. To assess the risk of disease transmission to wild fish populations, it is necessary to examine the pathogens' specific properties. When it comes to infectious diseases, viruses, bacteria, fungi, and parasites all are the centre of attention. They are frequently referred to as infectious disease agents.

The presence and abundance of biological agents are affected by variables such as the system's temperature, pH, dissolved gases, and food supply. Infectious agents have two primary characteristics: (a) the potential for direct transmission and b) the capacity to proliferate within the host tissue.

7.4.3 Environment

Water and its holding system, including tanks, ponds, cages, and pens, constitute the habitat of farmed fish. The stability of the environment, especially the water's physico-chemical properties, which may be influenced by the fish culture activities or natural factors, directly affects the fish population's health. Variations in environmental conditions including temperature, pH, salinity, and dissolved oxygen that are too great or too small for the host can potentially cause stress and illness. Fish malnutrition, water quality issues, overpopulation, and the growth of unwanted aquatic plants are all possible outcomes of these environmental stresses.

Moreover, the landscape is constantly shifting its role in supporting the delicate equilibrium between pathogen and host. In several situations, the culture animals live a healthy life in the presence of pathogens; but as environmental stresses arise and the balance tips in favour of the disease, the pathogen gets the upper hand and disease conditions succeed.

7.4.4 Poor Water Quality

Tolerability by aquatic species depends on the water's chemical, biological, and physical properties, as these organisms are in constant contact with the water. A wide variety of physicochemical factors fall outside the species' tolerance range. For aquatic species to continue to thrive within their optimal range, changes to water quality parameters must be gradual. Otherwise, the fish may experience stress, develop a weakened immune system, and possibly die.

7.4.5 Nutrition

Good body composition/flesh quality and overall fish health can be attributed in large part to a fish's proper nutrition. As a result, intensive aquaculture relies heavily on carefully planned diets and feeding schedules. Nutritional techniques that boost the resistance and immunity of cultured organisms have received more attention in recent years as a means of lowering economic losses caused by disease.

The six nutrient groups (water, proteins and other nitrogenous molecules, lipids, carbohydrates, minerals, and vitamins) are required for reproduction, growth, and health maintenance of animals (Morris and Mohiuddin, 2020). The macronutrients (protein, lipids, and carbohydrates) produce the bulk energy required for the fish metabolic system to function and formation and repair tissues (Adewumi, 2018). Minerals including calcium, phosphorus, magnesium, chloride, sodium, potassium, and sulphur as well as trace minerals like cobalt, chromium, copper, iodine, iron, manganese, selenium, and zinc provide are cofactors needed for the metabolic steps. Both types of nutrients can be obtained from the environment, but in order to stimulate growth, they must be offered purposefully through designed meals (Adewumi, 2018).

7.4.6 Stress

Stress is a situation in which an animal is unable to sustain a normal physiologic state due to a variety of conditions that are detrimental to its health. Overcrowding, poor handling, poor environmental conditions, or malnutrition induces stress in the cultured fish. Importantly, placing a fish in a scenario that is above its typical tolerance level causes stress. Specific examples of things that can cause stress (stressors) are listed next (Figure 7.3 and Table 7.1).

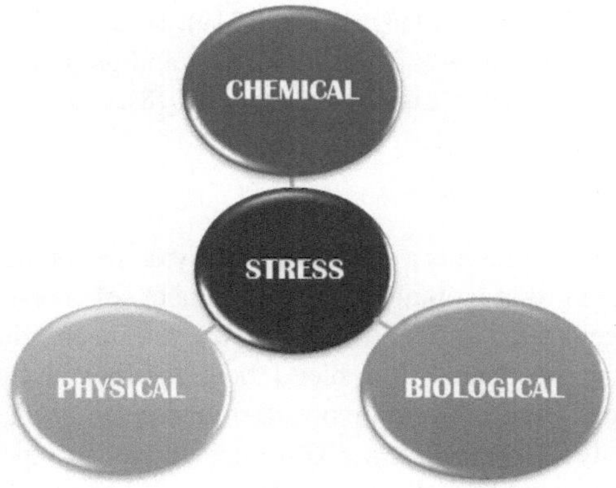

FIGURE 7.3 Types of stressors.

7.5 Principle of Disease Diagnosis

The most important aspect of any animal's health management against any infectious disease is the identification of the etiological agent. Aside from looking for the causative agents, some diagnostic techniques are used to screen healthy animals so as to ensure that they do not have any sub-clinical infections. International agencies like FAO and OIE recognize that disease diagnosis can be undertaken at principally three levels as given:

- Changes in feeding or swimming behaviour (Level I) are frequently used to initiate an investigation into potential disease problems, which is then followed by appropriate sampling and analysis. Level I diagnosis includes farm site observation for clinical signs and symptoms, examination of pond conditions that may be sufficient to make a management decision. Field-level kits may be used to evaluate sediment and water quality, and they may be useful in more accurately identifying the environmental conditions that may precipitate the disease. It is critical to begin management measures as soon as possible in order to on-time effectively cover the disease in a pond or farm and reduce mortalities.
- As many symptoms are shared by several diseases, extensive laboratory analysis is usually required, which may include diagnostic techniques such as histopathology, bacteriology, virology, and water analysis, which constitutes Level II diagnosis, although in case of some viral diseases viral isolation is the golden test of disease detection. So, this would be required to identify disease-causing agents, especially in the case of bacterial diseases, to determine the antimicrobial susceptibility of the organism involved. It regulates the use of antimicrobial agents in aquaculture and to prevent the advent and spread of microorganisms resistant to most commercially available antibiotics. At this level accurate diagnosis may take up to 2–3 days which may result in delayed management action, thereby leading to the major loss.
- Level III diagnosis is critical for confirming the diagnosis of viral infections. It utilises advanced techniques, such as polymerase chain reaction (PCR) and other modifications such as reverse transcriptase PCR and real-time PCR for rapid and confirmatory infection diagnosis. The pattern of mortality may indicate whether the problem is caused by an infectious agent, poor water quality, or toxicity.

7.6 Types and Sources of Immunostimulants

In aquaculture, the probability of fish survival can be enhanced by boosting the efficacy of immune system comprising both the innate (non-specific) and acquired (specific) immunity that protect fish against pathogenic infections. Such immunity boosters (immunostimulants) are established as supplements having the capability to modulate the immune system by increasing the host resistance to diseases. These can either be a suppressive or as a prophylactic treatment for pathogens (latent or sub-lethal) and seasonal outbreaks of certain known endemic diseases. Broadly, these can be categorised into biological and synthetic immunostimulants based on chemical characteristics and origin. Whereas based on kinds of sources, it can be classified into groups such as algae-derived, animal-derived, bacterial-derived, cytokines, hormones, and nutritional factors.

7.6.1 Biological Immunostimulants

These refer to the molecular (macro) products that are derived from microbes, algae, plants, or animals. Certain examples are glucans, peptidoglycans, polypeptides, lactoferrins, vitamins, hormones, etc.

TABLE 7.1

Types of Stressors

Biological	Chemical	Physical
• Population density (crowding) • Other species of fish (aggression, territoriality, lateral swimming space requirements) • Microorganisms (pathogenic and nonpathogenic) • Macroorganisms (internal and external parasites)	• Poor water quality (low dissolved oxygen, improper pH) • Intentional pollution (chemical treatments) • Accidental pollution (insect spray, spills) • Diet composition (type of protein, amino acids) • Nitrogenous and other metabolic wastes (accumulation of ammonia or nitrite)	• Temperature • Light • Sounds • Dissolved gases • Ice

7.6.1.1 Cytokines

Cytokines act as immunomodulating agents (substances that suppress or stimulate the immune system) as these have behavioural effects on adjacent cells (Guo and Li, 2021). These are the class of polypeptides with low molecular weight (>30 kDa) and are referred to as glycoproteins. Mostly produced by activated immune cells at the point of entry of the pathogen. However, some of them can be secreted by the cells of the endocrine or central nervous system (astrocytes and microglial) and may produce interferons (INF- α, INF-1, INF-2, and INF-6), and tumour necrosis factor (TNF). Cytokines act as regulators or informants, connecting various components of the immune system to properly drive an immune response (Urbinati et al., 2020). These also regulate the antigen-presenting cells (APCs) along with their migration to the lymph nodes for the initiation of adaptive immune responses (Wu et al., 2020). The cytokines possess pleiotropic effects as these can act both in autocrine or paracrine ways for regulating the host defence (Sakai et al., 2021). Cytokines are interferons, interleukins, transforming growth factor, tumour necrosis factor, colony stimulating factor, and chemokine.

7.6.1.2 Interferons

An interferon is the most demonstrated group of cytokines and has the ability to interfere with the viral replication. There are three major classes of interferon-type I IFN, type II IFN, and type III IFN. Type I IFN becomes abundant during an infection with a potent anti-viral effect. It has an effect via inducing the production of Th1 cells, which in turn suppresses viral replication by boosting expression of MHC I molecules on cells (infected by the virus) and destruction of NK cells. Type I IFN includes IFN α which is of 13 sub-types and IFN β. Unlike type I IFN, IFN γ is the single representative of type II IFN and has a major role in the activation of macrophages both in adaptive and innate immune responses. Type III IFN plays a major role in host defence against infections and biological effects are similar to type I IFN. Type III IFN includes IL-28/29 (Gan et al., 2020).

7.6.1.3 Interleukins

An interleukin represents a group of secreted proteins that help in the communication between leukocytes (Altan-Bonnet and Mukherjee, 2019). Currently, 40 interleukins have been identified and some of them are further divided into subtypes (Ferreira et al., 2018). IL 1 was first named as lymphocyte activating factor and is formed of 11 cytokines-1 anti-inflammatory cytokine (IL-37), 3 receptor antagonists (IL-1Ra, IL-36Ra, and IL-38), and 7 agonist ligands (IL-1α, IL-β, IL-33, IL-36α, IL-36β and IL-36 γ).

IL 1α helps in initiating the signalling cascade after cell death due to necrosis as in case of stroke and myocardial infraction (Abbate et al., 2020). IL 1β, on the other hand, is involved in degenerative, infectious, autoimmune, and especially with auto-inflammatory diseases (Kany et al., 2019; Wang et al., 2020). This cytokine was one among the first to be identified, and it plays a key role in the inflammatory response to infection in fish (Gallenga et al., 2019). IL 2 family acts as proliferation and growth factor for mature and progenitor cells (Ferreira et al., 2018; Lin and Leonard, 2018).

7.6.1.4 Chemokines

A chemokine represents a superfamily of cytokines (structurally homologous) that stimulate the movement of homeostatic immune cells such as B cells, monocytes, and neutrophils and regulate their migration from blood, bone marrow, and peripheral tissues. These are also involved in inflammatory reactions. The activity of chemokines is co-ordinated by the binding of seven trans-membrane domains to G- protein linked receptors. Chemokines are further divided into four families based on the position of N-terminal cysteine residues, i.e., C, CC, CXC and CX_3C, where CC and CXC form two major families.

7.6.1.5 Tumour Necrosis Factor (TNF)

It was first discovered in 1975 as the molecule that causes the *in-vitro* necrosis of tumours cell, thus giving it the name tumour necrosis factor. TNF family consists of more than 40 members, out of which TNF α (TNF) and TNF β (lymphotoxin) are the most outstanding ones and the rest includes ligands (FAS, CD40, FAS, GITR, OX40) and membrane proteins (Dostert et al., 2019). TNF α is a pro-inflammatory cytokine; it plays a crucial role in a wide range of host responses, including cell distinctiveness, growth, death, and disintegration (Jahandideh et al., 2020). The binding of TNFs to their respective receptors triggers the inflammatory reactions for destruction of pathogens (Dostert et al., 2019). It was also observed that the cells of host defence system (macrophages, monocytes, NK cells, and T cells) promote the expressions of TNF. The stimulus for the expression of TNF includes the binding of toll like receptors (TLRs) to the lipopolysaccharide or other parts of pathogens (Dostert et al., 2019). It is often referred to as an endogenic pyrogen as it increases infection by exciting the cells of hypothalamus to create prostaglandins.

7.6.1.6 Transforming Growth Factor (TGF-β)

The cytokine transforming growth factor is widely distributed. Different leukocytic lineages, including dendritic cells, lymphocytes, macrophages, granulocytes, and natural killer (NK) cells, are all affected by its regulation of cell growth, differentiation, multiplication, survival, and motility. TGF- β also helps in promoting the immune response by inducing Th17 cells generation. In grass carp, TGF- β1 stimulates the proliferation of blood lymphocytes (peripheral) and also down-regulates the lipopolysaccharide/phytohemagglutinin (PS/PHA) (Dahmani and Delisle, 2018).

7.6.1.7 Colony Stimulating Factors (CSFs)

These stimulate haematopoiesis that contributes to the activation of macrophages and growth of progenitors of eosinophils, basophils, neutrophils, and monocytes. Macrophage colony stimulating factor (M-CSF) and granulocyte-macrophage colony stimulating factor (GM-CSF) have pro-inflammatory action.

In fish, the most characterized and functionally described cytokine in jawed fish is of IL-1 family and has the most potent pro-inflammatory action (Secombes, 2022). Anti-inflammatory cytokines like IL-10 and TGF-β inhibit immunological responses. Where, IL10 inhibits the gene expression of major histocompatibility complex (MHC) while, TGF-β prevents the up-regulation of IL-1β and TNF-α along with the nitric oxide response. Chemokines such as IL-8 whose main function is to promote the activation of monocyte and neutrophils and stimulate their respiratory activity (Xiao et al., 2019). Thus, chemokine also acts by inhibiting bacterial growth, inducing lymphocytic proliferation, and regulating the immune response of the host against pathogenic infection (Sun et al., 2019).

7.6.1.8 β-glucan

It is among widely used immunostimulant in aquaculture. The β-glucans are long-chain polysaccharides derived from yeast (mostly from *Saccharomyces cerevisiae*), and are a heterogeneous group of polymers of glucose having β-(1,6)-linked side chains with a backbone of β-(1,3)-linked β-d-glucopyranosyl (Hansen et al., 2021). Studies have reported that treatment with β-glucan increases lysosomal, phagocytic, and bactericidal activities of macrophages, along with improved intestinal immune responses indicating the enhanced activity of defence mechanisms (Dawood et al., 2020; Hasan et al., 2019; Yamamoto et al., 2020). In a study, β-glucan (0.09%) dietary supplementation has also been found to increase the growth of large yellow croakers (*Pseudosciaena crocea).* This growth promotion can be accredited to the activity of glucanase enzyme, produced by digestive gland which decomposes β-glucans for generation of the energy which can be ultimately used in the production of proteins. Instances of improved intestinal immune responses have been observed by supplementation of β-glucans and supporting the growth in aquaculture. Moreover, the oral administration of β-glucan to Nile tilapia *(Oreochromis niloticus)* for 5 days has stimulated the production of cytokines (TNF-α, IL-10, IL-12, IL-1β) (Chansue et al., 2000). Similarly, when fry of rainbow trout were exposed to β- glucans through bath treatment for 45 min, there was observed an enhancement in the gene expression of anti-inflammatory cytokines (TGF-β, IL-10) and pro-inflammatory cytokines (TNF-α, IL-6, IL-1β).

Additionally, β-glucans have also displayed their role in maintenance of the ion balance during fish transport (Vanderzwalmen et al., 2019). Supplementary feeding of β-glucan (0.2%) with mannanoligosaccharide in fish diet was also found to improve feeding efficiency, intestinal morphology, and growth response. In *Aeromonas hydrophila*–infected fish, dietary exposure of β-glucan along with *Bacillus licheniformis* has shown to enhance immune reactions, growth rate, and survival rate (Anjugam et al., 2022).

Therefore, β-glucan is supposed to act in two different ways: 1) directly stimulates the immune system by getting absorbed in the small intestine (de Oliveira et al., 2019) and 2) stimulates the gastrointestinal bacteria (typical prebiotic). As β-glucan works in two different aspects so can be classified both as an immunostimulant and a prebiotic.

7.6.1.8.1 Levan

It is a natural polymer that belongs to a class of fructo-oligosaccharides composed of D-fructosyl residues linked through β (2, 6) linkages. It is derived mostly from bacteria and other sources such as yeasts, plants, and fungal species that feeds on sucrose-based substrates. Microbial levan is mostly used as a nutraceutical feed additive in aquaculture as it can effectively augment the non–specific immune response, growth performance, maintain the liver homeostasis, and improves the thermal tolerance. Immunological aspect when inspected, displayed that upon nutritional supplement of bacterial levan in the *Labeo rohita* infected with *A. hydrophila*, resulted in the down-regulation of monitoring genes (TGF-β) and HSP70, while up-regulation of immune responsive genes (IFN-γ, β-2 M, and TLR22), decreased blood glucose, and cortisol (Gupta, 2021). Microbial levan at concentration of 0.5% had shown to induce maximum immunomodulatory effects in common carp by raising haemoglobin count, total erythrocyte count, respiratory burst, lysosomal, and phagocytic activity.

7.6.1.8.2 Lactoferrin

Lactoferrin is a glycoprotein belonging to transferrin family (proteins that are capable of transferring and binding Fe^{3-} ions) (Rascón-Cruz et al., 2021). Due to its localization on the surface of mucosa, it forms the first line of defence against the agents that invade organisms. It also exhibits antimicrobial activities and influences the proliferation of pathogens like protozoa, bacteria, and fungi. Dietary supplementation of lactoferrin has been found to enhance the tolerance to increased temperature stress in Japanese flounder and non-specific immunity in Nile tilapia (Chandan et al., 2020; Hashem et al., 2022). Therefore, by virtue of its anti-neoplastic, immunomodulatory, and antimicrobial properties, it has great potential to be used in aquaculture.

7.6.2 Bacterial Derivatives

Bacterial derivatives such as muramyl dipeptide (MDP) and peptidoglycan shares intrinsic properties of parent cells like ability to load specific antigens and immune modulation. Nanoscale size makes them easy to be taken and are highly biocompatible. They provide better safety profiles than live bacteria within proper range of dose because the non-viability of the derivatives reduces the risk of bacterial proliferation and transmission to non-target sites.

7.6.2.1 Muramyl Dipeptide (MDP)

Muramyl dipeptide (MDP) is a simple glycoprotein and is a purified form of mycobacteria. It has been found to enhance antibody activity, stimulate the activation of macrophages, and polyclonal activation of lymphocytes. In rainbow trout, muramyl dipeptide induced macrophages through nitric oxide (HNO_3) synthesis.

7.6.2.2 Lipopolysaccharides

Lipopolysaccharides are the major component of cell wall of Gram-negative bacteria. It is a glycolipid consisting of three structurally and biosynthetically different moieties: lipid A, core oligosaccharide, and O- polysaccharide. Lipopolysaccharides exerts potent effects on the immune system as these can stimulate phagocytosis, B-cell proliferation, and production of superoxide anions (Zhu and Su, 2022). Lipopolysaccharides toxicity is a known phenomenon in mammals, but the same has no potential in fish due to the absence of several co-stimulatory molecules like (TICAM2, LY96, and CD14) and intracellular mediators that are present in mammals. Enhanced non-specific parameters and growth performance have also been reported by the dietary supplementation of lipopolysaccharide (20 μg/kg of body weight day^{-1}) in fish. Even the lower concentration (3 mg LPS/kg) of lipopolysaccharides provides protection against various diseases. When the striped catfish (*Pangasianodon hypophthalmus*) was fed with lipopolysaccharides derived from *Escherichia coli,* it enhanced immune response, white blood cells and protection from *Edwardsiella ictaluri.*

7.6.2.3 Outer-membrane Proteins

Outer membrane proteins are the surface protein present on specific outer membranes of Gram-negative bacteria. These proteins can stimulate the immunity (both humoral and cellular) and can induce a protection in the host. The outer membrane proteins of *Aeromonas sobria* GC-2 and *Bacillus* subtilis JB-1 strains provide protection against *Yersinia ruckeri* in salmonids. When the common carp was vaccinated with the recombinant protein of *Edwardsiella tarda* i.e., outer membrane protein-A, it elicited production of antibodies and higher survival percentage against the infection by *Edwardsiella tarda.*

7.6.2.4 Peptidoglycans

Peptidoglycan is an important and unique constituent of the cell wall of bacteria. It is a polymer of β (1, 4) linked N-acetylglucosamine and N-acetylmuramic acid. Peptidoglycan has been found to modify the innate immune response by forming a network with gut-associated lymphoid tissue and intestinal microbiota. Oral administration of peptidoglycan enhances the immune system (innate) of rainbow trout and can induces a resistance in Japanese flounder and yellowtail kingfish against *Vibrio anguillarum* and *Lactococcus garvieae,* respectively so can be used to boost the fish immunity.

7.6.2.5 Freund's Complete Adjuvant (FCA)

Freund's complete adjuvant (FCA) is the adjuvant of mineral oil having killed *Mycobacterium butyricum.* It increases the NK cell activity, phagocytic activity, respiratory burst, and immune response of fish. Injection of FCA shows an increase in resistance against *A. hydrophila, A. salominicida, V. ordalli,* furunculosis, vibriosis and red mouth disease in fish.

7.6.2.6 Achromobacter Stenohalis

Achromobacter stenohalis is a Gram-negative aerobic bacterium isolated from sea water. Its lipopolysaccharide is capable of activating the B-lymphocytes and macrophages in mouse. Inactivated *Achromobacter stenohalis* may also enhance the complement activity, immune response by kidney cells and protection against *A. salmonicida,* the cause of fish furunculosis.

7.6.2.7 Clostridium Butyricum

Clostridium butyricum is a Gram-positive obligate anaerobic bacterium. It has a role in regulating the healthy intestinal environment by repairing and regenerating the intestinal epithelium. *C. butyricum* produces short-chain fatty acids, especially butyric acid. It was found that the live *Clostridium butyricum* or lipoteichoic acid extracted from it is quite effective in suppressing the pathogenic growth. Its oral administration can enhance *vibriosis* resistance, superoxide anion production, leucocyte activation, and phagocytosis in rainbow trout, along with stimulation of NK cells, macrophages and improved resistance against various infection in fish (Li et al., 2019).

7.6.3 Nutritional Factors

For the well-being of aquatic organisms, both micro- and macro-nutrients are required. The health of organisms depends on what they eat or what they are fed in aquaculture. The nutritional factors are aimed to provide the aquatic organisms with an improved immune response and protection of tissues from collateral damage.

7.6.3.1 Vitamins

At appropriate concentrations, vitamins help in improving the adaptive as well as innate immunity, organism's performance, and reduce mortality.

7.6.3.1.1 Vitamin C/Ascorbic Acid

An intracellular anti-oxidant micronutrient (Abiri and Vafa, 2020). It is not synthesised in the body so it needs to be supplemented to improve the growth, immunity, wound healing, disease resistance, and lipid metabolism *via* carnitine synthesis (Dawood and Koshio, 2018). It is required for various hydroxylation reactions, thus aiding in removal of waste out of body. It also plays an important role in maintaining the integrity of blood vessels, bone tissue and connective tissue because of its involvement in the production of proline and hydroxyproline (Gasco et al., 2018). Other functions of vitamin C include catecholamine bio-synthesis, conversion of tryptophan and folic acid to serotonin, and metabolically active form, respectively. Deficiency of vitamin C in fish is characterized by reduced growth, skeletal damage, anorexia, fin erosion, fin haemorrhage, distorted gill filaments, and increased mortality. Vitamin C–supplemented diets (250 mg/kg and 500 mg/kg) when fed to pacu for 60 days resulted in an increased number of thrombocytes in a dose-dependent manner. Reduction in the stress response and bioaccumulation has been reported in *Channa punctatus* when fed with vitamin C. Vitamin C can also act as a co-factor in many processes like synthesis of collage, neuromodulation -related cellular functions, and immune and hormone system.

7.6.3.1.2 Vitamin E

Tocopherol, also known as vitamin E, is a family of phenol molecules with antioxidant properties. It can enhance cytokine production, phagocytosis, antibody generation, and improves cytotoxicity (Gombart et al., 2020; Wu and Meydani, 2019). The most well-known effect of this vitamins is the defence it provides against oxidative stress, reactive chemicals, and unsaturated fatty acids contained in biological membranes. Because of its propensity to transfer the phenolic hydrogen to radicals, vitamin E has antioxidant properties (lipid-free). Vitamin E's role in regulating PUFA peroxidation is facilitated by their placement as a structural component of cell membranes (polyunsaturated fatty acids). Signalling within cells, maintenance of membrane homeostasis, and regulation of gene expression are just a few of its many metabolic roles. This vitamin is mostly required under stressful (acute/chronic) conditions. Supplementation of vitamin E has shown to enhance antibody production and serum complement against *Y. ruckeri* infestation in rainbow trout and *Ictalurus punctatus,* respectively (Gasco et al., 2018). Moreover, it has shown to influence the cell mediated and specific immunity against the infection of *Edwardseiella tarda* in Japanese flounder.

7.6.3.1.3 Vitamin B6

Pyridoxine is an essential nutritional factor for fish as it promotes the leukocyte count, albumin level, globulin level, total plasma proteins, macrophage activity, lysozyme activity, and respiratory burst activity of fish (Akhtar and Ciji, 2020). Vitamin B6 has shown to enhance protection against the fungal infection (*Saprolegnia parasitica)* in *Labeo rohita* by improving the total globulin level, lysozyme activity, and respiratory burst activity.

7.6.3.2 Hormones

Hormones play a vital role in the proper function, nutrient intake, enzyme synthesis, and growth of aquatic organisms. These are secretions regulated by feedback mechanisms.

7.6.3.2.1 Prolactin

Prolactin is a single-chain polypeptide hormone generally secreted by the anterior pituitary. Other than anterior pituitary, prolactin is also released from other locations including ovary, placenta, mammary epithelium, skin cells, spleen, adipose tissue and bone marrow (Borba et al., 2019). It has been found to increase the phagocytic activity of macrophages, mitotic activity of leukocytes and affect the NK cells, lymphocytes, and macrophages directly (Villalba et al., 2020).

7.6.3.2.2 Growth Hormone

Growth hormone is a peptide hormone which stimulates the cell reproduction, growth in animals and directly affects the immune competent cells like lymphocytes, macrophages, and natural killer cells. In fish, growth hormone activates the natural killer cells, produces superoxide anions, and also has mitogenic activity on the lymphocytes.

7.6.3.2.3 Thyroxine

Thyroxine helps in the growth, development, and reproduction of fish (Abdollahpour et al., 2019; Abdollahpour et al., 2018; Alinezhad et al., 2020). Supplementation of thyroxin hormone (1-10 μ g BW^{-1}) to the larvae of Nile tilapia indicated larval viability and faster development of fish digestive system.

A combination of growth hormone and thyroxine hormone when provided to the larvae of stripped catfish resulted in a faster development and an increased growth rate.

7.6.4 Animal Derived Immunostimulants

7.6.4.1 Hde

Hde is glycoprotein fraction extracted from the *Haliotis discus hannai* (abalone) extract. When rainbow trout were injected with Hde, it improved phagocytosis and increased survival rate in the fish infected with *V. anguillarum.* Moreover, Hde can enhance the tumour cells death in *in vitro* and inhibits tumour growth under *in vivo* condition (Gong et al., 2019).

7.6.4.2 Ete

Ete is an extract from a marine tunicate, *Ecteinascida turbinate.* Ete has the potential to increase the survival and enhance phagocytosis against infection by *A. hydrophila.*

7.6.4.3 Propolis

Propolis is derived from honeybees and has antitumor, anti-inflammatory, and immunomodulatory properties (Yosri et al., 2021). It can induce antibodies production and immune cells production which involves both acquired and innate responses (Wolska et al., 2019).

7.6.4.4 Polysaccharides

Polysaccharides refer to a group of biological macromolecules that are composed of repetitive residues of monosaccharide polymer linked to each other by glycosidic bonds. They boost the growth performance, immune responses, and resistance to pathogens in fish (Mohan et al., 2019; Wu et al., 2019).

7.6.4.5 Chitin

Chitin is a type of alkaline polysaccharide that is the major constituent of the exoskeleton of insects and crustaceans and is also found in the cell walls of fungi. Its supplementation can enhance the haemolytic complement activity, cytotoxicity, macrophage activity, and respiratory burst activity of leukocytes. It has also shown an increased resistance towards the infections of *A. salmonicida, P. piscicida,* and *V. anguillarum* in brook trout, yellow tail, and rainbow trout, respectively (Kamilya and Khan, 2020).

7.6.4.6 Chitosan

Chitosan is the deacetylation product of chitin. It is a natural polymer that is edible, non-toxic, bio-degradable, and biocompatible. In aquaculture, chitosan improves immunity, promotes growth, purifies water, inhibits the growth of pathogens, and enhances the infection resistance in aquatic organisms.

7.6.5 Algal Derivatives

Algae contain~~s~~ pigments, minerals, proteins, and oils that accelerate the growth, immunomodulatory response, and disease resistance; however, it decreases the lipid content of serum, muscle, and liver in fish (Camacho et al., 2019). It is among the significant bases of sugar polymers (alginate, carrageenan, and agarose) (Xie and Cheong, 2021). Although algae are a rich source of bioactive compounds yet their use as an immunostimulant is poorly studied. Brown algae (*Hizika fusiformis*) and microalgae (*Chlorella sp.*) are widely used in aquaculture. Dietary administration of *Chlorella* (5%, 10%, and 15%) and *Hizika fusuformis* (2%, 4%, and 6%) have been observed to increase~~s~~ the weight of olive flounder, whereas microalgae has been found to down regulate the glutamic oxaloacetic transaminase, serum glucose, and glutamic pyruvic transaminase; and the brown algae developed the *Streptococcus* resistance, indicating that both these algae are safe as additive (Hasan et al., 2019).

Alginate is another polysaccharide (anionic) present in brown algae that originated to have immune-stimulating possessions in fish. N-oxide-quaternary alkaloid fraction isolated from a marine *Chlorophyceae macroalga* has been shown to have immunostimulatory and disease tolerance potential.

7.6.6 Plant-Derived Immunostimulant

Many immunostimulants have limitations like limited efficacy and high cost, which can be solved by using plants as alternatives. Medicinal plants may represent a fresh opportunity in aquaculture as these can act as appetite stimulators, anti-stressor, and growth stimulator against a wide range of pathogens and as replacement of animal protein in fish meal. Various parts of the plant as well as its derivatives possess medicinal properties and can be used in aquaculture for disease prevention by enhancing the immunity and controlling the microbes that are pathogenic in aquaculture spceis (Awad and Awaad, 2017; Hoseinifar et al., 2020; Doan et al. 2020). Compounds like terpenoids, phenols, polyphenols, quinones, flavonoids, lectins, proteoglycans, polypeptides and alkaloids have found to be quite effective alternatives of antibiotics, vaccines and medical drugs (Nair et al., 2019). The fruit pulp of *Phyllanthus emblica* contains large amount of vitamin C and so is considered as an immunostimulant (Mohanan et al., 2020). When the fruit extract of *Phyllanthus emblica* was fed to Nile tilapia at various concentrations (0, 5, 10, 20 and 40 mg kg^{-1}) for 8 weeks, maximum activity was observed at 20 mg/kg, with enhanced the peroxidase activities, skin mucus lysozyme, growth parameters, and exhibited high resistance towards *S. agalactiae* infection (Van Doan et al., 2022). Tulsi, a medicinal plant known for its leaves rich in various water-soluble phenolic compounds, as well as constituents such as caryophyllene, eugenol, and methyl eugenol, demonstrated significant effects when administered to Tilapia (Oreochromis mossambicus) in terms of enhancing protection against Aeromonas hydrophila infection and altering antibody response patterns. Furthermore, dietary exposure of Tulsi extract to fingerlings of Rohu (Labeo rohita) resulted in increased total RBC count, WBC count, globulin levels, superoxide anion production, total plasma immunoglobulin, and lysozyme activity in fish.

7.6.6.1 *Resveratrol*

A natural polyphenol extracted from roots of white hellebore, berries, peanuts, and grapes has shown to behave as a reversible and non-competitive inhibitor of myeloperoxidase (MPO). In a study evaluating the effect of an extract composed of 14 herbs against Aeromonas hydrophila infection in Nile Tilapia, Psidium guajava (at a ratio of 1:4 w/w of commercialized feed) demonstrated the most potent antimicrobial activity. Enhanced total immunoglobulin and serum protein level was evidenced when rainbow trout was fed with the extract of black cumin seed for three weeks. Aloe vera has also exhibited many properties like immunomodulatory, wound healing, antiviral, antifungal, antibacterial, and anti-diabetic (Gao et al., 2019). During the transport of Matrinxã/Sabalo (*Prochilodus lineatus*), the addition of aloe-vera powder to water at 0.02, 0.2, and 2 mg of the powder per litre increased the respiratory activity of leucocytes in a concentration-dependent manner. In another study, infection with heat killed *A. hydrophila* (24 h prior), along with dietary exposure of aloe vera for 10 days before transportation exhibited an improvement in fish immunity response, including serum lysozyme and complement activities plus an enhancement in respiratory burst activity of leukocytes in *Piaractus mesopotamicus*. Raja Rajeswari et al., (2012) reported increased total haemocyte count, phenol oxidase activity, phagocytosis, haemagglutinin activity, bacterial clearance, and growth rate at varying rates in *Fenneropenaeus indicus* by supplementing five herbs including *Zingiber officinale, Tinospora cordifolia, Picrorhiza kurora, Hygrophila spinosa,* and *Acatyphya indica* against *Vibrio harveyi* infection. In a proportional research of the potential of four essential oils including leech lime (*Citrus hystrix*), cinnamon (*Cinnamomum verum*), turmeric (*Curcuma longa*), and lemon grass (*Cymbopogon citratus*) against *Streptococcus iniae* infection in Nile tilapia, cinnamon oil (40 µg/ml) showed the highest inhibitory capacity.

Inclusion (at 0.1 and 1%) of the plant materials (mistletoe, nettle, and ginger) in a fish species diet has been observed to upsurge the extracellular respirational burst action. *Clarias gariepinus,* when fed for 12 weeks with clove basil (*Ocimum gratissimum*), was shown to be inversely proportional to the dose required to produce a dose-dependent increase in intestinal villi height, breadth, and uptake region (Abdel-Tawwab et al., 2018). Additionally, total protein, albumin, and globulin all increased dramatically in fish fed the clove basil extract, while blood glucose and cholesterol both reduced significantly (Abdel-Tawwab et al., 2018; Adeshina et al., 2019).

7.6.7 Synthetic Immunostimulants

Synthetic immunostimulants refer to those products that are chemically synthesised. These include substances like levamisole and quaternary ammonium compounds.

7.6.7.1 *Levamisole*

Levamisole is a synthetic anthelmintic compound that has strong restorative feat on the innate immune system of fish. It is a derivative of imidazothiazole having an imidazole ring and a thiol group (Kumar et al., 2022). In fish, levamisole is used against various skin and gill parasites (Tabari et al., 2020). It mainly acts by increasing the level of cGMP by interacting with thymopoietin receptor sites. Higher concentration of c-AMP induces the augmentation of chemotactic responses and lymphocytic proliferation leading to enhanced phagocytosis, antibody production, and lymphokine production. Immunomodulating responses of oral levamisole treatment on rainbow trout include elevated basophil, eosinophil, monocyte, lysozyme, and phagocytic indices (Amiri and Bahrekazemi, 2017).

Diets containing levamisole have positive effects on the lysozyme concentration with no differences in the growth parameter of *Pseudoplatystoma reticulatum.* In another study where levamisole-containing diets (100 mg/kg) were fed to Pacu (*Piaractus mesopotamicus*) for 15 days and followed by inoculation with *Aeromonas hydrophilia* through injection, an increase in the leukocyte respiratory burst was observed. For treating the monogenean (*Anacanthorus spatulatus, Linguadactyloides, Mymarothecium boegeri,* and *Notozothecium janauachensis*) infestations in *Colossoma macropomum*, therapeutic baths (125 mg/L) of levamisole have been found to be effective (88.2%) (Alves et al., 2019).

7.6.7.2 Quaternary Ammonium Compound (QAC)

QACs are polymeric cationic compounds used in aquaculture because of their limited haemolytic activity and potent antibacterial activity. QACs include some disinfectants like dodecyl dimethyl ammonium chloride, dioctyl dimethyl ammonium chloride and benzalkonium chloride (alkyl benzyl dimethyl ammonium chloride) that are used to improve the water quality by making them bacteria (*Streptococcus agalactiae)* free. QACs also include antiseptic agents such as cetylpyridinium chloride and conditioning agents like hexadecyltrimethylammonium (Droge et al., 2021).

7.7 Efficacy of Immunostimulants

Due to the practice of accelerated fish farming, infectious diseases are a major problem in the aquaculture industry (Assefa and Abunna, 2018). Diseases in fish are an association between host fish, pathogen, and environmental problems like deteriorated water quality or certain other stressors. The disease issue can be overcome through better management practices. As there is always certain mismanagement leading to establishment of pathogens in fish~~es~~ and produce diseases. In order to combat diseases, chemotherapeutics, vaccines, and immunostimulants are used. The benefits of using immuno-stimulants in aquaculture are numerous. They help in up-regulating the innate defence of aquatic organisms, hence making them more prepared towards overcoming the invading pathogens (Jinendiran et al., 2019; Kumar et al., 2022). Immunostimulants are found to be safer than chemotherapeutics and their range of efficacy is more than that of vaccination.

For evaluating the efficacy of immunostimulants, there are two main procedures (i) *in vitro-* measurement of efficiency of humoral and cellular immune mechanisms and (ii) *in vivo* challenge test using pathogens of fish. The parameters for *in-vitro*-based evaluation are complement activation, total leukocyte/ erythrocyte counts, serum lysozyme activity, chemokinesis, chemotaxis, lymphocyte proliferation, and phagocytosis. The range of techniques that are involved in the evaluations varies from inexpensive and simple methods to the use of flow cytometry, immunoassays, and some other biological approaches. Till date, a large number of immunostimulants like fungi, mycobacterium, glucans, polysaccharides, muramyl dipeptide, peptidoglycan, hormones, plant and animal extracts, and nutritional factors have been tested in different fishes for their immune response. They stimulate the complement system, lysozyme, antibodies, and natural killer cells and facilitate the function of phagocytic cells. So, immunostimulants can compensate the limitations of both chemotherapeutics and vaccines.

7.8 Immunostimulation Methods

Immunostimulants are most commonly given orally, rather than by injection or immersion, for the purpose of boosting the body's resistance to disease. The least potent but non-stressful form of administering bulk immunity is through oral administration. Because of this, it is widely accepted as the best method for most types of fish and shrimp. In most research, immunostimulants have been administered orally to Indian main carps. Carp culture immunostimulants have only been administered intraperitoneally in a few experiments. The longer the length of immunostimulant administration (beyond 40 days), however, several research predicts higher results. Carps' immune systems are affected directly and indirectly by the concentration of a drug's dosage. Comparatively, results demonstrate that the optimal dose of immunostimulants has a greater growth rate and survival rate than the higher dose of immunostimulants (Sattanathan et al., 2020; Devi et al., 2019). In addition to this, research on Indian major carps demonstrated that the immune response was directly stimulated in a dose-dependent manner.

7.9 Future Perspectives and Conclusion

Novel nanotechnological approaches (therapeutic agents encapsulated inside nanocarriers) are promising strategies that require more study and research. These nano strategies will provide opportunities to reconsider the compounds for treatment that were once considered toxic. So, coupling the novel aspects with nanotechnology may prove promising in the development of bioactive compounds with accurate disease treatment potential.

In conclusion, successful and ecologically responsible fish farming will not be possible without the immunostimulants' benefits. However, the efficacy of these ingredients is dependent on the dosages, duration, and the method of administration; research is required to establish the ideal strategy to maximize benefits while avoiding immunosuppression. Concentration and dose-dependent stimulatory effects of medicinal plants have been demonstrated in several research, but the mechanism of ligand recognition, extract composition, and activation of fish immune responses is still poorly understood.

REFERENCES

Abbate, Antonio, Stefano Toldo, Carlo Marchetti, Jordana Kron, Benjamin W. Van Tassell, and Charles A. Dinarello. "Interleukin-1 and the inflammasome as therapeutic targets in cardiovascular disease." *Circulation Research* 126, no. 9 (2020): 1260–1280.

Abdel-Tawwab, Mohsen, Ibrahim Adeshina, Adetola Jenyo-Oni, Emmanuel K. Ajani, and Benjami O. Emikpe. "Growth, physiological, antioxidants, and immune response of African catfish, Clarias gariepinus (B.), to dietary clove basil, Ocimum gratissimum, leaf extract and its susceptibility to Listeria monocytogenes infection." *Fish & Shellfish Immunology* 78 (2018): 346–354.

Abdollahpour, Hamed, Bahram Falahatkar, Iraj Efatpanah, and Bahman Meknatkhah. "Effect of thyroxine injection on female growth and reproductive performance of sterlet sturgeon (Acipenser ruthenus)." *Journal of Marine Biology* (2019): 63–78.

Abdollahpour, Hamed, Bahram Falahatkar, Iraj Efatpanah, Bahman Meknatkhah, and Glen Van Der Kraak. "Influence of thyroxine on spawning performance and larval development of Sterlet sturgeon Acipenser ruthenus." *Aquaculture* 497 (2018): 134–139.

Abiri, Behnaz, and Mohammadreza Vafa. "Micronutrients that affect immunosenescence." *Reviews on New Drug Targets in Age-Related Disorders* (2020): 13–31.

Adeshina, Ibrahim, Adetola Jenyo-Oni, Benjamin O. Emikpe, Emmanuel K. Ajani, and Mohsen Abdel-Tawwab. "Stimulatory effect of dietary clove, Eugenia caryophyllata, bud extract on growth performance, nutrient utilization, antioxidant capacity, and tolerance of African catfish, Clarias gariepinus (B.), to Aeromonas hydrophila infection." *Journal of the World Aquaculture Society* 50, no. 2 (2019): 390–405.

Adewumi, A.A. (2018). The impact of nutrition on fish development, growth and health. *International Journal of Scientific and Research Publications (IJSRP)* 8 (6), 147–153.

Akhtar, M. S., and A. Ciji. "Pyridoxine and its biological functions in fish: current knowledge and perspectives in aquaculture." *Reviews in Fisheries Science & Aquaculture* 29, no. 2 (2020): 260–278.

Alinezhad, Soheil, Hamed Abdollahpour, Naghmeh Jafari, and Bahram Falahatkar. "Effects of thyroxine immersion on sterlet sturgeon (Acipenser ruthenus) embryos and larvae: Variations in thyroid hormone levels during development." *Aquaculture* 519 (2020): 734745.

Altan-Bonnet, Grégoire, and Ratnadeep Mukherjee. "Cytokine-mediated communication: a quantitative appraisal of immune complexity." *Nature Reviews Immunology* 19, no. 4 (2019): 205–217.

Alves, Carliane Maria Guimarães, Joziele Neves Nogueira, Isaac Belo Barriga, Joilson Rodrigues Dos Santos, Gracienhe Gomes Santos, and Marcos Tavares-Dias. "Albendazole, levamisole and ivermectin are effective against monogeneans of Colossoma macropomum (Pisces: Serrasalmidae)." *Journal of Fish Diseases* 42, no. 3 (2019): 405–412.

Amiri, Z., and M. Bahrekazemi. "Effect of oral administration of Levamisole, Quil-A and Cinnamon in growth amount, hematological and immune parameters of Marmalade cichlid, Labeotrophus fuelleborni (Ahl, 1926)." *Iranian Journal of Aquatic Animal Health* 3, no. 2 (2017): 86–97.

Anjugam, Mahalingam, Arokiadhas Iswarya, Ashokkumar Sibiya, Chandrabose Selvaraj, Sanjeev Kumar Singh, Marimuthu Govindarajan, Naiyf S. Alharbi *et al.* "Molecular interaction analysis of β-1, 3 glucan binding protein with Bacillus licheniformis and evaluation of its immunostimulant property in Oreochromis mossambicus." *Fish & Shellfish Immunology* 121 (2022): 183–196.

Assefa, Ayalew, and Fufa Abunna. "Maintenance of fish health in aquaculture: review of epidemiological approaches for prevention and control of infectious disease of fish." *Veterinary Medicine International* 2018 (2018): 5432497.

Awad, Elham, and Amani Awaad. "Role of medicinal plants on growth performance and immune status in fish." *Fish & Shellfish Immunology* 67 (2017): 40–54.

Bais, B. "Fish scenario in India with emphasis on Indian major carps." *International International Journal of Avian & Wildlife Biology* 3, no. 6 (2018): 409–411.

Barrow, Dylan R., Abbate, Lauren M., Paquette, Max R., Driban, Jeffrey B., Vincent, Heather K., Newman, Connie, Messier, Stephen P., Ambrose, Kirsten R., & Shultz, Sarah P. (2019). Exercise prescription for weight management in obese adults at risk for osteoarthritis: synthesis from a systematic review. *BMC Musculoskeletal Disorders* 20, 1–9.

Borba, Vânia Vieira, Gisele Zandman-Goddard, and Yehuda Shoenfeld. "Prolactin and autoimmunity: The hormone as an inflammatory cytokine." *Best Practice & Research Clinical Endocrinology & Metabolism* 33, no. 6 (2019): 101324.

Camacho, Franciele, Angela Macedo, and Francisco Malcata. "Potential industrial applications and commercialization of microalgae in the functional food and feed industries: A short review." *Marine Drugs* 17, no. 6 (2019): 312.

Chandan, Nitish Kumar, Rakhi Kumari, and G. M. Siddaiah. "Role of nutraceuticals in fish feed." In *Fish nutrition and its relevance to human health*, pp. 229–243. CRC Press, 2020.

Chansue, N., M. Endo, T. Kono, and M. Sakai. "The stimulation of cytokine-like proteins in Tilapia (Oreochromis niloticus) orally treated with beta-1, 3-glucan." *Asian Fisheries Science* 13, no. 3 (2000): 271–278.

Dahmani, Amina, and Jean-Sébastien Delisle. "TGF-β in T cell biology: implications for cancer immunotherapy." *Cancers* 10, no. 6 (2018): 194.

Dawood, Mahmoud AO, Abd El-Salam Metwally, Mohamed E. El-Sharawy, Ahmed M. Atta, Zizy I. Elbialy, Hany Mr Abdel-Latif, and Bilal Ahamad Paray. "The role of β-glucan in the growth, intestinal morphometry, and immune-related gene and heat shock protein expressions of Nile tilapia (Oreochromis niloticus) under different stocking densities." *Aquaculture* 523 (2020): 735205.

Dawood, Mahmoud AO, and Shunsuke Koshio. "Vitamin C supplementation to optimize growth, health and stress resistance in aquatic animals." *Reviews in Aquaculture* 10, no. 2 (2018): 334–350.

de Oliveira, Carlos AF, Vaclav Vetvicka, and Fábio S. Zanuzzo. "β-Glucan successfully stimulated the immune system in different jawed vertebrate species." *Comparative Immunology, Microbiology and Infectious Diseases* 62 (2019): 1–6.

Devi, Gunapathy, Ramasamy Harikrishnan, Bilal Ahmad Paray, Mohammad K. Al-Sadoon, Seyed Hossein Hoseinifar, and Chellam Balasundaram. "Effects of aloe-emodin on innate immunity, antioxidant and immune cytokines mechanisms in the head kidney leucocytes of Labeo rohita against

Aphanomyces invadans." *Fish & Shellfish Immunology* 87 (2019): 669–678.

Dostert, Catherine, Melanie Grusdat, Elisabeth Letellier, and Dirk Brenner. "The TNF family of ligands and receptors: communication modules in the immune system and beyond." *Physiological Reviews* 99, no. 1 (2019): 115–160.

Droge, Steven TJ, James M. Armitage, Jon A. Arnot, Patrick N. Fitzsimmons, and John W. Nichols. "Biotransformation potential of cationic surfactants in fish assessed with rainbow trout liver S9 fractions." *Environmental Toxicology and Chemistry* 40, no. 11 (2021): 3123–3136.

Ferreira, Vinicius L., Helena HL Borba, Aline de F. Bonetti, L. Leonart, and Roberto Pontarolo. "Cytokines and interferons: types and functions." *Autoantibodies and Cytokines* 13 (2018).

Gallenga, C. E., F. Pandolfi, Al Caraffa, S. K. Kritas, G. Ronconi, E. Toniato, S. Martinotti, and P. Conti. "Interleukin-1 family cytokines and mast cells: activation and inhibition." *Journal of Biological Regulatory Homeostatic Agents* 33, no. 1 (2019): 1–6.

Gan, Zhen, Shan Nan Chen, Bei Huang, Jun Zou, and Pin Nie. "Fish type I and type II interferons: composition, receptor usage, production and function." *Reviews in Aquaculture* 12, no. 2 (2020): 773–804.

Gao, Yan, Kit Ieng Kuok, Ying Jin, and Ruibing Wang. "Biomedical applications of Aloe vera." *Critical Reviews in Food Science and Nutrition* 59, no. supp 1 (2019): S244–S256.

Gasco, Laura, Francesco Gai, Giulia Maricchiolo, Lucrezia Genovese, Sergio Ragonese, Teresa Bottari, and Gabriella Caruso. "Supplementation of vitamins, minerals, enzymes and antioxidants in fish feeds." In *Feeds for the Aquaculture Sector*, pp. 63–103. Springer, Cham, 2018.

Gombart, Adrian F., Adeline Pierre, and Silvia Maggini. "A review of micronutrients and the immune system–working in harmony to reduce the risk of infection." *Nutrients* 12, no. 1 (2020): 236.

Gong, Fang, Mei-Fang Chen, Yuan-Yuan Zhang, Cheng-Yong Li, Chun-Xia Zhou, Peng-Zhi Hong, Sheng-Li Sun, and Zhong-Ji Qian. "A novel peptide from abalone (Haliotis discus hannai) to suppress metastasis and vasculogenic mimicry of tumor cells and enhance anti-tumor effect in vitro." *Marine Drugs* 17, no. 4 (2019): 244.

Guo, Ming, and Chenghua Li. "An overview of cytokine used as adjuvants in fish: current state and future trends." *Reviews in Aquaculture* 13, no. 2 (2021): 996–1014.

Gupta, Sanjay Kumar. "Ameliorative and protective effects of prebiotic, microbial levan in common carp,(Cyprinus carpio) fry under experimental exposure to fipronil." *International Journal of Aquatic Biology* 9, no. 2 (2021): 134–147.

Hansen, Jon Øvrum, Leidy Lagos, Peng Lei, Felipe Eduardo Reveco-Urzua, Byron Morales-Lange, Line Degn Hansen, Marion Schiavone *et al.* "Down-stream processing of baker's yeast (Saccharomyces cerevisiae)–Effect on nutrient digestibility and immune response in Atlantic salmon (Salmo salar)." *Aquaculture* 530 (2021): 735707.

Hasan, Md Tawheed, Won Je Jang, Jong Min Lee, Bong-Joo Lee, Sang Woo Hur, Sang Gu Lim, Kang Woong Kim, Hyon-Sob Han, and In-Soo Kong. "Effects of immunostimulants, prebiotics, probiotics, synbiotics, and potentially immunoreactive feed additives on olive flounder (Paralichthys olivaceus): a review." *Reviews in Fisheries Science & Aquaculture* 27, no. 4 (2019): 417–437.

Hashem, Nada MA, Mai AM El-Son, Ahmed I. Ateya, and Rasha M. Saleh. "Impact of lactoferrin supplementation on oxidative stress, gene expression and immunity dysfunction induced by Aeromonas veronii in Nile tilapia (Oreochromis niloticus)." *Aquaculture Research* 53, no. 6 (2022): 2392–2407.

Hoseinifar, Seyed Hossein, Yun-Zhang Sun, Zhigzhang Zhou, Hien Van Doan, Simon J. Davies, and R. Harikrishnan. "Boosting immune function and disease bio-control through environment-friendly and sustainable approaches in finfish aquaculture: herbal therapy scenarios." *Reviews in Fisheries Science & Aquaculture* 28, no. 3 (2020): 303–321.

Jahandideh, Bahman, Mehdi Derakhshani, Hossein Abbaszadeh, Ali Akbar Movassaghpour, Amir Mehdizadeh, Mehdi Talebi, and Mehdi Yousefi. "The pro-Inflammatory cytokines effects on mobilization, self-renewal and differentiation of hematopoietic stem cells." *Human Immunology* 81, no. 5 (2020): 206–217.

Jinendiran, Sekar, Abel Arul Nathan, Dharmaraj Ramesh, Baskaralingam Vaseeharan, and Natesan Sivakumar. "Modulation of innate immunity, expression of cytokine genes and disease resistance against Aeromonas hydrophila infection in goldfish (Carassius auratus) by dietary supplementation with Exiguobacterium acetylicum S01." *Fish & Shellfish Immunology* 84 (2019): 458–469.

Kaleem, Oliver, and Abudou-Fadel Bio Singou Sabi. "Overview of aquaculture systems in Egypt and Nigeria, prospects, potentials, and constraints." *Aquaculture and Fisheries* 6, no. 6 (2021): 535–547.

Kamilya, Dibyendu, and Md Idrish Raja Khan. "Chitin and chitosan as promising immunostimulant for aquaculture." In *Handbook of Chitin and Chitosan*, pp. 761–771. Elsevier, 2020.

Kany, Shinwan, Jan Tilmann Vollrath, and Borna Relja. "Cytokines in inflammatory disease." *International Journal of Molecular Sciences* 20, no. 23 (2019): 6008.

Kumar, Sudhir, Abhay Kumar Choubey, and Praveen Kumar Srivastava. "The effects of dietary immunostimulants on the innate immune response of Indian major carp: A review." *Fish & Shellfish Immunology* (2022).

Li, Tao, Fei Ke, Jian-Fang Gui, Li Zhou, Xiao-Juan Zhang, and Qi-Ya Zhang. "Protective effect of Clostridium butyricum against Carassius auratus herpesvirus in gibel carp." *Aquaculture International* 27, no. 3 (2019): 905–914.

Lin, Jian-Xin, and Warren J. Leonard. "The common cytokine receptor γ chain family of cytokines." *Cold Spring Harbor Perspectives in Biology* 10, no. 9 (2018): a028449.

Mohan, Kannan, Samuthirapandian Ravichandran, Thirunavukkarasu Muralisankar, Venkatachalam Uthayakumar, Ramachandran Chandirasekar, Palaniappan Seedevi, and Durairaj Karthick Rajan. "Potential uses of fungal polysaccharides as immunostimulants in fish and shrimp aquaculture: a review." *Aquaculture* 500 (2019): 250–263.

Mohanan, Salini, B. Nidhin, M. Harikrishnan, and M. P. Prabhakaran. "Effect of the Natural Herb Amla (Phyllanthus emblica) on Growth, Survival and Disease Resistance in Macrobrachium rosenbergii Juveniles." *Fishery Technology* 57, no. 4 (2020).

Mondal, Hemanta Kumar, Usha Jyoti Maji, Sriprakash Mohanty, Pramoda Kumar Sahoo, and Nikhil Kumar Maiti. "Alteration of gut microbiota composition and function of Indian major carp, rohu (Labeo rohita) infected with Argulus siamensis." *Microbial Pathogenesis* 164 (2022): 105420.

Morris, Alyssa L., and Shamim S. Mohiuddin. "Biochemistry, nutrients." (2020).

Nair, A., D. Chattopadhyay, and B. Saha. "New Look to Phytomedicine." (2019): 435–499.

Pastoret, P. P., P. Griebel, H. Bazin, & A. Govaerts (Eds.). (1998). Handbook of Vertebrate Immunology. Academic Press.

Raja Rajeswari, P., S. Velmurugan, M. Michael Babu, S. Albin Dhas, K. Kesavan, and T. Citarasu. "A study on the influence of selected Indian herbal active principles on enhancing the immune system in Fenneropenaeus indicus against Vibrio harveyi infection." *Aquaculture International* 20, no. 5 (2012): 1009–1020.

Rascón-Cruz, Quintín, Edward A. Espinoza-Sánchez, Tania S. Siqueiros-Cendón, Sayuri I. Nakamura-Bencomo, Sigifredo Arévalo-Gallegos, and Blanca F. Iglesias-Figueroa. "Lactoferrin: A glycoprotein involved in immunomodulation, anticancer, and antimicrobial processes." *Molecules* 26, no. 1 (2021): 205.

Sakai, Masahiro, Jun-ichi Hikima, and Tomoya Kono. "Fish cytokines: current research and applications." *Fisheries Science* 87, no. 1 (2021): 1–9.

Sattanathan, Govindharajan, Thanapal Palanisamy, Swaminathan Padmapriya, Vijaya Anand Arumugam, Sungkwon Park, In Ho Kim, and Balamuralikrishnan Balasubramanian. "Influences of dietary inclusion of algae Chaetomorpha aerea enhanced growth performance, immunity, haematological response and disease resistance of Labeo rohita challenged with Aeromonas hydrophila." *Aquaculture Reports* 17 (2020): 100353.

Secombes, Christopher J. "Cytokines and Immunity." In *Principles of Fish Immunology*, pp. 301–353. Springer, Cham, 2022.

Sun, Baiming, Yang Lei, Zhenjie Cao, Yongcan Zhou, Yun Sun, Ying Wu, Shifeng Wang, Weiliang Guo, and Chunsheng Liu. "TroCCL4, a CC chemokine of Trachinotus ovatus, is involved in the antimicrobial immune response." *Fish & Shellfish Immunology* 86 (2019): 525–535.

Tabari, Mohaddeseh A., Mohammad Reza Youssefi, Seyed Mehdi Hosseinifard, Ali Akbar Moghaddamnia, Sohrab Kazemi, NadAli Yousefi Sadati, Azadeh Jalali Mothahari, and Mario Giorgi. "Pharmacokinetics of levamisole after intramuscular and oral administrations to Caspian salmon (Salmo trutta caspius)." *Journal of Veterinary Pharmacology and Therapeutics* 43, no. 3 (2020): 276–281.

Urbinati, Elisabeth Criscuolo, Fábio Sabbadin Zanuzzo, and Jaqueline Dalbello Biller. "Stress and immune system in fish." In *Biology and physiology of freshwater neotropical fish*, pp. 93–114. Academic Press, 2020.

Van Doan, Hien, Chompunut Lumsangkul, Korawan Sringarm, Seyed Hossein Hoseinifar, Mahmoud AO Dawood, Ehab El-Haroun, Ramasamy Harikrishnan, Sanchai Jaturasitha, and Marina Paolucci. "Impacts of Amla (Phyllanthus emblica) fruit extract on growth, skin mucosal and serum immunities, and disease resistance of Nile tilapia (Oreochromis niloticus) raised under biofloc system." *Aquaculture Reports* 22 (2022): 100953.

Van Doan, Hien, Hoseinifar, Seyed Hossein, Hung, Tran Quang, Lumsangkul, Chompunut, Jaturasitha, Sanchai, Ehab El-Haroun, & Paolucci, Marina. Dietary inclusion of chestnut (Castanea sativa) polyphenols to Nile tilapia reared in biofloc technology: Impacts on growth, immunity, and disease resistance against Streptococcus agalactiae. *Fish & Shellfish Immunology*, 105 (2020): 319–326.

Vanderzwalmen, Myriam, Lewis Eaton, Carrie Mullen, Fiona Henriquez, Peter Carey, Donna Snellgrove, and Katherine A. Sloman. "The use of feed and water additives for live fish transport." *Reviews in Aquaculture* 11, no. 1 (2019): 263–278.

Villalba, Melina, Gabriel Gómez, Lidia Torres, Nicolas Maldonado, Constanza Espiñeira, Gardenia Payne, Luis Vargas-Chacoff, Jaime Figueroa, Alejandro Yáñez, and Víctor H. Olavarría. "Prolactin peptide (pPRL) induces anti-prolactin antibodies, ROS and cortisol but suppresses specific immune responses in rainbow trout." *Molecular Immunology* 127 (2020): 87–94.

Wang, Yongjie, Mingxue Che, Jingguo Xin, Zhi Zheng, Jiangbi Li, and Shaokun Zhang. "The role of IL-1β and TNF-α in intervertebral disc degeneration." *Biomedicine & Pharmacotherapy* 131 (2020): 110660.

Wieczorek, Marek, Esam T. Abualrous, Jana Sticht, Miguel Álvaro-Benito, Sebastian Stolzenberg, Frank Noé, and Christian Freund. "Major histocompatibility complex (MHC) class I and MHC class II proteins: conformational plasticity in antigen presentation." *Frontiers in Immunology* 8 (2017): 292.

Wolska, K., A. Gorska, K. Antosik, and K. Lugowska. "Immunomodulatory effects of propolis and its components on basic immune cell functions." *Indian Journal of Pharmaceutical Sciences* 81, no. 4 (2019): 575–588.

Wu, Chun, Jinfeng Shan, Junchang Feng, Junli Wang, Chaobin Qin, Guoxing Nie, and Chenlong Ding. "Effects of dietary Radix Rehmanniae Preparata polysaccharides on the growth performance, immune response and disease resistance of Luciobarbus capito." *Fish & Shellfish Immunology* 89 (2019): 641–646.

Wu, Dayong, and Simin Nikbin Meydani. "Vitamin E, immune function, and protection against infection." In *Vitamin E in Human Health*, pp. 371–384. Humana Press, Cham, 2019.

Wu, Liting, Zhendong Qin, Haipeng Liu, Li Lin, Jianmin Ye, and Jun Li. "Recent advances on phagocytic B cells in teleost fish." *Frontiers in Immunology* 11 (2020): 824.

Xiao, Yingping, Lintian Yu, Guohong Gui, Yujie Gong, Xueting Wen, Wenrui Xia, Hua Yang, and Long Zhang. "Molecular cloning and expression analysis of interleukin-8 and-10 in yellow catfish and in response to bacterial pathogen infection." *BioMed Research International* 2019 (2019).

Xie, Xu-Ting, and Kit-Leong Cheong. "Recent advances in marine algae oligosaccharides: Structure, analysis, and potential prebiotic activities." *Critical Reviews in Food Science and Nutrition* (2021): 1–16.

Yamamoto, Fernando Y., Sergio Castillo, Clement R. de Cruz, Kequan Chen, Michael E. Hume, and Delbert M. Gatlin III. "Synergistic effects of the β-1, 3 glucan paramylon and vitamin C on immunological responses of hybrid striped bass (Morone chrysops× M. saxatilis) were pronounced in vitro but more moderate in vivo." *Aquaculture* 526 (2020): 735394.

Yosri, Nermeen, Aida A. Abd El-Wahed, Reem Ghonaim, Omar M. Khattab, Aya Sabry, Mahmoud AA Ibrahim, Mahmoud F. Moustafa *et al.* "Anti-viral and immunomodulatory properties of propolis: Chemical diversity, pharmacological properties, preclinical and clinical applications, and in silico potential against SARS-CoV-2." *Foods* 10, no. 8 (2021): 1776.

Zhu, Wentao, and Jianguo Su. "Immune functions of phagocytic blood cells in teleost." *Reviews in Aquaculture* 14, no. 2 (2022): 630–646.

8

Application of Immunostimulants for Aquaculture Health Management

Femi John Fawole[1], Shamna Nazeemashahul[2], Thongam Ibemcha Chanu[3], Arun Sharma[4], Gbadamosi Oluyemi Kazeem[5], S. Ferosekhan[6], and Tejaswini Kinnera[2]

[1]*Fish Nutrition and Biochemistry Unit, Department of Aquaculture and Fisheries, University of Ilorin, Ilorin, Nigeria*

[2]*Fish Nutrition, Biochemistry and Physiology Division, ICAR – Central Institute of Fisheries Education, Mumbai, Maharashtra, India*

[3]*Aquaculture Division, ICAR-Central Institute of Fisheries Education, Mumbai, Maharashtra, India*

[4]*Aquatic Animal Health Management Division, ICAR-Central Institute of Fisheries Education, Mumbai, Maharashtra, India*

[5]*Department of Fisheries and Aquaculture Technology, Federal University of Technology, Akure, Ondo State, Nigeria*

[6]*Department of Aquaculture, ICAR-Central Institute of Freshwater Aquaculture, Bhubaneswar, Odisha, India*

8.1 Introduction

Aquaculture is the potent sector to achieve a blue economy for a fast-growing global population. To meet the protein needs of the ever-increasing population, intensification of aquaculture production systems is one of the keys. However, problems associated with intensification adversely affect fish in terms of disease resistance and immune responses. The mushrooming intensive aquaculture systems in most developing countries is challenged by the disease outbreak caused by the various pathogenic organisms. To overcome this challenge, farmers are using more and more chemical disinfectants, antibiotics, and vaccines. Antibiotics are further constraints due to the developing of antibiotic-resistant bacteria[1,2] and antimicrobial residual in the carcasses that is harmful for the consumer health status. Also, there is not adequate vaccines to cover all infectious diseases in aquaculture sector and in some cases vaccination towards some economically important infectious diseases warranted further investigations. Immunostimulants, as they are known, comprise a group of natural or synthetic compounds that stimulate the innate immune response of target animals, and increasing disease resistance. Different types of immunostimulants have been discovered and used in aquaculture from various sources including plant and animal extracts, microbial derivatives, and antimicrobial components[3] with specific mechanisms of action such as anti-microbial, anti-fungal, anti-parasite, and anti-stress. Immunomodulators can modulate or change the response of the immune system. They help modulate or normalize the immune system by stimulating or suppressing the animal immune system.[4] Nowadays, the role of immunostimulants is considered very important and is gaining popularity as an alternative to antibiotics in aquaculture.[5] Common immunostimulants used in aquaculture include Levamisole, β-glucan, peptidoglucan, lipopolysaccharides from bacteria, chitin, chitosan, vitamins C and E, hormones, and nucleic acids. This chapter aims to discuss the types of immunomodulators and future perspectives of immunomodulators in aquaculture.

8.2 Immunomodulators

Immunomodulators are substances that change the immune response by increasing or decreasing the ability of the immune system. As a result, production of antibodies or cells get sensitized and react with the materials that initiated their production. Immunomodulators can be any substance like polyphenols, corticosteroids, tannins, and immunoglobulins. Immunomodulators are naturally present in the body of all organisms. However, the pharmacological preparations and nutraceuticals are the main source of immunostimulants used in aquaculture sector. Both terrestrial and aquatic plants are rich in immunostimulants.

DOI: 10.1201/9781003361183-10

8.3 Immunomodulation in Fish

Immunological studies showed that the invasion of pathogens and spread of infectious diseases in aquaculture systems are mainly due to exposure to stressors. The experimental and clinical immunological trials revealed that varying environmental conditions can induce stress conditions in animals which suppress their immune systems. Fish are able to respond to stress conditions by changing their immunological, physiological, and metabolic conditions to confer protection against the pathogens. A reduction in the synthesis process and diverting energy for vasodialation, antioxidant defence, and immune responses are commonly observed in stress-exposed fish. The understanding of components of immune system and their functions, especially the specific functions of nonspecific and specific immune systems, clear the way for interventions. The action of natural or synthetic immune-modulators can be easily understood if the mechanisms of immune-modulations in fish are known.

8.4 Fish Immune System

The biological structures and processes that protect an organism against pathogens, foreign materials, or diseases are considered as the immune system. A healthy immune system identifies pathogens like bacteria, viruses, parasites, or chemical substances and distinguishes them from its own healthy tissues. Though the immune system is divided into innate (nonspecific) and acquired (specific) systems, it has been observed that the action of these is combinational. Generally, the adaptive immune response succeeds the innate response.

8.4.1 Innate Immune System in Fish

The innate immune system is the first level of immune defence in fish, and it targets toxins produced by foreign cells that successfully enter the body of an organism. Pattern recognition is a mode of identification in innate immune response, which has a broad sensitivity.[6] Injured cells, or any alarm signals from cells due to stress, damage, or injury are even recognized by the same receptors.[7] Hence, there is very less specificity in this system, and their response is in a generic way.[8] However, this system is dominant with less or no memory.[9]

8.4.2 Adaptive Immune System in Fish

The adaptive immune system in fish bears specificity and are considered as advanced than innate system. The cellular and humoral responses are the major components of adaptive immune system. Like other higher vertebrates, the adaptive responses of fish also have memory;[10] however, as they are cold-blooded animals and the reactions change with changes in environmental temperature.[11,12] Fish pathogens include various types of organisms like viral particles, parasites, fungus, and bacteria and the host organism gets stimulated by contact with these microbes. Once these microbes or antigens enter the bloodstream of the host, they bind with lymphocytes. The white blood cell, lymphocyte, undergoes a conformational change as a result of antigen interaction and produce antibodies against that pathogen/foreign particle. These antibodies will be circulated in the bloodstream and neutralise the antigens present in it by antigen-antibody complex production. The lymphocytes present in fish have two variants: B-lymphocytes (B-cells) and T-lymphocytes (T-cells). The thymus gland produces the T-cells and kidney act a source of the B-cells. The joint action of T- and B-cells gives adaptive response in fish when exposed to antigens. The B-cells have the capacity to produce memory by producing antibody generating plasma cells and cells that can produce plasma cells on subsequent exposure to antigens. Three types of cellular immunity is provided by T-cells through three different cells, namely the T-helper cells (TH1) that provide supports to phagocytes, TH2 cells that support the division and differentiation of B cells, and T-cytotoxic cells that destroy viruses and infected cells.[13]

8.5 Classification of Immunomodulators

The immune system is an interwoven network of various biochemical reactions which defend against the invading pathogens. Immuno-modulation using natural and chemical substances is an age-old practice even described in Ayurveda and Chinese medical records.[14] There are several compounds that can either stimulate or suppress the immune system of fish (Figure 8.1). Immunostimulants are envisaged to stimulate the resistance to pathogens or foreign particles which invade the organism.

8.5.1 Immunosuppressants

Immunosuppressants can control the pathological reactions in auto-immune disorders by suppressing the immune response. Very often these compounds are used for preventing graft rejections or treating autoimmune disorders. Suppressor cells, a subset of T lymphocytes that exist in the immune system, help in regulating the antigen-driven responses and enhance the tolerance to self-antigens. An inbuilt self-regulating mechanism of the immune system modulates these reactions, especially in cells and tissues. In addition to its own immune suppression, many situations may cause unwanted immunosuppression and may lead to disease outbreaks. Several external factors that cause this suppression include the following.

8.5.1.1 Stress

Stress can be defined as "the nonspecific response of the body to any demand made upon it".[15] Stress can originate from improper diet and feeding schemes, poor water quality, handling, crowding, or confinement. Exposure to stress depresses certain cellular and humoral immune systems; hence, lowering immunity. On stress exposure, a cascade of neuroendocrine changes occurs that elevate the level of the steroid hormone, cortisol, in circulation. Cortisol and other stress hormones such as catecholamines reduce the count and distribution of circulating leukocytes and antibody-producing cells and depress macrophage activity in various parts of the body.

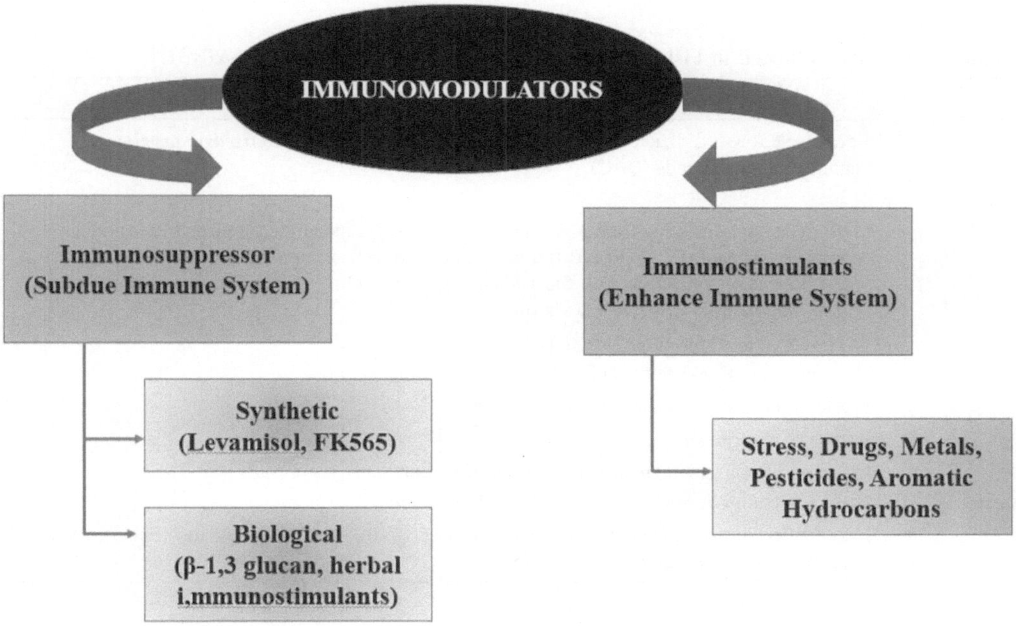

FIGURE 8.1 Classification of immunomodulators.

8.5.1.2 Metals

Metals at a higher required dosage also act as immune suppressing agents and lower the activity of specific and non-specific systems. Examples include mercury, cadmium, and zinc.

8.5.1.3 Aromatic Hydrocarbons

Polychlorinated biphenyls (PCBs) are known for reducing antibody-producing cells and increase disease susceptibility.

8.5.1.4 Pesticides

Pesticides such as endrin, malathion, dichlorodiphenyltrichloroethane (DDT), bayluscide, and tributylin have exerted their immunosuppressive nature in terms of reducing the number of lymphocytes and the activity of phagocytes, inducing necrosis of the thymus.

8.5.1.5 Drugs

An antibiotic like oxytetracycline, used in aquaculture to treat bacterial diseases, can have an immunosuppressive property and reduce the number of antibody-producing cells.

8.5.2 Immunostimulants

An immunostimulant is "anything that stimulates the defence mechanisms or immune response,[16] thus subjecting the animal more resistant to diseases". The definition includes "chemical, drug, stressor, or action". These agents are widely used to improve immune function and stabilize improved immune status. They are used in aquaculture as a prophylactic treatment for seasonal or endemic infections or treatment for latent or sublethal pathogens or conquered the immunosuppressive effects due to stressors.[17] The use of immunostimulants for the prevention of disease in fish is considered as an attractive tool and promising area in the field of aquaculture.[16,18] To confer high protection, the immunostimulants have a role in facilitating phagocytosis and bactericidal activities and stimulating the activity of natural killer cells, the complement system, lysozyme activity, and antibody responses in fish and shellfish, which confer enhanced protection from infectious diseases.[19] Many substances such as glucan, chitin, prolactin, lactoferrin, and levamisole, and nutrients like vitamin B and C, and a few hormones (Table 8.1) have been studied as prospective immunostimulants for fish.[17,20,21] All these compounds are found to influence non-specific immune system, especially activities of phagocytic cell, lysozyme, and natural killer cells. In farmed fish, a few studies showed that complement levels and total immunoglobulin (I_g) levels are also enhanced when supplemented with immunostimulants. In addition, phytobiotics have great promise in curing aquatic animals.

The use of chemicals to control diseases in aquaculture systems brings lots of issues like chemical bioaccumulation, antibiotic resistance, trophic level carries over, and aquatic pollution. Moreover, most of these chemicals are expensive and created immune suppression to aquatic animals. The widespread use of antibiotics leads to development of antibiotic resistant bacteria, immunosuppression, and destabilisation of beneficial bacterial populations. In addition, the application of antibiotics and other chemicals to aquaculture farms is quite expensive; hence, from the ecological point of view, it is undesirable to use chemicals and antibiotics in aquaculture sector. The food and agriculture organization of the United Nations (FAO) encourage the development of affordable vaccines, the use of immune modulators, and integrated systems for protecting the aquatic environments and controlling disease outbreaks in aquaculture.[22]

Immunostimulants play a dual role, both as a prophylactic agent and also as an alarm molecule that activate the immune system.[32] The immune systems of fish are mainly dependent

TABLE 8.1
The Types of Immunostimulants Evaluated in Fish and Shrimp (Modified from Barman et al., 2013)[23]

Type	Substances/Products/Compounds	References
Synthetic chemical	Levamisole, FK-565 (Lactoyltetrapeptide from *Streptomyces olvaceogriseus*), quaternary ammonium compounds (QAC).	24
Biological substances		
Bacterial derivatives	MDP (Muramyl dipeptide from *Mycobacterium* species), Lipopolysaccharide (LPS), Freund's complete adjuvant (FCA), EF203 (fermented egg product), Peptidoglucan (from *Brevibacterium lactofermentum* and *Vibrio* sp.), *Clostridium butyricum* cells, *Achromobacter stenohalis*cells, *Vibrio anguillarum* cells (*Vibrio* vaccine)	25
Yeast	derivatives β-1, 3 glucan, β-1, 6 glucan	26
Nutritional factors	Vitamin C and E, n-3 fatty acid	27
Hormones	Prolactin, Growth hormone, tri-iodothyronine	28
Cytokines	Interferon, Interleukin	25
Polysaccharides	Chitosan, Chitin, *Schizophyllan*, Oligosaccharide, Lentinan	29
Animal and plant extracts	Hde (Abalone), *Glycyrrhizin*, Ete (tunicate) Firefly squid	30
Others	Lactoferrin, Quil A, *Spirulina*, *Achyranthes aspera* (Herb), Mucorcircinelloides (Fungi), Quillaja, Saponin (Soap tree), Soybean protein	31

on non-specific defence mechanisms compared to mammals. Therefore, the immunostimulants in fish play a crucial role in the health management strategies of aquatic animals. In the current scenario, there are more than 20 different compounds that are used as immunostimulants, adjuvants, and fish vaccine carriers, including levamisole, lipopolysaccharides, glucan, and vitamins C and E. Glucan is one of the most promising ones. It is the immunostimulant that has been more studied for evaluation of nonspecific defence mechanisms of aquatic animals. For instance, administration of glucan in several fish species such as common carp, Atlantic salmon, rainbow trout, yellow tail, and African catfish resulted in an enhanced disease resistance against bacterial pathogens such as *Vibrio anguillarum*, *Aeromonas salmonicida*, *A. hydrophila*, and *Yersinia ruckeri*.[33] Administration of β-1,3 glucan in rainbow trout, Atlantic salmon, catfish, and carp improve nonspecific cellular defence mechanisms of fish by raising the number of phagocytes and the bactericidal activity of fish macrophages.[21]

Vitamin C is one of the most studied vitamins, especially in relation to its immunomodulatory role, and it has been widely used in aquaculture as a functional feed additive to prevent nutritionally related diseases such as cracked-skull disease in African catfish or serve as an antioxidant to prevent the incidence of oxidative stress. Vitamin C has been established to confer immunocompetence in both healthy and immunosuppressed fish. Sahoo and Mukherjee (1999)[34] exhibited that dietary vitamin C could be used to restore immunocompetency in fish exposed to aflatoxin B1 (AFB1) by enhancing lysozyme and bactericidal activities, neutrophil, and phagocytic ratio, and showed improved resistance to *A. hydrophila* infection. Liu et al. (2008)[35] reported that use of dietary vitamin C in stressed Chinese longsnout catfish can alleviates the effects of chronic stress. Both studies emphasised that the inclusion of vitamin C reversed the immunosuppressive action of AFB1 and ammonia and restored the health status of the fish. Recently, Ibrahim et al. (2020)[36] reported that vitamin C supplementation stimulated the activities of lysozyme, nitric oxide level, and phagocytic percentage and enhanced the survival of the fish after being challenged with virulent *A. sobria*. Other studies have reported similar results[37–40] and it was clearly noted that the effectiveness of vitamin C could be related to its ability to stimulate neutrophil migration to the site of infection, increase phagocytosis, improve oxidant production, and facilitate microbial killing activity.

Nucleotides and their metabolites play a key role in numerous essential physiological functions, like mediating energy metabolisms, genetic information encoding, and signal transduction. Sakai et al. (2001)[41] reported the beneficial influences of oral administration of nucleotides in Atlantic salmon, coho salmon, rainbow trout, common carp, hybrid tilapia, and hybrid striped bass in terms of immune functions or resistance to disease. The diet containing yeast ribonucleic acid administered to *Labeo rohita* juveniles at 0.4% improves the phagocyte respiratory burst and protection against *A. hydrophila* challenge.[42] Kajita et al. (1990)[43] observed that levamisole enhanced nonspecific immune activity and resistance to an experimental challenge with pathogenic bacteria in carp and rainbow trout. Whole microorganisms, chitin particles, lactoferrin, sodium alginates, and dietary carbohydrates also demonstrated an encouraging result as an immunostimulant in fish and shellfish health management.[44]

Several phytobiotics with immunomodulatory properties have also been tested in fish and are reported to improve aquatic animal immunity and resistance to diseases. Plants have variants of secondary metabolites that serve as a natural defense against phytophagous insects and herbivores. However, at low concentrations, these inherent bioactive substances have medicinal functions and are used in folklore medicine to treat diseases.[45] Phytobased prophylactic strategies are becoming popular in aquafarming for modulation of the innate immune system, conferring immunocompetency in fish and increasing resistance against diseases or stressors. They have been shown scientifically and practically to be environmentally friendly and sustainable. In aquatic animals, herbs and herbal products can elicit an immune response through stimulating the development and migration of immune-related cells. Interestingly, the antioxidant and antibacterial capacities of phytobiotics are already known[46–49] and have been reported to improve resistance against pathogens in aquatic species.[50–53] Specifically, modulatory effects of extracts of *Psidium guajava* (PG) and *Phyllanthus amarus* (PA) on cellular and humoral immune responses were reported in striped catfish, *Pangasianodon hyphophthalmus*. The study showed that both the PG and PA modulate the activities of lysozyme, alternative complement and respiratory burst, nitric oxide synthase, total immunoglobulin, and improve fish survival following *Edwardsiella ictaluri* infection.[54] Furthermore, they went deeper to examine the impacts of the extract on the pathways related to immune responses and antioxidants and noted that the combination of both extracts synergistically up-regulates the expression of proteins involved in leukocyte trans-endothelial migration, NK cell-mediated cytotoxicity, antigen processing and presentation, glutathione metabolism, oxidative phosphorylation, NOD-like receptor signalling, and apoptosis pathways.[54] Respiratory burst activity is dependent on the production of reactive oxygen species (ROS) by phagocytes during the respiratory burst process,[55,56] and activated macrophages produce lysozyme. Hence, the higher survival recorded in striped catfish fed PA and PG could result from the activation of these immune cells in engulfing and destroying the injected bacteria. Hoseinifar et al. (2020)[57] reported similar responses in *Cyprinus carpio* in which *Adiantum capillus-veneris* leaf powder was used as a phyto-immunostimulant to enhance carp immunity. The study showed that the tested leave powder at 1–2% remarkably increases serum and mucosal immunities by enhancing the lysozyme, total immunoglobulin, superoxide dismutase, and serum bactericidal activities against *Aeromonas hydrophila* infection. Thus, it is clear that using a phyto-based immunomodulators in fish diets would help to ensure immunocompetency in fish and enhance resistance against pathogens. Table 8.2 summarizes some immunomodulators used in aquaculture and their biological functions.

TABLE 8.2

Immunomodulators Used in Aquaculture and Their Biological Functions

Immunomodulator	Aquatic Species	Conditions	Main Biological Effects on Immune Response	References
β-glucan	*Oncorhynchus mykiss*	Normal experimental conditions (NEC)	↑ WBC, lysozyme, neutrophil, IgM, C3, C4.	58
	Oreochromis niloticus	Crowding stress	↑TP, lysozyme, PA, PI, IFN-γ, TNF-α, IL-1β ↓GLU, HSP70	59
Vitamin C	*Oreochromis niloticus*	NEC	↑ lysozyme, phagocytic activity of leucocytes, IELs, goblet cells, IL-1β	60
WS root extract + vitamin C (1:1)	*Labeo rohita*	Low pH and waterborne iron toxicity	↑RBA, WBC, TI, lysozyme, TP. ↓GLU, ALP, ALT, AST	61
Chitosan	*Oreochromis niloticus*	Lead toxicity	↑ PA, lysozyme, SOD, CAT, GPx	62
	Labeo rohita	*Aeromonas hydrophila* challenge	↑RBA, lysozyme, CAT, myeloperoxidase, high relative percentage survival	63
Mannan oligosaccharides	*Sparus aurata*	Thermal stress	**Before stress** ↑WBC, lymphocytes, monocytes, TI, IL-1β, TNF-α. **After stress** ↑ monocytes, lysozyme ↓Cortisol, IL-1β, TNF-α	64
Fructooligosaccharide	*Rutilus rutilus*	NEC and salinity stress challenge	↑Ig levels, lysozyme activity, ACH50, survival rate.	65
	Cyprinus carpio	NEC and salinity stress challenge	↑WBC, respiratory burst activity, modulated cultivable autochthonous gut microbiota levels, resistance to salinity stress	66
Astaxanthin	*Litopenaeus vannamei*	*Vibrio alginolyticus* challenge	↑ serum T-SOD, lysozyme, AKP. ↑↑ antimicrobial peptide gene: (*Cru, Pen-3α); SOD, LZM,* LGBP ↓*proPO,* mortality	67
Astaxanthin-rich *Haematococcus pluvialis* powder	*Plectropomus leopardus*	*Vibrio harveyi* challenge	↑Serum SOD, CAT, GPx, ACP, lysozyme, C3, C4. ↑↑*sod-1, sod-2, CAT, acp6, akp, lz-c, IgM, c3, c4-b.* High survival rate	68
Fucoidan	*Clarias gariepinus*	NEC and cadmium toxicity	**NEC group** ↑oxidative burst, PA (%), LTI, lysozyme, NO, bactericidal activity. **Cadmium toxicity group** oxidative burst, PA (%), LTI, lysozyme, NO, absolute lymphocyte count, bactericidal activity, and higher survival.	69
Polysaccharide *Agaricus bisporus*	*Procambarus clarkii*	*Aeromonas hydrophila*	↑hemolymph ACP, AKP, T-SOD, lysozyme, survival rate. ↑↑*gpx, lec, HSP70* mRNA, *ctsb, alf, mtMnsod expression* ↓mortality	70

(*Continued*)

TABLE 8.2 (Continued)

Immunomodulators Used in Aquaculture and Their Biological Functions

Immunomodulator	Aquatic Species	Conditions	Main Biological Effects on Immune Response	References
Houttuynia cordata leaf extract	*Labeo rohita*	NEC	↑TP, WBC, haematocrit, RBA, lysozyme ↑↑IFN-γ, TNF-α.	71
Tea polyphenols	Hybrid grouper (*Epinephelus fuscoguttatus* ♀ × *E. lanceolatus* ♂)	NEC	↑ACP, AKP, lysozyme, IgM, CAT, T-SOD, GPx. *Liver* ↑↑*sod, cat, gpx, nrf2, keap1, hsp70, hsp90, myd88, tnfα.* ↓↓*tlr22, il8, il1β, il10 expression* *Head kidney* ↑↑*myd88, il1β, tnfα, il6.* ↓↓*tlr22, il10 expression*	72
Turmeric powder	*Oreochromis niloticus*	NEC	↑Lysozyme, NO, interleukin 4 (IL 4), WBC, lymphocyte count, improvement in lymphocyte proliferation in response to PWM which is a potent B-cell stimulant	73

Increase (↑); decrease (↓); upregulation(↑↑); downregulation (↓↓); IELs, intraepithelial lymphocytes; IL-1β, interleukin-1beta; POD, peroxidase; TP, total protein; T-SOD, total superoxide dismutase; CAT, catalase; IgM, immunoglobulin M; WBC, white blood cell; PA, phagocytic activity; PI, phagocytic index; TNF-α, tumor necrosis factor-alpha; IFN-γ, interferon-gamma; HSP, heat shock protein; WS, *Withania somnifera;* RBA, respiratory burst activity; TI, total immunoglobulin; GLU, glucose, ALT, alanine transferase, AST, aspartate transferase; ALP, alkaline phosphatase; C, Complement component; ACH50, alternative complement activity; Cru, Crustin gene; proPO, Prophenoloxidase gene; *LZM,* Lysozyme gene; Pen-3α, Penaiedin-3α gene; LGBP, Lipopolysaccharide- and β-1,3-glucan-binding protein; AKP, alkaline phosphatase; ACP, Acid phosphatase, ctsb, Cathepsin B; *gpx:* glutathione peroxidase; mtMnsod: mitochondrial manganese superoxide dismutase; alf, anti-lipopolysaccharide factor; sod-1, copper-zinc superoxide dismutase; sod-1, manganese superoxide dismutase; NO, nitric oxide; LTI, lymphocytes transformation index; PWM, pokeweed mitogen; *tlr,* Toll-Like Receptor; *il,* interleukin; *myd88,* myeloid differentiation factor 88; *keap1,* Kelch-like ECH-associated protein 1; *nrf2,* Nuclear factor-erythroid 2 related factor 2.

8.6 Antimicrobial Functions of Immunostimulants

In the last decade, several studies have evaluated a wide range of herbal extracts for their antimicrobial attributes and have shown promising benefits attributed mainly to their phytochemical constituents, including phenolic acids, tannins, alkaloids, saponin, essential oils, terpenoids, and flavonoids. The antimicrobial activity exhibited by herbal extracts against fish or human pathogens has been demonstrated in many studies.[74–77] For example, the antimicrobial activity of *Adiantum capillus-veneris* against fish and human bacteria was tested in fish serum and skin mucus after the plant extract was fed. The study revealed that serum had inhibitory and bactericidal activities against *A. hydrophila, Pseudomonas aeruginosa, Escherichia coli, Staphylococcus aureus, Klebsiella pneumonia,* and *Yersinia ruckeri*, while mucosal bactericidal and inhibitory activities were observed against *K. pneumonia, E. coli, A. hydrophila,* and *Y. ruckeri*, but there were no effects on *P. aeruginosa* and *S. epidermidis*. The most effective concentration was found in body fluid samples from the group of fish fed 2% *Adiantum capillus-veneris* leaf powder.[57] In a related study, Pereira-Torres et al. (2016)[78] investigated the effect of *Drimys winteri* extract on the growth of *Saprolegnia parasitica*, and the report indicated that the extract suppresses the growth of *S. parasitica.* To further examine the stimulatory effect, the head kidney cell line of Atlantic salmon was incubated with *D. winteri* extract and exposed to *S. parasitica.* The result showed immunomodulation by up-regulating the expression of Interleukin 1β and interferon gamma genes, hence confirming the potency of *D. winteri* extract against the tested bacteria and its immunomodulatory power. Additionally, Fawole et al. (2003)[47] reported the antimicrobial activity of five ethanolic plant extracts (*Mangifera indica, Psidium guajava, Allium cepa, Thespesia populnea,* and *Ixora coccinea*) alone or in combination against *A. hydrophila, Micrococcus luteus,* and *Vibrio vulnificus* and the result revealed that all extracts inhibited the growth of all pathogenic bacteria tested.

8.7 Anti-Stress Function of Immunostimulants

Positive influences and anti-stress functions of immunostimulant supplementation in aquafeed have been attributed to the role of the functional components in stimulating the immune system, modulating the activity of metabolic and antioxidative stress enzymes, and ultimately enhancing the disease resistance and growth performance of fish. For instance, Abo-State et al. (2014)[79] observed that supplementing Moringa leaf meal in the diet of Nile tilapia, *O. niloticus*, improved FCR and growth rates at lower supplementation levels of 0.5 to 0.75 g/kg after *A. hydrophila* stress challenge. The reason can be the activation of antioxidant activity, enhanced disease resistance and growth performance of fish.[79] Flavonoids like quercetin and kaempferol neutralise the reactive oxygen species (ROS) generated by supporting antioxidant enzymes.[80] Tekle and Sahu (2015)[81] further attributed the positive influence of immunostimulant supplementation to the role of the antioxidants in the plant, which helped in the suppression of reactive oxygen species (ROS) formation either by inhibition of enzymes or by chelating trace elements involved in free radical generation.

Immunostimulants also function as anti-stress by decreasing the level of glucose in fish. According to Barton and Iwama (2005),[82] this observation may be due to the hypoglycaemic activity of some functional components in the enhancement of peripheral metabolism of glucose. Tekle et al. (2015)[83] reported that the supplementation of Moringa flower blocked the activation of the hypothalamic-pituitary-interrenal (HPI) axis associated with stressors and, thus decreased or inhibited the release of the stress hormone cortisol into the bloodstream of Nile tilapia subjected to *A. hydrophila*-induced stress. In concordance, Kaleeswarana et al. (2011)[84] reported positive effects of *Cynodon dactylon* (L.) on innate immunity and disease resistance of Indian major carp, *Catla catla*, subjected to *A. hydrophila* induced stress.

8.8 Role of Immunostimulants in Modulating Heat Shock Protein 70 (HSP70) in Fish

In many anti-stress studies, exposure of fish to harmful stressors provided an opportunity to evaluate the effectiveness of dietary treatments in terms of protection against pathogens, while the use of hyperthermia stress coupled with a nutrigenomic method is effective in the expression of heat shoch proteins (HSPs) of fish. After *A. hydrophila* and hyperthermia-induced stress, dietary supplementation with aqueous extracts of some phytogens resulted in an increased accumulation of HSPs in Nile tilapia.[85] Similarly, a *Labeo rohita* diet supplemented with aqueous extract of *Psidium guajava* leaves had significantly an elevated level of HSPs genes after challenge with a virulent strain of *A. hydrophila* for two weeks.[86] The increase in the HSPs level was attributed to the increased expression and accumulation of the stress proteins, which facilitates the ability of cells to both repair and synthesize new proteins to replace those that were damaged after a particular metabolic insult.[87] The tissue or organ affected by disease is meant to have a robust stress response for its survival and recovery.[86] In infectious diseases, stress proteins are major targets of the fish immune system.[88,89] Oral administration of similar phytogenic products like *Achyranthes aspera* leaves augmented expression of IL-1b and TNF immune response genes in *Catla catla*.[90] Yang et al. (2015)[91] observed a strong up-regulation of both IL-1b and TNF-b genes in grass carp (*Ctenopharyngodon idella*) fed polysaccharides extracted from *Ficus carica* leaves after challenge with a virulent strain of *A. hydrophila* for three weeks and attributed the up-regulation to the anti-microbial, anti-inflammatory, and antioxidant activities in the plant extract. Ojima et al. (2005)[87] reported that during stress, protein folding/assembly events may be affected, and the high expression of stress proteins facilitates the ability of fish cells to both repair and replace those that were damaged.

8.9 Role of Immunostimulants in Controlling Reproduction

Immunostimulants along with immunomodulatory properties also affect the reproductive performance of the animal. Elias and Abdel-latif (2005)[92] reported in female *C. gariepinus* that levamisole treated *Y. ruckeri* injected brooders showed significantly a higher fecundity with fully condensed, mature ova in their ovary compared to *Y. ruckeri* injected brood stock. Dietary supplementation of β-mannan oligosaccharide ameliorated gonadal degeneration by stress mitigating and antioxidant effects and improving the levels of sex-steroids like testosterone and estradiol.[93] Several immunostimulants have been tested in fish to enhance reproductive performance, including those described below and schematically represented in Figure 8.2.

8.9.1 Role of Probiotics

Probiotics are a class of immunostimulants that have gained a great deal of importance in aquaculture due to their ability to secrete antimicrobial substances, compete with pathogenic

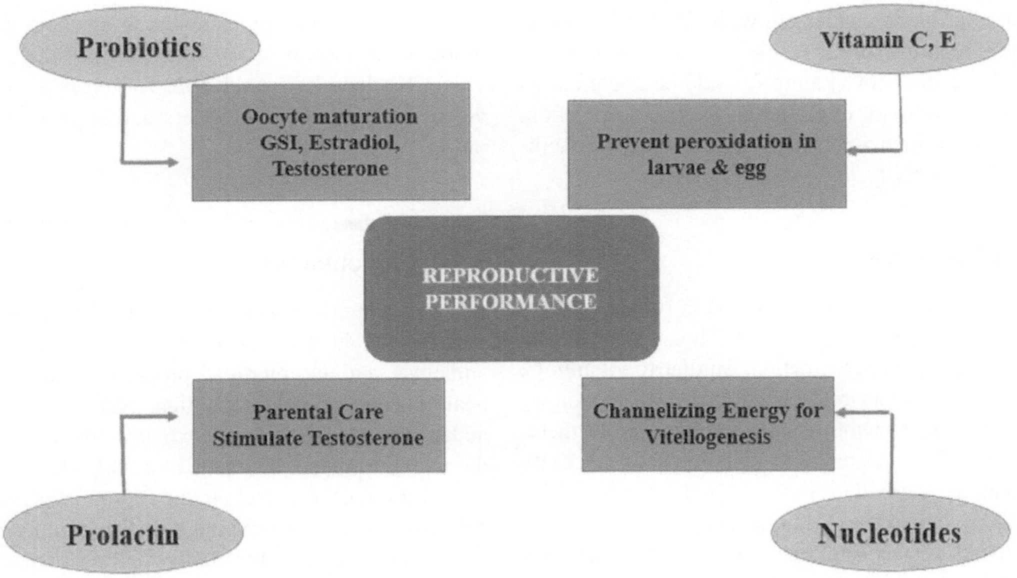

FIGURE 8.2 Schematic representation of the role of selected immunostimulants on fish reproductive performance.

bacteria, and have antitoxin effects. Recently, effect of probiotics on egg and sperm production gained importance. Administration of probiotic *L. rhamnosus* brought changes in oocyte maturation, chemical composition of oocytes, increased the number of vitellogenic follicles along with increased gonado-somatic index concomitant with increased expression of cytochrome P 450 aromatase (cyp 19a) gene expression in ovary and vitellogenin (vtg), estrogenic receptor alpha (erα) in liver.[94,95] Valcarce et al. (2015)[96] in their studies have observed *Pediococcus acidilactici* had role in spermatogenesis by increasing the leptin, brain derived neurotrophic factor (bdnf), and doublesex and dmrt1 gene expressions.

8.9.2 Role of Vitamin C

Vitamin C, an important water-soluble vitamin is highly essential for fish as they lack the enzyme machinery necessary for its synthesis, hence must be provided in their diets and also vit C being water-soluble quickly leaches out, thereby requiring higher amounts of inclusion. Vitamin C is important for growth, development, and reproduction because of its role in the synthesis of collagen, the absorption of iron, and its most important antioxidant role. There are many studies in fish where the role of vitamin C in reproductive performance and immunity has been elucidated. Dietary vitamin C is believed to be transferred from female brooder into eggs for growth and development until first feeding of larvae; it is also important for reproduction, gonadal maturation, and larval metamorphosis.[97,98] Shahkar et al. (2015)[99] reported in male Japanese eel broodstock that dietary increment (410.8 and 911.8 mg/kg diet) in the supplementation of ascorbic acid improved the number of spermatogonia. Shao et al. (2018)[100] obtained results in ongoing female gibel carp, where L-ascorbic acid at 444.1 mg/kg diet decreased ovarian development by reducing GnRH levels. Furthermore, as the role of vitamin C in fish immunity is critically important, Ibrahim et al. (2020)[36] reported that vitamin C supplementation stimulated the activities of lysozyme, nitric oxide level, and phagocytic percentage and enhanced the survival of the fish after being challenged with virulent *A. sobria.* Other studies have reported similar results[37–40] and it was clearly noted that the effectiveness of vitamin C could be related to its ability to stimulate neutrophil migration to the site of infection, increase phagocytosis, improve oxidant production, and facilitate microbial killing activity.

8.9.3 Role of Vitamin E

Vitamin E is a fat-soluble micronutrient and is found as structural component in cell membranes where it has important role in protecting the cells from peroxidation. Similar to vitamin C, vitamin E executes important role in reproductive function by protecting egg and fast-growing tissues of larvae, from metabolic free radicals. The tissues with high metabolic activity showed an enhanced antioxidant defence, and vitamin E also plays a significant role by acting synergistically.[101] Vitamin E is transferred to gonads during vitellogenesis along with lipoproteins, and thus has a significance in reproduction.[102] Guerriero et al. (2004)[103] reported low levels of vitamin E in dead eggs of sea bass and lowest survival in the embryos. Furthermore, it was found that feeding zebrafish and goldfish brood stock with vitamin E deficient diets leads to poor fecundity, spawning, hatchability, and mortality of the embryos compared to the vitamin E fed group.[104,105] In addition to this, recent studies have revealed the role of vitamin E in stimulating the expression of pituitary gonadotropins such as FSH and LH in turbot.[106] Similarly, dietary E was reported to improve innate immunity in Caspian trout (*Salmo caspius*) and decreases the activities of liver function enzyme such as alanine aminotransferase (ALT) and aspartate aminotransferase (AST).[107]

8.9.4 Role of Prolactin

Prolactin is a polypeptide hormone found in almost all vertebrate groups, except in the agnatha such as hag fishes; however, a recent study has discovered the presence of prolactin-like peptide in lamprey, citing its evolutionary importance.[108] Prolactin is involved in osmoregulation, growth and development, metabolism, immunoregulation, reproduction, and reproductive behaviour. Very early studies have reported the effect of prolactin on nest-building behaviour of fish, by promoting mucus production and increasing the number of mucus producing cells.[109] Prolactin is directly involved in stimulation of testosterone, thereby was found to be involved in gonadogenesis.[110] In mouth brooding cichlids, it was found that prolactin and oestrogen have promoted the mouth breeding behaviour and reported that prolactin lineage genes are involved in complex parental care behaviour.[111]

8.9.5 Role of Nucleotides

Nucleotides are the building blocks of genetic material the DNA and RNA. They are made of three components: nitrogenous base, sugar, and phosphate group. They are involved in a variety of functions like energy metabolism, growth and development, cell signalling, cofactors for enzymes, etc. Not much literature could be found on the effects of nucleotides on reproductive performance. In a study, Vecino (2004)[112] reported that dietary supplementation of nucleotides in haddock brood stock improved fecundity, egg quality, and larval survival. Nucleotides could enhance growth and reproductive vigour in Nile tilapia brooders and supported gonadal maturation[113] (Figure 8.2).

8.10 Conclusions

Immunostimulants have been reported to have a multitude of applications in aquaculture with their immune modulatory, antistress, and antimicrobial properties, which help to achieve healthy and reliable production cycles. Recent studies have added knowledge to the existing literature and expanded the use of immunostimulants in a multi-dimensional way. The role of immunostimulants in the reproduction and reproductive behaviour of fish gives great scope to explore. Further studies that would facilitate the understanding of the mechanism of action of these substances on reproduction and could help in the development of multifunctional products are warranted.

8.11 Future Perspectives

Immunomodulators appear to be the most promising and beneficial tools for the prophylactic and therapeutic treatment of farmed fish and shellfish. Although immunomodulators will not replace vaccines, proper nutrition, or good management strategies, the strength of these compounds appears to lie in their ability to enhance larval culture before the specific immune system matures. It has immense potential for aquatic health management, but a series of scientific investigations is necessary to understand the actual mode of action. An emphasis should be given to determining the specific dosage and efficacy of different immunostimulants for a variety of fish species and their pathogens and also to decreasing the cost of the immunomodulators. To study the mode of action of different immunomodulators more precisely, transcriptomics and proteomics tools may provide more scientific insights. New immunostimulants industrialised using genetic engineering could provide a safer alternative to antibiotic use and also improve the safety and reliability of aquaculture products. Furthermore, research aimed at explaining the effects of dietary phytogenic products on mucosal immunity, mucosal microbiota, and immunomodulation of mucosal surfaces will provide more knowledge as a means of enhancing the health of farmed fish. This could reduce the current dependence on the use of antibiotics and other chemicals as stress reducers in aquaculture. It will also help to provide necessary information as the allochthonous (invasive) and autochthonous (indigenous) microbial balances in fish intestines due to dietary influences and stress-induced responses are still mostly unclear. In future research on immunomodulators, the immunostimulatory potential of a particular substance should be clearly established, followed by studies on its efficiency and potency in the context of immunostimulants. Nevertheless, due to their environmentally friendly, lower cost, biodegradability, and easy availability, plant-based immunostimulants and their extracts may play a key role in aquafarming. However, the world of gut microbiota and probiotics may be a more promising immunomodulatory tool for aquaculture industry.

Acknowledgement

The authors express immense gratitude to director/vice-chancellor, ICAR-CIFE, India and vice-chancellor, University of Ilorin, Nigeria for providing permission to write this chapter on application of immunostimulants for aquaculture health management.

Disclosure statement: The authors declare that they have no conflicts of interest.

REFERENCES

1. Nathan, C. and Cars, O., 2014. Antibiotic resistance—problems, progress, and prospects. *New England Journal of Medicine*, 371(19), pp. 1761–1763.
2. Xian, J.A., Wang, A.L., Tian, J.X., Huang, J.W., Ye, C.X., Wang, W.N. and Sun, R.Y., 2009. Morphologic, physiological and immunological changes of haemocytes from *Litopenaeus vannamei* treated by lipopolysaccharide. *Aquaculture*, 298(1-2), pp. 139–145.
3. Thangaraj, S.K., Poornima, M. and Alavandi, S., 2017. Immunostimulants in Aquaculture. pp. 175.
4. Goldsby, T.J., Griffis, S.E. and Roath, A.S., 2006. Modeling lean, agile, and leagile supply chain strategies. *Journal of Business Logistics*, 27(1), pp. 57–80.
5. Ching, J.J., Shuib, A.S., Abdul Majid, N. and Mohd Taufek, N., 2021. Immunomodulatory activity of β-glucans in fish: Relationship between β-glucan administration parameters and immune response induced. *Aquaculture Research*, 52(5), pp. 1824–1845.
6. Medzhitov, R., 2007. Recognition of microorganisms and activation of the immune response. *Nature*, 449 (7164), pp. 819–826.
7. Matzinger, P., 2002. The danger model: a renewed sense of self. *Science*, 296(5566), pp. 301–305.
8. Alberts, B., Alexander, J., Julian, L., Martin, R., Keith, R. and Peter, W., 2002. *Molecular Biology of the Cell*; Fourth Edition. New York and London: Garland Science. ISBN 0-8153-3218-1.
9. Litman, G.W., Cannon, J.P. and Dishaw, L.J., 2005. Reconstructing immune phylogeny: new perspectives. *Nature Review in Immunology*, 5 (11), pp. 866–879.
10. Van Muiswinkel, W.B., 1992. Fish immunology and health. *Netherlands Journal of Zoology*, 42(2-3), pp. 494–499.
11. Bly, J.E. and Clem, L.W., 1992. Temperature and teleost immune functions. *Fish & Shellfish Immunology*, 2, pp. 159–171.
12. Carlson, R.E., Baker, E.P. and Fuller, R.E., 1995. Immunological assessment of hybrid striped bass at three culture temperature. *Fish & Shellfish Immunology*, 5, pp. 359–373.
13. Ellis, A.E., 1988. General principles of fish vaccination. In: Ellis A.E., editor. *Fish vaccination*. London: Academic Press; pp. 1–19.
14. Vaghasiya, J., Datani, M., Nandkumar, K., Malaviya, S. and Jivani, N., 2010. Comparative evaluation of alcoholic and aqueous extracts of Ocimum sanctum for immunomodulatory activity. *International Journal on Pharmaceutical and Biological Research*, 1(1), pp. 25–29.
15. Selye, H., 1973. The Evolution of the Stress Concept: The originator of the concept traces its development from the discovery in 1936 of the alarm reaction to modern therapeutic applications of syntoxic and catatoxic hormones. *American scientist*, 61(6), pp. 692–699.
16. Anderson, D.P., 1992. Immunostimulants, adjuvants and vaccine carriers in fish: applications to aquaculture. *Annual review of fish diseases*, 2, pp. 281–307.
17. Sakai, M., 1999. Current research status of fish immunostimulants. *Aquaculture*, 172, pp. 63–92.
18. Secombes C.J., 1994. Enhancement of fish phagocyte activity. *Fish & Shellfish Immunology*, 4, pp. 421–436.
19. Harikrishnan, R., Balasundaram, C. and Heo, M.S., 2011a. Impact of plant products on innate and adaptive immune system of cultured finfish and shellfish. *Aquaculture*, 317, pp. 1–15.

20. Dugenci, S.K., Arda, N. and Candan, A., 2003. Some medicinal plants as immunostimulant for fish. *Journal of Ethnopharmacology*, 88, pp. 99–106.
21. Rao, Y.V., Das, B.K., Jyotyrmayee, P. and Chakrabarti, R., 2006. Effect of Achyranthes aspera on the immunity and survival of Labeo rohita infected with Aeromonas hydrophila. *Fish & Shellfish Immunology*, 20, pp. 263–273.
22. Subasinghe, R., Soto, D. and Jia, J., 2009. Global aquaculture and its role in sustainable development. *Reviews in Aquaculture*, 1(1), pp. 2–9.
23. Barman, D., Nen, P., Mandal, S.C. and Kumar, V., 2013. Immunostimulants for aquaculture health management. *J Marine Sci Res Dev*, 3(3), pp. 1–11.
24. Cuesta, A., Meseguer, J. and Esteban, M.A., 2004. Total serum immunoglobulin M levels are affected by immunomodulators in seabream (Sparus aurata L.). *Veterinary Immunology and Immunopathology*, 101(3-4), pp. 203–210.
25. Bricknell, I. and Dalmo, R.A., 2005. The use of immunostimulants in fish larval aquaculture. *Fish & Shellfish Immunology*, 19(5), pp. 457–472.
26. Wang, W., Sun, J., Liu, C. and Xue, Z., 2017. Application of immunostimulants in aquaculture: current knowledge and future perspectives. *Aquaculture Research*, 48(1), pp. 1–23.
27. Zhou, X., Wang, Y., Yao, J. and Li, W., 2010. Inhibition ability of probiotic, *Lactococcus lactis*, against *A. hydrophila* and study of its immunostimulatory effect in tilapia (*Oreochromis niloticus*). *International Journal of Engineering, Science and Technology*, 2(7).
28. Iromo, H., Zairin Junior, M., Agus Suprayudi, M. and Manalu, W., 2015. The optimum dose of thyroxine hormone supplementation in broodstock mud crab (*Scylla serrata*) to accelerate ovarian maturation. *Journal of Aquaculture Research and Development*, 6(3).
29. Zhou, J., Song, X.L., Huang, J. and Wang, X.H., 2006. Effects of dietary supplementation of A3α-peptidoglycan on innate immune responses and defense activity of Japanese flounder (*Paralichthys olivaceus*). *Aquaculture*, 251(2-4), pp. 172–181.
30. Davis, J.F. and Hayasaka, S.S., 1984. The enhancement of resistance of the American eel, Anguilla rostrata Le Sueur, to a pathogenic bacterium, *Aeromonas hydrophila*, by an extract of the tunicate, *Ecteinascidia turbinata*. *Journal of Fish Diseases*, 7(4), pp. 311–316.
31. Bindhu, M.R. and Umadevi, M., 2014. Antibacterial activities of green synthesized gold nanoparticles. *Materials Letters*, 120, pp. 122–125.
32. López, N., Cuzon, G., Gaxiola, G., Taboada, G., Valenzuela, M., Pascual, C., Sánchez, A. and Rosas, C., 2003. Physiological, nutritional, and immunological role of dietary β 1-3 glucan and ascorbic acid 2-monophosphate in *Litopenaeus vannamei* juveniles. *Aquaculture*, 224(1-4), pp. 223–243.
33. Matsuyama, H., Mangindaan, R.E. and Yano, T., 1992. Protective effect of schizophyllan and scleroglucan against Streptococcus sp. infection in yellowtail (*Seriola quinqueradiata*). *Aquaculture*, 101(3-4), pp. 197–203.
34. Sahoo, P.K. and. Mukherjee, S.C., 1999. Influence of the immunostimulant, chitosan on immune responses of healthy and cortisol- treated rohu (Labeo rohita). *Journal of Aquaculture in the Tropics*, 14 (3), pp. 209–215.
35. Liu, H., Xie, S., Zhu, X., Lei, W., Han, D. and Yang, Y., 2008. Effects of dietary ascorbic acid supplementation on the growth performance, immune and stress response in juvenile *Leiocassis longirostris* Günther exposed to ammonia. *Aquaculture Research*, 39(15), pp. 1628–1638.
36. Ibrahim, R.E., Ahmed, S.A., Amer, S.A., Al-Gabri, N.A., Ahmed, A.I., Abdel-Warith, A.W.A.... ... . and Metwally, A.E., 2020. Influence of vitamin C feed supplementation on the growth, antioxidant activity, immune status, tissue histomorphology, and disease resistance in Nile tilapia, Oreochromis niloticus. *Aquaculture Reports*, 18, pp. 100545.
37. Narra, M.R., Rajender, K., Reddy, R.R., Rao, J.V. and Begum, G. (2015). The role of vitamin C as antioxidant in protection of biochemical and haematological stress induced by chlorpyrifos in freshwater fish Clarias batrachus. *Chemosphere*, 132, pp. 172–178.
38. Imanpoor, M., Imanpoor, M.R. and Roohi, Z., 2017. Effects of dietary vitamin C on skeleton abnormalities, blood biochemical factors, haematocrit, growth, survival and stress response of Cyprinus carpio fry. *Aquaculture International*, 25(2), pp. 793–803.
39. Yusuf, A., Huang, X., Chen, N., Apraku, A., Wang, W., Cornel, A. and Rahman, M.M., 2020. Impact of dietary vitamin C on plasma metabolites, antioxidant capacity and innate immunocompetence in juvenile largemouth bass, Micropterus salmoides. *Aquaculture Reports*, 17, pp. 100383.
40. Gbadamosi, O.K., 2021. Effects of ascorbic acid and iron nanoparticles supplements on hyperthermia-induced stress in African catfish, *Clarias gariepinus* (Burchell, 1822). *The Journal of Basic and Applied Zoology*, 82(1), pp. 1–9.
41. Kono, T. and Sakai, M., 2001. The analysis of expressed genes in the kidney of Japanese flounder, *Paralichthys olivaceus*, injected with the immunostimulant peptidoglycan. *Fish & Shellfish Immunology*, 11(4), pp. 357–366.
42. Choudhury, D., Pal, A.K., Sahu, N.P., Kumar, S., Das, S.S. and Mukherjee, S.C., 2005. Dietary yeast RNA supplementation reduces mortality by *Aeromonas hydrophila* in rohu (*Labeo rohita* L.) juveniles. *Fish & Shellfish Immunology*, 19(3), pp. 281–291.
43. Kajita, Y., Sakai, M., Atsuta, S. and Kobayashi, M., 1990. The immunomodulatory effects of levamisole on rainbow trout, *Oncorhynchus mykiss*. *Fish Pathology*, 25(2), pp. 93–98.
44. Cheng, A.C., Tu, C.W., Chen, Y.Y., Nan, F.H. and Chen, J.C., 2007. The immunostimulatory effects of sodium alginate and iota-carrageenan on orange-spotted grouper *Epinephelus coicoides* and its resistance against *Vibrio alginolyticus*. *Fish & Shellfish Immunology*, 22(3), pp. 197–205.
45. Fawole, F.J., Yisa, R.O., Jayeoba, O.O., Adeshina, I., Ahmed, A.O. and Emikpe, B.O., 2022. Effect of Dietary Polyherbal Mixture on Growth Performance, Haemato-Immunological Indices, Antioxidant Responses, and Intestinal Morphometry of African Catfish, *Clarias gariepinus*. *Aquaculture Nutrition*, 2022.
46. Zheng, W. and Wang, S.Y., 2001. Antioxidant activity and phenolic compounds in selected herbs. *Journal of Agricultural and Food Chemistry*, 49(11), pp. 5165–5170.

47. Fawole, F.J., Sahu, N.P., Pal, A.K. and Lakra, W.S., 2013. Evaluation of antioxidant and antimicrobial properties of selected Indian medicinal plants. *International Journal of Medicinal and Aromatic Plants*, 3(1), pp. 69–77.
48. Aliyu, A.B., Achika, J.I., Adewuyi, J.A., Gangas, P., Ibrahim, H. and Oyewale, A.O., 2019. Antioxidants from Nigerian Medicinal Plants: What Are the Evidence?. *In Lipid Peroxidation Research*, pp. 43. IntechOpen.
49. Chaves, N., Santiago, A. and Alías, J.C., 2020. Quantification of the antioxidant activity of plant extracts: Analysis of sensitivity and hierarchization based on the method used. *Antioxidants*, 9(1), pp. 76.
50. Giri, S.S., Sen, S.S., Chi, C., Kim, H.J., Yun, S., Park, S.C. and Sukumaran, V., 2015. Effect of guava leaves on the growth performance and cytokine gene expression of Labeo rohita and its susceptibility to Aeromonas hydrophila infection. *Fish & Shellfish Immunology*, 46(2), pp. 217–224.
51. Fawole, F.J., Sahu, N.P., Pal, A.K. and Ravindran, A., 2016. Haemato-immunological response of L abeo rohita (Hamilton) fingerlings fed leaf extracts and challenged by *Aeromonas hydrophila*. *Aquaculture Research*, 47(12), pp. 3788–3799.
52. Nhu, T.Q., Hang, B.T.B., Hue, B.T.B., Quetin-Leclercq, J., Scippo, M.L., Phuong, N.T. and Kestemont, P., 2019. Plant extract-based diets differently modulate immune responses and resistance to bacterial infection in striped catfish (Pangasianodon hypophthalmus). *Fish & Shellfish Immunology*, 92, pp. 913–924.
53. Adeshina, I., Tiamiyu, L.O., Akpoilih, B.U., Jenyo-Oni, A. and Ajani, E.K., 2021. Dietary Mitracarpus scaber leaves extract improved growth, antioxidants, non-specific immunity, and resistance of Nile tilapia, *Oreochromis niloticus* to *Gyrodactylus malalai* infestation. *Aquaculture*, 535, pp. 736377.
54. Nhu, T.Q., Bich Hang, B.T., Cornet, V., Oger, M., Bach, L.T., Anh Dao, N.L.,… … and Kestemont, P., 2020. Single or combined dietary supply of Psidium guajava and *Phyllanthus amarus* extracts differentially modulate immune responses and liver proteome in striped catfish (*Pangasianodon hyphophthalmus*). *Frontiers in Immunology*, 11, pp. 797.
55. Magnadóttir, B. (2006). Innate immunity of fish (overview). *Fish & Shellfish Immunology*, 20(2), pp. 137–151.
56. Harikrishnan, R., Balasundaram, C., Kim, M.C., Kim, J.S., Han, Y.J. and Heo, M.S., 2009. Innate immune response and disease resistance in *Carassius auratus* by triherbal solvent extracts. *Fish & Shellfish Immunology*, 27(3), pp. 508–515.
57. Hoseinifar, S.H., Jahazi, M.A., Mohseni, R., Raeisi, M., Bayani, M., Mazandarani, M.,… … and Mozanzadeh, M.T., 2020. Effects of dietary fern (Adiantum capillus-veneris) leaves powder on serum and mucus antioxidant defence, immunological responses, antimicrobial activity and growth performance of common carp (Cyprinus carpio) juveniles. *Fish & Shellfish Immunology*, 106, pp. 959–966.
58. Khanjani, M.H., Ghaedi, G. and Sharifinia, M., 2022. Effects of diets containing β-glucan on survival, growth performance, haematological, immunity and biochemical parameters of rainbow trout (Oncorhynchus mykiss) fingerlings. *Aquaculture Research*, 53(5), pp. 1842–1850.
59. Dawood, M.A., Metwally, A.E.S., El-Sharawy, M.E., Atta, A.M., Elbialy, Z.I., Abdel-Latif, H.M. and Paray, B.A., 2020. The role of β-glucan in the growth, intestinal morphometry, and immune-related gene and heat shock protein expressions of Nile tilapia (Oreochromis niloticus) under different stocking densities. *Aquaculture*, 523, pp. 735205.
60. Abdel Rahman, A.N., Khalil, A.A., Abdallah, H.M. and ElHady, M., 2018. The effects of the dietary supplementation of Echinacea purpurea extract and/or vitamin C on the intestinal histomorphology, phagocytic activity, and gene expression of the Nile tilapia. *Fish & Shellfish Immunology*, 82, pp. 312–318.
61. Laltlanmawia, C., Saha, R.K., Saha, H. and Biswas, P., 2019. Ameliorating effects of dietary mixture of Withania somnifera root extract and vitamin C in Labeo rohita against low pH and waterborne iron stresses. *Fish & Shellfish Immunology*, 88, pp. 170–178.
62. Salaah, S.M., Dalia, M. and Gaber, H.S., 2022. Potential effects of dietary chitosan against lead-induced innate immunotoxicity and oxidative stress in Nile tilapia (Oreochromis niloticus). *The Egyptian Journal of Aquatic Research*, 48(2), pp. 123–129.
63. Kumar, N., Thakur, N., Sharma, C., Shanthanagouda, A.H., Taygi, A. and Singh, A., 2022. Effect of dietary chitosan nanoparticles on immune response and disease resistance against *Aeromonas hydrophila* infection in tropical herbivore fish (rohu, Labeo rohita). *Aquaculture International*, pp. 1–14.
64. Khosravi-Katuli, K., Mohammadi, Y., Ranjbaran, M., Ghanaatian, H., Khazaali, A., Paknejad, H. and Santander, J., 2021. Effects of mannan oligosaccharide and synbiotic supplementation on growth performance and immune response of Gilthead Sea Bream (*Sparus aurata*) before and after thermal stress. *Aquaculture Research*, 52(8), pp. 3745–3756.
65. Soleimani, N., Hoseinifar, S.H., Merrifield, D.L., Barati, M. and Abadi, Z.H., 2012. Dietary supplementation of fructooligosaccharide (FOS) improves the innate immune response, stress resistance, digestive enzyme activities and growth performance of Caspian roach (Rutilus rutilus) fry. *Fish & Shellfish Immunology*, 32(2), pp. 316–321.
66. Hoseinifar, S.H., Soleimani, N., and Ringø, E., 2014. Effects of dietary fructo-oligosaccharide supplementation on the growth performance, haemato-immunological parameters, gut microbiota and stress resistance of common carp (Cyprinus carpio) fry. *British Journal of Nutrition*, 112(8), pp. 1296–1302.
67. Liu, L., Li, J., Cai, X., Ai, Y., Long, H., Ren, W.… … . and Xie, Z.Y., 2022. Dietary supplementation of astaxanthin is superior to its combination with Lactococcus lactis in improving the growth performance, antioxidant capacity, immunity and disease resistance of white shrimp (*Litopenaeus vannamei*). *Aquaculture Reports*, 24, pp. 101124.
68. Zhu, X., Hao, R., Zhang, J., Tian, C., Hong, Y., Zhu, C. and Li, G., 2022. Dietary astaxanthin improves the antioxidant capacity, immunity and disease resistance of coral trout (*Plectropomus leopardus*). *Fish & Shellfish Immunology*, 122, pp. 38–47.

69. El-Boshy, M., El-Ashram, A., Risha, E., Abdelhamid, F., Zahran, E. and Gab-Alla, A., 2014. Dietary fucoidan enhance the non-specific immune response and disease resistance in African catfish, Clarias gariepinus, immunosuppressed by cadmium chloride. *Veterinary Immunology and Immunopathology*, 162(3-4), pp. 168–173.
70. Zhu, X., Xu, N., Liu, Y., Ai, X. and Yang, Y., 2022. The effects of *Agaricus bisporus* polysaccharides enriched diet on growth, nonspecific immunity and disease resistance in crayfish (*Procambarus clarkii*). *Aquaculture Reports*, 24, pp. 101168.
71. Garg, C.K., Sahu, N.P., Maiti, M.K., Shamna, N., Deo, A.D. and Sardar, P., 2021. Dietary Houttuynia cordata leaf extract and meal enhances the immunity and expression of immune genes in *Labeo rohita* (Hamilton, 1822). *Aquaculture Research*, 52(1), pp. 381–394.
72. Pan, S., Yan, X., Li, T., Suo, X., Liu, H., Tan, B.... ... . and Dong, X., 2022. Impacts of tea polyphenols on growth, antioxidant capacity and immunity in juvenile hybrid grouper (*Epinephelus fuscoguttatus*♀× *E. lanceolatus*♂) fed high-lipid diets. *Fish & Shellfish Immunology*, 128, pp. 348–359.
73. Abdelrazek, H.M.A., Tag, H.M., Kilany, O.E., Reddy, P.G. and Hassan, A.M., 2017. Immuomodulatory effect of dietary turmeric supplementation on Nile tilapia (*Oreochromis niloticus*). *Aquaculture Nutrition*, 23(5), pp. 1048–1054.
74. Castro, S.B.R., Leal, C.A.G., Freire, F.R., Carvalho, D.A., Oliveira, D.F. ... and Figueiredo, H.C.P., 2008. Antibacterial activity of plant extracts from Brazil against fish pathogenic bacteria. *Brazilian Journal of Microbiology*, 39, pp. 756–760.
75. Akinpelu, D.A., Aiyegoro, O.A., Akinpelu, O.F. and Okoh, A.I. (2014). Stem bark extract and fraction of Persea americana (Mill.) exhibits bactericidal activities against strains of Bacillus cereus associated with food poisoning. *Molecules*, 20(1), pp. 416–429.
76. Gonelimali, F.D., Lin, J., Miao, W., Xuan, J., Charles, F., Chen, M. and Hatab, S.R., 2018. Antimicrobial properties and mechanism of action of some plant extracts against food pathogens and spoilage microorganisms. *Frontiers in Microbiology*, 9, pp. 1639.
77. Mostafa, A.A., Al-Askar, A.A., Almaary, K.S., Dawoud, T.M., Sholkamy, E.N. and Bakri, M.M., 2018. Antimicrobial activity of some plant extracts against bacterial strains causing food poisoning diseases. *Saudi Journal of Biological Sciences*, 25(2), pp. 361–366.
78. Pereira-Torres, D., Gonçalves, A.T., Ulloa, V., Martínez, R., Carrasco, H., Olea, A.F.... ... . and Astuya, A. (2016). In vitro modulation of Drimys winteri bark extract and the active compound polygodial on Salmo salar immune genes after exposure to Saprolegnia parasitica. *Fish & Shellfish Immunology*, 59, pp. 103–108.
79. Abo-State, H., Yasser Hammouda, H.Y., El-Nadi, A. and AboZaid, H., 2014. Evaluation of feeding raw Moringa (Moringa oleifera Lam.) leaf meal in Nile tilapia fingerlings (Oreochromis niloticus) diets. *Global Veterinaria*, 13 (1), pp. 105–111.
80. Kumar, S., Abhay K. and Pandey, A.K., 2013. Chemistry and Biological Activities of Flavonoids: An Overview. *The Scientific World Journal*, 13, pp. 1–16.
81. Tekle, E.W. and Sahu, N.P., 2015. Growth and immunodulatory response of Nile tilapia, Oreochromis niloticus fingerlings to ethanolic extract of Moringa oleifera flower. *International Journal of Sustainable and Renewable Products*, 5(7), pp. 285–296.
82. Barton, B.A. and Iwama, G.K., 2005. Simple field methods for monitoring stress and general condition of fish. *Aquaculture Research*, 26, pp. 273–282.
83. Tekle, E.W., Sahu, N.P. and Makesh, M., 2015. Antioxidative and antimicrobial activities of different solvent extracts of Moringa oleifera: an in vitro evaluation. *International Journal of Sustainable and Renewable Products*, 5(5), pp. 255–266.
84. Kaleeswarana B., Ilavenilb, S. and Ravikumara, 2011. Dietary supplementation with Cynodon dactylon (L.) enhances innate immunity and disease resistance of Indian major carp, Catla catla (Ham.). *Fish & Shellfish Immunology*, 31, pp. 953–962.
85. Gbadamosi, O.K. and Olanipekun, O.S., 2020. Dietary supplementation with goat weed leaf (Ageratum conyzoides) improves growth performance, haematological parameters and attenuates pathological-induced stress in African catfish (Clarias gariepinus) fingerlings. *Livestock Research for Rural Development*, 32 (7), pp. 1–9.
86. Giri, S.S., Sen, S.S., Chi, C., Kim, H.J., Yun, S., Park, S.C. and Sukumaran, V., 2015. Effect of guava leaves on the growth performance and cytokine gene expression of Labeo rohita and its susceptibility to Aeromonas hydrophila infection. *Fish & Shellfish Immunology*, 46(2), pp. 217–224.
87. Ojima, N., Yamashita, M. and Watanabe, S., 2005. Quantitative mRNA expression profiling of heat shock protein families in rainbow trout and Nile tilapia cells. *Biochemical and Biophysical Resources Communication*, 329, pp. 51–57.
88. Li, L., Kobayashi, M., Kaneko, H., Nakajima-Takagi, Y., Nakayama, Y. and Yamamoto, M., 2008. Molecular evolution of Keap1: Two Keap1 molecules with distinctive intervening region structures are conserved among fish. *Journal of Biological Chemistry*, 283(6), pp. 3248–3255.
89. Liu, F., Shi, H.Z., Guo, Q.S., Yu, Y.B., Wang, A.M., Lv, F. and Shen, W.B., 2016. Effects of astaxanthin and emodin on the growth, stress resistance and disease resistance of yellow catfish (Pelteobagrus fulvidraco). *Fish & Shellfish Immunology*, 51, pp. 125–135.
90. Chakrabarti, R., Srivastava, P.K., Verma, N. and Sharma, J.G., 2014. Effect of seeds of Achyranthes aspera on the immune responses and expression of some immune-related genes in carp *Catla catla, Fish & Shellfish Immunology*, 4, pp. 164–169.
91. Yang, X., Guo, J.L., Ye, J.Y., Zhang, Y.X. and Wang, W., 2015. The effects of Ficus carica polysaccharide on immune response and expression of some immune-related genes in grass carp, *Ctenopharyngodon idella. Fish & Shellfish Immunology*, 42(1), pp. 132–137.
92. Elias, N.S. and Abdel-latif, J.I., 2009. Impairment of fish fecundity by ecps yersinia ruckeri with levamisole treatment. Presented at the Proceedings of the 2nd Global Fisheries and Aquaculture Research Conference, Cairo International Convention Center, 24-26 October 2009, Massive Conferences and Trade Fairs, pp. 315–328.

93. El-Fahla, N.A., Khalil, K.A., Dessouki, A.A., Abdelrazek, H.M., Mohallal, M.E. and El-Hak, H.N.G., 2022. Dietary Beta-MOS® ameliorated lead induced reproductive toxicity and stress in Nile tilapia. *Aquaculture*, 548, pp. 737711.
94. Carnevali, O., Maradonna, F. and Gioacchini, G., 2017. Integrated control of fish metabolism, wellbeing and reproduction: the role of probiotic. *Aquaculture*, 472, pp. 144–155.
95. Gioacchini, G., Giorgini, E., Merrifield, D.L., Hardiman, G., Borini, A., Vaccari, L. and Carnevali, O., 2012. Probiotics can induce follicle maturational competence: the Danio rerio case. *Biology of Reproduction*, 86, pp. 65–1.
96. Valcarce, D.G., Pardo, M., Riesco, M., Cruz, Z. and Robles, V., 2015. Effect of diet supplementation with a commercial probiotic containing *Pediococcus acidilactici* (Lindner, 1887) on the expression of five quality markers in zebrafish (Danio rerio (Hamilton, 1822)) testis. *Journal of Applied Ichthyology*, 31, 18–21.
97. Dabrowski, K. and Ciereszko, A., 2001. Ascorbic acid and reproduction in fish: endocrine regulation and gamete quality. *Aquaculture Research*, 32, pp. 623–638.
98. Adebayo, O. and Fawole, F., 2012. Growth and reproductive performance of African giant catfish, *Heterobranchus longifilis* Valenciennes 1840 broodstock on ascorbic acid supplementation. *Indian Journal Fisheries*, 59(2), pp. 135–140.
99. Shahkar, E., Yun, H., Kim, D.-J., Kim, S.-K., Lee, B.I. and Bai, S.C., 2015. Effects of dietary vitamin C levels on tissue ascorbic acid concentration, hematology, non-specific immune response and gonad histology in broodstock Japanese eel, *Anguilla japonica*. *Aquaculture*, 438, pp. 115–121.
100. Shao, L., Han, D., Yang, Y., Jin, J., Liu, H., Zhu, X. and Xie, S., 2018. Effects of dietary vitamin C on growth, gonad development and antioxidant ability of on-growing gibel carp (Carassius auratus gibelio var. CAS III). *Aquaculture Research*, 49, pp. 1242–1249.
101. El-Sayed, A.M., Izquierdo, M., 2022. The importance of vitamin E for farmed fish—A review. *Reviews in Aquaculture*, 14, pp. 688–703.
102. Tokuda, M., Yamaguchi, T., Wakui, K., Sato, T., Ito, M. and Takeuchi, M., 2000. Tocopherol affinity for serum lipoproteins of Japanese flounder Paralichthys olivaceus during the reproduction period. *Fisheries Science*, 66, pp. 619–624.
103. Guerriero, G., Ferro, R., Russo, G. and Ciarcia, G., 2004. Vitamin E in early stages of sea bass (*Dicentrarchus labrax*) development. *Comparative Biochemistry and Physiology Part a: Molecular and Integrative Physiology*, 138, pp. 435–439.
104. James, R., Vasudhevan, I. and Sampath, K. (2008). Effect of dietary vitamin E on growth, fecundity, and leukocyte count in goldfish (Carassius auratus). *The Israeli Journal of Aquaculture – Bamidgeh*, 60(2), pp. 121–127
105. McDougall, M., Choi, J., Truong, L., Tanguay, R. and Traber, M.G., 2017. Vitamin E deficiency during embryogenesis in zebrafish causes lasting metabolic and cognitive impairments despite refeeding adequate diets. *Free Radical Biology and Medicine*, 110, pp. 250–260.
106. Huang, B., Wang, N., Wang, L., Jia, Y., Liu, Bin, Gao, X., Liu, Baoliang, and Wang, W., 2019. Vitamin E stimulates the expression of gonadotropin hormones in primary pituitary cells of turbot (*Scophthalmus maximus*). *Aquaculture*, 509, 47–51.
107. Saheli, M., Islami, H.R., Mohseni, M. and Soltani, M., 2021. Effects of dietary vitamin E on growth performance, body composition, antioxidant capacity, and some immune responses in Caspian trout (Salmo caspius). *Aquaculture Reports*, 21, 100857.
108. Gong, N., Ferreira-Martins, D., Norstog, J.L., McCormick, S.D., Sheridan, M.A., 2022. Discovery of prolactin-like in lamprey: Role in osmoregulation and new insight into the evolution of the growth hormone/prolactin family. *Proceedings of the National Academy of Sciences* 119, e2212196119
109. Machemer, L. and Lorke, D., 1981. Embryotoxic effect of cadmium on rats upon oral administration. *Toxicology and Applied Pharmacology*, 58(3), pp. 438–443. Rubin and Specker, 1992
110. Summers, K. and Zhu, Y., 2008. Positive selection on a prolactin paralog following gene duplication in cichlids: adaptive evolution in the context of parental care. *Copeia*, 2008(4), pp. 872–876.
111. Vecino, J.L.G., 2004. Nucleotide Enhancement of Diets, Fish Reproduction and Egg Quality. Open University (United Kingdom).
112. Arshadi, A., Yavari, V., Oujifard, A., Mousavi, S.M., Gisbert, E. and Mozanzadeh, M.T., 2018. Dietary nucleotide mixture effects on reproductive and performance, ovary fatty acid profile and biochemical parameters of female Pacific shrimp Litopenaeus vannamei. *Aquaculture Nutrition*, 24, pp. 515–523.
113. de Lima, S.A., de Oliveira Pedreira, A.C., de Freitas, J.M.A., Dalmaso, A.C.S., Chiella, R.J., Meurer, F., Romão, S. and Bombardelli, R.A., 2020. Diets containing purified nucleotides reduce oxidative stress, interfere with reproduction, and promote growth in Nile tilapia females. *Aquaculture*, 528, pp. 735509.

Section III

Immunomodulators in aquaculture health management

9

Herbal Immunomodulators for Aquaculture

Shamna Nazeemashahul[1], Femi John Fawole[2], Babitha Rani A.M.[3], Manish Jayant[1], Neha Qureshi[4], Hussain Nottanalan[1], Ashutosh D. Deo[1], and Parimal Sardar[1]

[1]*Fish Nutrition, Biochemistry and Physiology Division, ICAR – Central Institute of Fisheries Education, Mumbai, Maharashtra, India*

[2]*Fish Nutrition and Biochemistry Unit, Department of Aquaculture and Fisheries, University of Ilorin, Ilorin, Nigeria*

[3]*Aquaculture Division, ICAR – Central Institute of Fisheries Education, Mumbai, Maharashtra, India*

[4]*Fisheries Economics, Extension and Statistics Division, ICAR-CIFE, Mumbai, Maharashtra, India*

9.1 Introduction

The growing demand for increased aquaculture production to meet the targeted protein requirements of a growing population is putting immense pressure on various aquaculture systems. The stagnation in fisheries sector and limited land and water resources directed the vertical expansion of aquafarming by increasing the stocking density. However, increased stocking numbers in a culture system cause stress in the animals. The multiple stressors like crowding and feed restriction can induce oxidative stress in animals which can damage the cellular and physiological functions.[1] The level of cortisol in the serum is widely accepted as a stress indicator in fish and it generally increases after stress exposure. The increase in the level of cortisol negatively influences the growth and reproduction in fish and suppresses the immune system.[2] Stressors of any kind are indicated by physiological responses. Responses to stress inside the body are often indicated as primary, secondary, and tertiary, and the activation of brain centers resulting in the secretion of cortisol and catecholamines is considered a primary response.[3] Secondary responses are observed in tissue and blood level due to the actions of these hormones, whereas tertiary responses, exemplified by immune-suppression and lower performance of the fish. As a result, stressors can weaken the immune system, making disease outbreaks in the culture more likely. With the intensification of the culture system, numerous disease outbreaks like viral haemorrhagic septicemia, white spot viral disease (WSVD), Tilapia Lake Virus (TiLV), and vibriosis are commonly observed in aquaculture facilities.[4] Drugs and antibiotics are frequently used in culture systems to prevent or cure diseases without any regulations or control. This has resulted in generating antibiotic resistance and bioaccumulation of chemicals and antibiotic residues at various trophic levels.[5] Hence, it is essential to find an alternative to these medications or drugs.

Several attempts were made to boost the immune system of fish using different kinds of biogenic compounds commonly known as "nutraceuticals". The ability of the nutraceuticals to boost fish immune systems and promote growth has been reported by many researchers.[6] According to research reports, they promote increased feed intake, improve the fish and shellfish's resistance to stress and pathogens, and accelerate maturation without having any negative effects. Some feed additives used in aquaculture are also listed as nutraceuticals.[7] Among these biogenic compounds, the most potent and effective ones turned out to be herbal extracts or phytobiotics. The potential herbal plants across the globe are enriched with biologically active compounds such as tannins, alkaloids, terpenoids, saponins, phenolics, steroids, and flavonoids, which will contribute to growth promotion, immune stimulation, antibacterial, antiviral, antiparasitic, antifungal, and antistress activities in fish. Many of these compounds are part of the natural defense system of plants, and at higher levels they act as antimetabolites in animals. The lower dose of flavonoids, alkaloids, terpenoids and tannins were found to be immune-modulators in fish, and they did not cause any bioaccumulation or detrimental effect to the environment.[8] Thus, initiatives on herbal-based products as therapeutic agents not only reduce the harmful effects of synthetic drugs and additives, but also ensure protection of the environment.

9.2 Nutraceuticals in Aquaculture

Nutraceuticals are nutritionally functional foods or food products that can promote growth and health benefits in fish. Several food-derived bioactive compounds are found to have potential as functional feed additives. Nutraceutical therapy can promote wellness and control disease symptoms.[9] They can be nutrients or herbal products like extracts or concentrates. They increase

DOI: 10.1201/9781003361183-12

the feed consumption, induce maturation, enhance antimicrobial capability and antistress characteristics that will be of immense use in the culture of shrimp and other fin fish without any environmental or hazardous problems.

9.3 Herbal Extracts/Phytobiotics as Nutraceuticals

Herbs are any plants that have potential use as food, medicine or flavouring agents from their parts or as a whole. Some of the herbs or herbal compounds are used as nutraceuticals in fish feeds. The extract prepared out of various parts of a herb can have several bioactive compounds like tannin, alkaloids, and phenolic compounds, which have the ability to neutralize the generation of oxygen anions and scavenge free radicals. Hence, phytobiotics are those compounds which are derived from plants or parts of plants given to fish through diet to enhance their performance.[10] Most of the phytobiotics act as antioxidants or antistress compounds, while some show properties like appetite stimulation, bile and digestive enzyme secretions, hepatoprotection, and growth promotion.[11]

9.3.1 Some Commonly Used Herbs and Their Importance in Aquaculture

The effect of many herbs was studied in different fish species. Some of the most studied phytobiotics include neem (*Azadirachta indica*), aloe (*Aloe vera*), picrorhiza (*Picrorhiza kurroa*), turmeric (*Curcuma longa*), garlic (*Allium sativum*), tulsi (*Ocimum sanctum*), ginger (*Zingiber officinale*), Indian ginseng (*Withania somnifera*), kalmegh (*Andrographis paniculata*), and bhringaraj (*Eclipta allea*).

9.3.1.1 Neem

Neem is popularly known as an Indian herbal doctor which contains many bioactive compounds, mainly nimbin, azadirachtin, meliantroil, vepol, and salannin, which have great potential against bacterial, viral, fungal, and parasitic diseases of fishes. The application of neem extract in the pond soil can effectively reduce the virulence of WSVD.[11] Aquaneem, one commercially available neem product, can suppress the growth of *Aeromonas hydrophila*, *Pseudomonas fluorescenes*, *E. coli*, and fish/shrimp pathogenic *Flavobacteria*. Neem seed oil and neem leaf extract showed better recovery against epizootic ulcerative syndrome (EUS) and red spot disease.[12] In brackish water, 90 ppm of neem oil reduced the total ammonia nitrogen (TAN) level by 0.45 mg/l in 96 hours, while neemazal and neemgold, commercial neem products, reduced the TAN level by 69% and 79%, respectively.[13]

9.3.1.2 Picrorhiza

Picrorhiza kurroa extract is a good example of an herbal extract with hepatoprotective functions and immune modulation in fish and shrimp.[14] The extract has a high antioxidant potential and showed a high glutathione restoration capacity in malaria infection.[15]

9.3.1.3 Aloe

It has excellent antifungal, antibacterial, anti-inflammatory, antimutagenic, wound healing, and antioxidant properties.[16] It promotes the innate and adaptive immune system by increasing the lysozyme activity, respiratory burst activity, and total globulin protein levels in many fish species and also it shows disease resistance against *A. hydrophila*.[17] Due to its broad-spectrum, antimicrobial property, aloe vera leaf extract is an excellent healer against EUS in striped murrel[18] and acts as an excellent protective agent against WSVD and *Vibrio parahaemolyticus* infection in cultured shrimps.[19] Dietary inclusion of aloe vera extract at 4% in the fish diet showed a higher sex reversal rate in GIFT tilapia[20,21] and killed ingested microbes.[22]

9.3.1.4 Turmeric (Curcuma Longa)

Turmeric, a rhizomatous perennial plant belonging to the family Zingiberaceae, is widely used as a traditional medicine for treating various ailments. Turmeric rhizome contained terpenes and terpenoids (3–6%),[23] as well as diferuloylmethane (curcumin), a polyphenolic compound with antimicrobial, antioxidant, and orexigenic properties.[24] Its main bioactive compound is curcumin, which has anti-inflammatory, antimicrobial, antioxidant, hepatoprotective, and wound-healing properties.[25] Curcumin's antioxidant potential is based on its ability to donate hydrogen and neutralize free radicals,[26] and it has been demonstrated that 20 mM curcumin can detoxify hydrogen peroxide better than synthetic antioxidants of equal concentration.[27] Given this, a turmeric-enriched diet has shown better antibacterial effect and survival rate against various fastidious bacteria like *Vibrio alginolyticus* in gaint freshwater prawn (*Macrobrachium rosenbergii*), *Vibrio harveyi* in black tiger shrimp (*Penaeus monodon*), and *Pseudomonas fluorescence* in Nile tilapia (*Oreochromis niloticus*).[28,29] Dietary curcumin elevated the red blood cells (RBC) and white blood cells (WBC) counts, heamoglobin levels, enzymatic antioxidant activities, specific growth rate, and weight gain in rainbow trout (*Oncorhynchus mykiss*).[30] The low absorption of curcumin can be improved by co-feeding of the same with piperine (bioactive compound from black pepper) and Patil et al. (2016) observed a 20-fold increase in the bioavailability of curcumin when fed with piperine.[31] The stability and bioavailability are found to be higher in the nanoform of curcumin.[32]

9.3.1.5 Garlic

It is commonly known as an 'all-healing' herb in the family Alliaceae. The principal bioactive compound, Allicin, has antimicrobial, antiviral, antibacterial, antifungal, and antiparasitic properties. When fed through dietary supplementation, it can enhance fish feeding efficiency, survival rate, and immune-haematological parameters.[33] In this line, significant changes in growth performance, antibacterial activity, total serum protein, and globulin in rainbow trout were observed.[34] Several studies have demonstrated that garlic can effectively eliminate principal pathogenic bacteria such as *Myxococcus piscicola*, *Pseudomonas fluorescens*, *Vibrio anguillarum*, *Aeromonas punctata*, *Edwardsiella tarda*, *Yersinia ruckeri*, and *Fibrobacter intestinalis* in freshwater fish. It also proved effective against different protozoans, including *Opalina ranarum*, *Opalina dimidicita*, *Balantidium entozoon*, *Entamoeba histolytica*, *Trypanosoma*, *Leishmania*, *Leptomonas*, and *Crithidia*.[35]

9.3.1.6 Tulsi

Tulsi belongs to the family of *Lamiaceae*, and has antibacterial, antiviral, antifungal, antioxidant, hypoglycemic, hypolipidemic, hepatoprotective, neuroprotective, and reno-protective properties.[36] Tulsi contains some chemicals that play a significant role in phagocytic activity in fish. Aquaculture primarily uses leaves, stems, flowers, seeds, and roots for therapeutic purpose. Tulsi leaves contain water-soluble phenolic compounds and other constituents, such as eugenol, methyl-eugenol, and caryophyllene, that act as immunostimulants, increasing specific and non-specific immune response, disease resistance against *Aeromonas hydrophila* in fish.[37] Greasy grouper (*Epinephelus tauvina*) fed with *O. sanctum* extract (200 mg/Kg) reduced mortality during *Vibrio* infection.[38] However, tulsi-fed tilapia exhibited resistance against *Aeromonas hydrophila* infection through a high antibody level.[39] Dietary supplementation of *O. sanctum* (0.2%) enhanced the serum total protein, globulin, heamoglobin content, and survival when Indian carp (*Labeo rohita*) fingerlings were challenged with *Aeromonas hydrophila*.[40] The essential oil extracted from *Ocimum americanum* did not alter the intestinal microbial composition and growth, while it improved the membrane fragility of RBC and supported innate immunity.[41]

9.3.1.7 Ginger

Ginger is an herbaceous perennial plant containing zingiberene as the main component[42] and a mixture of zingerone and shogaols, gingerols, and gingerols sesquiterpenoids. A diet enriched with ginger has been shown to promote growth, immunity, and antiparasitic, antibacterial, antiviral, and antifungal properties of fish in aquaculture.[43,44] Feeding of ginger can improve fish immune responses against fish pathogens,[45] whereas bath treatment with ginger extract reduced the *Gyrodactylus* infection in guppy (*Poecilia reticulata*).[46] Dried ginger at 0.8% diet enhanced the survival percentage in *Labeo rohita* fingerlings when challenged with *Aeromonas hydrophila*.[47]

9.3.1.8 Indian Ginseng

It has been widely used for millennia in the field of Ayurveda. It also contains essential biologically active constitutes such as withanolides, withanone, rutinosides, alkaloids, steroidal lactones, and saponins[48] to ameliorate serum antibacterial, antiviral properties, immunity, and growth of fish in aquaculture.[49] Feeding of Ashwagandha root powder at 2% enhanced growth in rohu fingerlings.[50] Similarly, another study by Sharma et al. (2017) reported that the erythrocytes, leukocytes and haemoglobin values, survival %, and growth rate were high in rohu challenged with *Aeromonas hydrophila* after feeding with ashwagandha.[51] Waterborne iron toxicity can be reduced in rohu fingerlings by supplementing L-ascorbic acid and Ashwagandha root extract (1%) in a 1:1 ratio.[52]

9.3.1.9 Kalmegh

Kalmegh is a plant that belongs to the family *Acanthaceae*, commonly used for the treatment of several diseases like fever, laryngitis, malaria, and diarrhea. It is rich in several bioactive compounds, especially diterpenoids and its glycosides like andrographolide, deoxyandrographolide andneo-andrographolide etc. The leaves of this plant are used as a nutraceutical in aquaculture. It is very effective against columnaris disease caused by *Flavobacter columnar* in fish and also helps in enhancing fish immunity.[53]

9.3.1.10 Bhringaraj

Bhringaraj leaves are mainly used in aquaculture to control parasites, and it is mainly administered through dietary supplementation. Direkbusarakom (1998) reported the antibacterial properties of Bhringaraj (*Eclipta erecta*) in fish.[54] Aqueous extract of *Eclipta alba* is found to be effective against Aeromonas infection in tilapaia.[55] Similarly, Rani (1999) also reported that *Eclipta alba* has high anti-stress and growth promoting properties.[56]

9.3.1.11 Others

There are several herbal and weed plants which can be used as potential nutraceuticals in aquafeed. Chameleon plant (*Houttunia cordata*) leaf meal, leaf extract, and root extract have been shown in studies to improve immunity (interferon and TNF-expresssion, phagocytosis, and lysozyme activity) and promote growth in *Labeo rohita* fingerlings.[57] The antibacterial effects of *Cinnamonum verum* (Ceylon cinnamon) and *Eupatorium odoratum* (Siam weed) is reported in fish.[58] The antiparasitic (against monogenean infection) effect of *Santalum album* was studied in goldfish and it was reported that sandalwood is a good antihelminth in aquaculture farms.[59] The plant extract prepared from *Cynodon dactylon* exhibited an antiviral property in *black tiger shrimp* infected with WSVD.[60] The herbs Devil's horsewhip (*Achrathes aspera*), Indian gooseberry, chaga mushroom, *Solanum trilobatum*, *Aucklandia lappa*, and red horn weed have proven to be potential antibacterial and antifungal agents in fish.[61–64] The growth promotion efficiency was shown by *Picorhiza kurooa, Tridax procumbens, Phyllanthus niruri, Eclipta alba, Piper longum,* and *Crataegi fructus* in different fish species.[65,66]

There are several attempts done to explore the potential of herbals as immune-modulators or adjuvants in aquaculture (Soltani et al., 2020)[67] and many herbs and plants at specific concentration provides immune-modulatory, stress mitigating and growth promoting effect in fish (Vijayaram et al., 2022; Doan et al., 2020; Gupta et al., 2021).[68-70]

9.4 Preparation of Phytobiotics from Different Parts of an Herb

It is well known that herbs are rich in various bioactive compounds such as tannins, alkaloids, terpenoids, saponins, phenolics, steroids, and flavonoids. Hence, several parts of medicinal plants are used to extract these active compounds, which include 37% leaves, 8% seeds, 6% fruits, 18% roots, 6% bark, 4% flowers, and 22% whole plant.[71] Polar or non-polar organic solvents like methanol, ethanol, petroleum ether, chloroform, acetone, hexane, etc., or a mixture of solvents are used to extract the bioactive compounds present in the plant. Organic solvents are more efficient in extracting active compounds with higher immunostimulatory and antimicrobial action than water.[72] It is always preferable to use a polar organic solvent with an aldehyde or hydroxyl group as they are less reactive than other groups.

9.5 Evaluation of Phytobiotics

Phytobiotics can be screened for their bioactive compounds in vitro and in vivo. There are several qualitative and quantitative tests for the in vitro evaluation of the various compounds present in herbs.

9.5.1 In Vitro Evaluation

Phytochemical screening of compounds such as flavonoids, phenol, tannin, coumarin, and saponin in phytobiotics can be carried out using different colorimetric tests described by Mishra et al. (2018).[73] The formation of a yellow colour after chemical reaction of phytobiotics with 10% NaOH indicates the presence of coumarin. Whereas, stirring of herbal extract in distilled water with heating results in frothing (appearance of creamy miss of small bubbles) indicates the presence of saponin. A reaction between the ferric chloride and the supernatant (heated herbal extract in distilled water) forms a dark green colour, which indicates the presence of tannin. Similarly, the addition of three drops of hydrochloric acid to the product of a chemical reaction of herbal extract with 0.1 N NaOH produces an intense yellow colour, which indicate the presence of flavonoids. If the reaction between one drop of ferric chloride and herbal extract (dissolved in alcohol) generates an intense colour, it suggests the presence of the phenolic compound.

Natural antioxidants, including polyphenolic compounds, alkaloids and vitamins are found to be effective in capturing free radicals produced during metabolic activities.[74] Hence, it is important to study the antioxidant capacity of dietary crude extracts before conducting the animal experiment.[75] Phenols and flavonoids are considered abundant secondary metabolites of plants, which are required for stress response. Total phenols can be determined by the Folin-Ciocalteu method[76] using gallic acid as standard while total flavonoids can be estimated by aluminium chloride colorimetric assay, described by Marinova et al. (2005) using rutin as standard.[77] The phenol assay by Folin Ciocalteu reagent allows measuring the reducing capacity of the sample, and the results exhibit a positive relation between antioxidant capacity and total phenol content.[78] The antioxidant activity of different herbal extracts can be quantified using different methods like 2,2-diphenyl-1-picrylhydrazyl (DPPH) inhibition,[79] ferric reducing antioxidant power (FRAP) assay,[75] phosphomolybdenum method,[80] and ferrous ion chelating assay.[81] Sanchez-Moreno (2002)[82] and Pulido et al. (2000)[83] suggested that DPPH inhibition (%) and FRAP activity are effective and simple methods to estimate the activities of polyphenolic compounds present in crude herbal extracts or phytobiotics. Different bioactive phytochemicals in herbal extracts can be standardized qualitatively and quantitatively using different chromatographic techniques like gas chromatography, column chromatography, thin layer chromatography (TLC), high-performance liquid chromatography (HPLC), and high-performance thin layer chromatography (HPTLC).[84,85] In vitro antibacterial and antifungal activities of different phytobiotics or herbal extracts can be determined by measuring the size of inhibitory zones against pathogenic bacteria and fungal species following agar-disk diffusion method.[86,87] For instance, antibacterial activity of methanol extracts of *Ficus benghalensis* (prop root) and *Leucaena leucocephala* (pod seed) was evaluated by measurement of zone of inhibition against pathogenic bacteria, *Escherichia coli,* and *Aeromonas hydrophila.*[88] Anti-amylase and hypoglycemic activities of herbal extracts can be determined using spectrophotometric method described by Giancarlo et al. (2006).[89]

9.5.2 In Vivo Evaluation

In aquaculture, it has been reported that herbal extracts have the potential to act as alternative growth boosters, anti-stressors, immunological stimulants, appetizers, and digestive stimulants.[90] Many studies have shown that plant immune stimulants can enhance specific and non-specific immune protective mechanisms and increase or decrease fish losses against a variety of pathogens[91-94]. In most studies, growth performance, digestive and stress enzyme activities, immunological parameters, serum biochemical parameters, histology of tissues, haematological parameters and survival (%) were used as response parameters to evaluate the dietary supplementation of different herbal extracts in fish.[95-99] Fazio et al. (2021)[100] reported that *Labeo rohita* fed with ginger, *Zingiber officinale*, showed greater efficacy against bacteria compared to ashwagandha, *Withania coagulans*, suggesting that ginger can be considered as a potential and functional ingredient for carp feed. Mansour et al. (2018)[101] recommended that *Moringa oleifera* leaf meal in gilthead seabream (*Sparus aurata*) can be included in the diets at 5% level for the best immune status or at 10% for the high growth performance and acceptable immune surveillance. Verma et al. (2015)[102] reported that *Ficus benghalensis* and *Leucaena leucocephala* are perennial trees commonly known for their wide medicinal use in India. In addition, the immunostimulatory effect of *F. benghalensis* (aerial root) or *L. leucocephala* (pod seeds) at 5% powder was evaluated in fish feed of Indian spotted snakehead, *Channa punctate*. The study highlighted the potential immunostimulatory response of *F. benghalensis* and *L. leucocephala* powder when added to the diet of *C. punctatus*. Chowdhury et al. (2021)[103] reported that a combination of turmeric with ginger or garlic with an equal proportion at 1% inclusion level can be useful to improve growth, digestive, metabolic, antioxidant enzyme activities, and health status of *Labeo rohita* fingerlings. Phyto-additives probably could enhance gut health by establishing microbes that are more beneficial. It probably could modulate digestive enzymes' activities in fish.[104] Garg et al. (2021)[55] suggested that the dietary supplementation of ethanolic extract of the *Houttuynia cordata* leaf at 10 g/kg level could enhance the immune response of *L. rohita* fingerlings.

Kaur et al. (2020)[105] investigated the therapeutic effects of turmeric on the health of rohu (*Labeo rohita*) when challenged with pathogenic *Aeromonas veronii*. The highest supplementation of turmeric (20 g/kg feed) showed reduced bacterial pathogenicity by maintaining nearly normal microarchitecture of tissues compared to other turmeric-supplemented groups. Dietary effects of Chinese herbal mixture (CHM) composed of astragalus, angelica, hawthorn, licorice root, and honeysuckle on immune responses and disease resistant of Nile tilapia, GIFT strain were investigated.[106] The author suggested that CHM can be applied as a tilapia feed supplement to elevate fish immunity and disease resistance against *Aeromonas hydrophila* infection. Dietary *Withania somnifera* (ashwagandha) root powder at the level of 2 g/kg showed significantly improved immunological parameters and protection against *A. hydrophila* infection in *Labeo rohita.*[107] Exposure of *L. rohita* to 75 mg/kg of *Chaetomorpha antennina* methanolic extract (MECA) showed significantly a higher survival percentage against *Edwardsiella tarda* infection. Results indicated that MECA could be used as an immuno-stimulant in *L. rohita* diet.[108]

9.6 Mode of Administration of Green Drugs in Aquaculture

Herbs or herbal products can be administered in aquaculture through various modes like dietary supplementation, immersion or bath treatment, and intraperitoneal or intramuscular injection. For therapeutic and prophylactic treatment, the dose should be determined as per the medicinal plant and the extent of the infection.[109] Though, injection is found to be an effective method of phytobiotic administration, causing rapid production of an antibacterial substance in blood and tissue levels, the oral method was found to be feasible in aquafarming.[110] Studies have suggested that oral supplementation of phytobiotics can promote growth and disease resistance against several contagious fish pathogens.[111] Administration of phytobiotics through the intramuscular or intra-venal method is a tedious and stressful process, and hence, for large scale production units, it is not advisable. Immersion or bath treatment also has similar limitation. Moreover, this method is very effective for ectoparasites rather than endoparasites. The time of exposure and dose of the phytobiotics are also very critical in bath treatments. In oral administration, the inclusion of herbal extract or powder is possible along with the vitamin-mineral mix or coating of these phytobiotics in gel in the final feed. Water stability and leaching issues are serious concerns in this mode of delivery. Furthermore, the gastrointestinal absorption of phytobiotics is slow,[112] whereas oral delivery is economically viable and less labour intensive. The co-administration of herbal extracts or powders with probiotics or trace elements is also found to be more efficient than the normal oral delivery.[113] There are attempts to improve the oral delivery of phytobiotics by preparing nano-formulations. The targeted delivery of phytobiotics using nanoforms can reduce gut losses and improve intestinal absorption.[114] However, all these studies showed that the combination of several extracts is found to be more beneficial than individual extracts. This might be due to the synergistic effect of various bioactive compounds present in different plants. Moreover, a few studies reported that the dosage of administration is observed to be lower in combinations of extracts than in individual ones.[115] However, before administration of a specific mixture of herbs in aquaculture, it is important to be aware of the herbs toxicity and their bioactive compounds, as some in some cases the bioactive compounds are toxic to target fish or have an antagonistic function.

9.7 Action of Phytobiotics on Fish

The action of phytobiotics on fish is by immune modulation, stimulation of feed intake, enhancement of digestive enzyme activities, antioxidant activities, growth promotion, and antimicrobial action (Figure 9.1).

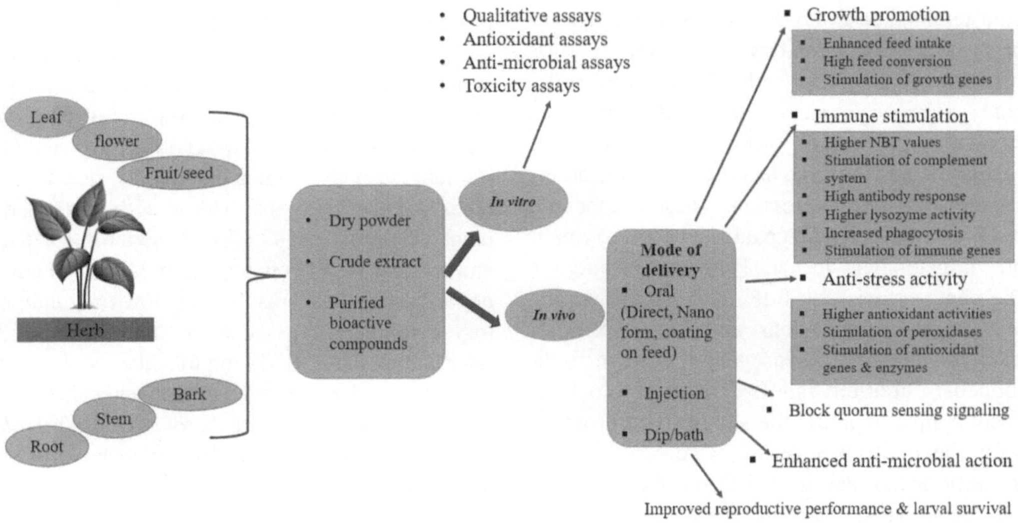

FIGURE 9.1 Phytobiotics, forms, and role in fish.

9.7.1 Immune Stimulation

The herbal extracts mainly stimulate the innate immune system of animals including fish, which has cellular, humoral, and epithelial barriers against pathogens. Immunostimulants trigger the specific and non-specific immune defense system in fish, which ensures and improve immunocompetency in fish. The first line of defense in fish is mainly through the innate immune system, comprising of monocytes, granulocytes, macrophages, and other humoral components like the complement system, lysozyme, etc.[116] Prophylactics through immunostimulants is possible through the stimulation of the innate elements like respiratory burst activity, phagocytosis, lysozyme activity, etc.[117] Several herbal meals and extracts also show specific antibody reaction. For example, *Cynodon dactylon* ethanolic extract at 5% level could improve the antigen-antibody reaction in *Catla catla*.[118] Similarly, *Achyranthes aspera* also showed enhanced protection in *Catla catla* and showed an enhanced bovine serum albumin (BSA)-specific antibodies.[119] Intraperitoneal as well as oral administration of phytobiotics exhibited increased innate immunity and antibody response in fish.[120] Moreover, administration of several herbal formulations showed enhanced lysozyme and phagocytic activities in fish.[121,122] Moreover, herbal extracts support haematological, antioxidant and biochemical responses in fish. Garlic supplementation increased mucus immune parameters in Caspian roach.[123] Compounds like curcumin present in turmeric reduces the chemokines and interleukines by binding on TNF-α and inhibiting NF-κB.[124]

The antioxidant property of herbal extract or meal is mainly because of free radical scavenging and electron donation.[125] Antimicrobial activity in fish is mainly estimated through lysozyme, peroxidase, and antiprotease activities.[126] During stress exposure, free radical production will be increased in the body, and the imbalance created between antioxidants and free radicals can generate oxidative stress at the cellular level.[127] The flavonoids, polyphenols, alkaloids, tannins, vitamins, and minerals are known to have free radical scavenging or electron donation properties at certain doses. Hence, they are known as non-enzymatic antioxidants.[128] A diet supplemented with 200 mg/Kg curcumin mitigated chromium toxicity in fish.[129] Similarly, Rajabiesterabadi et al. (2020) also reported that the deleterious effects of heavy metal toxicity (chromium exposure) are alleviated in common carp fed with 10 mg turmeric/Kg diet.[130] In this study, turmeric powder (10 mg/Kg diet) increased enzymatic antioxidants (superoxide dismutase, catalase, and glutathione peroxidase) and lysozyme activity while decreasing hepatic interleukines and TNF-expression. These compounds can work in parallel with the body enzymatic and non-enzymatic antioxidants and mitigate stress in fish. However, in microbicidal actions like phagocytosis, the free radicals that can kill the bacterial cells are used as weapons, and this is promoted by the herbal extracts. The antiprotease usually obstructs the proteolytic pathogens. Curcumin and its derivatives block the shikimate pathway, which supports the synthesis of aromatic rings required for the multiplication of bacteria.[131] Tilapia fed *Astragallus* polysaccharides at 1,500 mg/Kg enhanced respiratory burst, phagocytosis, and bactericidal activities.[132] The blood and serum chemistry are also influenced by the phytobiotics administration. Dietary supplementation of peppermint extract elevated the blood leukocyte count and respiratory burst activity in Capian trout (Salmo *trutta)*.[133] A quercetin glycoside presents in some herbal plants, rutin showed strong antioxidant and anti-stress properties in fish. Rutin extracted from *Toona sinensis* when given to *Vibrio* exposed white leg shrimp improved biochemical and haematological indicators. Increase in WBC, RBC and Hb values due to *Aeromonas* infection was brought back to normal level in goldfish by the supplementation of herbal mixture (*Ocimum sanctum*, *Curcuma longa* and *Azaridacta indica)* at 400 or 800 mg/Kg feed.[134]

The growth promotion in fish is facilitated by the enhanced digestive enzyme secretion and activity in fish. The herbal extracts stimulate the gut wall and pancreatic secretion of digestive enzymes.[67] *Mucuna pruriens* and *Cucurbita mixta* seed meal supplementation at 4 g/Kg feed improved weight gain and feed conversion ratio in *Oreochromis mossambicus*.[135,136] Similarly, Ali et al. (2017) reported that Ajwain and marjoram extract at 1–2% level improved growth and immunity in rainbow trout.[137] Another observation by Abdel-Tawwab et al (2018)[138] denoted that clove basil leaf extract could enhance the intestinal villi length and absorption area, thereby promoting the growth of *Clarias gariepinus*. Similar results were observed by Meena et al. (2022)[139] when rohu was fed with *Terminalia arjuna* bark extract.

9.8 Research Progress in Aquaculture

Incorporating herbal material or phytobiotics sourced from the leaves, barks, flowers, roots, seeds, or stems of plants as functional additives is one of the practical solutions to the management of diseases in aquaculture.[140] Herbal products have been reported to have immunomodulating potential attributed to the abundant content of bioactive compounds such as alkaloids, terpenoid, quinone, polyphenols, and flavonoids.[141-143] They are used in aquaculture as antibacterial, antiparasitic, antioxidant, and antifungal agents to improve fish resistance against pathogenic invaders.[144] Several studies have shown that herbal supplements possess the inherent potential to promote growth, improve feed palatability, gut functions and antioxidant responses, activate the immune system, and enhance resistance against pathogenic bacteria in fish.[145-148] A recent study that investigated the use of *Flos populi* extract as an immunostimulant showed an improvement in the associated immune response indicators mainly due to increased serum lysozyme and myeloperoxidase activities, as well as complement C3, immunoglobulin M, superoxide dismutase (SOD), and catalase (CAT) enzyme activities.[149] The inclusion of *Flos populi* extract also resulted in down-regulation of inflammatory cytokines/genes such as tumor necrosis factor (TNF-), interleukin (IL)-8, IL-1, and toll-like receptor 2 (TLR2), which all culminated in improved growth and enhanced survival rate in gibel carp challenged with *Aeromonas hydrophila*.[149] Similarly, *Ocimum gratissimum* administration greatly increased growth performance, innate immunity, and survival of *Clarias gariepinus* exposed to *Listeria monocytogenes*.[150] Combinations of herbal supplements have also been investigated for their synergistic effect in fish. Fawole et al. (2016)[144] investigated the additive effect of mango and guava leaf extracts in *Labeo rohita* and found that their use enhances lysozyme and respiratory burst activity, WBC, and protects the fish against *Aeromonas hydrophila* infection. A combination of herbal dietary supplements (*Coriandrum sativum*, *Malva sylvestris*, and *Quercus brantii*) has been shown to promote growth, immunological response, and resistance to *A. hydrophila* challenge in common carp[147] (Figure 9.2).

In African catfish, Fawole et al. (2022) confirmed the additive effect of a polyherbal formulation and stated that the immune response indices such as respiratory burst activity, heterophil count, total immunoglobulin, SOD, and CAT activity were improved.[148] The polyherbal formulation was also found to significantly improve intestinal villus height, absorption area, and muscle thickness, thus helping optimal absorption of nutrients and improving brush border integrity as well as growth. The presence of several bioactive substances or phytochemicals in these plants explains their medicinal potential in the form of immunostimulation, oxyradical scavengers, and growth enhancement. Furthermore, these herbal supplements were found to stimulate skin and mucosal immunities in response to pathogen infection in various species of fish, including *Nile tilapia,*[96] great sturgeon (*Huso huso),*[151] rainbow trout,[152] and *Acipenser baerii.*[153] Table 9.1 summarizes the most recent herbal supplements, alone or in combination, that have been studied for efficacy in various fish species.

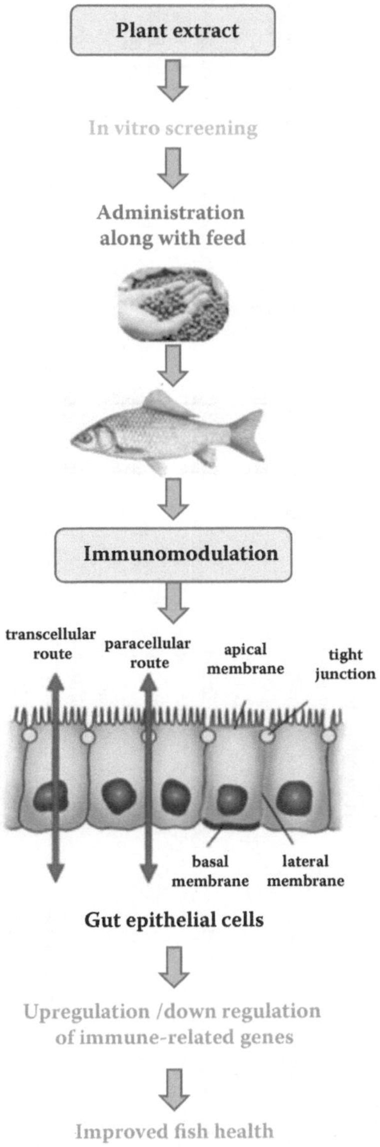

FIGURE 9.2 Overview of the possible outcome of the action of herbal immunostimulants on fish immunity.

Modified from Elumalai et al. (2020).

9.9 Market Scenario of Herbal Medicine

The global market for medical plants stood at US$151.91 billion in 2021 and is expected to increase at an annual growth rate of 11.6% to US$347.50 billion by 2029. The total world herbal trade is currently assessed at US$120 billion. India's share in the global export of herbs and herbal products is low due to unsophisticated agricultural and quality control procedures, lack of processing, research and development, standardisation in products and regulatory framework in trade of medicinal plants. These bottlenecks make also it convenient for the meat industry to continue using synthetic feed additives, which not only pose a threat to the environment but to health. Most of the commonly used feed additives in India are banned in many countries owing to the threat they pose to the health of their consumers. Considering the economic loss which cutting down on the use of these additives might accrue, producers do not consider this as an option until and unless an economically feasible option is available (Table 9.2).

Looking at the cost and perils of using synthetic feed additives, it is imperative to switch to an alternative that will ensure sustainable production and the health of consumers. Hence, in this section we are discussing the economic and health implications of using phytobiotics in aquaculture.

9.9.1 Phytobiotics Preparation and Cost

Most of the phytobiotics are plant derivatives; hence, the cost of raw materials is negligible. The most common ingredients used are fruits, vegetable wastes, agri-leaf waste, or very low-cost industry by-products, like leaf meal, etc. The formulation of any phytobiotic or plant-based additive includes the preparation of crude extract, which is usually done using organic solvents like industry grade ethanol, which costs less than $1.14/litre. Many studies have provided empirical evidence to prove that mostly ethanolic and aqueous extraction gives maximum dry matter recovery and enhances antioxidant properties,[133,164] which makes it a potent source for protection against free radicals and potential utilisation in aquafeed as a nutraceutical. Extraction follows simple procedures. The basic requirement is a solvent which can be re-collected and used again (70–80% is reused), thereby minimising wastage. Instruments like a rotary evaporator and a lyophiliser (Cryodessicator) are the major instruments required for the extract preparation. If the industrial preparation is standardised, then the overall cost of production for herbal extracts will be equivalent to or less than that of their synthetic counterparts.

TABLE 9.1

Recent Herbal Supplements Used in Aquaculture

Herbal Supplements	Plant Part	Bioactive Compound/ Molecule	Dose (%)	Species	Mode of Action	Pathogen Challenge	Survival Rate (%)	References
Hybanthus enneaspermus	Leaf extract	Alkaloids, flavonols, steroids, terpenoids	0.0, 0.1, 0.2, 0.3*, 0.4	*Labeo rohita*	Growth and SGR. lysozyme, phagocytic activity, respiratory burst activity, RBC, WBC, TP, SOD, ALP↑. AST, ALT, BG, cholesterol↓	*A. hydrophila*	69.5%	154
Allium stipitatum	Powder	thiosulfinates, polysulfanes, polyphenols, tannins, flavonoids, alkaloids, saponins, fructans, fructo-oligosaccharides, essential oils, amino acids, vitamins, pigments	0, 0.5, 1.0, 2.0 Recommended 1.27–1.35%	*Oncorhynchus mykiss*	Improved FW, WG, SGR, and FCR. protease, amylase, and LPS. TP, LZM, Ig, and complement C4 and C3↑. BG, LDH↓.	–	–	152
Mitracarpus scaber	Leaf extract	psychorubrin, ursolic acid, rutin, kaempferol-3-O-rutinoside, kaempferol, stigmasterol, and quercetin.	0, 0.2, 0.4, 0.6 (recommended 0.5–0.6%)	*Oreochromis niloticus*	Improved growth, Hb, Hct, RBC, WBC, lymphocyte, lysozyme, RBA,TP, SOD, CAT, GPx↑ AST, ALT, ALP, GLU, Chol., MDA↓	*Gyrodactylus malalai*	95%	155
Cymbopogon citratus	leaf powder	phytosterols, anthocyanin, amino acids, organic acid, phenolic compounds, volatile components, fatty acids, flavonoids, p-coumaric acid	0, 0.5, 1.0*, 1.5, 2.0	*Clarias gariepinus*	Hb, RBC, Hct, WBC, lymphocyte↑	*A. salmonicida*	75-95	156
Houttuynia cordata	leaf powder & extract	flavonoids, alkaloids, volatile oils, sterols, fatty acids, and polyphenolic acids	0, 0.25, 0.5, 1.0*, 2.0.	*Labeo rohita*	IGF-I, WG (%), SGR, protease enzyme↑	–	–	55
Momordica charantia	Extract	saponins, flavonoids, phenolic acids, cucurbitane triterpenoids, and triterpene glycosides	0, 0.25, 0.5, 1.0	*Cyprinus carpio*	No effect on growth, SOD, CAT, T-AOC, IL-6 and IL-10↑. IL-1β, IL-8 expression, MDA↓.	–	–	157
Centella asiatica	Powder	rutin, kaempferol, quercetin, gallic acid, luteolin and catechin	0, 0.5, 1.0*, 2.0	*Oreochromis niloticus*	No effect on growth, lysozyme, PA, ACH50, phagocytosis, RBA↑.	–	–	158
Scrophularia striata	Powder	salicylic acid, ferulic acid, cinnamic acid, coumaric acid, caffeic acid (phenolic acids), genistein, myricetin (flavonoids), and resveratrol (stilbenoid).	0, 1.0, 2.0, 3.0*	*Cyprinus carpio*	Improved growth and FCR, WBC, Hct,TP, ALB, globulin, lymp, monocyte, ACH50, BA, lysozyme, Ig M↑.	*Aeromonas hydrophila*	61.76	159
	Powder		0, 1.0, 2.0, 4.0*, 8.0			-	-	160

Nephelium lappaceum		geraniin, ellagic acid, corilagin, phenolics, ellagitannins		*Pangasianodon hypophthalmus*	Improved growth and FCR, lysozyme, PA, IL-1, IFN2a, IFN2b, and MHCII↑.			
Olea europaea	leaf extract	oleuropein, hydroxytyrosol, flavonoids, hydroxycinnamic acids, triterpenic acids, coumarin	0, 1.0*, 1.5, 2.0	*Oreochromis niloticus*	WBC, RBC, Hb, Hct, TP, AG ratio↑. CHOL, TRYGL↓.	–	–	161
Allium cepa	Peel powder	flavonoids (quercetin), vitamins, saponins, sulphur and phenolics compounds	0, 2.0, 4.0, 6.0	*Clarias gariepinus*	Improve WG, SGR, and FCR. WBC, neutrophils, SOD, CAT, GSH↑. CHOL, cortisol, GLU↓.	–	–	162
Flos populi	Extract	flavonoids and their glucopyranosides (quercetin, kaempferol, luteolin, apigenin, pinocembrin, chrysin), phenolics, and cardiac glycosides	0, 0.05, 0.1*, 0.15*, 0.2	*Carassius Auratus*	Improved FW, WG, and SGR. SOD, CAT, GPx, GSH, LZM, MPO, IgM, complement C3↑. PCC, MDA↓. *Keap1*, *TNF-α*, *IL-8*, *IL-1β*, *TLR2*↓↓	*A. hydrophila*	79.60	149
Camellia sinensis	Leaf extract	Catechins, epicatechins, theaflavins, flavonol glycosides, L-theanine, caffeine, theobromine, & volatile organic substances	0, 0.1, 0.2*, 0.4, 0.8	*Oreochromis niloticus*	Improved FW, WG, SGR, and FCR. LZM, peroxidase, ACH50, phagocytosis, RBA↑.	*Streptococcus agalactiae*	83.33	163
Berberis vulgaris	Fruit extract	acanthine, bargustanine, beriambine, flavonoids, ascorbic acid, triterpenes, and palmatine	0, 150, 300, 600, 750 mg/kg	*Acipenser baerii*	WBC, Lymp, complement C4, LZM, ACH50, protease, esterase, AKP, GPx, SOD↑. MDA↓. *HSP70*, cytochrome*P450* mRNA ↓↓	*Streptococcus iniae, Yersinia ruckeri, Escherichia coli, Listeria monocytogenes, Aeromonas hydrophila, Lactococcus garvieae*	–	153
Polyherbal formulation								
CHMM (*Ferula sinkiangensis, Medicago falcata, Allium sativum*)	Powder	mixed constituent	0, 0.5, 1.0, 1.5, 2.0*, 2.5	*Lateolabrax japonicus*	Improve growth and FCR, LZM SOD, TP, LPS, TA↑.	–	–	164
PHB (7 traditional herbal plant)	Powder	mixed constituent	0, 0.05, 0.1*, 0.2	*Clarias gariepinus*	Improve WG, SGR, Hb, RBC, Hct, heterophil, RBA, SOD, CAT, VH, AA, MT↑. ALT↓.	–	–	148

(*Continued*)

TABLE 9.1 (Continued)

Recent Herbal Supplements Used in Aquaculture

Herbal Supplements	Plant Part	Bioactive Compound/ Molecule	Dose (%)	Species	Mode of Action	Pathogen Challenge	Survival Rate (%)	References
Mangifera indica & Psidium guajava	Leaf extract	mangiferin phenolic acids, benzophenones, flavonoids, carotenoids, quercetin, isoquercetin, ascorbic acid, and tocopherols. rutin, naringenin, gallic acid, catechin, epicatechin, kaemp-ferol, isoflavonoids, and flavonoids.	0.5*, 1.0, (0.5 mango extract + 0.5 guava extract)	*Labeo rohita*	RBA, RBC, WBC, LA, Hb, TP, albumin, globulin ↑ SOD, CAT, A/G ratio ↓	*A. hydrophila*	80% (mango(, 69% (guava)	165
CHMM (12 traditional herbal plant)	Leaf powder	mixed constituent	0, 0.25, 0.5, 0.75, 1.0, 1.25. Recommended 0.9%	*Anguilla Anguilla*	Improve WG and SGR. enteric TA, amylase, TP, SOD, LZM, ALP/AKP↑.	–	–	166
CMO (*Coriandrum sativum, Malva sylvestris, Quercus Brantii*)	Extract	mixed constituent	0, 0.5, 1.0, 3.0*, 5.0	*Cyprinus carpio.*	Improve FW, WG, SGR, and FCR. RBC, WBC, Hct, Hb, TP, ALB, GLO, LZM, Ig, RBA, ACH50, protease, AKP, SOD, CAT, Gpx, GR↑. TRYGL, CHOL, BG, cortisol, MDA↓.	*A. hydrophila*	75.0	167

Note:

* Represents the optimal dosage with the most significant results. Increase (↑), Decrease (↓), Downregulation (↓↓)

Abbreviations: PA: peroxidase activities, ACH50: alternative hemolytic complement activity, MPO: myeloperoxidase activities, RBA: respiratory burst activities, lymp: lymphocyte, Hb: haemoglobin, RBC: red blood cell, WBC: white blood cell, Hct: haematocrit, TP: total protein, ALB: albumin, GLO: Globulin, RBA: Respiratory Burst Activity, LZM: lysozyme, bactericidal activities, Ig M: immunoglobulin M, Ig: total immunoglobulin, MHC II: major histocompatibility complex (class II), IL-I: interleukin I, IFN2a: Interferon alpha-2a, IFN2b:Interferon alpha-2b, SOD: superoxide dismutase, GPx: Glutathione peroxidase, CAT: Catalase, GSH: Glutathione, LPS: Lipase, TA: Trypsin activities, CHMM: Chinese herbal medicine, PHB: Polyherbal mixture, FW: final weight, WG: weight gain, SGR: specific growth rate, VH: villus height, AA: area of absorption, MT: muscle thickness, LDH: lactate dehydrogenase, BG: blood glucose, AG ratio: Albumin-globulin ratio, CHOL: Cholesterol, TRYGL: triglyceride, LCA: leucine aminopeptidase, PMA: metabolic activity of splenic macrophages, PKA: phagocytes activity, TAS: Total antioxidant status, GLURED: glutathione reductase, GR: Glutathione reductase, ALP/AKP: Alkaline phosphatase, AST: aspartate aminotransferase, ALT: alanine aminotransferase, PCC; Protein carbonyl, MDA: Malondialdehyde, Keap1: Kelch-like-ECH-associated protein 1, TNF-α: Tumor necrosis factor-α, IL-8: Interleukin, TLR2: toll-like receptor 2, HSP70: Heat shock protein 70.

TABLE 9.2

Prices of Commonly Used Synthetic Feed Additives

S. No	Synthetic Additives	Price (in US$)
1.	Butylated Hydroxyanisole, BHA	3.5/kg
2.	Powder Butylated Hydroxy Toluene – BHT, Pack Size: 25 Kg	8–10/kg
3.	Choline chloride	25–30/kg
4.	Taurine	4–8/kg
5.	Betaine	8–15/kg
6.	Ractopamine (Banned in many countries)	415–450/kg

9.10 Conclusion

Phytobiotics are versatile compounds with potent antiviral, antibacterial, antioxidant, and antifungal properties. Moreover, these substances are cheap, environmentally eco-friendly, biodegradable, and biocompatible, having less impact on the environment. Nevertheless, the application of phytobiotics in aquaculture is still at an incipient stage and confined to laboratory conditions only. The application of various research on the phytobiotics as a green approach to disease control can open a new avenue to commercialise herbal-based products.

Acknowledgement

The authors express immense gratitude to director/vice-chancellor, ICAR-CIFE, India, and vice-chancellor, University of Ilorin, Nigeria for providing permission to write this chapter on herbal immunoceuticals.

Disclosure statement: The authors declare that they have no conflicts of interest.

REFERENCES

1. Shamna, N., Sahu, N.P., Sardar, P., Fawole F.J., Kumar S. Changes in weight gain, digestive and metabolic enzyme activities in Labeo rohita fingerlings in response to multiple stress exposure and dietary nutraceutical. *Tropical Animal Health and Production*, *53*, pp.509.
2. Wendelaar Bonga, S.E., 1997. The stress response in fish. *Physiological Reviews*, *77*(3), pp.591–625.
3. Barton, B.A. and Iwama, G.K., 1991. Physiological changes in fish from stress in aquaculture with emphasis on the response and effects of corticosteroids. *Annual Review of Fish Diseases*, *1*, pp.3–26.
4. Rehman, S., Gora, A.H., Ahmad, I. and Rasool, S.I., 2017. Stress in aquaculture hatcheries: source, impact and mitigation. *International Journal of Current Microbiology and Applied Sciences*, *6*(10), pp.3030–3045.
5. Serrano, P.H., 2005. *Responsible use of antibiotics in aquaculture* (*Vol. 469*). Food & Agriculture Org.
6. Mutalipassi, M., Esposito, R., Ruocco, N., Viel, T., Costantini, M. and Zupo, V., 2021. Bioactive Compounds of Nutraceutical Value from Fishery and Aquaculture Discards. *Foods*, *10*(7), p.1495.
7. Brower, V., 1998. Nutraceuticals: poised for a healthy slice of the healthcare market?. *Nature Biotechnology*, *16*(8), pp.728–731.
8. Sahoo, P.K., 2007. Role of immunostimulants in disease resistance of fish. *CABI Reviews*, (2007), p.18.
9. Mutalipassi, M., Esposito, R., Ruocco, N., Viel, T., Costantini, M. and Zupo, V., 2021. Bioactive Compounds of Nutraceutical Value from Fishery and Aquaculture Discards. *Foods*, *10*(7), p.1495.
10. Cristea, I.A., Kok, R.N. and Cuijpers, P., 2016. The effectiveness of cognitive bias modification interventions for substance addictions: a meta-analysis. *PLoS One*, *11*(9), p.e0162226.
11. Mukherjee, A.K., Doley, R. and Saikia, D., 2008. Isolation of a snake venom phospholipase A2 (PLA2) inhibitor (AIPLAI) from leaves of Azadirachta indica (Neem): Mechanism of PLA2 inhibition by AIPLAI in vitro condition. *Toxicon*, *51*(8), pp.1548–1553.
12. Alam, M.N., Ahmed, G.U. and Chowdhury, M.B.R., 2014. Performance of herbal extracts on diseased fish. *Bangladesh Journal of Veterinary Medicine*, *12*(2), pp.225–230.
13. Krishnani, K.K., Gupta, B.P., Joseph, K.O., Muralidhar, M. and Nagavel, A., 2002. Studies on the use of neem products for removal of ammonia from brackishwater. *Journal of Environmental Science and Health, Part A*, *37*(5), pp.893–904.
14. Citarasu, T., 2010. Herbal biomedicines: a new opportunity for aquaculture industry. *Aquaculture International*, *18*(3), pp.403–414.
15. Chandra, H., Kumari, P., Prasad, R., Gupta, S.C. and Yadav, S., 2021. Antioxidant and antimicrobial activity displayed by a fungal endophyte Alternaria alternata isolated from Picrorhiza kurroa from Garhwal Himalayas, India. *Biocatalysis and Agricultural Biotechnology*, *33*, p.101955.
16. Joseph, B. and Justin Raj, S. 2010. Pharmacogno`stic and phytochemical properties of Aloe vera linn an overview. *International Journal of Pharmacuetical Science Reviews and Research*, *4*(2), pp.106–110.
17. Ranjbar, M., Ghorbanpoor, M., Peyghan, R., Mesbah, M. and Razi Jalali, M., 2010. Effects of dietary Aloe vera on some specific and nonspecific immunity in the common carp (Cyprinus carpio). *Iranian Journal of Veterinary Medicine*, *4*(3).
18. Trejo-Flores, J.V., Luna-González, A., Álvarez-Ruíz, P., Escamilla-Montes, R., Peraza-Gómez, V., Diarte-Plata, G., Esparza-Leal, H.M., Campa-Córdova, Á.I., Gámez-Jiménez, C. and Rubio-Castro, A., 2016. Protective effect of Aloe vera in Litopenaeus vannamei challenged with Vibrio parahaemolyticus and white spot syndrome virus. *Aquaculture*, *465*, pp.60–64.
19. Haniffa, M.A., Bharathi, B.K., Margaret, I.V. and Paray, B.A., 2013. Effect of a Probiotic and Herbal Additives on Growth, Survival and Disease Resistance of Striped Murrel. World Aquaculture June, pp.64–67.
20. Gabriel, N.N., Qiang, J., Ma, X.Y., Xu, P. and Nakwaya, D.N., 2017. Effects of dietary Aloe vera crude extracts on digestive enzyme activities and muscle proximate composition of GIFT tilapia juveniles. *South African Journal of Animal Science*, *47*(6), pp.904–913.

21. Upadhyay, R.K., 2018. Nutraceutical, therapeutic, and pharmaceutical potential of Aloe vera: A review. *International Journal of Green Pharmacy (IJGP)*, *12*(01).
22. Gabriel, N.N., Qiang, J., Ma, X.Y., He, J., Xu, P., & Omoregie, E. (2017). Sex-Reversal Effect of Dietary Aloe vera (Liliaceae) on Genetically Improved Farmed Nile Tilapia Fry. North American Journal of Aquaculture, *79*(1), pp.100–105.
23. Prasad, S., Tyagi, A.K. and Aggarwal, B.B., 2014. Recent developments in delivery, bioavailability, absorption and metabolism of curcumin: the golden pigment from golden spice. *Cancer research and treatment: official journal of Korean Cancer Association*, *46*(1), pp.2–18.
24. Alagawany, M., Ashour, E.A. and Reda, F.M., 2016. Effect of dietary supplementation of garlic (Allium sativum) and turmeric (Curcuma longa) on growth performance, carcass traits, blood profile and oxidative status in growing rabbits. *Annals of Animal Science*, *16*(2), pp.489–505.
25. Alagawany, M., Farag, M.R., Abdelnour, S.A., Dawood, M.A., Elnesr, S.S. and Dhama, K., 2021. Curcumin and its different forms: A review on fish nutrition. *Aquaculture*, *532*, p.736030.
26. Wei, Q.Y., Chen, W.F., Zhou, B., Yang, L. and Liu, Z.L., 2006. Inhibition of lipid peroxidation and protein oxidation in rat liver mitochondria by curcumin and its analogues. *Biochimica et Biophysica Acta (BBA)-General Subjects*, *1760*(1), pp.70–77.
27. Ak, T. and Gülçin, İ., 2008. Antioxidant and radical scavenging properties of curcumin. *Chemico-Biological Interactions*, *174*(1), pp.27–37.
28. Alambra, J.R., Alenton, R.R.R., Gulpeo, P.C.R., Mecenas, C.L., Miranda, A.P., Thomas, R.C., Velando, M.K.S., Vitug, L.D. and Maningas, M.B.B., 2012. Immunomodulatory effects of turmeric, Curcuma longa (Magnoliophyta, Zingiberaceae) on macrobrachium rosenbergii (Crustacea, Palaemonidae) against Vibrio alginolyticus (Proteobacteria, Vibrionaceae). *Aquaculture, Aquarium, Conservation & Legislation*, *5*(1), pp.13–17.
29. Mahmoud, H.K., Al-Sagheer, A.A., Reda, F.M., Mahgoub, S.A. and Ayyat, M.S., 2017. Dietary curcumin supplement influence on growth, immunity, antioxidant status, and resistance to Aeromonas hydrophila in Oreochromis niloticus. *Aquaculture*, *475*, pp.16–23.
30. Yonar, M.E., Yonar, S.M., İspir, Ü. and Ural, M.Ş., 2019. Effects of curcumin on haematological values, immunity, antioxidant status and resistance of rainbow trout (Oncorhynchus mykiss) against Aeromonas salmonicida subsp. achromogenes. *Fish & Shellfish Immunology*, *89*, pp.83–90.
31. Patil, V.M., Das, S. and Balasubramanian, K., 2016. Quantum chemical and docking insights into bioavailability enhancement of curcumin by piperine in pepper. *The Journal of Physical Chemistry A*, *120*(20), pp.3643–3653.
32. Ghalandarlaki, N., Alizadeh, A.M., and Ashkani-Esfahani, S. 2014. Nanotechnology-applied curcumin for different diseases therapy. *BioMed Research International 2014*, Article ID 394264, 23 pages, 2014. https://doi.org/10.1155/2014/394264
33. Martin, K.W. and Ernst, E., 2003. Herbal medicines for treatment of bacterial infections: a review of controlled clinical trials. *Journal of Antimicrobial Chemotherapy*, *51*(2), pp.241–246.
34. Nya, E.J. and Austin, B., 2009. Use of garlic, Allium sativum, to control Aeromonas hydrophila infection in rainbow trout, Oncorhynchus mykiss (Walbaum). *Journal of fish diseases*, *32*(11), pp.963–970.
35. Shakya, S.R. and Labh, S.N., 2014. Medicinal uses of garlic (Allium sativum) improves fish health and acts as an immunostimulant in aquaculture. *European Journal of Biotechnology and Bioscience*, *2*(4), pp.44–47.
36. Mohan, L., Amberkar, M.V. and Kumari, M., 2011. Ocimum sanctum linn.(TULSI)-an overview. *Int J Pharm Sci Rev Res*, *7*(1), pp.51–53.
37. Shankar, K. and Kiran, B.R., 2013. Review on usage of medicinal plants in fish diseases. International Journal of Pharma and Bio Sciences, *4*(3), pp.975–986.
38. Sivaram, V., Babu, M.M., Immanuel, G., Murugadass, S., Citarasu, T. and Marian, M.P., 2004. Growth and immune response of juvenile greasy groupers (Epinephelus tauvina) fed with herbal antibacterial active principle supplemented diets against Vibrio harveyi infections. *Aquaculture*, *237*(1-4), pp.9–20.
39. Logambal, S.M., Venkatalakshmi, S. and Dinakaran Michael, R., 2000. Immunostimulatory effect of leaf extract of Ocimum sanctum Linn. in Oreochromis mossambicus (Peters). *Hydrobiologia*, *430*(1), pp.113–120.
40. Das, R., Raman, R.P., Saha, H. and Singh, R., 2015. Effect of Ocimum sanctum Linn.(Tulsi) extract on the immunity and survival of Labeo rohita (Hamilton) infected with Aeromonas hydrophila. *Aquaculture Research*, *46*(5), pp.1111–1121.
41. Sutili, F.J., Velasquez, A., Pinheiro, C.G., Heinzmann, B.M., Gatlin III, D.M. and Baldisserotto, B., 2016. Evaluation of Ocimum americanum essential oil as an additive in red drum (Sciaenops ocellatus) diets. *Fish & Shellfish Immunology*, *56*, pp.155–161.
42. Ali, B.H., Blunden, G., Tanira, M.O. and Nemmar, A., 2008. Some phytochemical, pharmacological and toxicological properties of ginger (Zingiber officinale Roscoe): a review of recent research. *Food and Chemical Toxicology*, *46*(2), pp.409–420.
43. Nya, E.J. and Austin, B., 2009. Use of garlic, Allium sativum, to control Aeromonas hydrophila infection in rainbow trout, Oncorhynchus mykiss (Walbaum). *Journal of Fish Diseases*, *32*(11), pp.963–970.
44. Gholipour Kanani, H., Nobahar, Z., Kakoolaki, S. and Jafarian, H., 2014. Effect of ginger-and garlic-supplemented diet on growth performance, some hematological parameters and immune responses in juvenile Huso huso. *Fish Physiology and Biochemistry*, *40*(2), pp.481–490.
45. Immanuel, G., Uma, R.P., Iyapparaj, P., Citarasu, T., Punitha Peter, S.M., Michael Babu, M. and Palavesam, A., 2009. Dietary medicinal plant extracts improve growth, immune activity and survival of tilapia Oreochromis mossambicus. *Journal of Fish Biology*, *74*(7), pp.1462–1475.
46. Levy, G., Zilberg, D., Paladini, G. and Fridman, S., 2015. Efficacy of ginger-based treatments against infection with Gyrodactylus turnbulli in the guppy (Poecilia reticulata (Peters)). *Veterinary Parasitology*, *209*(3-4), pp.235–241.

47. Sukumaran, V., Park, S.C. and Giri, S.S., 2016. Role of dietary ginger Zingiber officinale in improving growth performances and immune functions of Labeo rohita fingerlings. *Fish & Shellfish Immunology*, *57*, pp.362–370.
48. Mishra, B. and Singh Sangwan, N., 2019. Amelioration of cadmium stress in Withania somnifera by ROS management: active participation of primary and secondary metabolism. *Plant Growth Regulation*, *87*(3), pp.403–412.
49. Yogeeswaran, A., Velmurugan, S., Punitha, S.M.J., Babu, M.M., Selvaraj, T., Kumaran, T. and Citarasu, T., 2012. Protection of Penaeus monodon against white spot syndrome virus by inactivated vaccine with herbal immunostimulants. *Fish & Shellfish Immunology*, *32*(6), pp.1058–1067.
50. Srivastava, A., Ansal, M.D. and Khairnar, S.O., 2020. Effect of ashwagandha (Withania somnifera) root powder supplementation on survival, growth and flesh quality of an indian major carp, Labeo rohita (Ham.) fingerlings. *Animal Nutrition and Feed Technology*, *20*(3), pp.515–524.
51. Sharma, A., Chanu, T.I. and Deo, A.D., 2017. Dietary ashwagandha, Withania somnifera (L. dunal) potentiates growth, haemato-biochemical response and disease resistance of Labeo rohita (Hamilton, 1822) against Aeromonas hydrophila infection. *Journal of Entomology and Zoology Studies*, *5*(5), pp.1113–1119.
52. Laltlanmawia, C., Saha, R.K., Saha, H. and Biswas, P., 2019. Ameliorating effects of dietary mixture of Withania somnifera root extract and vitamin C in Labeo rohita against low pH and waterborne iron stresses. *Fish & Shellfish Immunology*, *88*, pp.170–178.
53. Maiti, S., Saha, S., Jana, P., Chowdhury, A., Khatua, S. and Ghosh, T.K., 2021. Effect of dietary Andrographis paniculata leaf extract on growth, immunity, and disease resistance against Aeromonas hydrophila in Pangasianodon hypopthalmus. *Journal of Applied Aquaculture*, pp.1–25.
54. Direkbusarakom, S., Ezura, Y., Yoshimizu, M. and Herunsalee, A., 1998. Efficacy of Thai traditional herb extracts against fish and shrimp pathogenic bacteria. *Fish Pathology*, *33*(4), pp.437–441.
55. Garg, C.K., Sahu, N.P., Shamna, N., Deo, A.D., Fawole, F.J., Kumar, S. and Maiti, M.K., 2019. Effect of dietary Houttuynia cordata leaf meal and leaf extract on the growth performance, nutrient utilization and expression of IGF-I gene in Labeo rohita. *Aquaculture Nutrition*, *25*(3), pp.702–711.
56. Pandey, G., Madhuri, S. and Mandloi, A.K., 2012. Medicinal plants useful in fish diseases. *Plant Archives*, *12*(1), pp.1–4.
57. Christybapita, D., Divyagnaneswari, M., Michael, D.R. 2007. Oral administration of Eclipta alba leaf aqueous extract enhances the non-specific immune responses and disease resistance of *Oreochromis mossambicus*. *Fish and Shellfish Immunology*, *23*(4), 840–852.
58. Rani, T.V.J. 1999. Fourth year annual report (CSIR Research Associateship) submitted to Council of Scientific and Industrial Research, New Delhi, 1999.
59. Tu, X., Ling, F., Huang, A., Zhang, Q. and Wang, G., 2013. Anthelmintic efficacy of Santalum album (Santalaceae) against monogenean infections in goldfish. *Parasitology Research*, *112*(8), pp.2839–2845.
60. Balasubramanian, G., Sarathi, M., Venkatesan, C., Thomas, J. and Hameed, A.S., 2008. Oral administration of antiviral plant extract of Cynodon dactylon on a large scale production against white spot syndrome virus (WSSV) in Penaeus monodon. *Aquaculture*, *279*(1-4), pp.2–5.
61. Rao, Y.V., Das, B.K., Jyotyrmayee, P. and Chakrabarti, R., 2006. Effect of Achyranthes aspera on the immunity and survival of Labeo rohita infected with Aeromonas hydrophila. *Fish & Shellfish Immunology*, *20*(3), pp.263–273.
62. Harikrishnan, R., Balasundaram, C. and Heo, M.S., 2012. Effect of *Inonotus obliquus* enriched diet on hematology, immune response, and disease protection in kelp grouper, *Epinephelus bruneus* against *Vibrio harveyi*. *Aquaculture*, *344*, pp.48–53.
63. Xue-Gang, H., Lei, L., Cheng, C., Kun, H., Xian-Le, Y. and Gao-Xue, W., 2013. In vitro screening of Chinese medicinal plants for antifungal activity against Saprolegnia sp. and Achlya klebsiana. *North American Journal of Aquaculture*, *75*(4), pp.468–473.
64. Dubber, D. and Harder, T., 2008. Extracts of *Ceramium rubrum*, *Mastocarpus stellatus* and *Laminaria digitata* inhibit growth of marine and fish pathogenic bacteria at ecologically realistic concentrations. *Aquaculture*, *274* (2-4), pp.196–200.
65. Gupta, N., Kar, S.R. and Chakraborty, A. 2021. A Review on Medicinal Plants and Immune Status of Fish. *Egyptian Journal of Aquatic Biology and Fisheries*, *25*(2), pp.897–912.
66. Elumalai, P., Kurian, A., Lakshmi, S., Faggio, C., Esteban, M.A. and Ringø, E., 2020. Herbal immunomodulators in aquaculture. *Reviews in Fisheries Science & Aquaculture*, *29*(1), pp.33–57.
67. Soltani, M., Lymbery, A., Song, S.K., Shekarabi, P.H. 2019.Adjuvant effects of medicinal herbs and probiotics for fish vaccines. *Reviews in Aquaculture*, *11*, pp.1325–1341.
68. Doan, H.V., Hoseinifar, S.Y., Jaturasitha, S., Dawood, M.O., Harikrishnan, R. 2020. The effects of berberine powder supplementation on growth performance, skin mucus immune response, serum immunity, and disease resistance of Nile tilapia (Oreochromis niloticus) fingerlings. *Aquaculture*, *520*, pp.734927.
69. Vijayaram, S., Sun, Y., Zuorro, A., Ghafarifarsani, H., Doan, H.V., Hoseinifar, S.H. 2022. Bioactive immunostimulants as health-promoting feed additives in aquaculture: A review. *Fish and Shellfish Immunology*, *130*, pp.294–308.
70. Gupta, N., Kar, S.R., Chakraborty, A. 2021. A Review on Medicinal Plants and Immune Status of Fish. *Egyptian Journal of Aquatic Biology & Fisheries*, *25*(2), pp.897 – 912.
71. Reverter, M., Tapissier-Bontemps, N., Sasal, P. and Saulnier, D., 2017. Use of medicinal plants in aquaculture. *Diagnosis and control of diseases of fish and shellfish*, pp.223–261.
72. Van Hai, N., 2015. The use of medicinal plants as immunostimulants in aquaculture: A review. *Aquaculture*, *446*, pp.88–96.
73. Mishra, R., Gupta, A.K., Kumar, A., Lal, R.K., Saikia, D. and Chanotiya, C.S., 2018. Genetic diversity, essential oil composition, and in vitro antioxidant and antimicrobial

activity of Curcuma longa L. germplasm collections. *Journal of Applied Research on Medicinal and Aromatic Plants*, *10*, pp.75–84.
74. Ames, B.N., Gold, L.S. and Willett, W.C., 1995. The causes and prevention of cancer. *Proceedings of the National Academy of Sciences*, *92*(12), pp.5258–5265.
75. Omidbaigi, R. and Nasiri, M.F., 2004. Quantitative distribution of hesperidin in Citrus species, during fruit maturation and optimal harvest time. *Natural Product Radiance*, *3*(1), pp.12–15.
76. Singleton, V.L. and Rossi, J.A., 1965. Colorimetry of total phenolics with phosphomolybdic-phosphotungstic acid reagents. *American journal of Enology and Viticulture*, *16*(3), pp.144–158.
77. Marinova, G. and Batchvarov, V., 2011. Evaluation of the methods for determination of the free radical scavenging activity by DPPH. *Bulgarian Journal of Agricultural Science*, *17*(1), pp.11–24.
78. Hoseinifar, S.H., Sun, Y.Z., Zhou, Z., Van Doan, H., Davies, S.J. and Harikrishnan, R., 2020. Boosting immune function and disease bio-control through environment-friendly and sustainable approaches in finfish aquaculture: herbal therapy scenarios. *Reviews in Fisheries Science & Aquaculture*, *28*(3), pp.303–321.
79. Brand-Williams, W., Cuvelier, M.E. and Berset, C.L.W.T., 1995. Use of a free radical method to evaluate antioxidant activity. *LWT-Food Science and Technology*, *28*(1), pp.25–30.
80. Benzie, I.F. and Strain, J.J., 1996. The ferric reducing ability of plasma (FRAP) as a measure of "antioxidant power": the FRAP assay. *Analytical Biochemistry*, *239*(1), pp.70–76.
81. Prieto, P., Pineda, M. and Aguilar, M., 1999. Spectrophotometric quantitation of antioxidant capacity through the formation of a phosphomolybdenum complex: specific application to the determination of vitamin E. *Analytical Biochemistry*, *269*(2), pp.337–341.
82. Singh, N. and Rajini, P.S., 2004. Free radical scavenging activity of an aqueous extract of potato peel. *Food Chemistry*, *85*(4), pp.611–616.
83. Sánchez-Moreno, C., 2002. Methods used to evaluate the free radical scavenging activity in foods and biological systems. *Food Science and Technology International*, *8*(3), pp.121–137.
84. Pulido, R., Bravo, L. and Saura-Calixto, F., 2000. Antioxidant activity of dietary polyphenols as determined by a modified ferric reducing/antioxidant power assay. *Journal of Agricultural and Food Chemistry*, *48*(8), pp.3396–3402.
85. Kushwaha, S.K., Kushwaha, N., Maurya, N. and Rai, A.K., 2010. Role of markers in the standardization of herbal drugs: a review. *Archives of Applied Science Research*, *2*(1), pp.225–229.
86. Farooqui, N.A., Dey, A., Gn, S. and Ts, E., 2014. Development and validation of reverse phase high performance liquid chromatography method for quantitative estimation of vasicine in bulk and pharmaceutical dosage form. *Development*, *7*(5).
87. Rios, J.L., Recio, M.C. and Villar, A., 1988. Screening methods for natural products with antimicrobial activity: a review of the literature. *Journal of Ethnopharmacology*, *23*(2-3), pp.127–149.
88. Alzoreky, N.S. and Nakahara, K., 2003. Antibacterial activity of extracts from some edible plants commonly consumed in Asia. *International Journal of Food Microbiology*, *80*(3), pp.223–230.
89. Verma, V.K., Sehgal, N. and Prakash, O., 2015. Characterization and screening of bioactive compounds in the extract prepared from aerial roots of Ficus benghalensis. *International Journal of Pharmacuetical Science and Research*, *6*(5056), pp.5056–5069.
90. Giancarlo §, S., Rosa §, L.M., Nadjafi, F. and Francesco, M., 2006. Hypoglycaemic activity of two spices extracts: Rhus coriaria L. and Bunium persicum Boiss. *Natural Product Research*, *20*(9), pp.882–886.
91. Heidarieh, M., Mirvaghefi, A.R., Akbari, M., Sheikhzadeh, N., Kamyabi-Moghaddam, Z., Askari, H. and Shahbazfar, A.A., 2013. Evaluations of H ilyses™, fermented S accharomyces cerevisiae, on rainbow trout (O ncorhynchus mykiss) growth performance, enzymatic activities and gastrointestinal structure. *Aquaculture Nutrition*, *19*(3), pp.343–348.
92. Citarasu, T., 2010. Herbal biomedicines: a new opportunity for aquaculture industry. *Aquaculture International*, *18*(3), pp.403–414.
93. Harikrishnan, R., Balasundaram, C. and Heo, M.S., 2011. Impact of plant products on innate and adaptive immune system of cultured finfish and shellfish. *Aquaculture*, *317*(1-4), pp.1–15.
94. Reverter, M., Bontemps, N., Lecchini, D., Banaigs, B. and Sasal, P., 2014. Use of plant extracts in fish aquaculture as an alternative to chemotherapy: current status and future perspectives. *Aquaculture*, *433*, pp.50–61.
95. Van Doan, D., Huynh, B.A., Nguyen, T.D., Cao, X.T., Nguyen, V.C., Nguyen, T.L.H., Nguyen, H.T. and Le, V.T., 2020. Biosynthesis of silver and gold nanoparticles using aqueous extract of Codonopsis pilosula roots for antibacterial and catalytic applications. *Journal of Nanomaterials*, *2020*.
96. Kumar, V., Akinleye, A.O., Makkar, H.P.S., Angulo-Escalante, M.A. and Becker, K., 2012. Growth performance and metabolic efficiency in Nile tilapia (Oreochromis niloticus L.) fed on a diet containing Jatropha platyphylla kernel meal as a protein source. *Journal of Animal Physiology and Animal Nutrition*, *96*(1), pp.37–46.
97. Thakur, N.S., Verma, K.S. and Rana, R.C., 2014. Growth and yield performance of ashwagandha (Withania somnifera) under agroforestry. *Indian Journal of Agricultural Sciences*, *84*(8), pp.937–941.
98. Sharma, A., Bachheti, A., Sharma, P., Bachheti, R.K. and Husen, A., 2020. Phytochemistry, pharmacological activities, nanoparticle fabrication, commercial products and waste utilization of Carica papaya L.: A comprehensive review. *Current Research in Biotechnology*, *2*, pp.145–160.
99. Chowdhury, M.A.K., Song, H., Liu, Y., Bunod, J.D. and Dong, X.H., 2021. Effects of Microencapsulated Organic Acid and Their Salts on Growth Performance, Immunity, and Disease Resistance of Pacific White Shrimp Litopenaeus vannamei. *Sustainability*, *13*(14), p.7791.
100. Syed, R., Masood, Z., Ul Hassan, H., Khan, W., Mushtaq, S., Ali, A., Gul, Y. et al. 2022. Growth performance, haematological assessment and chemical composition of

Nile tilapia, Oreochromis niloticus (Linnaeus, 1758) fed different levels of Aloe vera extract as feed additives in a closed aquaculture system. *Saudi Journal of Biological Sciences*, *29*(1), pp. 296–303.
101. Fazio, F., Naz, S., Habib, S.S., Hashmi, M.A.H., Ali, M., Saoca, C. and Ullah, M., 2021. Effect of Fortified Feed with Phyto-Extract on the First Physical Barrier (Mucus) of Labeo rohita. *Animals*, *11*(5), p.1308.
102. Mansour, A.T., Miao, L., Espinosa, C., García-Beltrán, J.M., Ceballos Francisco, D.C. and Esteban, M., 2018. Effects of dietary inclusion of Moringa oleifera leaves on growth and some systemic and mucosal immune parameters of seabream. *Fish Physiology and Biochemistry*, *44*(4), pp.1223–1240.
103. Verma, V.K., Rani, K.V., Sehgal, N. and Prakash, O., 2015. Enhanced disease resistance in the Indian snakehead, Channa punctata against Aeromonas hydrophila, through 5% feed supplementation with F. benghalensis (aerial root) and L. leucocephala (pod seed). *Aquaculture International*, *23*(5), pp.1127–1140.
104. Chowdhury, D.K., Sahu, N.P., Sardar, P., Deo, A.D., Bedekar, M.K., Singha, K.P. and Maiti, M.K., 2021. Feeding turmeric in combination with ginger or garlic enhances the digestive enzyme activities, growth and immunity in Labeo rohita fingerlings. *Animal Feed Science and Technology*, *277*, p.114964.
105. Hashemi, S.R. and Davoodi, H., 2011. Herbal plants and their derivatives as growth and health promoters in animal nutrition. *Veterinary Research Communications*, *35*(3), pp.169–180.
106. Kaur, A., Shanthanagouda, A.H., Kaur, V.I., Bansal, N. and Billekallu Thammegowda, N.K., 2020. Biochemical and histomorphological associated in vivo responses of turmeric supplemented diets in Rohu, Labeo rohita (Linn.). *Aquaculture Research*, *51*(9), pp.3915–3923.
107. Tang, J., Cai, J., Liu, R., Wang, J., Lu, Y., Wu, Z. and Jian, J., 2014. Immunostimulatory effects of artificial feed supplemented with a Chinese herbal mixture on Oreochromis niloticus against Aeromonas hydrophila. *Fish & Shellfish Immunology*, *39*(2), pp.401–406.
108. Sharma, A., Deo, A.D., Riteshkumar, S.T., Chanu, T.I. and Das, A., 2010. Effect of Withania somnifera (L. Dunal) root as a feed additive on immunological parameters and disease resistance to *Aeromonas hydrophila* in *Labeo rohita* (Hamilton) fingerlings. *Fish & Shellfish Immunology*, *29*(3), pp.508–512.
109. Sattanathan, G., Tamizhazhagan, V., Padmapriya, S., Liu, W.C. and Balasubramanian, B., 2020. Effect of green algae Chaetomorpha antennina extract on growth, modulate immunity, and defenses against Edwardsiella tarda infection in Labeo rohita. *Animals*, *10*(11), p.2033.
110. Direkbusarakom, S., 2004. Application of medicinal herbs to aquaculture in Asia. *Walailak Journal of Science and Technology (WJST)*, *1*(1), pp.7–14.
111. Kum, C. and Sekkin, S., 2011. The immune system drugs in fish: immune function, immunoassay, drugs. *Recent Advances in Fish Farms*, pp.169–210.
112. Adeeyo, A.O., Edokpayi, J.N., Alabi, M.A., Msagati, T.A. and Odiyo, J.O., 2021. Plant active products and emerging interventions in water potabilisation: disinfection and multi-drug resistant pathogen treatment. *Clinical Phytoscience*, *7*(1), pp.1–16.
113. Bulfon, C., Volpatti, D. and Galeotti, M., 2015. Current research on the use of plant-derived products in farmed fish. *Aquaculture Research*, *46*(3), pp.513–551.
114. Harikrishnan, R., Balasundaram, C. and Heo, M.S., 2011. Impact of plant products on innate and adaptive immune system of cultured finfish and shellfish. *Aquaculture*, *317*(1-4), pp.1–15.
115. Seethalakshmi, P.S., Rajeev, R., Kiran, G.S. and Selvin, J., 2021. Shrimp disease management for sustainable aquaculture: innovations from nanotechnology and biotechnology. *Aquaculture International*, *29*(4), pp.1591–1620.
116. Gupta, S. and Bharalee, R., 2021. Genetic Diversity and Population Structure of a Medicinal Herb *Houttuynia cordata* Thunb. of North-East India. *Plant Molecular Biology Reporter*, *39*(2), pp.434–442.
117. Magnadóttir, B., 2006. Innate immunity of fish (overview). *Fish & Shellfish Immunology*, *20*(2), pp.137–151.
118. Bilen, S., Bulut, M. and Bilen, A.M., 2011. Immunostimulant effects of Cotinus coggyria on rainbow trout (Oncorhynchus mykiss). *Fish & Shellfish Immunology*, *30*(2), pp.451–455.
119. Kaleeswaran, B., Ilavenil, S. and Ravikumar, S., 2012. Changes in biochemical, histological and specific immune parameters in Catla catla (Ham.) by Cynodon dactylon (L.). *Journal of King Saud University-Science*, *24*(2), pp.139–152.
120. Chakrabarti, R. and Vasudeva, R.Y., 2006. Achyranthes aspera stimulates the immunity and enhances the antigen clearance in Catla catla. *International Immunopharmacology*, *6*(5), pp.782–790.
121. Harikrishnan, R. and Balasundaram, C., 2008. In vitro and in vivo studies of the use of some medicinal herbals against the pathogen Aeromonas hydrophila in goldfish. *Journal of Aquatic Animal Health*, *20*(3), pp.165–176.
122. Wu, N., Song, Y.L., Wang, B., Zhang, X.Y., Zhang, X.J., Wang, Y.L., Cheng, Y.Y., Chen, D.D., Xia, X.Q., Lu, Y.S. and Zhang, Y.A., 2016. Fish gut-liver immunity during homeostasis or inflammation revealed by integrative transcriptome and proteome studies. *Scientific Reports*, *6*(1), pp.1–17.
123. Bilen, S., Özkan, O., Alagöz, K. and Özdemir, K.Y., 2018. Effect of dill (Anethum graveolens) and garden cress (Lepidium sativum) dietary supplementation on growth performance, digestive enzyme activities and immune responses of juvenile common carp (Cyprinus carpio). *Aquaculture*, *495*, pp.611–616.
124. Ghehdarijani, M.S., Hajimoradloo, A., Ghorbani, R. and Roohi, Z., 2016. The effects of garlic-supplemented diets on skin mucosal immune responses, stress resistance and growth performance of the Caspian roach (Rutilus rutilus) fry. *Fish & Shellfish Immunology*, *49*, pp.79–83.
125. Anthwal, A., Thakur, B.K., Rawat, M.S.M., Rawat, D.S., Tyagi, A.K. and Aggarwal, B.B., 2014. Synthesis, characterization and in vitro anticancer activity of C-5 curcumin analogues with potential to inhibit TNF-α-induced NF-κB activation. *BioMed Research International*, *2014*.
126. Alok, S., Jain, S.K., Verma, A., Kumar, M., Mahor, A. and Sabharwal, M., 2014. Herbal antioxidant in clinical practice: A review. *Asian Pacific Journal of Tropical Biomedicine*, *4*(1), pp.78–84.

127. Klūga, A., Terentjeva, M., Vukovic, N.L. and Kačániová, M., 2021. Antimicrobial activity and chemical composition of essential oils against pathogenic microorganisms of freshwater fish. *Plants*, *10*(7), p.1265.
128. Herrera, M., Mancera, J.M. and Costas, B., 2019. The use of dietary additives in fish stress mitigation: comparative endocrine and physiological responses. *Frontiers in Endocrinology*, *10*, p.447.
129. Mohamed, A.A.R., El-Houseiny, W., Abd Elhakeem, E.M., Ebraheim, L.L., Ahmed, A.I. and Abd El-Hakim, Y.M., 2020. Effect of hexavalent chromium exposure on the liver and kidney tissues related to the expression of CYP450 and GST genes of Oreochromis niloticus fish: Role of curcumin supplemented diet. *Ecotoxicology and Environmental Safety*, *188*, p.109890.
130. Rajabiesterabadi, H., Yousefi, M. and Hoseini, S.M., 2020. Enhanced haematological and immune responses in common carp Cyprinus carpio fed with olive leaf extract-supplemented diets and subjected to ambient ammonia. *Aquaculture Nutrition*, *26*(3), pp.763–771.
131. de Oliveira, E.F., Tikekar, R. and Nitin, N., 2018. Combination of aerosolized curcumin and UV-A light for the inactivation of bacteria on fresh produce surfaces. *Food Research International*, *114*, pp.133–139.
132. Zahran, E., Risha, E., AbdelHamid, F., Mahgoub, H.A. and Ibrahim, T., 2014. Effects of dietary Astragalus polysaccharides (APS) on growth performance, immunological parameters, digestive enzymes, and intestinal morphology of Nile tilapia (Oreochromis niloticus). *Fish & Shellfish Immunology*, *38*(1), pp.149–157.
133. Adel, M., Pourgholam, R., Zorriehzahra, J. and Ghiasi, M., 2016. Hemato–Immunological and biochemical parameters, skin antibacterial activity, and survival in rainbow trout (Oncorhynchus mykiss) following the diet supplemented with Mentha piperita against Yersinia ruckeri. *Fish & Shellfish Immunology*, *55*, pp.267–273.
134. Harikrishnan, R. and Balasundaram, C., 2008. In vitro and in vivo studies of the use of some medicinal herbals against the pathogen *Aeromonas hydrophila* in goldfish. *Journal of Aquatic Animal Health*, *20*(3), pp.165–176.
135. Musthafa, M.S., Ali, A.R.J., Kumar, M.S.A., Paray, B.A., Al-Sadoon, M.K., Balasundaram, C. and Harikrishnan, R., 2017. Effect of *Cucurbita mixta* (L.) seed meal enrichment diet on growth, immune response and disease resistance in Oreochromis mossambicus. *Fish & Shellfish Immunology*, *68*, pp.509–515.
136. Musthafa, M.S., Asgari, S.M., Elumalai, P., Hoseinifar, S.H. and Van Doan, H., 2018. Protective efficacy of Shilajit enriched diet on growth performance and immune resistance against Aeromonas hydrophila in Oreochromis mossambicus. *Fish & Shellfish Immunology*, *82*, pp.147–152.
137. Ali, M., Soltanian, S., Akbary, P. and Gholamhosseini, A., 2018. Growth performance and lysozyme activity of rainbow trout fingerlings fed with vitamin E and selenium, marjoram (Origanum spp.), and ajwain (Trachyspermum ammi) extracts. *Journal of Applied Animal Research*, *46*(1), pp.650–660.
138. Meena, D.K., Sahoo, A.K., Jayant, M., Sahu, N.P., Srivastava, P.P., Swain, H.S., Behera, B.K., Satvik, K. and Das, B.K., 2022. Bioconversion of Terminalia arjuna bark powder into a herbal feed for Labeo rohita: Can it be a sustainability paradigm for Green Fish production?. Animal Feed Science and Technology, *284*, p.115132.
139. Abdel-Tawwab, M., Adeshina, I., Jenyo-Oni, A., Ajani, E.K. and Emikpe, B.O., 2018. Growth, physiological, antioxidants, and immune response of African catfish, *Clarias gariepinus* (B.), to dietary clove basil, *Ocimum gratissimum*, leaf extract and its susceptibility to Listeria monocytogenes infection. *Fish & Shellfish Immunology*, *78*, pp.346–354.
140. Tan, X., Sun, Z., Liu, Q., Ye, H., Zou, C., Ye, C., Wang, A., and Lin, H. 2018. Effects of dietary ginkgo biloba leaf extract on growth performance, plasma biochemical parameters, fish composition, immune responses, liver histology, and immune and apoptosis-related genes expression of hybrid grouper (Epinephelus lanceolatus♂× Epinephelus fuscoguttatus♀) fed high lipid diets. *Fish & Shellfish Immunology*, *72*, pp. 399–409.
141. Fawole, F.J., Adeoye, A.A., Tiamiyu, L.O., Ajala, K.I., Obadara, S.O. and Ganiyu, I.O., 2020. Substituting fishmeal with Hermetia illucens in the diets of African catfish (Clarias gariepinus): Effects on growth, nutrient utilization, haemato-physiological response, and oxidative stress biomarker. *Aquaculture*, *518*, p.734849.
142. Zheng, Y., Zhao, Z., Wu, W., Song, C., Meng, S., Fan, L., Bing, X. and Chen, J., 2017. Effects of dietary resveratrol supplementation on hepatic and serum pro-/anti-inflammatory activity in juvenile GIFT tilapia, Oreochromis niloticus. *Developmental & Comparative Immunology*, *73*, pp.220–228.
143. Yang, Q.H., Tan, B.P., Dong, X.H., Chi, S.Y. and Liu, H.Y., 2015. Effects of different levels of Yucca schidigera extract on the growth and nonspecific immunity of Pacific white shrimp (Litopenaeus vannamei) and on culture water quality. *Aquaculture*, *439*, pp.39–44.
144. Fawole, F.J., Sahu, N.P., Pal, A.K. and Ravindran, A., 2016. Haemato-immunological response of L abeo rohita (H amilton) fingerlings fed leaf extracts and challenged by A eromonas hydrophila. *Aquaculture Research*, *47*(12), pp.3788–3799.
145. Huang, Z., Aweya, J.J., Zhu, C., Tran, N.T., Hong, Y., Li, S., Yao, D. and Zhang, Y., 2020. Modulation of crustacean innate immune response by amino acids and their metabolites: inferences from other species. *Frontiers in Immunology*, *11*, p.574721.
146. Kurian, A., Lakshmi, S., Fawole, F.J., Faggio, C. and Elumalai, P., 2021. Combined effects of Leucas aspera, oxy-cyclodextrin and bentonite on the growth, serum biochemistry, and the expression of immune-related gene in Nile tilapia (Oreochromis niloticus). *Turkish Journal of Fisheries and Aquatic Sciences*, *21*(3), pp.147–158.
147. Raissy, M., Ghafarifarsani, H., Hossein Hoseinifar, S., El-Haroun, E.R., Shahbazi Naserabad, S., and Van Doan, H. 2022. The effect of dietary combined herbs extracts (oak acorn, coriander, and common mallow) on growth, digestive enzymes, antioxidant and immune response, and resistance against Aeromonas hydrophila infection in common carp, Cyprinus carpio. *Aquaculture 546*, p.737287
148. Fawole, F.J., Yisa, R.O., Jayeoba, O.O., Adeshina, I., Ahmed, A.O. and Emikpe, B.O., 2022. Effect of Dietary Polyherbal Mixture on Growth Performance, Haemato-Immunological Indices, Antioxidant Responses, and

Intestinal Morphometry of African Catfish, Clarias gariepinus. *Aquaculture Nutrition, 2022.*

149. Zhang, X., Sun, Z., Wang, Y., Cao, Y., Wang, G. and Cao, F., 2022. Enhancement of growth, antioxidative status, nonspecific immunity, and disease resistance in gibel carp (Carassius auratus) in response to dietary Flos populi extract. *Fish Physiology and Biochemistry*, *48*(1), pp.67–83.
150. Abdel-Tawwab, M., Adeshina, I., Jenyo-Oni, A., Ajani, E.K. and Emikpe, B.O., 2018. Growth, physiological, antioxidants, and immune response of African catfish, Clarias gariepinus (B.), to dietary clove basil, Ocimum gratissimum, leaf extract and its susceptibility to Listeria monocytogenes infection. *Fish & Shellfish Immunology*, *78*, pp.346–354.
151. Safari, R., Hoseinifar, S.H., Imanpour, M.R., Mazandarani, M., Sanchouli, H. and Paolucci, M., 2020. Effects of dietary polyphenols on mucosal and humoral immune responses, antioxidant defense and growth gene expression in beluga sturgeon (Huso huso). *Aquaculture*, *528*, p.735494.
152. Shekarabi, S.P.H., Javarsiani, L., Mehrgan, M.S., Dawood, M.A. and Adel, M., 2022. Growth performance, blood biochemistry profile, and immune response of rainbow trout (Oncorhynchus mykiss) fed dietary Persian shallot (Allium stipitatum) powder. *Aquaculture*, *548*, p.737627.
153. Shekarabi, S.P.H., Mehrgan, M.S., Ramezani, F., Dawood, M.A., Van Doan, H., Moonmanee, T., Hamid, N.K.A. and Kari, Z.A., 2022. Effect of dietary barberry fruit (Berberis vulgaris) extract on immune function, antioxidant capacity, antibacterial activity, and stress-related gene expression of Siberian sturgeon (Acipenser baerii). *Aquaculture Reports*, *23*, p.101041.
154. Giri, S.S., Jun, J.W., Sukumaran, V. and Park, S.C., 2017. Evaluation of dietary Hybanthus enneaspermus (Linn F. Muell.) as a growth and haemato-immunological modulator in Labeo rohita. *Fish & Shellfish Immunology*, *68*, pp.310–317.
155. Adeshina, I., Tiamiyu, L.O., Akpoilih, B.U., Jenyo-Oni, A. and Ajani, E.K., 2021. Dietary Mitracarpus scaber leaves extract improved growth, antioxidants, non-specific immunity, and resistance of Nile tilapia, Oreochromis niloticus to Gyrodactylus malalai infestation. *Aquaculture*, *535*, p.736377.
156. Adebayo, A., Oke, I.O. and Dada, A.A., 2020. Evaluation of lemon grass (Cymbopogon citratus) as phyto-additive in the diet of African catfish (Clarias gariepinus) fingerlings. *Applied Tropical Agriculture*, *25*(1), pp.29–36.
157. Qin, C., Wang, J., Zhao, W., Pi, D., Yan, X. and Nie, G., 2022. Effects of dietary bitter melon extract on growth performance, antioxidant capacity, inflammatory cytokines expression, and intestinal microbiota in common carp (Cyprinus carpio L.). Aquaculture Nutrition, 2022.
158. Hamid, N.K.A., Somdare, P.O., Harashid, K.A.M., Othman, N.A., Kari, Z.A., Wei, L.S. and Dawood, M.A., 2022. Effect of papaya (Carica papaya) leaf extract as dietary growth promoter supplement in red hybrid tilapia (Oreochromis mossambicus× Oreochromis niloticus) diet. *Saudi Journal of Biological Sciences*, *29*(5), pp.3911–3917.
159. Hosseini, H., Pooyanmehr, M., Foroughi, A., Esmaeili, M., Ghiasi, F. and Lorestany, R., 2022. Remarkable positive effects of figwort (Scrophularia striata) on improving growth performance, and immunohematological parameters of fish. *Fish & Shellfish Immunology*, *120*, pp.111–121.
160. Le Xuan, C., Wannavijit, S., Outama, P., Montha, N., Lumsangkul, C., Tongsiri, S.,... ... & Van Doan, H. (2022). Effects of dietary rambutan (Nephelium lappaceum L.) peel powder on growth performance, immune response and immune-related gene expressions of striped catfish (Pangasianodon hypophthalmus) raised in biofloc system. *Fish & Shellfish Immunology*, *124*, pp.134–141.
161. Srichaiyo, N., Tongsiri, S., Hoseinifar, S.H., Dawood, M.A., Jaturasitha, S., Esteban, M.Á., Ringø, E. and Van Doan, H., 2020. The effects gotu kola (Centella asiatica) powder on growth performance, skin mucus, and serum immunity of Nile tilapia (Oreochromis niloticus) fingerlings. *Aquaculture Reports*, *16*, p.100239.
162. Aluta, U.P., Aderolu, A.Z., Lawal, M.O. and Olutola, A.A., 2021. Inclusion effect of onion peel powder in the diet of African catfish, Clarias gariepinus: Growth, blood chemistry, hepatic antioxidant enzymes activities and SOD mRNA responses. *Scientific African*, *12*, p.e00780.
163. Van Doan, H., Hoseinifar, S.H., Sringarm, K., Jaturasitha, S., Yuangsoi, B., Dawood, M.A., Esteban, M.Á., Ringø, E. and Faggio, C., 2019. Effects of Assam tea extract on growth, skin mucus, serum immunity and disease resistance of Nile tilapia (*Oreochromis niloticus*) against Streptococcus agalactiae. *Fish & Shellfish Immunology*, *93*, pp.428–435.
164. Xu, A., Shang-Guan, J., Li, Z., Gao, Z., Huang, Y.C. and Chen, Q., 2020. Effects of dietary Chinese herbal medicines mixture on feeding attraction activity, growth performance, nonspecific immunity and digestive enzyme activity of Japanese seabass (Lateo labrax japonicus). *Aquaculture Reports*, *17*, p.100304.
165. Fawole, F.J., Sahu, N.P., Nazeemashahul, S. and Adeoye, A.A., 2018. Effect of Psidium guajava and Mangifera indica leaves extracts on growth, antioxidant and metabolic enzymes activities of Labeo rohita fingerlings. *Journal of the Association of Nigerian Fisheries Scientists*, *1*, pp.40–49.
166. Huang, Z., Lu, J., Ye, Y., Xu, A. and Li, Z., 2020. Effects of dietary Chinese herbal medicines mixture on growth performance, digestive enzyme activity and serum biochemical parameters of European eel, Anguilla anguilla. *Aquaculture Reports*, *18*, p.100510.
167. Raissy, M., Ghafarifarsani, H., Hoseinifar, S.H., El-Haroun, E.R., Naserabad, S.S. and Van Doan, H. 2022. The effect of dietary combined herbs extracts (oak acorn, coriander, and common mallow) on growth, digestive enzymes, antioxidant and immune response, and resistance against *Aeromonas hydrophila* infection in common carp, *Cyprinus carpio*. *Aquaculture*, *546*, p.737287.

10

Prebiotics and Probiotics as Effective Immunomodulators in Aquaculture

Mehdi Soltani[1,2], Koushik Ghosh[3], Dipanjan Dutta[3,4], and Einar Ringø[5]
[1]*Centre for Sustainable Aquatic Ecosystems, Harry Butler Institute, Murdoch University, Perth, Australia*
[2]*Department of Aquatic Animal Health Faculty of Veterinary Medicine, University of Tehran, Tehran, Iran*
[3]*Aquaculture Laboratory, Department of Zoology, The University of Burdwan, Golapbag, Burdwan, West Bengal, India*
[4]*Post Graduate Department of Zoology, Hooghly Mohsin College, Chinsurah, Hooghly, West Bengal, India*
[5]*Norwegian College of Fishery Science, Faculty of Biosciences, Fisheries and Economics, UiT, The Arctic University of Norway, Tromsø, Norway*

10.1 Introduction

Immunological and physiological functions of fish and shellfish species are affected under in intensive and super-intensive systems of aquacultural development, thus, under such growing systems the target host become vulnerable to infectious agents (Merrifield *et al.*, 2014). Our understanding on immune systems in fish or shrimp is scarce and it is lagging behind the higher vertebrates. The fish immune system consists of the humoral [complement system, lysozyme, antimicrobial peptides, interferon (IFN), lectins, proteases, protease inhibitors, or eicosanoids] and cellular [monocyte-macrophages, granulocytes, natural killer (NK), and non-specific cytotoxic (NCC) cells] components of the innate immune system (see Manning, 1998), which gets activated after coming into contact with pathogen-associated molecular patterns (PAMPs) viz., viral dsRNA, bacterial lipopolysaccharide (LPS), and certain sugars. The antigen is then processed and presented to the adaptive immune system components (B and T lymphocytes). Nevertheless, the response achieved is not so effective unlike the mammalian response due to lack of thermoregulation. The control and integration of the immune response is carried out by cytokines, derived from lymphocytes and/or macrophages after stimulation.

Apart from the fish immune organs (head-kidney, spleen, and thymus), the fish immune system also consists of physical barriers, like epithelia, and their mucus secretions, called mucosa-associated lymphoid tissue (MALT) (Rombout *et al.*, 2011) composed of T and B lymphocytes, macrophages, plasma cells, granulocytes, andmast cells. The MALT in teleost fish is categorized into gut-associated lymphoid tissue (GALT), gill-associated lymphoid tissue (GIALT), and skin-associated lymphoid tissue (SALT) (Salinas *et al.*, 2011). It releases IgT and IgM against several pathogens (Zhang *et al.*, 2010; Salinas *et al.*, 2011; Xu *et al.*, 2013). Unlike mammals, fish lack lymph nodes, M cells, or IgA secretion in the gut.

The outbreaks of infectious by bacterial agents are generally more dominated in aquaculture farming as majority of potential pathogenic bacteria are ubiquitous more prevalent in gastrointestinal (GI) tracts of aquatic organisms and in their environments (Austin *et al.*, 2016). Thus, frequent treatments of infectious diseases such as bacterial infections can increase various problems including development of residual of antibiotics, bacterial resistance to antibiotics, and disruption of the aquatic ecosystems. Use of prebiotics or probiotics as an alternative tool either as a preventive or as a protective tool can reduce the impacts of chemotherapy in aquaculture industry (Chauhan *et al.*, 2019; Van Doan *et al.*, 2021). Prebiotics as complex carbohydrates are a good source of energy for beneficial gut microbiota or probiotics to enhance animal immunity and animal tolerance to the environmental stressors (Song *et al.*, 2014). The prebiotics are also capable of resisting the host's digestive enzymes and can be fermented by the gut microorganisms with an overall target of promoting microbial growth to improve host health, and thereby increasing production of aquaculture species. Edible mushroom, yeast, animal dairy products, and products with the plant-based indulging fruits, microalgae, vegetables beans, and seaweed are major prebiotic sources that can increase the efficiency of feed utilization in various levels depending on the type, dosage, and duration use of prebiotic (Wee *et al.*, 2022). β-glucan, inulin, arabinoxylan oligosaccharide (AXOS), mannan oligosaccharide (MOS), galacto-oligosaccharide (GOS), and fructo-oligosaccharides (FOS) are mostly used as prebiotics in aquaculture. The usual mode of

DOI: 10.1201/9781003361183-13

action of the prebiotics involves preventing the attachment of harmful microbiota and eliminating them from gut epithelial cells acting as a receptor or modulator in the host immune system to control inflammation (Al-Sheraji *et al.,* 2013). However, host microbial flora is selective to the prebiotic, as the microorganisms intended to be benefitted by applied prebiotics should possess required enzymes to carry out the fermentation process. Probiotics are microorganisms particularly of bacterial members and yeast that stimulate immune-physiological variables of the target host to provide a better growth and health status. Like prebiotics, although immunomodulatory effect by probiotics is well studied in finfish and shellfish, the details of the mechanisms still required more research. This chapter addressed the immunomodulatory effects of prebiotics and probiotics in aquaculture species.

10.2 Sources of Probiotics in Aquaculture

In the aquaculture sector, many microorganisms from a wide range of bacteria especially members of Gram-positive, yeasts, and microalgae have been characterized and tested for their potentials as probiotics in different finfish and shellfish species (Soltani *et al.*, 2019a; Van Doan *et al.*, 2021; Amenyogbe *et al.*, 2020; Ringø *et al.*, 2020; Caipang et al, 2020), but still no adequate commercial probiotics are available for usage in the aquaculture industry. The source of these probiotics varies greatly, including the gastrointestinal tracts of fish and shellfish, water of rearing of aquatic organisms, sediments of culture ponds/tanks, fermented food products, and terrestrial animals (Shefat *et al.*, 2018). As the key principle of a microorganism as a probiotic is to create a relationship with the permanent or temporary harmful and beneficial bacteria present in fish/shellfish intestine, then the gastrointestinal tract and the surface mucus of aquatic organisms are usually the most common sources for the isolation of potential autochthonous probiotics (Lazado *et al.*, 2015). It is important to say that the host-associated microorganisms are preferred as potential probiotics because the microbiota living in healthy hosts are considered as a part of the natural defense system, being advantageous to the host in several ways. In addition, indigenous or autochthonous probiotics can survive naturally with a physiological function at an optimum level, exhibiting a longer-lasting effect on the host (Ringø *et al.*, 2020). In contrast to terrestrial animals, the gastrointestinal microbiota of aquatic animals is highly dependent on their surrounding environments due to the continuous water flow through the digestive tract; thus, part of bacteria populations that could colonize the gastrointestinal tract or external surfaces (skin, gill, fin) are temporary and could be varied if the animal environmental conditions are changed (Martınez-Cruz *et al.*, 2012).

10.3 Selection Criteria of Probiotics in Aquaculture

There are numerous published data on the screening, selection, and characterization of fish and shellfish probiotic bacterial strains, and the potential candidates obtained from different origins are screened through several steps of *in vitro* and *in vivo* assays to evaluate their potential as ideal probiotics (Alonso et al., 2019). The safety and lack of pathogenicity of isolated microorganisms are the first essential steps before the probiotic candidate being successfully assessed for the next criteria. Some of these criteria, classifying them as either essential or favorable, are proposed by Merrifield *et al.* (2010). The essential criteria to consider a microorganism as a suitable probiotic are (1) being not pathogenic to the host, other aquatic animals and humans as the consumers; (2) being free of plasmid-encoded antibiotic resistance genes; (3) being able to resist a wide range of pH conditions and high concentration (>2.5%) of bile salts; (4) being able to adhere to and/or grow well within the intestinal mucus; (5) being able to colonize the intestinal epithelial surface; (6) being registered for use as a feed additive; (7) exhibiting beneficial growth characters such as having a short lag period, having a short generation time of growth, grow well at the host rearing temperature; (8) displaying a wide range of antagonistic activity towards at least one of the economically important pathogens; (9) be able to secrete adequate relevant extracellular digestive enzymes or vitamins; (10) being autochthonous to the host or the culturing environment; (11) be able to stay viable under normal storage conditions and being sufficiently strong to survive the industrial processes (mass production); (12) having good sensorial properties such as fermentative action, tolerance to freeze-drying and viability in feed during packaging and storing process; and (13) having advantage effects on the growth, stimulation of immunity, and protection of the host towards different pathogenic bacteria. There may be other criteria for selection a more effective probiotic for instance, evaluation of a synergistic or antagonistic efficacy of two or more probiotics when they are used in a mixture form. Up to date it is almost improbable to find a probiotic candidate that can fulfill all the above criteria thus, the more of these criteria achievable, the more probable it will be a favorable probiotic (Hai *et al.*, 2015; Simón *et al.*, 2021). However, the *in vitro* inhibitory activity or *in vivo* efficacy towards the target pathogens is the principal criteria to select potential probiotics among different candidates.

10.4 Probiotics Immunomodulatory Mode of Action

In aquaculture, it is well known that administration of some indigenous or exogenous bacteria particularly of Gram-positive members including bacilli and lactobacilli and, members of Gram-negative bacteria such as *Aeromonas*, *Pseudomonas,* and *Vibrio*, can reduce or prevent the process of adhesion, colonization, and multiplication of pathogenic agents in the GI tract of finfish and shellfish (Cheng *et al.*, 2014; Soltani *et al.*, 2018; Soltani *et al.*, 2019a; Ringø *et al.*, 2022; Van Doan *et al.*, 2021; Simon *et al.*, 2021). Probiotic bacteria can inhibit or elongate the generation time of pathogenic microorganisms through the structural cell surfaces such as the cell receptors and adhesive molecules, stimulate the immunity of the target host, enhance digestibility of feed, and improve the quality of rearing water (Zokaeifar *et al.*, 2012; Zheng *et al.*, 2017; Soltani *et al.*, 2019a; Kewcharoen *et al.*, 2019; Ringø *et al.*, 2020). Although immunomodulatory role of probiotics in aquaculture species have been extensively studied, more research works are required to elucidate the actual detail mechanisms of such immunostimulatory functions (Soltani *et al.*, 2018; Soltani *et al.*, 2019a, 2019b; Kuebutoqrnyea *et al.*, 2019; Ringø *et al.*, 2020).

In fish and shellfish, the innate immunity with various functions is the first line of defence of the host against disease agents, and potential probiotics can stimulate their immune responses by enhancing the populations of immune-competent cells, and such functions have been shown in various species of commercial interest such as tilapia, catfish, rainbow trout, carps, and Atlantic salmon (*Salmo salar*) (Soltani *et al.*, 2019a; Ringø *et al.*, 2020; Van Doan *et al.*, 2021). Probiotics particularly Gram-positive members also conserve molecular patterns of microorganisms including lipoteichoic acids, exopolysaccharides, lipopolysaccharides, peptidoglycan, flagellin, the proteins of S and A layers, and the nucleic acids, which are recognized by recognition pattern of receptors. This can persuade a signalling of theimmunological cascades followed by a secretion of cytokines/interleukins, and other molecular effectors which are activators of animal immune functions resulting in induction of immunostimulatory or immunomodulatory effects (Remus *et al.*, 2012; Soltani *et al.*, 2019a; Zhou *et al.*, 2019; Mariman *et al.*, 2014). In addition, there are relations between tool-like cell receptor signalling-mediated recognition of probiotics and stimulation of the animal mucosal immunity (Vinderola *et al.*, 2019; Galdeano *et al.*, 2019). The heat shock protein 70 is also stimulated by some probiotics of lactobacilli such as *Lactobacillus fructivorans* and *Lactobacillus plantarum;* thus, signs of an improvement in the animal tolerance to the stressors, and an up-regulation of gene expression of cytokines including interleukins 1β and10 and tumour necrose factor (TNF-α) (Perez-Sanchez *et al.*, 2011; Soltani *et al.*, 2019a, 2019b; Van Doan *et al.*, 2021). Despite such data associated with modulation of fish and shellfish immunity by probiotics, especially members of Gram-positive groups, various aspects of immune systems of aquaculture species merits future studies.

10.5 *Bacillus* for Use as Probiotics

Genus of *Bacillus* are thermophilic, psychrophilic, acidophilic, alkaliphilic, halotolerant, and halophilic representatives (Slepecky *et al.*, 2006) with unusual physiological characters, enabling them to survive in various environmental conditions such as temperature, salinity, and pH. They can degrade various substrates such as cellulose, starch, proteins, agar, and hydrocarbons (Slepecky *et al.*, 2006; Ray *et al.*, 2012). Bacilli are also antibiotic producers, heterotrophic nitrifiers, denitrifiers, nitrogen fixers, iron precipitators, selenium oxidizers, oxidizers, manganese reducers, and facultative chemolithotrophs. Details of biological functions of bacilli are discussed somewhere else (e.g., Slepecky *et al.*, 2006; Sella *et al.*, 2014). *Bacillus* bacteria are a part of dominated primary inhabitants of the marine as well as freshwater environments and generally exist more frequently in sediments than in the water column, and their spores may account for population size of the total heterotrophic microbial community mainly in the sediments, and therefore, they can naturally be ingested by aquatic animals e.g., shrimps and some demersal fish that feed in or on the sediments (Rengpipat *et al.*, 1998). Numerous studies have revealed the presence of bacilli in GI tract of finfish and shellfish (e.g., Soltani *et al.*, 2018; 2019a; Zhou *et al.*, 2019; Kewcharoen *et al.*, 2019; Daroonpunt *et al.*, 2019; Anyanwu *et al.*, 2019; Peng *et al.*, 2019). Some *Bacillus* species are autochthonous (adherent to mucosal surface) and can contribute to nutrition, inhibit pathogen adherence and colonization, and may positively affect the animal immune system (Soltani *et al.*, 2019a). The intestinal microbiota including bacilli is involved in a wide range of biological processes that benefit the host, and so far, many bacilli species have been reported from the digestive tract of different fish and shellfish. Readers with special interest are referred to the review of Soltani *et al.* (2019a).

Bacillus with the aquatic organism's source produce many exo-enzymes including protease, amylase, lipase, non-starch polysaccharide-degrading enzymes, phytase, tannase, and chitinase that have a vital role in the host digestive functions (Meidong *et al.*, 2018; Soltani *et al.*, 2018; 2019a; Di *et al.*, 2019; Kuebutoqrnyea *et al.*, 2019). Various gut-associated bacilli such as *Bacillus coagulans*, *Bacillus cereus*, *Bacillus subtilis*, *Bacillus thuringiensis,* and *Bacillus licheniformis* reported in various fish species can produce digestive enzymes including amylase, protease, and lipase (Mukherjee *et al.*, 2017; Nandi *et al.*, 2017; Soltani *et al.*, 2019a). Because of cellulose and/or xylanase-degrading bacteria such as *B. subtilis*, *Bacillus circulans*, *Bacillus pumilus*, *Bacillus tequilensis*, *Bacillus megaterium,* and *Bacillus altitudinis* in the fish gut (Dey *et al.*, 2016; Nandi *et al.*, 2017; Soltani *et al.*, 2019a), they may play a role in the fermentative degradation of cellulose and hemicelluloses for the host fish. Further, phytase, tannase, and chitinase activities have been demonstrated in different bacilli species e.g., *B. licheniformis B. subtilis*, *B. thuringiensis*, *B. cereus*, *B. atrophaeus*, *Brevibacillus agri*, *B. pumilus,* and *B. flexus* isolated from the GI tract of various fish species (Talukdar *et al.*, 2016; Mukherjee *et al.*, 2017; Soltani *et al.*, 2019a, Di *et al.*, 2019). A more detailed overview on the exo-enzymes produced by bacilli and their beneficial attributes in aquaculture is given by e.g., Soltani *et al.* (2019a), Kuebutoqrnyea *et al.* (2019), and Ghosh *et al.* (2019).

Probiotic *Bacillus* also produce many bacteriocins including subtilin, cerein, haloduracin, thuricin, subtilosin and megacin that exhibit a broader spectrum of inhibition against different Gram-negative and Gram-positive bacterial finfish and shellfish pathogens, e.g., *Aeromonas*, *Edwardsiella*, *Streptococcus*, *Pseudomonas* and *Vibrio* (Sumathi *et al.*, 2017; Mukherjee *et al.*, 2019a; Soltani *et al.*, 2019a). Under *in vivo* condition, some probiotic bacilli such as *B. subtilis*, *B. velezensis*, *B. amyloliquefaciens*, *B. circulans*, *B. thuringiensis,* and *B. aerius* could increase resistance of finfish and shrimp towards some pathogenic bacteria including *Streptococcus*, *Aeromonas*, *Vibrio*, *Enterococcus,* and *Lactococcus* (e.g., Soltani *et al.*, 2019a; Ringø 2020; Mukherjee *et al.*, 2019b; Vogeley *et al.*, 2019; Peng *et al.*, 2019; Ringø *et al.*, 2020). Some *Bacillus* species such as *B. subtilis* are a natural resource for screening new quorum quenching bacteria and are commonly regarded as safe bacteria for use in aquaculture as agents for improving water quality and disease control (Chen *et al.*, 2019).

Use of *Bacillus* as the bioremediatory tool in the rearing water of aquaculture species and in soil of aquaculture ponds exhibited a feasible way of improving water quality through removing of toxic gases e.g., ammonia, nitrite, nitrate and carbon dioxide that are harmful for aquatic organisms (Gupta *et al.*, 2016; Zokaeifar *et al.*, 2014; Dash *et al.*, 2018; Soltani *et al.*, 2019a; Kewcharoen *et al.*, 2019). Some bacilli such as *B. subtilis*, *B. licheniformis*, *B.cereus,* and *B. coagulans* are suitable bioremediatory tools for removing of organic detritus (Divya *et al.*, 2015), but may not be naturally present in a high enough population level in aquatic ecosystems i.e., water column and sediment. However, *B. subtilis* and *B. licheniformis* may be more suitable candidates for bioremediation of aquaculture rearing water (Soltani *et al.*, 2019a). Use of *Bacillus* supplementation in the rearing water can make a balance between the microorganisms in the water column or in the pond soil through a bacterial competition with a consequence in the decreasing in load of secondary bacterial pathogens (Kumar *et al.*, 2016).

Bacillus can also provide a suitable condition in the GI tracts of finfish and shellfish, giving enhanced digestion and absorption of the feed nutrients by aquatic organisms with an ultimate improvement in the animal growth performance and health status (e.g., Meidong *et al.*, 2018; Soltani *et al.*, 2018; 2019a; Mukherjee *et al.*, 2019b; Li *et al.*, 2019; Zhou *et al.*, 2019; Vogeley *et al.*, 2019; Tsai *et al.*, 2019). The detail mode of actions by these bacilli as the major producers of proteinaceous substances is however not fully understood and merits further research. Bacilli can modulate the gut microbiota of recipient aquatic animals via a bacterial competition that can result in inhibition of pathogen attachment and colonization to intestinal mucosa, ultimately enhancing of health and growth status of the target aquatic organism (Meidong *et al.*, 2018; Soltani *et al.*, 2018; 2019a; Vogeley *et al.*, 2019; Kuebutoqrnyea *et al.*, 2019). Modulation of finfish and shellfish innate immune responses e.g., phagocytic and lysozyme activity, respiratory burst, antiprotease and peroxidase, superoxide dismutase and myeloperoxidase by *Bacillus* administration is well studied (Yi *et al.*, 2018; Yi *et al.*, 2019; Zhou *et al.*, 2019). Bacilli can change in animal cell physiology e.g., neutrophil migration, plasma bactericidal activity, and increasing of neutrophil adherence ability, that can eventually result in the improving of immune responses e.g., increase in complement activity, immunoglobulin production and cell cytotoxicity (Soltani *et al.*, 2018; 2019a; Li *et al.*, 2019; Di *et al.*, 2019). These immune- stimulatory effects by *Bacillus* occur in the gut-associated lymphoid tissue of fish, although the detail mechanisms required further research.

10.5.1 Immunomodulatory Effects of *Bacillus* Probiotics in Fish Culture

Among probiotic *Bacillus,* more attention has been focussed on *Bacillus subtilis* than other bacilli members. The *in vivo* immunostimulatory effects and disease resistance of *B. subtilis* in different fish species including rainbow trout (*Oncorhynchus mykiss*), Nile tilapia (*Oreochromis niloticus*), yellow croaker (*Larimichthys crocea*), grouper (*Epinephelus coioides*), olive flounder (*Paralichthys olivaceus*), and grass carp (*Ctenopharyngodon idellus*) are well known, and for more details, readers are referred to the reviews by Soltani *et al.* (2019a) and Ringø *et al.* (2020).

The immunomodulatory effect of *B. subtilis* in aquaculture was studied for the first time by Newaj-Fyzul *et al.* (2007), who isolated *B. subtilis* from the digestive tract of rainbow trout and fed the fish with the bacilli for two weeks and revealed significant enhancement in immune responses including serum and gut lysozyme, respiratory burst, phagocytosis, peroxidase activity, and antiprotease activity. The probiotic-fed fish also displayed a higher resistance against *Aeromonas* sp. infection compared to control fish. In salmonids, some probiotic bacilli are effective in stimulating the fish immune responses and health status but only a few studies evaluated immunomodulatory effect of probiotic *Bacillus* in commercial salmonid species (Merrifield *et al.*, 2010). Administration of *B. subtilis* individually or in combination with *B. licheniformis*, or lactic acid bacteria (LAB) including *Pediococcus acidilactici*, *Lactobacillus rhamnosus,* and *Enterococcus faecium* is potent to modulate the fish immunity (Merrifield *et al.*, 2010). Rainbow trout fed with *B. subtilis* with or without trans-cinnamic acid presents a better immunostimulatory effect of granulocyte population, antibody level, and activities of respiratory burst, phagocytosis, phagocytic index, total antiprotease, and myeloperoxidase as well as a higher resistance to *Y. ruckeri* infection (Yılmaz *et al.*, 2019).

Tilapia fed *B. subtilis* individually or in combination with other bacilli species such as *B. licheniformis* or yeast induce a well immunostimulatory effect detectable by bioassays of lysozyme, neutrophil adherence, leucocytes population, respiratory burst, phagocytosis, catalase, myeloperoxidase IgM, protease, antiprotease and superoxide dismutase, cytokine, and complement activity (Aly *et al.*, 2008; Zhou *et al.*, 2010; Telli *et al.*, 2014; Iwashita *et al.*, 2015; Liu *et al.*, 2017; Abarike *et al.*, 2018; Won *et al.*, 2020a). Such immunostimulatory effects can be confirmed by a higher resistance in the treated fish towards different pathogenic bacteria, including *Aeromonas hydrophila*, *Streptococcus agalactiae,* and *Streptococcus iniae* (Abarike *et al.*, 2018). The immunomodulatory effects of the bacilli probiotics are also detectable at molecular levels e.g., an up-regulation in immune genes of heat shock protein, β-defensin, transforming growth factor (TGFα), C-lysozyme, IL-1β, interferon-gamma, and tumour necrosis factor (Abarike *et al.*, 2018; Galagarza *et al.*, 2018; Won *et al.*, 2020a).

Carp species including common carp (*Cyprinus carpio*), grass carp, koi carp, and rohu (*Labeo rohita*) fed with *B. subtilis* strains induced immunomodulatory effects by enhancing fish immune responses, including up-regulation of immune genes (heat shock protein 70, IL-1β, IL-8, IL-1β IL-10); tumour necrosis factor (TNF-α); TNF-α protein claudin b; transforming growth factor (TGF-β); pro-inflammation cytokine; adaptor protein myeloid differentiation factor 88; antioxidant enzymes of superoxide dismutase; catalase and glutathione peroxidase; and activity of malondialdehyde, catalase, superoxide dismutase, glutathione, and also a disease resistance to pathogenic agents such as *A. hydrophila* infection (He *et al.*, 2011; Banerjee *et al.*, 2017; Kong *et al.*, 2017). Such an immunostimulatory effects could decrease infections by e.g., *A. hydrophila* through bacterial competition by reducing the adhesion, colonization, and multiplication of the pathogen

(Tang *et al.*, 2019). It is worth noticing that dosage optimization of the probiotic is crucially important to be determined prior to probiotic application as lower or higher dosages can result in nil to an over-immunomodulatory effect. For instance, orally used *B. subtilis* at 10^6–10^{10}CFU/g feed in rohudiets exhibited significantly a better immunostimulatory function than lower dosages even when fish were challenged with *A. hydrophila* infection, a higher disease resistance was seen at these dosages of bacilli administration.

Interestingly, intraperitoneal or bath immersion used of whole cell or extracellular proteins of the probiotics such as bacilli can stimulate the fish immunity at both innate and adaptive responses and causes an increase in fish resistance to the bacterial diseases, suggesting differences in the immunostimulatory effects of probiotics compounds. For instance, rohu subjected to whole cell or extracellular proteins of a strain of *B. subtilis* displayed an improvement in immune response and higher resistance towards *A. hydrophila* infection than control fish (Ramesh *et al.*, 2018).

In eels, oral administration of *B. subtilis* in Japanese eel significantly enhanced fish innate immune responses, including an up-regulation in IgM and heat shock protein and activity of lysozyme, superoxide dismutase, myeloperoxidase (Lee *et al.*, 2017), and this immunomodulatory effect was stronger than that of *Lb. plantarum* administration, indicating the host-specific effect. If *B. subtilis* is orally used in combination with the prebiotics such as mannan-oligosaccharide, a better immunostimulatory effect is noticed in eel (Lee *et al.*, 2018), suggesting a positive synergistic immunostimulatory function which increased fish resistance against *Vibrio anguillarum* infection.

Immunomodulatory effects of *B. subtilis* in other commercial fish species including sea bream (*Sparus aurata* or *Pagrus major*), sea bass (*Dicentrarchus labrax*), yellow croaker (*Larimichthys polyactis*), and grouper is also promising through the enhancement of growth performance and immunity of the target animals. For instance, feeding gilthead sea bream with *B. subtilis* improves fish immune responses (e.g., IgM, respiratory burst, phagocytic, lysozyme, complement, catalase, peroxidase, and bactericidal activities, and a higher up-regulation in immune genes of extension factor 1α, IgM heavy-chain, T-cell receptor b, major histocompatibility complex, b-defensin 1, colony-stimulating factor receptor-1, a higher resistance to *Photobacterium damselae* sub sp. *piscicida* infection, and tolerate oxidative stress conditions (Cerezuela *et al.*, 2012; Zaineldin *et al.*, 2018).

Bacilli can stimulate the fish immunity even after short time feeding asa short time feeding of Brazilian catfish with *B. subtilis* provides a higher immunostimulatory effect by increasing the fish immunocompetent cell population i.e., neutrophils, lymphocytes, thrombocytes, and leukocytes (Veiga *et al.*, 2020), suggesting optimization of duration administration of the probiotics is an important criterion to achieve an optimal immunostimulatory effect. Yellow croaker (*Larimichthys polyactis*) orally fed *B. subtilis* exhibits a significant improvement in immune variables and higher disease resistance against *V. harveyi* infection (Ai *et al.*, 2011). Grouper fed *B. subtilis*, a probiotic isolated from fermented boiled soybeans, displayed significant immunostimulatory effect including an increase in lysozyme, phagocytosis, respiratory burst and complement activities (Liu *et al.*, 2012), and a higher resistance to *Streptococcus* sp. infection, particularly at higher administration dosage of the probiotic. Similarly, dietary *B. subtilis* can cause a remarkable protection to *S. iniae* challenged in olive flounder fed B. *subtilis* due to probiotic immunostimulatory effects i.e., higher levels of lysozyme, respiratory burst, and superoxide dismutase activities (Cha *et al.*, 2013). However, less data is available on efficacy of bacilli on surgeon health status (Hosseinifar et al., 2016). Oral use of *B. licheniformis*, *B. subtilis*, and *B. circulans* in Persian sturgeon (*Acipenser persicus*) could be effective in stimulating fish performance, resistance to environmental stressors including fluctuations in water quality parameters, and survival rate. Mixture of *B. subtilis*, *B. licheniformis,* and yeast *S. cerevisiae* may be more effective in stimulating Persian sturgeon immune functions including population of immunocompetent cells, lysozyme, immunoglobulins, complement, IL-1, and antioxidant activity than the bacilli alone (Darafsh *et al.*, 2020). However, further research is needed on sturgeons.

In addition to *B. subtilis*, individual or combination immunostimulatory effects of other species of *Bacillus* of both indigenous and exogenous sources are investigated in some commercial finfish species. For instance, orange spotted grouper fed *B. clausii* and *B. pumilus* (isolated from orange spotted grouper) individually significantly enhanced fish innate immune responses i.e., phagocytosis, complement C3, and lysozyme activity (Sun *et al.*, 2010).

Positive immunomodulatory and disease resistance effects by *B. amyloliquefaciens* in various fish species such as catla (*Catla catla*), tilapia, turbot (*Scophthalmus maximus*), eel (*Anguilla anguilla*), and zebrafish (*Danio rerio*) are promising, and have been proved by fish immune responses both at paraclinical and molecular levels (e.g., immunocompetent cell, erythrocyte populations, total protein and globulin, superoxide dismutase, respiratory burst, nitric oxide, bactericidal, and lysozyme activities) (Das *et al.*, 2013; Selim & Reda, 2015; Reda *et al.*, 2015; Chen *et al.*, 2016; Lin *et al.*, 2019; Lu *et al.*, 2011; Lin *et al.*, 2019). Such immunomodulations were confirmed by demonstration of a disease resistance of the fish to bacterial infections including *Edwardsiella* septicaemia caused by *E. tarda*, vibriosis caused by *V. anguillarum*, motile *Aeromonas* septicaemia caused by *A. hydrophila* and, streptococcosis caused by *S. agalactiae.* The probiotic also can improve fish tolerance to hepatic stress condition.

Administration of *B. coagulans*, *B. licheniformis*, *Paenibacillus polymyxa*, *B. pumilus*, and *B. velezensis* individually or in combination to common carp, catfish, grouper, Atlantic salmon (*Salmo salar*), rohu, and rainbow trout revealed improved immunomodulatory effects and displayed higher survival to infections by *V. parahaemolyticus*, *A. hydrophila*, *A. salmonicida,* and *Yersinia ruckeri.* However, the route of administration (oral in feed or bath immersion), the dosage, duration of administration and water quality parameters such as water temperature are important determinants of level of the immunostimulatory effect. For instance, bath administration of *B. licheniformis* at a lower dosage was more effective by stimulating of catfish immune parameters than the higher dosage, suggesting dosage optimization of probiotic is necessary before being applied in farm fish (Gobi *et al.*, 2016; Gupta *et al.*, 2014).

The application of bacilli probiotics in combination with LAB or yeast may have a better immunostimulatory effect in finfish, although the detail mode of action of such synergistic immunomodulatory effect merits further research. Oral feeding of Atlantic salmon with *B. velezensis* in combination witha yeast (*Rhodotorula mucilaginosa*) revealed higher immunostimulatory effects and a higher disease resistance than individual usage of the probiotics after challenging fish with *A. salmonicida* (Wang *et al.*, 2019). Oral use of *B. velezensis* in combination with *Lb. plantarum* to Nile tilapia showed a better immunostimulatory effects and higher resistance against *S. agalactiae* challenge than individual usage of the bacilli (Van Doan *et al.*, 2018). In addition, feeding rohu with mixture of *B. subtilis*, *P. aeruginosa* and *Lb. plantarum* may exhibit a remarkable efficacy in stimulating fish immune responses and provide a higher disease resistance towards *A. hydrophila* challenge than individual use of the bacilli or lactobacilli (Giri *et al.*, 2014). Immunostimulatory effect of a combination of *B. subtilis* and *B. licheniformis* spores may be more effective than individual usage in rainbow trout in term of increasing antibody titer, lymphocyte and total protein and a higher disease resistance to *Y. ruckeri* infection (Raida *et al.*, 2003). It is probable that *Bacillus* and *Yersinia* may share antigens, with clones of lymphocytes (B and T cells) induced by the bacilli and inducing protection against yersinia septicaemia. Feeding Asian sea bass (*Lates calcarifer*) with a mixture of *B. subtilis* and *B. licheniformis* can improve the population of fish immunocompetent cells (Adorian *et al.*, 2019) that may increase fish resistance to pathogenic agents.

10.5.2 Immunomodulatory Effects of *Bacillus* Probiotics in Shellfish Culture

Shrimp can defend themselves against infectious agents by various immune responses including physical barriers, model formation, encapsulation, phagocytosis, antimicrobial activity, pro-phenol-oxidase system, clotting reactions, and reactive oxygen intermediates (Kulkarni *et al.*, 2021). The induction of shrimp innate immune response has been demonstrated by a diversity of pattern recognition receptors on the membranes of haemocytes to show pathogen-associated molecule pattern and decrease signals during the pathogen invasion. By activation of haemocyte cells, antimicrobial molecules are degranulated to inactive target pathogens (Song *et al.*, 2014).

In recent years, the probiotic immunostimulatory in shrimps have received much attention (Ringø, 2020; Van Doan *et al.*, 2021). Individual or combination of various members of autochthonous or exogenous *Bacillus* species including B. *subtilis*, *B. megaterium*, *B. cereus*, *B. infanti*, *B. licheniformis*, *B. flexus*, *B. aryabhattai*, *B. pumilus*, *Paenibacillus polymyxa*, *B.* polymyxa, *B. thuringiensis*, *B. coagulans*, *B. firmus*, *B. coagulans,* and *B. amyloliquefaciens* as probiotics can influence an immunostimulatory effect in shrimp aquaculture by enhancing the stimulating of innate immune responses and improving disease resistance against pathogenic agents (Laranja *et al.*, 2017; Soltani *et al.*, 2019a; Zhao *et al.*, 2019; Sumon *et al.*, 2022). In the earlier studies, a positive immunostimulatory effect and a higher disease resistance to *Vibrio harveyi* infection were seen in black tiger shrimp (*P. monodon*) fed bacilli (Rengpipat *et al.*, 1998). In following studies by other researchers, some details of mode of action of probiotics such as *Bacillus* were studied in shrimps (Li *et al.*, 2009; Chandran *et al.*, 2014; Zokaeifar *et al.*, 2014). In a previous study by Rengpipat *et al.* (2000), *Bacillus* administration induced an immunostimulatory effect by improving functions of phagocytosis, phenoloxidase, and antimicrobial activity of black tiger shrimphemolymph, and also increased the shrimp resistance to *V. harveyi* challenge. In the following work by Rengpipat *et al.* (2003), black tiger shrimp fed with *Bacillus* sp. exhibited a higher resistance to *V. harveyi* challenge, and a significant enhancement was seen in survival of black tiger shrimp after a bath-treatment with cell-free extracts of a *B. subtilis* strain and challenged with *V. harveyi* infection (Vaseeharan *et al.*, 2003). *B. subtilis* isolated from shrimpintestine demonstrated an immunostimulatory effect in back tiger shrimp through increasing hemocyte population, phenoloxidase activity, superoxide anion, and bactericidal activity (Purivrirojkul *et al.*, 2005). Black tiger shrimp fed *B. cereus* showed higher values of lysozyme, respiratory burst, phenoloxidase, and bactericidal activity than the control (Chandran *et al.*, 2014), and such immunostimulatory effect was supported by a higher survival of treated shrimp to *V. harveyi* challenge. When black tiger shrimp were fed *B. subtilis*, improvement in the shrimp immune response and resistance to *V. harveyi* infection were noticed (Utiswannakul *et al.*, 2011).

Oral use of *B. subtilis* in white leg shrimp may be superior to other bacilli probiotics in stimulating the animal growth and immunity. For instance, higher immunostimulatory effects in terms of growth and immunity was seen in white shrimp fed *B. subtilis* than other bacilli probiotics including *B. megaterium*, *B. cereus,* and *B. infantis* (Tamilarasu *et al.*, 2021). It is however, important to emphasize that a favorable probiotic should fulfill more the probiotic selection criteria; thus, more studies are required to confirm the superiority of a specific probiotic to others. Daily or day internal bath immersion of *B. fusiformimis* displayed an increase in survival and the farm production of white leg shrimp larvae (Guo *et al.*, 2016). Administration of exogeneous *B. subtilis* with protease-producing capacity (isolated from fermented soyabeans) in a white leg shrimp diet revealed an immunostimulatory effect through enhancement in activity of lysozyme and prophenoloxidase I and II as well as an increase in stress tolerance to-water quality fluctuations such as temperature, salinity, and nitrite, and in another study using the same bacilli induced an increased disease resistance of white leg shrimp to *V. alginolyticus* infection due to the stimulation of shrimp immunity such as phenoloxidase and phagocytic activities (Tseng *et al.*, 2009). Oral use of *Paenibacillus polymyxa* improves the growth, immune and antioxidant activities, intestinal health, and disease resistance in white leg shrimp challenged with *Vibrio parahaemolyticus* (Amoah *et al.*, 2020). The protective and immunomodulatory effects of *Bacillus* sp. was seen in juveniles of kuruma shrimp (*Marsupenaeus japonicus*) through increasing the immune responses and tolerance to temperature stress (Dong *et al.*, 2014).

In giant freshwater prawns (*Macrobrachium rosenbergii*), several bacillar probiotics including *B. pumilus*, *B. subtilis B. coagulans B. licheniformis*, and *B. vireti* are effective in stimulating the animal immune responses and increasing towards the infections (Zhao *et al.*, 2019; Frozza *et al.*, 2021). For instance, oral administration of *B. pumilus* improves activity of respiratory burst, catalase and phenoleoxidase, and digestive enzymes and with a better growth performance than control of giant freshwater prawns (Zhao *et al.*, 2019). It seems that the application of combined probiotics of *Bacillus* and LAB may be superior in stimulating of shrimp immunity. For instance, oral use of *B. subtilis* in combination with LAB exhibited a great enhancement in white shrimp immunity and disease resistance to *V. parahaemolyticus* infection (Lee *et al.*, 2022). However, more *in vivo* works are required to confirm such synergistic efficacy.

In carbs, bacilli probiotics including *B. subtilis*, *B. pumilus*, and *B. cereus* (Wu *et al.*, 2014) were effective in stimulating mud crab (*Scylla paramamosain*) immune responses such as upregulation of immune genes i.e., catalase, prophenoloxidase and superoxide dismutase, and displayed resistance to *V. parahaemolyticus* infection. The same species of mud crab fed *B. subtilis* exhibited a significant improvement in phenoloxidase activity, phagocytic activity, and disease resistance (Yeh *et al.*, 2014).

Feeding lobster (*Panulirus ornatus*) with *B. pumilus*, *B. pumilus,* and *B. cereus* for 60 days increased growth, and reduced mortality towards *Vibrio owensii* (Nguyen *et al.*, 2014a). In addition, *Bacillus* spp. administrated via *Artemia* nauplii in European lobster (*Homarus gammarus*) induced a higher growth and survival than control (Daniels *et al.*, 2010, 2013), but was ineffective in bath immersion route (Middlemiss *et al.*, 2015).

Even use of bacilli administration to the rearing water can modulate the shrimp immunity. For instance, administration of *B. subtilis* and *B. licheniformis* in water rearing of white leg shrimp increased the levels of lysozyme, serine protein, prophenoloxidase, peroxinectin, and lipopolysaccharide and β-1,3-glucan-binding protein, and improved shrimp resistance to *V. harveyi* infection (Zokaeifar *et al.*, 2012, 2014; Sadat Hoseini Madani *et al.*, 2018).

Studies have demonstrated that *Bacillus* strains isolated from the same environment where they will be applied, can enhance immune responses of shrimp species. For instance, feeding black tiger shrimp with endogenous *B. cereus* enhanced lysozyme, respiratory burst, phenoloxidase, and bactericidal activity (Chandran *et al.*, 2014), and such immunostimulatory effect was supported by a higher disease resistance than control shrimp in a challenge with *V. harveyi* challenge. A *Bacillus* sp. isolated from giant freshwater prawn and used in giant freshwater prawn enhanced haemocyte population and activities of phenoloxidase and respiratory burst, suggesting a possible of probiotic species-specific effect (Rahiman *et al.*, 2010). Another *Bacillus* strain isolated from the intestine of Chinese shrimp (*Fenneropenaeus chinensis*), induced higher immunostimulatory effect in white leg shrimp than the control through enhancing the shrimp immune responses including prophenoloxidase, peroxinectin, penaeidin, thioredoxin, lectins, haemocyanin, and crustin (Chai *et al.*, 2016). The treated shrimp also displayed enhanced resistance to white spot viral disease than the control group. In addition, *B. aryabhattai* isolated from the sediment demonstrated an improvement in shrimp innate immunity, antioxidant activity, and disease resistance to *V. harveyi* challenge (Tepaamorndech *et al.*, 2019).

Bacillus may also increase disease resistance to viral infections in shrimp. For instance, Li *et al.* (2009) revealed that application of a *Bacillus* strain to white leg shrimp increased protection towards white spot viral disease probably due to enhancement of phenoloxidase, phagocytosis functions. The probiotic mixture of *B. licheniformis*, *B. subtilis*, and *B. subtilis* subsp. *subtilis* was able to decrease the morbidity and mortality of white leg shrimp infected with white spot viral disease and infectious hypodermal and hematopoietic necrosis viral disease that could be due to enhancing the animal immunity such as up-regulation of immune-related genes i.e., prophenoloxidase, LvToll1 gene, and superoxide dismutase (Sánchez-Ortiz *et al*, 2016).

Bacilli may modulate the activities of innate immune enzymes and digestive enzymes and improve the animal tolerance to stress condition. For instance, tolerance to environmental stress and disease resistance to *V. harveyi* were shown in white leg shrimp after feeding with *B. licheniformis* and *B. flexus* either in single or in a combined form (Cai *et al.*, 2019). Use of *B. subtilis* in white leg shrimp induced a significant immunostimulatory effect shown by an increase in expression of the immune genes including lipopolysaccharide and β-1,3-glucan-binding protein, hemocyanin, and prophenoloxidase before and after challenging with *V. parahaemolyticus* (Interaminensea *et al.*, 2019; Chien *et al.*, 2020). These results demonstrated an enhancement of white leg shrimp immunity via glutamine metabolism and hexosamine biosynthesis when the shrimp were fed a diet containing *B. subtilis*. The authors showed that the transcriptome analysis revealed a higher expression of ten genes associated with antioxidant enzymes, pattern recognition proteins, and antimicrobial molecules in shrimp fed a diet supplemented with the probiotic compared to control group. Also, the expressions of four genes involved with hexosamine biosynthesis pathway and UDP-N-acetylglucosamine-peptide N-acetylglucosaminyltransferase for protein O-glycosylation were higher in hepatopancreas of probiotic-treated shrimp than in shrimp fed the control diet. Additionally, the authors showed a higher activity of heat shock factor 1, and protein O-glycosylation in hepatopancreas of shrimp fed probiotic than the control shrimp. In addition, superoxide dismutase lysozyme activity and immune-related gene expression of serine protease, prophenoloxidase and actin in white leg shrimp fed *B. subtilis* were higher than control shrimp (Won *et al.*, 2020b). Such immunostimulatory effect was supported by a higher disease resistance in treated shrimp against *V. parahaemolyticus* challenge.

Some *Bacillus* species such as *B. thuringiensis* carry poly-beta-hydroxybutyrate (PHB), a well-known bacterial storage compound and a polymer of the short chain fatty acid β-hydroxybutyrate, that exerts its beneficial effect by stimulating both the specific and non-specific immune mechanism in aquatic animals (Suguna *et al.*, 2014; Defoirdt *et al.*, 2007; Soltani *et al.*, 2019a). In the study by Laranja *et al.* (2014) the immunostimulatory effect of PHB-accumulating *Bacillus* spp. on immune status and survival of shrimp species towards pathogenic agents was promising. In the subsequent study by Laranja *et al.* (2017) *Bacillus* strains with PHB stimulated black tiger shrimp innate immune responses (innate immune genes for transglutaminase, heat shock protein 70 and prophenoloxidase) after shrimp were fed *Bacillus* sp. enriched

Artemia larvae and challenged with *V. campbellii*. Therefore, like fish application of bacilli probiotics individually or in a combination form can modulate the shrimp immune function, but still details mode of action warranted future works. From a practical aquaculture point of view, it is worth mentioning that optimization of the probiotic dosage and duration of utilization are two essential factors to be considered probiotic administration in aquaculture farm as over-dosage may induce a revere immunostimulatory effect in the target animal.

10.6 Lactic Acid Bacteria (LAB) for Use as Probiotics

Lactic acid bacteria (LAB) are rod-shaped or coccid Gram-positive, non-endosporing bacteria of the Latobacillales order. They are oxidase- and catalase-negative bacteria, are mostly non-motile, and are mesophile with an optimum growth at pH 5.5–5.8 but with a complex nutritional requirement. The homofermentative LAB group secrete lactic acid from sugars, while the heterofermentative group secrete lactic acid, carbon dioxide, alcohol, or acetic acid (Ringø *et al.*, 2018). The secretory compounds such as hydrogen peroxide, bacteriocins, and diacyls by LAB are inhibitory to adhesion, colonization, and proliferation of pathogenic microorganisms in the host intestine (Li *et al.*, 2018; Simon *et al.*, 2021).

In aquaculture, among the LAB genera more attention has been paid to study the potential of *Lactobacillus*, *Lactococcus*, *Pediococcus* genera, and less data is available on other genera such as *Aerococcus*, *Enterococcus*, *Leuconostoc*, *Streptococcus*, *Carnobacterium*, *Vagococcus*, and *Weissella* (Fusco *et al.*, 2015; Ringø *et al.*, 2020; Van Doan *et al.*, 2021). Beneficial LAB can be isolated from various sources including finfish GI tract, plants, fruits, dairy products, and humans (Merrifield *et al.*, 2014; Ringø *et al.*, 2010; 2018). Probiotic LAB, especially the endogenous strains, play a major role in improving of stimulating the aquatic animal immune functions, disease resistance to infections and food digestibility (Piazzon *et al.*, 2017; Tarnecki *et al.*, 2017; Li *et al.*, 2018; Wang *et al.*, 2018). Many studies have evaluated the potential of LAB as probiotics particularly the autochthonous strains in both finfish and shellfish, and today several of them are currently used as commercial probiotics for use in the aquaculture sector (Merrifield *et al.*, 2014; Ringø *et al.*, 2010). However, it is important to notice that some members of LAB may be pathogenic for aquaculture species (Austin *et al.*, 2016); thus, careful considerations must be undertaken when one is trying to use endogenous or exogenous LAB strains in the aquaculture industry. As numerous reports are available on the immunomodulatory effects of LAB on finfish and shellfish (Ringø *et al.*, 2018), thus, in this section an overview of LAB immunostimulatory effect in aquaculture species is discussed, focusing on recently published studies.

10.6.1 Immunomodulatory Effects of LAB Probiotics in Fish Culture

Immunomodulatory effects of different LAB species including *Lb. plantarum*, *Lb. rhamnosus*, *Lb. acidophilus*, *Lb. sakei*, *Lb. casei*, *Enterococcus faecium*, *Bifidobacterium bifidium*, *Pediococcus* (*P.*) *pentosaceus*, and *P. acidilactici* are revealed in several fish species such as carps, tilapia, grouper, and sturgeons and in most cases these LAB improved the fish immune responses and had a higher resistance against infectious agents (Naiel *et al.*, 2020).

P. pentosaceus is immunomodulator in grass carp improving up-regulation of fish immune genes, IgM, and C3 complement protein in both the liver and spleen tissues and induce fish resistance to bacterial infections including motile *Aeromonas* septicaemia caused by *A. hydrophila* (Gong *et al.*, 2019). Oral use of *Lb. acidophilus* in common carp increased immunocompetent cell population and improved fish resistance to bacterial diseases including pseudomoniosis caused by *Pseudomonas aeruginosa* and motile *Aeromonas* septicaemia caused by *A. hydrophila*, but the level of the probiotic immunostimulatory was dose-dependent (Adeshina, 2018). Bath immersion of various LAB including *Ent. faecalis* and *P. acidilactici* induced a higher resistance in common carp against *A. hydrophila* infection through enhancing fish immunity e.g., total immunoglobulin, protease activity, and up-regulation of immune genes of lysozyme and both proinflammatory (TNF-a, IL-1b, IL-6, IL-12) and anti-inflammatory cytokines (IL-10, TGF-b) (Shabirah *et al.*, 2019; Mulyani *et al.*, 2018; Feng *et al.*, 2019; Hoseinifar *et al.*, 2019). Oral administration of an autochthonous *Lb. casei* strain is immunomodulatory to Shabuoth juveniles (*Tor grypus*) through improving values of leukocytes population, haemoglobin, and expression of immune genes of IL-1b, TNF-a and IL-8 (Mohammadian *et al.*, 2018). The innate immune responses of Caspian white fish (*Rutilus frisii kutum*) were improved through oral administration of *Lb. acidophilus*, *Lb. casei*, *Enterococcus faecium*, and *Bifidobacterium bifidium* (Mirghaed et al., 2018). Oral use of *Lb. fermentum* in common carp led to an increase in hematological parameters, and activity of antioxidant enzymes, lysozyme, and serum respiratory burst and higher resistance towards *A. hydrophila* infection (Ahmadifar *et al.*, 2019). Feeding common carp *Lb. plantarum* increased B-lymphocytes and phagocytic activity, enhanced γ-globulins and total protein, and exhibited a higher resistance to challenges with *A. hydrophila* (Kazu´n *et al.*, 2018). Common carp fed with a LAB exhibited enhancement immunomodulatory properties through an increase in the expression of cytokines and resistance to *A. hydrophila* infection (Feng *et al.*, 2019).

In tilapia, oral application of *Lb. plantarum* or *Lb. rhamnosus* individually, or in combination with Jerusalem artichoke, improved the innate immune variables of fish including complement, lysozyme, and peroxidase activities; total Ig concentration; numbers of mucin-secreting goblet cells; heat shock protein 70, IL-1b, and IFN-c lysozyme; and provide a significant protection to streptococcosis caused by *S. agalactiae* and *A. veronii* infection (Van Doan *et al.*, 2018; Sewaka *et al.*, 2019; Xia *et al.*, 2018).

In salmonids, the immunomodulatory effects of several LAB species including *Lactococcus lactis*, *Leu. mesenteroides*, *Lb. sakei*, *Lb. plantarum Lb. rhamnosus*, *Lb. delbrueckii*, *Ent. faecium*, *Ent. faecalis*, *P. acidilactici*, *Carnobacterium divergens*, *Carnobacterium maltaromaticum*, and *Carnobacterium inhibens* in rainbow trout and Atlantic salmon have been confirmed by assessment of different immune responses and disease resistance of treated fish (Merrifield et al., 2010, Ringø *et al.*, 2018). For instance, oral application of *Lb. plantarum* increased functions of respiratory burst, complement, lysozyme and serum bactericidal activities and a higher resistance to *A. hydrophila* challenge (Soltani *et al.*, 2017).

Individual or mixture of either exogenous or autochthonous LAB including *Lac. Lactis*, *Ent. faecalis*, *P. pentosaceus*, *Weissella cibaria*, *Lb. plantarum*, *Weissella confuse*, *Lactococcus garvieae*, *Lb. curvatus*, *Leuconostoc mesenteroides*, *Lb. acidophilus*, *Lb. delbrueckii*, *Lb rhamnosus*, and *Pediococcus acidilactici* are revealed to be beneficial in Persian sturgeon (*Acipenser persicus*), great sturgeon (*Huso huso*), and Siberian sturgeon (*Acipenser baerii*) through improved growth performance, digestive enzymes activities, innate immunity, and survival to some bacterial diseases, but the level of the immunomodulatory effect can be varied depending on the type and source of LAB, concentration, and duration of probiotic administration (Jafaryan *et al.*, 2012; Askarian et al., 2012; Rezvani Gil Kolaei *et al.*, 2019; Rastekenari *et al.*, 2021; Bazari Moghaddam *et al.*, 2021; Hosseini Shekarabi *et al.*, 2022). However, probiotic therapy in the sturgeon species needs more investigations.

Olive flounder fed with endogenous strains of *Lac. lactis*, *Lb. sakei*, and *Lb. plantarum* (strains isolated from olive flounder) improved resistance towards streptococcosis caused by *S. iniae* by enhancement in the innate immune responses including respiratory burst, values of superoxide dismutase, serum lysozyme, myeloperoxidase and antiprotease, and upregulation of immune genes of pro-inflammatory cytokines (TNF-a, IL-1b and IL-6), MHC-II, IgM, and TCR-b (Hasan et al., 2018; Feng et al., 2019). The same LAB strains fed to great sturgeon increased fish total serum Ig, lysozyme activity, and respiratory burst activity that may be in a dose-dependent manner (Ghiasi *et al.*, 2018). Humpback grouper (*C. altivelis*) fed with endogenous strains of *Lac. lactis* (isolated from humpback grouper) enhanced protection of vibriosis caused by *V. harveyi* through an enhancement in innate immunity including lysozyme, respiratory burst, and serum acid phosphatase (Sun *et al.*, 2018). Application of *P. acidilactici* to Asian sea bass (*Late calcarifer*) diet significantly stimulated respiratory burst, lysozyme, and haemolysis activities (Ashouri *et al.*, 2018). Use of *Lb. rhamnosus* in zebrafish (*Danio rerio*) revealed an enhancement in fish innate immune responses and improved hepatic stress tolerance (Gioacchini *et al.*, 2014).

In aquaculture, generally the parasitic or fungal infections cause a lower mortality than viral and bacterial pathogens, but they can affect the host health status causing an economic loss. Despite this, little attention has been paid to assess the efficacy of probiotics towards pathogenic parasites of aquaculture species. In a study by Pieters *et al.* (2008), oral use of *Brochothrix thermosphacta* and *Aeromonas sobria* in rainbow trout exhibited an increase resistance towards *Ichthyophthirius multifiliis,* the cause of white spot disease (Ich), but *A. sobria* was more efficacious in protecting the fish towards this ciliated protozoan. Catfish exposed to *Lb. plantarum* exhibited a decrease to infection with *Saprolegnia parasitica* (Nurhajati *et al.*, 2012). In their study by Yanuhar *et al.* (2019), feeding Koi carp with *Lactobacillus* sp. in combination with *Bacillus* spp., and *Nitrosomonas* sp. at different doses exhibited an improvement in disease resistance towards *Myxobolus* sp., in terms of decreasing level of gill tissue damage. More studies are required to show additional immunomodulatory effects of probiotics to parasitic and fungal infections.

10.6.2 Immunomodulatory Effects of LAB Probiotics in Shellfish Culture

Several LAB species including *Lb. plantarum*, *Lb. acidophilus*, *Lb. bulgaricus*, and *Lb. bulgaricus* revealed immunomodulatory properties and are currently used as probiotics in shrimp aquaculture. The immunomodulatory effects of LAB by oral usage are more effective than bath immersion in improving the shrimp immune responses, disease resistance and growth performance. In an earlier study, Phianphak *et al.* (1999) revealed the probiotic potential of *Lactobacillus* sp. towards control of *Vibrio harveyi* infection in the black tiger shrimp culture. Oral application of *Lb.plantarum* to white leg shrimp (*Litopenaeus vannamei*) improved the immune responses including phenoloxidase, and superoxide dismutase, respiratory burst, phagocytosis, alkaline phosphatase, acid phosphatase, total nitric oxide synthase, total antioxidant capacity, and lysozyme; and upregulated immune relative genes of prophenoloxidase I, serine proteinase, and peroxinectin, lysozyme, superoxide dismutase, prophenoloxidase, heat shock protein 70, Imd, Toll, Relish, TOR, 4E-BP, eIF4E1α, eIF4E2, and LGBP; and also enhanced the shrimp metabolites including inosine monophosphate, valine, and betaine (Chiu *et al.*, 2007; Wei *et al.*, 2022). Even the heat-killed (paraprobiotic) or cell-free extract (postbiotic) of *Lactobacillus* species can induce a tolerance stress condition in white shrimp or kuruma shrimp towards poor health management such as water quality fluctuations e.g., a low salinity stress condition through enhancing in animal immune status e.g., upregulation of the immune genes of prophenoloxidase, superoxide dismutaseand lysozyme, thus application of the probiotic can result in a higher survival during outbreaks of vibriosis caused by various species of *Vibrio* pathogens especially *V. parahaemolyticus* and *V. harveyi* infections (Zheng *et al.*, 2017; Thanh Tung *et al.*, 2010; Nguyen *et al.*, 2018; Wei *et al.*, 2022). Therefore, in shrimp aquaculture practice, a continuous application of the probiotics at an optimum dosage can significantly decrease morbidity and mortality through both immunomodulatory effect and modification of the shrimp intestinal microbiota. In addition, *Lb. plantarum* in combination with galactooligosaccharide may be more effective in stimulating the shrimp immune parameters including phenoloxidase, respiratory bursts, phagocytosis, and expressing of immune genes e.g., prophenoloxidase I, serine proteinase, and peroxinectin (Chiu *et al.*, 2007; Huynh *et al.*, 2018). Interestingly, inclusion of autochthonous *Lb. plantarum* in combination with other autochthonous LAB strains or with yeast can be more effective by stimulating shrimp immunity and induce a higher resistance to the infectious diseases including acute hepatopancreas necrosis disease caused by *V. parahaemolyticus* (Chomwong *et al.*, 2018). For instance, survival of 93% was revealed in white leg shrimp feda mixture of *Lb. plantarum* and *Lac. lactis* prior to challenge with *V. parahaemolyticus*, while individual use of these probiotics revealed 50–60% survival, but only 10% survival in the control. In addition, dietary supplementation of a mixture of *Lb. acidophilus*, *Lb. bulgaricus*, *Lb. bulgaricus* and *Streptococcus cremoris* displayed enhanced disease resistance against *Vibrio alginolyticus* that could explained by the enhancement of the shrimp nonspecific immune responses (Ajitha *et al.*, 2004). Further, oral administration of the mixed probiotics containing *Lb. pentosus*, *L. fermentum*, *B. subtilis*, and *S. cerevisiae* in white shrimp was more efficient than using the single probiotics in terms of growth and health status (higher resistance to *V. alginolyticus* challenge) that was due to higher immunostimulatory effects including phenoloxidase activity, respiratory bursts, and lysozyme activity of hemocytes by the mixed probiotics (Wang *et al.*, 2019). Enriched diets with *Lb. acidophilus* in combination with *Saccharomyces cerevisiae* improved shrimp survival against *V. parahaemolyticus* infection via improvement in haemocyte parameters, including the total haemocyte count, granular haemocytes percentage, and phenoloxidase activity as well as highly upregulation of haemocyte genes (*Crustin* and *penaeidin*) and resulted in a high survival rate (Pooljun *et al.,* 2020). Therefore, it could be concluded that administration of a mixture of several *Lactobacillus* strains are important in the management of disease outbreaks in shrimp aquaculture. However, the synergistic immunomodulatory effect of LAB administration needs to be carefully evaluated prior to their application in the animal farm. For instance, no significant difference was noticed in disease resistance in white shrimp fed *Lb. pentosus* or *Lb. fermentum* individually or in a mixture that may in part be due to lack of dosage optimization (Wang *et al.*, 2019), or combination use of *Lb. plantarum* with *B. subtilis* in mud crab (*Scylla paramamosian*) displayed a reduction in efficiency of probiotics in terms of immune responses and disease resistance that was probably due to an antagonistic effect of growth between two probiotics (Yeh *et al.*, 2014). *E. faecalis* and *P. pentosaceus* administered to Mud crab (*Scylla paramamosain*) (Yang *et al.*, 2019) were effective in stimulating the animal immune responses such as up-regulation of immune genes i.e., catalase, prophenoloxidase and superoxide dismutase, and displayed resistance to *V. parahaemolyticus* infection. Feeding lobster (*Panulirus ornatus*) *Lb. plantarum* for two months increased growth and survival of the lobster towards *Vibrio owensii* infection (Nguyen *et al.*, 2014a).

In addition, the source and dosage of the LAB are important in evaluation of their immunostimulatory effect as strains/species of a specific LAB isolated from different sources may have different immunomodulatory effect. For instance, strains of *Lb. sakei* isolated from fermented and salted shrimp was more effective in stimulating shrimp immunity than other strains or species isolated from different origins, and the immunostimulatory effect by *Lb. bulgaricus* in white leg shrimp may be dose dependent as a better immunostimulatory effect and a higher survival were noticed at higher dosage (10^9CFU/g feed) than lower one (10^7CFU/g feed) (Roomiani *et al.*, 2018; Le *et al.*, 2018).

Furthermore, application of diets containing *Lactobacillus* increase the abundance of *Lactobacillus* in the shrimp GI tract and subsequently decrease the population level of pathogenic bacteria including *Vibrio;* thus, oral administration of *Lactobacillus* could enhanced the biodiversity of the GI microbiota resulting in enhanced immune status, digestive enzymes activities, feed utilization, growth performance, and increase disease resistance towards pathogenic infections (Kumar *et al.*, 2016; Li *et al.*, 2018; Zheng *et al.*, 2017).

Apart from the commercial shrimp species, white leg shrimp, and black tiger shrimp LAB can also induce immunomodulatory effects in other shrimp species including *Penaus indicus*, giant freshwater prawn and kuram shrimp (Sumon *et al.*, 2022). For instance, administration of *Lb. rhamnosus* and *Lac. lactis* induced an immunostimulatory effect in *Penaeus indicus* and kuruma shrimp by enhancing the immunity and improving higher disease resistance (Maeda *et al.*, 2014; Naresh *et al.*, 2014). Even in other crustaceans or molluscs the immunomodulation of LAB is studied, and the results are promising for the development of the sustainability of invertebrate aquaculture. For instance, Japanese abalone (*Haliotis discus hannai*) fed endogenous *Lb. pentosus*, narrow clawed crayfish fed *Lb. plantarum* and sea cucumber fed strains of *Lb. plantarum*, *Weissella confuse*, *Lac. lactis,* and *Ent. faecalis* can develop various innate immune responses (e.g., lysozyme, acid phosphatase, catalase, superoxide dismutase, alkaline phosphatase, phenoloxidase, total plasma proteins) and improve the resistance to bacterial diseases, *Vibrio splendidus* and *V. parahaemolyticus* infections (Xiaolong *et al.*, 2018; Li *et al.*, 2018; Valipour *et al.*, 2019).

It is also revealed that probiotic LAB have antiviral activities (Arena *et al.*, 2018); thus, their application in the prevention and control of viral diseases in shrimp aquacultures could be an efficient novel tool. The antiviral activity of LAB mainly depends on bacterial strain, and the antiviral effects may be in part due to antimicrobial secretions by the probiotic bacteria that are effective as antiviral bacterial derivatives, or to the immunomodulatory role of probiotic towards activation of the immune system. Probiotic LAB produce hydrogen peroxide, lactic acid, bacteriocins and bacteriocin-like substances, short-chain fatty acids, and polysaccharides that assist in detecting viral infections and enhanced the antiviral activity (Arena *et al.*, 2018). Peraza-Gómez *et al.* (2009) demonstrated an increase in survival of white shrimp towards white spot viral disease after feeding the shrimp with LAB strains, isolated from tilapia, and one yeast, suggesting LAB and yeast as potential microbial adjunct in the shrimp culture against white spot viral disease. In addition, Jiravanichpaisal *et al.* (1997) demonstrated *Lactobacillus* sp. as an effective probiotic additive for black tiger shrimp against white spot viral infection and vibriosis after the shrimp being treated with the probiotic.

It is worth mentioning that isolation of probiotics such as LAB from the commensal microbiota may be a more useful approach to enrich the pool of probiotics, and available data shows beneficial immunostimulatory effects when these autochthonous LAB are used to the hosts (Merrifield *et al.*, 2014; Van Doan *et al.*, 2018). When the native hosts, however, are fed the probiotics at a high or a low dosage, no effects or adverse effects may display that could be in part due to the imbalance of the gut microbiota induced by the excessive feeding of autochthonous bacteria; thus, this topic warranted future studies.

10.7 Gram-Negative Bacteria for Use as Probiotics

Despite many data on the immunostimulatory effects of Gram-positive probiotics in finfish and shellfish, less attention has been paid to assess the immunomodulatory effects of Gram-negative probiotics in aquaculture species and this may in part be due to less safety of Gram-negatives in aquaculture systems than Gram-negatives. However, in addition to Gram-positive probiotics, particularly members of LAB and bacilli, strains of different Gram-negatives, including members of *Shewanella*, *Vibrio*, *Aeromonas*, *Alteromonas*, *Pseudomonas*, *Phaeobacter*, *Pseudoalteromonas*, *Rhodococcus*, *Rhodopseudomonas*, *Arthrobacter,* and *Roseobacter* have been assessed as potential probiotics for use in aquaculture.

10.7.1 Immunomodulatory Effects of Gram-Negative Probiotics in Fish Culture

In a recent publication by Ringø *et al.* (2022), attention has been paid to some members of Gram-negatives including *Aeromonas*, *Vibrio*, *Aliivibrio*, *Enterovibrio*, *Alteromonas*, *Arthrobacter*, *Phaeobacter*, *Pseudoalteromonas*, *Pseudomonas*, *Rhodococcus*, *Rhodopseudomonas*, and *Shewanella* as potential probiotics in finfish aquaculture. In most cases the reported strains of these Gram-negatives were effective in immunomodulation of fish immune responses or increased disease resistance to the pathogenic bacterial infections in finfish.

From the available literature it seems that members of *Shewanella,* particularly *Shewanella putrefaciens* strain Pdp11 are the most frequently used Gram-negative probiotics in finfish aquaculture (Ringø *et al.*, 2022). For instance, feeding gilthead seabream *Sh. putrefaciens* Pdp11, isolated from seabream skin, induced an upregulation of immune related genes (antioxidant, INFγ and IL-1β genes), a protection against intracellular reactive oxygen species and stimulated fish growth (Esteban *et al.*, 2014; Guzmán-Villanueva *et al.*, 2014). Administration of the same probiotic strain (Pdp11) was effective in skin wound healing in gilthead seabream, and increased immune variables including ratio of albumin/globulin, protease and peroxidase activities, anti-inflammatory cytokines (IL-10 and TGF-b), up-regulation of the expression of the genes; antioxidant enzymes, the intestinal inflammatory responses, and decreased pro-inflammatory cytokines (IL-1 b, IL-6, IL-8 and TN-a), suggesting the wound healing and skin damage by the probiotic (Chen *et al.*, 2020a, Chen *et al.*, 2020b). Endogenous *Sh. xiamenensis* strains A-1 and A-2 administrated to grass carp improved activity of phagocytosis, lysozyme and respiratory burst, and induced an upregulation of immune related genes (IL-8, IL-1β, lysozyme-C and TNF-a), and a resistance towards *A. hydrophila* challenge (Wu *et al.*, 2015).

In addition, the probiotic potentials of other members of Gram-negatives have been assessed on finfish immune systems with mostly positive results. For instance, administration of *A. sobria* individually in form of live cells, disrupted cells, or cell free supernatant, or in combination with *Bacillus* spp. was able to enhance rainbow trout survival towards streptococcosis, lactococcosis and furunculosis caused by *S. iniae*, *L. garvieae*, and *Aeromonas salmonicida*, respectively (Brunt *et al.*, 2005; 2007). Cellular components (whole cell proteins, outer membrane proteins, cell wall proteins/and or lipopolysaccharides) of a strain of *Aeromonas sobria* increased resistance of rainbow trout towards *Yersinia ruckeri* infection (Abbass *et al.*, 2010). Feeding grass carp (*Ctenopharyngodon idella*) with *A. veronii* was isolated from the same fish was immune promoter inducing a significant increase of respiratory burst, phagocytic and lysozyme activities, and induced an upregulation of immune genes (IL-8, IL-1b, lysozyme-C and TNF-a), and resistance towards *A. hydrophila* infection (Wu *et al.*, 2015). Use of probiotics *A. hydrophila*, *Vibrio fluvialis,* and *Vibrio* spp. reduced fry and fingerling mortalities in rainbow trout towards furunculosis (Irianto *et al.*, 2002; 2003). Further, administration of *A. sobria*, *Pseudomonas spp.,* and *Pseudomonas fluorescens* could increase survival of rainbow trout or Atlantic salmon against *Vibrio anguillarum*, *Vibrio ordalii,* and *Vibrio harveyi,* the causative agents of vibriosis (Gram *et al.*, 1999; Spanggaard *et al.*, 2001; Brunt *et al.*, 2007; Arijo *et al.*, 2008). Two-month feedings to major carp/rohu (*Labeo rohita*) of *Pseudomonas aeruginosa* induced an enhancement in fish immune responses, including activity of lysozyme, phagocytosis, alternative complement pathway, respiratory burst, superoxide dismutase, serum IgM level, and increased disease resistance to *A. hydrophila* infection (Giri *et al.*, 2012). Biovars I, II, and III of *P. fluorscens* that were inhibitory against the growth of *Pseudomonas angulliseptica* and *Strepcoccus faecium,* significantly reduced mortality of tilapia challenged with the pathogens and effectively stimulated fish immunity such as immunocompetent cell population (Eissa *et al.*, 2014). *Pseudomonas monteilii* administered to grass carp increased fish resistance to *A. hydrophila* infection (Qi *et al.*, 2020).

Furthermore, two strains of Gram-negative probiotics belonging to *Vibrionaceae* significantly enhanced immune responses (serum peroxidase, complement, phagocytic, and cytotoxic activities) of gilthead seabream (*Sparus aurata*) (Díaz-Rosales *et al.*, 2006). Feeding sea bass (*Dicentrarchus labrax*) larvae with *Vibrio lentus* was effective in stimulation of upregulation of gene expression related to cell proliferation, cell adhesion, reactive oxygen species, iron transport, and immune functions (Schaeck *et al.*, 2017). Inclusion of endogenous strains of *Aliivibrio* in rearing water of Atlantic salmon and lumpfish (*Cyclopterus lumpus*) improved fish growth and ulcer prevalence, suggesting the efficacy of *Aliivibrio* on fish health status (Klakegg *et al.*, 2020a; 2020b).

Additionally, *Alteromonas* sp. or *Pseudoalteromonas* sp. fed to European sea bass (*Dicentrarchus labrax*) was immunomodulator inducing an enhancement in respiratory burst, phagocytic activity, and gene expression of lysozyme, Mx protein, caspase 3, TNF-α, and IL-10 (Mladineo *et al.*, 2016), but administration of *Enterovibrio coralii* to the same fish species exhibited no significant stimulation of fish immune responses (Mladineo *et al.*, 2016). Strain of *Chromobacterium aquaticum* (isolated from lake water) administered to zebrafish (*Danio rerio*) displayed an enhancement in the innate immune-related genes, increased disease resistance towards infections by *A. hydrophila* and *Streptococcus iniae* (Yi *et al.*, 2019). Use of endogenous *Enterobacter* strain and *Enterobacter amnigenus* in rainbow trout induced disease resistance to

Flavobacterium psychrophilum the cause of bacterial cold-water disease (Burbank *et al.*, 2011), and use of *Enterobacter cloacae* in Kenyi cichlid (*Maylandia lombardoi*) improved growth and immunity (respiratory burst activity) of fish (Girijakumari *et al.*, 2018). Strains of *Phaeobacter gallaeciensis* (isolated from seawater of scallop cultures) administrated to Atlantic cod (*Gadus morhua*) increased resistance to *Vibrio anguillarum* infection (D'Alvise *et al.*, 2013). Administration of strains of *Rhodopseudomonas palustris* to tilapia or yellow catfish (*Pelteobagrus vachelli*) was effective in enhancing of respiratory burst, superoxide dismutase, catalase and myeloperoxidase, alkaline phosphatase, acid phosphatase, and digestive enzymes activities (protease, amylase, lipase activities), and increased the catfish resistance towards *A. hydrophila* infection (Zhou *et al.*, 2010; Liu *et al.* (2020). Use of strains of *Psychrobacter namhaensis* and *Psychrobacter maritimus* in Nile tilapia exhibited an enhancement in immunoglobulin, complement, phagocytic and lysozyme activities, IL4, erythrocytes and leucocytes populations, and up-regulation of several immune genes (Makled *et al.*, 2017; Makled *et al.*, 2020). Bighead catfish (*Clarias macrocephalus*) fed a strain of *Acinetobacter* sp. (isolated from skin mucus of bighead catfish) exhibited an enhancement in phagocytosis, and activity of lysozyme, respiratory burst activity, and alternative complement pathway, and the up-regulation of immune genes as well as an increase in disease resistance towards *A. hydrophila* infection (Bunnoy *et al.*, 2019).

There is very little data concerning anti-fungal activity of probiotics in aquatic animals. Inclusion of *Aeromonas media* to the rearing water of silver perch (*Bidyanus bidyanus*) could reduce infection by *Saprolegina paracitica* the cause of saprolegniosis (Lategan *et al.*, 2004). In a study by González-Palacios *et al.*, (2020), bath treatment of rainbow trout with two strains of *Pseudomonas fluorescens* displayed a decrease in *S. parasitica* infection through enhancing fish innate immune response (phagocytosis and serum proteins), and the production of siderophores, suggesting *P. fluorescens* as a biocontrol tool of fish saprolegniosis probably through the production of siderophores.

10.7.2 Immunomodulatory Effects of Gram-Negative Probiotic in Shellfish Culture

Like finfish, compared to Gram-positive probiotics fewer studies have been directed to evaluate the immunomodulatory effects of Gram-negative probiotics in shellfish (Butt *et al.*, 2021). When *Vibrio diabolicus* was added into the rearing water of white leg shrimp it was very effective in controlling vibriosis affecting the shrimp at larval stages and juveniles through enhancing animal immunity (Gullian *et al.*, 2004). Feeding *Roseobacter* strains including *Rhodobacter sphaeroides* and *Roseobacter gallaeciensis* isolated from marine environment, shrimp or fish intestines were effective in stimulating the growth, intestinal enzymes (protease, amylase, lipase) of white shrimp and induced disease resistance to *V. parahaemolyticus* infection (Wang 2007; Ruiz-Ponte *et al.*, 1999; Balcázar *et al.*, 2007). Oral use of *V. alginolyticus* isolated from intestine of white shrimp stimulated the shrimp immunity and induced a higher growth and disease resistance against *V. parahaemolyticus* (Balcázar *et al.*, 2007). Bath application of *Arthrobacter* sp. to white leg shrimp breeding tanks improved animal immune responses (phenoloxidase activity, phagocytic activity and clearance efficiency) and induced a higher survival and growth of larvae (Xia *et al.*, 2014). Bath treatment of white leg shrimp with *Rhodopseudomonas palustris* (isolated from shrimps and common carp ponds) enhanced the shrimp growth and improved immune responses including activity of prophenoloxidase, sodium peroxidase dismutase and antibacterial activity (Wang *et al.*, 2018). Some strains of *Pseudomonas* and *V. fluvialis* as a bath were ineffective in stimulating immune system of tiger shrimp (Alavandi *et al.*, 2004) that may in part due to a low dosage (10^3 cfu/ml) used. Application of *Pseudomonas* sp. and *V. fluvialis* in the form of a bath for 45 days did not change heamolymph bactericidal activity, prophenoloxidase, and total haemocyte count in tiger shrimp (Alavandi *et al.*, 2004), but *Pseudomonas aestumarina* isolated from intestine of white shrimp induced disease resistance against *V. parahaemolyticus* infection in white shrimp (Balcázar *et al.*, 2007). Use of different strains of marine *Peudoalteromonas* sp. in white shrimp and *Litopenaeus stylirostris* displayed a disease resistance against pathogenic *Vibrio* including *Vibrio* sp., *V. parahaemolyticus,* and *V. nigripulchritudo*, and enhanced the growth, digestive enzymes (protease, amylase, and lipase), and immune responses including immune gene expression of lysozyme and penaeidin (Longeon *et al.*, 2004; Sorieul *et al.*, 2018; Wang *et al.*, 2018). Administration of commercial *Pseudomonas synxantha* and *Pseudomonas aeruginosa* for 84 days in Western king prawn (*Penaeus latisulcatus*) enhanced the survival and specific growth rate (van Hai *et al.*, 2009), that could be due to immunomodulatory effect induced by the probiotics. Two-month feeding *Vibrio* to giant freshwater prawn (*M. rosenbergii*) stimulated the animal immunity in terms of increasing total haemocyte counts, phenoloxidase, and respiratory burst and growth (Rahiman *et al.*, 2010).

In bivalve mollusks, a five-day bath administration of *Aromonas media* in Pacific oyster (*Crassostrea gigas*) induced a disease resistance (Gibson *et al.*, 1998). Larvae and juveniles of eastern oyster (*C. virginica*) bath treated with either a strain of *Phaeobacter* (Karim *et al.*, 2013; Sohn *et al.*, 2016a), or in combination with *B. pumilus* exhibited an increase in survival rate in challenge with *Vibrio tubiashii* and *Roseovarius crassostreae* infections (Karim *et al.*, 2013; Sohn *et al.*, 2016a, Sohn *et al.*, 2016b). In addition, a 24-hour bath treatment of bay scallop (*Argopecten irradians*), hard clam (*Mercenaria mercenaria*), blue mussel (*Mytilus edulis*), and razor clam (*Ensis directus*) with these probiotics individually or in mixture form displayed an increase resistance to *Vibrio coralliilyticus* infection only in bay scallop, suggesting a species-specific action of the probiotics (Sohn *et al.*, 2016b). Red abalone (*Haliotis rufescens*) fed probiotics *Agarivorans albus* and two strains of *Vibrio* spp. for seven months, and Japanese abalone (*H. discus*) fed *Sh. colwelliana* and *Sh. olleyana* for 28 days displayed an improvement in some immune functions such as total hemocyte population and activity of phagocytosis, lysozyme, and respiratory burst and an exhibited a disease resistance towards *V. parahaemolyticus* and *V. harveyi* infections (Silva-Aciares *et al.*, 2013; Jiang *et al.*, 2013).

10.8 Probiotic Immunomodulatory and Fish Haematology

Haematological profiles including leucocytes, erythrocytes, haematocrit, and hemoglobulin as the essential diagnostic tools for evaluation of the health condition of fish are depends on the host species and age, nutritional status, and environmental conditions. Detection of these blood indices is, however, very important to assess the immune-physiological functions induced bythe probiotic bacteria in the aquaculture species. From the available data it is clear that the evaluation of probiotics effects on fish haematology is mostly based on standard aquaculture conditions plus the fish responses following the challenging with pathogenic infectious agents. However, many environmental distresses such as fluctuations in water temperature, level of pollutants, and nutrition conditions can cause a physiological stress in fish (e.g., Bly *et al.*, 1997; Alfonso *et al.*, 2021). Thus, it is crucially important to determine the influence of probiotics on the animal haematological profiles under the optimal cultural conditions as well as under the above stressor conditions.

Numerous studies and comprehensive reviews on the immunostimulatory effects of probiotics on finfish health are available, but many interferences have been made by researchers concerning the use of probiotics on improving fish blood indices (e.g., Kiron *et al.*, 2010; Dahiya *et al.*, 2012; Da Paixão *et al.*, 2017; Michael *et al.*, 2019) reviewed the immunostimulatory effects of probiotics on fish haematological values, but there is less data available on the actual mechanisms of action. The function of leucocytes that are available in both bloodstream and tissues of thymus, kidney, and spleen of finfish is important in cellular defence and humoral immunity and can be influenced by the toxicants and stressors (Farrell, 2011). The application of individual probiotics or in combination in fish diets have shown to increase the immunocompetent cell population higher than the controls. For instance, use of combined *Lb. acidophilus* and *B. subtilis* in Nile tilapia (Aly *et al.*, 2008), a mixture of *Lb. sporogenes*, *Lb. acidophilus*, *B subtilis*, *B. licheniformis,* and *Saccharomyces cerevisiae* inmrigal carp (*Cirrhinus cirrhosus*) (Sharma *et al.*, 2013); a mixture of *B. subtilis* and *S. cerevisiae* in mrigal carp (Ullah *et al.*, 2018); a mixture of *B subtilis*, and *B. licheniformis* in Persian sturgeon (Darafsh et al., 2020); and *B. subtilis* in rainbow trout (Kamgar *et al.*, 2014) all displayed a higher enhancement in value of leucocyte population and a better immune responses than control fish. In addition, erythrocyte level in fish species subjected to probiotic administrations can be stimulated (Azarin *et al.*, 2015). For instance, the application of *Lb. acidophilus* in rainbow trout (Faramaz *et al.*, 2011), B. *licheniformis* and *B. subtilis* in Caspian white fish (Azarin *et al.*, 2015), and a combined dosage of *B. subtilis*, *B. licheniformis*, *Lb. sporogenes*, *Lb. acidophilus*, and *S. cerevisiae* in mrigal carp (Sharma *et al.*, 2013) significantly enhanced the erythrocyte population *vs.* fish feed basal diets.

10.9 Probiotics as a Vector for Fish Vaccines

From a aquaculture practice point of view, the oral route of vaccination is much easier and less cost- and time effective than the other of vaccination routes i.e., bath immersion and injection. The oral route of immunization is critically important in cases where initial infection of some infectious diseases occurs at the mucosa of the susceptible fish. Therefore, developing of the vaccines to induce protective immunity via developing mucosal immune responses is feasible. However, in the oral vaccination route the antigens must be protected from unsuitable gut condition such as low pH and proteases in GI tract.In recent years, application of *Bacillus* spores as a highly efficient oral vaccine delivery tool has received much attention that is due to features of *Bacillus* spores e.g., tolerance specialty, operability of the genes, safety, and adjuvant possessions. This topic was reviewed by Soltani *et al.* (2018) and revealed that vaccinated fish fed the probiotics or the immunized fish the vaccine delivered by the bacilli spores displayed higher survival after being challenged with pathogens. In a previous study by Venkatalakshmi *et al.* (2015), vaccinated tilapia with *A. hydrophila* bacterin and fed *Lactobacillus* probiotic significantly stimulated immunocompetent cell populations comparedto the vaccinated group fed the basal diet. Five-month feeding of tilapia immunized with of *A. hydrophila* bacterin followed by feeding with a combination of *B. subtilis*, *Lb. acidophilus*, *S. cervisiae*, and *Aspergillus oryzae* exhibited a significant increase in the specific antibody level and improved disease resistance to *A. hydrophila* infection (Aly *et al.*, 2016). Common carp and koi carp orally immunized with the *Lb. plantarum* expressing two antigens including a glycoprotein of rhabdovirus (the causative agent of spring viremia of carp) and ORF81 protein of herpesvirus (the aetiology of koi herpesvirus disease in koi carp) increased anti-specific IgM and higher survivals (71% and 53% survivals in immunized carp and koi, respectively) than control fish 65 days post-vaccination (Cui *et al.*, 2015), suggesting a poly valent protection way of carps to these two economically important viral diseases using a recombinant *Lb. plantarum*. Feeding carp with *Lb. plantarum* expressing viral glycoprotein of spring viremia of carp (SVC) and encapsulated with chitosan-alginate stimulateda specific immune responses and provided protection against SVC (Jia *et al.*, 2020). Rainbow trout orally vaccinated with a recombinant *L. lactis* NZ3900 expressing the G gene of viral hemorrhagic septicemia virus (VHSV) displayed significantly a higher relative expression of IFN-1 and MX-1 genes in the head-kidney, enhanced anti-VHS antibody titer, significant survival towards VHSV challenge, and a significant decrease in the viral loads in the vaccinated fish compared to the controls, suggesting a protective immunity and efficacy of recombinant *L. lactis* vaccine towards VHSV in trout (Naderi-Samani *et al.*, 2020). A *Lactobacilli/Escherichia coli* shuttle vector pPG1 (surface-displayed) and a secretory vector (pPG2) with the capsid of a VP2 gene (a virulent gene in infectious pancreatic necrosis virus, IPNV) inserted and transformed into *Lb. casei* to produce two recombinant strains (Lbc:PG1-VP2 and Lbc:PG2-VP2) were designed by Min *et al.*, 2012), and used by

Zhao *et al.*, 2012) in a study on rainbow trout orally vaccinated with Lbc:PG1-VP2, Lbc:PG2-VP2, Lbc:PG1-VP3 and Lbc:PG2-VP3, displayed a high anti-specific IgM to IPNV. In addition, the Lc:PG2-VP2 was significantly more immunogenic than Lbc:PG1-VP2, Lbc:PG1-VP3, and Lbc:PG2-VP3 by assay of serum IgM titer, suggesting the Lc:PG2-VP2 as a more suitable candidate for the immunization of trout towards IPNV infection.

Flounder fed *Lac. lactis* BFE920 and adapted to express the *S. iniae* SiMA antigen twice a week exhibited a significant enhancement in CD4 and CD8 T cell populations, antibody titer, and IFNɣ level, and in three-month post-vaccination, the immunized fish exhibited a significant protection towards *S. iniae* infection by 84% and 82% survivals in intraperitoneal and bath challenges, respectively. Interestingly, the weight gain and feed conversion ratio in the immunized fish were a 21% higher and 16% less than the control fish, respectively (Kim *et al.*, 2016). Bath-vaccinated rainbow trout with a streptococcosis/lactococosis vaccine (a bivalent vaccine developed by Professor M. Soltani) followed by a two-month feeding with *Lb. plantarum* revealed a higher values of immuno-competent cell population, IgM, total protein, lysozyme, complement activity, specific antibody than vaccinated fish fed basal diet (Kani *et al.*, 2016). Bath-vaccinated rainbow trout with a *Yersinia ruckeri* vaccine (Antiyersin vaccine, Bojan Ltd, Iran) and fed with the same strain of *Lb. plantarum* for 72 days demonstrated higher immune resposnes including IgM level and activities of complement and lysozyme than vaccine-only group. In the studies by Kani *et al.*, (2016) and Soltani *et al.*, (2019c), immunized rainbow trout fed the probiotic exhibited a higher growth preformance than only vaccination. Forty-one days of feeding surubim hybrid (*P. corruscans x P. reticulatum*) with *Weissella cibaria* exhibited higher levels of lysozyme, antibody titre, and antimicrobial activity. Such an immunostimulatory effect was supported by a higher protection to *A. hydrophila* infection than the control fish (Pereira *et al.*, 2016). Jiang et al. (2017) showed that oral feeding of grass carp with *B. subtilis* spores expressed the *Clonorchis sinensisis* enolase could induce a high level of immunostimulatory effects through improving both innate and adaptive immune responses. In the study by Tang et al. (2017), the recombinant *B. subtilis* spores expressing cysteine protease of *C. sinensis* displayed a high specific immune response i.e., a high level of IgM in both serum and intestine mucus that may be in part due to the high count of the bacilli spores in intestine of vaccinated fish in which can present high level of antigen particles for absorption and stimulation of animal immunity. In Nile tilapia, an efficacious in humoral and cellular immune responses as well as significantly a higher survival (41.7% *vs.* 24.2% in control) in a challenge test with *S. agalactiae* and immune gene expression of fish intestine and spleen were shown after the fish were subjected to *B. subtilis* spores expressed by the *S. agalactiae* Sip antigen via gavage leads (Yao et al., 2019). In addition, an eight-week-oral use of *B. subtilis* spores expressing viral VP4 protein, the major outer capsid protein of grass carp reovirus responsible for haemorrhagic disease in grass, demonstrated significantly high levels of serum IgM and intestinal mucus IgZ as well as 47% survival after challenging the fish with the virus (Jiang *et al.*, 2019). The vaccinated fish also showed an increase in expression of various innate immune genes (IL-4/13 A, IL-4/13B and CSF1R) and adaptive immune genes (CD8 and CD4L cells, MHC-II, IL1b cytokines, tumour necrose factor-α, TGF-β and interferon type I-related signalling pathway genes).

Chinese liver fluke (*Clonorchis sinensis*) is an important fish borne zoonotic parasite threatening public health and, the parasite can be treated with paramyosin. Feeding grass carp with paramyosin expressed on the surface of *B. subtilis* spores increased specific IgM in sera, skin mucus, bile, and intestinal mucus, as well as mRNA levels of IgM and IgZ in the spleen and head kidney of treated fish (Sun *et al.*, 2020). The transcripts levels of IL-8 and TNF-α in the spleen and head kidney of treated fish were also significantly higher than the control group. These findings suggest a promising effective, safe, and needle-free candidate vaccine against *C. sinensis* infection for grass carp. In a study by Yanuhar *et al.* (2019), feeding Koi carp with a mixture of *Bacillus* spp., *Lactobacillus* sp., and *Nitrosomonas* sp. at different doses exhibited an improvement in disease resistance towards *Myxobolus* sp., in terms of gill tissue damage reduction. It is also worth mentioning that the topic of the regulatory mechanisms of *B. subtilis* on the immune systems of aquatic organisms remains unclear and merits further studies. Zhou *et al.* (2019) suggested the molecular mechanism of the dendritic cells-mediated immune response was triggered by *B. subtilis* administration in grass carp.

Even in the shrimp model, oral use of *B. subtilis* expressing VP28 antigen (a major chimeric envelope protein of white spot virus) conferred protection in Chinese white shrimp (*F. chinensis*) (Fu *et al.*, 2010), crayfish (*Cambarus clarkia*) (Ning *et al.*, 2011), white leg shrimp (Fu *et al.*, 2011; Nguyen *et al.*, 2014b; Valdez *et al.*, 2014), and black tiger shrimp (Pham *et al.*, 2017). It is, however, worth mentioning that factors such as dosage optimization and duration of administrationare topics merits future investigations. Oral use of *B. subtilis* spores expressing the VP28 antigen of white spot virus could be a potential treatment route to protect white leg shrimp against the disease that may in part be due to improving the shrimp immune responses and disease resistance observed in the treated shrimp (Nguyen *et al.*, 2014b), and a 75% protection was obtained in black tiger shrimp fed *B. subtilis* expressing the VP28 antigen spores on black tiger shrimp against white spot viral disease (Pham *et al.*, 2017).

10.10 Prebiotics

10.10.1 Prebiotics and Immunity

In general, the dietary intake of prebiotics shifts the balance of gut-flora towards increasing the number of "good" bacteria, thereby indicating an indirect effect of prebiotics on host immunity (Hoseinifar *et al.*, 2015). Prebiotics may also attain the form of microbe-associated molecular patterns (MAMPs) such as teichoic acid, peptidoglycan, glycosylated protein, or the capsular polysaccharide of bacteria to interact with fish pattern recognition receptors (PRRs) in (Bron *et al.*, 2012). It needs to be mentioned here that the PRRs have been identified in teleost fish, including toll-like receptors (TLRs), NOD-like receptors (NLR), C-type lectin receptors (CLRs), and peptidoglycan recognition proteins (PGRPs) (Boltanã *et al.*, 2011).

In addition, prebiotics administration may lead to higher production of metabolites such as propionate, butyrate, or short-chain fatty acids (SCFA) by the gut microbiota, which can be utilized by the immune cells of the GALT, as reported in mammals (Knudsen *et al.*, 2003), and be activated by specific SCFA-receptors (Petit and Wiegertjes, 2022). Application of prebiotics also affects gut histo-morphology thereby affecting the ability of pathogens to adhere and translocate the intestinal barrier and colonize epithelial tissues. Enhanced mucus productionin inulin fed gilthead seabream (Cerezuela *et al.*, 2013a) and MOS-fed rainbow trout (Rodriguez-Estrada et al., 2009), despite a decrease in the number of goblet cells, proved efficacy of the tested prebiotics. Moreover, in seabream fed inulin, up-regulation of the gut transcripts of occludin, a protein involved in the TJs was demonstrated (Cerezuela *et al.*, 2013b). This in turn causes disease resistance by hindering the pathogen entry and colonization.

The exact mechanism used by prebiotics to augment the gut and systemic immunity is largely unknown, but several hypotheses have been proposed: (i) luminal prebiotics may activate GALT leucocytes by directly contact, (ii) immune activation may be facilitated by enterocytes metabolism, (iii) transformed gut microbiota is responsible for the immune activation, and (iv) prebiotics may encounter leucocytes by crossing the epithelia. All of the above hypotheses are in a nascent state and further studies are warranted especially in teleost model. However, the most explored hypothesis states that prebiotics increases the local and systemic immunity by mediating the microbiota or their products. Nevertheless, it is unidentified if GALT leucocytes can translocate epithelial membrane and directly recognize the MAMPs or they are stimulated by microbial products.

10.10.2 Types of Prebiotics

The initial investigation on the effect of Mannan oligosaccharide as prebiotics in hybrid red tilapia, was reported by Hanley *et al.* (1995). After that, several studies were carried out regarding the efficacy of prebiotics and the following compounds were established as prebiotics in aquaculture industry.

10.10.2.1 β-glucan

β-glucans are a set of β-D-glucose polysaccharides possessing a linear backbone with 1–3 β-glycosidic bonds and typically available in the cell walls of cereals, bacteria, and fungi. They have an array of diverse physiological effects in animals depending upon molecular mass, solubility, viscosity, branching structure, and gelation properties. They are commercially manufactured from gums of *Hordeum* sp. by bacterial hydrolysis using *Bacillus amyloliquefaciens* as a by-product of wine production in breweries.

Filho et al (2019) applied β-1,3/1,6 glucan at 1 g/kg diet in Atlantic salmon (*Salmo salar*) and reported increased resistance to *Moritella viscosa* and infectious salmon anaemia virus that augmented protective effect of the vaccine against the pathogens. In another study, of Nile tilapia (*Orechromis niloticus*) were applied with β-glucans at a much lesser dose that significantly improved the fish immune status with an overall decreased mortality while challenged with *Aeromonas hydrophila* (Sherif and Mahfouz, 2019). However, application of a commercial probiotic, Algamune^{TM} at various doses in Nile tilapia exhibited contrasting results. Although there was no improvement in disease resistance of the fish, macrophage and superoxide anion production was increased (Yamamoto *et al.*, 2018). Nevertheless, significantly higher complement activity with increased disease resistance against *A. hydrophila* was observed in silver catfish (*Rhamdia quelen*) on application of prebiotic (Domenico *et al.*, 2017). Further, application of the probiotic was noticed with improved resistance against white muscle disease in *Macrobrachium rosenberghii* juveniles, where insignificantly improved respiratory burst and phenoloxidase activities were recorded (Pavadi *et al.*, 2018). In another study, application of β-glucan in *Litopenaeus vannamei* was noticed with significantly improved superoxide dismutase activity (Xu *et al.*, 2021). Petit *et al.* (2022) reported immune modulation in European common carp (*Cyprinus carpio carpio*) by activating C-type lectin receptor (CLR), a protein similar to Dectin-1.

10.10.2.2 Mannanoligosaccharides (MOS)

It is indigestible linear polymer of the sugar mannose. It can be produced from hydrolysis of either insoluble galactomannan or soluble glucomannan. Glucomannan MOS is applied as prebiotics in animal husbandry and nutritional supplements due to its bioactivity. It is commercially derived from yeast (*Saccharomyces cerevisiae*) cell wall. It is one of the most widely applied prebiotics in aquaculture (Merrifield *et al.*, 2010). In several studies, dietary supplementation of MOS at varying dose and duration was noticed to increase serum lysozyme activity. Similar results were reported by Talpur *et al.* (2014) in snakehead (*Channa striata*) (2 g kg^{-1} for 12 weeks) or Zhou *et al.* (2010) and Buentello *et al.* (2010) in red drum (*Sciaenops ocellatus*) (10 g kg^{-1} for 8 weeks or 6 weeks, respectively). Although contrasting results were described by Torrecillas *et al.* (2007, 2011) in European sea bass (*Dicentrarchus labrax*) and Welker *et al.* (2012) in channel catfish (*Ictalurus punctatus*), where dietary supplementation of MOS had no significant impact on the fish immune system Azevedo *et al.* (2016) reported that dietary supplementation of MOS at 2 g kg^{-1} was capable of significantly improving lysozyme activity in Siamese fighting fish (*Betta splendens* Regan, 1910) larvae. Dietary MOS supplementation through *Artemia* enrichment at 12 mg L^{-1} significantly improved Pacific white shrimp post larval immunity with augmentation of phagocytic, phenoloxidase and respiratory burst activities (Widanarni *et al.*, 2019). Furthermore, inclusion of MOS at 0.5% w/w in the diet of *Litopenaeus vannamei* blocked the adhesion sites of potential pathogens to decrease disease infestation (Gainza and Romero, 2020). In another study, *L. vannamei* exhibited an upregulation of expression levels of catalase (CAT), glutathione peroxidase (GPX), superoxide dismutase (SOD), penaeidin-3a (Pen-3a), and heat shock protein (Hsp-70) on 0.2% dietary administration of MOS (Chen *et al.*, 2020). Nevertheless, further studies are necessary to determine the proper inclusion level of MOS and period of administration in fish.

10.10.2.3 Fructo-Oligosaccharides (FOS)

It denotes short and medium chains of β-D-fructans in which fructosyl units are bound by β-(2–1) glycosidic linkages and attached to a terminal glucose unit. It can be found naturally in *Allium*, *Musa*, *Asparagas,* and from some other sources. A second type of prebiotic under FOS group, known as the short chain FOS (scFOC) is derived from *Hordeum* and *Triticum.* Two different classes of fructooligosaccharides (FOS) are produced commercially by transfructosylation action of a β-fructosidase of *Aspergillus niger* on sucrose. Addition of FOS in fish diet could modulate the fish GI tract to elevate the immune system of fish. Although no vertebrates possess specific receptors for FOS uptake (Mahious *et al.*, 2006; Ringø *et al.*, 2010). Enhanced lysozyme activity was noted in red drum fed the diet containing 10 g kg^{-1} FOS for six weeks (Buentello *et al.*, 2010) or eight weeks (Zhou *et al.*, 2010), respectively. Soleimani *et al.* (2012) included different doses of FOS in the diet of Caspian roach (*Rutilus rutilus*) fry for seven weeks and reported higher immune parameters viz., Ig level, lysozyme activity and ACH_{50}. They also pointed out that dietary inclusion of lower dose of FOS improved lesser number of immune parameters and vice versa. Similar findings were reported by Akrami *et al.* (2013) in stellate sturgeon (*Acipenser stellatus*). However, contrasting results were depicted by Ai *et al.* (2011) and Ye *et al.* (2011) where dietary supplementation of FOS had no significant impact on immune system of large yellow croaker, *Larimichthys crocea* and Japanese flounder, *Paralichthys olivaceus*, respectively. On the other hand, FOS had no significant effects on the hematological parameters in common carp (*Cyprinus carpio*), although immunological parameters like respiratory burst activity was augmented (Hoseinifar *et al.*, 2014). In a recent study by Varela-Granados *et al.* (2021), inclusion of MOS and FOS in the diet of *Macrobrachium rosenberghii* had no significant impact on the immune parameters of prawn. In contrast, Paz *et al.* (2019) documented that supplementation of FOS @ 0.1–2% w/w in the diet of tambaqui (*Colossoma macropomum*) significantly influenced the haematological parameters (WBC, PAS-GL^{+}) and immune parameters like phagocytic activity. Similar findings were exhibited in pacu (*Piaractus mesopotamicus*) on dietary inclusion of FOS @ 2.0 g kg^{-1} (Deon *et al.*, 2021).

10.10.2.4 Inulin

Inulin is a heterogeneous collection of fructose polymers in which chain-terminating glucosyl moieties and a repetitive fructosyl moiety are linked by β(2,1) glycosidic bonds. They are a group of naturally occurring polysaccharides produced by many types of plants like *Allium* sp. and *Hordeum* sp. However, industrially it is extracted from *Cichorium* sp.

An interesting observation on application of inulin in fish diet was reported by Cerezuela et al., 2012a, b). Inulin incorporated at 10 g kg^{-1} in the diet of gilthead sea bream (*Sparus aurata* L.) had no significant impact on the complement and peroxidase activity when the initial body weight was 150 g. While, inulin could significantly improve the complement activity and IgM level when the initial body weight was 50 g. Ibrahem *et al.* (2010), Mouriño *et al.* (2012) and Reyes-Becerril *et al.* (2014) reported similar findings where dietary inclusion of inulin at 5 g kg^{-1} significantly elevated the lysozyme activity in Nile tilapia (*Oreochromis niloticus*), total immunoglobulin in hybrid surubim (*Pseudoplatystoma* sp.) and lysozyme activity, IgM level, and myeloperoxidase activity in leopard grouper (*Mycteroperca rosacea*), respectively. Yones *et al.* (2019) suggested a low-level inclusion of inulin at 2.5 g/kg diet in the diet of Nile Tilapia, *Oreochromis niloticus* significantly affects alternate complement pathway and superoxide dismutase activity. In another study, inulin-fortified basal diet of rainbow trout (*Oncorhynchus mykiss*) was found to elevate diffent immunological parameters, e.g., lysozyme, alkaline phosphatise, and total Ig (Ghafarifarsani *et al.*, 2021). In a recent study, dietary supplementation of inulin enhanced immune parameters like total Ig, ACH_{50} and lysozyme along with haematological and biochemical parameters in common carp, *Cyprinus carpio* (Ajdari *et al.*, 2022).

10.10.2.5 Galactooligosaccharides (GOS)

It consists of 2–20 molecules of galactose and glucose derived by enzymatic treatments of lactose (Kiron, 2012). Dietary administration of GOS at 10 g kg^{-1} for eight weeks significantly enhanced the lysozyme activity of red drum (Zhou *et al.*, 2010). On the other hand, contrasting results were recorded in Atlantic salmon, when GOS were administered for four months (Grisdale-Helland *et al.*, 2008). Mustafa *et al.* (2019) supplemented GOS in the diet of *L. vannamei* at 0.25 and 0.40% w/w and recorded a rise of phagocytic activity after prebiotic administration. Further, dietary supplementation of GOS (2%) was noticed to upregulate cytokines and lysozyme activity in common carp (*Cyprinus carpio*) indicating its role in immunomodulation (Pietrzak *et al.*, 2020).

10.10.2.6 Arabinoxylan-Oligosaccharides (AXOS)

It is produced commercially from wheat bran by hydrolysis using *Bacillus subtilis*. It may be mentioned that wheat bran is a low-cost by-product of traditional wheat milling. Geraylou *et al.* (2012) assessed the effects of two differentstructures of AXOS, namely AXOS-32-0.30 (average degree of polymerization: 32, average degree of substitution: 0.30) and AXOS-3-0.25, on the juveniles of Siberian sturgeon (*Acipenser baerii*). AXOS-32-0.30 performed better than its counterpart as it significantly influenced ACH_{50} and total peroxidase enzyme concentration, although lysozyme activity was not at all affected by AXOS. Another study by Geraylou *et al.* (2013) in the same model organism depicted higher ACH_{50} but no significant impact on theserum peroxidase and total immunoglobulin. Combined application of AXOS and inulin were capable of reducing the abundance of pathogens like *Vibrio*, *Rhodococcus,* and *Photobacterium* and increasing the abundance of beneficial species (Li *et al.*, 2021).

10.10.2.7 Levan

It is a naturally occurring fructan made up of fructose monomer connected by 2,6 beta glycosidic linkages. The polymer possesses both branched and linear structures of relatively low molecular weight. The terminal end of levangenerally contains a glucosyl residue.

Levan incorporated at 5 and 10 g kg^{-1} inthe feeds of juvenile common carp, *Cyprinus carpio* significantly affected the serum lysozyme activity (Rairakhwada *et al.*, 2007). In another study, levan-fortified diet increased the serum lysozyme activity and relative survival percentage against *Aeromonas hydrophila* in *L. rohita* indicating its immunomodulatory role (Gupta *et al.*, 2008). Huang *et al.* (2014) also reported a similar finding when lysozyme activity and relative survival percentage seemed to increase in orange spotted grouper (*Epinephelus coioides* H.) after challenge with *Vibrio harveyi*.

10.10.2.8 Chitooligosaccharide (COS)

It is obtained by chemical and enzymatic hydrolysis of chitosan, which is a linear polysaccharide composed of D-glucosamine and *N*-acetyl-D-glucosamine linked by β-(1-4)-glycosidic bond. It is derived from chitin shells of shrimp and other crustaceans treating with sodium hydroxide.

COS has a lower molecular weight and is readily soluble in water that indicates its higher activity and more physiological functions than chitosan. The COS significantly improved the serum SOD and lysozyme activity in koi (*Cyprinus carpio koi*) when incorporated at 2 g kg^{-1} (Lin *et al.*, 2012).

10.10.2.9 Isomaltooligosaccharide (IMO)

It is commercially derived from corn starch. It is a mixture of short-chain carbohydrates, which has a digestion-resistant property. The starch is first converted, by means of simple enzymatic hydrolysis, into high maltose syrup which is further converted into α(1,6)-glycosidic linkages, creating "iso" linkages between glucose moieties and forming isomalto-oligosaccharide. IMO fortified (0.2% w/w) diet elevated different immunological parameters (e.g., phenoloxidase, respiratory burst, phosphatise) and increased survival percentage in *L. vannamei* when challenged with potential pathogenic white spot syndrome virus (WSSV) (Li *et al.*, 2009).

10.10.2.10 Xylooligosaccharide (XOS)

It is commercially derived from defatted rice bran. Sun *et al.* (2018) reported that XOS administered at 1-6 g kg^{-1} in the diet of *L. vannamei* improved different immune parameters like superoxide dismutase, phenoloxidase, and lysozyme activity increasing the overall resistance against *Vibrio alginolyticus*. In another study by Morshedi *et al.* (2018), addition of 0.5% XOS in the diet of sobaity seabream, *Sparidentex hasta*, significantly affected the alternate complement pathway, although no marked variation was observed on total immunoglobulin and lysozyme activity.

10.10.2.11 Honey

Honey contains 17.2% water, 38.2% fructose, 31.3% glucose, 0.7% sucrose, and 3.1% oligosaccharide (Bogdanov *et al.*, 2008). Honey can function as potential source of prebiotic because of its ability to increase the population of beneficial bacteria (*Bifidobacteria* and Lactobacilli) in the GI tract (Sanz *et al.*, 2005). Moreover, oligosaccharide is fermentable in the colon and induces growth of the beneficial flora (Merrifield and Ringø, 2014). In various studies done in recent years, honey was noticed to increase the growth parameters in *L. vannamei* and *O. niloticus* (Widanarni *et al.*, 2019; Aryati *et al.*, 2020), but whether honey can influence the immune system of the model organism is still unknown and requires further investigation.

10.10.2.12 Commercial Prebiotics

Li and Gatlin (2003) applied commercial prebiotic (Brewtech), manufactured from brewer's yeast by partial autolysis, in the diet of hybrid striped bass (*Morone chrysops* × *M. saxatilis*) and reported improved serum peroxidase activity when the initial weight of the fish is >50 g, while no change was observed when the initial weight is <50 g.

Another commonly used commercial prebiotic, containing MOS and β-glucans derived from the cells elevated the serum lysozyme and complement activities in rainbow trout (Ahmadi *et al.*, 2014a). However, Ta'ati *et al.* (2011) indicated no significant impact on the serum lysozyme activity and IgM level of great Sturgeon (*Huso huso*). Similar findings were reported by Yu *et al.* (2014) in Japanese seabass (*Lateolabrax japonicus*).

Grobiotic-A, a commercial prebiotic, composed of a mixture of partially autolyzed brewer's yeast, dairy ingredient components, and dried fermentation productshave been extensively studied (Ringø *et al.*, 2010). However, no significant impact on the humoral immune response in different fish species, such as hybrid striped bass (Li and Gatlin, 2004, 2005), Golden shiners (*Notemigonus crysoleucas*) (Lochmann *et al.*, 2009) and Nile tilapia (Zheng *et al.*, 2011) were detected.

Vitacel (a commercial prebiotic primarily composed of cellulose and hemicellulose) was found to remarkably elevate serum lysozyme, ACH_{50}, bactericidal activity, and agglutination antibody titer in rainbow trout (Yarahmadi et al., 2014b).

10.11 Conclusions

Application of the functional feed additives like probiotics and prebiotics as natural therapeutic agents has created growing interest in view of an alternate and environment-friendly approach of disease management in aquatic animals (Hoseinifar et al., 2015; Mukherjee et al., 2019). The issue of health management is one the most critical factors, for development of a sustainable aquaculture in the globe. Sustainable aquaculture development, therefore, requires an advanced disease and health management due to the susceptibility of aquatic organisms to ubiquitous pathogens in the aquatic environments where the application of chemotherapy such as antibiotic therapy can seriously affect the human health and aquatic environments;

therefore, prophylactic alternatives such as immunostimulants are highly desirable. The application of probiotics and prebiotics as prefect immunomodulator tool is a promising for development of global aquaculture industry.

B. subtilis is one of the most frequently used probiotics in aquaculture due to its efficacy for stimulating growth performance, immunity, and disease resistance to pathogens. In addition, *in vivo* data shows that administration of other bacilli, *B. licheniformis, B. velezensis, B. amyloliquefaciens, B. circulans, B. thuringiensis,* and *B. aerius* also stimulate fish and shellfish immune systems with subsequent increase in diseases resistance to pathogenic bacteria and viruses. *Bacillus* can also adhere and colonize the host GI tract, where they can improvedigestion and absorption of the nutrients, and enhance growth performance, modulating the gut microbiota through bacterial competition, resulting in inhibition of pathogen to adhere and colonize the intestinal mucosa. The modulations of fish and shellfish innate immune responses including antiprotease and peroxidase, superoxide dismutase and myeloperoxidase, phagocytosis, lysozyme activity, and respiratory burst by *Bacillus* have been confirmed. In addition, *Bacillus* have immunostimulatory effects on fish and shellfish immune-competent cells population such as neutrophil migration, plasma bactericidal activity and increasing of neutrophil adherence ability, that can eventually result in the enhancing of animal immune functions such as an increase in complement activity, immunoglobulin production and cell cytotoxicity (e.g., Soltani *et al.*, 2019a; Li *et al.*, 2019; Di *et al.*, 2019). These immune-stimulatory effects by *Bacillus* occur in the gut-associated lymphoid tissue of fish, but the detail mechanisms merit further studies. In addition, bacilli have ability to produce several antimicrobial substances inhibiting growth of many pathogenic agents. The sporulation capacity of *Bacillus* gives them a double advantage due to their heat-tolerance and longer shelf-life in various environmental conditions compared to other probiotics e.g., the LAB.

On the other hand, several prebiotics [e.g., β-glucan, galacto-oligosaccharide (GOS), mannan-oligosaccharide (MOS) and fructo-oligosaccharide (FOS)] were shown to improve immune response, survival, and protection against pathogens in fish and shellfish species (Hoseinifar *et al.,* 2015; Ai *et al.,* 2011). Among the prebiotics evaluated in fishes, the functional roles of FOS and related prebiotics are fairly established, while functional mechanism of others need to be evaluated.

Further studies are required to clarify several issues including (a) the detail mechanisms of immunomodulation effects of the probiotics and prebiotics; (b) the optimum duration of functional feed additive application regarding the host age and species, and its environment; (c) the possibility of the immune memory of fish and shellfish; (d) the evaluation of probiotic and prebiotic components responsible for the beneficial effects on the host immunity; (e) validation of immunomodulatory data between the efficacy of a single and a combination use of probiotics and prebiotics; (f) the detail mechanisms when probiotics are used in combination with prebiotics, oral vaccines and non-specific immunostimulants; (g) the duration effectiveness of a probiotic culture; (h) the major differences between probiotics and para-probiotics in immunomodulation of the host immunity; and (i) the relationship between the life cycle of the host (fry, fingerling, fattening, brood stock) and the probiotic as well as probiotic application.

Disclosure statement: The authors declare that they have no conflicts of interest.

Authors contributions: MS by data collection, data management, writing the draft and final editing. ER helped by writing and final editing, KG and DD by data collection and writing prebiotic section.

REFERENCES

Abarike, E.D., Cai, J., Lu, Y., Yu, H., Chen, L., Jian, J., Tang, J., Jun, L., & Kuebutornye, F.K. (2018) Effects of a commercial probiotic BS containing *Bacillus subtilis* and *Bacillus licheniformis* on growth, immune response and disease resistance in Nile tilapia, *Oreochromis niloticus. Fish & Shellfish Immunology* 82, 229–238. doi: 10.1016/j.fsi.2018.08.037

Abbass A., Sharifuzzaman S.M., & Austin B. (2010) Cellular components of probiotics control *Yersinia ruckeri* infection in rainbow trout, *Oncorhynchus mykiss* (Walbaum). *Journal of Fish Diseases* 33, 31–37. 10.1111/j.1365-2761.2009.01086.x

Adeshina, I. (2018) The effect of *Lactobacillus acidophilus* as a dietary supplement on nonspecific immune response and disease resistance in juvenile common carp, Cyprinus carpio. *International Food Research Journal* 25, 2345–2351.

Adorian, T.J., Jamali, H., Ghafari Farsani, H., Darvishi, P., Hasanpour, S., Bagheri, T., & Roozbehfar R. (2019) Effects of probiotic bacteria *Bacillus* on growth performance, digestive enzyme activity, and hematological parameters of Asian sea bass, *Lates calcarifer* (Bloch). *Probiotics and Antimicrobial Proteins* 11, 248–255. doi: 10.1007/s12602-018-9393-z

Ahmadi, P.Y., Farahmand, H., Miandare, H.K., Mirvaghefi, A., & Hoseinifar, S.H. (2014a) The effects of dietary Immunogenon innate immune response, immune related genes expression, and disease resistance of rainbow trout (*Oncorhynchus mykiss*). *Fish & Shellfish Immunology* 37, 209–214. doi: 10. 1016/j.fsi.2014.02.006

Ahmadifar, E., Moghadam, M.S., Dawood, M.A.O., & Hoseinifar, S.H. (2019) *Lactobacillus fermentum* and/or ferulic acid improved the immune responses, antioxidative defence and resistance against *Aeromonas hydrophila* in common carp (*Cyprinus carpio*) fingerlings. *Fish & Shellfish Immunology* 94, 916–923. 10.1016/j.fsi.2019.10.019

Ai, Q., Xu, H., Mai, K., Xu, W., Wang, J., & Zhang, W. (2011) Effects of dietary supplementation of *Bacillus subtilis* and fructooligosaccharide on growth performance, survival, non-specific immune response and disease resistance of juvenile large yellow croaker, *Larimichthys crocea. Aquaculture* 317, 155–161.

Ai, Q., Xu, H., Mai, K., Xu, W., Wang, J., & Zhang, W. (2011) Effects of dietary supplementation of *Bacillus subtilis* and fructooligosaccharide on growth performance, survival, non-specific immune response and disease resistance of juvenile large yellow croaker, *Larimichthys crocea. Aquaculture* 317, 155–161.doi: 10.1016/j.aquaculture.2011.04.036

Ajdari, A., Ghafarifarsani, H., Hoseinifar, S.H., Javahery, S., Narimanizad, F., Gatphayak, K., & Doan, H.V. (2022) Effects of dietary supplementation of PrimaLac, inulin, and Biomin Imbo on growth performance, antioxidant, and innate immune responses of Common Carp (*Cyprinus carpio*). *Aquaculture Nutrition* 2022, 8297479. doi:10.1155/2022/8297479

Ajitha, S., Sridhar, M., Sridhar, N., Singh, I.S.B., & Varghese, V. (2004) Probiotic effects of lactic acid bacteria against *Vibrio alginolyticus* in *Penaeus* (*Fenneropenaeus*) Indicus (H.Milne Edwards). *Asian Fisheries Sci* 17, 71–80.

Akrami, R., Iri, Y., Rostami, H.K., & Mansour, M.R. (2013) Effect of dietary supplementation of fructooligosaccharide (FOS) on growth performance, survival, lactobacillus bacterial population and hemato-immunological parameters of stellate sturgeon (*Acipenser stellatus*) juvenile. *Fish & Shellfish Immunology* 35, 1235–1239.doi: 10.1016/j.fsi.2013.07.039

Alavandi, S.V., Vijayan, K.K., Santiago, T.C., Poornima, M., Jithendran, K.P., Ali, S.A., & Rajan, J.J.S. (2004) Evaluation of *Pseudomonas* sp. PM 11 and *Vibrio fluvialis* PM 17 on immune indices of tiger shrimp, *Penaeus monodon*. *Fish & Shellfish Immunology*& Shellfish Immunology 17, 115–120. 10.1016/j.fsi.2003.11.007.

Alfonso, S., Gesto, M.I., & Sadoul, B. (2021) Temperature increases and its effects on fish stress physiology in the context of global warming. DOI: 10.1111/jfb.14599

Alonso, S., Castro, M.C., Berdasco, M., de la Banda, I.G., Moreno-Ventas, X., & de Rojas, A.H. (2019) Isolation and partial characterization of lactic acid bacteria from the gut microbiota of marine fishes for potential application as probiotics in aquaculture. *Probiotics Antimicrob Proteins* 11(2): 569–579. doi: 10.1007/s12602-018-9439-2

Al-Sheraji, S.H., Ismail, A., Manap, M.Y., Mustafa, S., Yusof, R.M., & Hassan, F.A. (2013). Prebiotics as functional foods: A review. *Journal of Functional Foods*, 5(4): 1542–1553. doi.org/10.1016/j.jff.2013.08.009

Aly S., Mohamed A.A.Z., Rahmani A.H., & Nashwa M.A.A. (2016) Trials to improve the response of Orechromis niloticus to *Aeromonas hydrophila* vaccine using immunostimulants (garlic, Echinacea) and probiotics (Organic Green TM and Vet-Yeast TM). *African Journal of Biotechnology* 2016; 15(21): 989–994.

Aly, S.M., Ahmed, Y.S.G., Ghareeb, A.A.A., & Mohamed, M.F. (2008) Studies on *Bacillus subtilis* and *Lactobacillus acidophilus*, as potential probiotics, on the immune response and resistance of Tilapia nilotica (*Oreochromis niloticus*) to challenge infections. *Fish & Shellfish Immunology* 25, 128–136. 10. 1016/j.fsi.2008.03.013.

Amenyogbe, E., Chen, G., Wang, Z., Huang, J., Huang, B., & Li, H. (2020) The exploitation of probiotics, prebiotics and synbiotics in aquaculture: present study, limitations and future directions: a review. *Aquac Int* 28, 1017–1041. 10.1007/s10499-020-00509-0

Amoah, K., Huang, Q., Dong, X., Tan, B., Zhang, S., Chi, S., Yang, Q., Liu, H., & Yang, Y. (2020) *Paenibacillus polymyxa* improves the growth, immune and antioxidant activity, intestinal health, and disease resistance in *Litopenaeus vannamei* challenged with Vibrio parahaemolyticus. *Aquaculture* 518, 734563. 10.1016/j. aquaculture.2019.734563.

Anyanwu, N.G., & Ariole, C.N. (2019) Probiotic potential of an indigenous marine *Bacillus thuringiensis* on shrimp (*Penaeus monodon*) culture infected with *Vibrio mimicus*. *Journal of Applied Science* 19, 173–179.

Arena, M.P., Elmastour, F., Sane, F., Drider, D., Fiocco, D., Spano, G., & Hober, D. (2018) Inhibition of *Coxsackievirus* B4 by *Lactobacillus plantarum*. *Microbiological Research* 210, 59–64.

Arijo, S., Brunt, J., Chabrillón, M., Díaz-Rosales, P., & Austin, B. (2008) Subcellular components of *Vibrio harveyi* and probiotics induce immune responses in rainbow trout, *Oncorhynchus mykiss* (Walbaum), against *V. harveyi*. *J. Fish Dis.* 31, 579–590

Aryati, Y., Widanarni, W., Wahjuningrum, D., Rusmana, I., & Lusiastuti, A.M. (2020) The effect of dietary honey prebiotic on microbiota diversity in the digestive tract of Nile tilapia (*Oreochromis niloticus*) and its growth performance. *Aquaculture Research* 52, 1215–1226. doi:10.1111/are.14980

Ashouri, G., Mahboobi Soofiani, N., Hoseinifar, S.H., Jalali, S.A.H., Morshedi, V., Van Doan, H., & Torfi Mozanzadeh, M. (2018). Combined effects of dietary low molecular weight sodium alginate and Pediococcus acidilactici MA18/5M on growth performance, haematological and innate immune responses of Asian sea bass (Lates calcalifer) juveniles. *Fish & Shellfish Immunology*, 79: 34–41. doi: 10.1016/j.fsi.2018.05.009.

Askarian, F., Kousha, A., Salma, W., & Ringø, E. (2012) The effect of lactic acid bacteria administration on growth, digestive enzyme activity and gut microbiota in Persian sturgeon (*Acipenser persicus*) and beluga (*Huso huso*) fry. *Aquaculture Nutrition* 17, 488–497.

Austin, B., & Austin, D.A. (2016) *Bacterial Fish Pathogens: Disease of Farmed and Wild Fish*, Springer, London, The UK, Pp. 643.

Azarin, H., Aramli, M.S., Imanpour, M.R., & Rajabpour, M. (2015) Effect of a probiotic containing *Bacillus licheniformis* and *Bacillus subtilis* and Ferroin solution on growth performance, body composition and haematological parameters in Kutum (*Rutilus frisii kutum*) fry. *Probiotics Antimicrobial* 7(1), 31–37.

Azevedo, R.V., Silva-Azevedo, D.K., Santos-Júnior, J.M., Filho, J.C.F., Andrade, D.R., Braga, L.G.T., & Junior, M.V. (2016) Effects of dietary mannan oligosaccharide on the growth, survival, intestinal morphometry and nonspecific immune response for Siamese fighting fish (*Betta splendens* Regan, 1910) larvae. *Latin American Journal of Aquatic Research* 44, 800–806. doi: 10.3856/vol44-issue4-fulltext-15

Balcázar, J.L., Rojas-Luna, T., & Cunningham, D.P. (2007) Effect of the addition of four potential probiotic strains on the survival of pacific white shrimp (*Litopenaeus vannamei*) following immersion challenge with *Vibrio parahaemolyticus*. *Journal of Invertebrate Pathology*, 96, 147–150. 10.1016/j.jip.2007.04.008.

Banerjee, G., Nandi, A., & Ray, A.K. (2017) Assessment of hemolytic activity, enzyme production and bacteriocin characterization of *Bacillus subtilis* LR1 isolated from the gastrointestinal tract of fish. *Archives of Microbiology* 199, 115–124. doi: 10.1007/s00203-016-1283-8

Bazari Moghaddam, S., & Pourjaafari, M. (2021) The effects of four types of specific probiotic on growth performance, liver enzymes and immune indices of juvenile Persian sturgeon (*Acipenser persicus*). *Iranian Journal of Fisheries Science*, 20 (4):1179–1191. Doi: 20.1001.1.15622916.2021.20.4.13.8

Bly, J.E., Quiniou, S.M., & Clem, L.W. (1997) Environmental effects on fish immune mechanisms. *Dev Biol Stand* 90, 33–43.

Bogdanov, S., Jurendic, T., Sieber, R., & Gallmann, P. (2008) Honey for nutrition and health: a review. *Journal of the American College of Nutrition* 27, 677–689. doi: 10.1080/07315724.2008.10719745

Boltanã, S., Roher, N., Goetz, F.W., & Mackenzie, S.A. (2011) PAMPs, PRRs and the genomics of gramnegative bacterial recognition in fish. *Developmental and Comparative Immunology* 35, 1195–1203. doi: 10.1016/j.dci.2011.02.010

Bron, P.A., van Baarlen, P., & Kleerebezem, M. (2012) Emerging molecular insights into the interaction between probiotics and the host intestinal mucosa. *Nature Reviews Microbiology* 10, 66–78. doi:10.1038/nrmicro2690

Brunt, J., Newaj-Fyzul, A., & Austin, B. (2007) The development of probiotics for the control of multiple bacterial diseases of rainbow trout, *Oncorhynchus mykiss* (Walbaum). *Journal of Fish Diseases* 30(10):573–579.

Brunt, J., & Austin, B. (2005) Use of a probiotic to control lactococcosis and streptococcosis in rainbow trout, *Oncorhynchus mykiss* (Walbaum). *Journal of Fish Diseases* 28, 693–701.

Buentello, J.A., Neill, W.H., & Gatlin III, D.M. (2010) Effects of dietary prebiotics on the growth, feed efficiency and non-specific immunity of juvenile red drum *Sciaenops ocellatus* fed soybean-based diets. *Aquaculture Research* 41, 411–418. doi: 10.1111/j.1365-2109.2009.02178.x

Bunnoy, A., Na-Nakorn, U., & Srisapoome, P. (2019) Probiotic effects of a novel strain, *Acinetobacter* KU011TH, on the growth performance, immune responses, and resistance against *Aeromonas hydrophila* of Bighead Catfish (*Clarias Macrocephalus* Günther 1864). *Microorganisms* 7, 613. doi: 10.3390/ microorganisms7120613

Burbank, D.R., Shah, D.H., LaPatra, S.E., Fornshell, G., & Cain, K.D. (2011) Enhanced ersistance to coldwater disease following feeding of probiotic bacterial strains to rainbow trout (*Oncorhynchus Mykiss*). *Aquaculture* 321, 185–190. doi: 10.1016/j.aquaculture.2011.09.004

Butt, U.D., Lin, N., Akhter, N., Siddiqui, T., & Li, S., Wu, B. (2021) Overview of the latest developments in the role of probiotics, prebiotics and synbiotics in shrimp aquaculture. *Fish and Shellfish Immunology* 114, 263–281. 10.1016/ j.fsi.2021.05.003

Cai, Y., Yuan, W., Wang, S., Guo, W., Li, A., & Wu, Y. (2019) *In vitro* screening of putative probiotics and their dual beneficial effects: to white shrimp (*Litopenaeus vannamei*) postlarvae and to the rearing water. *Aquaculture*, 498, 61–71. doi:10.1016/j.aquaculture.2018.08.024

Caipang C., Suharman I., Avillanosa A., & Bargoyo V. (2020) Host-derived probiotics for finfish aquaculture. In: *IOP Conf Ser: Earth Environ Sci*. Bristol, UK: IOP Publishing. 430, 012026. doi: 10.1088/1755-1315/430/1/012026

Cerezuela, R., Cuesta, A., J. Meseguer, J., & Esteban, M. (2012) Effects of dietary inulin and heat-inactivated *Bacillus subtilis* on gilthead seabream (*Sparus aurata* L.) innate immune parameters. *Beneficial Microbes* 3(1), 77–81. Top of Form. Bottom of Form 10.3920/BM2011.0028

Cerezuela, R., Cuesta, A., Meseguer, J., & Esteban, M.A. (2012a) Effects of dietary inulin and heat-inactivated *Bacillus subtilis*on gilthead seabream (*Sparus aurata*L.) innate immune parameters. *Beneficial Microbes* 3, 77–81. doi: 10.3920/ BM2011.0028

Cerezuela, R., Fumanal, M., Tapia-Paniagua, S.T., Meseguer, J., Moriñigo, M.A., M. _A.Esteban, M.Á. (2013a) Changes in intestinal morphology and microbiota caused by dietary administration of inulin and *Bacillus subtilis*in gilthead sea bream (*Sparus aurata*L.) specimens. *Fish & Shellfish Immunology* 34, 1063–1070. doi: 10.1016/j.fsi.2013.01.015

Cerezuela, R., Guardiola, F.A., Meseguer, J., & Esteban, M.Á. (2012b) Increases in immune parameters by inulin and *Bacillus subtilis* dietary administration to gilthead seabream (*Sparus aurata* L.) did not correlate with disease resistance to *Photobacterium damselae*. *Fish & Shellfish Immunology* 32, 1032–1040. doi: 10.1016/j.fsi.2012.02.025

Cerezuela, R., Meseguer, J., & Esteban, M.Á. (2013b) Effects of dietary inulin, *Bacillus subtilis* and microalgae on intestinal gene expression in gilthead seabream (*Sparus aurata* L.). *Fish & Shellfish Immunology* 34, 843–848. doi:10.1016/ j.fsi.2012.12.026

Cha, J.-H., Rahimnejad, S., Yang, S.-Y., Kim, K.-W., & Lee, K.-J. (2013) Evaluations of Bacillus spp. as dietary additives on growth performance, innate immunity and disease resistance of olive flounder (*Paralichthys olivaceus*) against *Streptococcus iniae* and as water additives. *Aquaculture*, 402, 50–57.

Chai, P.C., Song, X.L., Chen, G.F., Xu, H., & Huang, J. (2016) Dietary supplementation of probiotic *Bacillus* PC465 isolated from the gut of *Fenneropenaeus chinensis* improves the health status and resistance of Litopenaeus vannamei against white spot syndrome virus. *Fish Shellfish Immunol.*, 54, 602–611. 10.1016/j.fsi.2016.05.011

Chandran, M.N., Iyapparaj, P., Moovendhan Ramasubburayan, R., Prakash, S., & Immanuel, G. (2014) Influence of probiotic bacterium *Bacillus cereus* isolated from the gut of wild shrimp *Penaeus monodon* in turn as a potent growth promoter and immune enhancer in *P. monodon*. *Fish & Shellfish Immunology* 36, 38–45.

Chauhan A., & Singh R. (2019) Probiotics in aquaculture: a promising emerging alternative approach. *Symbiosis* 77(2):99–113. doi: 10.1007/s13199- 018-0580-1

Chen, B., Peng, M., Tong, W., Zhang, Q., & Song, Z. (2019) The quorum quenching bacterium *Bacillus licheniformis* T-1 protects Zebrafish against *Aeromonas hydrophila* infection. *Probiotics and Antimicrobial Proteins* 1–12.

Chen, M., Chen, X.-Q., Tian, L.-X., Liu, Y.-J., & Niu, J. (2020) Beneficial impacts on growth, intestinal health, immune responses and ammonia resistance of pacific white shrimp (*Litopenaeus vannamei*) fed dietary synbiotic (mannan oligosaccharide and *Bacillus licheniformis*). *Aquaculture Reports* 17, 100408. doi: 10.1016/j.aqrep.2020.100408

Chen, Y., Li, J., Xiao, P., Li, G.Y., Yue, S., Huang, J., Zhu, W.Y., & Mo, Z.L. (2016) Isolation and characterization of *Bacillus* spp. M001 for potential application in turbot (*Scophthalmus maximus* L.) against *Vibrio anguillarum*. *Aquacult. Nutr.*, 22, 374–381.

Chen, Z., Ceballos-Francisco, D., Guardiola, F.A., & Esteban, M.Á. (2020a) Dietary administration of the probiotic *Shewanella Putrefaciens* to experimentally wounded gilthead seabream (*Sparus Aurata* L.) facilitates the skin wound healing. *Scientific Reports* 10, 11029. doi: 10.1038/s41598-020-68024-z

Chen, Z., Ceballos-Francisco, D., Guardiola, F.A., & Esteban, M.Á. (2020b) Influence of skin wounds on the intestinal inflammatory response and barrier function. Protective role of dietary *Shewanella Putrefaciens* SpPdp11 administration to gilthead seabream (*Sparus Aurata* L.). *Fish & Shellfish Immunology* 99, 414–423. doi: 10.1016/j.fsi.2020.02.022.

Cheng, G., Hao, H., Xie, S., Wang, X., Dai, M., Huang, L., *et al.* (2014) Antibiotic alternatives: the substitution of antibiotics in animal husbandry? *Front. Microbiol.* 5, 217. 10.3389/ fmicb.2014.00217.

Chien, C-C, Lin, T-Y, Chin, C-C, & Liu, C-H (2020) Probiotic, *Bacillus subtilis* E20 alters the immunity of white shrimp, *Litopenaeus vannamei* via glutamine metabolism and hexosamine biosynthetic. *Fish & Shellfish Immunology*. doi: 10.1016/j.fsi.2020.01.014.

Chiu, C-H, Guu, Y-K, Liu, C-H, Pan, T-M, Cheng, W.J.F., & Immunology, S. (2007) Immune responses and gene expression in white shrimp, *Litopenaeus vannamei*, induced by *Lactobacillus plantarum*. *Fish & Shellfish Immunology* 23, 364–377.

Chomwong, S., Charoensapsri, W., Amparyup, P., & Tassanakajon, A. (2018). Two host gut-derived lactic acid bacteria activate the proPO system and increase resistance to an AHPND-causing strain of Vibrio parahaemolyticus in the shrimp Litopenaeus vannamei. *Developmental & Comparative Immunology* 89, 54–65. doi: 10.1016/j.dci.2018.08.002.

Cui, L.C., Guan, X.T., Liu, Z.M., Tian, C.Y., & Xu, Y.G. (2015) Recombinant lactobacillus expressing G protein of spring viremia of carp virus (SVCV) combined with ORF81 protein of koi herpesvirus (KHV): A promising way to induce protective immunity against SVCV and KHV infection in cyprinid fish via oral vaccination. *Vaccine* 33, 3092–3099.

D'Alvise, P.W., Lillebø, S., Wergeland, H.I., Gram, L., & Bergh, Ø. (2013) Protection of codlarvae from vibriosis by *Phaeobacter* Spp.: A comparison of strains and introduction times. *Aquaculture* 384-387, 82–85. doi: 10.1016/j.aquaculture.2012.12.013

Da Paixão, A., Dos Santos, J., Pinto, M., Pereira, D., De Oliveira, R., *et al.* (2017) Effect of commercial probiotics (*Bacillus subtilis* and *Saccharomyces cerevisiae*) on growth performance, body composition, hematology parameters, and disease resistance against Streptococcus agalactiae in tambaqui (*Colossoma macropomum*). *Aquacul Int* 25(6), 2035–2045.

Dahiya, T., Sihag, R.C., & Gahlawat, S. (2012) Effect of probiotics on the haematological parameters of Indian magur (*Clarius batrachus* L.). *J Fish Aquat Sci* 7(4), 279–290.

Daniels, C.L., Merrifield, D.L., Boothroyd, D.P., Davies, S.J., Factor, J.R., & Arnold, K.E. (2010) Effect of dietary *Bacillus* spp. and mannan oligosaccharides (MOS) on European lobster (*Homarus gammarus* L.) larvae growth performance, gut morphology and gut microbiota. *Aquaculture* 304, 49–57.

Daniels, C.L., Merrifield, D.L., Ringø, E., & Davies, S.J. (2013) Probiotic, prebiotic and synbiotic applications for the improvement of larval European lobster (*Homarus gammarus*) culture. *Aquaculture* 416, 396–406.

Darafsh, F., Soltani, M., Abdolhay, H.A., & Shamsaei Mehrejan, M. (2020) Efficacy of dietary supplementation of *Bacillus licheniformis* and *Bacillus subtilis* probiotics and *Saccharomyces cerevisiae* (yeast) on the hematological, immune response, and biochemical features of Persian sturgeon (*Acipenser persicus*) fingerlings. *Iranian Journal of Fisheries Sciences* 19(4) 2024–2038 2020 DOI: 10.22092/ijfs.2018.117847

Daroonpunt, R., Yiamsombut, S., Sitdhipol, J., & Tanasupawat, S. (2019) Bacillus salacetis sp. nov., a slightly halophilic bacterium from Thai shrimp paste (Ka-pi). *Int J Syst Eval Microbiol*, 69, doi.org/10.1099/ijsem.0.003286

Das, A., Nakhro, K., Chowdhury, S., & Kamilya, D. (2013) Effects of potential probiotic *Bacillus amyloliquefaciens* [corrected] FPTB16 on systemic and cutaneous mucosal immune responses and disease resistance of catla (*Catla catla*). *Fish & Shellfish Immunology*, 35, 1547–1553.

Dash, P., Tandel, R.S., Bhat, R.A.H., Mallik, S., Singh, A.K., & Sarma, D. (2018) The addition of probiotic bacteria to microbial floc: Water quality, growth, non-specific immune response and disease resistance of *Cyprinus carpio* in mid-Himalayan altitude. *Aquaculture*, 495, 961–969.

Defoirdt, T., Boon, N., Sorgeloos, P., Verstraete, W., & Bossier, P. (2007) Alternatives to antibiotics to control bacterial infections: luminescent vibriosis in aquaculture as an example. *Trends Biotechnol.*, 25, 472–479.

Deon, M.P.P., Bicudo, Á.J.A., & Sado, R.Y. (2021) Performance, hematology, and immunology of pacu in response to dietary supplementation with fructooligosaccharides. *Pesquisa Agropecuária Brasileira* 56, e02460. doi:10.1590/S1678-3921.pab2021.v56.02460

Dey, A., Ghosh, K., & Hazra, N. (2016) Evaluation of extracellular enzyme-producing autochthonous gut bacteria in walking catfish, *Clarias batrachus* (L.). *Journal of Fish* 4, 345–352.

Di, J., Chu, Z., Zhang, S., Huang, J., Du, H., & Wei, Q. (2019) Evaluation of the potential probiotic *Bacillus subtilis* isolated from two ancient sturgeons on growth performance, serum immunity and disease resistance of *Acipenser dabryanus*. *Fish & Shellfish Immunology*, 711–719. doi: 10.1016/j.fsi.2019.08.020.

Díaz-Rosales, P., Salinas, I., Rodríguez, A. *et al.* (2006) Gilthead seabream (*Sparus aurata* L.) innate immune response after dietary administration of heat-inactivated potential probiotics. *Fish & Shellfish Immunology* 20, 482–492. 10.1016/j.fsi.2005.06.007

Divya, M., Aanand, S., Srinivasan, A., & Ahilan, B. (2015) Bioremediation – An eco-friendly tool for effluenttreatment: A Review. *International Journal of Applied Research* 1(12): 530–537.

Domenico, J.D., Canova, R., Soveral, L.F., Nied, C.O., Costa, M.M., Frandoloso, R., & Kreutz, L.C. (2017) Immunomodulatory effects of dietary β-glucan in silver catfish (*Rhamdia quelen*). *Pesquisa Veterinária Brasileira* 37, 73–78. doi: 10.1590/S0100-736X2017000100012

Dong, H-B, Su, Y-Q, Mao, Y., & Wang, J. (2014) Dietary supplementation with *Bacillus* can improve the growth and survival of the kuruma shrimp *Marsupenaeus japonicus* in high-temperature environments. *Aquaculture International*, 22, 10.1007/s10499-013-9688-8

Eissa, El-Gheit, E.A., & Shaheen, A.A. (2014) Protective effect of *Pseudomonas fluorescens* as a probiotic in controlling fish pathogens. *American Journal of BioScience* 2014; 2(5): 175–181. doi: 10.11648/j.ajbio.20140205.12

Esteban, M.A., Cordero, H., Martínez-Tomé, M. et al (2014) Effect of dietary supplementation of probiotics and palm fruits extracts on the antioxidant enzyme gene expression in the mucosae of gilthead seabream (*Sparus aurata* L.). *Fish Shellfish Immunol* 39, 532–540. 10. 1016/j.fsi.2014.06.012

Faramaz, I.M., Kiaalvandi, S., Lashkarbolooki, M., & Iranshahi, F. (2011) The investigations of *Lactobacillus acidophilis* as Probiotics Trout, grown performance and disease resistance of rainbow trout (*Oncorhynchus mykiss*). *American Eurasian Journal of Science* 6(1), 32–38.

Farrell, A.P. (2011) Cellular Composition of the Blood. *Encyclopedia of Fish Physiology: From Genome to Environment* 2, 984–991.

Feng J., Chang X., Zhang Y. et al (2019) Effects of *Lactococcus lactis* from *Cyprinus carpio* L. as probiotics on growth performance, innate immune response and disease resistance against *Aeromonas hydrophila*. *Fish & Shellfish Immunology* 93, 73–81. 10.1016/j.fsi.2019.07.028

Filho, F.O.R., Koch, J.F.A., Wallace, C., & Leal, M.C. (2019) Dietary β-1,3/1,6-glucans improve the effect of a multivalent vaccine in Atlantic salmon infected with *Moritella viscosa* or infectious salmon anemia virus. *Aquaculture International* 27, 1825–1834. doi: 10.1007/s10499-019-00436-9

Frozza, A., Fiorini, A., Vendruscolo, E.C.G., Rosado, F.R., Konrad, D., Rodrigues, M.C.G., & Ballester, E.L.C. (2021) Probiotics in the rearing of freshwater prawn *Macrobrachium rosenbergii* (de Man, 1879) in a biofloc system. *Aquatic Research. are* 15265. 10.1111/are.15265.

Fu, L.L., Shuai, J.B., Xu, Z.R., Li, J.R., & Li, W.F. (2010) Immune responses of *Fenneropenaeus chinensis* against white spot syndrome virus after oral delivery of VP28 using *Bacillus subtilis* as vehicles. *Fish & Shellfish Immunology* 28, 49–55. doi: 10.1016/j.fsi.2009.09.016

Fu, L.L., Wang, Y., Wu, Z.C., & Li, W.F. (2011) *In vivo* assessment for oral delivery of *Bacillus subtilis* harboring a viral protein (VP28) against white spot syndrome virus in *Litopenaeusvannamei*. *Aquaculture* 322, 33–38. doi: 10.1016/j.aquaculture.2011.09.036

Fusco, V., Quero, G.M., Cho, G.-S., Kabisch, J., Meske, D., Neve, H., *et al.*(2015) The genus Weissella: taxonomy, ecology and biotechnological potential. *Front. Microbiol.* 6, 155. doi: 10.3389/fmicb.2015.00155

Gainza, O., & Romero, J. (2020) Effect of mannan oligosaccharides on the microbiota and productivity parameters of *Litopenaeus vannamei* shrimp under intensive cultivation in Ecuador. *Scientific Reports* 10, 2719. doi: 10.1038/s41598-020-59587-y

Galagarza, O.A., Smith, S.A., Drahos, D.J.,Robert, J.D., Williams, C., & Kuhn, D.D. (2018) Modulation of innate immunity in Nile tilapia (Oreochromis niloticus) by dietary supplementation of Bacillus subtilis endospores. *Fish & Shellfish Immunology* 83, 171–179. doi: 10.1016/j.fsi.2018.08.062

Galdeano, M.C., Cazorla, S.I., Lemme Dumit, J.M., V´elez, E., & Perdigon, G. (2019) Beneficial effects of probiotic consumption on the immune system. *Ann. Nutr. Metab.* 74, 115–124. 10.1159/000496426.

Geraylou, Z., Souffreau, C., Rurangwa, E., D'Hondt, S., Callewaert, L., Courtin, C.M., Delcour, J.A., Buyse, J., & Ollevier, F. (2012) Effects of arabinoxylan-oligosaccharides (AXOS) on juvenile Siberian sturgeon (*Acipenser baerii*) performance, immune responses, and gastrointestinal microbial community. *Fish & Shellfish Immunology* 33, 718–724.doi: 10.1016/j.fsi.2012.06.010

Geraylou, Z., Souffreau, C., Rurangwa, E., Meester, L.D., Courtin, C.M., Delcour, J.A., Buyse, J., & Ollevier, F. (2013) Effects of dietary arabinoxylan-oligosaccharides (AXOS) and endogenous probiotics on the growth performance, nonspecific immunity and gut microbiota of juvenile Siberian sturgeon (*Acipenser baerii*). *Fish & Shellfish Immunology* 35, 766–775. doi: 10.1016/j.fsi.2013.06.014

Ghafarifarsani, H., Rashidian, G., Bagheri, T., Hoseinifar, S.H., & Doan, H.V. (2021) Study on growth enhancement and the protective effects of dietary prebiotic inulin on immunity responses of rainbow trout (*Oncorhynchus mykiss*) fry infected with *Aeromonas hydrophila*. *Annals of Animal Science* 21, 543–559. doi: 10.2478/aoas-2020-0074

Ghiasi, M., Binaii, M., Naghavi, A., Rostami, H.K., Nori, H., & Amerizadeh, A. (2018) Inclusion of *Pediococcus acidilactici* as probiotic candidate in diets for beluga (*Huso huso*) modifies biochemical parameters and improves immune functions. *Fish Physiol Biochem* 44, 1099–1107.

Gibson, L.F., Woodworth, J., & George, A.M. (1998) Probiotic activity of Aeromonas media when challenged with *Vibrio tubiashii*. *Aquaculture* 169, 111–120.

Gioacchini, G., Giorgini, E., Olivotto, I. et al (2014) The influence of probiotics on zebrafish Innate Immunity and hepatic stress. *Zebrafish* 11, 98–106. 10.1089/zeb.2013.0932

Giri, S.S., Sen, S.S., & Venkatachalam, S. (2012) Effects of dietary supplementation of potential probiotic *Pseudomonas aeruginosa* VSG-2 on the innate immunity and disease resistance of tropical freshwater fish, *Labeo rohita*. *Fish & Shellfish Immunology* 32, 1135–1140. 10.1016/j.fsi.2012.03.019.

Giri, S.S., Sukumaran, V., Sen, S.S., & Jena, P.K. (2014) Effects of dietary supplementation of potential probiotic *Bacillus subtilis* VSG1 singularly or in combination with *Lactobacillus plantarum* VSG3 or/and *Pseudomonas aeruginosa* VSG2 on the growth, immunity and disease resistance of *Labeo rohita*. *Aquatic Nutrition*, 20, 163–171.

Girijakumari, N.R., Ethiraja, K., & Marimuthu, P.N. (2018) In Vitro and In Vivo Evaluation of Probiotic Properties of Enterobacter Cloacae in Kenyi Cichlid, *Maylandia Lombardoi*. *Aquacult. Inter.* 26, 959–980. doi: 10.1007/s10499-018-0262-2

Gobi, N., Malaikozhundan, B., Sekar, V., Shanthi, S., Vaseeharan, B., Jayakumar, R., & Nazar, A.K. (2016) GFP tagged *Vibrio parahaemolyticus* Dahv2 infection and the protective effects of probiotic *Bacillus licheniformis* Dahb1 on the growth, immune and antioxidant responses in *Pangasius hypophthalmus*. *Fish & Shellfish Immunology* 52, 230–238.

Gong, L., He, H., Li, D., Cao, L., Khan, T.A., Li, Y., Pan, L., Yan, L. *et al.* (2019) A new isolate of *Pediococcus pentosaceus* (SL001) with antibacterial activity against fish pathogens and potency in facilitating the immunity and growth performance of grass carps. *Front Microbiol* 10, 1384.

González-Palacios, G., Fregeneda-Grandes, J-M., & Aller-Gancedo, J-M. (2020) Possible Mechanisms of Action of Two *Pseudomonas fluorescens* Isolates as Probiotics on Saprolegniosis Control in Rainbow Trout (*Oncorhynchus mykiss* Walbaum). *Animals* 2020, 10, 1507; doi:10.3390/ani10091507

Gram, L., Melchiorsen, J., Spanggaard, B., Huber, I., & Nielsen, T.F. (1999) Inhibition of *Vibrio anguillarum* by *Pseudomonas fluorescens* AH2, a possible probiotic treatment of fish. *Applied Environmental Microbiology* 65, 969–973.

Grisdale-Helland, B., Helland, S.J., & Gatlin III, D.M. (2008) The effects of dietary supplementation with mannanoligosaccharide, fructooligosaccharide, or galactooligosaccharide on the growth and feed utilization of Atlantic salmon (*Salmo salar*). *Aquaculture* 283, 163–167.doi: 10.1016/j.aquaculture.2008.07.012

Gullian, M., Thompson, F., & Rodriguez, J. (2004) Selection of probiotic bacteria and study of their immunostimulatory effect in *Penaeus vannamei*. *Aquaculture* 233, 1–14.

Guo, X., Chen, D-D, Peng, K-S, Cui, Z-W, Zhang, X-J, Li, S., & Zhang Y-A (2016) Identification and characterization of *Bacillus subtilis* from grass carp (*Ctenopharyngodon idellus*) for use as probiotic additives in aquatic feed. *Fish and Shellfish Immunology* 52, 74–84. doi: 10.1016/j.fsi.2016.03.017

Gupta, A., Gupta, P., & Dhawan, A. (2014) Dietary supplementation of probiotics affects growth, immune response and disease resistance of *Cyprinus carpio* fry. *Fish & Shellfish Immunology*, 41, 113–119.

Gupta, A., Gupta, P., & Dhawan, A. (2016) *Paenibacillus polymyxa* as a water additive improved immune response of *Cyprinus carpio* and disease resistance against *Aeromonas hydrophila*. *Aquatic Reports*, 4, 86–92.

Gupta, S.K., Pal, A.K., Sahu, N.P., Dalvi, R., Kumar, V., & Mukherjee, S.C. (2008) Microbial levan in the diet of *Labeo rohita*Hamilton juveniles: effect on non-specific immunity and histopathological changes after challenge with *Aeromonas hydrophila*. *Journal of Fish Diseases* 31, 649–657. doi: 10.1111/j.1365-2761.2008.00939.x

Guzmán-Villanueva, L.T., Tovar-Ramírez, D., Gisbert, E. et al (2014) Dietary administration of β-1,3/1,6- glucan and probiotic strain *Shewanella putrefaciens*, single or combined, on gilthead seabream growth, immune responses and gene expression. *Fish & Shellfish Immunology* 39, 34–41. 10.1016/j.fsi.2014.04.024

Hai, N.V. (2015) The use of probiotics in aquaculture. *Journal of Applied Microbiology* 119 (4):917–935. doi: 10.1111/jam.12886

Hanley, F., Brown, H., & Carbery, J. (1995) First observations on the effects of mannan oligosaccharide added to hatchery diets for warmwater hybrid red tilapia. Poster at the 11th Annual Symposium on Biotechnology in the Feed Industry, Lexington, KY, USA.

Hasan, K.N., & Banerjee, G. (2018) Recent studies on probiotics as beneficial mediator in aquaculture: review. *J Basic Appl Zool* 81, 53. doi: 10.1186/s41936-020-00190-y

He, S., Liu, W., Zhou, Z., Mao, W., Ren, P., Marubashi, T., & Ringø, E. (2011) Evaluation of probiotic strain *Bacillus subtilis* C-3102 as a feed supplement for koi carp (*Cyprinus carpio*). *Journal of Aquaculture Research and Development* S1–005, 1–7. doi: 10.4172/2155-9546.S1-005

Hoseinifar, S.H., Ringø, E., Shenavar Masouleh, A., & Esteban, M. Á. (2016). Probiotic, prebiotic and synbiotic supplements in sturgeon aquaculture: a review. *Rev. Aquacult.* 8, 89–102, doi: 10.1111/raq.12082

Hoseinifar, S.H., Esteban, M.Á., Cuesta, A., & Sun, Y-Z. (2015) Prebiotics and fish immune response: a review of current knowledge and future perspectives. *Reviews in Fisheries Science* 23, 315–328. doi:10.1080/23308249.2015.1052365

Hoseinifar, S.H., Hossein, M., Paknejad, H., Safari, R., Jafar, A., Yousefi, M., Doan, H.V., & Mozanzadeh, M.T. (2019) Enhanced mucosal immune responses, immune related genes and growth performance in common carp (*Cyprinus carpio*) juveniles fed dietary *Pediococcus acidilactici* MA18/5M and raffinose. *Dev Comp Immunol* 94, 59–65.

Hoseinifar, S.H., Soleimani, N., & Ringø, E. (2014) Effects of dietary fructo-oligosaccharide supplementation on the growth performance, haemato-immunological parameters, gut microbiota, and stress resistance of common carp (*Cyprinus carpio*) fry. *British Journal of Nutrition* 112, 1296–1302. doi: 10.1017/S0007114514002037

Hosseini Shekarabi, S.P., Ghodrati, M., Dawood, M.A.O., & Roudbaraki, A.F. (2022) The multi-enzymes and probiotics mixture improves the growth performance, digestibility, intestinal health, and immune response of Siberian sturgeon (*Acipenser baerii*). *Annals of Animal Science*, DOI: 10.2478/aoas-2022-0006

Huang, M.-Y., Chang, C.-I., Chang, C.-C., Tseng, L.-W., & Pan, C.-L. (2014) Effects of dietary levan on growth performance, nonspecific immunity, pathogen resistance, and body composition of orange-spotted grouper (*Epinephelus coioides*H.). *Aquaculture Research* 46, 2752–2767. doi: 10.1111/are.12430

Huynh, T-G, Cheng, A-C, Chi, C-C, Chiu, K-H, & Liu, C-H (2018) A synbiotic improves the immunity of white shrimp, *Litopenaeus vannamei*: Metabolomic analysis reveal compelling evidence. *Fish Shellfish Immunol* 79, 284–293.

Ibrahem, M.D., Fathi, M., Mesalhy, S., & Abd El-Aty, A.M. (2010) Effect of dietary supplementation of inulin and vitamin C on the growth, hematology, innate immunity, and resistance of Nile tilapia (*Oreochromis niloticus*). *Fish &Shellfish Immunology* 29, 241–246. doi:10.1016/j.fsi.2010.03.004

Interaminensea, J.A., Vogeleyb, J.L., Gouveiaa, C.K., Portelaa, R.S., Oliveirac, J.P., & Silvad, S.M.B.C. (2019) Effects of dietary *Bacillus subtilis* and *Shewanella* algae in expression profile of immune-related genes from hemolymph of *Litopenaeus vannamei* challenged with *Vibrio parahaemolyticus*. *Fish & Shellfish Immunology* 86, 253–259. doi: 10.1016/j.fsi.2018.11.051

Irianto, A., & Austin, B. (2002) Use of probiotics to control furunculosis in rainbow trout, Oncorhynchus mykiss (Walbaum). *J. Fish Dis.* 25, 333–342.

Irianto, A., & Austin, B. (2003) Use of dead probiotic cells to control furunculosis in rainbow trout, *Oncorhynchus mykiss* (Walbaum). *J. Fish Dis.* 26, 59–62.

Iwashita, M.K.P., Nakandakare, I.B., Terhune, J.S., Wood, T., & Ranzani-Paiva, M.J.T. (2015) Dietary supplementation with *Bacillus subtilis*, *Saccharomyces cerevisiae* and *Aspergillus oryzae* enhance immunity and disease resistance against *Aeromonashydrophila* and *Streptococcus iniae* infection in juvenile tilapia *Oreochromis niloticus*. *Fish & Shellfish Immunology* 43, 60–66. doi: 10.1016/j.fsi.2014.12.008

Jafaryan, H., & Soltani, M. (2012) Effects of bioencapsulated Daphnia magna with *Saccharomyces cerevisiae* on the growth and feeding performance of Persian sturgeon (*Acipenser persicus*) larvae. *Iranian Journal of Veterinary Medicine*, 6, 13–18. Doi. 10.22059/IJVM.2012.24619

Jia, S., Zhou, K., Pan, R. et al (2020) Oral immunization of carps with chitosan–alginate microcapsule containing probiotic expressing spring viremia of carp virus (SVCV) G protein provides effective protection against SVCV infection. *Fish Shellfish Immunol* 105, 327–329. 10.1016/j.fsi.2020.07.052

Jiang, H.F., Liu, X.L., Chang, Y.Q., Liu, M.T., & Wang, G.X. (2013) Effects of dietary supplementation of probiotic *Shewanella colwelliana* WA64, *Shewanella olleyana* WA65 on the innate immunity and disease resistance of abalone, *Haliotis discus hannai* Ino. *Fish Shellfish Immunol* 35, 86–91.

Jiang, H., Chen, T., Sun, H., Tang, Z., Yu, J., Lin, Z., Ren, P., Zhou, X., Huang, Y., Li, X., & Yu, X. (2017). Immune response induced by oral delivery of Bacillus subtilis spores expressing enolase of Clonorchis sinensis in grass carps (Ctenopharyngodon idellus). *Fish & Shellfish Immunology*, 60, 318–325, doi.org/10.1016/j.fsi.2016.10.011

Jiang, H., Bian, Q., Zeng, W., Ren, P., Sun Z., Lin, Z., et al (2019) Oral delivery of *Bacillus subtilis* spores expressing grass carp reovirus VP4 protein produces protection against grass carp reovirus infection. *Fish & Shellfish Immunology* 84, 768–780.

Jiravanichpaisal, P., Chuaychuwong, P., & Menasveta, P. (1997) The use of Lactobacillus sp. as the probiotic bacteria in the giant tiger shrimp (*Penaeus monodon* Fabricius), poster session of the second Asia-Pacific marine biotechnology conference and third Asia-pacific conference on algal Biotechnology, pp. 7–10.

Kamgar, M., & Ghane, M. (2014) Studies on Bacillus subtilis, as Potential Probiotics, on the Hematological and Biochemical Parameters of Rainbow trout, Oncorhynchus mykiss (Walbaum). *Journal of Applied Environmental Microbiology* 2(5), 203–207.

Kani, A.M., Soltani, M., Ebrahimzahe-Mousavi, H.A., & Pakzad K.. (2016) Influence of probiotic, *Lactobacillus plantarum* on serum biochemical and immune parameters in vaccinated rainbow trout (*Oncorhynchus mykiss*) against streptococcosis/lactococosis. *International Journal of Aquatic Biology* 4(4): 285–294.

Karim, M., Zhao, W., Rowley, D., Nelson, D., & Gomez-Chiarri, M. (2013) Probiotic strains for shellfish aquaculture: protection of eastern oyster, Crassostrea virginica, larvae and juveniles againsl bacterial challenge. *Journal of Shellfish Research* 32, 401–408.

Kazuń, B., Małaczewska, J., Kazuń, K. *et al.* (2018) Immune-enhancing activity of potential probiotic strains of Lactobacillus plantarum in the common carp (*Cyprinus carpio*) fingerling. *J Vet Res* 62, 485–492. 10.2478/jvetres-2018-0062

Kewcharoen, W., & Srisapoome, P. (2019) Probiotic effects of Bacillus spp. from Pacific white shrimp (Litopenaeus vannamei) on water quality and shrimp growth, immune responses, and resistance to Vibrio parahaemolyticus (AHPND strains). *Fish & Shellfish Immunology*, 10.1016/j.fsi.2019.09.013

Kim, D., Beek, B.B., Lee, S.U., Jeon, J., Lee, D.W., Lee, J.I. *et al.* (2016) Pellet feed adsorbed with the recombinant *Lactococcus lactis* BFE920 expressing SiMA antigen induced strong recall vaccine effects against *Streptococcus iniae* infection in olive flounder (P. olivaceus). *Fish & Shellfish Immunology* 55, 374–38.

Kiron, A.P.V., & Watanabe, S.S.T. (2010) Probiotic bacteria *Lactobacillus rhamnosus* influences the blood profile in rainbow trout *Oncorhynchus mykiss* (Walbaum). *Fish Physiol Biochem* 36(4), 969–977.

Kiron, V. (2012) Fish immune system and its nutritional modulation for preventive health care. *Animal Feed Science and Technology* 173, 111–133.doi: 10.1016/j.anifeedsci.2011.12.015

Klakegg, Ø, Salonius, K., Nilsen, A. et al (2020a) Enhanced growth and decreased mortality in Atlantic salmon (Salmo salar) after probiotic bath. *Journal of Applied Microbiology* 129, 146–160. 10.1111/jam.14649

Klakegg, Ø., Salonius, K., Nilsen, A., Fülberth, M., & Sørum, H. (2020b). Improved Health and Better Survival of Farmed Lumpfish (*Cyclopterus Lumpus*) After a Probiotic Bath With Two Strains of *Aliivibrio*. *Aquaculture* 518, 734810. doi: 10.1016/aquaculture.2019.734810

Knudsen, K.E.B., Serena, A., Canibe, N., & Juntunen, K.S. (2003) New insight into butyrate metabolism. *Proceedings of the Nutrition Society* 62, 81–86. doi: 10.1079/PNS2002212

Kong, W., Huang, C., Tang, Y., Zhang, D., Wu, Z., & Chen, X. (2017) Effect of *Bacillus subtilis* on *Aeromonas hydrophila*-induced intestinal mucosal barrier function damage and inflammation in grass carp (*Ctenopharyngodon idella*). *Science Reports*, DOI: 10.1038/s41598-017-01336-9

Kuebutoqrnyea, F.K.A., Abarikea, E.D., & Lua, Y. (2019) A review on the application of *Bacillus* as probioitcs in aquaculture. *Fish & Shellfish Immunology*, 87, 820–828.

Kulkarni, A., Krishnan, S., Anand, D., Uthaman, S.K., Otta, S.K., Karunasagar, I., & Valappil, R.K. (2021) Immune responses and immunoprotection in crustaceans with special reference to shrimp. 13, 431–459. 10.1111/raq.12482

Kumar, V., Roy, S., Meena, D.K., & Sarkar, U.K. (2016) Application of probiotics in shrimp aquaculture: Importance, mechanisms of action, and methods of administration. *Rev Fish Sci Aquacult* 24, 342–368.

Laranja, J.L.Q., Ludevese-Pascual, G.L., Amar, E.C., Sorgeloos, P., Bossier, P., & De Schryver, P. (2014) Poly-βhydroxybutyrate (PHB) accumulating *Bacillus* spp. improve the survival, growth and robustness of Penaeus monodon (Fabricius, 1798) postlarvae. *Vet Microbiol* 173, 310–317.

Laranja, J.L.Q., Amar, E.C., Ludevese-Pascual, G.L., Niu, Y., Geaga, M.J., De Schryver, P., & Bossier, P.A. (2017) Probiotic *Bacillus* strain containing amorphous poly-betahydroxybutyrate (PHB) stimulates the innate immune response of Penaeus monodon postlarvae. *Fish & Shellfish Immunology*, 68, 202–210. 10.1016/j.fsi.2017.07.023

Lategan, M.J., Torpy, F.R., & Gibson, L.F. (2004) Biocontrol of Saprolegniosis in Silver Perch *Bidyanus Bidyanus* (Mitchell) by *Aeromonas media* Strain A199. *Aquaculture* 235, 77–88. doi: 10.1016/j.aquaculture.2003.09.014

Lazado, C.C., Caipang, C.M.A., & Estante, E.G. (2015) Prospects of host-associated microorganisms in fish and penaeids as probiotics with immunomodulatory functions. *Fish Shellfish Immunol* (2015) 45(1):2–12. doi: 10.1016/j.fsi.2015.02.023

Le, B., & Yang, S.H. (2018) Probiotic potential of novel *Lactobacillus* strains isolated from salted-fermented shrimp as antagonists for *Vibrio parahaemolyticus*. *J Microbiol* 56, 138–144.

Lee, J.W., Chiu, S-T, Wang, S-T, Liao, Y-C, Chang, H-T, Ballantyne, R., Lin, J-S., & Liu, C-H. (2022) Dietary SYNSEA probiotic improves the growth of white shrimp, *Litopenaeus vannamei* and reduces the risk of *Vibrio* infection via improving immunity and intestinal microbiota of shrimp. *Fish and Shellfish Immunology* 127 (2022), 482–491. 10.1016/j.fsi.2022.06.071

Lee, S., Katya, K., Hamidoghli, A., Hong, J., Kim, D-J, & Bai, S.C. (2018) Synergistic effects of dietary supplementation of *Bacillus subtilis* WB60 and mannanoligosaccharide (MOS) on growth performance, immunity and disease resistance in Japanese eel, *Anguilla japonica*. *Fish & Shellfish Immunology* 83, 283–291. doi: 10.1016/j.fsi.2018.09.031

Lee, S., Katya, K., Park, Y., Won, S., Seong, M., Hamidoghli, A., & Bai. S.C. (2017) Comparative evaluation of dietary probiotics *Bacillus subtilis* WB60 and *Lactobacillus plantarum* KCTC3928 on the growth performance, immunological parameters, gut morphology and disease resistance in Japanese eel, *Anguilla japonica*. *Fish & Shellfish Immunology* 61, 201–210. doi: 10.1016/j.fsi.2016.12.035

Lee, S., Kumar Katya, K., Hamidoghli, A., Hong, J., Kim, D-J., & Bai, S.C. (2018) Synergistic effects of dietary supplementation of *Bacillus subtilis* WB60 and mannanoligosaccharide (MOS) on growth performance, immunity and disease resistance in Japanese eel, *Anguilla japonica*. *Fish & Shellfish Immunology* 83, 283–291. doi: 10.1016/j.fsi.2018.09.031

Li, E., Xu, C., Wang, X., Wang, S., Zhao, Q., Zhang, M., Qin, J.G., & Chen, L. (2018) Gut microbiota and its modulation for healthy farming of Pacific white shrimp Litopenaeus vannamei. *Rev Fish Sci Aquac* 26, 381–399.

Li, J., Wu, Z.B., Zhang, Z., Zha, J.W., Qu, S.Y., Qi, X.Z., Wang, G.X., & Ling, F. (2019) Effects of potential probiotic *Bacillus velezensis* K2 on growth, immunity and resistance to *Vibrio harveyi* infection of hybrid grouper (*Epinephelus lanceolatus*♂×*E. fuscoguttatus*♀). *Fish & Shellfish Immunology*, doi: 10.1016/j.fsi.2019.08.047

Li, J., Tan, B., & Mai, K. (2009) Dietary probiotic *Bacillus* OJ and isomaltooligosaccharides influence the intestine microbial populations, immune responses and resistance to white spot syndrome virus in shrimp (*Litopenaeus vannamei*). *Aquaculture* 291, 35–40. doi: 10.1016/j.aquaculture.2009.03.005

Li, P., & Gatlin III, D.M. (2003) Evaluation of brewer's yeast (*Saccharomyces cerevisiae*) as a feed supplement for hybrid striped bass (*Morone chrysops*×*M.saxatilis*). *Aquaculture* 219, 681–692. doi:10.1016/S0044-8486(02)00653-1

Li, P., & Gatlin III, D.M. (2004) Dietary brewer's yeast and the prebiotic Grobiotic™ AE influence growth performance, immune responses and resistance of hybrid striped bass (*Morone chrysops*×*M. saxatilis*) to *Streptococcus iniae*infection. *Aquaculture* 231, 445–456. doi: 10.1016/j.aquaculture.2003.08.021

Li, P., & Gatlin III, D.M. (2005) Evaluation of the prebiotic GroBiotic-A and brewer's yeast as dietary supplements for sub-adult hybrid striped bass (*Morone chrysops*×*M. saxatilis*) challenged in situ with *Mycobacterium marinum*. *Aquaculture* 248, 197–205. doi: 10.1016/j.aquaculture.2005.03.005

Li, Y., Yuan, W., Zhang, Y., Liu, H., & Dai, X. (2021) Single or combined effects of dietary arabinoxylan-oligosaccharide and inulin on growth performance, gut microbiota, and immune response in Pacific white shrimp *Litopenaeus vannamei*. *Journal of Oceanology and Limnology* 39, 741–754. doi: 10.1007/s00343-020-9083-z

Lin, S., Mao, S., Guan, Y., Luo, L., & Pan, Y. (2012) Effects of dietary chitosan oligosaccharides and *Bacillus coagulans*on the growth, innate immunity, and resistance of koi (*Cyprinus carpio koi*). *Aquaculture* 342, 36–41.doi: 10.1016/j.aquaculture.2012.02.009

Lin, Y-S, Saputra, F., Chen, Y-C, & Hu, S-Y (2019) Dietary administration of *Bacillus amyloliquefaciens* R8 reduces hepatic oxidative stress and enhances nutrient metabolism and immunity against *Aeromonas hydrophila* and *Streptococcus agalactiae* in zebrafish *(Danio rerio)*. *Fish Shellfish Immunol*, 10.1016/j.fsi.2018.11.047

Liu, C.-H., Chiu, C.-H., Wang, S.-W., & Cheng, W. (2012) Dietary administration of the probiotic, Bacillus subtilis E20, enhances the growth, innate immune responses, and disease resistance of the grouper, *Epinephelus coioides*. *Fish & Shellfish Immunology*, 33, 699–706.

Liu, H., Wang, S., Cai, Y., Guo, X., Cao, Z., & Zhang, Y., *et al.* (2017) Dietary administration of *Bacillus subtilis* enhances growth, digestive enzyme activities, innate immune responses and disease resistance of tilapia, *Oreochromis niloticus*. *Fish & Shellfish Immunology* 60, 326–333.

Liu, R.J., Wu, W.S., Xu, X.H., Wang, Y.L., Yu, T., Wang, J.H., *et al.* (2020) *Rhodopseudomonas Palustris* in effluent enhances the disease resistance, TOR and NF-Kappa B signalling pathway, intestinal microbiota and aquaculture water quality of *Pelteobagrus Vachelli*. *Aquaculture Research* 51 (10), 3959–3971. doi: 10.1111/are.14736.

Lochmann, R.T., Sink, T.D., & Phillips, H. (2009) Effects of dietary lipid concentration, a dairy–yeast prebiotic, and fish and nonfish protein sources on growth, survival, and nonspecific immune response of golden shiners in indoor tanks and outdoor pools. *North American Journal of Aquaculture* 71, 16–23. doi: 10.1577/A07-093.1

Longeon, A., Peduzzi, J., Barthelemy, M., Corre, S., Nicolas, J.L. & Guyot, M. (2004) Purification and partial identification of novel antimicrobial protein from marine bacterium Pseudoalteromonas species strain X153. *Mar. Biotechnol.*, 6, 633–641.

Lu, L., Cao, H., He, S., Wei, R., & Diong, M. (2011) Bacillus amyloliquefaciens G1: a potential antagonistic bacterium against eel-pathogenic *Aeromonas hydrophila*. *Evid. Based Complement. Alternat. Med.*, 2011, 824104.

Maeda, M., Shibata, A., Biswas, G., Korenga, H., Kono, T., Itami, T., & Sakai, M. (2014) Isolation of lactic acid bacteria from kuruma shrimp (*Marsupenaeus japonicus*) intestine and assessment of immunomodulatory role of a selected strain as probiotic. *Marine Biotechnology* 16, 181–192.

Mahious, A.S., Gatesoupe, F.J., Hervi, M., Metailler, R., & Ollevier, F. (2006) Effect of dietary inulin and oligosaccharides as prebiotics for weaning turbot, *Psetta maxima* (Linnaeus, C. 1758). *Aquaculture International* 14, 219–229. doi:10.1007/s10499-005-9003-4

Makled, S.O., Hamdan, A.M., & El-Sayed, A.F.M. (2020) Growth Promotion and Immune Stimulation in Nile Tilapia, *Oreochromis Niloticus*, Fingerlings Following Dietary Administration of a Novel Marine Probiotic, Psychrobacter Maritimus S. *Prob. Antimicrob. Prot.* 12 (2), 365–374. doi: 10.1007/s12602-019-09575-0

Makled, S.O., Hamdan, A.M., El-Sayed, A.F.M., & Hafez, E.E. (2017) Evaluation of marine psychrophile, *Psychrobacter Namhaensis* SO89, as a probiotic in Nile tilapia (*Oreochromis Niloticus*) diets. *Fish Shellfish Immunol.* 61, 194–200. doi: 10.1016/j.fsi.2017.01.001

Manning, M.J. (1998) Immune defence systems. In: *Biology of farmed fish*, Black, K.D., Pickering, A.D. (eds.), 180–221, Sheffield Academic Press, Sheffield, UK

Mariman, R., Tielen, F., Koning, F., & Nagelkerken, L. (2014) The probiotic mixture VSL#3 dampens LPS-induced chemokine expression in human dendritic cells by inhibition of STAT-1 phosphorylation. *PLoS One* 9, e115676. 10.1371/journal.pone.0115676.

Martınez-Cruz, P., Iba´ ñez, A.L., Monroy-Hermosillo, O.A., & Ramırez-Saad, H.C.´ (2012) Use of probiotics in aquaculture. *ISRN Microbiology* 916845, 1–13. doi: 10.5402/2012/916845

Meidong, R., Khotchanalekha, K., Doolgindachbaporn, S., Nagasawa, T., Nakao, M., Sakai, K., & Tongpim, S. (2018) Evaluation of probiotic *Bacillus aerius* B81e isolated from healthy hybrid catfish on growth, disease resistance and innate immunity of Pla-mong *Pangasius bocourti*. *Fish & Shellfish Immunology*, 73, 1–10.

Merrifield, D.L., Dimitroglou, A., Foey, A., Davies, S.J., Baker, R.T., Bøgwald, J., *et al.* The current status and future focus of probiotic and prebiotic applications for salmonids. *Aquaculture* (2010) 302(1-2):1–18. doi: 10.1016/ j.aquaculture.2010.02.007

Merrifield, D.L., Balcazar, J.L., Daniels, C.L., Zhou, Z., Carnevali, O., Sun, Y.Z., Hoseinifar, S.H., & Ringø, E. (2014) Indigenous lactic acid bacteria in fish and crustaceans. In: Merrifield, D.L., Ringø, E. (Eds.), *Aquaculture Nutrition: Gut Health, Probiotics and Prebiotics*. West Sussex, Wiley-Blackwell, pp. 128–168.

Merrifield, D.L., Dimitroglou, A., Foey, A., Davies, S.J., Baker, R.T.M., Bøgwald, J. Castex, M., & Ringø, E. (2010) The current status and future focus of probiotic and prebiotic applications for salmonids. *Aquaculture* 302, 1–18. doi:10. 1016/j.aquaculture.2010.02.007

Merrifield, D., & Ringø,E., (eds) (2014) *Aquaculture nutrition: gut health, probiotics and prebiotics*. Wiley-Blackwell, Oxford

Michael, S.E., Abarike, E.D., & Cai, J. (2019) A Review on the Probiotic Effects on Haematological Parameters in Fish. *Journal of Fisheries Science*, 13, 25–31.

Middlemiss, K.L., Daniels, C.L., Urbina, M.A., & Wilson, R.W. (2015) Combined effects of UV irradiation, ozonation, and the probiotic *Bacillus* spp. on growth, survival, and general fitness in European lobster (*Homarus gammarus*). *Aquaculture* 444, 99–10.

Min, L., Li-Lia, Z., Jun-Weia, G., Xin-Yuana, Q., Yi-Jinga, L., & Di-Qiub, L. (2012) Immunogenicity of *Lactobacillus*-expressing VP2 and VP3 of the infectious pancreatic necrosis virus (IPNV) in rainbow trout. *Fish and Shellfish Immunology*, 32, 196–203.

Mirghaed, A.T., Yarahmadi, P., Hosseinifar, S.H., Tahmasebi, D., Gheisvandi, N., & Ghaedi, A. (2018) The effects singular or combined administration of fermentable fiber and probiotic on mucosal immune parameters, digestive enzyme activity, gut microbiota and growth performance of Caspian white fish (*Rutilus frisii kutum*) fingerlings. *Fish & Shellfish Immunology* 77, 194–199.

Mladineo, I., Buselic, I., Hrabar, J., Radonic, I., Vrbatovic, A., Jozic, S., *et al.* (2016) Autochthonous bacterial isolates successfully stimulate *in vitro* peripheral blood leukocytes of the European sea bass (*Dicentrarchus Labrax*). *Frontiers in Microbiology* 7. doi: 10.3389/fmic.2018.01244

Mohammadian, T., Alishahi, M., Tabandeh, M.R., Nejad, A.J., Karami, E., & Zarea, M. (2018) Effects of autochthonous probiotics, isolated from Tor grypus (Karaman, 1971) intestine and *Lactobacillus casei* (PTCC 1608) on expression of immune-related genes. *Aquaculture International* 27, 239.

Morshedi, V., Agh, N., Noori, F., Jafari, F., Tukmechi, A., Marammazi, J., & Pagheh, E. (2018) Effects of dietary xylooligosaccharide on growth and feeding performance, body composition and physiological responses of sobaity seabream (*Sparidentex hasta*) juvenile. *Aquaculture Nutrition* 24, 1796–1803. doi: 10.1111/anu.12818

Mouriño, J.L.P., Vieira, F.D.N., Jatobá, A.B., Da Silva, B.C., Jesus, G.F.A., Seiffert, W.Q., & Martins, M.L. (2012) Effect of dietary supplementation of inulin and *W. cibaria* on haematoimmunological parameters of hybrid surubim (*Pseudoplatystoma* sp.). *Aquaculture Nutrition*18, 73–80.doi: 10.1111/j.1365-2095.2011.00879.x

Mukherjee, A., Banerjee,G., Mukherjee, P., Ray, A.K., Chandra, G., & Ghosh, K. (2019a). Antibacterial substances produced by pathogen inhibitory gut bacteria in *Labeo rohita*: physico-chemical characterization, purification and identification through MALDI-TOF Mass Spectrometry. *Microbial Pathology* 130, 146–155.

Mukherjee, A., Chandra, G., & Ghosh, K. (2019b). Single or conjoint application of autochthonous *Bacillus* strains as potential probiotics: effects on growth, feed utilization, immunity and disease resistance in Rohu, *Labeo rohita* (Hamilton). *Aquaculture* 512: 10.1016/j.aquaculture.2019.734302

Mukherjee, A., Dutta, D., Banerjee, S., Ringø, E., Breines, E.M., Hareide, E., Chandra, G., & Ghosh, K. (2017) Culturable autochthonous gut bacteria in rohu, *Labeo rohita*. *In vitro* growth inhibition against pathogenic *Aeromonas* spp., stability in gut, bio-safety and identification by 16S rRNA gene sequencing. *Symbiosis* 73, 165–177.

Mulyani, Y., Aryantha, I.N.P., Suhandono, S., & Pancoro, A. (2018) Intestinal bacteria of common carp (*Cyprinus carpio* L.) as a biological control agent for *Aeromonas*. *Journal of Pure and Applied Microbiology* 12, 601–610.

Mustafa, A., Buentello, A., Gatlin, D., Lightner, D., Hume, M., & Lawrence, A. (2019) Dietary supplementation of galactooligosaccharides (GOS) in Pacific white shrimp, *Litopenaeus vannamei*, cultured in a recirculating system and its effects on gut microflora, growth, stress, and immune response. *Journal of Immunoassay and Immunochemistry* 40, 662–675. doi: 10.1080/15321819.2019.1675694

Naderi-Samani, M., Soltani, M., Dadar, M., Taheri-Mirghaed, A., Zargar, A., & Ahmadivand, S. *et al.* (2020) Oral immunization of trout fry with recombinant *Lactococcus lactis* NZ3900 expressing G gene of viral hemorrhagic septicaemia virus (VHSV). *Fish Shellfish Immunol* 105, 62–70.

Naiel, M.A.E., Farag, M.R., Gewida, A.G.A., Elnakeeb, M.A., Amer, M.S., & Alagawany, M. (2020) Using lactic acid bacteria as an immunostimulants in cultured shrimp with special reference to *Lactobacillus* spp. *Aquaculture International* 10.1007/s10499-020-00620-2

Nandi, A., Dan, S.K., Banerjee, G., Ghosh, P., Ghosh, K., Ringø, E., & Ray, A.K. (2017) Probiotic potential of autochthonous bacteria isolated from the gastrointestinal tract of four freshwater teleost. *Probiotics and Antimicrobial Proteins* 9, 12–21.

Naresh, S., Suneetha, Y., & Srinivasulu, M. (2014) Reddy effect of *Lactobacillus rhamnosus* and *Bacillus subtilis* supplemented probiotic diets on the growth patterns and anti oxidant enzyme activities in *Penaeus monodon* and *Penaeus indicus*. *American International Journal of Research in Formal, Applied & Natural Sciences*, 8, 76–80, http://www.iasir.net

Newaj-Fyzul, A., Adesiyun, A.A., Mutani, A., Ramsubhag, A., Brunt, J., & Austin, B. (2007) *Bacillus subtilis* AB1 controls *Aeromonas* infection in rainbow trout (*Oncorhynchus mykiss*, Walbaum). *Journal of Applied Microbiology* 103, 1699–1706.

Nguyen, T.T.G., Nguyen, T.C., Leelakriangsak, M., Pham, T.T., Pham, Q.H., & Lueangthuwapranit, CJTM (2018) Promotion of *Lactobacillus plantarum* on growth and resistance against acute hepatopancreatic necrosis disease pathogens in white-leg shrimp (*Litopenaeus vannamei*). *Thi J Sci* 48, 19–28.

Nguyen, V.D., Pham, T.T., Nguyen, T.H.T., Nguyen, T.T.X., & Hoj, L. (2014a) Screening of marine bacteria with bacteriocin-like activities and probiotic potential for ornate spiny lobster (*Panulirus ornatus*) juveniles. *Fish & Shellfish Immunology* 40, 49–60. doi: 10.1016/j.fsi.2014.06.017.

Nguyen, A.T.V., Pham, C.K., Pham, H.L., Nguyen, A.H., Dang, L.T., Huynh, H.A., Cutting, S.M., & Phan, T.N. (2014b). *Bacillus subtilis* spores expressing the VP28 antigen: A potential oral treatment to protect *Litopenaeus vannamei* against white spot syndrome. *FEMS Microbiology Letters* 358(2). doi: 10.1111/1574-6968.12546

Ning, D., Leng, X., Li, Q., & Xu, X. (2011) Surface-displayed VP28 on *Bacillus subtilis* spores induce protection against white spot syndrome virus in crayfish by oral administration. *Journal of Applied Microbiology* 111, 1327–1336. doi: 10.1111/j.1365-2672.2011.05156.x

Nurhajati, J., Atira, Aryantha, I.N.P., & Kadek Indah, D.G. (2012) The curative action of Lactobacillus plantarum FNCC 226 to Saprolegnia parasitica A3 on catfish (*Pangasius hypophthalamus Sauvage*). *International Food Research Journal*, 19 (4):1723–1727.

Pavadi, P., Murthy, H.S., Honnananda, B.R., & Choudhary, B.K. (2018) Enhancement of growth, immunity, resistance and survival of freshwater prawn, *Macrobrachium rosenbergii* against White Muscle Disease (WMD) due to dietary administration of probiotic and biogut. *International Journal of Current Microbiology and Applied Sciences* 7, 569–579. doi: 10.20546/ijcmas.2018.704.067

Paz, A.L., da Silva, J.M., da Silva, K.M.M., & Val, A.L. (2019) Protective effects of the fructooligosaccharide on the growth performance, hematology, immunology indicators and survival of tambaqui (*Colossoma macropomum*, Characiformes: Serrasalmidae) infected by *Aeromonas hydrophila*. *Aquaculture Reports* 15, 100222. doi:10.1016/j.aqrep.2019.100222

Peng, M., Zhang, Y., & Song, Z. (2019) Isolation and characterization of a *Bacillus* spp. against *Vibrio Parahaemolyticus* from shrimp culture ponds. *International Journal of Microbiology and Biotechnology*, 4, 29–37.

Peraza-Gómez, V., Luna-González, A., Campa-Córdova, ÁI, López-Meyer, M., Fierro-Coronado, J.A., & Álvarez-Ruiz, P. (2009) Probiotic microorganisms and antiviral plants reduce mortality and prevalence of WSSV in shrimp (*Litopenaeus vannamei*) cultured under laboratory conditions. *Aquatic Research* 40, 1481–1489.

Pereira, G.V., Jesus, G.F.A., Vieira, F.N., Pereira, S.A., Ushizima, T.T., Mouriño, J.L.P., & Martins, M.L. (2016) Probiotic supplementation in diet and vaccination of hybrid surubim (*Pseudoplatystoma reticulatum*♀ x *P. corruscans*♂). *Ciência Rural, Rural, Santa Maria* 46 (2): 343–353. 10.1590/0103-8478cr20150543

Perez-Sanchez, T., Balcazar, J.L., Merrifield, D.L., Carnevali, O., Gioacchini, G., de Blas, I., & Ruiz-Zarzuela, I. (2011) Expression of immune-related genes in rainbow trout (Oncorhynchus mykiss) induced by probiotic bacteria during Lactococcus garvieae infection. *Fish Shellfish Immunol.* 31, 196–201. 10.1016/j.fsi.2011.05.005.

Petit, J., de Bruijn, I., Goldman, M.R.G., van den Brink, E., Pellikaan, W.F., Forlenza, M., & Wiegertjes, G.F. (2022) β-Glucan-induced immuno-modulation: a role for the intestinal microbiota and short-chain fatty acids in common carp. *Frontiers in Immunology* 12, 761820. doi: 10.3389/fimmu.2021.761820

Petit, J., & Wiegertjes, G.F. (2022) Conservation of members of the free fatty acid receptor gene family in common carp. *Developmental & Comparative Immunology* 126, 104240. doi: 10.1016/j.dci.2021.104240

Pham, K.C., Tran, H.T., Van Doan, H., Le, P.H., Van Nguyen, A.T., Nguyen, H.A., Hong, H.A., Cutting, S.M., & Pham, T.-N. (2017) Protection of *Penaeus monodon* against white spot syndrome by continuous oral administration of a low concentration of *Bacillus subtilis* spores expressing the VP 28 antigen. *Letters in Applied Microbiology* 64, 184–191. doi: 10.1111/lam.12708

Phianphak, W., Rengpipat, S., Piyatiratitivorakul, S., & Menasveta, P. (1999) Probiotic use of *Lactobacillus* spp. for black tiger shrimp, *Penaeus monodon. J. Sci. Res. Chula. Univ.*, 42–51.

Piazzon, M.C., Calduch-Giner, J.A., Fouz, B., Estensoro, I., Simó-Mirabet, P., Puyalto, M., *et al.* (2017) Under control: how a dietary additive can restore the gut microbiome and proteomic profile, and improve disease resilience in a marine teleostean fish fed vegetable diets. *Microbiome* 5, 164. doi: 10.1186/s40168-017-0390-3

Pieters, N., Brunt, J., Austin, B., & Lyndon, A.R. (2008) Efficacy of in-feed probiotics against *Aeromonas bestiarum* and *Ichthyophthirius multifiliis* skin infections in rainbow trout (*Oncorhynchus mykiss*, Walbaum). *Journal of Applied Microbiology*, 105(3):723–732. doi: 10.1111/j.1365-2672.2008.03817.x

Pietrzak, E., Mazurkiewicz, J., & Slawinska, A. (2020) Innate immune responses of skin mucosa in Common Carp (*Cyprinus carpio*) fed a diet supplemented with galactooligosaccharides. *Animals (Basel)* 10, 438. doi: 10.3390/ani10030438

Pooljun, C., Daorueang, S., Weerachatyanukul, W., Direkbusarakom, S., & Jariyapong, P. (2020). Enhancement of shrimp health and immunity with diets supplemented with combined probiotics: application to Vibrio parahaemolyticus infections. *Diseases of Aquatic Organisms*, 140: 37–46. doi: 10.3354/dao03491

Purivrirojkul, W., Maketon, M., & Areechon, N. (2005) Probiotic properties of *Bacillus pumilus*, *Bacillus sphaericus* and *Bacillus subtilis* in black tiger shrimp (*Penaeus monodon* Fabricius) culture. *Food and Agriculture Organization of United Nations*, pp. 263–271.

Qi, X., Xue, M.Y., Cui, H.B., Yang, K.C., Song, K.G., Zha, J.W., *et al.* (2020) Antimicrobial Activity of *Pseudomonas Monteilii* JK-1 Isolated From Fish Gut and its Major Metabolite, 1-Hydroxyphenazine, Against *Aeromonas Hydrophila*. *Aquaculture* 526, 735366. doi: 10.1016/j.aquaculture.2020.735366

Rahiman, K.M.M., Jesmi, Y., Thomas, A.P., & Hatha, A.A.M. (2010) Probiotic effect of Bacillus NL110 and Vibrio NE17 on the survival, growth performance and immune response of Macrobrachium rosenbergii (de Man). *Aquatic Research*, 41, 120–134, doi:10.1111/j.1365-2109.2009.02473.x

Raida, M.K., Larsen, J.L., Nielsen, M.E., & Buchmann, K. (2003) Enhanced resistance of rainbow trout, *Oncorhynchus mykiss* (Walbaum), against *Yersinia ruckeri* challenge following oral administration of *Bacillus subtilis* and *B. licheniformis* (BioPlus2B). *Journal of Fish Diseases*, 26, 495–498.

Rairakhwada, D., Pal, A.K., Bhathena, Z.P., Sahu, N.P., Jha, A., & Mukherjee, S.C. (2007) Dietary microbial levan enhances cellular non-specific immunity and survival of common carp

(*Cyprinus carpio*) juveniles. *Fish & Shellfish Immunology* 22, 477–486.doi: 10.1016/j.fsi.2006.06.005

Ramesh, D., & Souissi, S. (2018) Effects of potential probiotic *Bacillus subtilis* KADR1 and its subcellular components on immune responses and disease resistance in *Labeo rohita*. *Aquaculture Research* 49, 367–377. doi: 10.1111/are.13467

Rastekenari, H.Y., Kazami, R., Shenavar Masouleh, A., Banavreh, A., Najjar Lashgari, S., Hassani, M.H.S., Ghorbani Vaghei, R., Alizadeh Roudposhti, M., & Hallajian, A. (2021) Autochthonous probiotics *Lactococcus lactis* and *Weissella confuse* in the diet of fingerlings great sturgeon, *Huso huso*: effects on growth performance, feed efficiency, haematological parameters, immune status and intestinal morphology. *Aquaculture Researach*, 52, 3687–3695. 10.1111/are.15213

Ray, A.K., Ghosh, K., & Ringø, E. (2012) Enzyme-producing bacteria isolated from fish gut: A review. *Aquacult. Nutr.* 18, 465–492.

Reda, R.M., & Selim, K.M. Evaluation of *Bacillus amyloliquefaciens* on the growth performance, intestinal morphology, hematology and body composition of Nile tilapia, *Oreochromis niloticus. Aquac. Int.*, 23, 203–217 (2015).

Remus, D.M., van Kranenburg, R., van Swam, I.I., Taverne, N., Bongers, R.S., Wels, M., Bron, P., & Kleerebezen, M. (2012) Impact of four*Lactobacillus plantarum* capsular polysaccharide clusters on surface glycan composition and host cell signaling. *Microbiol. Cell. Fact.* 11, 149. 10.1186/1475-2859-11-149.

Rengpipat, S., Rukpratanporn, S., Piyatiratitivorakul, S., & Menasaveta, P. (2000) Immunity enhancement in black tiger shrimp (*Penaeus monodon*) by a probiont bacterium (*Bacillus* S11). *Aquaculture*, 191, 271–288. 10.1016/S0044-8486(00)00440-3

Rengpipat, S., Tunyanun, A., Fast, A.W., Piyatiratitivorakul, P., & Menasveta, P. (2003) Enhanced growth and resistance to *Vibrio* challenge in pond-reared black tiger srhimp *Penaeus monodon* fed a *Bacillus* probiotic. *Dis Aquat. Org.*, 55, 169–173. Doi. 10.3354/dao055169.

Rengpipat, S., Phianphak, W., Piyatiratitivorakul, S., & Menasveta, P. (1998) Eeffcts of a probiotic bacterium on black tiger shrimp *Penaeus monodon* survival and growth. *Aquaculture* 167, 301–313.

Reyes-Becerril, M., Ascencio, F., Gracia-Lopez, V., Macias, M.E., Roa, M.C., & Esteban, M.Á. (2014) Single or combined effects of *Lactobacillus sakei*and inulin on growth, non-specific immunity, and IgM expression in leopard grouper (*Mycteroperca rosacea*). *Fish Physiologyand Biochemistry* 40, 1169–1180.doi: 10.1007/s10695-014-9913-z

Rezvani Gil Kolaei, A., Shoaibi Omrani, B., & Afraei Bandpey, M.A. (2019) The Effect of dietary probiotic *Lactobacillus Plantarum* on nutrition performance in Siberian sturgeon (*Acipenser Baerii*). *Journal of Aquaculture Development*, 13, 79–88.

Ringø, E., Hoseinifar, S.H., Ghosh, K., Doan, H.V., Beck, B.R., & Song, S.K. (2018) Lactic Acid Bacteria in Finfish—An Update. *Frontiers in Microbiology* 9, 1818. doi: 10.3389/fmicb.2018.01818

Ringø, E., Li, X., Van Doan, Hv, & Ghosh, K. (2022) Interesting probiotic bacteria other than the more widely used lactic acid bacteria and bacilli in finfish. *Frontiers in Marine Science* 9, 848037. doi: 10.3389/fmars.2022.848037

Ringø, E. (2020) Probiotics in shellfish aquaculture. *Aquatic Fish* 5 (2), 1–27. 10.1016/j.aaf.2019.12.001.

Ringø, E., Løvmo, L., Kristiansen, M., Bakken, Y., Salinas, I., Myklebust, R., Olsen, R.E., & Mayhew, T.M. (2010) Lactic acid bacteria vs. pathogens in the gastrointestinal tract of fish: a review. *Aquac. Res.* 41, 451–467. 10.1111/j.1365-2109.2009. 02339.x.

Ringø, E., Olsen, R.E., Gifstad, T.Ø., Dalmo, R.A., Amlund, H., Hemre, G.-I., & Bakke, A.M. (2010) Prebiotics in aquaculture: a review. *Aquaculture Nutrition* 16, 117–136. doi: 10.1111/j.1365-2095.2009.00731.x

Ringø, E., van Doan, H., Lee, S.H., Soltani, M., Hoseinifar, S.H., Ramasamy, H., & Song, S.K. (2020) Probiotics, lactic acid bacteria and bacilli: interesting supplementation for aquaculture. *Journal of Applied Microbiology* 129, 116–136. doi: 10.1111/jam.14628

Rodriguez–Estrada, U., Satoh, S., Haga, Y., Fushimi, H., & Sweetman, J. (2009) Effects of single and combined supplementation of *Enterococcus faecalis*, mannan oligosaccharide and polyhydroxybutyrate acid on growth performance and immune response of rainbow trout *Oncorhynchus mykiss*. *Aquaculture Science* 57, 609–617. doi: 10.11233/AQUACULTURESCI.57.609

Rombout, J.H.W.M., Abelli, L., Picchietti, S., Scapigliati, G., & Kiron, V. (2011) Teleost intestinal immunology. *Fish & Shellfish Immunology* 31, 616–626. doi:10.1016/j.fsi.2010.09.001

Roomiani, L., Ahmadi, S., & Ghaeni, M.J.A.Ü.V.F.D. (2018) Immune response and disease resistance in the white shrimp, *Litopenaeus vannamei* induced by potential probiotic *Lactobacillus*. *Unk Uni Vet Fak Der* 65, 323–329.

Ruiz-Ponte, C., Samain, J.F., Sanchez, J.L., & Nicolas, J.L. (1999) The benefit of a *Roseobacter* species on the survival of scallop larvae. *Marine Biotechnology*, 1, 52–59.

Sadat Hoseini Madani, N., Adorian, T.J., Ghafari Farsani, H., & Hoseinifar, S.H. (2018) The effects of dietary probiotic bacilli (*Bacillus subtilis* and *Bacillus licheniformis*) on growth performance, feed efficiency, body composition and immune parameters of whiteleg shrimp (*Litopenaeus vannamei*) postlarvae. *Aquacultre Research* 49, 1926–1933. doi: 10.1111/are.13648

Salinas, I., Zhang, Y-A., & Sunyer, J.O. (2011) Mucosal immunoglobulins and B cells of teleost fish. *Developmental & Comparative Immunology* 35, 1346–1365. doi:10.1016/j.dci.2011.11.009

Sánchez-Ortiz, A.C., Angulo, C., Luna-González, A., Álvarez-Ruiz, P., Mazón-Suástegui, J.M., & Campa Córdova, ÁI (2016) Effect of mixed-Bacillus spp. isolated from pustulose *ark Anadara tuberculosa* on growth, survival, viral prevalence and immune-related gene expression in shrimp *Litopenaeus vannamei*. *Fish & Shellfish Immunology* 59, 95–102.

Sanz, M.L., Polemis, N., Morales, V., Corzo, N., Drakoularakou, A., Gibson, G.R., & Rastall, R.A. (2005) In vitro investigation into the potential prebiotic activity of honey oligosaccharides. *Journal of Agricultural and Food Chemistry* 53, 2914–2921. doi:10.1021/jf0500684

Schaeck, M., Reyes-López, F.E., Vallejos-Vidal, E. et al (2017) Cellular and transcriptomic response to treatment with the probiotic candidate *Vibrio lentus* in gnotobiotic sea bass (*Dicentrarchus labrax*) larvae. *Fish & Shellfish Immunology* 63, 147–156. doi.org/10.1016/j.fsi.2017.01.028

Selim, K.M., & Reda, R.M. (2015) Improvement of immunity and disease resistance in the Nile tilapia, *Oreochromis niloticus*, by dietary supplementation with *Bacillus amyloliquefaciens*. *Fish & Shellfish Immunology*, 44, 496–503.

Sella S.R.B.R., Vandenberghe L.P.S., & Socol C.R. (2014) Life cycle and spore resistance of spore-forming *Bacillus atrophaeus*. *Microbiology Research* 169, 931–939.

Sewaka, M., Trullas, C., Chotiko, A., Rodkhum, C., Chansue, N., Boonanuntanasam, S., & Pirarat, N. (2019) Efficacy of synbiotic Jerusalem artichoke and *Lactobacillus rhamnosus* GG-supplemented diets on growth performance, serum biochemical parameters, intestinal morphology, immune parameters and protection against *Aeromonas veronii* in juvenile red tilapia (*Oreochromis* spp.). *Fish Shellfish Immunol* 86, 260–268.

Shabirah, A., Mulyani, R.Y., & Lili, W. (2019) Effect of types isolated lactic acid bacteria on hematocrit and differential leukocytes fingerling common carp (*Cyprinus carpio* L.) infected with Aeromonas hydrophila bacteria. *WNOFNS* 24, 25–35.

Sharma, P., Sihag, R.C., & Gahlawat, S.K. (2013) Effect of probiotic on haematogical paramaters of diseased fish (*Cirrihinus mrigal*). *Journal of Fish Science* 7(4), 323–328.

Shefat S.H.T. (2018) Probiotic strains used in aquaculture. *International Research Journal on Microbiology* 7, 43–55. doi: 10.14303/irjm.2018.023

Sherif, A.H., & Mahfouz, M.E. (2019) Immune status of *Oreochromis niloticus* experimentally infected with *Aeromonas hydrophila* following feeding with 1, 3 β-glucan and levamisole immunostimulants. *Aquaculture* 509, 40–46. doi:10.1016/j.aquaculture.2019.05.016

Silva-Aciares, F., Moraga, D., Auffret, M., Tanguy, A., & Riquelme, C. (2013) Transcriptomic and cellular response to bacterial challenge (pathogenic *Vibrio parahaemolyticus*) in farmed juvenile *Haliotis rufescens* fed with or without probiotic diet. *Journal of Invertebrate Pathology* 113, 163–176.

Simón, R., Docando, F., Nuñez-Ortiz, N., Tafalla, C., & Díaz-Rosales, P. (2021) Mechanisms used by probiotics to confer pathogen resistance to teleost fish. *Frontiers in Immunology* 12, 653025. doi: 10.3389/fimmu.2021.653025

Simon, R., Docando, F., Nuñez-Ortiz, N., Tafalla, C., & Díaz-Rosales, P. (2021) Mechanisms used by probiotics to confer pathogen resistance to teleost fish. *Frontiers in Immunology* 12, 653025. doi: 10.3389/fimmu.2021.653025

Slepecky, R., & Hemphill, E. (2006) The genus *Bacillus*. Nonmedical. The Prokaryotes, (Dworkin, M., Falkow, S., Rosenberg, E., Schleifer, K-H & Stackebrandt, E., eds), 4:530–562. Springer, New York.

Sohn, S., Lundgren, K.M., Tammi, K., Karim, M., Smolowitz, R., Nelson, D.R., Rowley, D.C., & Gómez-Chiarri, M. (2016a) Probiotic strains for disease management in hatchery larviculture of the eastern oyster *Crassostrea virginica*. *Journal of Shellfish Research* 35, 307–317.

Sohn, S., Lundgren, K.M., Tammi, K., Smolowitz, R., Nelson, D.R., Rowley, D.C., & Gómez-Chiarri, M. (2016b) Efficacy of probiotics in preventing vibriosis in the larviculture of different species of bivalve shellfish. *Journal of Shellfish Research* 35, 319–328.

Soleimani, N., Hoseinifar, S.H., Merrifield, D.L., Barati, M., & Abadi, Z.H. (2012) Dietary supplementation of fructooligosaccharide (FOS) improves the innate immune response, stress resistance, digestive enzyme activities, and growth performance of Caspian roach (*Rutilus rutilus*) fry. *Fish &Shellfish Immunology* 32, 316–321. doi: 10.1016/j.fsi.2011.11.023

Soltani, M., Abdy, E., Alishahi, M. *et al.* (2017) Growth performance, immune-physiological variables and disease resistance of common carp (*Cyprinus carpio*) orally subjected to different concentrations of *Lactobacillus plantarum*. *Aquacult Int*, 25, 1913–1933. 10.1007/s10499-017-0164-8

Soltani, M., Kane, A.M., Taherimirghaed, A., Pakzad, K., & Hosseini Shekarabi, P. (2019b). Effect of the probiotic, *Lactobacillus plantarum* on growth performance and haematological indices of rainbow trout (*Oncorhynchus mykiss*) immunized with bivalent streptococcosis /lactococcosis vaccine. *Iranian Journal of Fisheries Sciences*. 18(2) 283–295. DOI: 10.22092/ijfs.2018.117757

Soltani, M., Lymbery, A., Song, S.K., & Hosseini Shekarabi, P. (2018) Adjuvant effects of medicinal herbs and probiotics for fish vaccines. *Rev Aquac* 11(4), 1325–1341. doi.org/10.1111/raq.12295

Soltani, M., Pakzad, K., Taheri-Mirghaed, A., Mirzargar, S., Shekarabi, S.P.H., Yosefi, P., & Soleymani, N. (2019c). Dietary application of the probiotic *Lactobacillus plantarum* 426951 enhances immune status and growth of rainbow trout (*Oncorhynchus mykiss)* vaccinated against *Yersinia ruckeri*. *Probiotics Antimicrobial Proteins* 11(1):207–219. doi: 10.1007/s12602-017-9376-5.

Soltani, M., Ghosh, K., Hoseinifar, S.H., Kumar, V., Lymbery, A.J., Roy, S., & Ringø, E. (2019a) Genus *Bacillus*, promising probiotics in aquaculture: Aquatic animal origin, bioactive components, bioremediation and efficacy in fish and shellfish. *Rev Fish Sci Aquacult*, DOI: 10.1080/23308249.2019.1597010

Song, S.K., Beck, B.R., Kim, D., Park, J., Kim, J., Kim, H.D., & Ringø, E. (2014) Prebiotics as immunostimulants in aquaculture: A review. 10.1016/j.fsi.2014.06.016

Sorieul, L., Wabete, N., Ansquer, D., Mailliez, J-M., Pallud, M., Zhang, C., Lindivat, C., Boulo, V., & Pham. D. (2018) Survival improvement conferred by the *Pseudoalteromonas* sp. NC201 probiotic in *Litopenaeus stylirostris* exposed to *Vibrio nigripulchritudo* infection and salinity stress. *Aquaculture*, 495, 888–898. 10.1016/j.aquaculture.2018.06.058

Spanggaard, B., Huber, I., Nielsen, J., Sick, E.B., Pipper, C.B., Martinussen, T., Slierendrecht, W.J., & Gram, L. (2001) The probiotic potential against vibriosis of the indigenous microflora of rainbow trout. *Environ. Microbiol.* 3, 755–765.

Suguna, P., Binuramesh, C., Abirami, P., Saranya, V., Poornima, K., Rajeswari, V., & Shenbagarathai, R. (2014) Immunostimulation by poly-b hydroxybutyrate-hydroxyvalerate (PHB-HV) from *Bacillus thuringiensis* in *Oreochromis mossambicus*. *Fish & Shellfish Immunology*, 36, 90–97. doi.org/10.1016/j.fsi.2013.10.012

Sumathi, C., Nandhini, A., & Padmanaban, J. (2017) Antagonistic activity of probiotic *Bacillus megaterium* against *Streptococcus mutans*. *Int J Pharmacol Biolo Sci* 8, 270–274.

Sumon, T.F., Hussain, M.A., Sumon, M.A.A., Jang, W.J., Abellan, F.G., Sharifuzzaman, S.M., Brown, C.L., Lee, E-W., Kim, C.h., & Hasan, M.T. (2022) Functionality and prophylactic role of probiotics in shellfish aquaculture. *Aquaculture Reports* 25, 101220. 10.1016/j.aqrep.2022.101220

Sun, H., Shang, M., Tang, Z., Jiang, H., Dong, H., Zhou, X., Lin, Z., Shi, C., Ren, P., Zhao, L., Shi, M., Zhou, L., Pan, H., Chang, O., Li, X., Huang, Y., & Yu, X. (2020) Oral delivery of *Bacillus subtilis* spores expressing *Clonorchis sinensis* paramyosin protects grass carp from cercaria infection. *Applied Microbiology and Biotechnology*. doi: 10.1007/s00253-019-10316-0.

Sun, Y., Wang, G., Peng, K., Huang, Y., Cao, J., Huang, W., Chen, B., & Hu, J. (2018) Effects of dietary xylooligosaccharides on growth performance, immunity and *Vibrio alginolyticus* resistance of juvenile *Litopenaeus vannamei*. *Aquaculture Research* 50, 358–365. doi: 10.1111/are.13911

Sun, Y.-Z., Yang, H.-L., Ma, R.-L., & Lin, W.-Y. (2010) Probiotic applications of two dominant gut Bacillus strains with antagonistic activity improved the growth performance and immune responses of grouper *Epinephelus coioides*. *Fish & Shellfish Immunology*, 29, 803–809.

Ta'ati, R., Soltani, M., Bahmani, M., & Zamini, A. (2011) Growth performance, carcass composition, and immunophysiological indices in juvenile great Sturgeon (*Huso huso*) fed on commercial prebiotic, Immunoster. *Iranian Journal of Fisheries Sciences* 10, 324–335.

Talpur, A.D., Munir, M.B., Mary, A., & Hashim, R. (2014) Dietary probiotics and prebiotics improved food acceptability, growth performance, haematology and immunological parameters, and disease resistance against *Aeromonas hydrophilain* snakehead (*Channa striata*) fingerlings. *Aquaculture* 426, 14–20. doi: 10.1016/j.aquaculture.2014.01.013

Talukdar, S., Ringø, E., & Ghosh, K. (2016) Extracellular tannase-producing bacteria detected in the digestive tracts of seven freshwater fishes. *Acta Ichthyologica et Piscatoria* 46, 201–210.

Tamilarasu, A., Ahilan, B., Gopalakannan, A., & Lingam, R.S.S. (2021) Evaluation of probiotic potential of *Bacillus* strains on growth performance and physiological responses in *Penaeus vannamei*. *Aquaculture Research*, DOI: 10.1111/are.15159

Tang, Y., Han, L., Chen, X., Xie, M., Kong, M., & Wu, Z. (2019) Dietary supplementation of probiotic *Bacillus subtilis* affects antioxidant defenses and immune response in grass carp under *Aeromonas hydrophila* challenge. *Probiotics and Antimicrobial Proteins* 11,545–558. doi: 10.1007/s12602-018-9409-8

Tarnecki, A.M., Burgos, F.A., Ray, C.L., & Arias, C.R. (2017) Fish intestinal microbiome: diversity and symbiosis unraveled by metagenomics. *Journal of Applied Microbiology* 123, 2–17. doi: 10.1111/jam.13415.

Telli, G.S., Ranzani-Paiva, M.J.T., de Carla Dias, D., Sussel, F.R., Ishikawa, C.M., & Tachibana, L. (2014) Dietary administration of *Bacillus subtilis* on hematology and non-specific immunity of Nile tilapia,*Oreochromis niloticus* raised at different stocking densities. *Fish & Shellfish Immunology*, 39, 305–311.

Tepaamorndech, S., Chantarasakha, K., Kingcha, Y., Chaiyapechara, S., Phromson, M., Sriariyanun, M., Kirschke, C.P., Huang, L., & Visessanguan, W. (2019) Effects of *Bacillus aryabhattai* TBRC8450 on vibriosis resistance and immune enhancement in Pacific white shrimp,*Litopenaeus vannamei*. *Fish & Shellfish Immunology*, 86, 4–13. DOI: 10.1016/j.fsi.2018.11.010

Thanh Tung, H., Koshio, S., Ferdinand Traifalgar, R., Ishikawa, M., & Yokoyama, S. (2010) Effects of dietary heat-killed *Lactobacillus plantarum* on larval and post-larval kuruma shrimp, *Marsupenaeus japonicus* Bate. *J World Aquacult Soc* 41, 16–27.

Torrecillas, S., Makol, A., Caballero, M.J., Montero, D., Ginés, R., Sweetman, J., & Izquierdo, M. (2011) Improved feed utilization, intestinal mucus production, and immune parameters in sea bass (*Dicentrarchus labrax*) fed mannan oligosaccharides (MOS). *Aquaculture Nutrition* 17, 223–233. doi: 10.1111/j.1365-2095.2009.00730.x

Torrecillas, S., Makol, A., Caballero, M.J., Montero, D., Robaina, L., Real, F., Sweetman, J., Tort, L., & Izquierdo, M.S. (2007) Immune stimulation and improved infection resistance in European sea bass (*Dicentrarchus labrax*) fed mannan oligosaccharides. *Fish &Shellfish Immunology* 23, 969–981. doi: 10.1016/j.fsi.2007.03.007

Tsai, C-Y., Chi,C-C., & Liu C-H. (2019) The growth and apparent digestibility of white shrimp, *Litopenaeus vannamei*, are increased with the probiotic, *Bacillus subtilis*. 10.1111/are.14022

Ullah, A., Zuberi, A., Ahmad, M., Bashir, A., Younus, N., *et al.* (2018) Dietary administration of the commercially available probiotics enhanced the survival, growth, and innate immune responses in Mori (*Cirrhinus mrigala*) in a natural earthen polyculture system. *Fish & Shellfish Immunology* 72, 266–272.

Utiswannakul, P., Sangchai, S., & Rengpipat, S. (2011) Enhanced growth of black tiger shrimp Penaeus monodon by dietary supplementation with Bacillus (BP11) as a probiotic. *Journal of Aquatic Research and Development* 3(4), 2155–9546.

Valdez, A., Yepiz-Plascencia, G., Ricca, E., & Olmos, J. (2014) First *Litopenaeus vannamei* WSSV 100% oral vaccination protection using CotC: Vp26 fusion protein displayed on *Bacillussubtilis* spore surface. *Journal of Applied Microbiology* 117, 347–357. doi: 10.1111/jam.12550

Valipour, A., Nedaei, S., Noori, A., Khanipour, A.A., & Hoseinifar, S.H. (2019) Dietary *Lactobacillus plantarum* affected on some immune parameters, air-exposure stress response, intestinal microbiota, digestive enzyme activity and performance of narrow clawed cray fish (*Astacus leptodactylus*, Eschscholtz). *Aquaculture* 504, 121–130.

Van Doan, H., Hoseinifar, S.H., Khanongnuch, C., Kanpiengjai, A., Unban, K., & Srichaiyo, S. (2018) Host-associated probiotics boosted mucosal and serum immunity, disease resistance and growth performance of Nile tilapia (*Oreochromis niloticus*). *Aquaculture* 491, 94–100. doi: 10.1016/ j.aquaculture.2018.03.019

Van Doan, H.V., Soltani, M., & Ringø, E. (2021) *In vitro* antagonistic effect and *in vivo* protective efficacy of Gram-positive probiotics versus Gram-negative bacterial pathogens in finfish and shellfish. *Aquaculture* 538, 10.1016/j.aquaculture.2021.736581

van Hai, N., Buller, N., & Fotedar, R. (2009) Effects of probiotics (*Pseudomonas synxantha* and *Pseudomonas aeruginosa*) on the growth, survival and immune parameters of juvenile western king prawns (*Penaeus latisulcatus Kishinouye*, 1896). *Aquaculture Research* 40, 590–602. 10.1111/j.1365-2109.2008.02135.x.

Varela-Granados, Y., Frías-Gómez, S.A., Harnández-Harnández, L.H., Powell, M.S., & Vega-Villasante, F. (2021) Effects of mannan oligosaccharides and fructooligosaccharides on the growth and nonspecific immune responses of juvenile freshwater prawn *Macrobrachium acanthurus*. *Latin American Journal of Aquatic Research* 49, 299–306. doi: 10.3856/vol49-issue2-fulltext-2586

Vaseeharan, B., & Ramasamy, P. (2003) Control of pathogenic *Vibrio* spp. by *Bacillus subtilis*, a possible probiotic treatment for black tiger shrimp *Penaeus monodon*. *Letters of Applied Microbiology* 36, 83–87.

Veiga, P.T.N., Owatari, M.S., Nunes, A.L., Rodrigues, R.A., Kasai, R.Y.D., Fernandes, C.E., & Campos, C.M. (2020) *Bacillus subtilis* C-3102 improves biomass gain, innate defense, and intestinal absorption surface of native Brazilian hybrid surubim (*Pseudoplatystoma corruscans x P. reticulatum*). *Aquaculture International*. doi: 10.1007/s10499-020-00519-y

Venkatalakshmi S., & Ebanasar J. (2015) Immunostimulatory effect of *Lactobacillus sporogenes* on the nonspecific defense mechanisms of *Oreochromismossambicus* (Peters). *International Journal of Fisheries and Aquatic Studies*, 2(4): 362–369.

Vinderola, G., Matar, C., & Perdigon, G. (2019) Role of intestinal epithelial cells in immune effects mediated by gram-positive probiotic bacteria: involvement of toll-like receptors. *Clinical Diagnostic Laboratory of Immunology* 12 (9), 1075–10784. doi.org/10.1128/ CDLI.12.9.1075-1084.2005.

Vogeley, J.L., Interaminense, J.A.,Buarque D.S., da Silva, S.M.B.C., Coimbra, M.R.M., Peixoto, S.M., & Soares, R.B. (2019) Growth and immune gene expression of Litopenaeus vannamei fed Bacillus subtilis and Bacillus circulans supplemented diets and challenged with Vibrio parahaemolyticus. *Aquaculture International*, 27,1451–1464.

Wang, C., Liu, Y., Sun, G., Li, X., & Liu, Z. Growth, immune response, antioxidant capability, and disease resistance of juvenile Atlantic salmon (*Salmo salar* L.) fed *Bacillus velezensis* V4 and *Rhodotorula mucilaginosa* compound. *Aquaculture*, 500, 65–74 (2019).

Wang, H., Wang, C., Tang, Y., Sun, B., Huang, J., & Song, X. (2018) *Pseudoalteromonas* probiotics as potential biocontrol agents improve the survival of Penaeus vannamei challenged with acute hepatopancreatic necrosis disease (AHPND)-causing *Vibrio parahaemolyticus*. *Aquaculture*, 494, 30–36.

Wang, Y.B. (2007) Effect of probiotics on growth performance and digestive enzyme activity of the shrimp *Penaeus vannamei*. *Aquaculture*, 269, 259–264.

Wee, W., Abdul Hamid, N.K., Mat, K., Raja Khalif, R.I.K., Rusli, N.D., Rahman, M.M., Kabir, M.A., & Wei, L.S. (2022) The effects of mixed prebiotics in aquaculture: A review. 10.1016/j.aaf.2022.02.005

Wei, C., Luo, K., Wang, M., Li, Y., Pan, M., Xie, Y., Qin, G., Liu, Y., Li, L., Liu, Q., & Tian, X. (2022) Evaluation of potential probiotic properties of a strain of *Lactobacillus plantarum* for shrimp farming: From beneficial functions to safety assessment. *Front. Microbiol.* 13, 854131. doi: 10.3389/fmicb.2022.854131

Welker, T.L., Lim, C., Yildirim-Aksoy, M., & Klesius, P.H. (2012) Effect of short-term feeding duration of diets containing commercial wholecell yeast or yeast subcomponents on immune function and disease resistance in channel catfish, *Ictalurus punctatus*. *Journal of Animal Physiology and Animal Nutrition* 96, 159–171. doi: 10.1111/j.1439-0396.2011.01127.x

Widanarni, W., Putri, F.N., & Rahman, R. (2019) Growth performance of white shrimp *Litopenaeus vannamei* fed with various dosages of prebiotic honey. *IOP Conference Series: Earth and Environmental Science* 278, 012079. doi: 10.1088/1755-1315/278/1/012079

Widanarni, W., Taufik, A., Yuhana, M., & Ekasari, J. (2019) Dietary mannan oligosaccharides positively affect the growth, digestive enzyme activity, immunity and resistance against*Vibrio harveyi* of Pacific White Shrimp (*Litopenaeusvannamei*) larvae. *Turkish Journal of Fisheries and Aquatic Sciences* 19, 271–278. doi: 10.4194/1303-2712-v19_4_01

Won, S., Hamidoghli, A., Choi, W., Bae, J., Jang, W.J., & Lee, B.S.C. (2020b). Evaluation of potential probiotics *Bacillus subtilis* WB60, *Pediococcus pentosaceus*, and *Lactococcus lactis* on growth performance, immune response, gut histology and immune-related genes in whiteleg shrimp, *Litopenaeus vannamei*. *Microorganisms* 8(2), 1–15. doi: 10.3390/microorganisms8020281.

Won, S., Hamidoghli, A., Choi, W., Park, Y., Jang, W.J., Kong, I.S., & Bai, S.C. (2020a). Effects of *Bacillus subtilis* WB60 and *Lactococcus lactis* on growth, immune responses, histology and gene expression in Nile tilapia, *Oreochromis niloticus*. *Microorganisms* 8(1), 1–15. doi: 10.3390/microorganisms8010067.

Wu, H-J, Sun, L-B, Li, C-B, Li, Z-Z, Zhang, Z., Wen, X-B, Hu, Z., Zhang, Y-L, & Li, S-K (2014) Enhancement of the immune response and protection against *Vibrio parahaemolyticus* by indigenous probiotic Bacillus strains in mud crab (*Scylla paramamosain*). *Fish & Shellfish Immunology* 41, 156–162.

Wu, Z.-Q., Jiang, C., Ling, F., & Wang, G.-X. (2015) Effects of Dietary Supplementation of Intestinal autochthonous bacteria on the innate immunity and disease resistance of grass carp (*Ctenopharyngodon Idellus*). *Aquaculture* 438, 105–114. doi: 10.1016/j.aquaculture.2014.12.041

Xia, Y., Lu, M., Chen, G. et al (2018) Effects of dietary *Lactobacillus rhamnosus* JCM1136 and *Lactococcus lactis* subsp. lactis JCM5805 on the growth, intestinal microbiota, morphology, immune response and disease resistance of juvenile Nile tilapia *Oreochromis niloticus*. *Fish & Shellfish Immunology* 76, 368–379. doi.org/10.1016/j.fsi.2018.03.020

Xia, Z., Zhu, M., & Zhang, Y. (2014) Effects of the probiotic *Arthrobacter* sp. CW9 on the survival and immune status of white shrimp (*Penaeus vannamei*). *Letters of Applied Microbiology* 58, 60–64.

Xiaolong, G., Mo, Z., Xian, L., Yin, H., Fucun, W., & Ying, L. (2018). The effects of feeding *Lactobacillus pentosus* on growth, immunity, and disease resistance in *Haliotis discus* hannai Ino. *Fish & Shellfish Immunology* 78, 42–51.

Xu, B., Zhang, G., Wang, L., Sagada, G., Zhang, J., Shao, Q. (2021) The influence of dietary β-1,3-glucan on growth performance, feed utilization, antioxidative and immune status of Pacific white shrimp, *Litopenaeus vannamei*. *Aquaculture Nutrition* 27, 1590–1601. doi: 10.1111/anu.13299

Xu, Z., Parra, D., Gomez, D., Salinas, I., Zhang, Y-A., Jørgensen, L.G., Heinecke, R.D., Buchmann, K., LaPatra, S., & Sunyer,

J.O. (2013) Teleost skin, an ancient mucosal surface that elicits gut-like immune responses. *The Proceedings of the National Academy of Sciences* 110, 13097–13102. doi: 10. 1073/pnas.1304319110

Yamamoto, F.Y., Sutili, F.J., Hume, M., & Gatlin III, D.M. (2018) The effect of β-1,3-glucan derived from *Euglena gracilis* (Algamune^{TM}) on the innate immunological responses of Nile tilapia (*Oreochromis niloticus* L.). *Journal of Fish Diseases* 41, 1579–1588. doi: 10.1111/jfd.12871

Yang, Q., Lü, Y., Zhang, M., Gong, Y., Li, Z., Tran, N.T., He, Y., Zhu, C., Lu, Y., Zhang, Y., & Li, S. (2019) Lactic acid bacteria, Enterococcus faecalis Y17 and *Pediococcus pentosaceus* G11, improved growth performance, and immunity of mud crab (*Scylla paramamosain*). *Fish Shellfish Immunol* 93, 135–143.

Yanuhar, U., Junirahma, N.S., Susilowati, K., Caesar, N.R., & Musa, M. (2019) Effects of probiotic treatment on histopathology of Koi carp (*Cyprinus carpio*) infected by *Myxobolus* sp. *J Phys: Conf Ser*, 1374, 12051. doi: 10.1088/ 1742-6596/1374/1/012051

Yao, Y.-Y., Chen, D.-D., Cui, Z.-W., Zhang, X.-Y., Zhou, Y.-Y., Guo, X., Li, A.-H., & Zhang, Y.-A. (2019) Oral vaccination of tilapia against Streptococcus agalactiae using Bacillus subtilis spores expressing Sip. *Fish & Shellfish Immunology* 86, 999–1008. doi: 10.1016/j.fsi.2018.12.060.

Yarahmadi, P., Miandare, H.K., Farahmand, H., Mirvaghefi, A., & Hoseinifar, S.H. (2014) Dietary fermentable fiber upregulated immune related genes expression, increased innate immune response, and resistance of rainbow trout (*Oncorhynchus mykiss*) against *Aeromonas hydrophila*. *Fish & Shellfish Immunology* 41, 326–331. doi: 10.1016/j.fsi. 2014.09.007

Ye, J.D., Wang, K., Li, F.D., & Sun, Y.Z. (2011) Single or combined effects of fructo- and mannan oligosaccharide supplements and *Bacillus clausii*on the growth, feed utilization, body composition, digestive enzyme activity, innate immune response, and lipid metabolism of the Japanese flounder *Paralichthys olivaceus*. *Aquaculture Nutrition* 17, e902–e911. doi: 10.1111/j.1365-2095.2011.00863.x

Yeh, S.p., Y.L. Shiu, Z.L. Huang, & C.H. Liu (2014) Effects of diets supplemented with either individual or combined probiotics, *Bacillus subtilis* E20 and *Lactobacillus plantarum* 7-40, on the immune response and disease resistance of the mud crab, *Scylla paramamosain* (Estampador). *Aquaculture Research* 45, 1164–1175 10.1111/are.12061.

Yi, C.-C., Liu, C.-H., Chuang, K.-P., Chang, Y.T., & Hu, S.-Y. (2019) A potential probiotic *Chromobacterium* aquaticum with bacteriocin-like activity enhances the expression of indicator genes associated with nutrient metabolism, growth performance and innate immunity against pathogen infections in Zebrafish (*Danio Rerio*). *Fish & Shellfish Immuno*logy 93, 124–134. doi: 10.1016/j.fsi.2019.07.042

Yi, Y., Zhang, Z., Zhao, F., Liu, H., Yu, L., Zha, J., & Wang, G. (2018) Probiotic potential of *Bacillus velezensis* JW: Antimicrobial activity against fish pathogenic bacteria and immune enhancement effects on *Carassius auratus*. *Fish Shellfish Immunol*, 78, 322–330.

Yılmaz, S., Ergun, S., Yigit, M., & Çelik, E.S. (2019) Effect of combination of dietary *Bacillus subtilis* and *trans*-cinnamic acid on innate immune responses and resistance of rainbow trout, *Oncorhynchus mykiss* to *Yersinia ruckeri*. *Aquaculture Research* 51(2), 441–454. doi: 10.1111/are.14379

Yones, A.M.A.S., Eissa, I.A.M.M., Ghobashy, M.A., & Marzok, S.S. (2019) Effects of dietary inulin as prebiotic on growth performance, immuno-haematological indices and ectoparasitic infection of fingerlings Nile Tilapia, *Oreochromis niloticus*. *The Egyptian Journal of Histology* 43, 88–103. doi: 10.21608/EJH.2019.15495.1152

Yu, H.H., Han, F., Xue, M., Wang, J., Tacon, P., Zheng, Y.H., Wu, X.F., & Zhang, Y.J. (2014) Efficacy and tolerance of yeast cell wall as an immunostimulant in the diet of Japanese seabass (*Lateolabrax japonicus*). *Aquaculture* 432, 217–224.doi: 10.1016/j.aquaculture.2014.04.043

Zaineldin, A.I., Hegazi, S., Koshio, S., Ishikawa, M., Bakr, A., El-Keredy, A.M., & Yukun, Z. (2018) *Bacillus subtilis* as probiotic candidate for red sea bream: growth performance, oxidative status, and immune response traits. *Fish & Shellfish Immunology* 79, 303–312. doi: 10.1016/j.fsi. 2018.05.035

Zhang, Y-A., Salinas, I., Li, J., Parra, D., Bjork, S., Xu, Z., LaPatra, S.E., Bartholomew, J., & Sunyer, J.O. (2010) IgT, a primitive immunoglobulin class specialized in mucosal immunity. *Nature Immunology* 11, 827–835. doi:10.1038/ ni.1913

Zhao, L.L., Liu, M., Ge, J.W., Qiao, X.Y., Li, Y.J., & Liu, D.Q. (2012) Expression of infectious pancreatic necrosis virus (IPNV) VP2-VP3 fusion protein in *Lactobacillus casei* and immunogenicity in rainbow trouts. *Vaccine*, 30(10), 1823–1829. doi:10.1016/j.vaccine.2011.12.132

Zhao, C., Zhu, J., Hu, J., Dong, X., Sun, L., Zhang, X., & Miao, S. (2019) Effects of dietary *Bacillus pumilus* on growth performance, innate immunity and digestive enzymes of giant freshwater prawns (*Macrobrachium rosenbergii*). *Aquatic Nutrition* 25, 712–720. doi.org/10.1111/anu. 12894.

Zheng, C.N., & Wang, W. (2017) Effects of *Lactobacillus pentosus* on the growth performance, digestive enzyme and disease resistance of white shrimp, *Litopenaeus vannamei* (Boone, 1931). *Aquatic Research* 48, 2767–2777. 10.1111/ are.13110.

Zheng, Z.L., Wang, K.Y., Gatlin III, D.M., & Ye, J.M. (2011) Evaluation of the ability of GroBiotic®-A to enhance growth, muscle composition, immune responses, and resistance against *Aeromonas hydrophila*in Nile tilapia, *Oreochromis niloticus*. *Journal of the World Aquaculture Society* 42, 549–557. doi: 10.1111/j.1749-7345.2011.00497.x

Zhou, Q-C., Buentello, J.A., & Gatlin III, D.M. (2010) Effects of dietary prebiotics on growth performance, immune response and intestinal morphology of red drum (*Sciaenops ocellatus*). *Aquaculture* 309, 253–257. doi: 10.1016/j.aquaculture. 2010.09.003

Zhou, S., Song, D., Zhou, X., Mao, X., Zhou, X., Wang, S., Wei, J., Huang, Y., Wang, W., Xiao, S.M., & Qin, Q. (2019) Characterization of *Bacillus subtilis* from gastrointestinal tract of hybrid Hulong grouper (*Epinephelus fuscoguttatus* × *E. lanceolatus*) and its effects as probiotic additives. *Fish & Shellfish Immunology* 84, 1115–1124. doi: 10.1016/j.fsi. 2018.10.058

Zhou, X.X., Tian, Z.Q., Wang, Y.B., & Li, W.F. (2010) Effect of Treatment With Probiotics as Water Additives on Tilapia (*Oreochromis Niloticus*) Growth performance and immune response. *Fish Physiology & Biochemistry* 36 (3), 501–509. doi: 10.1007/s10695-009-9320-z

Zokaeifar, H., Babaei, N., Saad, C.R., Kamarudin, M.S., Sijam, K., & Balcazar, J.L. (2014) Administration of *Bacillus subtilis* strains in the rearing water enhances the water quality, growth performance, immune response, and resistance against *Vibrio harveyi* infection in juvenile white shrimp, *Litopenaeus vannamei*. *Fish & Shellfish Immunology*, 36, 68–74.

Zokaeifar, H., Balcazar, J.L., Saad, C.R., Kamarudin, M.S., Sijam, K., Arshad, A., & Nejat, N. (2012) Effects of *Bacillus subtilis* on the growth performance, digestive enzymes, immune gene expression and disease resistance of white shrimp, *Litopenaeus vannamei*. *Fish & Shellfish Immunol*ogy 33, 683–689. 10.1016/j.fsi.2012.05.027.

11

Immunomodulation in Fish Through Nutrients, Antioxidants and Hormones

Chiranjiv Pradhan[1], Nikhila Peter[1], and Sweta Das[2]
[1]*Fish Nutrition and Feed Technology Laboratory, Department of Aquaculture, Kerala University of Fisheries and Ocean Studies, Panangad, Kochi, Kerala*
[2]*Department of Aquatic Animal Health Management, Kerala University of Fisheries and Ocean Studies, Panangad, Kochi, Kerala*

11.1 Introduction

Aquaculture is a sunrise industry growing with an annual growth rate of 5.8%, and provides about half of the global fish supply. Aquaculture is developing, expanding and intensifying in almost all regions of the world and aquatic food products are playing a critical role in food security and nutrition. The growing demand of sea food has been closely associated with intensification of production conditions. In the intensive mode of aquaculture, high stocking density of fish is reared in unit area of water body with provision of artificial feed, aeration and water exchange. The intensification is very risky in the absence of proper management, often leads to stress among the fish and become cause of high mortality primarily due to infectious diseases in the early production stages. A lot of prevention majors in terms of application of chemicals and antibiotics are taken in order to stop the infectious disease and to reduce the mortality in aquaculture practice. These types of preventive majors has its own disadvantages, often increases chemical and antibiotic load in the environment. Presently, a lot of emphasis is given on the alternative disease prevention methods by adopting several techniques. One among them is improving aquatic animal health with proper nutritional strategy. There is a significant role of nutrition on immunity and disease resistance in fish (Kiron, 2012). Adequate nutrition modulates the immune system of fish and gives innate and adaptive responses to pathogens and often referred as immunomodulation through nutrients. The term "immunomodulation" is the activation or suppress of immune system of body by some causative agents and through which body fights with infections and diseases. In this process macrophage, granulocyte, natural killer and lymphocytes get activated as part of nonspecific stimulation and numerous effector chemicals get synthesized by the activated cells which take part in the defence towards the pathogens. Immunomodulation through nutrition is an effective strategy to focus on prevention of disease rather than treatment for a better health. It is well known that inadequate or excess supply of nutrients induces physiological disorder in animals including fish and causes nutritional diseases. The nutritional deficiency related disease is very difficult to diagnose and is found associated with secondary infections by microbes and parasites in fish due to compromised immunity (Lavilla-Pitogo and Amar, 2010). The immune boosting nutrition can influence the immune system in various ways, and the specific effects depend on the type, quality and quantity of the nutrient consumed. Adequate and appropriate nutrition nourishes all cells including immune cells for optimal function in the body. A well balanced nutrition, especially in terms of adequate protein and amino acids; lipids and fatty acids; vitamins, minerals, antioxidants, other additives and hormones modulate immune system for a better health and protection against the pathogens.

11.2 Role of Nutrients, Antioxidants and Hormones in Immunity

Nutritional imbalance can disturb systemic homeostasis and disrupt normal physiology of living organisms. Nutrients acts on immune cells and through metabolic, neurological, or endocrine pathways alter immune responses in animals (Reddy and Frey, 1990). The dietary macronutrients (proteins, carbohydrates and lipids), micronutrients (vitamins and minerals), non-nutrient immunostimulatory compounds (phytochemicals, antioxidants and probiotics) and even feeding strategies take part in modulating immune homeostasis in fish. Macronutrients are necessary in substantial amounts to sustain body functions and to carry out daily activities. Macronutrients are also called energy providing nutrients. The three important macronutrients viz., carbohydrates, proteins and lipids are the very basis of sustaining health, growth, reproduction and immunity in living organisms. Micronutrients are needed by the body in low doses and holds essential part in fish nutrition. Micronutrients are very essential in enhancing innate and adaptive immune responses and alleviating stress responses in fish. Almost all micronutrients such as

DOI: 10.1201/9781003361183-14

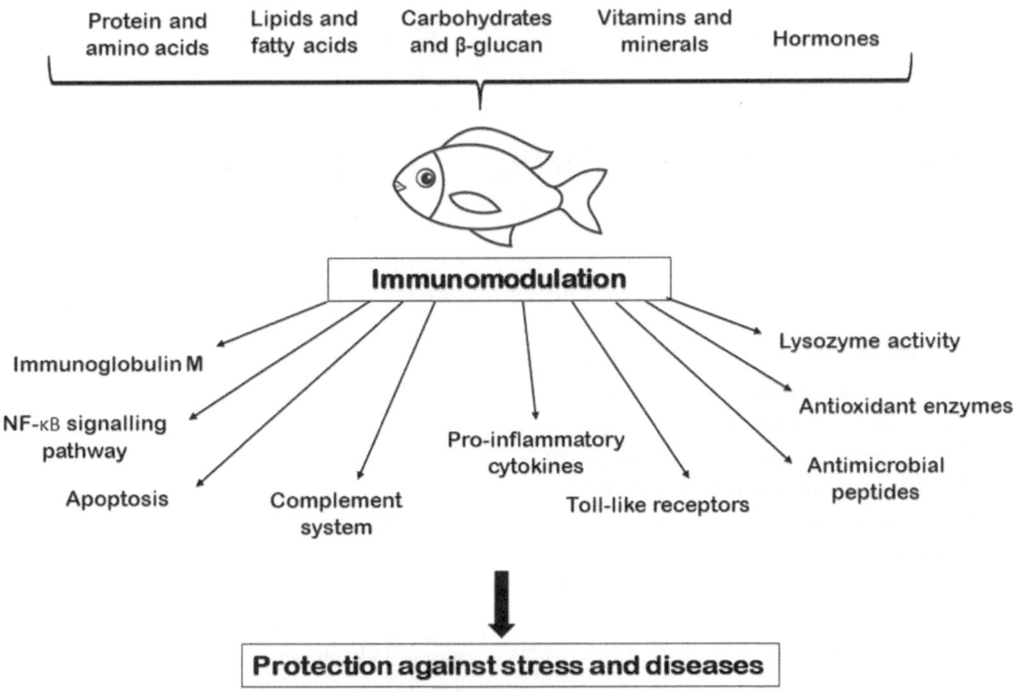

FIGURE 11.1 Effect of nutrients and hormones in immunomodulation of fish.

vitamins A (including beta-carotene), vitamin B6, folate, vitamin B12, vitamin C and vitamin D, and minerals such as copper, iron, selenium and zinc are important for some aspects of immune function in fish. They act in different ways. Some act as co-factors for enzymes that drive the biochemical reactions and some are required for the manufacture of DNA and RNA and in the production of immune cells. In terms of micronutrients, this is accomplished by those that act as antioxidants. The processes involved in innate and adaptive immune function include the identification and elimination of pathogens, cell development, the generation of antibodies and cytokines and the so-called "respiratory or antioxidant burst" (Figure 11.1). All of these processes require adequate amounts of macronutrients and micronutrients.

11.3 Immunomodulation Through Nutrients

11.3.1 The Role of Proteins and Amino Acids in Immunomodulation

Dietary protein or amino acids has been associated with immune function of animal and human being and upon deficiency increase the susceptibility to infectious disease (Li et al., 2007). Protein is very essential nutrient for fish and the requirement is two to four times more in comparison to mammals. The diet containing optimal protein level helps growth, immunity and disease resistance in fish. The effect of protein on immunity of fish is always discussed in terms of the amino acid composition of the diet (Li et al., 2007). In fish feed development, replacement of fishmeal is a major research thrust. The replacement of fishmeal with plant ingredients or with alternative protein sources often lead to improper amino acid balance and increase in the concentration of antinutritional factors. A study conducted by Jalili et al. (2013) reported that when plant protein inclusion increased up to 70–100% in the diet of rainbow trout (*Oncorhynchus mykiss*) resulted in undesirable effects on serum total immunoglobulin and alternative complement activity including growth. Otherwise, 40% fishmeal replacement with plant counterparts did not cause negative effects on growth and humeral immune parameters (lysozyme activity and total antibody). At higher inclusion levels (over 75%) of plant ingredients, gilthead seabream (*Sparus aurata*) showed liver steatosis, accompanied by a decrease in complement activity (Sitjà-Bobadilla et al., 2005). However, Ding et al. (2015) observed that at moderate replacement of fishmeal by fermented soybean meal (FSM) did not affect the immune response of prawn but total replacement significantly reduced immunity. Similarly, at moderate replacement levels (56.25%) of fishmeal with plant protein in the diet of red sea bream (*Pagrus major*) did not show change in immune defence system (Dossou et al., 2018). It is indicated from the above studies that moderate replacement levels of fishmeal with plant ingredients did not influence the immune status in fish could be due to no drastic change in amino acid profile and antinutritional factors in the feed. Though the studies related to effect of antinutritional factors on immune status of fish is less reported. Soybean protein is often associated with enteritis in Atlantic salmon (Booman et al., 2018). The distal intestine of salmonids is immunologically sensitive to common antinutritional factors could be the reason for disturbed digestive process and a reduced health status. However, FSM has been effectively used to improve immune responses in fish. The *Aspergillus oryzae* FSM improved phosphorous utilization as well as non-specific immune responses of parrot fish (*Oplegnathus fasciatus*), which was evidenced through increased superoxide dismutase (SOD) activity (Kim et al., 2009). Similar observation has also reported earlier in parrot fish when fed with traditional Korean FSM with rice straw showed higher DPPH (2,2-diphenyl-1-picryl-hydrazyl-hydrate) scavenging activity and liver SOD compared to raw soybean meal (Pham and Lee, 2007). The antioxidant defense system is associated with health status and immune system in fish (Martínez-Álvarez et al., 2005). A study conducted by Dossou et al. (2018) observed that incorporation of fermented rape seed meal (FRM) up to 28% in the diet of red sea bream *P. major* was suitable for higher antioxidant activity and the reason could be the beneficial effects of the increased level of rape seed meal antinutrient degradation products present in FRM. The degraded products of antinutritional factors such as glucosinolates and phenolic compounds which present in some plant protein sources are known to possess antioxidative properties (Adom and Liu, 2002; Mazumder et al., 2016). Dietary glucosinolate is also very effective against lice infection in Atlantic salmon. The antioxidant properties of glucosinolates and its role in increasing the expression of interferon, and higher expression profiles of Type 1 pro-inflammatory factors, are beneficial against skin lice (Holm et al., 2016). Another attractive alternative protein source in aquafeed is insect meal which is increasingly gaining popularity. The black soldier fly (*Hermetia illucens*) larval (BSFL) meal is one of the proteins rich insect meals has been commonly used as alternative of fish meal (FM) in fish feed. The BSFL meal regulates survival and growth of aquatic animals and increases the specific and non-specific immunity in aquatic animals. The role of BSFL meal in disease resistance of aquatic animals against pathogens and up regulating the growth and immune genes in aquatic animals has also been confirmed in many studies (Mohan et al., 2022). Yellow catfish (*Pelteobagrus fulvidraco*) fed with 25% FM protein replaced by BSFL meal protein showed higher growth performances and increased serum lysozyme activity significantly (Xiao et al., 2018). Similarly, European seabass (*Dicentrarchus labrax*) fed with BSFL meal showed increased serum antioxidant biomarkers such as malondialdehyde levels, and catalase, superoxide dismutase and glutathione peroxidase enzyme activities in a dose-dependent manner (replacement of fishmeal 25% to 50%) compared to the control group. The nonspecific immune indices, including phagocytic activity, phagocytic index, serum lysozyme and respiratory burst activities were significantly elevated in BSFL meal groups compared to those in the control group. Additionally, after the challenge with *Vibrio alginolyticus*, the relative percent of survival was significantly elevated in fish groups fed on diets containing graded levels of BSFLM over the control group (Abdel-Latif et al., 2021). A study on the potential effect of yellow mealworm (*Tenebrio molitor*) substitution for fish meal revealed that at 4% incorporation level growth, immune and antioxidant capacity in juvenile large-mouth bass (*Micropterus salmoides*) can be enhanced (Gu et al., 2022). Sankian et al. (2018) found that 30% yellow meal worm inclusion significantly enhanced lysozyme and glutathione peroxidase activities in mandarin fish (*Siniperca scherzeri*) compared to the control group.

Amino acids (AA) plays array of roles such as protein synthesis, cell signaling, osmoregulation, growth and development, antioxidant, energy substrate, survival and modulation of immunity (Wu, 2010). Emerging evidence from studies shows that many AA regulate immune responses in different fish species (Figure 11.2). Lysine is one the most essential amino acids for fish. The lysine deficiency causes growth retardation in fish. However, there is limited information regarding the role of

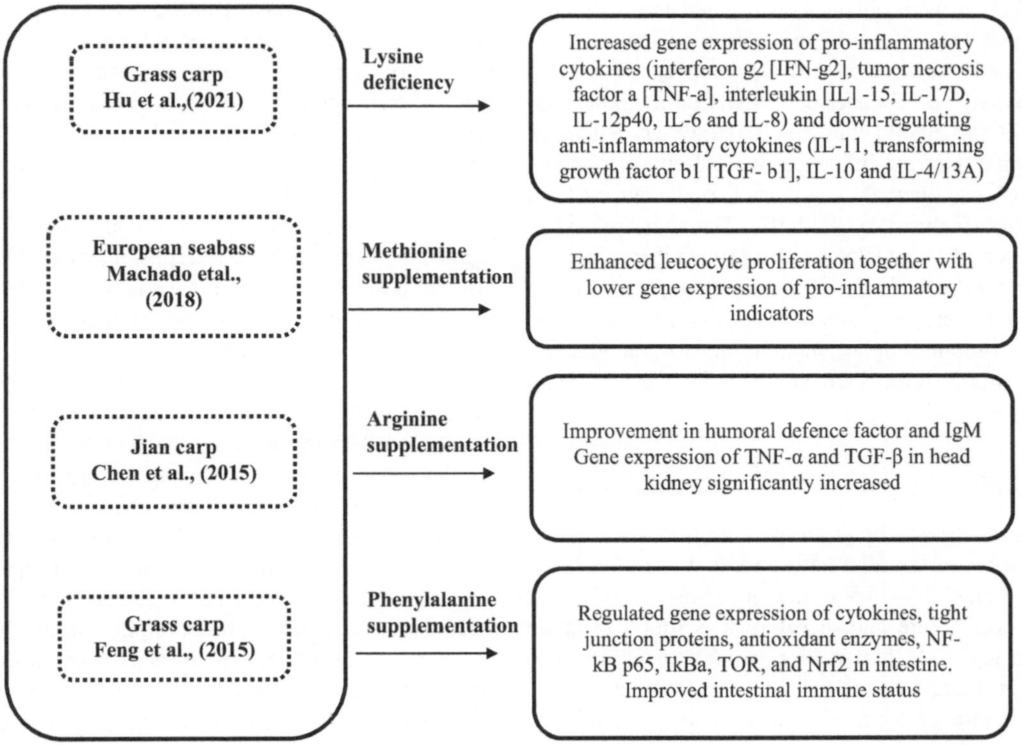

FIGURE 11.2 Effect of amino acids in immunomodulation in fish.

lysine in fish immunity. The study conducted by Hu et al. (2021) reported that lysine deficiency reduced immune response of immune organs in grass carp (*Ctenopharyngodon idella*). Lysine deficiency aggravated the inflammatory response of immune organs in the fish by increasing the gene expressions of pro-inflammatory cytokines (interferon g2 [IFN-g2], tumor necrosis factor a [TNF-a], interleukin [IL]-15, IL-17D, IL-12p40, IL-6 and IL-8) and down-regulating anti-inflammatory cytokines (IL-11, transforming growth factor b1 [TGF-b1], IL-10 and IL-4/13 A). The role of methionine on immune system of European seabass has been studied by Machado et al. (2018) and they found that highest methionine supplementation level enhanced leucocyte proliferation together with lower gene expression of pro-inflammatory indicators. This indicated that increasing methionine as dietary content may improve immune status of fish without triggering an inflammatory response. Kuang et al. (2012) observed an increased leucocytes concentration and survival rate after injection with *A. hydrophila* in juvenile Jian carp (*Cyprinus carpio var. Jian*) fed graded levels of methionine hydroxyl analog, a synthetic methionine source. The essential amino acid arginine has been widely studied for its immunomodulating roles in animal models including fish. Arginine through its metabolic product nitric oxide inactivates enzymes of the mitochondrial respiratory chain cease the energy production for the proliferation of bacteria and also oxidizes bacterial cell membrane directly. Dietary graded level of arginine for nine weeks to Jian carp followed by challenge test injecting *A. hydrophila* influenced head kidney and spleen weights, as well as erythrocyte and leukocyte counts (Chen et al., 2015). There was improvement in humoral defence factor and IgM. Gene expression of TNF-α and TGF-β in head kidney significantly increased with dietary arginine levels up to 21.9 g/kg diet, and decreased thereafter. The highest survival rates following *A. hydrophila* infection were obtained in fish fed the diets containing arginine at 16.1–21.9 g/kg diet. Arginine supplementation is associated with increased survivability of shrimp (*Penaeus monodon*) and channel catfish (*Ictalurus punctatus*) (Chen et al., 1992; Buentello et al., 2000). However, deficiency of dietary arginine does not influence survivability of rainbow trout, turbot (*Psetta maxima*), gilthead seabream and in European seabass (Fournier et al., 2002). The study related to dietary application of phenylalanine also proved that it has immunomodulating effects. Grass carp fed with phenylalanine showed improved intestinal immune status, and regulated gene expression of cytokines, tight junction proteins, antioxidant enzymes, NF-kB p65, IkBa, TOR and Nrf2 in intestine (Feng et al., 2015).

11.3.2 The Role of Lipids in Immunomodulation

Among the macronutrients, lipids are the energy dense nutrient. It is the good source of essential fatty acids and vitamins. Lipid is required in a range of 4–15% in the diet of fish. Lipids have diverse chemical structure and in nature it is found in many forms such as triglyceride, fatty acid, phospholipid, cholesterol, spingolipids, progesterone, testosterone, and vitamin A. Triglyceride is a storage form of lipid whereas phospholipids and cholesterol are the major component in the cell membrane, responsible for normal function of cellular membranes.

The n-3 polyunsaturated fatty acids (PUFA), particularly docosahexaenoic acid (DHA) and eicosapentaenoic acid (EPA) are known for their immunomodulatory functions. These fatty acids are biological regulators and upon dietary supplementation increases IgM level and stimulate immune function. The dietary treatment of DHA elevates immunoglobulins production by B cells and interferon-g (IFN-g) by interlukin-2 (IL-2) in comparison to EPA-based diet (Sugano et al., 2000). Traditionally, fish oil (FO) has been used as the source of long chain PUFA in aquafeed. However, due to limited availability of FO, the use of vegetable oils (VOs) as replacer of FO is increasing in aquafeeds. The incorporation of VO influences the fish immune system and antioxidant status differently in comparison to FO. Nile tilapia (*niloticus*) fed with palm oil (PO) as substitute of FO for eight weeks did not show altered lysozyme activity. This indicated that replacing FO with PO in diets of tilapia did not deplete disease resistance as lysozyme provides a defense line in preventing proliferation and colonization of pathogenic microbes (Larbi Ayisi et al., 2018). However, a contradictory report was presented by Xueqin et al. (2013) that replacement FO with VO had significant effect on serum lysozyme activity in juvenile dark barbel catfish (*Pelteobagrus vachelli*). The possible explanation for the differences in results could be due to the differences in dietary fatty acid composition and the ratio of n-6/n-3 fatty acids of the diets which are known to have effects on immunity and disease resistance in fish (Montero et al., 2008). A study conducted in 2010 by Montero and colleagues reported that replacement of dietary FO by VOs affects humoral immunity and expression of pro-inflammatory cytokines genes in gilthead sea bream. The study concluded that the substitution of FO by VOs containing high levels of n-6 fatty acids may induce imbalances in fish immune response, leading to a lower potential response against infections. The proper balance of EPA and DHA also plays very important role in maintaining immune status and disease resistance in fish. In a recent study, Kumar et al. (2022) found that combinatorial mixture of EPA + DHA at the rate of 0.2 and 0.4% in the diet improved growth performance, anti-oxidative status, immunity of the pangasius (*pangasianodon hypopthalmus*) and reduced the infection against *A. hydrophila*. The replacement of FO also equally affects immune function in shrimp species. In a study when black tiger shrimp (*P. monodon*) was fed, a low FO diet showed the lowest expression levels of inhibitor of apoptosis proteins, Toll and tumor necrosis factor receptor associated factor 6 (TRAF6) in the hepatopancreas (Xie et al., 2022). It is concluded from the study that low dietary FO can impair the immune response of shrimp through the immune deficiency and Toll signaling pathways.

The role of phospholipid (PL) and cholesterol for survival rate, non-specific immunity and antioxidant capacity in fish has been reported in several studies (Zheng et al., 2018; Lin et al., 2020; Hou et al., 2022). Phospholipid, phosphatidyl serine (PS) plays an important role in immunity of shrimp. Leg white shrimp (*Litopenaeus vannamei*) showed elevated exocytosis of pro-phenol oxidase activity, hemocyanin-derived phenol oxidase activity, and antibacterial and bacteriolytic activities when injected with PS (Yang and Pan, 2013). Reports also confirm that dietary treatment of PS can enhance total hemocyte count and phenoloxidase values of hemocyte in leg white shrimp. Dietary inclusion of 3.29% PL increased lysozyme (LZ), acid phosphatase (ACP) activities and complement component 3 (C3) content, up-regulated the mRNA relative expression levels of interleukin 10, transforming growth factor β1 (TGF-β1), inhibitor protein κBα (IκBα), target of rapamycin (TOR) and casein kinase 2 (CK2), and down-regulated tumor necrosis factor α (TNF-α), interleukin 1β, nuclear factor κB p65 (NF-κB p65), IκB kinase β (IKKβ) and IκB kinase γ (IKKγ) mRNA relative expression levels in the grass carp intestine, suggesting that optimum PL could improve fish intestinal immunity (Chen et al., 2015). In another study, inclusion of PL in diet were checked in the aspects of gill health by Feng et al. (2016) and found that PL improved innate immune response through increased activities of lysozyme, complement component 3 (C3) and several antioxidant enzymes. At the same time, expression of TNFα, IL1β, NF-κB p65 decreased. This study concluded the dietary PL is associated with the improvement of the immunity, antioxidant status and tight junction barrier of fish gills. Cholesterol (CH) plays a very significant role in innate as well as adaptive immunity in animals (Aguilar-Ballester et al., 2020). As such, there is no dietary CH requirement in fish but a low level of cholesterol affects immunity and disease resistance and the same has been reported in yellow tail (Maita et al., 2006). Deng et al. (2013) also reported lower respiratory burst activity of macrophages and phagocytic activity of leucocytes in the head kidney of rainbow trout fed with low level of CH. The excess dietary cholesterol has negative health effects both in higher and lower doses. Grass carp fed with lower CH exhibited lower innate and adaptive immune components along with increased skin hemorrhage and lesion morbidity. Excess dietary CH impaired immunity and aggravated inflammation response in fish (Wang et al., 2018). However, at appropriate doses of CH (0.721–0.826%) the ability against skin hemorrhage and lesions, lysozyme activity in the head kidney and the acid phosphatase activity in the spleen of juvenile grass carp was enhanced. Some studies were performed to investigate the effects of the interaction between dietary PL and CH on immune response of crustaceans. The interaction between dietary PL and CH has significant influence on the activities of glutathione peroxidase and lysozyme in the hepatopancrease of juvenile swimming crab (*Portunus trituberculatus*) (Sun et al., 2017). Similarly, Yan et al. (2020) observed that the interaction of PL and CH enhanced the SOD activity, and tolerance against *Vibrio alginolyticus* in with *leg shrimp.* The gut microbiota consisting of healthy microbes directly antagonizes the colonization or over growth of pathogens and enhances fish health by suppression of virulence factors. There is a direct relationship with diet type with gut microbiome and nutrient assimilation in fish (Parata et al., 2020). In this line, Hou et al. (2022) reported that optimal dietary PL (60 g/kg) and CH levels (5 g/kg) could significantly improve the abundance of intestinal microflora, change and optimize the intestinal flora composition and decrease pathogenic bacteria relative abundance in juvenile female crayfish (*Procambarus clarkii*).

11.3.3 The Role of Carbohydrates in Immunomodulation

Fish can survive without carbohydrate as a dietary source (Kamalam et al., 2017). However, optimal carbohydrate helps for the protein sparing and good growth of fish. There are polysaccharides that can play as immunomodulators when fed as dietary components. β-glucans are the most important polysaccharides that take part in immunomodulation because of their higher biological activity, conformational complexity, molecular weight and number of branches. β-glucans triggers innate immune response through pattern recognition receptors (PRRs) including TLRs, Dectin-1 (β-glucan receptor (βGR)), mannose receptor, complement receptors CR3, scavenger receptors and lactosylceramide (Gantner et al., 2003, Herre et al., 2004). There are soluble and insoluble β-glucans and research has shown that insoluble (1,3/1,6) β-glucan has greater biological activity than that of a soluble one. The immunomodulatory β-glucan is a dietary fiber and is commonly obtained from oat (*Avena sativa*), barley (*Hordeum vulgare*), yeast (*Saccharomyces cerevisiae*), bacteria (*Alcaligenes faecalis*), seaweed (*Laminaria digitata*) and mushroom (*Lentinula edodes*) (Mudgil, 2017). Oat and barley β-glucans are linear with Beta (1,4) and Beta (1,3) linkages (Du et al., 2019). Mushrooms β-glucans have short Beta (1,6) linked branches from Beta (1,3) backbone (Han et al., 2020). Yeast β-glucans have Beta (1,6) branches further with additional Beta (1,3) regions. Dietary supplementation of β-glucans has shown improvement in health, growth and general performance of farmed shrimp and fish. The mechanism of action of β-glucan starts when it reaches intestine of fish through dietary means where it comes in contact with apolipoprotein A-IV (apoa4) synthesized by the epithelial enterocyte before being secreted into systemic circulation. There it is recognized by innate immune cell receptors, such as lectin (lectin a, b and c) which with the spleen tyrosine kinase (Syk) generate intracellular signal transduction downstream by the mitogen-activated protein kinase (mapkin2) and nuclear factor kappa B (NF-κB) pathways. The NF-κB signaling pathway has an important role in inflammation, immune response, apoptosis, cell growth and differentiation (Liu et al., 2017). Similarly, the interaction between β-glucan and lactosylceramide receptor stimulates macrophage inflammatory protein (MIP)-2 and also activate NFκB which leads to the enhancement of neutrophil oxidative burst and antimicrobial functions (Meena et al., 2013). The receptor of β-glucan attaches with it and triggers immune responses and activates phagocytosis. In this process, certain cytokines such as IL-1, IL-6, GM-CSF and interferons are generated and antigens are processed. These cytokines are very much helpful for the synthesis of new white blood cells, providing immunity to β-glucan binding receptors present in fish (Rodrigues et al., 2020).

There are several studies conducted in different fish species to check the efficacy of β-glucan either injecting into fish body or through dietary sources. Effect of β-glycan in carp (*Cyprinus carpio*) is depicted in Table 11.1. Rohu (*Labeo rohita*) injected with 10 mg/kg barley β-glucan showed an increased activity in phagocytic activity, leukocytes count, lysozyme activity, complement activity and resistance to *A. hydrphilla* and

TABLE 11.1

Immunomodulatory Effects of β-Glucan in Common Carp (*Cyprinus Carpio*)

Source	Dosage	Effects	References
Injected with β-glucan from *Saccharomyces cerevisiae* along with lipopolysaccharide (LPS)	100 μg β-glucan + 10 μg lipopolysaccharide/fish	Increased activity in superoxide anion and resistance to *A. hydrophilla* infection	Selvaraj et al., 2006.
b-1,3-glucan	25 μg/mL (in vitro study, macrophage stimulation)	Increased oxidative burst Increased lysozyme activity Increased protection against *Aeromonashydrophila* challenge.	Gopalakannan and Arul, 2010
MacroGard®	6 mg/kg feed	Decreased the transcript levels of several pro-inflammatory cytokines in gut and head kidney tissues. Levels of TNFα1, TNFα2, IL-1β, and IL-6 became significantly higher in fish fed β-glucan supplemented diet at 6 h post-infection. Such differential effects may reflect the complex interactions between the bacterium and the immunostimulant relationship with the inflammatory response of the host.	
Macrogard®	1% of feed	Boost the host innate immune defense by inducing neutrophil extracellular trap formation Stabilize neutrophil extracellular traps against bacterial nuclease degradation, and thereby reduce the severity of an infection of *A. hydrophila.*	Brogden et al., 2012
β-glucan preparations of baker's yeast such as zymosan, composed of insoluble β-glucan and mannan, and *MacroGard®*	100 μg/mL (*in vitro* assay)	Stimulate macrophages to increase the production of reactive oxygen and nitrogen radicals. Affect the expression patterns of cytokine genes.	Pietretti et al., 2013
β-glucan	20 mg/mL in in vitro head-kidney cells	Did not enhance wound recovery.	Vera-Jimenez and Nielsen, 2013
0·1, 1 and 2% w/w MacroGard® a β-(1,3)(1,6)-D-glucan derived from the yeast *Saccharomyces cerevisiae,*	0.1; 1.0; 2.0% of feed	Affect the composition of the carp intestinal microbial communities. Positive effects on intestinal microvilli length and density.	Kühlwein et al., 2013
Macrogard®	0.1% in vivo	Significant increase in classical complement activity when compared to fish fed the control diet. β-glucan enriched diet enhances the immune defense parameters of juvenile carp.	Sych et al., 2013
MacroGard	0.1 μg/mL (a stock solution was prepared (0.5 g MacroGard/500 mL Milli-Q water)	β-glucan enriched bath promotes the closure of wounds in common carp and induced a local change in cytokine expression.	Przybylska-Diaz et al., 2013
MacroGard®	1% *MacroGard®*	β-glucan stimulated the kidney derived neutrophil to produce more neutrophil extracellular traps and entrapped a significantly higher percentage of bacteria than the head kidney derived neutrophil extracellular traps.	Brogden et al., 2014
MacroGard®	1% in diet or 10 mg glucan per kg body weight.	Increased antiviral activity of macrophages.	Falco et al., 2014
MacroGard®	6 mg/kg live weight	Induced and enhanced an immune response.	Pionnier et al., 2014
MacroGard®	0% (control), 0.1%, 1%, or 2% MacroGard	Enhance localized intestinal leucocyte infiltration in the anterior intestine of mirror carp.	Kühlwein et al., 2014
MacroGard®	0–1,000 μg incubated for 6, 24, and 48 h (in vitro)	Higher than 500 μg, MacroGard induces to a higher percentage of apoptosis in vitro.	Miest and Hoole 2015
Macrogard®	10 g kg–1 diet	Protection against bacteria.	Jung-Schroers et al., 2018

*MacroGard, commercial β glucan derived from baker's yeast (*Saccharomyces cerevisiae*).

Edwardsiella tarda infections (Misra et al., 2006). Channel catfish injected with 5 and 75 μg/kg body weight of yeast β-glucan showed an increase in antibody concentration, phagocytosis and resistance to *Edwardsiella ictaluri* challenge (Chen and Ainsworth, 1992). Rodríguez et al. (2009) injected β-glucan to *A. hydrophilla* challenged zebrafish and found that the bactericidal activity was increased, whereas there was no influence on TNF-α and IL1β gene expression. Oral administration of β-1,3/1,6-glucan or laminaran to Nile tilapia increased phagocytic activity and index in immunocompromised (IMC) fish due to mercuric chloride exposure. Similarly, oral administration increased oxidative burst and neutrophil adhesion cells in IMC fish fed β-glucan or laminaran. Increased survival after challenge with *A. hydrophila* infection was recorded in normal and IMC fish fed β-glucan (Petit et al., 2016).

11.4 Immunomodulation Through Micronutrients and Antioxidants

Oxidative stress is a very common phenomenon occurs in the living organism with the imbalance between levels of reactive oxygen species (ROS) and their antioxidants. Increased oxidative stress by a dietary deficiency in antioxidant nutrients affects host immune responses, leading to impairment of host resistance to infectious microorganisms (Iddir et al., 2020). Dietary micronutrients are essential elements of the antioxidant network. Therefore, adequate levels of micronutrients are required for a proper immune response against infectious microorganisms, whereas micronutrient deficiency leads to an increased incidence of infection. In general, trace elements do not act directly as antioxidants, but are critical components of the antioxidant enzymes. Dietary antioxidants of substantial importance for immune system functions include vitamin A, vitamin C (ascorbic acid), vitamin E and carotene. In addition, minerals such as selenium and zinc also act in the antioxidant network, primarily by serving as cofactors for enzymes with antioxidant activity.

11.4.1 Role of Vitamin A in Immunomodulation

Vitamin A (VA) is a liposoluble vitamin and it is essential nutrient for growth of fish (Lall and Lewis-McCrea LM, 2007). VA plays a significant role in the immune system (both innate and adaptive immunity), which helps the body fight infections in a variety of ways. High dietary levels of VA enhanced certain immune functions such as kidney leucocyte migration and serum bactericidal activity in Atlantic salmon (Thompson et al., 1994), serum complement, antiprotease activity, leukocyte migration, serum complement activity and phagocytic activity in rainbow trout (Thompson et al., 1995; Amar et al., 2000), respiratory burst activity in gilthead seabream (Cuesta et al., 2002) and serum antibacterial activity in Japanese flounder (Hernandez et al. (2007). Fernández et al. (2015) demonstrated that feeding flat fish (*Senegalese sole)* larvae a diet high in VA (203,000 IU/kg) prior to exposing them to *Photobacterium damselae* challenge enhanced fish survival. They claimed that this impact was brought on by the C1 inhibitor's excessive activation of the C1 complex and a decrease in the amount of iron available for the growth of the bacteria, which was primarily controlled by transferrin and Steap4. Intestinal health which is associated with intestinal immune function is very much important for fish growth. The role of dietary VA on intestinal health was assessed by Zhang et al. (2017) and found deficiency decreased growth of grass carp and immune responses in the posterior intestine (PI), middle intestine (MI) and distal intestine (DI) were mediated partly by NF-κB canonical signaling and p38MAPK signaling pathways. Based on percentage of weight gain, to protect fish against enteritis morbidity and acid phosphatase activity, the optimum dietary VA levels were estimated to be 0·664, 0·707 and 0·722 mg /kg, respectively. Liang et al. (2020) observed that VA deficiency reduced intestinal activities of ACP, alkaline phosphatase, lysozyme, complement 3, complement 4 contents, and activities of digestive enzymes in juvenile hybrid grouper (*Epinephelus fuscoguttatus* ♀ × *Epinephelus lanceolatus* ♂). Along with this, grouper also exhibited down-regulated antimicrobial peptides (β-defensin and Hepcidin), anti-inflammatory cytokines (interleukin 10 and transforming growth factor β1 [not in DI]), tight junction proteins (occluding and claudin3) mRNA levels in the PI, MI and DI, and up-regulated pro-inflammatory cytokines (tumor necrosis factor α [not in MI] and interleukin 1β [not in MI]), signaling molecules c-Rel and p65 due to VA deficiency. The authors concluded that VA deficiency could reduce growth performance because of a negative effect on intestinal health by depressing digestive abilities, intestinal morphology, immunity and tight junction function in the intestine.

11.4.2 Role of Vitamin C in Immunomodulation

Vitamin C (VC) is one of the powerful antioxidant defence systems in fish. Except sturgeons, other fish species lack L-gulonolactone oxidase for which unable to synthesize VC. As a necessary micronutrient, VC must be provided in the feed for growth, health and maintenance. Reduced growth, scoliosis, lordosis, internal and fin haemorrhage, deformed gill filaments, fin erosion, anorexia and increased mortality are the main symptoms of ascorbate deficiency (Verlhac and Gabaudan, 1994; Boglione et al., 2013). Through its antioxidant properties, VC plays as an important part in fish immune response and resistance to infectious diseases. VC influences immunological functions by protecting cells from auto-oxidation, specifically in the case of the initiation of the oxidative burst of macrophages (Carr and Maggini, 2017). Phagocytes, which play a key role in the innate immune response of fish, have large levels of VC in their cytoplasm, which offers good defence against the massive generation of reactive oxygen species during pathogen defence (Wu et al., 2020). The high level of VC is found in thymus, brain, and leukocytes confirm the significance of it in preserving vital tissues from oxidation processes (Omoniyi and Ovie, 2018). Liver and head kidney are important storage organs for VC in fish (Omoniyi and Ovie, 2018). The elevated level found in the head kidney is likely to be related to the presence of lymphopoietic tissues. Rohu fed with VC showed enhanced non-specific immune parameters, such as bactericidal activity, phagocytic ratio and respiratory burst activity, compared to control fish (Tewary and Patra, 2008).

The phagocytic activity is often enhanced by dietary VC due to the antioxidative properties (Leal et al., 2017). A study conducted by Rahimnejad et al. (2021) exhibited that rainbow trout fed with oxidized fish oil (OFO) at peroxide value of 182 meq kg^{-1} with/without VC influenced no effect on the growth and tissue VC concentration, but supplementation of VC at 500 mg kg^{-1} reversed the adverse effects of OFO on non-specific immune response and antioxidant capacity. Vitamin C also plays very important role in stress modulation which in turn helps health and immunity of fish. For instance, a dose of 133.7–251.5 mg/kg VC in the diet of Wuchang bream (*Megalobrama amblycephala*) increased non-specific immune and anti-oxidation capacities, and enhanced resistance against high pH stress and pathogenic bacterial infection (Liu et al., 2016). In a study, dietary VC supplementation at the level of 350 to 700 mg ascorbic acid kg^{-1} diet in pangasius juveniles enhanced fish immunity and stress responses towards *A. hyrdophila* challenge (Daniel et al., 2021). The effects of dietary VC on the growth, and head kidney, spleen and skin immunity, structural integrity and related signaling molecules mRNA expression levels of young grass carp revealed that under VC deficiency and challenging fish with *A. hydrophila* an induction of inflammation in tissues of head kidney, spleen and skin plus an oxidative injury in these tissues were seen due to a disruption in the antioxidant system (Xu et al., 2016). VC also regulated NF-κB, TOR, Nrf2, apoptosis and MLCK signaling (Xu et al., 2016). A study on the role of VC to maintain the immunophysical barriers, and regulation of relative mRNA levels of signaling molecules in grass carp gill depicted that upon deficiency gill of the fish was susceptible to oxidative injury. The fish also showed disturbed tight junctional complexes in the gill, decreased immunity and aggravated inflammation in gills. The VC requirement for against gill rot morbidity of grass carp (264–1,031 g) was estimated to be 156.0 mg/kg diet (Xu et al., 2016). In addition, based on the gill biochemical indices (antioxidant indices MDA, PC and VC contents, and immune indices lysozyme activity and ACP activity) the VC requirements for grass carp at 264–1,031 g were estimated to be in range 110.8–156.6 mg/kg diet (Xu et al., 2016).

11.4.3 Role of Vitamin E in Immunomodulation

Vitamin E (VE) is the lipophilic compound plays an important role as antioxidant. It protects the macromolecules such as fatty acids, nucleic acids and membrane phospholipids from singlet oxygen produced during oxidation process. Dietary VE is known to improve growth, helps in reproductive performance, prevention of muscle degeneration, meat life and improvement of immune response in fish (El-Sayed and Izquierdo, 2022). Generally, the dietary VE requirement reported for fish species such as Rohu (Sau et al., 2004), mrigal, (*Cirrhinus mrigala;* Paul et al., 2004), Caspian trout (*Salmo caspius;* Saheli et al., 2021) hybrid snakehead (*Channa argus* × *Channa maculate*; Zhao et al., 2018) and Japanese eel (*Anguilla japonica*; Shahkar et al., 2018), varies from 6 to 200 mg kg^{-1} α-tocopherol. Galaz et al., (2010) conducted a study to find out an optimum dose of VE for improved immune function in parrot fish and reported that a mega dose of VE 500 mg/kg was required for the improvement of nonspecific immune response based on the results in NBT and MPO activities and disease resistance against *V. anguillarum* challenge. However, studies suggest that effect of smaller dose of VE is capable to trigger immune response and ensure health status in fish. Pan et al. (2017) demonstrated that VE deficiency in grass carp caused oxidative damage, decreased SOD, glutathione peroxidase (GPx), catalase (CAT) and glutathione reductase (GR) activities, and down-regulated the mRNA levels of antioxidant enzymes and signaling molecules Nrf2 significantly. Deficiency of VE also can induce apoptosis by up-regulating capase–2, –3, –7 and –8 mRNA levels in grass carp tissues including head kidney, spleen and skin. The authors concluded that, based on immune indicator (LA activity in head kidney) and antioxidant indicator (protection of spleen against MDA), the optimal VE requirements for young grass carp were estimated to be 123.8 and 136.4 mg/kg diet, respectively. Zhou et al. (2013) reported that juvenile cobia (*Rachycentron canadum)* fed diets supplemented with 61.2 and 124 mg kg^{-1} VE showed higher immunoglobulin concentration, lysozyme and SOD activity than control fish. The dietary VE requirement of juvenile cobia was established based on second-order polynomial regression of weight gain and lysozyme to be 78 or 111 mg all-rac-α-tocopheryl acetate kg^{-1} diet, respectively. The role of VE/VC ratio on fish growth and health status has also been assessed in several studies and the results exhibited a synergetic effect on fish health and growth. The supplementation of VC has been shown to spare VE (El-Sayed and Izquierdo, 2022). A study conducted by Li et al. (2016) on role of dietary linolenic acid (LN), VE and VC in regulating fish growth and immune response in juvenile dark barbel catfish reported that dietary linolenic acid and VC combination improved the growth of the fish. However, the higher levels of dietary VE addition enhanced immune responses and the resistance to *A. hydrophila* challenge. VE and selenium (Se) also exhibits a positive interaction and even can spare metabolic requirement of each other. In a study, dietary Nano Se or Nano Se and VE significantly increased the antioxidant enzyme (GPX, SOD, CAT) activity, NBT, lysozyme and phagocytosis values and a decrease in MDA value. Liver and spleen TNF-α and IL-1β expressions were significantly up-regulated in tilapia fed on Nano Se or Nano Se and VE (Dawood et al., 2020). The authors concluded that Nano Se or/and VE can be used effectively in tilapia diets for improving the growth, oxidative status, immune-related gene expression and health condition of intestine and blood indices.

11.4.4 Role of Carotenoids in Immunomodulation

Carotenoids are lipid soluble pigments exhibit strong antioxidant property. These pigments are available in nature in a huge quantity and also they are of different kinds. Carotenoids influence the colour, survivability, growth, egg quality and provide disease resistance to fish (Nakano and Wiegertjes, 2020). Carotenoids protect the cells against oxidative injury and ensure optimal cellular functions including apoptosis, cell signaling and gene regulation (Chew and Park, 2004). Carotenoids are very effective in reducing the effect of ROS by quenching the singlet oxygen species and protect the cell membranes. Carotenoids maintain the balance between intracellular and extracellular milieu and ensure the immunity of the cells (Chew and Park, 2004). In aquafeeds the carotenoids such as astaxanthin is provided at 50 to 100 mg/kg for muscle pigmentation. It is semi essential for fish and promotes tremendous benefits in terms of growth and health at relatively lower inclusion level (Nakano and Wiegertjes, 2020). Larvae of Japanese parrotfish and spotted parrotfish (*O. punctatus*) that received b-carotene-enriched rotifers indicated better proliferation of lymphocytes and improved survival (Tachibana et al., 1997). Activities of lysozyme, complement and phagocytes and non-specific cytotoxicity were elevated upon b-carotene and astaxanthin supplementation in rainbow trout (Amar et al., 2001). In aquaculture, fish often comes across stress conditions due to sudden change in environmental factors. In this line a study conducted by Cheng et al. (2018) to investigate the effects of astaxanthin on growth performance, biochemical parameters, ROS production and immune-related gene expressions of the pufferfish (*Takifugu obscurus*) under high temperature stress found that the basal diet supplemented with 80–320 mg kg^{-1} astaxanthin could enhance growth, nonspecific immune responses and antioxidant defense system and improve resistance against high-temperature stress. In another study Liu et al. (2019) to find out the effect of carotenoid supplement in yellow catfish under high-temperature stress reported that when yellow catfish fed 150–200 mg·kg^{-1} carotenoids exhibited significantly an increase in values of serum lysozyme activity, total protein (TP), thyroxin (T4), immunoglobulin (Ig), complement 3 (C3), complement 4 (C4), liver total antioxidant capacity (TAC), CAT, SOD, liver hepcidin, lysozyme G (LysG) and heat shock protein 70 (HSP70) gene expression but decreased serum aminotransferase (ALT) and liver malondialdehyde (MDA) levels prior to stress exposure compared to control fish. The study recommended supplementing a basal diet with 150–200 mg·kg^{-1} carotenoids can enhance immune responses to disease condition and imparts resistance to high-temperature stress in yellow catfish.

11.4.5 Role of Zinc in Immunomodulation

Zinc is an important inorganic trace element required by animals including fish. It influences various immunophysiological functions including growth performance, immune system and defense mechanism against free ions and radicals (Watanabe et al., 1997). Zinc may improve the antioxidant status and reduce the lipid peroxidation level in fish (Anderson et al., 2001). Zinc has been demonstrated to be immunomodulatory for both cellular and humoral parameters, including inflammatory responses (transcriptome profiling) and complement C3-1 protein production (Sanchez-Dardon et al., 1999; Hogstrand et al., 2002). Zn is a very potent mineral to fight against stress condition caused heavy metals, temperature fluctuations and microbial infection. A study conducted by Kumar et al. (2017) found that zinc at 10 and 20 mg/kg feed restored the biochemical and immunological parameters against concurrent exposure to lead (Pb), temperature and pathogenic infection in common name here, *P. hypophthalmus*. The immunomodulating effect of zinc oxide nanoparticles (ZnO NPs) was studied in Nile tilapia and found that fish received 30 and 60 μg/g ZnO NPs showed a better immune response in terms of phagocytic activity, serum antibacterial activity, lysozymes, respiratory burst activity, gene expression of immunoglobin M-2, TNF-α, interleukin (IL)-1β, heat shock proteins, IL-10, insulin growth factor 1 (IGF 1), transforming growth factor-β2, SOD and CAT genes. Sherif et al. (2022) provided ZnO NPs *in* Nile tilapia feed and observed an improvement in both fish immune response and disease resistance against *A. hydrophila*. The effects of dietary supplementation of different zinc sources on growth performance, anti-oxidative status, immune response, and ammonia stress resistance of marbled spinefoot rabbitfish (*Siganus rivulatus*) revealed that the tolerance of fish against an ammonia challenge was increased with Zn-methionine and ZnO-NPs supplementation. Simultaneously, the innate immune response (lysozyme and phenoloxidase activities) significantly improved with organic and nanoparticles zinc forms (Sallam et al., 2020).

11.4.6 Role of Selenium in Immunomodulation

Selenium (Se) is known as antioxidant mineral. It is part of glutathione peroxidase and thioredoxine reductase, which degrade the ROS, protect the cells and play an important role in maintaining normal immune response (Zhang et al., 2020). Selenium is provided in fish feed and other animal feed in two forms: organic selenium (Se-glycinate, Se-proteinate, selenomethionine, seleno yeast) and inorganic selenium (sodium selenite, selenite) (Wang et al., 2022). The organic Se is known to be more effective than inorganic as it is assimilated more effectively. The dietary selenium yeast increased CAT and SOD activities and total antioxidant status as well as thiobarbituric reactive substances in liver homogenate in meagre (*Argyrosomus regius*) in a dose-dependent manner compared to control (Mansour et al., 2017). Mansour et al. (2017) also exhibited an increase in hematological and innate immune parameters (immunoglobulin, lysozyme, myeloperoxidase, ACH50 and respiratory burst activity) in fish fed the Se-yeast supplemented diets compared to control group and concluded that supplementation of 3.98 mg Se-yeast kg^{-1} diet can improve growth performance, antioxidant balance and innate immune status of meagre juveniles. In addition, Se modulates the immune functions such as inflammation and virulence development by regulating HSPs and TLRs in liver of many animals including various fish species (Rayman, 2000; Gao et al., 2019). Dietary Se deficiency increased the TNF-α, IL-1β and IL-6 productions and decreased TGF-β by HSP60 augment in carp (Gao et al., 2019). TNF-α and IL-1β are pro inflammatory cytokines and regulates host immune system (Kany et al., 2019), and an increase in these two factors indicates high inflammatory response which can cause liver damage. It is clearly understood that dietary Se controls the production of TNF-α and IL-1β and effectively protects cells from damage. Dietary Se deficiency induced the TLR2-MAPKs pathway active with HSP60 in carp and this clearly indicates the capacity of dietary Se to produce inflammatory response. Seleno-protein requires Se for synthesis as it is involved in the array of biological functions such as cell maintenance, thyroid hormone metabolism, calcium homeostasis and immune response (Castets et al., 2011). The seleno-proteins such as glutathione peroxidase and thioredoxine reductase play a crucial role in controlling of ROS. There are other seleno-proteins such as seleno-protein K (SelK) and S (SelS) are related to inflammation and immunity (Gao et al., 2004; Verma et al., 2011). In a study supplementation of diets with 4 mg Se Kg^{-1} diet using Sel-Plex remarkably improved fish response to viral pathogen-associated molecular pattern (PAMP) stimulation. Sel-Plex, being a highly bioavailable supplement of organic Se, is a very potent additive in fish feed for improving fish fitness and resistance against immune challenges (Pacitti et al., 2016).

11.5 Immunomodulation Through Hormones

The endocrine system, central nervous system and immune system are very closely connected with each other. Now it is very well established that hormonal signals from the endocrine system influence the immune systems with differential effects based on stress and environmental changes (Yada and Nakanishi, 2002). Immune cells can communicate by either cytokine or neurotransmitter release to signal the endocrine, gastrointestinal, and nervous systems. There is a bidirectional connection in how the endocrine and immune system exchange signals and molecules to bring normal homeostasis. There are hormones such as cortisol, growth hormones, steroid hormones, prolactin and reproductive hormones which are immunomodulatory in nature.

11.5.1 The Role of Cortisol in Immunomodulation

It is well known that the stress condition negatively influences the immune function of the cell and causes disease condition. When fish come across with any stressor first that is perceived by the central nervous system and many hormones release in to circulation as an indicator of stress. Cortisol has proven to be a reliable indicator of stress and is considered the major stress hormone (Sadoul B, and Geffroy, 2019). Cortisol is a glucocorticoids (GC) hormone that is produced by an adrenal gland. Cortisol has a proven role as immunosuppressive agentmediated through hypothalamo-pituitary-interrenal axis. It influences the immune cells in vertebrates through induction of apoptosis, change of differentiation patterns, inhibition of cytokine release and inhibition of immunocyte migration (Baschant and Tuckermann, 2010). The immune competence is seriously affected when high level of cortisol is secreted. Studies also show when cortisol is administered it reduces the resistance of fish to fungal and bacterial pathogens (Wojtaszek et al., 2002). The glucocorticoids are known to inhibit NF-kB pathway and the synthesis of IL-12 and IFN so the production of pro inflammatory cytokines decreases and this how they modulate the immune system (Sternberg, 2006; Borghetti et al., 2009). Cortisol treatment had a negative effect on leukocyte number in circulation was observed in juvenile coho salmon (*Oncorhynchus kisutch*), but number of these cells increased in the thymus and head kidney (Maule and Schreck, 1990). Similarly, intraperitoneal injection of cortisol (200 mg/kg) modified the percentages of circulating blood leukocytes subtypes in carp (Wojtaszek et al., 2002). The leukocytes carry the glucocorticoid receptors and this is how the effects of cortisol are mediated through this channel. A study conducted by Cortés et al. (2013) demonstrated the effect of doses of cortisol corresponding to acute and chronic levels on the complement and lysozyme activity in plasma of the rainbow trout and the results showed that there is a relationship between the complement and lysozyme activity in plasma and mRNA levels in liver, supporting the important role of this organ in producing these immune system proteins after a rise of cortisol in the fish plasma.

11.5.2 The Role of Growth Hormone in Immunomodulation

Growth hormone (GH) is produced from the lymphoid organs such as thymus, spleen and peripheral blood and they stimulate proliferation of leucocytes (Jeay et al., 2002). GH helps in immunoglobulin (IgG and IgM) synthesis and stimulates production of T and B cells. Non-specific immune defence mechanism such as cytotoxic, phagocytic, haemolytic and lysozyme activities are also activated by the administration of GH (Franz et al., 2016). The role of GH as an immunomodulatory cytokine is also well known (Harris and Bird, 2000). It promotes cell cycle progression of lymphoid cells and prevents apoptosis. The removal of pituitary (hypophysectomy) decreases fish immune function has been reported in several fish species indicates the importance of hypophyseal hormones in the regulation of the fish immune system (Yada and Nakanishi, 2002). Administration of carp pituitary extract which is the source of GH into a characin fish (*Astyanax mexicanus*) increased the number of leucocytes in the head kidney and the spleen (Yada, 2007). The effect of bovine GH as heterologous GH also increased somatic growth and antibody production in Mozambique tilapia (*Oreochromis mossambicus*) and rainbow trout due to the endocrine effect (Leedom et al., 2002; Biga et al., 2005). Some studies also suggest GH increases the phagocytic activity with increased activity of superoxide anion and this how pathogens are controlled (Yada et al., 2001). The increased lysozyme activity and haematolytic activity due to the administration of GH has also been noticed in some fish species (Sakai et al., 1996; Yada et al., 2001). GH plays an important role in the maintenance of basal antibody production and it has been proved when GH was administered to hypophysectomized fish, circulating Ig were restored to the levels in the intact fish (Yada et al., 2001). In rainbow trout, administration of GH enhanced leukocytes, phagocytes and lysozyme activity (Kajita et al., 1992; Sakai et al., 1996; Yada et al., 2001). Growth hormones stimulate the secretion of IGF1 that plays important roles in cell proliferation, differentiation and functional modulation of the immune system. The GH and IGF-I stimulated the proliferation of gilthead sea bream head kidney leukocytes *in vitro* Calduch-Giner et al., 1995). Yada (2007) reviewed GH and fish immune system and reported that GH gene is expressed in many extra-pituitary tissues of fish, especially in lymphoid organs and cells and immune system of fish is regulated by paracrine manner by GH and IGF-I.

11.5.3 The Role of Sex Hormone in Immunomodulation

Monosex fish population is preferred for aquaculture so that uncontrolled reproduction can be checked. In this way the energy for reproduction is diverted for higher growth performance of fish. There are several ways to produce monosex fish species however, hormonal treatment is known to very effective for the production of monosex fish. There are several hormones such as 19-norethyltestosterone, fluoxymesterone, ethyltestosterone and 17 α-methyltestosterone (MT), dihydrotestosterone (DHT), androstenedione, trenbolone acetate, mesterolone, testosterone and 17α-ethynyltestosterone and 17α-methyldihydrotestosterone are used for all male fish production (Abo-Al-Ela et al., 2017). These hormones are provided through oral route by feed. The exposures of sex hormone in early developmental stages influence the development of immune organs in several fish species. Sayed and Moneeb (2015) reported that the monosex Nile tilapia cultivated using methyltestosterone (MT) showed low levels of RBCs and lymphocytes. MT perturbs the hormonal system of living beings, and increases the risk of various diseases. In a study when Nile tilapia fry were fed with MT exhibited disturbed antioxidant enzyme activity and gene transcription (Zheng et al., 2016). In rainbow trout, 11-Ketotestosterone (11KT) decreased the number and capability of IgM-secreting cells of spleen, head kidney, blood and skin (Hou et al., 1999). Buchtíková et al. (2011) also found negative effects in terms of depressed innate immune responses and respiratory bursts in common carp due to 11KT. The phagocytosis process is also affected by MT treatment and that has been reported in carp both *in vitro* and *in vivo* conditions when treated with progesterone or 11-KT injections (Yamaguchi et al., 2001). Estrogens are also used in aquaculture for the production of YY super male in eels. There are also many reports which claim the effect of estrogen on immunity of fish (Burgos-Aceves et al., 2016). The treatment of 17β-estradiol enhanced inflammation by increasing the production of a pro-inflammatory cytokine, the interleukin-1β (IL1β), in head kidney phagocytes in gilthead seabream (Chaves-Pozo et al., 2003). Japanese medaka exposed to estrogen showed down regulated complement components, lysozyme and ceruloplasmin, while il-21, ifn, novel immune-type receptor-18 and Ikaros were up-regulated (Sun et al., 2011).

11.6 Conclusion

The interplay between nutrition and immune system is well established since time immemorial. In the case of fish, it is more crucial because fishes have different feeding habits and diverse immune components. It is reflected in the patterns of their response to different stimuli such as stress and diseases. However, the influence of dietary nutrients and hormone on fish immunity has not been studied in detail. So far the research undertaken elucidates the effect of nutrients on some immune indicators such as lysozyme, IgM, pro-inflammatory cytokines and antioxidant enzymes. This relationship becomes even more complex with different source and combination of nutrients. With the advancement of omics techniques such as micro-array, RNA-seq and proteomics, the effect of individual components on the overall physiology and immune system will become transparent. These approaches will redefine the nutrient requirement studies, which will include immune response to different stressors or disease that otherwise are so far studied mostly for growth performance. Recent studies on the effect of nutrients on gut microbiome and transcriptome is gaining momentum and more studies are needed to develop a feed specifically formulated to combat a specific stress and disease.

REFERENCES

Abdel-Latif, H. M., Abdel-Tawwab, M., Khalil, R. H., Metwally, A. A., Shakweer, M. S., Ghetas, H. A., & Khallaf, M. A. (2021). Black soldier fly (*Hermetia illucens*) larvae meal in diets of European seabass: Effects on antioxidative capacity, non-specific immunity, transcriptomic responses, and resistance to the challenge with *Vibrio alginolyticus*. *Fish & Shellfish Immunology*, *111*, 111–118.

Abo-Al-Ela, H. G., El-Nahas, A. F., Mahmoud, S., & Ibrahim, E. M. (2017). The extent to which immunity, apoptosis and detoxification gene expression interact with 17 alpha-methyltestosterone. *Fish & Shellfish Immunology*, *60*, 289–298.

Adom, K.K., & Liu, R.H. (2002). Antioxidant activity of grains. *Journal of Agricultural and Food Chemistry*, *50*, 6182–6187

Aguilar-Ballester, M., Herrero-Cervera, A., Vinué, Á., Martínez-Hervás, S., & González-Navarro, H. (2020). Impact of cholesterol metabolism in immune cell function and atherosclerosis. *Nutrients*, *12*(7), 2021.

Amar, E. C., Kiron, V., Satoh, S., & Watanabe, T. (2001). Influence of various dietary synthetic carotenoids on bio-defence mechanisms in rainbow trout, *Oncorhynchus mykiss* (Walbaum). *Aquaculture Research*, *32*, 162–173.

Amar, E. C., Kiron, V., Satoh, S., Okamoto, N., & Watanabe, T. (2000). Effects of dietary βcarotene on the immune response of rainbow trout *Oncorhynchus mykiss*. *Fisheries Science*, *66*(6), 1068–1075.

Anderson, R. A., Roussel, A. M., Zouari, N., Mahjoub, S., Matheau, J. M., & Kerkeni, A. (2001). Potential antioxidant effects of zinc and chromium supplementation in people with type 2 diabetes mellitus. *Journal of the American College of Nutrition*, *20*(3), 212–218.

Baschant, U., & Tuckermann, J. (2010). The role of the glucocorticoid receptor in inflammation and immunity. *The Journal of Steroid Biochemistry and Molecular Biology*, *120*(2–3), 69–75.

Biga, P. R., Peterson, B. C., Schelling, G. T., Hardy, R. W., Cain, K. D., Overturf, K., & Ott, T. L. (2005). Bovine growth hormone treatment increased IGF-I in circulation and induced the production of a specific immune response in rainbow trout (*Oncorhynchus mykiss*). *Aquaculture*, *246*(1–4), 437–445.

Boglione, C., Gisbert, E., Gavaia, P., Witten, P. E., Moren, M., Fontagné, S., & Koumoundouros, G. (2013). Skeletal anomalies in reared European fish larvae and juveniles. Part 2: main typologies, occurrences and causative factors. *Reviews in Aquaculture*, *5*, S121–S167. 10.1111/raq.12016

Booman, M., Forster, I., Vederas, J. C., Groman, D. B., & Jones, S. R.M. (2018). Soybean meal-induced enteritis in Atlantic salmon (*Salmo salar*) and Chinook salmon (*Oncorhynchus tshawytscha*) but not in pink salmon (*O. gorbuscha*), *Aquaculture*, *483*, 2018, 238–243.

Borghetti, P., Saleri, R., Mocchegiani, E., Corradi, A., & Martelli, P. (2009). Infection, immunity and the neuroendocrine response. *Veterinary Immunology and Immunopathology*, *130*(3–4), 141–162.

Brogden, G., Krimmling, T., Adamek, M., Naim, H. Y., Steinhagen, D., & von Köckritz-Blickwede, M. (2014). The effect of β-glucan on formation and functionality of neutrophil extracellular traps in carp (*Cyprinus carpio L.*). *Developmental & Comparative Immunology*, *44*(2), 280–285.

Brogden, G., von Köckritz-Blickwede, M., Adamek, M., Reuner, F., Jung-Schroers, V., Naim, H. Y., & Steinhagen, D. (2012). β-Glucan protects neutrophil extracellular traps against degradation by *Aeromonas hydrophila* in carp (*Cyprinus carpio*). *Fish & Shellfish Immunology*, *33*(4), 1060–1064.

Buchtíková, S., Šimková, A., Rohlenová, K., Flajšhans, M., Lojek, A., Lilius, E. M., & Hyršl, P. (2011). The seasonal changes in innate immunity of the common carp (*Cyprinus carpio*). *Aquaculture*, *318*(1–2), 169–175.

Buentello, J. A., & Gatlin III, D. M. (2000). The dietary arginine requirement of channel catfish (*Ictalurus punctatus*) is influenced by endogenous synthesis of arginine from glutamic acid. *Aquaculture*, *188*(3–4), 311–321.

Burgos-Aceves, M. A., Cohen, A., Smith, Y., & Faggio, C. (2016). Estrogen regulation of gene expression in the teleost fish immune system. *Fish & Shellfish Immunology*, *58*, 42–49.

Calduch-Giner, J. A., Sitjà-Bobadilla, A., Alvarez-Pellitero, P., & Pérez-Sánchez, J. (1995). Evidence for a direct action of GH on haemopoietic cells of a marine fish, the gilthead sea bream (*Sparus aurata*). *Journal of Endocrinology*, *146*(3), 459–467.

Carr, A. C., & Maggini, S. (2017). Vitamin C and immune function. *Nutrients*, *9*(11), 1211.

Castets, P., Bertrand, A. T., Beuvin, M., Ferry, A., Le Grand, F., Castets, M., Chazot, G., Rederstorff, M., Krol, A., Lescure, A., Romero, N. B., Guicheney, P., & Allamand, V. (2011). Satellite cell loss and impaired muscle regeneration in selenoprotein N deficiency. *Human Molecular Genetics*, *20*(4), 694–704.

Chaves-Pozo, E., Pelegrín, P., Mulero, V., Meseguer, J., & García Ayala, A. (2003). A role for acidophilic granulocytes in the testis of the gilthead seabream (*Sparus aurata L., Teleostei*). *Journal of Endocrinology*, *179*(2), 165–174.

Chen, D., & Ainsworth, A. J. (1992). Glucan administration potentiates immune defence mechanisms of channel catfish, *Ictalurus punctatus Rafinesque*. *Journal of Fish Diseases*, *15*(4), 295–304.

Chen, G., Liu, Y., Jiang, J., Jiang, W., Kuang, S., Tang, L., Zhang, Y. A., Zhou, X., & Feng, L. (2015). Effect of dietary arginine on the immune response and gene expression in head kidney and spleen following infection of Jian carp with *Aeromonas hydrophila*. *Fish & Shellfish Immunology*, *44*(1), 195–202.

Chen, H. Y., Leu, Y. T., & Roelants, I. (1992). Quantification of arginine requirements of juvenile marine shrimp, *Penaeus monodon*, using microencapsulated arginine. *Marine Biology*, *114*(2), 229–233.

Chen, Y. P., Jiang, W. D., Liu, Y., Jiang, J., Wu, P., Zhao, J., Kuang, X.Y., Tang, L., Tang, W. N., Zhang, Y. A., Zhou, X. Q., & Feng, L. (2015). Exogenous phospholipids supplementation improves growth and modulates immune response and physical barrier referring to NF-κB, TOR, MLCK and Nrf2 signaling factors in the intestine of juvenile grass carp (*Ctenopharyngodon idella*). *Fish & Shellfish Immunology*, *47*(1), 46–62.

Cheng, C. H., Guo, Z. X., Ye, C. X., & Wang, A. L. (2018). Effect of dietary astaxanthin on the growth performance, non-specific immunity, and antioxidant capacity of pufferfish (*Takifugu obscurus*) under high temperature stress. *Fish physiology and Biochemistry*, *44*(1), 209–218.

Chew, B. P., & Park, J. S. (2004). Carotenoid action on the immune response. *The Journal of Nutrition*, *134*(1), 257S–261S. 10.1093/jn/134.1.257S

Cortés, R., Teles, M., Trídico, R., Acerete, L., & Tort, L. (2013). Effects of cortisol administered through slow-release implants on innate immune responses in rainbow trout (*Oncorhynchus mykiss*). *International Journal of Genomics*, *2013*.

Cuesta, A., Ortuño, J., Rodriguez, A., Esteban, M. A., & Meseguer, J. (2002). Changes in some innate defence parameters of seabream (*Sparus aurata L.*) induced by retinol acetate. *Fish & Shellfish Immunology*, *13*(4), 279–291.

Daniel, N., Muralidhar, A. P., Srivastava, P. P., Jain, K. K., Prasad, K. P., & Ranjan, A. (2021). Effect of vitamin C on immune and stress responses in striped catfish, *Pangasianodon hypophthalmus* juveniles under pre-and post-challenge with *Aeromonas hydrophila*. *Aquaculture Research*, *52*(12), 6444–6452.

Dawood, M. A., Zommara, M., Eweedah, N. M., Helal, A. I., & Aboel-Darag, M. A. (2020). The potential role of nano-selenium and vitamin C on the performances of Nile tilapia (*Oreochromis niloticus*). *Environmental Science and Pollution Research*, *27*(9), 9843–9852.

Deng, J., Kang, B., Tao, L., Rong, H., & Zhang, X. (2013). Effects of dietary cholesterol on antioxidant capacity, non-specific immune response, and resistance to Aeromonas hydrophila in rainbow trout (*Oncorhynchus mykiss*) fed soybean meal-based diets. *Fish & Shellfish Immunology*, *34*(1), 324–331.

Ding, Z., Zhang, Y., Ye, J., Du, Z., & Kong, Y. (2015). An evaluation of replacing fish meal with fermented soybean meal in the diet of *Macrobrachium nipponense*: Growth, nonspecific immunity, and resistance to *Aeromonas hydrophila*. *Fish & Shellfish Immunology*, *44*(1), 295–301.

Dossou, S., Koshio, S., Ishikawa, M., Yokoyama, S., Dawood, M. A., El Basuini, M. F., El Hais, A. M., & Olivier, A. (2018). Effect of partial replacement of fish meal by fermented rapeseed meal on growth, immune response and oxidative condition of red sea bream juvenile, *Pagrus major*. *Aquaculture*, *490*, 228–235.

Du, B., Meenu, M., Liu, H., & Xu, B. (2019). A Concise Review on the Molecular Structure and Function Relationship of β-Glucan. *International Journal of Molecular Sciences*, *20*(16), 4032. 10.3390/ijms20164032

El-Sayed, A. F. M., & Izquierdo, M. (2022). The importance of vitamin E for farmed fish—A review. *Reviews in Aquaculture*, *14*(2), 688–703.

Falco, A., Miest, J. J., Pionnier, N., Pietretti, D., Forlenza, M., Wiegertjes, G. F., & Hoole, D. (2014). Β-Glucan-supplemented diets increase poly (I: C)-induced gene expression of Mx, possibly via Tlr3-mediated recognition mechanism in common carp (*Cyprinus carpio*). *Fish & Shellfish Immunology*, *36*(2), 494–502.

Feng, L., Chen, Y. P., Jiang, W. D., Liu, Y., Jiang, J., Wu, P.,… … & Zhou, X. Q. (2016). Modulation of immune response, physical barrier and related signaling factors in the gills of juvenile grass carp (*Ctenopharyngodon idella*) fed supplemented diet with phospholipids. *Fish & Shellfish Immunology*, *48*, 79–93.

Feng, L., Li, W., Liu, Y., Jiang, W. D., Kuang, S. Y., Jiang, J., Tang, L., Wu, P., Tang, W. N., Zhang, Y.A., & Zhou, X. Q. (2015). Dietary phenylalanine-improved intestinal barrier health in young grass carp (*Ctenopharyngodon idella*) is associated with increased immune status and regulated gene expression of cytokines, tight junction proteins, antioxidant enzymes and related signalling molecules. *Fish & Shellfish Immunology*, *45*(2), 495–509.

Fernández, I., López-Joven, C., Andree, K. B., Roque, A., & Gisbert, E. (2015). Vitamin A supplementation enhances Senegalese sole (*Solea senegalensis*) early juvenile's immunocompetence: new insights on potential underlying pathways. *Fish & Shellfish Immunology*, *46*(2), 703–709.

Fournier, V., Gouillou-Coustans, M. F., Metailler, R., Vachot, C., Guedes, M. J., Tulli, F. Olivia-Teles, A., Tibaldit, E., & Kaushik, S. J. (2002). Protein and arginine requirements for maintenance and nitrogen gain in four teleosts. *British Journal of Nutrition*, *87*(5), 459–469.

Franz, A. C., Faass, O., Köllner, B., Shved, N., Link, K., Casanova, A., Wenger, M., D'Cotta, H., Baroiller, J. F., Ullrich, O., Reinecke, M., & Eppler, E. (2016). Endocrine and local IGF-I in the bony fish immune system. *Biology*, *5*(1), 9.

Galaz, G. B., Kim, S. S., & Lee, K. J. (2010). Effects of different dietary vitamin E levels on growth performance, non-specific immune responses, and disease resistance against Vibrio anguillarum in parrot fish (*Oplegnathus fasciatus*). *Asian-Australasian Journal of Animal Sciences*, *23*(7), 916–923.

Gantner, B. N., Simmons, R. M., Canavera, S. J., Akira, S., & Underhill, D. M. (2003). Collaborative induction of inflammatory responses by dectin-1 and Toll-like receptor 2. *The Journal of Experimental Medicine*, *197*(9), 1107–1117.

Gao, X. J., Tang, B., Liang, H. H., Yi, L., & Wei, Z. G. (2019). Selenium deficiency induced an inflammatory response by the HSP60-TLR2-MAPKs signalling pathway in the liver of carp. *Fish & Shellfish Immunology*, *87*, 688–694.

Gao, Y., Feng, H. C., Walder, K., Bolton, K., Sunderland, T., Bishara, N., Quick, M., Kantham, L., & Collier, G. R. (2004). Regulation of the selenoprotein SelS by glucose deprivation and endoplasmic reticulum stress–SelS is a novel glucose-regulated protein. *FEBS Letters*, *563*(1—3), 185–190.

Gopalakannan, A., & Arul, V. (2010). Enhancement of the innate immune system and disease-resistant activity in *Cyprinus carpio* by oral administration of β-glucan and whole cell yeast. *Aquaculture Research*, *41*(6), 884–892.

Gu, J., Liang, H., Ge, X., Xia, D., Pan, L., Mi, H., & Ren, M. (2022). A study of the potential effect of yellow mealworm (*Tenebrio molitor*) substitution for fish meal on growth, immune and antioxidant capacity in juvenile largemouth bass (*Micropterus salmoides*). *Fish & Shellfish Immunology*, *120*, 214–221.

Han, B., Baruah, K., Cox, E., Vanrompay, D., & Bossier, P. (2020). Structure-Functional Activity Relationship of β-Glucans From the Perspective of Immunomodulation: A Mini-Review. *Frontiers in Immunology*, *11*, 658. 10.3389/fimmu.2020.00658

Harris, J., & Bird, D. J. (2000). Modulation of the fish immune system by hormones. *Veterinary Immunology and Immunopathology*, *77*, 163–176.

Hernandez, L. H. H., Teshima, S. I., Koshio, S., Ishikawa, M., Tanaka, Y., & Alam, M. S. (2007). Effects of vitamin A on growth, serum anti-bacterial activity and transaminase activities in the juvenile Japanese flounder, *Paralichthys olivaceus*. *Aquaculture*, *262*(2–4), 444–450.

Herre, J., Gordon, S., & Brown, G. D. (2004). Dectin-1 and its role in the recognition of β-glucans by macrophages. *Molecular Immunology*, *40*(12), 869–876.

Hogstrand, C., Balesaria, S., & Glover, C. N. (2002). Application of genomics and proteomics for study of the integrated response to zinc exposure in a non-model fish species, the rainbow trout. *Comparative Biochemistry and Physiology Part B: Biochemistry and Molecular Biology*, *133*(4), 523–535.

Holm, H. J., Wadsworth, S., Bjelland, A. K., Krasnov, A., Evensen, Ø., & Skugor, S. (2016). Dietary phytochemicals modulate skin gene expression profiles and result in reduced lice counts after experimental infection in Atlantic salmon. *Parasites & Vectors*, *9*(1), 1–14.

Hou, S., Li, J., Huang, J., & Cheng, Y. (2022). Effects of dietary phospholipid and cholesterol levels on antioxidant capacity, nonspecial immune response and intestinal microflora of juvenile female crayfish, *Procambarus clarkii*. *Aquaculture Reports*, *25*, 101245.

Hou, Y., Suzuki, Y., & Aida, K. (1999). Effects of steroids on the antibody producing activity of lymphocytes in rainbow trout. *Fisheries Science*, *65*(6), 850–855.

Hu, Y., Feng, L., Jiang, W., Wu, P., Liu, Y., Kuang, S., Tang, L., & Zhou, X. (2021). Lysine deficiency impaired growth performance and immune response and aggravated inflammatory response of the skin, spleen and head kidney in grown-up grass carp (*Ctenopharyngodon idella*). *Animal Nutrition*, *7*(2), 556–568.

Iddir, M., Brito, A., Dingeo, G., Fernandez Del Campo, S. S., Samouda, H., La Frano, M. R., & Bohn, T. (2020). Strengthening the immune system and reducing inflammation and oxidative stress through diet and nutrition: considerations during the COVID-19 crisis. *Nutrients*, *12*(6), 1562. 10.3390/nu12061562

Jalili, R., Tukmechi, A., Agh, N., Noori, F., & Ghasemi, A. (2013). Replacement of dietary fish meal with plant sources in rainbow trout (*Oncorhynchus mykiss*) effect on growth performance, immune responses, blood indices and disease resistance. *Iranian Journal of Fisheries Science*, *12* (3), 577–591

Jeay, S., Sonenshein, G. E., Postel-Vinay, M. C., Kelly, P. A., & Baixeras, E. (2002). Growth hormone can act as a cytokine controlling survival and proliferation of immune cells: new insights into signaling pathways. *Molecular and Cellular Endocrinology*, *188*(1–2), 1–7.

Jung-Schroers, V., Adamek, M., Harris, S., Syakuri, H., Jung, A., Irnazarow, I., & Steinhagen, D. (2018). Response of the intestinal mucosal barrier of carp (*Cyprinus carpio*) to a bacterial challenge by *Aeromonas hydrophila* intubation after feeding with β-1, 3/1, 6-glucan. *Journal of Fish Diseases*, *41*(7), 1077–1092.

Kajita, Y., Sakai, M., Kobayashi, M., & Kawauchi, H. (1992). Enhancement of non-specific cytotoxic activity of leucocytes in rainbow trout *Oncorhynchus mykiss* injected with growth hormone. *Fish & Shellfish Immunology*, *2*(2), 155–157.

Kamalam, B.S., Medale, F., & Panserat, S. (2017). Utilisation of dietary carbohydrates in farmed fishes: New insights on influencing factors, biological limitations and future strategies. *Aquaculture*, *467*, 3–27. 10.1016/j.aquaculture.2016.02.007

Kany, S., Vollrath, J. T., & Relja, B. (2019). Cytokines in Inflammatory Disease. *International Journal of Molecular Sciences*, *20*(23), 6008. 10.3390/ijms20236008

Kim, S.-S., Galaz, G.-B., Pham, M.-A., Jang, J.-W., Oh, D.-H., Yeo, I.-K., Lee, K.-J., (2009). Effects of dietary supplementation of a meju, fermented soybean meal, and *Aspergillus oryzae* for juvenile parrot fish (*Oplegnathus fasciatus*). *Asian-Australasian Journal of Animal Sciences*, *22*, 849–856.

Kiron, V. (2012). Fish immune system and its nutritional modulation for preventive health care. *Animal Feed Science and Technology*, *173*(1–2), 111–133.

Kuang, S. Y., Xiao, W. W., Feng, L., Liu, Y., Jiang, J., Jiang, W. D., Hu, H., Li, S. H., Tang, L., & Zhou, X. Q. (2012). Effects of graded levels of dietary methionine hydroxy analogue on immune response and antioxidant status of immune organs in juvenile Jian carp (*Cyprinus carpio var. Jian*). *Fish & Shellfish Immunology*, *32*(5), 629–636.

Kühlwein, H., Emery, M. J., Rawling, M. D., Harper, G. M., Merrifield, D. L., & Davies, S. J. (2013). Effects of a dietary β-(1, 3)(1, 6)-D-glucan supplementation on intestinal microbial communities and intestinal ultrastructure of mirror carp (*Cyprinus carpio L.*). *Journal of Applied Microbiology*, *115*(5), 1091–1106.

Kühlwein, H., Merrifield, D. L., Rawling, M. D., Foey, A. D., & Davies, S. J. (2014). Effects of dietary β-(1, 3)(1, 6)-D-glucan supplementation on growth performance, intestinal morphology and haemato-immunological profile of mirror carp (*Cyprinus carpio L.*). *Journal of Animal Physiology and Animal Nutrition*, *98*(2), 279–289.

Kumar, N., Chandan, N. K., Gupta, S. K., Bhushan, S., & Patole, P. B. (2022). Omega-3 fatty acids effectively modulate growth performance, immune response, and disease resistance in fish against multiple stresses. *Aquaculture*, *547*, 737506.

Kumar, N., Krishnani, K. K., Kumar, P., Jha, A. K., Gupta, S. K., & Singh, N. P. (2017). Dietary zinc promotes immunobiochemical plasticity and protects fish against multiple stresses. *Fish & Shellfish Immunology*, *62*, 184–194.

Lall, S. P., & Lewis-McCrea, L. M. (2007). Role of nutrients in skeletal metabolism and pathology in fish—an overview. *Aquaculture*, *267*(1–4), 3–19.

Larbi Ayisi, C., Zhao, J., & Wu, J. W. (2018). Replacement of fish oil with palm oil: Effects on growth performance, innate immune response, antioxidant capacity and disease resistance in Nile tilapia (*Oreochromis niloticus*). *PLoS One*, *13*(4), e0196100.

Lavilla-Pitogo, C. R., & Amar, E. C. (2010). Nutritional diseases. In G. D. Lio-Po & Y. Inui (Eds.), *Health Management in Aquaculture* (2nd ed., pp. 157–169). Tigbauan, Iloilo, Philippines: Aquaculture Department, Southeast Asian Fisheries Development Center

Leal, E., Zarza, C., & Tafalla, C. (2017). Effect of vitamin C on innate immune responses of rainbow trout (*Oncorhynchus mykiss*) leukocytes. *Fish & Shellfish Immunology*, *67*, 179–188.

Leedom, T. A., Uchida, K., Yada, T., Richman III, N. H., Byatt, J. C., Collier, R. J., Hirano, T., & Grau, E. G. (2002). Recombinant bovine growth hormone treatment of tilapia: growth response, metabolic clearance, receptor binding and immunoglobulin production. *Aquaculture*, *207*(3–4), 359–380.

Li, M., Chen, L., Qin, J., Yu, N., Chen, Y., Ding, Z., & Li, E. (2016). Growth, immune response and resistance to *Aeromonas hydrophila* of darkbarbel catfish *Pelteobagrus vachelli* fed diets with different linolenic acids, vitamins C and E levels. *Aquaculture Nutrition*, *22*: 664–674. 10.1111/anu.12287

Li, P., Yin, Y. L., Li, D., Kim, S. W., & Wu, G. (2007). Amino acids and immune function. *British Journal of Nutrition*, *98*(2), 237–252.

Liang, D., Yang, Q., Tan, B., Dong, X., Chi, S., Liu, H., & Zhang, S. (2020). Dietary vitamin A deficiency reduces growth performance, immune function of intestine, and alters tight junction proteins of intestine for juvenile hybrid grouper (*Epinephelus fuscoguttatus*♀× *Epinephelus lanceolatus*♂). *Fish & Shellfish Immunology*, *107*, 346–356.

Lin, Z., Han, F., Lu, J., Guo, J., Qi, C., Wang, C., Xiao, S., Bu, X., Wang, X., Qin, J., & Chen, L. (2020). Influence of dietary phospholipid on growth performance, body composition, antioxidant capacity and lipid metabolism of Chinese mitten crab, *Eriocheir sinensis*. *Aquaculture*, *516*, 734653.

Liu, B., Wan, J., Ge, X., Xiea, J., Zhou, Q., Miao, L., Ren, M., & Pan, L. (2016). Effects of dietary Vitamin C on the physiological responses and disease resistance to pH stress and *Aeromonas hydrophila* infection of *Megalobrama amblycephala*. *Turkish Journal of Fisheries and Aquatic Sciences*, *16*(2), 421–433.

Liu, F., Qu, Y. K., Wang, A. M., Yu, Y. B., Yang, W. P., Lv, F., & Nie, Q. (2019). Effects of carotenoids on the growth performance, biochemical parameters, immune responses and disease resistance of yellow catfish (*Pelteobagrus fulvidraco*) under high-temperature stress. *Aquaculture*, *503*, 293–303.

Liu, T., Zhang, L., Joo, D., & Sun, S. C. (2017). NF-κB signaling in inflammation. *Signal transduction and targeted therapy*, *2*, 17023–. 10.1038/sigtrans.2017.23

Machado, M., Azeredo, R., Fontinha, F., Fernández-Boo, S., Conceição, L. E., Dias, J., & Costas, B. (2018). Dietary methionine improves the European seabass (*Dicentrarchus labrax*) immune status, inflammatory response, and disease resistance. *Frontiers in Immunology*, *9*, 2672.

Machuca, C., Méndez-Martínez, Y., Reyes-Becerril, M., & Angulo, C. (2022). Yeast β-Glucans as Fish Immunomodulators: A Review. *Animals*, *12*(16), 2154.

Maita, M., Maekawa, J., Satoh, K. I., Futami, K., & Satoh, S. (2006). Disease resistance and hypocholesterolemia in yellowtail *Seriola quinqueradiata* fed a non-fishmeal diet. *Fisheries Science*, *72*(3), 513–519.

Mansour, A. T. E., Goda, A. A., Omar, E. A., Khalil, H. S., & Esteban, M. Á. (2017). Dietary supplementation of organic selenium improves growth, survival, antioxidant and immune status of meagre, *Argyrosomus regius*, juveniles. *Fish & Shellfish Immunology*, *68*, 516–524.

Martínez-Álvarez, R.M., Morales, A.E., & Sanz, A., (2005). Antioxidant defenses in fish: Biotic and abiotic factors. *Reviews in fish biology and fisheries*, *15*, 75–88.

Maule, A. G., & Schreck, C. B. (1990). Changes in numbers of leukocytes in immune organs of juvenile coho salmon after acute stress or cortisol treatment. *Journal of Aquatic Animal Health*, *2*(4), 298–304.

Mazumder, A., Dwivedi, A., Plessis, J. Du. (2016). Sinigrin and its therapeutic benefits. *Molecules 21*, 1–11

Meena, D. K., Das, P., Kumar, S., Mandal, S. C., Prusty, A. K., Singh, S. K.,… … & Mukherjee, S. C. (2013). Beta-glucan: an ideal immunostimulant in aquaculture (a review). *Fish Physiology and Biochemistry*, *39*(3), 431–457.

Miest, J. J., & Hoole, D. (2015). Time and concentration dependency of MacroGard® induced apoptosis. *Fish & Shellfish Immunology*, *42*(2), 363–366.

Misra, C. K., Das, B. K., Mukherjee, S. C., & Pattnaik, P. (2006). Effect of long term administration of dietary β-glucan on immunity, growth and survival of *Labeo rohita* fingerlings. *Aquaculture*, *255*(1–4), 82–94.

Mohan, K., Rajan, D. K., Muralisankar, T., Ganesan, A. R., Sathishkumar, P., & Revathi, N. (2022). Use of black soldier fly (*Hermetia illucens L.*) larvae meal in aquafeeds for a sustainable aquaculture industry: A review of past and future needs. *Aquaculture*, 738095.

Montero, D., Grasso, V., Izquierdo, M. S., Ganga, R., Real, F., Tort, L., Cabellaro, M.J., & Acosta, F. (2008). Total substitution of fish oil by vegetable oils in gilthead sea bream (*Sparus aurata*) diets: effects on hepatic Mx expression and some immune parameters. *Fish & Shellfish Immunology*, *24*(2), 147–155.

Mudgil, D. (2017). The interaction between insoluble and soluble fiber. In *Dietary Fiber for the Prevention of Cardiovascular Disease* (pp. 35–59). Academic Press.

Nakano, T., & Wiegertjes, G. (2020). Properties of Carotenoids in Fish Fitness: A Review. *Marine Drugs*, *18*(11), 568. 10.3390/md18110568

Omoniyi, A. D., & Ovie, I. A. (2018). Vitamin C: An important nutritional factor in fish diets. *Journal of Agriculture and Ecology Research International*, *16*, 1–7.

Pacitti, D., Lawan, M. M., Feldmann, J., Sweetman, J., Wang, T., Martin, S. A. M., & Secombes, C. J. (2016). Impact of selenium supplementation on fish antiviral responses: a whole transcriptomic analysis in rainbow trout (*Oncorhynchus mykiss*) fed supranutritional levels of Sel-Plex®. *BMC Genomics*, *17*(1), 1–26.

Pan, J. H., Feng, L., Jiang, W. D., Wu, P., Kuang, S. Y., Tang, L.,… … & Liu, Y. (2017). Vitamin E deficiency depressed fish growth, disease resistance, and the immunity and structural integrity of immune organs in grass carp (*Ctenopharyngodon idella*): Referring to NF-κB, TOR and Nrf2 signaling. *Fish & Shellfish Immunology*, *60*, 219–236.

Parata, L., Mazumder, D., Sammut, J., & Egan, S. (2020). Diet type influences the gut microbiome and nutrient assimilation of Genetically Improved Farmed Tilapia (*Oreochromis niloticus*). *PLoS One*, *15*(8), e0237775.

Paul, B. N., Sarkar, S., & Mohanty, S. N. (2004). Dietary vitamin E requirement of mrigal, *Cirrhinus mrigala* fry. *Aquaculture*, *242*(1–4), 529–536.

Petit, J., & Wiegertjes, G. F. (2016). Long-lived effects of administering β-glucans: Indications for trained immunity in fish. *Developmental & Comparative Immunology*, *64*, 93–102.

Pham, M. A., & Lee, K. J. (2007). Effects of dietary cheongkukjang on liver superoxide dismutase Activity of Parrotfish *Oplegnathus fasciatus*. *Journal of Aquaculture*, *20*(2), 132–139.

Pietretti, D., Vera-Jimenez, N. I., Hoole, D., & Wiegertjes, G. F. (2013). Oxidative burst and nitric oxide responses in carp macrophages induced by zymosan, MacroGard® and selective dectin-1 agonists suggest recognition by multiple pattern recognition receptors. *Fish & Shellfish Immunology*, *35*(3), 847–857.

Pionnier, N., Falco, A., Miest, J. J., Shrive, A. K., & Hoole, D. (2014). Feeding common carp *Cyprinus carpio* with β-glucan supplemented diet stimulates C-reactive protein and complement immune acute phase responses following PAMPs injection. *Fish & Shellfish Immunology*, *39*(2), 285–295.

Przybylska-Diaz, D. A., Schmidt, J. G., Vera-Jimenez, N. I., Steinhagen, D., & Nielsen, M. E. (2013). β-glucan enriched bath directly stimulates the wound healing process in common carp (*Cyprinus carpio L.*). *Fish & Shellfish Immunology*, *35*(3), 998–1006.

Rahimnejad, S., Dabrowski, K., Izquierdo, M., Hematyar, N., Imentai, A., Steinbach, C., & Policar, T. (2021). Effects of Vitamin C and E Supplementation on Growth, Fatty Acid Composition, Innate Immunity, and Antioxidant Capacity of Rainbow Trout (*Oncorhynchus mykiss*) Fed Oxidized Fish Oil. *Frontiers in Marine Science*, *1471*.

Rayman, M. P. (2000). The importance of selenium to human health. *The Lancet*, *356*(9225), 233–241.

Reddy, P. G., & Frey, R. A. (1990). Nutritional modulation of immunity in domestic food animals. *Advances in Veterinary Science and Comparative Medicine*, *35*, 255–281.

Rodrigues, M. V., Zanuzzo, F. S., Koch, J. F. A., de Oliveira, C. A. F., Sima, P., & Vetvicka, V. (2020). Development of fish immunity and the role of β-glucan in immune responses. *Molecules*, *25*(22), 5378.

Rodríguez, I., Chamorro, R., Novoa, B., & Figueras, A. (2009). β-Glucan administration enhances disease resistance and some innate immune responses in zebrafish (*Danio rerio*). *Fish & Shellfish Immunology*, *27*(2), 369–373.

Sadoul, B., & Geffroy, B. (2019). Measuring cortisol, the major stress hormone in fishes. *Journal of Fish Biology*. 2019 Apr; *94*(4), 540–555. doi: 10.1111/jfb.13904. Epub 2019 Mar 8. PMID: 30667059.

Saheli, M., Islami, H. R., Mohseni, M., & Soltani, M. (2021). Effects of dietary vitamin E on growth performance, body composition, antioxidant capacity, and some immune responses in Caspian trout (*Salmo caspius*). *Aquaculture Reports*, *21*, 100857.

Sakai, M., Kobayashi, M., & Kawauchi, H. (1996). In vitro activation of fish phagocytic cells by GH, prolactin and somatolactin. *Journal of Endocrinology*, *151*(1), 113–118.

Sallam, A. E., Mansour, A. T., Alsaqufi, A. S., Salem, M. E. S., & El-Feky, M. M. (2020). Growth performance, anti-oxidative status, innate immunity, and ammonia stress resistance of *Siganus rivulatus* fed diet supplemented with zinc and zinc nanoparticles. *Aquaculture Reports*, *18*, 100410.

Sanchez-Dardon, J., Voccia, I., Hontela, A., Chilmonczyk, S., Dunier, M., Boermans, H., Blakley, B., & Fournier, M. (1999). Immunomodulation by heavy metals tested individually or in mixtures in rainbow trout (*Oncorhynchus mykiss*) exposed in vivo. *Environmental Toxicology and Chemistry: An International Journal*, *18*(7), 1492–1497.

Sankian, Z., Khosravi, S., Kim, Y. O., & Lee, S. M. (2018). Effects of dietary inclusion of yellow mealworm (*Tenebrio molitor*) meal on growth performance, feed utilization, body composition, plasma biochemical indices, selected immune parameters and antioxidant enzyme activities of mandarin fish (*Siniperca scherzeri*) juveniles. *Aquaculture, 496*, 79–87.

Sau, S. K., Paul, B. N., Mohanta, K. N., & Mohanty, S. N. (2004). Dietary vitamin E requirement, fish performance and carcass composition of rohu (*Labeo rohita*) fry. *Aquaculture, 240*(1–4), 359–368.

Sayed, A. E. D. H., & Moneeb, R. H. (2015). Hematological and biochemical characters of monosex tilapia (*Oreochromis niloticus, Linnaeus, 1758*) cultivated using methyltestosterone. *The Journal of Basic & Applied Zoology, 72*, 36–42.

Selvaraj, V., Sampath, K., & Sekar, V. (2006). Adjuvant and immunostimulatory effects of β-glucan administration in combination with lipopolysaccharide enhances survival and some immune parameters in carp challenged *with Aeromonas hydrophila. Veterinary Immunology and Immunopathology, 114*(1–2), 15–24.

Shahkar, E., Hamidoghli, A., Yun, H., Kim, D. J., & Bai, S. C. (2018). Effects of dietary vitamin E on hematology, tissue α-tocopherol concentration and non-specific immune responses of Japanese eel, *Anguilla japonica. Aquaculture, 484*, 51–57.

Sherif, A. H., Abdelsalam, M., Ali, N. G., & Mahrous, K. F. (2022). Zinc Oxide Nanoparticles Boost the Immune Responses in *Oreochromis niloticus* and Improve Disease Resistance to *Aeromonas hydrophila* Infection. *Biological Trace Element Research*, 1–10.

Sitjà-Bobadilla, A., Peña-Llopis, S., Gómez-Requeni, P., Médale, F., Kaushik, S., & Pérez-Sánchez, J. (2005). Effect of fish meal replacement by plant protein sources on non-specific defence mechanisms and oxidative stress in gilthead sea bream (*Sparus aurata*). *Aquaculture, 249*(1–4), 387–400.

Sternberg, E. M. (2006). Neural regulation of innate immunity: a coordinated nonspecific host response to pathogens. *Nature Reviews Immunology, 6*(4), 318–328.

Sugano, M., Koga, T., & Yamada, K. (2000). Lipids and immunology. *Asia Pacific Journal of Clinical Nutrition, 9*(2), 146–152.

Sun, L., Shao, X., Hu, X., Chi, J., Jin, Y., Ye, W., & Fu, Z. (2011). Transcriptional responses in Japanese medaka (*Oryzias latipes*) exposed to binary mixtures of an estrogen and anti-estrogens. *Aquatic Toxicology, 105*(3–4), 629–639.

Sun, P., Jin, M., Ding, L., Lu, Y., Yuan, Y., Ma, H., & Zhoua, Q. (2017). Effect of dietary soybean lecithin and cholesterol on growth, antioxidant status and fatty acid composition of juvenile swimming crab, *Portunus trituberculatus. The Israeli Journal of Aquaculture –Bamidgeh*, IJA_69.2017.1421.

Sych, G., Frost, P., & Irnazarow, I. (2013). Influence of β-Glucan (MACROGARD®) on innate immunity of carp fry. *Journal of Veterinary Research, 57*(2), 219–223.

Tachibana, K., Yagi, M., Hara, K., Mishima, T., & Tsuchimoto, M. (1997). Effects of feeding of β-carotene-supplemented rotifers on survival and lymphocyte proliferation reaction of fish larvae (Japanese parrotfish (*Oplegnathus fasciatus*) and Spotted parrotfish (*Oplegnathus punctatus*): preliminary trials. *Hydrobiologia, 358*(1), 313–316.

Tewary, A., & Patra, B. C. (2008). Use of vitamin C as an immunostimulant. Effect on growth, nutritional quality, and immune response of *Labeo rohita (Ham.). Fish Physiology and Biochemistry, 34*(3), 251–259.

Thompson, I., Choubert, G., Houlihan, D. F., & Secombes, C. J. (1995). The effect of dietary vitamin A and astaxanthin on the immunocompetence of rainbow trout. *Aquaculture, 133*(2), 91–102.

Thompson, I., Fletcher, T. C., Houlihan, D. F., & Secombes, C. J. (1994). The effect of dietary vitamin A on the immunocompetence of Atlantic salmon (*Salmo salar* L.). *Fish Physiology and Biochemistry, 12*(6), 513–523.

Vera-Jimenez, N. I., Pietretti, D., Wiegertjes, G. F., & Nielsen, M. E. (2013). Comparative study of β-glucan induced respiratory burst measured by nitroblue tetrazolium assay and real-time luminol-enhanced chemiluminescence assay in common carp (*Cyprinus carpio* L.). *Fish & Shellfish Immunology, 34*(5), 1216–1222.

Verlhac, V., & Gabaudan, J. (1994). Influence of vitamin C on the immune system of salmonids. *Aquaculture Research, 25*(1), 21–36.

Verma, S., Hoffmann, F. W., Kumar, M., Huang, Z., Roe, K., Nguyen-Wu, E., Hashimoto, A. S., & Hoffmann, P. R. (2011). Selenoprotein K knockout mice exhibit deficient calcium flux in immune cells and impaired immune responses. *The Journal of Immunology, 186*(4), 2127–2137.

Wang, L., Sagada, G., Wang, R., Li, P., Xu, B., Zhang, C., Qiao, J., & Yan, Y. (2021). Different forms of selenium supplementation in fish feed: The bioavailability, nutritional functions, and potential toxicity. *Aquaculture*, 737819.

Wang, X. Z., Jiang, W. D., Feng, L., Wu, P., Liu, Y., Zeng, Y. Y., Jiang, J., Kuang, S. Y., Tang, L., Tang, W. N., & Zhou, X. Q. (2018). Low or excess levels of dietary cholesterol impaired immunity and aggravated inflammation response in young grass carp (*Ctenopharyngodon idella*). *Fish & Shellfish Immunology, 78*, 202–221.

Wang, Y., Yu, Y., Duan, Y., Wang, Q., Cong, X., He, Y., Gao, C., Hafeez, M., Jan, S., Rasheed, S. M., Cheng, S., & Wang, Z. (2022). Enhancing the activity of carboxymethyl cellulase enzyme using highly stable selenium nanoparticles biosynthesized by *Bacillus paralicheniformis* Y4. *Molecules, 27*(14), 4585.

Watanabe, T., Kiron, V., & Satoh, S. (1997). Trace minerals in fish nutrition. *Aquaculture, 151*(1–4), 185–207.

Wojtaszek, J., Dziewulska-Szwajkowska, D., Łozińska-Gabska, M., Adamowicz, A., & Dżugaj, A. (2002). Hematological effects of high dose of cortisol on the carp (*Cyprinus carpio L.*): cortisol effect on the carp blood. *General and Comparative Endocrinology, 125*(2), 176–183.

Wu, G. (2010). Functional amino acids in growth, reproduction, and health. *Advances in Nutrition, 1*(1), 31–37.

Wu, L., Qin, Z., Liu, H., Lin, L., Ye, J., & Li, J. (2020). Recent Advances on Phagocytic B Cells in Teleost Fish. *Frontiers in Immunology, 11*, 824. 10.3389/fimmu.2020.00824

Xiao, X., Jin, P., Zheng, L., Cai, M., Yu, Z., Yu, J., & Zhang, J. (2018). Effects of black soldier fly (*Hermetia illucens*) larvae meal protein as a fishmeal replacement on the growth and immune index of yellow catfish (*Pelteobagrus fulvidraco*). *Aquaculture Research, 49*(4), 1569–1577.

Xie, S., Wei, D., Liu, Y., Tian, L., & Niu, J. (2022). Dietary fish oil levels modulated lipid metabolism, immune response, intestinal health and salinity stress resistance of juvenile *Penaeus monodon* fed a low fish-meal diet. *Animal Feed Science and Technology, 289*, 115321.

Xu, H. J., Jiang, W. D., Feng, L., Liu, Y., Wu, P., Jiang, J., Kuang, S. Y., Tang, L., Tang, W. N., Zhang, Y. A., & Zhou, X. Q.

(2016). Dietary vitamin C deficiency depresses the growth, head kidney and spleen immunity and structural integrity by regulating NF-κB, TOR, Nrf2, apoptosis and MLCK signaling in young grass carp (*Ctenopharyngodon idella*). *Fish & Shellfish Immunology*, *52*, 111–138.

Xueqin, J., Liqiao, C., Jianguang, Q., Chuanjie, Q., Erchao, L., & Haibo, J. (2013). Effects of dietary soybean oil inclusion to replace fish oil on growth, muscle fatty acid composition, and immune responses of juvenile dark barbel catfish, *Pelteobagrus vachelli*. *African Journal of Agricultural Research*, *8*(16), 1492–1499.

Yada, T. (2007). Growth hormone and fish immune system. *General and Comparative Endocrinology*, *152*(2-3), 353–358.

Yada, T., & Nakanishi, T. (2002). Interaction between endocrine and immune systems in fish. *International Review of Cytology*, *220*, 35–92. doi: 10.1016/s0074-7696(02)20003-0. PMID: 12224552.

Yada, T., Azuma, T., & Takagi, Y. (2001). Stimulation of non-specific immune functions in seawater-acclimated rainbow trout, *Oncorhynchus mykiss,* with reference to the role of growth hormone. *Comparative Biochemistry and Physiology Part B: Biochemistry and Molecular Biology*, *129*(2-3), 695–701.

Yamaguchi, T. & Watanuki, H. & Sakai, M. (2001). Effects of estradiol, progesterone and testosterone on the function of carp, *Cyprinus carpio,* phagocytes in vitro. Comparative biochemistry and physiology. *Toxicology & Pharmacology: CBP*. *129*, 49–55. 10.1016/S1532-0456(01)00176-4.

Yan, M., Wang, W., Huang, X., Wang, X., & Wang, Y. (2020). Interactive effects of dietary cholesterol and phospholipids on the growth performance, expression of immune-related genes and resistance against *Vibrio alginolyticus* in white shrimp (*Litopenaeus vannamei*). *Fish & Shellfish Immunology*, *97*, 100–107.

Yang, L., & Pan, L. (2013). Effects of phosphatidyl serine on immune response in the shrimp *Litopenaeus vannamei*. *Central European Journal of Biology*, *8*(11), 1135–1144.

Zhang, L., Feng, L., Jiang, W. D., Liu, Y., Wu, P., Kuang, S. Y., Tang, L., Tang, W. N., Zhang, Y. A., & Zhou, X. Q. (2017). Vitamin A deficiency suppresses fish immune function with differences in different intestinal segments: the role of transcriptional factor NF-κB and p38 mitogen-activated protein kinase signalling pathways. *British Journal of Nutrition*, *117*(1), 67–82.

Zhang, Y., Roh, Y. J., Han, S. J., Park, I., Lee, H. M., Ok, Y. S., Lee, B. C., & Lee, S. R. (2020). Role of Selenoproteins in Redox Regulation of Signaling and the Antioxidant System: A Review. *Antioxidants (Basel, Switzerland)*, *9*(5), 383. 10.3390/antiox9050383

Zhao, H., Ma, H. J., Gao, S. N., Chen, X. R., Chen, Y. J., Zhao, P. F., & Lin, S. M. (2018). Evaluation of dietary vitamin E supplementation on growth performance and antioxidant status in hybrid snakehead (*Channa argus*× *Channa maculata*). *Aquaculture Nutrition*, *24*(1), 625–632.

Zheng, P., Wang, J., Han, T., Yang, M., Li, X., & Wang, C. (2018). Effect of dietary cholesterol levels on growth performance, body composition and gene expression of juvenile mud crab *Scylla paramamosain*. *Aquaculture Research*, *49*(10), 3434–3441.

Zheng, Y., Qu, J., Qiu, L., Fan, L., Meng, S., Song, C., Bing, X., & Chen, J. (2016). Effect of 17α-methyltestosterone (MT) on oxidation stress in the liver of juvenile GIFT tilapia, Oreochromis niloticus. *Springerplus*, *5*(1), 1–8.

Zhou, Q. C., Wang, L. G., Wang, H. L., Wang, T., Elmada, C. Z., & Xie, F. J. (2013). Dietary vitamin E could improve growth performance, lipid peroxidation and non-specific immune responses for juvenile cobia (*Rachycentron canadum*). *Aquaculture Nutrition*, *19*(3), 421–429.

12

Cytokines and Fish Health

Aifa Fathima[1], Yaser Arafath[1], Saqib Hassan[2,3], George Seghal Kiran[4], and Joseph Selvin[1]
[1]*Department of Microbiology, School of Life Sciences, Pondicherry University, Puducherry, India*
[2]*Department of Biotechnology, School of Bio and Chemical Engineering, Sathyabhama Institute of Science and Technology, Chennai, Tamilnadu, India*
[3]*FutureLeaders Mentoring Fellow, American Society for Microbiology, Washington, USA*
[4]*Department of Food Science and Technology, School of Life Sciences, Pondicherry University, Puducherry, India*

12.1 Introduction

Humans rely heavily on fish as a source for animal protein. The most essential source of high-quality protein for humans, accounting for 16% of global animal protein consumption. Fisheries contribute significantly to food security and livelihood, as well as in economic and social development in developing countries (FAO, 2016; Assefa and Abunna, 2018). Aquaculture sector has gained a lot of attention, and it is expanding quickly due to the advancement in aquaculture techniques (Kubečka et al., 2016). New approaches and increased demand for animal protein (fish) are the primary drivers for growth of aquaculture industries. This has led to the development in culture methods, and these methods have become more intensive in order to produce higher yields with a better quality (Rico et al., 2012). Aquaculture is experiencing a massive loss of production due to a variety of factors and disease outbreaks by many pathogens is one of the most serious constraints, causing damage to farmers by affecting their livelihood and also leads to food insecurity. To reduce losses in aquaculture owing to infectious diseases, scientifically validated and recommended as well as locally applicable techniques must be employed to control such diseases outbreaks (Assefa and Abunna, 2018). Previously, most of the diseases were controlled by using antibacterial drugs. But antibacterial medications should not be used to treat fish infections because it leads to the formation of superbugs (multi-drug resistant microbes) and also affects the food quality for consumption e.g., drug residuals (Sakai et al., 2021).

Higher vertebrate immune systems possess both innate and acquired immunity. The recognition of foreign antigens by humoral immune cells is referred to as acquired immunity. Antigen stimulation activates these antigen receptors, which have increased affinity for a particular antigen. The development of acquired immunity takes days to develope, against a bacterial infection, for example, is not always initiated (Nakanishi et al., 2018). An innate immune response, on the other hand, has a crucial role at the begining stages of the immune system, which is achieved by phagocytes and humoral factors (Sakai et al., 2021). In the prevention of infectious diseases in fish, innate immunity is thought to be important as much as the acquired immunity. During immune response the cytokines are secreted in larger amounts. These molecules cause signal transmission between cells and minor physiological changes take place through particular receptors that are present on specific cells (Zou and Secombes, 2016). Cytokines are important components of innate immune response. This chapter discusses fish cytokines and their role in fish health and discusses the present gaps.

A fish is a multi-shaped organism; their body is streamlined in shape and the external layer of the body is covered with scales where the fins helps in the swimming (Firdaus-Nawi and Zamri-Saad, 2016). However, this is not always the case, as some fish species, such as clingfish, lack scales while others, like eels, lack fins. Furthermore, some species, such as the lungfish, can survive outside water for several hours. Fish are classified as poikilothermic chordates with gills as the primary respiratory organ, according to Berra (2001). The entire body is protected by scales and/or mucus. Fish are members of a paraphyletic category that includes all gill-bearing aquatic animals without digits on limbs. As a result, hagfish, lamprey, cartilaginous and bony fish, as well as several extinct associated groups, fall in this category. Fish body temperature follows the ambient temperature because they are cold-blooded animals, with the exception of some huge fish like tuna and white sharks, which can maintain their temperature (Carey and Lawson, 1973; Goldman, 1997; Firdaus-Nawi and Zamri-Saad, 2016). Fish can be found in almost every water body on the planet, from highland streams (gudgeon and char) and in deepest oceans (anglerfish and gulpers) (Firdaus-Nawi and Zamri-Saad, 2016). In comparison to other vertebrate classes on the planet, fish are highly diverse. Over 32,000 fish species have been recognized thus far (Firdaus Nawi and Zamri-Saad, 2016).

DOI: 10.1201/9781003361183-15

12.2 Fish Immune System

The immune system protects against infection and maintains internal homoeostasis. As a result, the immune system functions as a barrier for the fish, protecting it from a wide range of invading microbes. Both immune systems in fishes have major role. Innate immunity recognizes the germ line-encoded molecules and responds to invading pathogens. Toll-like receptors (TLRs) and phagocytosis are important components of innate immunity that defend the host from foreign invading pathogens (Silva et al., 2002). Pathogens are recognized by adaptive immunity through the substances produced by somatic processes, which are then checked by B- and T-lymphocytes (Firdaus-Nawi and Zamri-Saad, 2016; Medzhitov and Janeway, 1997; Dixon and Stet, 2001).

Depending on the type of fish the immune organ varies (Firdaus-Nawi and Zamri-Saad, 2016). A real lymphoid organ is absent in lamprey, hagfish, and other jawless fish. They are solely reliant on lymphoid tissues present within non-lymphoid organs. Plasma cells, macrophages, and erythrocytes are produced in the pronephros or anterior kidney, as well as in some areas of the gut, where granulocyte maturation takes place. Rays and sharks, for example, have a more evolved immune system than cartilaginous fish. Leydig's organ, the spiral valve of the gut, and the epigonal organs all host plasma cells, lymphocytes, and granulocytes, are only found in Chondrichthyes and are not found in other fish. The critical immune organs of the thymus and spleen, which store and mature lymphocytes, plasma cells, and macrophages, are also found in cartilaginous fish. The principal site for granulocyte manufacture in paddlefish, sturgeon, and bichir of the chondrostean sub-class is located within the meninges. A tissue made up of reticular cells, lymphocytes, and some macrophages cover the chondrostean's heart. The development of macrophages, granulocytes, erythrocytes, and lymphocytes takes place ins inside the kidney, a vital chondrostean hemopoietic organ (Firdaus-Nawi and Zamri-Saad, 2016).

Similarly, the kidney, specifically the anterior kidney, is the essential hemopoietic organ for bony fish or teleost, where numerous types of immune cells originate. The teleost fish consists of spleen, thymus, gonads containing lymphoid cells, skin stomach, and gills are the lymphoid organs in the teleost. Lymphocytes, neutrophils, erythrocytes, and granulocytes are the main immune cells in the thymus (Chilmonczyk, 1992; Firdaus Nawi and Zamri-Saad, 2016). Zebrafish are one of the teleost animals with a lymphatic system similar to mammals (Küchler et al., 2006) (Table 12.1).

TABLE 12.1

Immune Response between Fish and Mammals with Varied Immune Diversity from Receptors to Antibody Activity (Rauta et al., 2012)

Immunological Response between Fish and Mammals			
Immunological Response	**Fish**	**Mammal**	**References**
Innate immune response TLR (toll like receptors)	TLR1, TLR2, TLR3, TLR4, TLR5, soluble TLR5, TLR7, TLR8, TLR9, TLR14, TLR21, TLR22, TLR23.	LR1, TLR2, TLR3, TLR4, TLR5,TLR6,TLR7, TLR8,TLR9,TLR10, TLR11, LTR12, TLR13	Aoki et al., 2008
NLRs (nod like receptors)	NOD1, NOD2	NOD1, NOD2, NALPs, IPAF, NAIP5	
Humoral variation Ig isotypes	IgM, IgD, IgT, IgW, IgH, IgZ, IgX/IgR, IgM	IgM, IgA, IgD, IgE, IgG	Tort et al., 2003.
Ig gene rearrangement	Multicluster	Translocon	
Antibody nature			
Affinity	Less	High	Tort et al., 2003.
Response	Comparatively slower	Fast	
Memory response	Weaker	Faster	
Maturation	Lower or absent	High	

12.3 Fish Cytokines

Cytokines are proteins that are secreted and influence the immune responses through their growth, differentiation and activation roles (Reyes-Cerpa et al., 2012a, 2012b). The activation of the innate immune response leads to the synthesis of cytotoxic T cells and antibody synthesis, cytokines are involved in many facets of the immune response. The mixture of cytokines produced by immunological stimulation increases the expression of immune-related genes via numerous signaling pathways in higher vertebrates, assisting in the initiation of the immune response (Reyes-Cerpa et al., 2012a, 2012b). When cytokines bind to their corresponding receptors, the modulation of immune response takes place through either autocrine or paracrine manner (Wang et al., 2011).

Pleiotropic actions of cytokines are overlapping and often contradicting, making classification challenging. Interferons (IFNs), interleukins (ILs), tumour necrosis factors (TNFs), colony stimulating factors, and chemokines are cytokines produced by macrophages, lymphocytes, granulocytes, dendritic cells, mast cells, and epithelial cells (Palanisamy et al., 2018). Activated immune cells secrete them after being infected by varied types of pathogens, like bacteria, parasites, or viruses (Salazar-Mather and Hokeness, 2006). Macrophages can produce cytokines like IL-1, IL-6, IL-12, TNF, as well as chemokines such as IL-8 and MCP-1, direct the immune cells towards the infected site and thereby activating the pathogen eliminators (Reyes-Cerpa et al., 2012a, 2012b). Moreover, cytokines generated by phagocytes in tissues can increase dendritic cells migration by activating acute phase proteins like mannose-binding lectin (MBL) and C-reactive protein (CRP) (DeVries et al., 1999). The two major cells involved in cytokine secretion are Th1 and Th2, first discovered in mice. Each cell has its specific receptor to regulate the response of T-cell and B-cell (Mosmann et al., 1986). The response to an antigen can be regulated by the secretion of cytokines through Th1 and Th2 helper cells. The stimulation and development of Th2 happen in the presence of interleukin-4 (IL-4) where interleukin–12, 18, 23, and 27 help in the stimulation of Th1 cells (Glimcher and Murphy, 2000). The Th1 and Th2 cells originate from the same precursor are differentiated based on the types of antigen it exposed (Hosken et al., 1995). The regulatory T-cells (TRs) are a special type of T-cell that helps in the inhibiting the unwanted immune response (Maloy et al., 2003). Majority of the cytokines are seen in fish that tend to express in a transcriptomic level (Secombes et al., 2011). The proinflammatory cytokines are IL-1β, TNF-α, and IL-6. The cytokines responsible for acquired immune responses are IL-2, IFN-γ, IL-4/13, IL-10, IL-17A/F, IL-21, IL-22, and TGF-β1 (Secombes et al., 2011). The mature trout interleukin one beta that was produced in *E. coli* was analyzed using polyclonal antibody against IL-1 and IL-1β extracted from macrophage cell line of rainbow trout (Hong et al., 2004). Apart from humans, they are found in chickens, frogs, and fish (Savan and Sakai, 2006). The cytokines in fish and humans are much more similar. Many numbers of cytokines were cloned in fish species (Reyes-Cerpa et al., 2012a, 2012b). TNF, IL-1, IL-6, and IFN are examples of cytokines found in fish.

12.4 Fish Cytokine Types

Fish and mammals have very identical cytokines, according to recent genetic studies (Savan and Sakai, 2006; Zou and Secombes, 2016). The key cytokine molecules of fish are discussed in below (Figure 12.1).

12.4.1 Interleukin-1 Family

IL-1 cytokine family plays a major role in innate immunity and also has pro-inflammatory activity. The first formed cytokine in fish is IL-1 (Zou et al., 1999). In mammals the same chromosome bears the gene encoding for IL-1α and IL-1β in adjacent to each other. In fish only IL-1β has been identified so far. The interleukin-1 activates lymphocytes and phagocytic cells in fish thus helps in fighting against *Aeromonas hydrophila* infection (Savan and Sakai, 2006). The recent discovery of new gene *il1b* in medaka (*Oryzias latipes*) shows 76% similarity to the *il1b* gene found on different chromosomes. There are 11 interleukin types found in mammals IL-1α (IL-1F1), IL-1β (IL-1F2), IL-1 antagonist receptor (IL-1ra/IL-1F3), IL-18 (IL-1F4), IL-1F5–10, and IL-33 (IL-1F11). They can act as antagonist or pro-inflammatory compounds acting upon specific cell lines. Only IL-1β and IL-18 homologue interleukins have been identified in fish (Secombes et al., 2011). The interleukin-1β activates a humoral immune response.

12.4.2 Interleukin-2 Family

In humans, seven types of interleukin-2 families have been identified: IL-2, IL-4, IL-7, IL-9, IL-15, and IL-21 and fish possess IL-2, IL-4/13 A, IL-4/13B, IL-7, IL-15, and IL-21 (Zou and Secombes, 2016). The interleukin IL-4/13 genes *il4/13a* and *il4/13b* show similarity with IL-4 and IL-13 found in mammals. The presence of Interleukin-15 was found in Japanese puffer fish *Takifugu rubripes*, whereas *il15* has similarity with *il15* of Japanese puffer fish and zebrafish (*Danio rerio*) (Gunimaladevi et al., 2007). The corresponding genes of IL-2, IL-15, and IL -21 are located in the same chromosome 4q.25–q35 and have similar antiparallel alpha helices with tertiary structure. In humans, the IL-2 gene has four exons with three interrupting introns; IL-21 has four introns and five exons; IL-15 has eight exons and seven introns (Fujita et al., 1983; Parrish-Novak et al., 2000; Krause et al., 1996).

The interleukin-2, which contains 149 amino acids in length, and Interleukin-21, which has conserved region (T-E-[LF]-x (2)-L-x-C- L-x(2)-E-L) was cloned from Japanese puffer fish (*Fugu*) for comparative analysis (Bird et al., 2005a). The IL-2 genes from other mammals show 24–34% lower amino acid identities. The Fugu IL-2 gene has exons (four) and introns (three) similar to humans. The IL-2 loci in *Fugu* and humans are conserved (Savan and Sakai, 2006).

12.4.3 Interleukin-6

Interleukin-6 is naturally pleomorphic and first to be isolated from humans and it plays a major role in formation of blood cells, inflammation, and immune function. IL-6 can be found

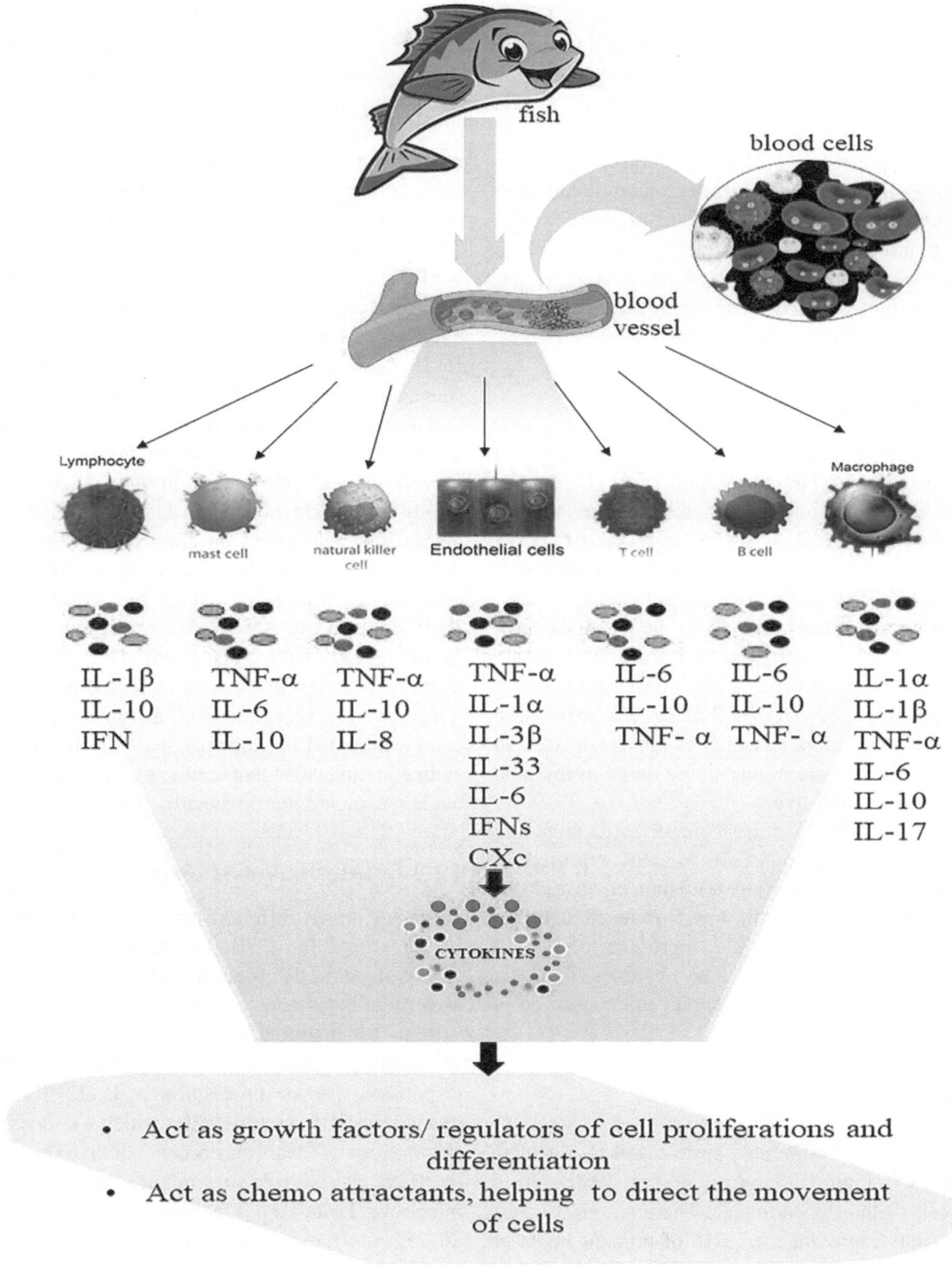

FIGURE 12.1 A schematic representation of fish immune response against pathogens via cytokine production.

in Japanese puffer fish (*Takifugu rubripes*) (Bird et al., 2005a), rainbow trout (Iliev et al., 2007), olive flounder (Nam et al., 2007), gilthead seabream (Castellana et al., 2008), zebrafish (Varela et al., 2012), orange spotted grouper (*Epinephelus coicoides*) (Chen et al., 2012), Atlantic halibut (Øvergård et al., 2012), and Nile tilapia (Wei et al., 2018). It helps in the differentiation of B-cells and T-cells. The same activity is also seen in aquatic animals like starfish, trout, and common carp (Beck and Habicht, 1996). Among teleost fish, the gene was identified from Fugu for the first time (Bird et al., 2005b). The c-DNA of *Fugu* IL-6 shows 2023 bp with a 182 bp 5VUTR and an 890 bp 3VUTR. The IL-6 consists of four cysteine residues among which two are conserved. The presence of third and fourth cysteine residues has higher biological activity found in fugu (Skelly et al., 1994). The IL-6 of Fugu is 1.317 kb in size having exons (five) and introns (four). Fugu IL-6 shows similarity with humans (16.5%), cows (15.6%), chickens (16.7%), and human herpes virus IL-6 gene (11.4%) (Savan and Sakai, 2006). Interleukin-6 along with IL-1β and tumor necrosis factor (TNFα) mediates inflammatory responses.

Interleukin-6 is composed of IL-11 and IL-31 other cytokines like ciliary neurotrophic factor, leukemia inhibitory factor, Oncostatin M, CT-1, and CT-2 (Secombes et al., 2011). The signaling pathway of IL-6 in fish is lesser known. The lipopolysaccharide poly I:C induces the macrophages to express IL-6 also IL-1β in RTS-11 macrophage cell line (Costa et al., 2011). IL-6 can act either through autocrine or paracrine mode for enhancing its expression (Reyes-Cerpa et al., 2012a, 2012b). To avoid the damage during inflammation, IL-6 downregulates the expression of TNFα1, TNFα2, and IL-1β (Costa et al., 2011). In rainbow trout, the recombinant IL-6 promotes phagocytic proliferation (Costa et al., 2011). The recombinant IL-6 induces transient expression of SOCS1–3 (Suppressor of Cytokine Signaling), IRF-1 (Interferon regulatory factor) and with increased gene expression has been observed in fresh macrophage cultures (Costa et al., 2011).

12.4.4 Interleukin- 11

The secretion of IL-11 in mammals is carried out by many cell types (Costa et al., 2011). The expression of IL-11 can be induced by the viral infection, bacterial infection, and by stimulation of cytokines like IL-1, TNFα, and TGF-β1. IL-11 acts on various cells like hematopoietic cells, hepatocytes, adipocytes, intestinal epithelial cells, tumor cells, macrophages, osteoblasts, and osteoclasts. IL-11 helps in the protection and restoration of gastrointestinal tract (Liu et al., 1996). In fish, IL-11 helps in protection against bacterial and viral infections (Huising et al., 2005a, 2005b). The expression of IL-11 is seen in trout kidney, which is also expressed in intestine and gills also they are highly upregulated in immune sites like spleen, kidney, and liver.

Interleukin-11 uses gp130 from its receptor as its signaling component. Apart from mammals, IL-11 was reported in puffer fish (green spotted) and trout (Huising et al., 2005a, 2005b). IL-11a present on the carp has a reduced level of expression on blood leucocytes and IL-11b present in Japanese flounder has a higher expression. The residues that are important for the binding to receptor complex are conserved in IL-11 (Wang et al., 2005).

12.4.5 Interleukin-10

Interleukin-10 suppresses the immune system and also acts as anti-inflammatory cytokine (Fiorentino et al., 1989). It is helical glycoprotein with 40 kDa in size. There are viral IL-10 homologs that closely represent the IL-10 of humans helps in suppressing the host immune response. The IL-10 family contains IL-10, IL-19, IL-20, IL-22, IL-24, IL-26, and IFN-γ. The IL-10 has been characterized in goldfish and carp (Piazzon et al., 2015). The genes for interleukins 10, 19, and 20 are seen in the same chromosome that codes for IL-22, IL-24, and IFN-γ in a tandem manner (Savan and Sakai, 2006). A recently identified gene IFN-γ gene (*ifngrel*) is located next to the *ifng* gene. This IFN-γ gene plays an important role against bacterial infections like *Aeromonas hydrophila*, *Edwardsiella tarda*, or *Shigella flexneri* in *Labeo rohita* Rohu (*Labeo rohita*) (Igawa et al., 2006). The primary interleukin to be cloned and identified through *in silico* aid are IL-10 genes isolated from pufferfish (green spotted), torafugu (tiger pufferfish), and common carp. The polypeptide obtained through cDNA synthesis codes for amino acids (180) (carp), 183 amino acid (torafugu), 184 amino acids (spotted green pufferfish and rainbow trout), and 181 amino acids (zebrafish). The sequence motif of IL-10 fish (cartilaginous) shows a conserved sequence. The IL-10 isolated from carp shows higher sequential identities with zebrafish (78.5%), trout (53.7%), torafugu (43.2%), and spotted green pufferfish (37.6%) (Savan and Sakai, 2006). When monocytes from gold fish were incubated with heat killed *Aeromonas salmonicidia* with IL-10, there was a decrease in the expression TNF-α1, TNF-α2, IL-1β1, IL-10, CXCL_8/IL-8, and the NADPH oxidase component p47(phox) (Grayfer et al., 2011). The pretreatment of monocytes with 1,000 ng/mL of IL-10 evades *A. salmonicida*. A similar reaction can be seen in Carp IL-10 in lipopolysaccharides, activated neutrophils, and macrophages (Piazzon et al., 2015). The IL-10 also inhibits MHC antigen-presenting gene in macrophages of carp. The human kidney cells taken from fish immunized with *Trypanoplasma borreli* antigen upon activation with IL-10 shows elevated antibody IgM production. The CD4+ and CD8+ of T cells of carp have a different effect on antibody IgM secretion and B cell differentiation. The Koi herpesvirus (Khv) has IL-10 viral homologue (KhvIL-10), which has moderate homology with IL-10 of carp but it not helpful in the viral replication (Van Beurden et al., 2011). The KhvIL-10 expresses its gene during acute reactivation phase with increased lysozyme positive cells when given to zebra fish similar to zebrafish with IL-10. This can be inhibited by injecting with KhvIL-10 morpholinos (the synthetic molecules that are the product of a redesign of natural nucleic acid structure) (Sunarto et al., 2012).

12.4.6 Interleukin-10 Family (IL-19, IL-20, IL-22)

There are seven different types of interleukins present in the IL-10 family: IL-10, IL-19, IL-20, IL-22, IL-24, IL-26, and IFN-ɣ. IL-10 in fish functions like a suppressing agent and acts as a reducing source at the time of inflammatory responses, similar to mammalian orthologs. Similar functions were studied in goldfish and carp (Piazzon et al., 2015). The genes responsible for the production of IL-22, IL-24, and IFN are located on il10 and il19/20, which is located in the same chromosome (Savan and Sakai, 2006). The genes stimulated by IFN play a role in immune defense against bacterial infections. Following infection with *A. hydrophila*, *E.tarda*, or *Sh. Flexner1i*, elevated expression of the IFN gene in the carp, rohu *Labeo rohita*, has been reported.

In spotted green pufferfish the first reported presence of single IL-10 commonly accepted as IL-20 located next to IL-10. Similar findings were observed in *Fugu* and zebrafish (Zou et al., 2004). These new genes show higher similarity to IL-19 and IL-20 of mammalian type. The presence if IL-20 gene can be seen in puffer fish, zebrafish, and trout (Wang et al., 2010). The IL-20-like gene shows higher expression in brain tissue, immune system, and nervous system. Mostly macrophages express IL-20L (Reyes-Cerpa et al., 2012a, 2012b). The heightened expression of IL-20L in RTS-11 can be seen at 24 h after lipopolysaccharide (LPS) stimulation and in spleen after 72 h of infection with *Yersinia ruckeri,* the cause of *Yersinia* septicemia in fish. This is due to the increased production of proinflammatory cytokines and other factors (Wang et al., 2010).

12.4.7 Interleukin-12

It is a heterodimeric cytokine (p35 and p40 subunits) that initiates the differentiation of Th1 cells and also assists T-cell and natural killer cells (NK) in the synthesis of interferon-γ (Savan and Sakai, 2006). Moreover, it plays important role in the governing T-cells into Th1 lineage by secreting lymphokines under cell-mediated response (Huang et al., 2000). The IL-12 subunits p35 and p40 were identified from Japanese puffer fish. Locus p35 is conserved and the expression of it can be activated by injecting the fish with poly I:C in the kidney and spleen. The p40 subunit is seen in Fugu tissues but not in the muscles and it is non-responsive to the poly (I:C) and LPS (Savan and Sakai, 2006). Apart from mammals, the subunits p35 and p40 were cloned in chickens later on the genes were also cloned from fish (Balu and Kaiser, 2003). The IL-12 encodes for 199 amino acids and consists of 359 amino acid polypeptides. The IL-12 p35 mRNA has higher antiviral function as well as it controls the IL-12 function (Savan and Sakai, 2006).

12.4.8 Interleukin-15

Interleukin-15 helps in activation of T-cells, dendritic cells, and NK cells also performs= a major part in disease and (autoimmune condition) (Savan and Sakai, 2006). Interleukin-15 has close activities of IL-2 and IL-2R beta and gamma units. The green spotted puffer fish has spliced IL-15 genes (CR704965 and AY374511). The IL-15 of zebrafish has genome size 7.7 kb with exons (four) and introns (three) similar to IL-2 genomic arrangement (Savan and Sakai, 2006). The protein sequence of IL-15 consists of 162 amino acids and has 42 amino acids as signal peptides. When IL-5 transcripts are cloned, it shows the presence of spliced genes (IL-15-2 and IL-15-3). When different exons were used, IL-15 shows three different IL transcripts IL-15-1, IL-15-2, and Il-15-3 (Savan and Sakai, 2006). The IL-15-1, IL-15-2, and IL-12-3 transcripts have four (E1, E2b, E3, and E4) has six (E1, E2a, E2b, E3, and E4) and three exons (E1a, E2b, and E4). The IL-15-1 transcript contains 162 amino acids along with 43 amino acids signal peptide and a 13.8 kDa mature peptide (Savan and Sakai, 2006).

12.4.9 Interleukin-17 Family

Humans have genes for six members of the IL-17 family: IL-17A, IL-17B, IL-17C, IL-17D, IL-17E, and IL-17F. In fish, IL-17A/F1, IL-17A/F2, IL-17A/F3, IL-17C, IL-17D, and a fish-specific IL-17 N have been identified (Korenaga et al., 2010). Using recombinant (r) protein, researchers investigated the uses of Japanese pufferfish IL-17A/F1 and discovered that it increased the expression of TNFα, *il1b*, and *il6* genes and activated phagocytes (Kono et al., 2011; Takahashi et al., 2020). The IL-17A/F1 has a greater ability than IL-17A/F2 and IL-17A/F3 to activate the transcriptional activity of NF-B in yellow croaker (*Larimicthys polyactis)* (Ding et al., 2016). In rainbow trout, rIL-17A/F2 can induce the expression of IL-6, the chemokine CXCL-8, and the antimicrobial peptide-defensin in splenocytes (Monte et al. 2013). Grass carp rIL-17A/F1 also induces several pro-inflammatory cytokines in kidney leukocytes, including *tnfa*, il1b, il6, and the chemokine cxcl8. Furthermore, when the il17a/f1 gene is knocked out by the clustered regularly interspaced short palindromic repeats (CRISPR)/CRISPR-associated protein (Cas) 9 system in Japanese medaka (*O. latipes*), which suppresses the digestive enzyme as well as antibacterial genes, indicating that medaka IL-17A/F1 may play a role in maintaining the fish health.

The IL-17 ligands have been found in zebrafish genome (Savan and Sakai, 2006). The IL-17 of zebrafish is around 1.9 to 3.8 kb in size where other IL-17 genes in zebrafish has shared three exons and two introns. The genes IL-17A and F were designated as IL-17A/F1 and 2 in zebrafish. Another set of homologue gene (IL-17A/F3) has been reported in zebrafish, rainbow trout, Atlantic salmon, puffer fish (IL-17A/F1, 2, and 3), and medaka (IL-17A/F1, 2, and 3) (Kono et al., 2011).

The level of IL-17D expression is high in Atlantic salmon and increases after the administration of *A. salmonicida.* When *A. salmonicida* is made to coexist with fish, it showed higher expression of IL-17D in skin and kidney. In trout, the IL-17C has IL-17C1 and IL-17C2, which can be where the IL-17C1 is seen more in gills and skin and IL-17C2 can be found more concentration in the head, kidney, and brain. Both interleukins' expression increases with the bacterial infection where the IL-17C2 has a higher degree on fold (Reyes-Cerpa et al., 2012a, 2012b).

12.4.10 Interleukin-22 and -26

The human chromosome no. 12 contains IL-22, IL-26, and IFNγ as clusters in the same locus. This was identified in the same locus of Fugu by Zou (2004). Similar locus was analyzed in zebrafish and green spotted puffer fish for identifying IL-22 and IL-26 genes (Igawa et al., 2006). The locus is 161.1 kb in size, which shows orthologus genes of IL-22, IL-26, and IFNγ. The flanking genes of flanking cytokines DYRK2 and MDM1 were conserved in both humans and fish. The IL-22 gene of zebrafish contains five exons and four introns with 4.5 kb in size. There is a gene present adjacent to IL-22 that encodes for 169 amino acids with similar transcriptional orientation, whereas in zebrafish it is commonly known as IL-22. Zebrafish have IL-26 with similar intron and exon as seen in humans (Savan and Sakai, 2006). The human IL-26 is 25 kbp with a comparatively longer third intron size. The similarity between humans and zebrafish IL-26 is 16.0% (Savan and Sakai, 2006).

The IL-22 present in mammals is produced by Th17 cells and NK22 (NK cells) (Cupedo et al., 2009). The IL-17 and IL-22 are secreted during infectious response (Feinen and Russell, 2012). The co-expression of IL-26 takes place along with interferon gamma and interleukin-22 through Th1 clones. It also has co-expression with IL-17 and IL-22 by Th17 cells (Wilson et al., 2007). In Fugu, IFNγ locus was first identified using gene synteny. This locus showed a homology to IL-22/26 which later on revealed the presence of two genes in zebrafish, one is IL-22 another with less homology with IL-26 (Igawa et al., 2006). The IL-22 gene expression can be seen in gills, intestine, kidney, and spleen. The IL-26 gene is expressed only in the intestine when treated with PolyI:C (Igawa et al., 2006). The disease resistance of haddock which was vaccinated against *Vibrio anguillarum* can show an elevated IL-22 expression in gills compared to controls (Corripio-Miyar et al., 2009) (Table 12.2).

TABLE 12.2
Shows the Developmental Duration of Kidney and Thymus in Fishes (Sahoo et al., 2021)

Species	Development of Fish Thymus and Kidney						References
	Formation of Lymphoid Organ		Formation of Hematopoietic Cell		Appearance of Lymphocytes		
	Kidney	Thymus	Kidney	Thymus	Kidney	Thymus	
Zebra fish	72 hrs (post fertilization)	54 hrs (post fertilization)	>96 hrs (post fertilization)	65 hrs (post fertilization)	3 weeks (post fertilization)	3 days (post fertilization)	Korzh et al., 1998
Rainbow trout	<8 days (before hatching)	5 days (before hatch)	5 days (before hatching)	1 day hatch	5 days (post hatching)	3 days (after hatch)	Grace and Manning et al., 1980
Seabream	<1 (days post hatch)	22–29 days (after hatch)	5 days (post hatch)	29–47 days (after hatch)	54 days (post hatch)	47 days (after hatch)	Jósefsson, and Tatner, 1993.
Channel catfish	unknown	unknown	<3 days (post hatch)	unknown	<7 days (post hatch)	5 (days after hatch)	Petrie-Hanson, Ainsworth, 2001.
Common carp	1 day (post hatch)	3 days (after fertilization)	unknown	unknown	6 days (post hatch)	4–5 days (after fertilization)	Botham and Manning, 1981

12.5 Against Bacterial Pathogens

The most commonly used organism for pathogenic studies is zebrafish (*Danio rerio*). They were exposed to *Salmonella typhimurium* to study the functioning of Traf6 (tumor necrosis factor receptor-associated factor 6), which is associated with innate immune response. During the course of infection, there is a change in the transcriptome level of the fish. When zebrafish and common carp *(Cyprinus carpio)* are infected with *Mycobacterium marinum,* the tissue of common carp express 39 TIR (Toll/interleukin-1 receptor) domains. Fish serve as a model organism in human disease studies. In catfish, infection causes *Enteric septicemia,* which drastically affects the economy of the catfish in the aquaculture industry. Studies show that actin cytoskeletal polymerization and remodeling, where *E. ictalurid* enters the fish by disrupting the intestinal barrier (Li et al., 2012). RNA-Seq data from *E. ictaluri* containing detailed information on nucleotide-binding oligomerization domain-like receptors (NLR), retinoic acid-inducible gene, and Toll-like receptors (TLR) give an insight of the infection. By the help of RNA-Seq analysis, the genetic biomarkers have resistance genes in yellow catfish *Pelteobagrus fulvidraco* (Zhu et al., 2017). The protein-protein interaction in Japanese flounder gills (*Paralichthys olivaceus*) with *Edwardsiella tarda* showed 24 hub genes during transcriptomic analysis (Li et al., 2012).

The invasive pathogen *Flavobacterium columnare* targets the fish mucosal organs like gill and skin. Where the channel catfish which are susceptible to *F. columnare* showed gills polarization leading to the pathogen entry (Peatman et al., 2013). The rhamnose-binding lectin (RBL) can be seen during the *F. columnare* infection in the channel catfish. RBL helps in bacterial attachment and aggregation. There are other sugars such as L-rhamnose and D-galactose containing RBL ligands, which has protective nature against the *F. columnare* infection (Beck et al., 2012). When the catfish from the channel was vaccinated against *F. columnare,* it showed stronger gene expression of deposition of collagen and remodeling of tissue when compared to non-vaccinated groups. The administration of mucosal vaccines helps in the innate immune response and tissue repair in fishes.

Aeromonas hydrophila affects brackish water and freshwater fishes. Studies using RNA-Seq have been conducted over large yellow croaker (Larimichthys crocea), blunt snout bream (*Megalobrama amblycephala*), common carp, grass carp (*Ctenopharyn godonidella*), golden mahseer (*Tor putitora*), and dark barrel catfish (*Pelteobagrus vachellii*) in order to understand the pathogenic mechanism of *A. hydrophila* (Qin et al., 2017). The infection caused by *Streptococcus agalactiae* in Nile tilapia activates toll-like receptor-mediated pathways and miRNAs in order to protect against the infection. The immune system of Nile tilapia activates the downstream immune system pathway to avoid destruction of tissues and cell apoptosis that was triggered by bacteria (Zhu et al., 2017). Due to climate change, the immune genes of Mozambique tilapia (*Oreochromis mossambicus*), which is responsible for *S. agalactiae* infection, get impaired.

The infectious study of *Vibrio harveyi* on Japanese sea bass (*Lateolabrax japonicus*) shows 8% specific transcripts that are unique to a fish, which was thought to be evolved during evolutionary period (Xiang et al., 2010). An infection caused by *Vibrio alginolyticus* on orange-spotted grouper (*Epinephelus coioides*) reveals the activation of the immune system (complement pathway) after four hours of bacterial infection along with hepcidin. When zebrafish were administered with live attenuated *Edwardsiella tarda* they showed up-regulation of the MHC-I pathway and down-regulation of MHC-II pathway at the initial stages of vaccination. Red mouth disease in fish was caused by *Yersinia ruckeri*, the protein of this specific bacteria was studied to virulence and pathogenicity.

12.6 Against Viral Pathogens

Viral infections has higher impact on the aquaculture industry. At the time of infection, specific and non-specific defensive mechanisms get activated inside the host in order to contain the virus from devastating effects (Smail and Munro, 2012). A detailed study has been conducted using RNA-Seq technology for a better understanding of viral infections in fish. In the aquaculture industry, the grass carp hold a huge economical part. In recent days, these species have been targeted by double-stranded grass carp reovirus (GCRV), which is an RNA virus that comes under the genus *Aquareovirus* of the Reoviridae family (Shi et al., 2014). This virus affects the fish at their early growth phase and causes hemorrhagic disease. Based on the findings from RNA-Seq technology, the virus does not affect any cells rather it distributes lipids and carbohydrates in fish organs (Shi et al., 2 014). The complement and coagulation cascade pathway can be identified using RNA-Seq along with RT-PCR validation in infection like GCRV.

The viral hemorrhagic septicemia virus (VHSV) affects both marine and fresh water fishes worldwide. It comes under *Novirhabdovirus* genus of the Rhabdoviridae family (Lenoir and De Kinkelin, 1975). The viral susceptibility test of VHSV was performed using rainbow trout (*Oncorhynchus mykiss*) clonal cell lines then with whole genome transcriptome analysis by RNA-Seq was carried on (Shi et al., 2 014). VHSV suppresses ATP synthesis and antioxidant systems resulting in the necrosis of the infected cells. The non-enveloped dsRNA Piscine orthoreovirus (PRV) causes heart and skeletal muscle inflammation in Atlantic salmon (Palacios et al., 2010). The North American, European, and Asian fish culture industry were greatly affected by hematopoietic necrosis virus (IHNV). Coinfection can occur in sockeye salmon (*Oncorhynchus nerka*) at the time of PRV infection the sockeye salmon is also affected by IHNV which can be seen by performing transcriptomic analysis of kidney tissue. The immune response against the IHNV has no effect on the PRV viral multiplication, which clearly state that neither of them dosent have any influence over the other (Polinski et al., 2016).

The infection in Atlantic salmon by salmonid alphavirus (SAV) results in the necrosis of pancreas, cardiomyopathy and skeletal myopathy. At the time of infection, the fish immune system showed stronger type-I interferon (IFN), which gets activated by the SAV-3 presented to the macrophage/dendritic like cell lines (Xu et al., 2015). The KHV (koi herpes virus) causes higher mortality rate in common carp and koi carp (*Cyprinus carpio koi*). At the time of CyHV-3 (Cyprinid herpesvirus-3), infection expresses various novel genes that help the innate immune system to fight against the virus (Lee et al., 2016), which was revealed during the spleen transcriptomic studies. At the time of administration of heat-inactivated vaccine against viral hemorrhagic septicemia virus (VHSV), it triggers numerous genes involved in the acute phase of infection along with TLR, IFN, and apoptosis pathways. The immunostimulant Poly (I:C) (Polyinosinic:polycytidylic acid) was used to study the genes and pathways in large yellow croaker spleen during innate and adaptive immunity (Mu et al., 2014). When the same immunostimulant is used in the rainbow trout, it helps in the production of heat labile molecules that activate macrophages. When Singapore grouper iridovirus (SGIV) was used in the orange-spotted grouper has profound action over mitogen-activated protein kinase, chemokines, TLRs, and RIG-I signaling pathways. During high temperatures, the megalocytivirus causes disease in the turbot (*Scophthalmus maximus*) and the infection was found to be low at lower temperatures (Zhang and Sun, 2017). Nodaviruses (RNA virus) have three groups: alphanodavirus (affects insects), betanodavirus (affects fishes), and gammanodavirus (infects prawns). The NNV (Nervous Necrosis Virus) causes viral encephalopathy and retinopathy, which has a broader host range, infecting 120 fish species, leading to 100% mortality during larval stages (Costa and Thompson, 2016). NNV virus escapes the host innate and adaptive immune response mechganism. The transcriptomic study on grouper cell lines showed the presence of capsid protein in the endoplasmic reticulum induces stress and eventually cell death. The NNV virus also escapes cell-mediated cytoxicity and elevated infection in European sea bass (Chaves-Pozo et al., 2017). The RNA-Seq-based analysis (transcriptomic) helps in the clear identification of host and viral interaction also the evading mechanism of virus from host immune system.

12.7 Against Parasites

Major parasitic infection were reported in fishes. Parasites can escape the immune system by avoding immune survillence, indirect life cycle, suppression of immune system, and manipulating the host neural system (Flegr, 2006). The parasite *Cryptocaryon irritans* (C. *irritans*) causes ich disease in large yellow croaker. Which activates toll pathway, signaling of chemokines, activation of complement system and initiation of coagulation cascade. Alternate pathway and complement system have a major role in the initial stages of infection. *C. irritans* infection can be fatal even at a lower microbial load. In orange-spotted grouper, the migration of leucocytes and TLR-associated cells were seen near the skin area (Hu et al., 2017). At lower temperatures numerous genes were expressed in the assisting the parasite to go into a dormant state. These genes serve as potential markers in the identification of cryptocaryoniasis, especially in tomont stage (Yin et al., 2016). The *C. irritans* has nine immobilization antigen transcripts and 161 protease transcripts that can be used in the drug and targeted vaccine development.

The intestinal myxozoan parasite *Enteromyxum scophthalmi* (*E. scophthalmi*) infects the intestine of the large yellow croaker. This leads to the activation of immune system (adaptive) in the spleen and kidney (Robledo et al., 2014), whereas the infection in turbot with *E. scophthalmi* activates the IFN-mediated immune response at the initial phase of the disease. *Tetracapsuloides bryosalmonae* (*T. bryosalmonae*) is a myxozoan parasite that infects salmonid fishes (farm and wild type) in Europe and North America resulting in loss of both farm animals and endangered species in the wild (El-Matbouli and Hoffmann, 2002).

12.8 Pro-inflammatory Cytokines

12.8.1 Tumor Necrosis Factor Alpha (TNFα)

TNFα is pro-inflammatory in nature helps in cell formation, differentiation, and programmed cell death, and induces other cytokines (Rahman and McFadden, 2006). Some viruses can neutralize TNFα by receptor-binding mechanism or by various signaling mechanisms. TNFα has been detected in species including Japanese flounder (*Paralichthys olivaceus*), rainbow trout, gilthead seabream (*Sparus aurata*), carp (*Cyprinus carpio*), tilapia (*Oreochromis*), turbot (*Scophthalmus maximus*), and goldfish (*Carassius auratus*) (Grayfer et al., 2008). There are various isoforms of TNFα seen among various species of teleost (Saeij et al., 2003). The major difference between the mammalian and fish TNFα is that their action over phagocyte activation is in fishes like goldfish, rainbow trout, turbot, and gilthead seabream under *in vitro* condition (Garcia-Castillo et al., 2004). When recombinant TNFα was injected into gilthead seabream, it showed a high phagocytic cell count near the site of injection with an increased respiratory burst activity (Garcia-Castillo et al., 2004). The TNFα assists the leukocytes to gather at site of inflammation. Ten TNF genes have been identified in puffer fish (Biswas et al., 2015). TNFα is an important immunological molecule among all the TNF family. There are three genes, *tnfa*, *tnfb,* and *ltb,* in humans arranged one in front of the other on the same chromosome (Savan et al., 2005). In fish, there is an additional *tnfa* located on a different chromosome (Kinoshita et al., 2014). The TNFα is expressed in trout thymus, which could promote thymocyte development. In addition, TNFα assists in pro-apoptotic activity (Hino et al., 2006), and helps in expressing and up-regulation of granzyme in cytotoxic cells (non-specific), protecting them against programmed cell death (Praveen et al., 2006).

12.8.2 Interleukin-1β

Interleukin-1 β is a pro-inflammatory cytokine helps the host to fight against infections in stepwise manner leading to inflammation. The up and down regulation of IL-1β is mediated by other cytokines and chemokines (Dinarello et al., 1984). In mammals IL-1β is synthesized by monocytes in blood and macrophages in tissues. The primary interleukin in fish can be seen in various species including rainbow trout, carp, sea bass (*Centropristis striata*), gilt head seabream and haddock (*Melanogrammus aeglefinus*), and tilapia (Lee et al., 2006). In trout, new IL-1β2 has been identified along with another interleukin (Pleguezuelos et al., 2000). The IL-1β genes contains IL-1 signature area, which is somewhat conserved (Zou et al., 1999). The arrangement of intron and exon of fish IL-1βis similar to mammalian arrangement. The IL-1β is induced in *vivo* by co-inhabitation with bacterial infections (Zou et al., 1999). The functional mechanism of carp IL-1β is studied by the over-expression technique by inserting it in a cytomegalo virus (CMV) vector. This significantly increased the phagocytosis, proliferation of lymphocytes, production of superoxide, and developed resistance against *Aeromonas hydrophila* infection better than controls (Kono and Sakai, 2001). There is an increase in antibody titer when carp IL-1β protein is used (Yin and Kwang, 2000).

12.8.3 Interferons

Interferons guide the cells towards viral pathogens and mediates both innate and acquired immunity. There are two types: type 1 and type 2 interferon. The type 2 contains a single gene called IFNγ (Savan and Sakai, 2006). Both interferons included under class II a-helical cytokine family are based on its structure with varied three-dimensional structural arrangement and are in contact with different receptors. The IFNα1 and IFNα2 from Atlantic salmon were cloned and characterized to study its gene structure and antiviral activity, which is induced by ISGs (Bergan et al., 2006). In salmon, IFNα1 induces both proteins (Mx and ISG1) present in TO cells, which have the same properties as mammalian IFNα/β and IFNλ (Robertsen et al., 2003). The *in vitro* analysis between IFNα 1 from salmon shows an antiviral activity against infectious pancreatic necrotic virus (Robertsen et al., 2003). However, the mechanism cannot be seen in *in vivo* conditions even after higher expression of IFNα in spleen and head kidney of fish (Reyes-Cerpa et al., 2012a, 2012b).

Three types of type I IFNs were discovered in rainbow trout. Atlantic salmon IFNα1 and IFNα2 show greater sequence similarity to IFN1 (rtIFN1) and (rtIFN2) (two cysteines). The combination of rainbow trout interferon 1 (rtIFN1) and rainbow trout interferon 2 (rtIFN2) assist in higher expression of Mx and inhibit the replication of viral hemorrhagic septicemia virus in RTG-2 cells. The rtIFN3 has weak Mx, inducing nature and lesser anti-viral properties.

12.8.4 Type I interferon

Type 1 interferons were first cloned and characterized from zebrafish, Atlantic salmon (*Salmo salar*), green spotted pufferfish, and catfish (*Siluriformes*) (Altmann et al., 2003). Among these, Fugu and zebrafish have been reported to possess two types of type I interferon (Zou et al., 2004). This newly identified type of interferon gene contains five exons and four introns, which is different from normal type 1 interferon (Savan and Sakai, 2006). The recombinant interferon obtained by combining catfish and zebrafish interferon shows an antiviral activity (Altmann et al., 2003).

12.8.5 Type II interferon (INFγ)

INFγ is seen in rainbow trout and Atlantic salmon. Using comparative genomics application, the interferon gamma of

Japanese puffer fish has been cloned (Zou et al., 2004). It is located on a genome with 3.7 kb in size, which encodes 189 amino acids. It has a similar structure like the human counterpart with more intron lengths (Savan and Sakai, 2006). In zebrafish and spotted green puffer fish, there are two similar interferon gamma-like genes (Igawa et al., 2006). The INFγ1 is 0.9 kb in size compared to IFNγ 2, which is 2.3 kb in size. Both of these IFNγ have been expressed in zebrafish (Savan and Sakai, 2006).

12.8.6 Chemokines

Chemokines are released in large quantities in areas of inflammation and cause leukocytes to migrate from blood vessels to inflamed tissues. More than 50 chemokines have been identified in humans to date (Liu et al., 2008). Chemokines have four highly conserved cysteine residues and are divided into four subfamilies based on the motif formed by the two N-terminal cysteine residues, namely CXC, CC, CX3C, and XC (Liu et al., 2008). For instance, more than 100 chemokines have been identified in zebrafish, but none of them belong to the CX3C subfamily (Nomiyama et al., 2008). Several CXC chemokines were found to show an increased expression in rainbow trout cell lines (i.e., RTG-2 and RTS-11 cells) after rIL-1β stimulation, including CXCLδ L1 (IL-δ), CXCL11 L1 (γIFN-inducible protein), CXCL F4, and CXCL F5 (Chen et al., 2013).

12.9 Conclusion

Fish's innate and adaptive immune responses to the numerous cytokines that have been documented are regulated by a complex arrangement. It will be intriguing to see whether human Th1, Th2, Th17, and regulatory T cells (Treg) responses are present in fish, regulating particular cell-mediated immunity, as research into the functional activity of these cytokines is now ongoing. The next hurdle in comprehending the equilibrium of such immune responses and aiding in the efficient design of therapeutic techniques to regulate the fish immune system will be the recombinant manufacture of these cytokines and antibodies against them.

Fish functional genomics research has produced a wealth of information about commercially valuable species. The main goal has been to discover transcripts in a whole organism or particular tissue that contribute to such reactions, as well as to functionally characterize the intensity of responses to different infections and the related molecular components. Complete mapping of a reaction is challenging and expensive due to the complicated biology of the immune response, in which various spatial-temporal expression occurs in many cell types at diverse body sites. In addition, arrays are only as useful as the transcripts that are displayed on them. As a result, the representation of transcripts important to the immune response is closely related to efforts to discover new genes through large-scale sequencing projects, where techniques like suppression subtractive hybridization SSH help to understand the transcriptomic response to particular pathogens as well as to discover new genes. High-throughput next-generation sequencing NGS technology has become more widely available in this field recently, and it has the potential to significantly advance our understanding of fish immune response. The main objective at hand is to define reactions common among fish species to particular pathogen groups and highly specialized responses through the meta-analysis of transcriptome data. This method will show host-specific expression profiles and make it easier to find disease prognostic indicators.

REFERENCES

Altmann, S. M., Mellon, M. T., Distel, D. L., & Kim, C. H. (2003). Molecular and functional analysis of an interferon gene from the zebrafish, Danio rerio. *Journal of Virology*, *77*(3), 1992–2002.

Aoki, T., Takano, T., Santos, M. D., Kondo, H., & Hirono, I. (2008, October). Molecular innate immunity in teleost fish: review and future perspectives. In *Fisheries for Global Welfare and Environment, Memorial Book of the 5th World Fisheries Congress* (pp. 263–276). Terrapub: Tokyo, Japan.

Assefa, A., & Abunna, F. (2018). Maintenance of fish health in aquaculture: review of epidemiological approaches for prevention and control of infectious disease of fish. *Veterinary Medicine International*, *2018*.

Balu, S., & Kaiser, P. (2003). Avian interleukin-12β (p40): cloning and characterization of the cDNA and gene. *Journal of Interferon & Cytokine Research*, *23*(12), 699–707.

Beck, B. H., Farmer, B. D., Straus, D. L., Li, C., & Peatman, E. (2012). Putative roles for a rhamnose binding lectin in *Flavobacterium columnare* pathogenesis in channel catfish *Ictalurus punctatus*. *Fish & Shellfish Immunology*, *33*(4), 1008–1015.

Beck, G., & Habicht, G. S. (1996). Characterization of an IL-6-like molecule from an echinoderm (Asterias forbesi). *Cytokine*, *8*(7), 507–512.

Bergan, V., Steinsvik, S., Xu, H., Kileng, Ø., & Robertsen, B. (2006). Promoters of type I interferon genes from Atlantic salmon contain two main regulatory regions. *The FEBS Journal*, *273*(17), 3893–3906.

Berra, T. M. (2001). *Freshwater fish distribution*. Academic Press.

Bird, S., Zou, J., Kono, T., Sakai, M., Dijkstra, J. M., & Secombes, C. (2005a). Characterisation and expression analysis of interleukin 2 (IL-2) and IL-21 homologues in the Japanese pufferfish, Fugu rubripes, following their discovery by synteny. *Immunogenetics*, *56*(12), 909–923.

Bird, S., Zou, J., Savan, R., Kono, T., Sakai, M., Woo, J., & Secombes, C. (2005b). Characterisation and expression analysis of an interleukin 6 homologue in the Japanese pufferfish, Fugu rubripes. *Developmental & Comparative Immunology*, *29*(9), 775–789.

Biswas, G., Kinoshita, S., Kono, T., Hikima, J. I., & Sakai, M. (2015). Evolutionary evidence of tumor necrosis factor super family members in the Japanese pufferfish (Takifugu rubripes): comprehensive genomic identification and expression analysis. *Marine Genomics*, *22*, 25–36.

Botham, J. W., & Manning, M. J. (1981). The histogenesis of the lymphoid organs in the carp Cyprinus carpio L. and the ontogenetic development of allograft reactivity. *Journal of Fish Biology*, *19*(4), 403–414.

Carey, F. G., & Lawson, K. D. (1973). Temperature regulation in free-swimming bluefin tuna. *Comparative Biochemistry and Physiology Part A: Physiology*, *44*(2), 375–392.

Castellana, B., Iliev, D. B., Sepulcre, M. P., MacKenzie, S., Goetz, F. W., Mulero, V., & Planas, J. V. (2008). Molecular characterization of interleukin-6 in the gilthead seabream (Sparus aurata). *Molecular Immunology*, *45*(12), 3363–3370.

Chaves-Pozo, E., Valero, Y., Esteve-Codina, A., Gómez-Garrido, J., Dabad, M., Alioto, T.,… … & Cuesta, A. (2017). Innate cell-mediated cytotoxic activity of European sea bass leucocytes against nodavirus-infected cells: a functional and RNA-seq study. *Scientific Reports*, *7*(1), 1–15.

Chen, H. H., Lin, H. T., Foung, Y. F., & Lin, J. H. Y. (2012). The bioactivity of teleost IL-6: IL-6 protein in orange-spotted grouper (*Epinephelus coioides*) induces Th2 cell differentiation pathway and antibody production. *Developmental & Comparative Immunology*, *38*(2), 285–294.

Chen, J., Xu, Q., Wang, T., Collet, B., Corripio-Miyar, Y., Bird, S.,… … & Zou, J. (2013). Phylogenetic analysis of vertebrate CXC chemokines reveals novel lineage specific groups in teleost fish. *Developmental & Comparative Immunology*, *41*(2), 137–152.

Chilmonczyk, S. (1992). The thymus in fish: development and possible function in the immune response. *Annual Review of Fish Diseases*, *2*, 181–200.

Corripio-Miyar, Y., Zou, J., Richmond, H., & Secombes, C. J. (2009). Identification of interleukin-22 in gadoids and examination of its expression level in vaccinated fish. *Molecular Immunology*, *46*(10), 2098–2106.

Costa, J. Z., & Thompson, K. D. (2016). Understanding the interaction between *Betanodavirus* and its host for the development of prophylactic measures for viral encephalopathy and retinopathy. *Fish & Shellfish Immunology*, *53*, 35–49.

Costa, M. M., Maehr, T., Diaz-Rosales, P., Secombes, C. J., & Wang, T. (2011). Bioactivity studies of rainbow trout (*Oncorhynchus mykiss*) interleukin-6: effects on macrophage growth and antimicrobial peptide gene expression. *Molecular Immunology*, *48*(15-16), 1903–1916.

Cupedo, T., Crellin, N. K., Papazian, N., Rombouts, E. J., Weijer, K., Grogan, J. L.,… … & Spits, H. (2009). Human fetal lymphoid tissue–inducer cells are interleukin 17–producing precursors to RORC+ CD127+ natural killer–like cells. *Nature Immunology*, *10*(1), 66–74.

DeVries, M. E., Ran, L., & Kelvin, D. J. (1999, April). On the edge: the physiological and pathophysiological role of chemokines during inflammatory and immunological responses. In *Seminars in Immunology* (*Vol. 11*, No. 2, pp. 95–104). Academic Press.

Dinarello, C. A. (1984). Interleukin-1. *Reviews of Infectious Diseases*, *6*(1), 51–95.

Ding, Y., Ao, J., Ai, C., & Chen, X. (2016). Molecular and functional identification of three interleukin-17A/F (IL-17A/F) homologues in large yellow croaker (*Larimichthys crocea*). *Developmental & Comparative Immunology*, *55*, 221–232.

Dixon, B., & Stet, R. J. M. (2001). The relationship between major histocompatibility receptors and innate immunity in teleost fish. *Developmental & Comparative Immunology*, *25*(8-9), 683–699.

El-Matbouli, M., & Hoffmann, R. W. (2002). Influence of water quality on the outbreak of proliferative kidney disease–field studies and exposure experiments. *Journal of Fish Diseases*, *25*(8), 459–467.

FAO, *The State of World Fisheries and Aquaculture, Contributing to food security and nutrition for all*, Food and Agriculture Organization of the United Nations, Rome, Italy, 2016.

Feinen, B., & Russell, M. W. (2012). Contrasting roles of IL-22 and IL-17 in murine genital tract infection by *Neisseria gonorrhoeae*. *Frontiers in Immunology*, *3*, 11.

Fiorentino, D. F., Bond, M. W., & Mosmann, T. R. (1989). Two types of mouse T helper cell. IV. Th2 clones secrete a factor that inhibits cytokine production by Th1 clones. *The Journal of Experimental Medicine*, *170*(6), 2081–2095.

Firdaus-Nawi, M., & Zamri-Saad, M. (2016). Major Components of Fish Immunity: A Review. *Pertanika Journal of Tropical Agricultural Science*, *39*(4).

Flegr, J. (2006). Evolutionary parasitology: the development of invasion, evasion, and survival mechanisms used by bacterial, viral, protozoan, and metazoan parasites. *Food Consumption and Disease Risk: Consumer-Pathogen Interactions*, 1st ed.; Potter, M., Ed, 251–270.

Fujita, T., Takaoka, C., Matsui, H., & Taniguchi, T. (1983). Structure of the human interleukin 2 gene. *Proceedings of the National Academy of Sciences*, *80*(24), 7437–7441.

Garcia-Castillo, J., Chaves-Pozo, E., Olivares, P., Pelegin, P., Meseguer, J., & Mulero, V. (2004). The tumor necrosis factor α of the bony fish seabream exhibits the in vivo proinflammatory and proliferative activities of its mammalian counterparts, yet it functions in a species-specific manner. *Cellular and Molecular Life Sciences CMLS*, *61*(11), 1331–1340.

Glimcher, L. H., & Murphy, K. M. (2000). Lineage commitment in the immune system: the T helper lymphocyte grows up. *Genes & Development*, *14*(14), 1693–1711.

Goldman, K. J. (1997). Regulation of body temperature in the white shark, *Carcharodon carcharias*. *Journal of Comparative Physiology B*, *167*(6), 423–429.

Grace, M. F., & Manning, M. J. (1980). Histogenesis of the lymphoid organs in rainbow trout, Salmo gairdneri Rich. 1836. *Developmental & Comparative Immunology*, *4*, 255–264.

Grayfer, L., Hodgkinson, J. W., Hitchen, S. J., & Belosevic, M. (2011). Characterization and functional analysis of goldfish (*Carassius auratus L.*) interleukin-10. *Molecular Immunology*, *48*(4), 563–571.

Grayfer, L., Walsh, J. G., & Belosevic, M. (2008). Characterization and functional analysis of goldfish (*Carassius auratus L.*) tumor necrosis factor-alpha. *Developmental & Comparative Immunology*, *32*(5), 532–543.

Gunimaladevi, I., Savan, R., Sato, K., Yamaguchi, R., & Sakai, M. (2007). Characterization of an interleukin-15 like (IL-15L) gene from zebrafish (*Danio rerio*). *Fish & Shellfish Immunology*, *22*(4), 351–362.

Hino, K., Nakamura, O., Yoshiura, Y., Suetake, H., Suzuki, Y., & Watanabe, T. (2006). TNF induces the growth of thymocytes in rainbow trout. *Developmental & Comparative Immunology*, *30*(7), 639–647.

Hong, S., Zou, J., Collet, B., Bols, N. C., & Secombes, C. J. (2004). Analysis and Characterisation of IL-1β processing in rainbow trout, *Oncorhynchus mykiss*. *Fish & Shellfish Immunology*, *16*(3), 453–459.

Hosken, N. A., Shibuya, K., Heath, A. W., Murphy, K. M., & O'Garra, A. (1995). The effect of antigen dose on CD4+ T helper cell phenotype development in a T cell receptor-alpha beta-transgenic model. *The Journal of Experimental Medicine*, *182*(5), 1579–1584.

Hu, Y., Li, A., Xu, Y., Jiang, B., Lu, G., & Luo, X. (2017). Transcriptomic variation of locally-infected skin of

Epinephelus coioides reveals the mucosal immune mechanism against *Cryptocaryon irritans*. *Fish & Shellfish Immunology*, *66*, 398–410.

Huang, D., Cancilla, M. R., & Morahan, G. (2000). Complete primary structure, chromosomal localisation, and definition of polymorphisms of the gene encoding the human interleukin-12 p40 subunit. *Genes & Immunity*, *1*(8), 515–520.

Huising, M. O., Kruiswijk, C. P., Van Schijndel, J. E., Savelkoul, H. F., Flik, G., & Verburg-van Kemenade, B. M. (2005a). Multiple and highly divergent IL-11 genes in teleost fish. *Immunogenetics*, *57*(6), 432–443.

Huising, M.O., Kruiswijk, C. P., van Schijndel, J. E., Savelkoul, H. F., Flik, G., Verburg-van Kemenade, B.M., (2005b). Multiple and highly divergent IL-11 genes in teleost fish. Immunogenetics, 1–12.

Igawa, D., Sakai, M., & Savan, R. (2006). An unexpected discovery of two interferon gamma-like genes along with interleukin (IL)-22 and-26 from teleost: IL-22 and-26 genes have been described for the first time outside mammals. *Molecular Immunology*, *43*(7), 999–1009.

Iliev, D. B., Castellana, B., MacKenzie, S., Planas, J. V., & Goetz, F. W. (2007). Cloning and expression analysis of an IL-6 homolog in rainbow trout (*Oncorhynchus mykiss*). *Molecular Immunology*, *44*(7), 1803–1807.

Jósefsson, S., & Tatner, M. F. (1993). Histogenesis of the lymphoid organs in sea bream (Sparus aurata L.). *Fish & Shellfish Immunology*, *3*(1), 35–49.

Kinoshita, S., Biswas, G., Kono, T., Hikima, J., & Sakai, M. (2014). Presence of two tumor necrosis factor (tnf)-α homologs on different chromosomes of zebrafish (*Danio rerio*) and medaka (*Oryzias latipes*). *Marine Genomics*, *13*, 1–9.

Kono, T., & Sakai, M. (2001). The analysis of expressed genes in the kidney of Japanese flounder, *Paralichthys olivaceus*, injected with the immunostimulant peptidoglycan. *Fish & Shellfish Immunology*, *11*(4), 357–366.

Kono, T., Korenaga, H., & Sakai, M. (2011). Genomics of fish IL-17 ligand and receptors: a review. *Fish & Shellfish Immunology*, *31*(5), 635–643.

Korenaga, H., Kono, T., & Sakai, M. (2010). Isolation of seven IL-17 family genes from the Japanese pufferfish *Takifugu rubripes*. *Fish & Shellfish Immunology*, *28*(5-6), 809–818.

Korzh, V., Sleptsova, I., Liao, J., He, J., & Gong, Z. (1998). Expression of zebrafish bHLH genes ngn1and nrd defines distinct stages of neural differentiation. *Developmental dynamics: an official publication of the American Association of Anatomists*, *213*(1), 92–104.

Krause, H., Jandrig, B., Wernicke, C., Bulfone-Paus, S., Pohl, T., & Diamantstein, T. (1996). Genomic Structure and Chromosomal Localization of The Human Interleukin 15 GENE (IL–15). *Cytokine*, *8*(9), 667–674.

Kubečka, J., Boukal, D. S., Čech, M., Hickley, P., Kitchell, J. F., Ricard, D.,… … & Welcomme, R. (2016). Ecology and ecological quality of fish in lakes and reservoirs. *Fisheries Research*, *173*, 1–3.

Küchler, A. M., Gjini, E., Peterson-Maduro, J., Cancilla, B., Wolburg, H., & Schulte-Merker, S. (2006). Development of the zebrafish lymphatic system requires VEGFC signaling. *Current Biology*, *16*(12), 1244–1248.

Lee, D. S., Hong, S. H., Lee, H. J., Jun, L. J., Chung, J. K., Kim, K. H., & Do Jeong, H. (2006). Molecular cDNA cloning and analysis of the organization and expression of the IL-1β gene in the Nile tilapia, *Oreochromis niloticus*. *Comparative Biochemistry and Physiology Part A: Molecular & Integrative Physiology*, *143*(3), 307–314.

Lee, X., Yi, Y., Weng, S., Zeng, J., Zhang, H., He, J., & Dong, C. (2016). Transcriptomic analysis of koi (*Cyprinus carpio*) spleen tissue upon *Cyprinid herpesvirus* 3 (CyHV3) infection using next generation sequencing. *Fish & Shellfish Immunology*, *49*, 213–224.

Lenoir, G., & De Kinkelin, P. (1975). Fish rhabdoviruses: comparative study of protein structure. *Journal of Virology*, *16*(2), 259–262.

Li, C., Zhang, Y., Wang, R., Lu, J., Nandi, S., Mohanty, S.,… … & Peatman, E. (2012). RNA-seq analysis of mucosal immune responses reveals signatures of intestinal barrier disruption and pathogen entry following *Edwardsiella ictaluri* infection in channel catfish, *Ictalurus punctatus*. *Fish & Shellfish Immunology*, *32*(5), 816–827.

Liu, Q., Du, X. X., Schindel, D. T., Yang, Z. X., Rescorla, F. J., Williams, D. A., & Grosfeld, J. L. (1996). Trophic effects of interleukin-11 in rats with experimental short bowel syndrome. *Journal of Pediatric Surgery*, *31*(8), 1047–1051.

Liu, Y., Chang, M. X., Wu, S. G., & Nie, P. (2009). Characterization of C–C chemokine receptor subfamily in teleost fish. *Molecular Immunology*, *46*(3), 498–504.

Maloy, K. J., Salaun, L., Cahill, R., Dougan, G., Saunders, N. J., & Powrie, F. (2003). CD4+ CD25+ TR cells suppress innate immune pathology through cytokine-dependent mechanisms. *The Journal of Experimental Medicine*, *197*(1), 111–119.

Medzhitov, R., & Janeway, C. A. (1997). Innate immunity: the virtues of a nonclonal system of recognition. *Cell*, *91*(3), 295–298.

Monte, M. M., Wang, T., Holland, J. W., Zou, J., & Secombes, C. J. (2013). Cloning and characterization of rainbow trout interleukin-17A/F2 (IL-17A/F2) and IL-17 receptor A: expression during infection and bioactivity of recombinant IL-17A/F2. *Infection and Immunity*, *81*(1), 340–353.

Mosmann, T. R., Cherwinski, H., Bond, M. W., Giedlin, M. A., & Coffman, R. L. (1986). Two types of murine helper T cell clone. I. Definition according to profiles of lymphokine activities and secreted proteins. *The Journal of Immunology*, *136*(7), 2348–2357.

Mu, Y., Li, M., Ding, F., Ding, Y., Ao, J., Hu, S., & Chen, X. (2014). De novo characterization of the spleen transcriptome of the large yellow croaker (*Pseudosciaena crocea*) and analysis of the immune relevant genes and pathways involved in the antiviral response. *PLoS One*, *9*(5), e97471.

Nakanishi, T., Hikima, J. I., & Yada, T. (2018). *Osteichthyes*: immune systems of teleosts (*Actinopterygii*). In *Advances in Comparative Immunology* (pp. 687–749). Springer, Cham.

Nam, B. H., Byon, J. Y., Kim, Y. O., Park, E. M., Cho, Y. C., & Cheong, J. (2007). Molecular cloning and characterisation of the flounder (*Paralichthys olivaceus*) interleukin-6 gene. *Fish & Shellfish Immunology*, *23*(1), 231–236.

Nomiyama, H., Hieshima, K., Osada, N., Kato-Unoki, Y., Otsuka-Ono, K., Takegawa, S.,… … & Yoshie, O. (2008). Extensive expansion and diversification of the chemokine gene family in zebrafish: identification of a novel chemokine subfamily CX. *BMC Genomics*, *9*(1), 1–19.

Øvergård, A. C., Nepstad, I., Nerland, A. H., & Patel, S. (2012). Characterisation and expression analysis of the Atlantic halibut (*Hippoglossus hippoglossus L.*) cytokines: IL-1β, IL-6, IL-11, IL-12β and IFNγ. *Molecular Biology Reports*, *39*(3), 2201–2213.

Palacios, G., Lovoll, M., Tengs, T., Hornig, M., Hutchison, S., Hui, J.,… … & Lipkin, W. I. (2010). Heart and skeletal muscle inflammation of farmed salmon is associated with infection with a novel reovirus. *PLoS one*, *5*(7), e11487.

Palanisamy, R., Bhatt, P., Kumaresan, V., Pasupuleti, M., & Arockiaraj, J. (2018). Innate and adaptive immune molecules of striped murrel *Channa striatus*. *Reviews in Aquaculture*, *10*(2), 296–319.

Parrish-Novak, J., Dillon, S. R., Nelson, A., Hammond, A., Sprecher, C., Gross, J. A.,… … & Foster, D. (2000). Interleukin 21 and its receptor are involved in NK cell expansion and regulation of lymphocyte function. *Nature*, *408*(6808), 57–63.

Peatman, E., Li, C., Peterson, B. C., Straus, D. L., Farmer, B. D., & Beck, B. H. (2013). Basal polarization of the mucosal compartment in *Flavobacterium columnare* susceptible and resistant channel catfish (*Ictalurus punctatus*). *Molecular Immunology*, *56*(4), 317–327.

Petrie-Hanson, L., & Ainsworth, A. J. (2001). Ontogeny of channel catfish lymphoid organs. *Veterinary Immunology and Immunopathology*, *81*(1-2), 113–127.

Piazzon, M. C., Savelkoul, H. F., Pietretti, D., Wiegertjes, G. F., & Forlenza, M. (2015). Carp Il10 has anti-inflammatory activities on phagocytes, promotes proliferation of memory T cells, and regulates B cell differentiation and antibody secretion. *The Journal of Immunology*, *194*(1), 187–199.

Pleguezuelos, O., Zou, J., Cunningham, C., & Secombes, C. J. (2000). Cloning, sequencing, and analysis of expression of a second IL-1 β gene in rainbow trout (*Oncorhynchus mykiss*). *Immunogenetics*, *51*(12), 1002–1011.

Polinski, M. P., Bradshaw, J. C., Inkpen, S. M., Richard, J., Fritsvold, C., Poppe, T. T.,… … & Johnson, S. C. (2016). De novo assembly of Sockeye salmon kidney transcriptomes reveal a limited early response to piscine reovirus with or without infectious hematopoietic necrosis virus superinfection. *BMC Genomics*, *17*(1), 1–22.

Praveen, K., Evans, D. L., & Jaso-Friedmann, L. (2006). Constitutive expression of tumor necrosis factor-alpha in cytotoxic cells of teleosts and its role in regulation of cell-mediated cytotoxicity. *Molecular Immunology*, *43*(3), 279–291.

Qin, C., Gong, Q., Wen, Z., Yuan, D., Shao, T., Wang, J., & Li, H. (2017). Transcriptome analysis of the spleen of the darkbarbel catfish *Pelteobagrus vachellii* in response to *Aeromonas hydrophila* infection. *Fish & Shellfish Immunology*, *70*, 498–506.

Rahman, M. M., & McFadden, G. (2006). Modulation of tumor necrosis factor by microbial pathogens. *PLoS Pathogens*, *2*(2), e4.

Rauta, P. R., Nayak, B., & Das, S. (2012). Immune system and immune responses in fish and their role in comparative immunity study: a model for higher organisms. *Immunology Letters*, *148*(1), 23–33.

Reyes-Cerpa, S., Maisey, K., Reyes-López, F., Toro-Ascuy, D., Sandino, A. M., & Imarai, M. (2012a). Fish cytokines and immune response. *New Advances and Contributions to Fish Biology*, 3–57.

Reyes-Cerpa, S., Reyes-López, F. E., Toro-Ascuy, D., Ibañez, J., Maisey, K., Sandino, A. M., & Imarai, M. (2012b). IPNV modulation of pro and anti-inflammatory cytokine expression in Atlantic salmon might help the establishment of infection and persistence. *Fish & Shellfish Immunology*, *32*(2), 291–300.

Rico, A., Satapornvanit, K., Haque, M. M., Min, J., Nguyen, P. T., Telfer, T. C., & Van Den Brink, P. J. (2012). Use of chemicals and biological products in Asian aquaculture and their potential environmental risks: a critical review. *Reviews in Aquaculture*, *4*(2), 75–93.

Robertsen, B., Bergan, V., Røkenes, T., Larsen, R., & Albuquerque, A. (2003). Atlantic salmon interferon genes: cloning, sequence analysis, expression, and biological activity. *Journal of Interferon & Cytokine Research*, *23*(10), 601–612.

Robledo, D., Ronza, P., Harrison, P. W., Losada, A. P., Bermúdez, R., Pardo, B. G.,… … & Martínez, P. (2014). RNA-seq analysis reveals significant transcriptome changes in turbot (*Scophthalmus maximus*) suffering severe enteromyxosis. *BMC Genomics*, *15*(1), 1–17.

Saeij, J. P., Stet, R. J., De Vries, B. J., Van Muiswinkel, W. B., & Wiegertjes, G. F. (2003). Molecular and functional characterization of carp TNF: a link between TNF polymorphism and trypanotolerance?. *Developmental & Comparative Immunology*, *27*(1), 29–41.

Sahoo, S., Banu, H., Prakash, A., & Tripathi, G. (2021). Immune system of fish: an evolutionary perspective. *Antimicrobial Immune Response*, *1*.

Sakai, M., Hikima, J. I., & Kono, T. (2021). Fish cytokines: Current research and applications. *Fisheries Science*, *87*(1), 1–9.

Salazar-Mather, T. P., & Hokeness, K. L. (2006). Cytokine and chemokine networks: pathways to antiviral defense. *Chemokines and Viral Infection*, 29–46.

Savan, R., & Sakai, M. (2006). Genomics of fish cytokines. *Comparative Biochemistry and Physiology Part D: Genomics and Proteomics*, *1*(1), 89–101.

Savan, R., Kono, T., Igawa, D., & Sakai, M. (2005). A novel tumor necrosis factor (TNF) gene present in tandem with theTNF-α gene on the same chromosome in teleosts. *Immunogenetics*, *57*(1), 140–150.

Secombes, C. J., Wang, T., & Bird, S. (2011). The interleukins of fish. *Developmental & Comparative Immunology*, *35*(12), 1336–1345.

Shi, M., Huang, R., Du, F., Pei, Y., Liao, L., Zhu, Z., & Wang, Y. (2014). RNA-seq profiles from grass carp tissues after reovirus (GCRV) infection based on singular and modular enrichment analyses. *Molecular Immunology*, *61*(1), 44–53.

Silva, J. R. M. C. D., Staines, N. A., Hernandez-Blazquez, F. J., Porto-Neto, L. R., & Borges, J. C. S. (2002). Phagocytosis and giant cell formation at 0 C by macrophage (MØ) of *Notothenia coriiceps*. *Journal of Fish Biology*, *60*(2), 466–478.

Skelly, S. M., Tackney, C., Hicklin, D., Tamkins, T., Goldstein, N., Waksal, H., & Dagan, S. (1994). High-level expression of a biologically active human interleukin-6 mutein. *Journal of Biotechnology*, *34*(1), 79–86.

Smail, D. A., & Munro, E. S. (2012). The virology of teleosts. *Fish Pathology*, 186–291.

Sunarto, A., Liongue, C., McColl, K. A., Adams, M. M., Bulach, D., Crane, M. S. J.,… … & Walker, P. J. (2012). Koi herpesvirus encodes and expresses a functional interleukin-10. *Journal of Virology*, *86*(21), 11512–11520.

Takahashi, Y., Okamura, Y., Morimoto, N., Mihara, K., Maekawa, S., Wang, H. C.,… … & Hikima, J. I. (2020). Interleukin-17A/F1 from Japanese pufferfish (*Takifugu rubripes*) stimulates the immune response in head kidney and intestinal cells. *Fish & Shellfish Immunology*, *103*, 143–149.

Tort, L., Balasch, J. C., & Mackenzie, S. (2003). Fish immune system. A crossroads between innate and adaptive responses. *Inmunología*, *22*(3), 277–286.

Van Beurden, S. J., Forlenza, M., Westphal, A. H., Wiegertjes, G. F., Haenen, O. L., & Engelsma, M. Y. (2011). The alloherpesviral counterparts of interleukin 10 in European eel and common carp. *Fish & Shellfish Immunology*, *31*(6), 1211–1217.

Varela, M., Dios, S., Novoa, B., & Figueras, A. (2012). Characterisation, expression and ontogeny of interleukin-6 and its receptors in zebrafish (*Danio rerio*). *Developmental & Comparative Immunology*, *37*(1), 97–106.

Wang, T., Díaz-Rosales, P., Martin, S. A., & Secombes, C. J. (2010). Cloning of a novel interleukin (IL)-20-like gene in rainbow trout *Oncorhynchus mykiss* gives an insight into the evolution of the IL-10 family. *Developmental & Comparative Immunology*, *34*(2), 158–167.

Wang, T., Holland, J. W., Bols, N., & Secombes, C. J. (2005). Cloning and expression of the first nonmammalian interleukin-11 gene in rainbow trout *Oncorhynchus mykiss*. *The FEBS Journal*, *272*(5), 1136–1147.

Wang, T., Huang, W., Costa, M. M., & Secombes, C. J. (2011). The gamma-chain cytokine/receptor system in fish: more ligands and receptors. *Fish & Shellfish Immunology*, *31*(5), 673–687.

Wei, X., Li, B., Wu, L., Yin, X., Zhong, X., Li, Y.,… … & Ye, J. (2018). Interleukin-6 gets involved in response to bacterial infection and promotes antibody production in Nile tilapia (*Oreochromis niloticus*). *Developmental & Comparative Immunology*, *89*, 141–151.

Wilson, N. J., Boniface, K., Chan, J. R., McKenzie, B. S., Blumenschein, W. M., Mattson, J. D.,… … & de Waal Malefyt, R. (2007). Development, cytokine profile and function of human interleukin 17–producing helper T cells. *Nature Immunology*, *8*(9), 950–957.

Xiang, L. X., He, D., Dong, W. R., Zhang, Y. W., & Shao, J. Z. (2010). Deep sequencing-based transcriptome profiling analysis of bacteria-challenged *Lateolabrax japonicus* reveals insight into the immune-relevant genes in marine fish. *BMC Genomics*, *11*(1), 1–21.

Xu, C., Evensen, Ø., & Munang'andu, H. M. (2015). De novo assembly and transcriptome analysis of Atlantic salmon macrophage/dendritic-like TO cells following type I IFN treatment and Salmonid alphavirus subtype-3 infection. *BMC Genomics*, *16*(1), 1–16.

Yin, F., Sun, P., Wang, J., & Gao, Q. (2016). Transcriptome analysis of dormant tomonts of the marine fish ectoparasitic ciliate *Cryptocaryon irritans* under low temperature. *Parasites & Vectors*, *9*(1), 1–12.

Yin, Z., & Kwang, J. (2000). Carp interleukin-1β in the role of an immuno-adjuvant. *Fish & Shellfish Immunology*, *10*(4), 375–378.

Zhang, J., & Sun, L. (2017). Transcriptome analysis reveals temperature-regulated antiviral response in turbot *Scophthalmus maximus*. *Fish & Shellfish Immunology*, *68*, 359–367.

Zhu, R., Liu, X. X., Lv, X., Li, S. Y., Li, Y. D., Yu, X. J., & Wang, X. G. (2017). Deciphering transcriptome profile of the yellow catfish (*Pelteobagrus fulvidraco*) in response to *Edwardsiella ictaluri*. *Fish & Shellfish Immunology*, *70*, 593–608.b

Zou, J., & Secombes, C. J. (2016). The function of fish cytokines. *Biology*, *5*(2), 23.

Zou, J., Bird, S., & Secombes, C. (2004). Fish cytokine gene discovery and linkage using genomic approaches. *Marine Biotechnology*, *6*, S533–S539.

Zou, J., Bird, S., Truckle, J., Bols, N., Horne, M., & Secombes, C. (2004). Identification and expression analysis of an IL-18 homologue and its alternatively spliced form in rainbow trout (*Oncorhynchus mykiss*). *European Journal of Biochemistry*, *271*(10), 1913–1923.

Zou, J., Grabowski, P. S., Cunningham, C., & Secombes, C. J. (1999). Molecular cloning of interleukin 1β from rainbow trout *Oncorhynchus mykiss* reveals no evidence of an ice cut site. *Cytokine*, *11*(8), 552–560.

13

Progressive Immunomodulation Through Nanotechnology

Heba Mahboub[1], Hiam Elabd[2], Mian Adnan Kakakhel[3], Gehad E. Elshopakey[4], Maram H. Abduljabbar[5], and Manal E. Alosaimi[6]

[1]*Fish Diseases and Management Department, Faculty of Veterinary Medicine, Zagazig University, Zagazig, Sharkia, Egypt*

[2]*Aquatic Animals' Medicine Department, Faculty of Veterinary Medicine, Benha University, Moshtohor, Toukh, Egypt*

[3]*MOE Key Laboratory of Cell Activities and Stress Adaptations, School of Life Sciences, Lanzhou University, Lanzhou, P.R. China*

[4]*Department of Clinical Pathology, Faculty of Veterinary Medicine, Mansoura University, Mansoura, Dakahlia, Egypt*

[5]*Department of Pharmacology and Toxicology, College of Pharmacy, Taif University, Taif, Saudi Arabia*

[6]*Department of Basic Science, College of Medicine, Princess Nourah Bint Abdulrahman University, Riyadh, Saudi Arabia*

13.1 Introduction

Fish represents almost 17% of the international human consumption of animal protein [1,2]. Fish display a very important role in food security as they secure food and job opportunities, besides a variety of essential nutrients particularly amino acids, fatty acids, vitamins, and minerals, for instance, selenium and iodine that is not present in other crops in such a rich manner [3]. The immune responses of fish plays a protective role against pathogens and diseases [4]. The fish immune response is categorized into cell-mediated and antibody-mediated immune responses in fish that is controlled by environmental factors.

The aquaculture industry is exposed to various diseases [5,6] that have their negative influences on the immune system. Use of antibiotics and other chemical remedies have many disadvantages, including difficulty in estimating concentration, increased bioburden of water, high cost, antibiotics resistance, and accumulation with residues in the fish muscles which has a hazardous effects on fish health, performance, growth, and can cause mortalities, beside the negative impacts of residues on fish consumers [7]. Several elements other than antibiotics are used, such as probiotics, oils, minerals, and vaccines to boost the immune response with various beneficial roles [2,8–10].

In aquaculture, nanotechnology has achieved great progress and success in the last decade, and currently they are used for multiple purposes, including immunomodulation, antioxidant activation, fish growth and digestibility enhancer, and treatment of anemia and treatment or control of other diseases such as fungal and bacterial diseases, besides their efficient use in drug delivery [11]. Recently, it has been reported that fish health and performance can also benefit from nanotechnology applications as nanoparticles prove their influence on fish growth, digestibility, blood indices, and on many essential biochemical parameters either immunological or antioxidant. This is due to their use in aquaculture systems, through the administration of different therapies in fish diets using nanomaterials with porous properties, in addition to nano-sensors to detect fish pathogens. They also achieved great success and promising immunological and antioxidant responses via direct administration of nanoparticles in the aqueous environment by mixing with water in an appropriate amount and they exert their beneficial action via absorption in the body of the aquatic organisms. Hence, in a wide range of pond ecosystem environments, nanoparticles have revealed great importance [11]. Nanoparticles can be considered immune-modulating substances as they have a potent influence on the immune system and reflect better immune response either directly by their compositions or indirectly using an intact carrier and their strength can be utilized for treatment of various diseases [11].

Several reports have clarified that smaller particles of nanoparticles (NPs) induce a better immune response compared to bigger parallels Figure 13.1 [11].

DOI: 10.1201/9781003361183-16

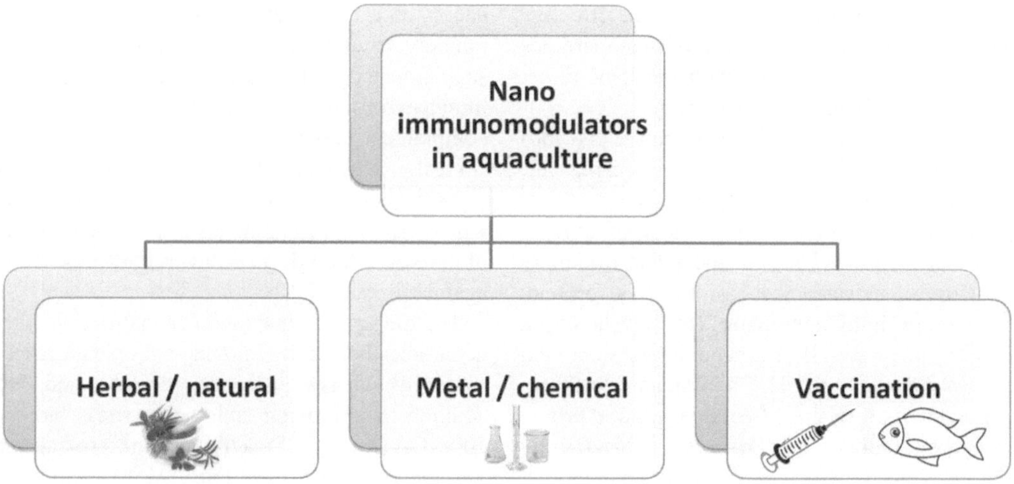

FIGURE 13.1 Types of nano immune-modulating materials in aquaculture.

13.2 Fish Immune Response

Fish immune response includes two main constituents (innate and adaptive) [4]. Non-specific (innate) immunity is the primary defense against pathogenic invasions and comprises non-specific humeral and cellular immune responses. The non-specific response consists of interferons, lysozyme, C-reactive proteins, the complement, transferrin, and lectins, while the non-specific cellular response involves the macrophages, neutrophils, toll-like receptors (TLRs), eosinophil, and cytotoxic cells. The innate immune response also includes the portions encoded by the disease-causing pathogen identification lineage. TLRs and phagocytes are the main constituents of this type of immune response that eliminates the foreign bodies through identifying and phagocytosis [12]. All components complement each other to protect against pathogens invasions [4].

Adaptive immune response, alternatively, comprises two main specific functional pathways called humeral and cellular [12]. Humeral immunity includes three main types of antibodies (IgM, IgD/W, and IgT/Z) that work to protect the body from invading pathogens and threats. The cellular immune response includes the cytotoxic T-lymphocyte cells that are responsible for eliminating cellular components infected by viruses and other pathogens [13]. Both immune response categories effectively synchronize working to protect the target host against diseases. The adaptive immune response includes firstly a process of pathogen recognition through somatic pathways monitored by both the humeral and cellular (B and T lymphocytes) responses [14].

13.3 Fish Immunomodulation

13.3.1 Fish Immunomodulation Through Nanotechnology in Aquaculture Practices

Nanotechnology is considered an encouraging and a successful key response that opens a new era for sustainable aquaculture. Nanotechnology encompasses the manipulation of nanometer-sized particles (1–100 nm) that elaborate innovative characteristics for generous applications [15]. Nanoparticles are prepared by different methods and the main marketable nanoparticles consist of iron, zinc oxides, silver, gold, calcium, titanium, and silica [16].

Nano-molecules are characterized by their smaller size, which provides a larger absorption surface area, improving solubility compared to traditional particle sizes. The nanometric size also allows the particles to move easier through biological barriers and thus be more effective. Ultimately, this will lead to the use of lower doses, lower costs, higher efficacy, and a shorter time frame [17].

Nano-materials could be also used as a delivery system by loading advantageous molecules or by coating such molecules, which appear to form a shell around them, facilitating their transport and release [11].

13.3.2 Fish Immunomodulation Through Natural and Herbal Nano-Materials

Currently, mixing both herbal natural plants and nanoparticles is extensively applied in aquaculture and is considered the best option for stimulating the fish immunological response [18].

Such mixture gathers all the benefits of both neutracuticals and nanoparticles. The natural and herbal plants are distinguished by many distinctive characteristics including safety, natural substances, inexpensive, higher bioavailability, and eco-friendly neutracutical based-products. Plant extracts are characterized by stress resistance promoters, antioxidant activators, and immuno-stimulators because they are abundance in many active ingredients including flavonoids (involve anthocyanins, flavonols, isoflavones, flavanones, and flavones), alkaloids, tannins (hydrolysable tannins and condensed tannins), stilbenes, and lignans [19]. Flavonoids are the most abundant polyphenols that have characterized by potent antioxidant activities [19]. Meanwhile, nanoparticles are distinguished by wide surface area and nano-sized particles that enable nanoparticles to exert their beneficial action. The herbal-based nano-materials are extensively used nowadays in the aquaculture industry as they verify an improved therapeutic value compared to the conventional plant extracts. Also, they are the best therapeutic option as they collect the advantages of both active constituents of plant extracts and the unique effective properties for nanoparticles (the nano-size of nanoparticles enables them for easily penetration and exerting their direct beneficial action besides their large surface area).

Chitosan (CS), a natural non-toxic, biodegradable, and biocompatible polymer obtained from the crustaceans carapaces, has been introduced in pharmacology and immune response improvement [20]. Nano-chitosan dietary incorporation in rainbow trout (*Oncorhynchus mykiss*) enhances fish immunity indicated by inducing an elevated level of the total protein, lysozyme profile, serum immunoglobulin (Ig) level, serum complement activity, increased IL-1β gene expression, and protection against *Yersinia ruckeri* [21]. Chitosan nanoparticles-dietary incorporation has a promising effect on the blood profile and immunity of grey mullet (*Liza ramada*) indicated by increasing red blood cells (RBCs), hematocrit, hemoglobin, globulin, lysozyme activity, and phagocytic index [22]. Furthermore, chitosan nanoparticles can enhance fish health, immune response, and antioxidant activity by improving catalase (CAT) and glutathione peroxidase (GPx) activities, and lessening malondialdehyde (MDA) level, demonstrating its efficacy as a functional supplement in aquaculture [22]. Nano-chitosans coated with folic acid (FA) are recorded with notable positive effects in rainbow trout fingerlings, and can be utilized as a growth promoter, anemic reliever, and antioxidant-immunological supplement [23]. They result in a significant increase in hemoglobin (Hb), white blood cells (WBCs), RBCs, hematocrit (Hct), CAT, superoxide dismutase (SOD), GPX, lysozyme (LA), immunoglobulin M (IgM), and complement component (C_3 and C_4) profiles [23].

As described by [24,25], the use of nano-chitosan increases growth performance and has an immune-antioxidant-stimulating activity (LA, respiratory burst, CAT, and SOD) in Nile tilapia (*Oreochromis niloticus*). Also, dietary chitosan-selenium nanoparticles (CTS-SeNP) can improve immunity and disease resistance in zebrafish by enhancing lysozyme activity, phagocytic respiratory burst, as well as splenocyte proliferation reflecting immunomodulation effect [26]. Recently, a report states that dietary nano-chitosan enhances the growth and immune responses of rainbow trout by improving the serum complement activity, antibacterial activity against *Yersinia ruckeri*, and inducible nitric oxide synthase (iNOS) gene expression [21]. A recent study uses vitamin C encapsulated in chitosan-based nanoparticles and detects elevated levels of vitamin C in the rainbow trout serum when chitosan-based nanoparticles are supplemented in the feed, which in turn increases the innate immune response due to elevated levels of hemolytic serum complement and lysozyme hemolytic activities, compared to control groups with vitamin C and chitosan [27].

Phytochemicals are also among those natural components which can boost immune response in fish through their content of tannins, alkaloids, and flavonoids, plus being able to protect against diseases [19].

The dietary administration of *Astragalus membranaceus* nanoparticles boosts the immune-antioxidant response as they significantly enhance lysozyme, nitrous oxide (NO) activities, in addition to improving antioxidant stress biomarkers, including GPx, CAT, and SOD activities, and provide protection against *Aeromonas veronii* infection. AM-NPs can also up-regulate the expression patterns of heat shock protein-70 and interleukin-1beta. Moreover, AM-NPs decrease glucose, cortisol, alanine transaminase, and aspartate aminotransferase levels in Nile tilapia (*O. niloticus*) challenged with hypoxia and cold stressors indicating its immunomodulatory properties. The immunomodulatory properties could be attributed to the *Astragalus* polysaccharides (APS) content as the main active ingredient, well known for its immunostimulant and hepato-protective activities [28].

Spirulina nanoparticles can significantly increase the immunological parameters (NO and LA), antioxidants (PGx, CAT, and SOD), biochemical (cortisol, glucose, alanine transaminase, and aspartate aminotransferase), and up-regulate liver's heat shock protein 70 (HSP70) and IL-1β expression in Nile tilapia challenged with cold, hypoxia, and *Aeromonas veronii* at 9×10^8 CFU/ml [29]. These positive characteristics reflect better growth rate and digestibility, and they could be due to their richness in protein and polyunsaturated fatty acids (PUFA) contents and flavonoids including chlorogenic, cinnamic, gallic, and pinostrobin [29].

13.3.3 Fish Immunomodulation through Metals/Chemical Nano-Materials

In spite of the established toxicological impacts of nanoparticles (NPs) in aquaculture [24,30–34], nanotechnology has gained much acclaim for its multiple applications in various fields, including drug/vaccine delivery and immunostimulants (Table 13.1) (Figure 13.2).

TABLE 13.1

Metals/Chemical Nano-Materials Modulating Fish Immune Response and Their Effect [65,66]

Nano-material	Effect	References
Metals/chemical		
Magnetite nanoparticles (Fe_3O_4 NPs)	Adsorb heavy metals from polluted water and improves growth, hematological profile, and immune-antioxidant biomarkers	[35]
Selenium nanoparticles (Se-NPs)	Increase fish growth performance and antioxidant capacity, as well as increase glutathione peroxidase activity	[39,40]
Zinc nanoparticles (Zn-NPs)	Improve oxidative stress and enhance the carbohydrate and protein metabolic enzymes	[39,40]
Silver NPs (nAg)	Inhibit bacterial protein regulation and synthesis	[39,40]
Iron nanoparticle	Improve the antioxidants (SOD and GSH), RBCs and hemoglobin levels, and innate immune response components	[39,40]
Silica nanoparticles (SiNPs)	Improve health and growth of aquatic organisms by strengthening the ionic exchange process	[39,40]

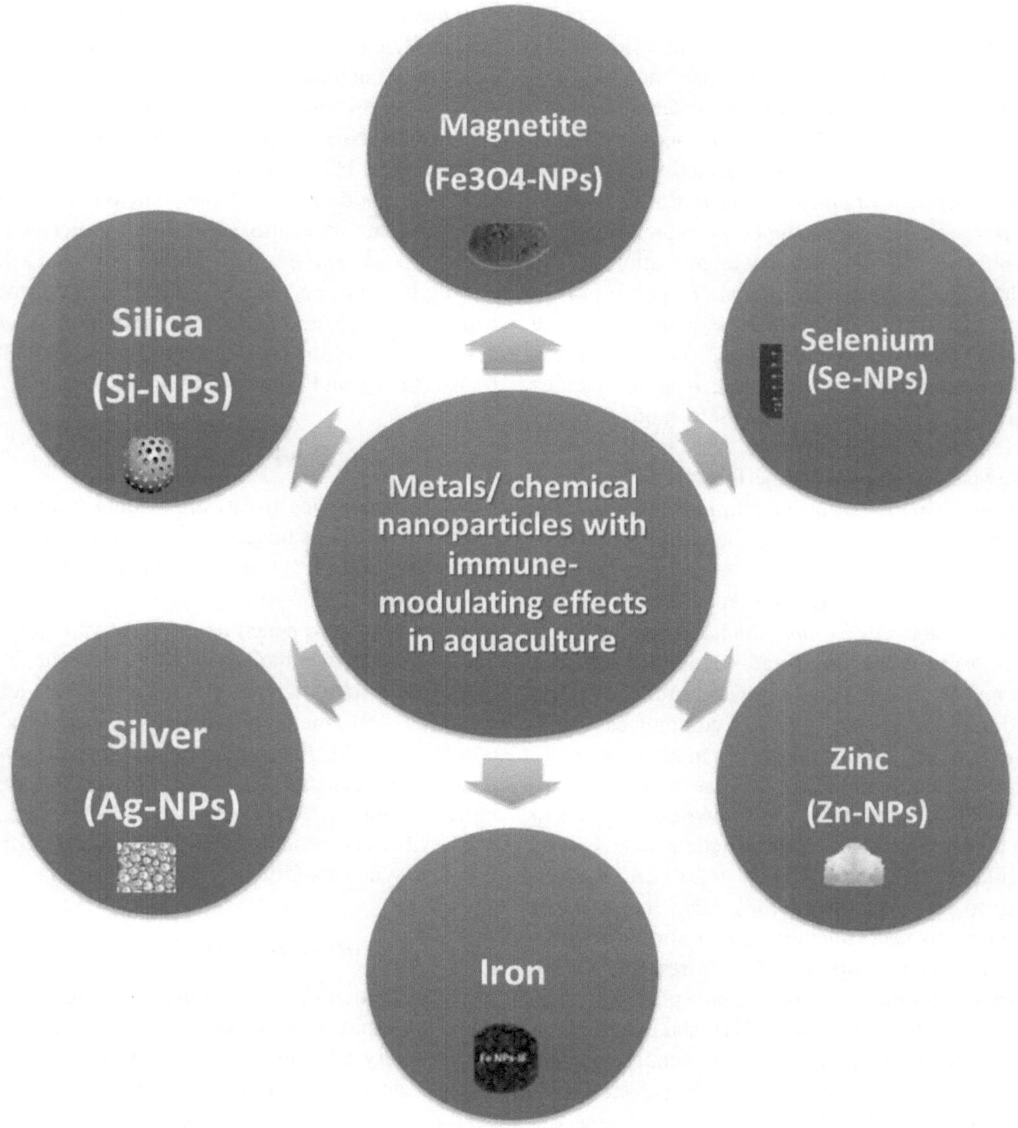

FIGURE 13.2 Types of metal NPs with immune-modulatory effects in aquaculture [65,66].

13.3.3.1 Magnetite Nanoparticles (Fe_3O_4 NPs)

Magnetite nanoparticles (Fe_3O_4 NPs) have a great ability to adsorb heavy metals (mercury) (Hg) from polluted water through their magnetic properties as the Fe_3O_4 NPs consist of crystal structure of free Fe^{2+} and Fe^{3+} electrons combined with oxygen. The detected free electrons in the crystal structure are responsible of the magnetic properties of Fe_3O_4 NPs and this electrostatic attraction can enhance the adsorption power of the magnetite [35]. Also, Fe_3O_4 NPs can enhance growth performance, hematological values, and immune-antioxidant biomarkers IgM, myeloperoxidase activities (MPO), and SOD in Nile tilapia [35]. Furthermore, Fe_3O_4 NPs are able to activate the melano-macrophage centers in the hepatic tissue and decrease the bioaccumulation of Hg and modulate immune-histo-reaction which is evidenced by computing the apoptotic marker (caspase 3) [35]. Nevertheless, it played a great role to enhance immunity immune-histo-reactivity for caspase-3 in the muscle and liver tissues. Another report by [35] concluded that NPs of Fe_3O_4 have impacts on the dendritic cell activity and improve the release of cytokines and interleukins. Also, MPO is an indicator of fish immunity, as it has been detected in neutrophils [36]; thus, the increase of MPO upon Fe_3O_4 NPs-supplementation reflects an immunostimulatory impact.

Few studies exist on the effects on marine fish species. It has been found that iron nanoparticles stimulate sturgeons and young carp to grow faster [37,38].

13.3.3.2 Selenium Nanoparticles (Se-NPs)

It has been reported that the nano-selenium-enriched diet could increase fish growth performance and immune-antioxidant capacity, as well as increase GPx activity in crucian carp (*Carassius Auratus Gibelio*) muscles [39,40].

Some reports have shown that nanoparticles interrelate with immune cells triggering an immune response [41]. *IL-10* is a vital indicator of fish immunological response, such as cytokine can reduce damage after pathogen invasion [42]. Interleukin *IL1-β* is a crucial cytokine that mediates the inflammatory response and is tangled in different cellular activities, including apoptosis [43]. Tumor necrosis factor *(TNF-α)* is an essential molecule that indicates the immunological reaction in different organisms, and is synthesized by various cell types and its major function represents in a modulation of phagocytosis and leukocyte migration in fish macrophages, therefore, reflects fish immune response [44].

Selenium nanoparticles (nano-Se) produce a strong immune response when supplemented in Nile tilapia diet and have a promising role in strengthening the immune response against cadmium toxicity [45]. Additionally, nano-Se is more effective in raising the selenium content in muscles than organic selenomethionine. Likewise, the growth and performance of ingested fish are assessed to be higher with the administration of these nano-level NPs because of augmenting the activity of GPxe enzyme in crucian carp [46].

Selenium nanoparticles can effectively enhance the thermal tolerance of catfish (*Pangasius hypophthalmus*) reared exposed to 34°C and lead (Pb) stressors [36]. Nano-selenium at a dose of 1 mg per kg diet reveals a significant improvement in the growth performance and immune defense system of common carp [47]. Besides its efficacy on composition of muscle, hemato-biochemical profile, and antioxidant defense mechanism of common carp [47]. Additionally, it shows a significant improvement in the blood profile including hemocyte levels and hemocyanin regulation of Chinese mitten crabs (*Eriocheir sinensis*) exposed to hypoxia stress [48], and the exposed Chinese mitten crabs to nano-selenium demonstrate improvement in the immunity and empower disease resistance [48]. Nano-Se can markedly improve the alkaline phosphatase, lactate dehydrogenase (ALT, AST, ALP, and LDH), and creatinine in Nile tilapia. Moreover, nano-Se shows more potent antioxidant activities than Se. Nano-Se can also effectively enhance IgM, lysozyme, GPx, NO, SOD, CAT, and their related gene expression [49]. The biosynthesized selenium nanoparticles verify a successful improvement of biochemical and hematological parameters of Nile tilapia in the elimination of Ag-NPs toxicity [50]. The dietary incorporation of nano-selenium in rainbow trout (5 mg/kg) down-regulates plenty of lipid molecules, including (3-beta-hydroxysteroid-Delta(8), Delta(7)-isomerase isoform X1 (EBP), Lanosterol 14-alpha demethylase (CYP51), Delta(24)-sterol reductase (DHCR24), 7-dehydrocholesterol reductase-like (DHCR7), and Apolipoprotein B-100-like (APOB) and pro-apoptotic proteins molecules (caspase-8 and Bad). An interaction analysis shows that nano-Se hinders apoptotic processes by up-regulating the expression of Hsp70, Hsp90a.1, Hspa8, Dnajb1, and Hspa8. Proving the novel properties of nano-Se includes tolerance of heat stress, enhancing protein repair, recovery of antioxidants, improvement the lipid metabolism, and apoptosis [51].

13.3.3.3 Zinc Nanoparticles (Zn-NPs)

Zinc nanoparticles (Zn-NPs) have a promising influence on fish immunity via mitigating oxidative stress and enhancing the carbohydrate and protein metabolic enzymes in catfish (*Pangasianodon hypophthalmus*) [52]. The biological zinc nanoparticles (BIO-ZnONPs) can enhance fish immunity by decreasing the total bacterial loads in water and fish tissues and concur the highest antibacterial activities against pathogenic bacteria in Nile tilapia [53]. Moreover, ZnONPs enhance performance, behavior, and immune response of Nile tilapia [53]. Zinc oxide nanoparticles (ZnONPs) have been found to boost the immune-antioxidant defense mechanism of African catfish (*C. gariepinus*). This finding has been achieved through elevating serum bactericidal killing activity, total antioxidant capacity (TAC), and antioxidant enzyme SOD and CAT plus main oxidative stress-associated genes, Glutathione S transferase (GST), and GPx [54].

13.3.3.4 Silver NPs (nAg)

Silver nanoparticles (nAg) are widely known for their antimicrobial activities including their antibacterial and antifungal properties because of their immune-stimulatory effect [16]. They are characterized by their strong ability for adhering in cell membrane and penetrate the cells without breaking the cell membrane and eventually inducing cell death. The mechanism of action of nAg includes a release of silver ions and bind to bacterial cell membrane causing their malfunction and destruction [55]. Its combination with chitosan (chitosan-Ag nanocomposites, CAgNCs) has also been tested as an antibacterial remedy for *Aliivibrio salmonicida*, the cause of cold-water vibriosis through inhibition of bacterial protein regulation and synthesis [56].

13.3.3.5 Iron Nanoparticles

Iron nanoparticles have been used in sturgeon and young carp, revealing rapid growth rates and improved immune response [37]. Dietary administration of nano-Fe in Indian major carp displays an immune-antioxidant impact evidenced by significantly improving the antioxidants (SOD and GSH), RBCs, hemoglobin levels, and innate immune response components like myeloperoxidase, respiratory burst, and bactericidal activities [38]. The incorporation of nano-Fe in the diet of Nile tilapia revealed recently stimulated immune response and compromised immune reaction evidenced by enhanced immune-antioxidant biomarkers (immune parameters: IgM, NO, and lysozyme; antioxidant parameters: SOD, CAT, and Gpx), plus an improvement in the associated gene transcripts (Insulin-Like Growth Factor-1 (*IGF-1*) and interleukin 1-*β* (*IL-1β*) genes) [38]. Also, the dietary intervention of nano-Fe in the Nile tilapia diet can enhance the growth performance, hematological profile, and the ionic exchange [57].

13.3.3.6 Silica Nanoparticles (SiNPs)

Among NPs, silica nanoparticles (SiNPs) are an exceptional class of inorganic nanoparticles with a diverse array of functional advantages [58]. They are currently used as a promising strategy, instead of antibiotics because of emergence of antibiotic resistance, for enhancing host immune response and thus, confronting many bacterial infections especially the mesoporous silica materials that are distinguished by ease of functionalization [58]. They are characterized by their unique advantages including wide surface area and biocompatibility for combating many bacterial infections [59]. They can also act as a potent tool and are considered one of the most frequently utilized nanoparticles in different applications of drug delivery especially the delivery of antimicrobials [59]. Silica nanoparticles are characterized by strong bactericidal efficiency owing to the released nitric oxide (NO), which is a more efficient molecule for reflecting immunity and correspondingly, kill the bacterium [59]. In Nile catfish, they are recently reported to enhance the immunological response via modulating immunological (NO and complement-3), and antioxidants (GPx, TAC, SOD, and CAT) indices and upregulation of *TGF-β*, *IL-1β*, *IL-8*, *NF-κβ*, *HSP70,* and *Hepcidin* genes, which are considered immune system regulators [60], in spite of their recorded toxicity [60] involving alteration in fish behavior and neuro-physiological disturbances of African catfish (*Clarias gariepinus*). They have been reported to improve health, immunity, and growth of aquatic organisms by strengthening the ionic exchange process plus controlling the apoptotic targeted genes in Nile tilapia [60]. They are characterized by a potent antioxidant activity that can be used in food packaging [61]. In aquaculture, SiNPs have been highlighted for their utility in fish nutrition, facilitating drug administration, and reducing the danger of disease outbreaks [62]. Furthermore, the nano-form of SiNPs is effective in treating wastewater through their magnetic properties in absorbing wastes [63,64], and controlling microbial load [55]. Recently, diets enriched with SiNPs have been appreciated in Nile tilapia as they proved a great influence in enhancing growth, immunological response, and protein.

13.3.4 Fish Immunomodulation Through Vaccination Using Nanoparticles

Fish vaccination has achieved a great promise in aquaculture as it is considered safe for the environment and animal, economic if it is applied for large-scale, ease of administration, and can exert potent immunity demonstrating minimal side effects, which offers great benefit for commercialization. The incorporation of nanoparticles in preparing a fish vaccine is the best option for better efficacy of vaccine. In drug administration, NPs have multiple privileges, such as administering vaccines to strengthen the resistance of farmed fish against pathogen-induced diseases. However, there is concern that the benefits of NPs administration may be accompanied by environmental and health risks [34]. The use of nanoparticles for the vaccination of farmed fish is a unique technique. In aquaculture, vaccination has a substantial impact on the control and prevention of infectious diseases despite evidence that there are many infectious diseases for which it is difficult to produce an efficient vaccine [2,67,68]. The use of NPs instead of bulk material in vaccine production is highly recommended as NPs, considered to be immune nano-stimulators, are particles of the size (20–100 nm) that can actively improve vaccine efficiency *in vivo* better than bulk molecules [69,70]. Some NPs have the ability to boost host immunity by stimulating animal immune cells. There are important factors that control the ability of NPs to activate the host immune response, including the size, surface chemistry, and shape of the NPs. Generally, NPs can produce immune system reactions by notably increasing inflammatory reactions and defense gene synthesis [70]. Nanoparticles could be taken up by host cells through the mechanism of endocytosis due to their nano-scale size [21,68], which increases the cellular uptake of antigens and promotes antigen presentation capacity [71]. Several studies have revealed that applied nanotechnology promote vaccine solubility, targeting, stability, permeability, and biocompatibility [2,39,72]. Nanoparticles are characterized by the presence of adjuvant that can efficiently promote the antigen. Nanoparticles enter living cells by cellular uptake through endocytosis due to their very small size [73,74]. Adjuvants help to enhance immune response and minimize the frequency of administration. To enable interfaces between macrophage receptors and NPs, many characters of NPs can be optimized, such as surface loading, hydrophobicity, thickness, surface topography, molecular weight, and material structure [75]. Adaptive immunity (B- and T-cells) has a set of receptors for the identification of different antigens. A variety of therapeutic treatments that depend on NPs have been used to enhance T-cells activity against bacterial, viral, or fungal infections [76]. DNA-nano vaccines (short DNA strands inside nano-capsules) are used in aquaculture to boost fish immune system [37]. Indeed, iron nanoparticles can enhance the development of fish immunitt by administering controlled-release drugs, which are transformed very rapidly through this approach. Nano-encapsulated vaccines have been efficaciously used to control *Listonella anguillarum* bacterium in rainbow trout [77,78] and Asian carp [79]. Also, iron nanoparticles could be used to improve fish growth, immunity, and probiotic counts in rainbow trout [80]. Metal oxides nanoparticles, as nano-ferric, manganese, aluminum, cerium, and titanium oxides, are characterized by a large surface area and hydrophilic properties that allow for a better availability [36].

Carbon nanotubes (CNTs) (1 nm in diameter) are able to absorb electrons through the fullerene core (an allotropes of carbon) and effectively scavenge reactive oxygen species (ROS) and possess anti-inflammatory properties due to the fullerene's steric bulk [81]. Carbon nanostructures could also be used for systematic and oral vaccine delivery [82].

Nanotechnology has essential key role in the vaccines development and there are substantial factors that control the success of nano-based vaccines including shape, size, hydrophobicity, and porosity of the nanoparticles [11,12]. The nanotechnology field is a very broad filed and we have tried to focus only on immunological applications of nanoparticles. The adjuvants can be used only in case of inactive vaccines that strengthen the quality and quantity of both humoral and cellular immune responses. The major role of nanoparticles in vaccination is to empower the antigens delivery to the immune system, and consequently, enhances animal immune responses [70]. The major common types of nanoparticles include virus-like particles (VLP) and squalene-based-oil emulsified in water such as MF59, this type has been used for decades [70,83]. The types of nano-based vaccines and their mechanism of action are recently documented by [83] and illustrated in Table 13.2.

TABLE 13.2

Types of Nano-Based Vaccines and Their Mechanism of Action

Nano-Based Vaccines	Range Size	Mechanism of Action
MF59[a]	150–200 nm	Neutrophil, antigen uptake, monocyte, and TH1-type immune responses
Virus-like particles	15–30 nm	Display of repetitive antigen, molecular or structural mimicry of the virus, penetration of tissue depending on the size of NPs, and reaching to lymphatics, and stimulation of TLR
Nanogel	30–40 nm	Entrapment of antigen in a hydrated nanogel matrix for slow release, and induction of antibody response and tumor-associated T cells
Cationic liposomes	200–1000 nm	Encapsulation and targeted antigen arrival and delivery of monocytes to the site of injection
W805EC[b]	200–400 nm	Uptake of antigen and stimulation of TLR2, TLR4, epithelial cells, and DCs, localized production of cytokine, and mucosal antibody responses
PLGA[c]	100–200 nm	Encapsulation for local antigens and release of co-mediator

Notes

[a] MF59 is an immunological adjuvant that utilizes the extract of shark liver oil named squalene

[b] W805EC is a nano-emulsion-dependent adjuvant (oil emulsified in water) that are applied intranasal

[c] PLGA is poly (lactic-co-glycolic acid).

The common mechanism of action for nanoparticles is via cytotoxic T lymphocyte (CTL) cell activation that targets virus-infected cells. Moreover, nanoparticles are surrounded by viral antigens to improve the activity of CTL. Also, they have the power to produce cytokines such as interleukin 12 (IL-12) and interleukin 15 (IL-15). The nano sensors are considered one of therapeutic essential applications for nanotechnology as they guide us to detect diseases associated with immune system and, consequently, discover emerging drugs [83].

Recently, vaccination is the most efficient way for enhancing the basic maintenance and the setup of a fish farm. Virus-like particles (VLPs) are considered the best option to owing to their structures look like that of the original virus. Therefore, VLPs have been considered the next-generation vaccines for fish. Current studies of VLP-linked fish vaccines reveal that these vaccines are potential and beneficial. The VLP-based vaccines are tested and achieve more success against many fish viruses involving salmonid alphavirus (SAV), nervous necrosis virus (NNV), infectious pancreatic necrosis virus (IPNV), and iridovirus [84].

13.4 Conclusion

Based on the review outcomes, nano-molecules have achieved a rapid development and are currently considered a successful emerging tool for boosting the fish's immune response, modulating the immune reaction, and controlling diseases owing to their distinctive properties. Also, the present report encompasses the major role of different types of nanoparticles for their attractive and efficient drug delivery with a potential role in the treatment of various diseases. They can be utilized in the aquaculture industry in various forms either via using metal NPs or plant-based NPs or recently, through vaccination for enhancing immunological response. Future studies are required to elucidate the behavior of nano-materials in the fish body, and the environment to lessen the issues for the public health. Moreover, further work is needed to address the mechanism of interaction between the immune system and nanoparticles.

Disclosure statement: The authors declare that they have no conflicts of interest.

REFERENCES

1. Nasr-Eldahan, S.; Nabil-Adam, A.; Shreadah, M. A.; Maher, A. M.; El-Sayed Ali, T. A Review Article on Nanotechnology in Aquaculture Sustainability as a Novel Tool in Fish Disease Control. *Aquac. Int.*, 2021, *29* (4), 1459–1480. 10.1007/s10499-021-00677-7.
2. Shah, B. R.; Mraz, J. Advances in Nanotechnology for Sustainable Aquaculture and Fisheries. *Rev. Aquac.*, 2020, *12* (2), 925–942. 10.1111/raq.12356.
3. Kwasek, K.; Thorne-Lyman, A. L.; Phillips, M. Can Human Nutrition Be Improved through Better Fish Feeding Practices? A Review Paper. *Crit. Rev. Food Sci. Nutr.*, 2020, *60* (22), 3822–3835. 10.1080/10408398.2019.1708698.
4. Firdaus-Nawi, M.; Saad, M. Major Components of Fish Immunity: A Review. *Pertanika J. Trop. Agric. Sci.*, 2016, *39*(4), 393–420.
5. Mahboub, H. H.; Shaheen, A. Prevalence, Diagnosis and Experimental Challenge of Dermocystidium Sp. Infection in Nile Tilapia (Oreochromis Niloticus) in Egypt. *Aquaculture*, 2020, *516*, 734556. 10.1016/j.aquaculture.2019.734556.
6. Mahboub, H. H.; Shaheen, A. A. Mycological and Histopathological Identification of Potential Fish Pathogens in Nile Tilapia. *Aquaculture*, 2021, *530*, 735849. 10.1016/j.aquaculture.2020.735849.
7. Jaime, R. Antibiotics in Aquaculture – Use, Abuse and Alternatives. In *Health and Environment in Aquaculture*; Carmen Gloria Feijoo, Ed.; Paola Navarrete ED1 - Edmir Daniel Carvalho ED2 - Gianmarco Silva David ED3 - Reinaldo J. Silva, Series Ed.; IntechOpen: Rijeka, 2012; p Ch. 6. 10.5772/28157.
8. Devi, K. P.; Nisha, S. A.; Sakthivel, R.; Pandian, S. K. Eugenol (an Essential Oil of Clove) Acts as an Antibacterial Agent against Salmonella Typhi by Disrupting the Cellular Membrane. *J. Ethnopharmacol.*, 2010, *130* (1), 107–115. 10.1016/j.jep.2010.04.025.
9. Kothari, D.; Patel, S.; Kim, S.-K. Probiotic Supplements Might Not Be Universally-Effective and Safe: A Review.

Biomed. Pharmacother., 2019, *111*, 537–547. 10.1016/j.biopha.2018.12.104.
10. Reverter, M.; Tapissier-Bontemps, N.; Sasal, P.; Saulnier, D. Use of Medicinal Plants in Aquaculture. In *Diagnosis and Control of Diseases of Fish and Shellfish*; John Wiley & Sons, Ltd, 2017; pp 223–261. 10.1002/9781119152125.ch9.
11. Kubackova, J.; Zbytovska, J.; Holas, O. Nanomaterials for Direct and Indirect Immunomodulation: A Review of Applications. *Eur. J. Pharm. Sci.*, 2020, *142*, 105139. 10.1016/j.ejps.2019.105139.
12. Silva, J. R. M. C.; Staines, N. A.; Hernandez-Blazquez, F. J.; Porto-Neto, L. R.; Borges, J. C. S. Phagocytosis and Giant Cell Formation at 0° C by Macrophage (MØ) of Notothenia Coriiceps. *J. Fish Biol.*, 2002, *60* (2), 466–478. 10.1111/j.1095-8649.2002.tb00294.x.
13. Sharma, M.; Pandey, G. Overviews of the Treatment and Control of Common Fish Diseases. *Int. Res. J. Pharm.*, 2012, *3*, 123–127.
14. Mugwanya, M.; Dawood, M. A. O.; Kimera, F.; Sewilam, H. Anthropogenic Temperature Fluctuations and Their Effect on Aquaculture: A Comprehensive Review. *Aquac. Fish.*, 2022, *7* (3), 223–243. 10.1016/j.aaf.2021.12.005.
15. Matteucci, F.; Giannantonio, R.; Calabi, F.; Agostiano, A.; Gigli, G.; Rossi, M. Deployment and Exploitation of Nanotechnology Nanomaterials and Nanomedicine. *AIP Conf. Proc.*, 2018, *1990* (1), 020001. 10.1063/1.5047755.
16. Khan, I.; Saeed, K.; Khan, I. Nanoparticles: Properties, Applications and Toxicities. *Arab. J. Chem.*, 2019, *12* (7), 908–931. 10.1016/j.arabjc.2017.05.011.
17. Zhang, L.; Gu, F. X.; Chan, J. M.; Wang, A. Z.; Langer, R. S.; Farokhzad, O. C. Nanoparticles in Medicine: Therapeutic Applications and Developments. *Clin. Pharmacol. Ther.*, 2008, *83* (5), 761–769. 10.1038/sj.clpt.6100400.
18. Doan, H. V.; Soltani, E.; Ingelbrecht, J.; Soltani, M. Medicinal Herbs and Plants: Potential Treatment of Monogenean Infections in Fish. *Rev. Fish. Sci. Aquac.*, 2020, *28* (2), 260–282. 10.1080/23308249.2020.1712325.
19. Stratev, D.; Zhelyazkov, G.; Noundou, X. S.; Krause, R. W. M. Beneficial Effects of Medicinal Plants in Fish Diseases. *Aquac. Int.*, 2018, *26* (1), 289–308. 10.1007/s10499-017-0219-x.
20. Wei, H.; Zhang, X.-Z.; Chen, W.-Q.; Cheng, S.-X.; Zhuo, R.-X. Self-Assembled Thermosensitive Micelles Based on Poly(L-Lactide-Star Block-N-Isopropylacrylamide) for Drug Delivery. *J. Biomed. Mater. Res. A*, 2007, *83A* (4), 980–989. 10.1002/jbm.a.31295.
21. Khani Oushani, A.; Soltani, M.; Sheikhzadeh, N.; Shamsaie Mehrgan, M.; Rajabi Islami, H. Effects of Dietary Chitosan and Nano-Chitosan Loaded Clinoptilolite on Growth and Immune Responses of Rainbow Trout (Oncorhynchus mykiss). *Fish Shellfish Immunol.*, 2020, *98*, 210–217. 10.1016/j.fsi.2020.01.016.
22. Dawood, M. A. O.; Gewaily, M. S.; Soliman, A. A.; Shukry, M.; Amer, A. A.; Younis, E. M.; Abdel-Warith, A.-W. A.; Van Doan, H.; Saad, A. H.; Aboubakr, M.; et al. Marine-Derived Chitosan Nanoparticles Improved the Intestinal Histo-Morphometrical Features in Association with the Health and Immune Response of Grey Mullet (Liza Ramada). *Mar. Drugs*, 2020, *18* (12), 611. 10.3390/md18120611.
23. Farahnak Roudsari, S.; Rajabi Islami, H.; Mousavi, S. A.; Shamsaie Mehrgan, M. Folic Acid-Coated Nanochitosan Ameliorated the Growth Performance, Hematological Parameters, Antioxidant Status, and Immune Responses of Rainbow Trout (Oncorhynchus mykiss). *Front. Vet. Sci.*, 2021, *8*, 647722.
24. Abdel Rahman, A. N.; Shakweer, M. S.; Algharib, S. A.; Abdelaty, A. I.; Kamel, S.; Ismail, T. A.; Daoush, W. M.; Ismail, S. H.; Mahboub, H. H. Silica Nanoparticles Acute Toxicity Alters Ethology, Neuro-Stress Indices, and Physiological Status of African Catfish (Clarias gariepinus). *Aquac. Rep.*, 2022, *23*, 101034. 10.1016/j.aqrep.2022.101034.
25. Abdel-Tawwab, M.; Razek, N. A.; Abdel-Rahman, A. M. Immunostimulatory Effect of Dietary Chitosan Nanoparticles on the Performance of Nile Tilapia, Oreochromis niloticus (L.). *Fish Shellfish Immunol.*, 2019, *88*, 254–258. 10.1016/j.fsi.2019.02.063.
26. Xia, I. F.; Cheung, J. S.; Wu, M.; Wong, K.-S.; Kong, H.-K.; Zheng, X.-T.; Wong, K.-H.; Kwok, K. W. Dietary Chitosan-Selenium Nanoparticle (CTS-SeNP) Enhance Immunity and Disease Resistance in Zebrafish. *Fish Shellfish Immunol.*, 2019, *87*, 449–459. 10.1016/j.fsi.2019.01.042.
27. Alishahi, A.; Mirvaghefi, A.; Tehrani, M. R.; Farahmand, H.; Koshio, S.; Dorkoosh, F. A.; Elsabee, M. Z. Chitosan Nanoparticle to Carry Vitamin C through the Gastrointestinal Tract and Induce the Non-Specific Immunity System of Rainbow Trout (Oncorhynchus mykiss). *Carbohydr. Polym.*, 2011, *86* (1), 142–146. 10.1016/j.carbpol.2011.04.028.
28. Elabd, H.; Wang, H.-P.; Shaheen, A.; Matter, A. Astragalus Membranaceus Nanoparticles Markedly Improve Immune and Anti-Oxidative Responses; and Protection against Aeromonas veronii in Nile Tilapia Oreochromis niloticus. *Fish Shellfish Immunol.*, 2020, *97*, 248–256. 10.1016/j.fsi.2019.12.025.
29. Elabd, H.; Wang, H.-P.; Shaheen, A.; Matter, A. Nano Spirulina Dietary Supplementation Augments Growth, Antioxidative and Immunological Reactions, Digestion, and Protection of Nile Tilapia, Oreochromis niloticus, against Aeromonas veronii and Some Physical Stressors. *Fish Physiol. Biochem.*, 2020, *46* (6), 2143–2155. 10.1007/s10695-020-00864-y.
30. Mahboub, H. H.; Khedr, M. H. E.; Elshopakey, G. E.; Shakweer, M. S.; Mohamed, D. I.; Ismail, T. A.; Ismail, S. H.; Abdel Rahman, A. N. Impact of Silver Nanoparticles Exposure on Neuro-Behavior, Hematology, and Oxidative Stress Biomarkers of African Catfish (Clarias gariepinus). *Aquaculture*, 2021, *544*, 737082. 10.1016/j.aquaculture.2021.737082.
31. Mahboub, H. H.; Rashidian, G.; Hoseinifar, S. H.; Kamel, S.; Zare, M.; Ghafarifarsani, H.; Algharib, S. A.; Moonmanee, T.; Van Doan, H. Protective Effects of Allium Hirtifolium Extract against Foodborne Toxicity of Zinc Oxide Nanoparticles in Common Carp (Cyprinus garpio). *Comp. Biochem. Physiol. Part C Toxicol. Pharmacol.*, 2022, *257*, 109345. 10.1016/j.cbpc.2022.109345.
32. Rashidian, G.; Lazado, C. C.; Mahboub, H. H.; Mohammadi-Aloucheh, R.; Prokić, M. D.; Nada, H. S.; Faggio, C. Chemically and Green Synthesized ZnO Nanoparticles Alter Key Immunological Molecules in

Common Carp (Cyprinus carpio) Skin Mucus. *Int. J. Mol. Sci.*, 2021, *22* (6), 3270. 10.3390/ijms22063270.
33. Mahboub, H. H.; Rashidian, G.; Hoseinifar, S. H.; Kamel, S.; Zare, M.; Ghafarifarsani, H.; Algharib, S. A.; Moonmanee, T.; Van Doan, H. Protective Effects of Allium Hirtifolium Extract against Foodborne Toxicity of Zinc Oxide Nanoparticles in Common Carp (Cyprinus carpio). *Comp. Biochem. Physiol. Part C Toxicol. Pharmacol.*, 2022, *257*, 109345. 10.1016/j.cbpc.2022.109345.
34. Almanaa, T. N.; Aref, M.; Kakakhel, M. A.; Elshopakey, G. E.; Mahboub, H. H.; Abdelazim, A. M.; Kamel, S.; Belali, T. M.; Abomughaid, M. M.; Alhujaily, M.; et al. Silica Nanoparticle Acute Toxicity on Male Rattus Norvegicus Domestica: Ethological Behavior, Hematological Disorders, Biochemical Analyses, Hepato-Renal Function, and Antioxidant-Immune Response. *Front. Bioeng. Biotechnol.*, 2022, *10*, 868111. 10.3389/fbioe.2022.868111.
35. Mahboub, H. H.; Beheiry, R. R.; Shahin, S. E.; Behairy, A.; Khedr, M. H. E.; Ibrahim, Seham. M.; Elshopakey, G. E.; Daoush, W. M.; Altohamy, D. E.; Ismail, T. A.; et al. Adsorptivity of Mercury on Magnetite Nano-Particles and Their Influences on Growth, Economical, Hemato-Biochemical, Histological Parameters and Bioaccumulation in Nile Tilapia (Oreochromis Niloticus). *Aquat. Toxicol.*, 2021, *235*, 105828. 10.1016/j.aquatox.2021.105828.
36. Kumar, N.; Krishnani, K. K.; Gupta, S. K.; Singh, N. P. Selenium Nanoparticles Enhanced Thermal Tolerance and Maintain Cellular Stress Protection of Pangasius Hypophthalmus Reared under Lead and High Temperature. *Respir. Physiol. Neurobiol.*, 2017, *246*, 107–116. 10.1016/j.resp.2017.09.006.
37. Fajardo, C.; Martinez-Rodriguez, G.; Blasco, J.; Mancera, J. M.; Thomas, B.; De Donato, M. Nanotechnology in Aquaculture: Applications, Perspectives and Regulatory Challenges. *Aquac. Fish.*, 2022, *7* (2), 185–200. 10.1016/j.aaf.2021.12.006.
38. Behera, T.; Swain, P.; Rangacharulu, P. V.; Samanta, M. Nano-Fe as Feed Additive Improves the Hematological and Immunological Parameters of Fish, Labeo Rohita H. *Appl. Nanosci.*, 2014, *4* (6), 687–694. 10.1007/s13204-013-0251-8.
39. Handy, R. D.; Cornelis, G.; Fernandes, T.; Tsyusko, O.; Decho, A.; Sabo-Attwood, T.; Metcalfe, C.; Steevens, J. A.; Klaine, S. J.; Koelmans, A. A.; et al. Ecotoxicity Test Methods for Engineered Nanomaterials: Practical Experiences and Recommendations from the Bench. *Environ. Toxicol. Chem.*, 2012, *31* (1), 15–31. 10.1002/etc.706.
40. Sekhon, B. S. Nanotechnology in Agri-Food Production: An Overview. *Nanotechnol. Sci. Appl.*, 2014, *7*, 31–53. 10.2147/NSA.S39406.
41. Elsabahy, M.; Wooley, K. L. Cytokines as Biomarkers of Nanoparticle Immunotoxicity. *Chem. Soc. Rev.*, 2013, *42* (12), 5552–5576. 10.1039/c3cs60064e.
42. Moore, K. W.; de Waal Malefyt, R.; Coffman, R. L.; O'Garra, A. Interleukin-10 and the Interleukin-10 Receptor. *Annu. Rev. Immunol.*, 2001, *19*, 683–765. 10.1146/annurev.immunol.19.1.683.
43. Ren, K.; Torres, R. Role of Interleukin-1β during Pain and Inflammation. *Brain Res. Rev.*, 2009, *60* (1), 57–64. 10.1016/j.brainresrev.2008.12.020.
44. Zou, J.; Peddie, S.; Scapigliati, G.; Zhang, Y.-A.; Bols, N.; Ellis, A. E.; Secombes, C. J. Functional Characterisation of the Recombinant Tumor Necrosis Factor in Rainbow Trout (Oncorhynchus mykiss). *Dev. Comp. Immunol.*, 2003, *27*, 813–822. 10.1016/S0145-305X(03)00077-6.
45. Abu-Elala, N. M.; Shaalan, M.; Ali, S. E.; Younis, N. A. Immune Responses and Protective Efficacy of Diet Supplementation with Selenium Nanoparticles against Cadmium Toxicity in Oreochromis niloticus. *Aquac. Res.*, 2021, *52* (8), 3677–3686. 10.1111/are.15212.
46. Zhou, X.; Wang, Y.; Gu, Q.; Li, W. Effects of Different Dietary Selenium Sources (Selenium Nanoparticle and Selenomethionine) on Growth Performance, Muscle Composition and Glutathione Peroxidase Enzyme Activity of Crucian Carp (Carassius Auratus gibelio). *Aquaculture*, 2009, *291* (1), 78–81. 10.1016/j.aquaculture.2009.03.007.
47. Ashouri, S.; Keyvanshokooh, S.; Salati, A. P.; Johari, S. A.; Pasha-Zanoosi, H. Effects of Different Levels of Dietary Selenium Nanoparticles on Growth Performance, Muscle Composition, Blood Biochemical Profiles and Antioxidant Status of Common Carp (Cyprinus carpio). *Aquaculture*, 2015, *446*, 25–29. 10.1016/j.aquaculture.2015.04.021.
48. Qin, F.; Shi, M.; Yuan, H.; Yuan, L.; Lu, W.; Zhang, J.; Tong, J.; Song, X. Dietary Nano-Selenium Relieves Hypoxia Stress and, Improves Immunity and Disease Resistance in the Chinese Mitten Crab (Eriocheir sinensis). *Fish Shellfish Immunol.*, 2016, *54*, 481–488. 10.1016/j.fsi.2016.04.131.
49. Neamat-Allah, A. N. F.; Mahmoud, E. A.; Abd El Hakim, Y. Efficacy of Dietary Nano-Selenium on Growth, Immune Response, Antioxidant, Transcriptomic Profile and Resistance of Nile Tilapia, Oreochromis niloticus against Streptococcus Iniae Infection. *Fish Shellfish Immunol.*, 2019, *94*, 280–287. 10.1016/j.fsi.2019.09.019.
50. Ibrahim, A. Th. A. Toxicological Impact of Green Synthesized Silver Nanoparticles and Protective Role of Different Selenium Type on Oreochromis Niloticus: Hematological and Biochemical Response. *J. Trace Elem. Med. Biol.*, 2020, *61*, 126507. 10.1016/j.jtemb.2020.126507.
51. Li, L.; Liu, Z.; Quan, J.; Sun, J.; Lu, J.; Zhao, G. Comprehensive Proteomic Analysis to Elucidate the Anti-Heat Stress Effects of Nano-Selenium in Rainbow Trout (Oncorhynchus mykiss). *Ecotoxicol. Environ. Saf.*, 2022, *241*, 113736. 10.1016/j.ecoenv.2022.113736.
52. Kumar, N.; Chandan, N. K.; Wakchaure, G.; Singh, N. P. Synergistic Effect of Zinc Nanoparticles and Temperature on Acute Toxicity with Response to Biochemical Markers and Histopathological Attributes in Fish. *Comp. Biochem. Physiol. Part C Toxicol. Pharmacol.*, 2020, *229*, 108678. 10.1016/j.cbpc.2019.108678.
53. El-Saadony, M. T.; Alkhatib, F. M.; Alzahrani, S. O.; Shafi, M. E.; El. Abdel-Hamid, S.; Taha, T. F.; Aboelenin, S. M.; Soliman, M. M.; Ahmed, N. H. Impact of Mycogenic Zinc Nanoparticles on Performance, Behavior, Immune Response, and Microbial Load in Oreochromis niloticus. *Saudi J. Biol. Sci.*, 2021, *28* (8), 4592–4604. 10.1016/j.sjbs.2021.04.066.
54. Mahboub, H. H.; Shahin, K.; Zaglool, A. W.; Roushdy, E. M.; Ahmed, S. A. A. Efficacy of Nano Zinc Oxide Dietary Supplements on Growth Performance, Immunomodulation

and Disease Resistance of African Catfish Clarias Gariepinus. *Dis. Aquat. Organ.*, 2020, *142*, 147–160. 10.3354/dao03531.
55. Huang, C.-M.; Chen, C.-H.; Pornpattananangkul, D.; Zhang, L.; Chan, M.; Hsieh, M.-F.; Zhang, L. Eradication of Drug Resistant Staphylococcus Aureus by Liposomal Oleic Acids. *Biomaterials*, 2011, *32* (1), 214–221. 10.1016/j.biomaterials.2010.08.076.
56. Dananjaya, S. H. S.; Godahewa, G. I.; Jayasooriya, R. G. P. T.; Lee, J.; De Zoysa, M. Antimicrobial Effects of Chitosan Silver Nano Composites (CAgNCs) on Fish Pathogenic Aliivibrio (Vibrio) Salmonicida. *Aquaculture*, 2016, *450*, 422–430. 10.1016/j.aquaculture.2015.08.023.
57. Elabd, H.; Youssuf, H.; Mahboub, H. H.; Salem, S. M. R.; Husseiny, W. A.; Khalid, A.; El-Desouky, H. S.; Faggio, C. Growth, Hemato-Biochemical, Immune-Antioxidant Response, and Gene Expression in Nile Tilapia (Oreochromis Niloticus) Received Nano Iron Oxide-Incorporated Diets. *Fish Shellfish Immunol.*, 2022. 10.1016/j.fsi.2022.07.051.
58. Bernardos, A.; Piacenza, E.; Sancenón, F.; Hamidi, M.; Maleki, A.; Turner, R. J.; Martínez-Máñez, R. Mesoporous Silica-Based Materials with Bactericidal Properties. *Small*, 2019, *15* (24), 1900669. 10.1002/smll.201900669.
59. Selvarajan, V.; Obuobi, S.; Ee, P. L. R. Silica Nanoparticles—A Versatile Tool for the Treatment of Bacterial Infections. *Front. Chem.*, 2020, *8*, 602.
60. Alandiyjany, M. N.; Kishawy, A. T. Y.; Abdelfattah-Hassan, A.; Eldoumani, H.; Elazab, S. T.; El-Mandrawy, S. A. M.; Saleh, A. A.; ElSawy, N. A.; Attia, Y. A.; Arisha, A. H.; et al. Nano-Silica and Magnetized-Silica Mitigated Lead Toxicity: Their Efficacy on Bioaccumulation Risk, Performance, and Apoptotic Targeted Genes in Nile Tilapia (Oreochromis niloticus). *Aquat. Toxicol.*, 2022, *242*, 106054. 10.1016/j.aquatox.2021.106054.
61. Dong, W.; Su, J.; Chen, Y.; Xu, D.; Cheng, L.; Mao, L.; Gao, Y.; Yuan, F. Characterization and Antioxidant Properties of Chitosan Film Incorporated with Modified Silica Nanoparticles as an Active Food Packaging. *Food Chem.*, 2022, *373*, 131414. 10.1016/j.foodchem.2021.131414.
62. Khosravi-Katuli, K.; Prato, E.; Lofrano, G.; Guida, M.; Vale, G.; Libralato, G. Effects of Nanoparticles in Species of Aquaculture Interest. *Environ. Sci. Pollut. Res.*, 2017, *24* (21), 17326–17346. 10.1007/s11356-017-9360-3.
63. Jarvie, H. P.; Al-Obaidi, H.; King, S. M.; Bowes, M. J.; Lawrence, M. J.; Drake, A. F.; Green, M. A.; Dobson, P. J. Fate of Silica Nanoparticles in Simulated Primary Wastewater Treatment. *Environ. Sci. Technol.*, 2009, *43* (22), 8622–8628. 10.1021/es901399q.
64. Bashar, A.; Hasan, N. A.; Haque, M. M.; Rohani, Md. F.; Hossain, Md. S. Effects of Dietary Silica Nanoparticle on Growth Performance, Protein Digestibility, Hematology, Digestive Morphology, and Muscle Composition of Nile Tilapia, Oreochromis niloticus. *Front. Mar. Sci.*, 2021, *8*, 706179.
65. Ismail, S. H.; Hamdy, A.; Ismail, T. A.; Mahboub, H. H.; Mahmoud, W. H.; Daoush, W. M. Synthesis and Characterization of Antibacterial Carbopol/ZnO Hybrid Nanoparticles Gel. *Crystals*, 2021, *11* (9), 1092. 10.3390/cryst11091092.
66. Selvaraj, V.; Mahboub, H. H.; Ganapathi, U.; Chandran, S. K.; Al-Onazi, W.; Al-Mohaimeed, A. M.; Chen, T.-W.; Faggio, C.; Paulraj, B. Enhanced Photodegradation of Methylene Blue from Aqueous Solution Using Al-Doped ZnS Nanoparticles. *Environ. Sci. Pollut. Res.*, 2022. 10.1007/s11356-022-20634-y.
67. Embregts, C. W. E.; Forlenza, M. Oral Vaccination of Fish: Lessons from Humans and Veterinary Species. *Dev. Comp. Immunol.*, 2016, *64*, 118–137. 10.1016/j.dci.2016.03.024.
68. Luis, A. I. S.; Campos, E. V. R.; de Oliveira, J. L.; Fraceto, L. F. Trends in Aquaculture Sciences: From Now to Use of Nanotechnology for Disease Control. *Rev. Aquac.*, 2019, *11* (1), 119–132. 10.1111/raq.12229.
69. Irvine, D. J.; Swartz, M. A.; Szeto, G. L. Engineering Synthetic Vaccines Using Cues from Natural Immunity. *Nat. Mater.*, 2013, *12* (11), 978–990. 10.1038/nmat3775.
70. Smith, D. M.; Simon, J. K.; Baker Jr, J. R. Applications of Nanotechnology for Immunology. *Nat. Rev. Immunol.*, 2013, *13* (8), 592–605. 10.1038/nri3488.
71. Shaalan, M.; Saleh, M.; El-Mahdy, M.; El-Matbouli, M. Recent Progress in Applications of Nanoparticles in Fish Medicine: A Review. *Nanomedicine Nanotechnol. Biol. Med.*, 2016, *12* (3), 701–710. 10.1016/j.nano.2015.11.005.
72. Doll, T. A. P. F.; Raman, S.; Dey, R.; Burkhard, P. Nanoscale Assemblies and Their Biomedical Applications. *J. R. Soc. Interface*, 2013, *10* (80), 20120740. 10.1098/rsif.2012.0740.
73. Treuel, L.; Jiang, X.; Nienhaus, G. U. New Views on Cellular Uptake and Trafficking of Manufactured Nanoparticles. *J. R. Soc. Interface*, 2013, *10* (82), 20120939. 10.1098/rsif.2012.0939.
74. Zhao, L.; Seth, A.; Wibowo, N.; Zhao, C.-X.; Mitter, N.; Yu, C.; Middelberg, A. P. J. Nanoparticle Vaccines. *Vaccine*, 2014, *32* (3), 327–337. 10.1016/j.vaccine.2013.11.069.
75. Nel, A. E.; Mädler, L.; Velegol, D.; Xia, T.; Hoek, E. M. V.; Somasundaran, P.; Klaessig, F.; Castranova, V.; Thompson, M. Understanding Biophysicochemical Interactions at the Nano–Bio Interface. *Nat. Mater.*, 2009, *8* (7), 543–557. 10.1038/nmat2442.
76. Fang, P.; Li, X.; Dai, J.; Cole, L.; Camacho, J. A.; Zhang, Y.; Ji, Y.; Wang, J.; Yang, X.-F.; Wang, H. Immune Cell Subset Differentiation and Tissue Inflammation. *J. Hematol. Oncol.J Hematol Oncol*, 2018, *11* (1), 97. 10.1186/s13045-018-0637-x.
77. Mongillo, J. F. *Nanotechnology 101*; Greenwood Press: Westport, Conn., 2007.
78. Ogunkalu, O. A. Utilization of Nanotechnology in Aquaculture and Seafood Sectors. *Eurasian J. Food Sci. Technol.*, 2019, *3* (1), 26–33.
79. Ghosal, S.; Singh, S.; Bhattacharya, S. K. Alkaloids of Mucuna Pruries Chemistry and Pharmacology. *Planta Med.*, 1971, *19* (3), 280–284.
80. Mohammadi, N.; Tukmechi, A. The Effects of Iron Nanoparticles in Combination with Lactobacillus Casei on Growth Parameters and Probiotic Counts in Rainbow Trout (Oncorhynchus mykiss) Intestine. 2015, *70*, 47–53.
81. Zhang, Y.; Wang, L.; Sun, Y.; Zhu, Y.; Zhong, Z.; Shi, J.; Fan, C.; Huang, Q. Conjugation of Dexamethasone to C60 for the Design of an Anti-Inflammatory

Nanomedicine with Reduced Cellular Apoptosis. *ACS Appl. Mater. Interfaces*, 2013, *5* (11), 5291–5297. 10.1021/am401153k.

82. Kim, M.-G.; Park, J. Y.; Shon, Y.; Kim, G.; Shim, G.; Oh, Y.-K. Nanotechnology and Vaccine Development. *Asian J. Pharm. Sci.*, 2014, *9* (5), 227–235. 10.1016/j.ajps.2014.06.002.
83. Keikha, R.; Daliri, K.; Jebali, A. The Use of Nanobiotechnology in Immunology and Vaccination. *Vaccines*, 2021, *9* (2), 74. 10.3390/vaccines9020074.
84. Jeong, K.-H.; Kim, H. J.; Kim, H.-J. Current Status and Future Directions of Fish Vaccines Employing Virus-like Particles. *Fish Shellfish Immunol.*, 2020, *100*, 49–57. 10.1016/j.fsi.2020.02.060.

Section IV

Current Status of Immunomodulators in Aquaculture

14

Efficacy and Limitations of Immunomodulators

Arathi Kollath[1], Lokesh Pawar[2], Ankeet Bhagat[3], Sunil Sharma[3], Owias Iqbal Dar[3,4], and Preetham Elumalai[1]
[1]*Department of Marine Biology, Microbiology and Biochemistry, Cochin University of Science and Technology (CUSAT), Cochin, Kerala, India*
[2]*College of Fisheries (Central Agricultural University - Imphal), Agartala, Tripura, India*
[3]*Aquatic Toxicology Lab, Department of Zoology, Guru Nanak Dev University, Amritsar, Punjab, India*
[4]*Department of Biosciences, Chandigarh University, Gharuan, Punjab, India*

14.1 Introduction

In 1796, Edward Jenner brought out the idea of immunomodulation (Porwal *et al.*, 2021). Since then, significant progress, research, and development have been made in our knowledge of the immunomodulation process. Immunomodulators are chemicals, either biological or synthetic, that help to control the immune system (Kumar *et al.*, 2022). Immunomodulators can also be referred to as biological response modifiers, immunological enhancers, or immune restoratives (Zia *et al.*, 2010). By enhancing or decreasing an immune system's function, which consists of innate and adaptive immunological responses, they are capable of changing the immunologic response. In immune system defence frameworks, immunomodulators operate as mixture operators (Siqueiros-Cendón *et al.*, 2014; Kumar *et al.*, 2022). Immune modulators are also widely used in chronic illness to restore immune system health in patients who have been on prolonged courses of antiviraltherapies or antibiotics because they are thought to modulate and restore weakened immune systems and control overactive immune systems, but they do not strengthen the immune system. They are also advised for people with auto-immune diseases (Zia *et al.*, 2010). Medicinal plant products, pharmaceuticals of natural and synthetic origins, a variety of immune system-derived proteins, a number of monoclonal antibodies, and microbiological products are some of the greatest examples of immunomodulators (Zia *et al.*, 2010, Ibrahim *et al.*, 2015). Immunomodulatory meds, which are recommended to improve immune response against tumours, infectious diseases, immunodeficiency, and changes in antibody transfer, among other conditions, alter immune response bydecreasing (immunosuppressives) or increasing (immunostimulators) the serum antibody production (Avorn., 2011; Asherson *et al.*,2008). Drugs that inhibit the immune system are used to treat autoimmune conditions such as lupus, allergies, and pemphigus as well as transplanted organ rejection (Bascones-Martinez *et al.*, 2014).

Immunomodulation alters the immune system's operations in detectable ways, which might have a particular or non-specific effect. Immunomodulator-specific activities alter the body's immune cells so they can react to a particular antigen or immunogen. Immunomodulators with non-specific actions help to either activate or inhibit the immune system without focusing the activity of stimulated cells on a specific antigen, similar to how therapeutic vaccines work. Selective immunomodulation occurs when the activation results in immunoreactions against antigens (Lee *et al.*, 2010; Bascones-Martinez *et al.*, 2014). They are additionally categorized into three types: type I, type II, and type III. Out of these, type I affects the immune system normally, type II affects the immunosuppressed body, and type III affects the immune system that is both immunosuppressed and normally functioning. Immunomodulator drugs also work to treat cardiovascular and autoimmune illnesses, as well as to slow the spread of cancer, rheumatoid arthritis, and many gastrointestinal ailments. Additionally, for a variety of disorders, immunostimulants basically serve as adjuvants to chemotherapy (Pujol *et al.*, 2015).

14.2 Classification of Immunomodulators

Immunosuppressants, immunostimulators, and immunoadjuvants are the three main categories of immunomodulators based on their actions. They are capable of mounting an immune action anddefense against infections or tumors, or both (Alamgir and Uddin., 2010).

Immunosuppressants, which are immune system-inhibiting molecules, are frequently employed to stop the rejection of transplanted organs. Immunosuppressants are also used to provide treatment for immunopathology, autoimmune diseases, and

DOI: 10.1201/9781003361183-18

hypersensitivity responses. Additionally, when combined, the immunosuppressants have substantial drug-drug interactions as well as significant interactions with medications intended to address comorbid diseases and adverse effects that frequently affect transplant recipients and those with autoimmune disorders (Fireman *et al.*, 2004). Immunosuppressants are classified into several categories, including lymphocyte gene expression and signaling inhibitors, cytotoxic agents, alkylating agents, and cytokine inhibitors. Examples of these agents are glucocorticoids, cyclosporine, cyclophosphamide, azathioprine, and sirolimus (Ibrahim *et al.*, 2015).

Immunostimulants are molecules that cause the immune system's messengers to become active or stimulated (Ibrahim *et al.*, 2015). These medications offer protection from several illnesses, including cancer, autoimmune diseases, and allergy disorders. These agents can operate through both the adaptive innate immune response immunological response, despite the fact that they are intrinsically non-specific in nature (Vikrant and Vivek., 2011; Porwal *et al.*, 2021). These molecules are classified into various categories, such as intracellular adhesion molecule inhibitors, recombinant cytokines, and hormonal analogues. Examples of these categories are efalizumb, isoprinosine, interferon alpha, and interferon gamma. Immunostimulators are also recognized as immunopotentiators and promoter agents as they activate the immune system's basic level (Porwal *et al.*, 2021). Immunostimulators are anticipated to operate as preventative or promotion agents in healthy individuals by raising the immune system's basal levels. In addition, these substances function as immunotherapeutic agents in individuals whose immune responses are compromised (Agarwal and Singh, 1999).

Immunoadjuvants are specialised immune stimulators that increase the effectiveness of vaccinations (Ibrahim *et al.*, 2015). An immunoadjuvant that is often employed is Freund's adjuvant. They frequently serve as selectors between humoral and cellular assistant/helper T1 cells and helper T2 cells, immune protective and destructive responses, and immunoglobulin G vs. immunoglobulin E, which presents a significant challenge to vaccine designers. These agents hold the potential to be the true modulators of the immune response (Agarwal and Singh, 1999; Arya and Gupta, 2011; Porwal *et al.*, 2021). These substances can produce monoclonal and polyclonal antibodies for use in vaccinations (Lee *et al.*, 2010).

14.3 Sources of Immunomodulators

Immunomodulators can be derived from various plants, from synthetic peptides and from some marine organisms. Among all these, the potential source of immunomodulators is medicinal plant products (Porwal *et al.*, 2021). In addition to these compounds, other phytochemicals such as terpenoids, flavonoids, alkaloids, saponins, essential oils, steroids, phenolic, and pigments have exhibit worth immunomodulatory effects (Ibrahim *et al.*, 2015). Most promising plants which have immunomodulatory properties are *Curcuma longa, Withaniasomnifera, Ocimum sanctum, Panax ginseng, Tinospora cordifolia, Picrorhizakurroa,* etc. (Agarwal and Singh., 1999). Research on immunomodulatory natural products has increased as a result of the immunomodulatory action of plant-based treatments. Plants and their extracts may now offer us new, useful entities to develop as innovative immunomodulatory drugs.

The following table describes the common plant-derived immunomodulators obtained from different parts of the plant with the mechanism or action (Table 14.1).

14.4 Immunostimulants and Immunosuppresent with Their Mode of Actions

Besides these plant-derived immunomodulatory agents, phytochemicals including alkaloids, terpenoids, essential oils, saponins, and pigments have immunomodulatory mechanisms, as shown in Table 14.2.

14.5 Marine-Derived Immunomodulators

Along with plant-derived immunomodulators, some research also shows marine-derived chemical constituents that have immunomodulatory actions. These agents provide invaluable leads for drug discovery. Various marine immunomodulators are given in Table 14.3 (Vikrant and Vivek 2011).

14.6 Immunomodulatory Drugs

Immunomodulators are substances that influence the pathways that control the functioning of the immune system. Immunomodulators are part of a family of pharmaceuticals called disease-modifying anti-rheumatic drugs (DMARDs), which alter and slow down or stop the progression of numerous autoimmune, inflammatory, and cancerous disorders (Timmermans *et al.*, 2019). Drugs that act as immunomodulators change how the immune system reacts by either promoting the production of serum antibodies (immunostimulators) or inhibiting it (immunosuppressives) (Avorn., 2011). Immunostimulants are used to boost the immune response to a variety of conditions, including malignancies, primary or secondary immunodeficiencies, viral infections, and alterations in antibody transfer (Asherson *et al.*, 2008). Drugs that inhibit the immune system are used to treat transplanted organ rejection as well as autoimmune diseases including lupus, allergies, and pemphigus (Carsons., 2008). Many disorders, including rheumatic diseases and other oral diseases, now have different treatment regimens thanks to immunomodulatory medications. Immunostimulants and immunosuppressants have been created to decrease or increase the activity of particular immune system populations and subpopulations like macrophages, B-cells, T-cells, neutrophils, NK cells, and dendritic cells due to the fact that action of immunomodulators is at various levels (Lee *et al.*, 2010). Some of the immunosuppressants compounds are described in Table 14.4.

Several cytokines, agonists, and adjuvants are used to stimulate the immune machinery through adaptive as well as

TABLE 14.1

The Common Plant-Derived Immunomodulators

Plant Name	Part Used	Mechanism	Other Biological Activities	References
Curcuma longa	Rhizome	Humoral antibody response	Anti-plasmodic, anti-inflammatory, gastrointestinal effects, inhibits the cancer growth	Agarwal and Singh 1999; Porwal *et al.*, 2021;
Withaniasomnifera	Roots	Enhance the production of WBC count	Immunomodulation, antibacterial, Sexual behavior and cardiovascular protection	Roy *et al.*, 2022
Ocimum sanctum	Whole plant	Increase the production of WBC, RBC, hemoglobin	Antiplasmodic, hepatoprotective and antiasthmatic	Porwal *et al.*, 2021; Dinesh *et al.*, 2011
Panax ginseng	Fruits and roots	Antibody plaque-forming cell response	Cardiovascular protection, antiarrhythmic, inhibits cancer growth	Porwal *et al.*, 2021; Dinesh *et al.*, 2011
Tinospora cordifolia	Whole herb	Bone marrow cellularity	Antipyretic, anti-diabetic, anti-malaria, anti-stress	Roy *et al.*, 2022; Porwal *et al.*, 2021; Dinesh *et al.*, 2011
Picrorhizakurroa	Leaf	Cell-mediated, humoral	Anti-oxidant, anti-inflammatory, hepatoprotective, anti-allergic	Agarwal and Singh, 1999; Porwal *et al.*, 2021
Piper longum	Fruits	Reduce level of pro inflammatory cytokines IL-1β, IL-6, and TNF-α. Down regulate expression of COX-2, NF-κB and NOS-2	Antioxidant	Porwal *et al.*, 2021; Dinesh *et al.*, 2011; Ibrahim *et al.*, 2015
Azadirachta indica	Flowers	Nitro blue tetrazolium dye, DTH reaction	Anti-inflammatory, hypoglycemia, antibacterial, antimalarial	Porwal *et al.*, 2021; Dinesh *et al.*, 2011
Moringaoleifera	Leaves	DTH test neutrophils	Anti-diabetic activity, cardiovascular protection, anti-oxidant activity	Dinesh *et al.*, 2011
Hibiscus rosasinensis	Flowers	Immunostimulatory, cell- mediated immunity	Anti-inflammatory, anti-diarrheal, to help wound healing	Porwal *et al.*, 2021; Dinesh *et al.*, 2011
Mangifera indica	Stem bark	Humoral antibody response	Antiseptic, anti-diarrheal	Dinesh *et al.*, 2011

TABLE 14.2

Phytochemicals and Their Immunomodulatory Mechanism

Phytoconstituents	Mechanisms	References
Alkaloids (Berberine, Piperine, Curcumin, Sinomenin, leonurine etc.)	Inhibit iNOS and COX-2 activity, down-regulate the T-helper cell cytokines (TNF-α, IL-6, IL-2, IL-4), and up-regulate IL-10 by activating the toll-like receptors and Inhibit T-cell proliferation	Vaibhav *et al.*, 2012; Ibrahim *et al.*, 2015; Xu *et al.*, 2012
Essential oils (Tetramethylpyra-zine, Z-ligustilide)	iNOS and COX-2 synthesis, macrophage chemotaxis, and the formation of ROS are all inhibited by controlling the MAPK and NF-κB signaling pathways	Chung *et al.*, 2012; Ibrahim *et al.*, 2015; Liu *et al.*, 2013
Flavonoids (Xanthohumol, Dihydroxanthohumol, Apigenin, Oroxylin A, Quercetin, Genistein, Thymoquinone, Apocynin)	Reduce the release of inflammatory mediators, and the expression of COX-2, VCAM-1, ICAM-1, and E-selectins, and inhibit NO generation, which is increased by LPS and INF-, reduced IL-1 and TNF production by PPAR activation, stop the generation of CD4+ and CD8+ T lymphocytes as well as inflammatory cytokines	Ibrahim *et al.*, 2015
Terpenoids (Oleanolic acid, Demethylzelasteral, Echinocystic acid, Asiaticoside, Celastrol)	Reduce level of IL-1α, TNF-α and IL-6 inhibit adenosine deaminase activity, enhance macrophages index in both humoral as well as in cell-mediated response; also inhibit pro-inflammatory cytokines gene includes IL-2, COX-2, iNOS, and also inhibit the proliferation of vascular endothelial cells	Kumar *et al.*, 2004; Ibrahim *et al.*, 2015; Brinker *et al.*, 2007

TABLE 14.3

Various Marine Immunomodulators and Their Immunomodulatory Mechanism

Marine source	Chemical constituent	Immunomodulatory mechanism
***Hyrrioserecta*, Thorectidae**	Puupehedione, dipuupehedione, bispuupehenone	Puupehedione exhibited immunomodulatory action
Litopenaeus vannamei, Penaeoidea	Polysaccharides	Immunomodulatory action of SOD (superoxide dismutase)
Hemifususpugilinus, Melongenidae	Fractions of ethyl acetate:petroleum ether (1:1) containing constituents	Type IV hypersentivity (Delayed type) reaction showed stimulation
Micrococcus luteus Micrococcaceae	Anti-*Micrococcus luteus* antibodies	Shows the immunosuppressive potential of Selective T-cells
Reniera species, Chalinidae	Cyclic tripeptide (Renieramide)	Activity in preliminary tests
Chlorella stigmatophora Syngnathidae	Polysaccharide	Immunosuppressant effects
Sargassum thunbergii, Sargassaceae	Polysaccharide (Fucoidan)	Increased the phagocytosis and chemiluminescence of phagocytotic cell i.e., macrophages

TABLE 14.4

Various Examples of Immunosuppressants and Their Modes of Action

No.	Immunosuppressants	Examples	Action	References
1.	Glucocorticoids	Betamethasone Dexamethasone Hydrocortisone	Induce redistribution of lymphocytes and down-regulation of cytokines, binds to promoter DNA of target genes	Almawi, and Melemedjian, 2002; Wikström, 2003
2.	Calcineurin inhibitors	Tacrolimus Cyclosporin	Blocks proinflammatory cytokines, inhibits T cell proliferation by blocking phosphatase activity, inhibition of signal 2 transduction in T-cells	Gupta *et al.*, 2002; Liddicoat and Lavelle, 2019; Barrett, 2022
3.	Antiproliferative/ antimetabolic agents	Azathioprine MycophenolateMofetil Sirolimus Everolimus Cyclophosphamide Chlorambucil Methotrexate	Binds toimmunoglobulin FKBP, inhibitscytolytic T cells, effects on B cell and humoral immunity	Johnson *et al.*, 2001; Allison and Eugui, 2005; Emadi *et al.*, 2009

TABLE 14.4 (Continued)

Various Examples of Immunosuppressants and Their Modes of Action

No.	Immunosuppressants	Examples	Action	References
4.	Antibodies	Anti-thymocyte globulin Anti CD3 monoclonal antibody (Murcmonab/OKT3) Anti -IL-2 receptor antibody (Daclizumab, basiliximab) Campath-1 H (Alemtuzumab) Anti TNF-α (infliximab, etanercept) Monoclonal anti-CD20 (rituximab)	Binds to T cells and inhibit them Binds to CD3 receptors and inhibits its function Binds to CD52 Depletion of B-cells Blocks IL-2 mediated T cell proliferation Binds to TNF-a and inhibits its function at the site of inflammation	Sageshima *et al.*, 2009; Mitoma *et al.*, 2018

TABLE 14.5

Various Examples of Immunostimulants and Their Modes of Action

No.	Immunostimulants	Examples	Action	References
1.	Levamisole	Tetramisole	Restores suppressed T cell function, Upregulates antibody, lymphokine production	Alvarez-Pellitero *et al.*, 2006
2.	Thalidomide	Distaval, Thalomid, lenaliodomide, and pomalidomide	Enhance production of IFN-Y, IL-12, and NK cell activity however comes with several side effects	Miller *et al.*, 2009
3.	BCG	–	Primarily regulate T-cell function, macrophages activity and phagocytosis Effect against bladder cancer	Kaufmann *et al.*, 2018.
4.	Recombinant Cytokines	IFN-alpha 2b IL-2 (aldesleukin)	Promote T-cell differentiation	Dummer *et al.*, 1998.

innate immune responses. Immunostimulants are expected to act as immunopotentiators in healthy humans, acting as both prophylactic and promoter agents (Juyal and Singla., 2001). Some immunostimulants and their action are mentioned in Table 14.5.

Wherever the immune system alone is incapable to fight off an infection or treat the growth of cancer, immunomodulator drugs are used. In Table 14.6, various immunomodulatory drugs are given as examples that have pharmacological effects and are useful for various therapeutic purposes (Preiss *et al.*, 2010).

TABLE 14.6

The Immunostimulants and Immunosuppressant Drugs and Their Pharmacological Effects

	Drugs	Pharmacological Effects
Immunostimulants	Specific monoclonal antibodies	Initiating the effector stage of a particular immune response to an antigen (e.g., tumor)
	Specific polyclonal antibodies	Triggering the effector phase of specific immunity against various antigens
	Interleukin 2	Activate the proliferation of lymphocytes
	Interferon gamma	Activation of lymphocytes, NK cells, macrophages, and also proliferation of monocytes
	Bacillus Calmette- Guerin (BCG)	Activation of phagocytosis cells includes NK cells, macrophages, and dendritic cells
	Some synthetic drugs (levamisole and isoprinosine)	T-lymphocyte development, growth, and activation, and the activation of NK cells, helper T cells, cytotoxic T cells, and chemotherapy taxis
Immunosuppressant	Anti-thymocyte	Triggering the effector phase of immunity against the lymphocytes
	Cyclosporine A, Tacrolimus	Block the proliferation of T-lymphocytes, Inhibits the gene expression of cytokines
	Azathioprine, Leflunomide, Cyclophosphamide	Inhibition of proliferation of B- and T-lymphocytes
	Dexamethasone and Prednisone	Suppression of cytokine transcription in macrophages and T lymphocytes
	Allergen medication	Reduced in reactivity of allergens, reverse the response from Th2 to Th1

14.7 Immunoadjuvants

Immunoadjuvants are immunostimulatory substances that activate or enhance non-specific immunity and acquired immunity. They are simply described as "helper chemicals" that improve the immune system's ability to respond to a vaccination and have the capacity to fend off illness. Through vaccination, adjuvants boost the strength of an immune response specific to an antigen.

14.7.1 Immunoadjuvants for Cancer Treatment

Many different approaches are used to treat tumors, including targeted nanoparticle systems, tumor microenvironment-responsive nanoparticle systems, phototherapy, nanoparticle-based chemotherapy, and immunotherapy. Among these, nanoparticle-based immunoadjuvant therapy has a long-term effect on treatment; it is used to affect tumor-avoiding agents in order to improve the patient's clinical outcome. Cytotoxic T cells are activated by immunotherapy; furthermore, it specifically targets immune cells and offers a chance in the healing of malignant tumors.

Immunotherapy works by rearranging the tumour micro-environment as a result of immunosuppressive proteins attaching to tumour cells, activated lymphocytes, macrophages, and regulatory T cells, which are often expressed on tumour cell surface. The chemicals known as immunoadjuvants employ tumours as antigens to elicit immune responses that are particular to the tumour. Troll-like receptors (Tregs), regulatory [TLCs] on antigen-presenting cells (APCs), function on immunoadjuvants; their molecular structure is similar to that of pathogens. Alum is a type of immunoadjuvant seen in the MHC [2] complex and preferably generates the Th2 responses and then leads to the stimulation of CD^{4+} T lymphocytesalso triggers the humoral immune response. Apart from this mode of action, surface receptors are present on the DCs affected by TLR agonist-based immunoadjuvants via the endosomal pathway and then perform lysosomal escape and present as cytoplasmic proteasomes. Adaptive immune responses evolve eventually due to the cross-presentation process, which directs the activation of cytotoxic T cells.

Cancer, infectious illnesses, and other immunological disorders are treated with bacterial cells and their by-products. The live cells of *Mycobacterium bovis* BCG and the immunotherapeutic activity of cell-wall skeletons [CWS] produced from the mycobacterial components, which are utilised as immunoadjuvants, are both widely used immunotherapeutic agents in the immunotherapy of human cancer cells. Mycobacterial cells that have been heat-killed and are suspended in mineral oil are the most effective immunoadjuvants for the start of cell-mediated immunity and humoral immunity (Freund's complete adjuvant, FCA). Mycolic acids, arabinogalactan, and mucopeptide make up BCG-CWS (Table 14.7).

Abbreviations and Descriptions: **TLRs:** [Toll-like receptors that mediate immune responses to a wide range of pathogens], **Pam3CSK4:** [N-palmitoyl-S- [2,3–bis (palmitoyloxy)-(2RS)-propyl]-CSKKKK], **MPL:** [Monophosphoryl lipid A, first non-alum vaccine adjuvant], **Saponin QS-21:** [Derived from the bark of the tree *Quillajasaponaria,* whose basic structure consists of quillaic acid triterpenes]; **HPSs:** [Heat shock proteins].

14.7.1.1 TLR Agonist-Based Immunoadjuvants

TLRs are group-recognition receptors that recognise damage-associated molecular pattern (DAMP) patterns. As an illustration, heat shock proteins are released by lipopolysaccharides, apoptotic, and necrotic cell bodies. TLRs are found on DCs, natural killer cells, macrophages, and epithelial cells. When they interact with a pathogen, they trigger an inflammatory response brought on by an adaptive immune response. TLRs can generate pro- or anti-tumor signals and are also present on the surfaces of tumour cells. It serves as a first line of defence against cancer. Pam3CSK4, a TLR1/2 complex generated from mycoplasma and bacterial lipoproteins, induces CD4+ T cell differentiation, and improves functional memory in CD8+ T cells.

14.7.1.2 Inorganic Immunoadjuvants

In 1926, Glenny regarded alum as a dominant immunoadjuvant. The alum brings about the activation of Carapace-1 and the secretion of IL- via the NALP3 inflammasome, and the activation of the innate immune system. Aluminum hydroxide and aluminium phosphate are responsible for the Th2 immune response, but they have limited effect on $CD8^{+}$ T cells. Alum works in tandem with IL-12 to maintain immunity against the melanoma antigen. The least naturally occurring inorganic immunoadjuvants present in the body are zinc phosphate and calcium phosphate (which initiates the activity of APCs). Mg (magnesium) is another immunoadjuvant that plays a vital role in cellular morphology, phagocytosis also the release of

TABLE 14.7

Details Regarding the Types of Immunoadjuvants Used for Cancer Treatment (Banstola et. al. 2020)

Types of Immunoadjuvants		References
TLR	Pam3CSK4 (TLR 1/2) MPL (TLR 2/4)	Chen *et al.*, 2014
Inorganic immunoadjuvants	Alum Magnesium	Wang *et al.*, 2019
Plant-derived Immunoadjuvants	Saponin QS-21 Polysaccharide	Fernandez-Tejada *et al.*, 2014
Exosomes	DC derived Exosomes	Besse *et al.*, 2016
Endogenous protein	HPSs	Tsan and Gao, 2004

lysosomal enzymes. GO (graphene oxide) aids in antigen loading and cytokine production, as well as activating cytotoxic T cells and inhibiting tumour growth.

14.7.1.3 Plant-Derived Immunoadjuvants

Plants also produce certain immunoadjuvants. Saponin is a plant-derived immunoadjuvant, and its combination with cholesterol and phospholipids forms immuno-stimulatory complexes (ISCOMs). Both adaptive andinnate immune responses can be triggered by saponin-based immunoadjuvants. The *Quillaja saponaria* tree's QS-21, a saponin-based natural substance, boosts the ASC/NALP3 inflammasome pathway's activity. The immune system is stimulated by pectic polysaccharides, commonly known as pectin and plant polysaccharides. Particularly pectin raises overall IgA and IgG levels in the blood while also increasing the quantity of circulating lymphocytes. It also promotes phagocytosis and causes macrophages to produce nitric oxide, regulating the immune system by enhancing spleen cell proliferation. The ripe fruit of *Physalis salkekengi* contains polysaccharides that increase CD4+ and CD8+ T lymphocytes.

14.7.1.4 Exosomes

Exosomes were developed in 1983, but since they can transmit antigens and activate T cell responses specific to certain antigens, they were recognised as immunoadjuvants in 1986. Engineered exosomes exhibited immunosuppressive qualities but were also utilised to cure cancer. They essentially function as cell-based nanovesicles that interact with MHC to form complexes and activate cytotoxic T cells. Exosomes on tumour cells or DCs stimulate the immune system. By subjecting DCs to cancer peptide treatment, one can produce DC-based exosomes.

14.7.1.5 Endogenous Proteins

Endogenous proteins include heat shock proteins (HSPs); examples of heat shock proteins are HSP40, HSP60, HSP70, and HSP90. Their immunogenic property upgrades the activity of inflammatory cytokines. HSPs aid in the release of TNF, the maturation of DC cells, and the release of nitric oxide. HSP40 helps bring out the anti-inflammatory response, and HSP90 functions in the activation of NK cells.

14.8 Adjuvants in Fish Vaccines

Adjuvants are widely used in fish vaccines. Adjuvant selection influences safety and potency, even though all adjuvants have different modulating capacities but their exact mode of action is still unknown. In spite of their side effects, oil-adjuvanted vaccines are widely used. Vaccines maintain continuity and slow antigen release at the injection site. Chronic peritonitis, adhesions, granulomas, and a delayed growth rate are the detrimental consequences of mineral oil usage (Midtlyng *et al.*, 1996; Poppe and Breck 1997).

However, herbal adjuvants and immunostimulants derived from plants are used in fish vaccines, due to anti-bacterial, anti-viral, immunostimulant, and anti-stress effect in fish and shellfish culture. Numerous research has demonstrated the beneficial benefits of herbal immunostimulants. For instance, Chu (2006) showed that the plant-derived substance Propolis increased antibody production and activity of leukocytein carp. *Ganoderma lucidum* and *Astragalus radix* boosts carp's immune response to an infection with *Aremonas hydrophila* (Yin *et al.*, 2009). Another immunostimulatory substance that also has anti-infective properties is *allium sativum*; it enhances lymphocyte formation, phagocytosis, and release of cytokines and activation of natural killer cells (Kyo *et al.*, 1998). In *Labeo rohita* and Rainbow trout, allium is thought to be an efficient immunostimulant against infection from *A. hydrophila* (Sahu *et al.*, 2007). Garlic and mineral oil mixed with a formalin-killed *A. hydrophila* cell (FKC) are frequently used as an adjuvant in fish vaccines to prevent the infection of Rohu, *L. rohita*.

As goldfish (*Carassius aurantus*) are highly susceptible to *A. hydrophila* infection and cause red fin diseases, motile aeromonad septicaemia, haemorrhagic septicaemia, etc. immunostimulants extracted from *Emblica officinalis, Cynodondactylon,* and *Adathodavasica* shown to reduce the microbial infection in goldfish (*C. aurantus*) and also enhance the immune system. Vibrio pathogenicity in Grouper larviculture can be controlled, according to Sivaram *et al.* (2004) by using methanolic extracts from plants. Magdelin (2005) used herbal stimulants to perform his research on the ornamental fish *Poeciliasphenops*. In the case of *Ixoracoccinea*, extractions from their flowers and leaves are believed to be anti-microbial, anti-inflammatory, and antioxidant in nature.

Because of their ability to defend and their positive action on the fish immune system, the cytokines are equally utilised as adjuvants on many forms of fish vaccines, according to *Invivo*. Pluripotent cytokines interfere with the formation of immunoglobulin, promote the cytotoxic activity of macrophages, activate T and B lymphocytes along with the stimulation of natural killer cells, and ultimately promote the expression of immune-related genes. IL1 immunoadjuvant is used in various fish vaccines to fight against pathogenic infection, but their mode of action is not clear. Recombinant IL1 was developed as an immunoadjuvant for fishes like carp, barramundi, sea bass, rainbow trout, and Japanese flounder. Wang *et al.* (2019) demonstrated that two cDNAs encoding inter-leukin-1 in channel catfish (*Ictalurus punctatus*) exhibit distinct expression patterns following bacterial infection.

14.9 Mode of Action of Adjuvant

Immunoadjuvants are often classified as either Signal 1 facilitators or Signal 2 facilitators depending on how they work. Both signal systems, which include an antigen presentation (Signal 1) and, in the case of Signal 2, an extra secondary signal, are important in the activation of certain T and B lymphocytes, which consequently activates the immune system. While Signal 2 facilitators aid in antigen detection through co-stimulation signals, Signal 1 facilitators control the vaccination antigen's time, position, and concentration, causing it to activate the immune system. They also offer a favourable setting for immunological reactions specific to an antigen. Table 14.8 describes different adjuvants with their actions.

TABLE 14.8

Signal 1 Adjuvants Used in Fish Vaccines, Oil Emulsion Adjuvants, Its Composition, and Their Mode of Action in Fish Vaccines (Taffalla *et al.* 2013)

Oil Emulsion Adjuvants			References
Name	**Composition**	**Action**	
Freud's Complete Adjuvant (FCA)	• Mineral oil and heat-killed mycobacteria	• Strong Th1 and Th17 response	• Opie EL and Freund J., 1937
Freund's Incomplete Adjuvant (FIA)	• W/O emulsion, lacks mycobacteria	• Significant reduction in toxicity and highly effective in vaccination	• Gjessing *et al.*, 2012
Montanide	• Mineral oil adjuvants (mostly) • Activity based on mineral oil or non-mineral oil or a mixture of both	• Improve efficacy and stability	• Lawrence *et al.*, 1997

Descriptions: **FCA** [composition of antigen emulsified in mineral oil, prepared from non-metabolizable oils, contains inactivated mycobacteria], **FIA** [it is being used in immunotherapy, less potent than FCA}, **Montanide** [contains natural metabolizable mineral oil, emulsifier from the mannide monooleate family, produce a strong and deep-rooted immunity].

An oil emulsion contains a dispersed phase (oil) and a continuous phase (antigenic media), and surfactants are additionally added to stabilize the emulsions. Freund's Complete Adjuvant (FCA) is made of mineral oil with surfactant and heat-killed mycobacteria that have been heated to death. Here, the FCA presence decreases the action of vaccine. The*peptococcus spp.* infection in rainbow trout is also depressed by FCA. FCA is replaced with FIA (Freund's Incomplete Adjuvant) as a consequence of its harmful effects. FIA-exempt vaccines used for *Edwards iellatarda* infection in Japanese flounder show high resistance against the virulence of the pathogen. AJ-oil (Alpha-Jet Oil 5200), PLGA (poly-(lactide-co-glycolide), and ISCOMs are further examples of immunoadjuvants used in fish vaccines. Mineral oil adjuvants are mainly categorised as "Montanide." These adjuvants are either mineral oil based or non-mineral oil based or a mixture of both. The vaccine against bacterial hemorrhagic ascites contains formalin-killed *Pseudomonas plecoglossicida* emulsified with montanide. Table 14.9 describes different adjuvants with their actions.

Descriptions: **Aluminium containing adjuvants** [main objective is to increase the adsorption of antigen on the adjuvant, Safe to use], **β–Glucans** – ligands for dectin – 1 [Protection against infection through Glucan intraperitoneal injection, effective against infection and tumor], **Flagellin** – Toll-like receptor 5 agonist [TLR5 is needed for the extracellular detection of bacterial flagellin], **CpG–Toll-like receptor 9 agonist** [Unmethylated cytidine phosphate guanosine – CpG, Oligodinucleotides-ODN], **Poly I:C – Toll-like receptor 3** agonist [Enhance chemotherapeutic efficacy in paclitaxel (PTX – chemotherapeutic agent used for treating breast cancer) resistant cell lines].

Immunoadjuvants containing aluminium are considered to besafe for use in human vaccinations but are rarely utilised in fish vaccines. Alum has a better capacity to elicit humoral immunity and Th2 responses, but they are ineffective against cell-mediated immunity. Few research have been conducted on fish containing

TABLE 14.9

Signal 2 Adjuvants in Fish Vaccines, Adjuvants, Composition, and Their Mode of Action in Fish Vaccines (Taffalla *et al.* 2013)

Adjuvants	Composition	Action	Reference
Aluminium containing Adjuvants	Aluminium phosphate Aluminium hydroxide	• Induces Th2 responses • Enhance humoral immunity	• Jiao *et al.*, 2010
β–Glucans – ligands for dectin-1	Beta-Glucans Yeast Glucans	• Activation of non-specific immunity	• Midtlyng *et al.*, 1996
Saponin	Natural glycosides of steroid Example: Quil A	• Stimulates both Th1 and Th2 responses	• Sun et al., 2009
Poly I:C - Toll-like receptor 3 agonist	Polyinosinicpolycytidilic acid (Poly ribonucleotide)	• Induces non-specific antiviral state	• Jensen et al., 2002
Lipopeptides	Lipoproteins and lipopeptides derived from mycobacteria and mycoplasms	• Long-lasting innate immunological response • Potent inflammatory response	• Hoel K and Lillehaugh A., 1997
Flagellin Toll-like receptor 5 agonist	Gram-negative flagella structural protein	• Promote the production of cytokines while acting as an effective activator of a variety of cell types (both adaptive immune and innate systems)	• Mizel *et al.*, 2010
CpG – Toll-like receptor 9 agonist	Unmethylated CpG present in bacterial DNA and synthetic oligodeoxynucleotides	• The development, differentiation, and proliferation of distinct immune cells, such as B and T cells	• Bode *et al.*, 2011

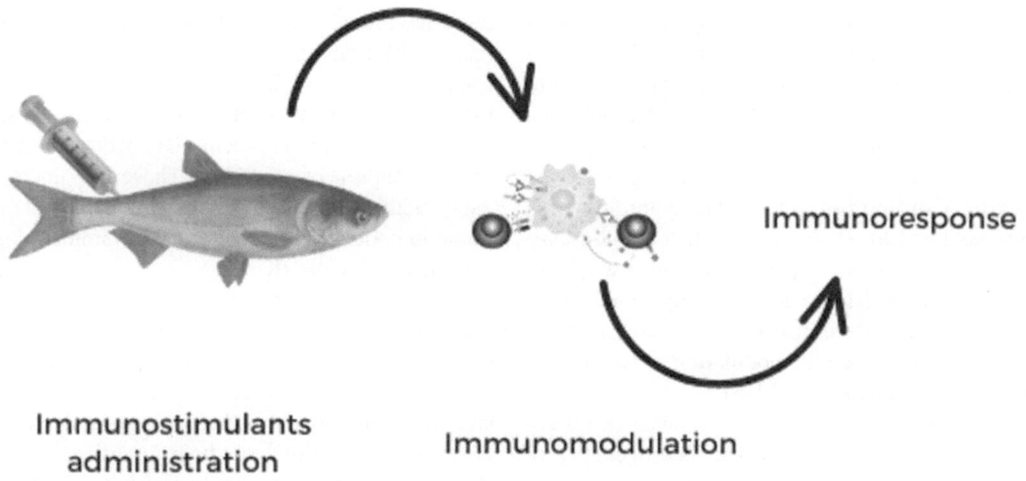

FIGURE 14.1 Represents general actions of immunomodulators.

lipopeptides, despite the fact that lipopeptides are present in many microbes, including mycobacteria and mycoplasms, and they have highly long-lasting adaptive immune responses.

polygonal troll I:C receptor3 agonist is utilised to induce Type I INF in various animals, including fish, by simply simulating viral infection using a double-stranded polyribonucleotide. A variety of cell types, particularly the adaptive and innate immune systems, are activated by flagellin, which is the structural protein of Gram-negative flagella.

Chitin and its derivatives as immunoadjuvants

Chitin, which is also found in fungus and bacteria's cell walls, makes up the exoskeleton of arthropods. It is bonded covalently to glucans and proteins in fungus and arthropods, respectively. Chitin and modified chitin have certain beneficial features. Tokura *et al.*, (1999) conducted a number of researches to comprehend the immunological properties of chitin and its derivatives. Chitin's immunostimulatory activity varies with particle size; the larger the particle (2–10 micron), the more active it is (Lee *et al.*, 2008; DaSilva *et al.*, 2009). Chitin and its derivatives induce macrophages to produce cytokines that have anti-tumor action and resistance to the infection of bacterial and viral in non-specific hosts, according to Suzuki *et al.* (1982) and Nishimura *et al.*, (1984). Heat-killed *Mycobacteriumbovis* induces Th1 immunity, which is favourably regulated by chitin. Th2 immunity is also influenced by chitin. According to Hamajima *et al.* (2003), the Th1 adjuvant action on chitin microparticles results in the induction of viral-specific immunity. It is useful to provide chitin because it serves as an immunoadjuvant and lowers the incidence of Th2 allergies. When given as chitin microparticles, shrimp chitin is an effective T and B cell immunoadjuvant that can convert a Th2 (T helper cells type 2) immune response into a Th1 immunological response.

14.10 Immunomodulator Efficacy

Immunosuppressive drugs are broadly used in organ transplantation and autoimmune disease like Rheumatoid arthritis, multiple sclerosis or SLE in human. Immunostimulants are promoted for preventive of respiratory diseases and other infectious diseases. Immunostimulants are proven to be efficient for antitumor treatment. In vaccine development, immunostimulants are used for the presentation of an immunogen to the immune system and to amplify immune response specific to the antigen; thus, immunostimulants are used in formulation of vaccines (A. M. Carmona-Ribeiro *et al.*,2017) (Figure 14.1).

14.11 Immunomodulator Efficacy in Fish

Due to their potential to replace vaccinations, serve as adjuvants, and increase vaccine effectiveness, immunostimulants are valuable in the prevention and management of fish diseases in aquaculture. They also have a role in enhancing the growth rate under stressful conditions (Heo *et al.*, 2004). Many animal and herbal extracts are used as immunomodulatory drugs in aquaculture. Polyherbal immunostimulants in fish significantly elevates ROS, macrophage phagocytic activity, levels of total protein and NOS activity, which indicates the potency of immunostimulants in aquaculture. It increases resistance against several strains of infectious bacteria such as *Aeromonas salmonicida, V. anguillarum, Streptococcus sp.* and *V. salmonicida,* viral infections like yellow head disease (YHV) and infectious hematopoietic necrosis (IHN). An herbal immunoregulation mixture (HIRM) comprising polyherbal formulation significantly elevates fish immunity. The macrophage phagocytic activity, levels of total protein, reactive oxygen species, albumin, globulin, and NOS activity in fish were all markedly improved by diets containing 0.5% and 1% HIRM. The HIRM of ginger (*Zingiber officinale*), nettle (*Urtica dioica*), and mistletoe (*Viscum album*) revealed increased respiratory burst activity in rainbow trout ($P < 0.001$) (Seung-Cheol *et al.*, 2007).

Significant decline in mortality rate was observed in fish given with immunostimulants. Administration of glucan with vaccine significantly enhanced vaccine efficacy. The glucan-fed vaccinated fish had a 5% death rate, while the control fish had a 30% mortality rate (Sahoo and Mukherjee., 2002). Beta-glucan dosages of 250 and 500 mg per kg of feed improved), feed conversion ratio (FCR), and specific growth rate (SGR) in *Aeromonas hydrophila*–challenged carps, but concentrations over 500 g glucan per fish increased the survival rate to 100%.

Following delivery of the glucan, the populations of neutrophils and monocytes as well as the overall number of leukocytes all rose. A little 0.09% of glucan might greatly increase the expression of IL-1ß mRNA by macrophages, phagocytic activity, and respiratory bursts of SOD (Selvaraj *et al.*, 2005). Administration of ß-1,3-glucan improved the survival of black tiger shrimp from white spot syndrome virus that cause white spot syndrome in shrimp. In common carp supplied with chitosan the relative percentage survival (RSP) increased to 80%, serum bactericidal activity, phagocytic index, and phagocytic ratio were also increased chitin stimulate immune system via non-specific modulation of haemolytic complement activity (Mehana *et al.*, 2015).

Immunostimulants are used to successfully manage a variety of fish diseases. Employing substances like oleuropein (Ole), which is derived from the leaves of the Olea europaea plant, it is possible to successfully manage viral infections such as viral hemorrhagic septicemia virus (VHSV) and Salmonid rhabdovirus (Micol *et al.*, 2005). Shrimp treated with butanolic extract of *Withania somnifera* efficiently suppressed *V. damsela* and *Vibrio parahaemolyticus*infection. Seaweed extracts in shrimp *Penaeus indicus* boosted immunity against *Vibrio parahaemolyticus* (Praseetha, 2005).

14.12 Aquaculture Immunomodulators

Aquaculture employs immunomodulators to lessen the immunosuppressive effects of stresses. Additionally, they can be utilized as preventative measures against seasonal endemic disease epidemics. Vaccines are important in preventing diseases in humans and animals. They are extensively used in cultured fish to enhance resistance against infectious pathogens. However, most vaccines increase resistance only against one or two pathogens and are also temporary in effect. But, in the case of immunostimulants, they increase resistance against a broad range of pathogens and are therefore nonspecific.

An algal extract of laminarin is also proven to elevate the respiratory burst activity in the leukocytes ofanterior kidneyand activate macrophages in Atlantic salmon, due to its increased solubility over yeast glucans. The epidermis and intestinal epithelium of the Atlantic halibut's (*Hippoglossus hippoglossus*) yolk-sac larvae, allow laminarin to be absorbed. This shows that laminarin may have the capacity to enhance immunity during the formative years of life, even before the development of acquired immunity (Barman *et al.* 2013). The impact of dietary pyridoxine on enzymes like LDH, MDH, AChE, etc. is extraordinary. LDH activity often rises in response to stress. Pyridoxine supplementation in the diet significantly reduced MDH activity. A decrease in the activity of LDH suggests that dietary pyridoxine may have helped in lowering the energy demands in *Labeorohita* fingerlings (Shahbaz Akhtar *et al.* 2010).

The chemical dichlorodiphenyldichloroethane raises blood glucose levels in rainbow trout (Benguira *et al.*, 2002) and aluminium concentrations in Atlantic salmon (Ytrestoyl *et al.*, 2001). Fish-fed beta-glucan orally express a high level of mucin, preserving the fish intestinal barrier's protective capability. Commoncarp-fedbeta-glucans have stimulated the growth of intestinal microflora to prevent bacterial invasion, lowering the risk of bacterial infection in the gut (Ching *et al.*, 2020). Immunostimulants can mitigate the loss caused by diseases in fish and crustaceans. Fish and crustaceans use several forms of beta-glucans to protect themselves from bacterial and viral illnesses (Paulsen *et al.*, 2001). These may increase the bactericidal activity of phagocytes in fish and shrimp. Other immunomodulators can increase both immunological responses and the activity of NK cells.

Antibiotics, vaccinations, and immunostimulants are the methods used to treat diseases in shrimp. The differences brought about by the use of chemotherapy and vaccinations are made up for by immunostimulants. They put forward a possible and reasonable solution regarding shrimp immunity. Hence, immunostimulants help in shrimp health management (Santosh Kumar *et al.*, 2022). In Chinese mitten crab, *riocheir sinensis,* the glycoprotein hemocyanin induces non-specific immunity to resist pathogens and also it has some immunomodulatory properties. A moderate level of iron in the body enhances the activity of hemocyanin. It has been demonstrated that E. sinensis's rate of growth and innate immune responses can be enhanced by consuming enough dietary iron supplements (Song *et al.*, 2021). The various other immunostimulants used in aquaculture include:

1. Synthetic chemicals: FK-565, Levamisole (an anti-helminthic used to treat nematode infection that helps in increasing the production of antibody-producing cells and enhances the phagocytic activity), Quaternary ammonium compounds (QAC), etc.
2. Biological derivatives: MDP (muramyl dipeptide)—increases phagocytic activity, respiratory bursts, etc. in rainbow trout. Lipopolysaccharides (LPS) stimulate superoxide anion production, phagocytosis, and macrophage production.

Freund's adjuvant (FCA) stimulates NK cell activity, respiratory burst, and phagocytosis.

- Yeast derivatives - Beta-1, 3 glucan, Beta-1,6 glucan, etc.
- Nutritional factors - Vitamin C, vitamin E, etc.
- Hormones - Prolactin, growth hormones, triiodothyronine, etc.
- Cytokines - Interferon and interleukin
- Polysaccharides - Chitin, chitosan, oligosaccharides, etc.
- Plant and animal extracts - HDE (abalone), ETE (tunicate), saponin (soap tree), etc (Figure 14.2).

14.12.1 Evaluation of the Efficacy of Immunomodulators in Aquaculture

The efficiency of humoral and cellular immune processes following immunomodulator administration may be measured *in vitro*, and an *in vivo* approach, such a protection test against fish infections, can be used to determine an immunomodulator's effectiveness. Total leucocyte and erythrocyte count, serum lysozyme, respiratory burst, complement, chemokinesis, phagocytosis, lymphocyte proliferation, and chemotaxis are the main

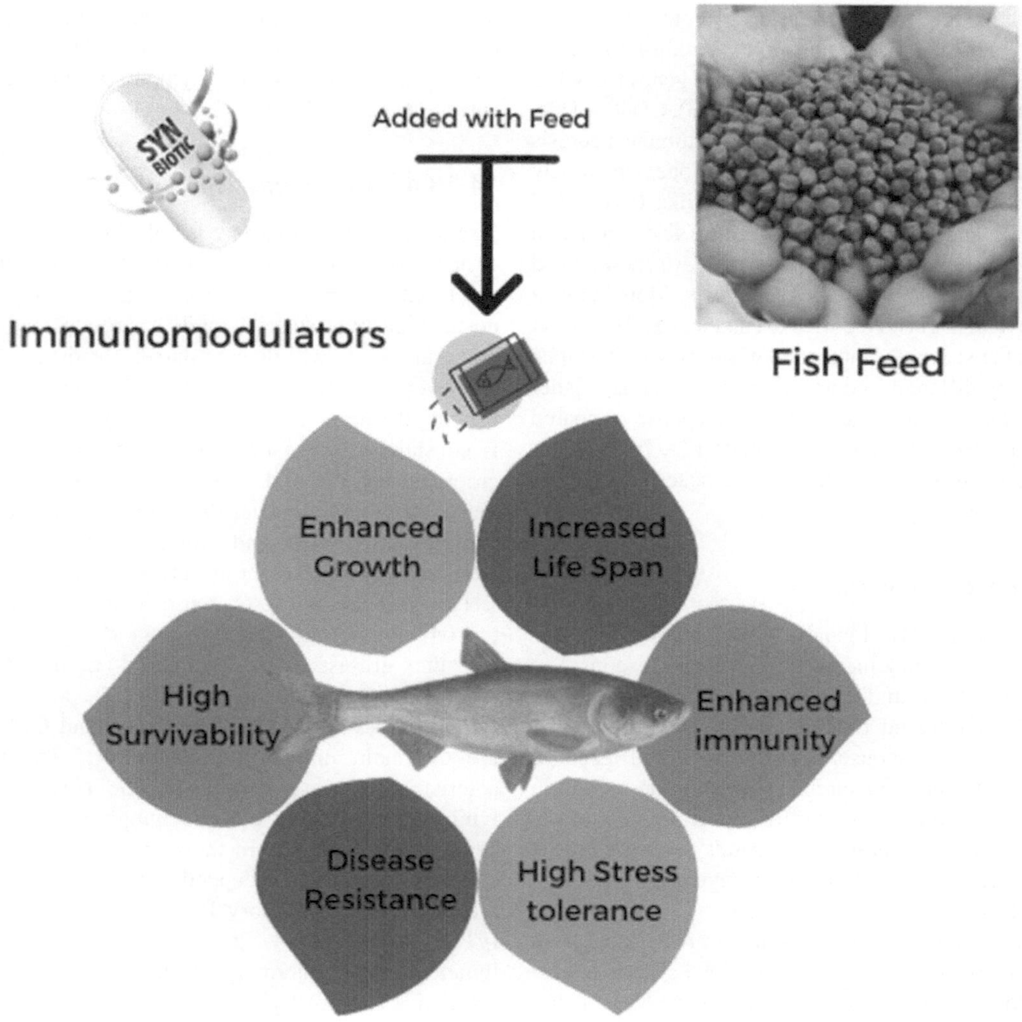

FIGURE 14.2 Illustrate the effects of immune modulators in finfish.

variables used *in vitro* research. Natural cytotoxic activity, C-reactive protein, and MAF (musculoaponeurotic fibrosarcoma) are some other indicators that are advised to be examined.

Currently, protection testing against fish diseases are utilised with great effectiveness. These studies have revealed a significant improvement in a fish's immune response to, or resistance to, a number of diseases, suggesting a significant role in aquaculture. When possible, *in vitro* testing should be combined with *in vivo* research to better understand the underlying processes of protection.

14.13 Immunoaduvants to Increase Vaccine Efficacy in Fish

The most effective way to control infectious illnesses is vaccination. To increase the effectiveness of a vaccination and the size of an adaptive response, adjuvants or immunostimulants are routinely utilised. The sort of adaptive response to a particular disease is also considered to be guided by them. The hunt for adjuvants that increase immunogenicity while inducing the optimal immune response has been going on for a long. The majority of the marketed bacterial vaccinations for fish include conventional adjuvants such as mineral oils.

14.13.1 Signal 1 Adjuvants

Freund's complete adjuvant (FCA) combined with the vaccine manifested the best protection against most of the pathogens compared to the vaccine in the absence of adjuvants used in fish culture. Coho salmon (*Oncorhynchuskisutch*) demonstrated enhanced vaccine efficacy when challenged with *Aeromonassalmonicida* (Olivier G., Evelyn TP *et al.*, 1985). Due of FCA's high level of toxicity, Freund's incomplete adjuvant has recently been used in place of it on a large scale. A water-oil immersion that is missing the mycobacterial components is Freund's incomplete adjuvant.

Enhanced efficacy of vaccines combined with adjuvants has been reported in the latest studies. Jiao XD *et al.* (2010) demonstrated enhanced efficacy of a vaccine encompasses a major antigenic protein from *Edwardsiellatarda* combined with FIA against the *Edwardsiellatarda* bacterium that infects many fish. Relative percent survival (RPS) attained by the vaccination with FIA, which measures protection against experimental challenge, improved from 34% to 81%. They claimed that, in contrast to vaccination immunization with the oil-adjuvanted antigen stimulated the expression of genes for MHC class I and class II, complement component 3 (C3), CD40, CD8a, interferon g (IFN-g), interleukin 6 (IL-6), and tumour necrosis factor a (TNF-a) in comparison to the antigen alone, which only increased the expression of some components like IgM, MHC class I and class II, and IgM. Similar trends were also reported in Catla (*Catlacatla*) challenged with *Aphanomyces invadans* and salmon injected with *Tenacibaculum maritimum.* Montamide, a mineral oil adjuvant, has been reported to enhance the RPS in fish when used with the vaccine compared to when the vaccine alone was given. Among different immunoadjuvants, FCA has higher efficacy and the highest and fastest antibody response compared to montamide or other adjuvants. Montanide ISA-76, ISA711, and ISA763A have proven to increase the efficacy of vaccines used in fish (Sitja-Bobadilla *et al.*, 2008).

14.13.2 Signal 2 Adjuvants

Toll-like receptor (TLR) ligands can function as immunoadjuvants as they may induce strong innate responses in fish, according to many studies, TLRs likely cause the synthesis of IL-12 after ligand binding, which favours the Th1 (Type 1 helper T cell) response. A cytotoxic T cell response may be facilitated under some circumstances by the activation of these TLRs, which may also cause the cross-presentation of antigens. Additionally, several TLR ligands promote type I IFN responses by activating interferon-regulating elements (Manicassamy S. and Pulendran B., 2009).

Glucans promote the non-specific immune response in fish and humans. In several researches today, beta-glucan is also employed as an adjuvant. In comparison to the vaccination alone, b-glucan was demonstrated to reduce mortality in fish exposed to *Aeromonas hydrophila.* The yeast glucan in the vaccination against tuberculosis has a comparable impact on the effectiveness of the shot. Toll-like receptor 3 agonists and Poly I:C trigger type I IFN and a general antiviral state after binding to TLR3. Fish inoculated with Poly I:C e Toll-like receptor 3 agonist displayed greater concentrations of certain antibodies in their systems.

IL-1b, IL-6, and IL-8 and TNF-a were shown to be upregulated in flagellin-immunized Atlantic salmon, according to Hynes *et al.* (2011). Additionally, flagellin increased the expression of genes such as IL-1b, Mx, IFN, MHC-Ia, CD8a, MHC-IIa, and IgM that are related to adaptive and innate immune responses. Interferon b (IFN b), Nitric oxide (NO), and the plasmid coding for IL-8 (pIL-8) were up-regulated by cytokines such as IRF-1 (interferon regulatory factor 1), which in combination with the VHSV vaccine improved the expression of interferon-inducible genes like Mx and the expression of TNF-a, IL-18, IL-11 and TGF-b in the spleen (Jimenez N, et. al., 2006).

14.13.3 Targets of Immunostimulant Diets in Fish

Fish external mucous, thymus, spleen, liver, and kidneys are the most immunocompetent organs and tissues affected by immunostimulants. Primary lymphoid organs that support the development and maturation of stem cells are the head, kidney, and thymus. The secondary lymphoid organs are the kidney, spleen, and MALT.

14.13.4 Fish Mucosal Immunity

Fish's mucosal body surface is linked to their adaptive immune system. Mucosal tissue is found on surfaces that come into touch with the outside environment, including the skin, gills, nose, digestive system, and gastrointestinal tract. These mucosal tissues will detect non-self, leading to local changes that eventually trigger the entire physiological response, enhancing the immune response at the systemic level. Because it is constantly thrown off, mucus prevents the growth of microorganisms. Second, after establishment has been made, mucus serves as a barrier that must be overcome. Lysozyme, complement, lectins, and proteolytic enzymes with antibacterial activity are present in mucus (Figure 14.3).

The immunological response in teleosts is controlled by the 04 mucosal-associated lymphoid tissues (MALT); they are the gill-associated lymphoid tissue (GIALT), gut-associated lymphoid tissue (GALT), and skin-associated lymphoid tissue (SALT). The SALT and GIALT molecules are the main immunological barriers. The primary characteristics of these tissues are the absence of organised lymphoid structures, such as lymphoid nodes or germinal centres, which results in the dispersion of their leukocytes. The existence of interdependency bacteria, some of which are coated with secretory Igs, a mucosal Ig class known as IgT/Z, and secretory Igs, which are transported into the lumen by a polymeric Ig receptor (pIgR) are further characteristics (Parra *et al.*, 2015).

14.13.5 Head Kidney

With its abundance of monocytes and macrophages, the head kidney functions as a generative and SLOs. Here, as opposed to the bone marrow where it happens in mammals, blood cell development takes place. The whole kidney is engaged in the early immune reaction and the formation of immune cells. The head kidney contains lymphoid cells that produce cytokines. As fish get older, the anterior kidney takes on a greater role in

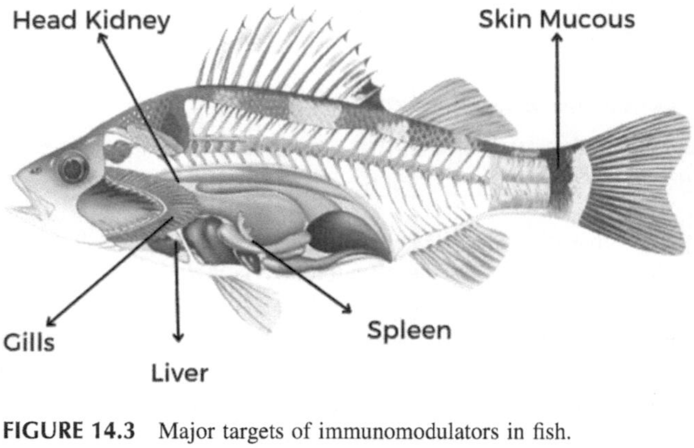

FIGURE 14.3 Major targets of immunomodulators in fish.

immunological and blood cell production, whereas the posterior kidney mainly handles filtration and/or urination duties. Fish kidney, liver, and spleen have a cluster of pigmented phagocytes called melanomacrophage centres, which are composed of aggregations of leukocytes and dendritic cell-associated macrophages with varied degrees of pigment. They are supposed to play a role in immunologic memory and allegedly involved in antigen trapping.

14.13.6 Spleen

Fish have a large secondary lymphoid organ called the spleen. The spleen is crucial for antigen presentation, lymphocyte activation, and the stimulation of T cell proliferation. The cooperation of dendritic cells (DC) at the cellular level makes this possible. The spleen contains hematopoietic and lymphoid cells. Fish spleens include lymphocytes and macrophages, which are found in ellipsoids and specialised capillary walls. The spleen's macrophages are organised into clusters called macrophage centres. It has been suggested that the rise in IL-10 in *Cyberlindner ajadinii* yeast subjected to autosylation, a downstream step, helps to reduce the inflammatory profile linked to Atlantic salmon challenged with soybean meal (Agboola *et al.*, 2021). Additionally, it has been reported that fish, fed with functional diets with shiitake mushroom (*Lentinula edodes*) can control the acute inflammatory profile in the spleen, minimising potentially detrimental reactions to an LPS challenge in salmonids like rainbow trout (*Oncorhynchus mykiss*) (Djordjevic B, Skugor S *et al.*, 2009).

14.13.7 Thymus

The primary lymphoid organ in fish, the thymus, is a paired bilateral structure located in the gill cavity behind the pharyngeal epithelium. T lymphocyte development takes place in the thymus. The function of the thymus supports the maturation and selection of the T cell repertoire, which is important for adaptive immunity and the T-cell progenitors go to the thymus from the bone marrow.

14.14 Factors Affecting the Efficacy of Immunomodulators in Fish

Several elements influence the action of an immunomodulatory substance on fish, such as

- Method of administration
    - Injection
    - Oral
    - Immersion
- Dosage of the drug
- Timing of administration and duration of drug
- Duration of drug

14.14.1 Method of administration

The three most common methods of administering immunomodulators to fish are injection, immersion, and oral, as shown in Figure 14.4. The most effective immunization route is injection.

14.14.1.1 Injection

This is one of the cost-effective methods when we consider large fish given that the size of the fish must be greater than 10–15 g. However, it is time-consuming compared to other methods. Only extensive aquaculture uses injection, and during processes the fish must be handled or at the very least confined to a restricted area. Injecting immunostimulants enhances leucocyte activity and offers protection against infections, according to several researches.

Injections can be administered intramuscularly, intraperitoneally, or into the dorso-median sinus. Broad protection against a number of infections has been demonstrated to be offered by injection. When channel catfish are exposed to *Edwardsiella ictaluri,* intraperitoneal glucan injections boost phagocytic activity, lowering fish mortality (Duncan, P.L., and

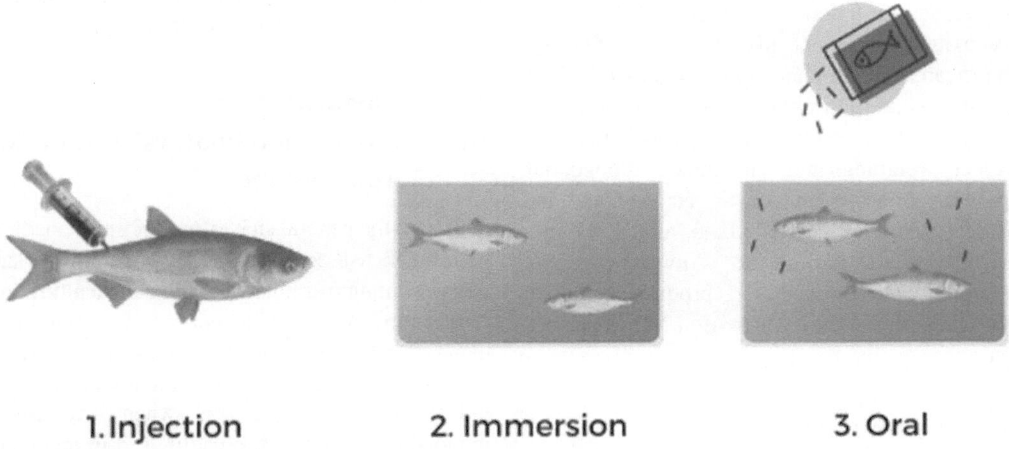

FIGURE 14.4 Different method of administration of immunomodulators in aquaculture.

Klesius, P.H., 1996). This approach can be used with adjuvants to improve the immunisation. Injecting a vaccine into little fish is not practicable. But with prolonged use, these techniques become less effective.

14.14.1.2 Oral

The concentration of the drug absorbed prior to digestion determines the effectiveness of oral delivery. Orally given immunostimulants, such as glucans, levamisole, chitosan, EF203, and lactoferrin, offer defence against contagious illnesses such furunculosis, vibriosis, and streptococcosis. Carp submerged in solution of levamisole for 24 hours showed activatedchemotactic behaviour, phagocytic activities, the production of active oxygen, and resistance to *A. hydrophila*, according to Masahiro Sakai's (1998) research. Feed is combined with immunostimulant granules using a coating of fish oil. To immunise fish larvae with live feeding organisms during their early larval stages, the bioencapsulation approach is now also used.

14.14.1.3 Immersion

Although administering immunostimulants through immersion is not as effective as injection, it is the most cost-effective method for small fish. The efficacy levels are not well defined in this method. It can be used to immunize small fishes, too. However, the oral method is the only non-stressful way of providing immunomodulators to fish. It has poor potency but is the less wasteful method. Medicated feed pellets are thrown upon the surface of the water, a process termed "broadcast feeding."

14.14.2 Drug Dosage

The dosage of a drug highly influences its efficacy as an immunomodulator; it is generally expressed in mg/kg b.w./day (kg b.w. means kg of bodyweight). According to research, the optimum dosage for an immunostimulant is 1,200 mg/kg. Lower than 600 mg/kg or higher than 800 mg/kg may induce an ineffective or altered pattern of immunostimulation. Fish fed a food supplemented with a modest dosage of levamisole (500 mg/kg) demonstrated greater weight gain and improved immunological activity as compared to fish fed the base diet. However, it has been demonstrated that fish given 1,000 mg of levamisole daily exhibit persistent toxicity. These findings highlight the significance of immunostimulant doses for fish health (Galindo-Villegas and Hosokawa, 2004). According to research, employing a premix as opposed to a 100% medicine will enable correct weighing and guarantee homogenous mixing. Glycoproteins boosted phagocytic activity and the generation of nitric oxide, interleukin-1, IL-6, and IL-12 in a dose-dependent manner, according to Anweicheng. At a dosage of 10 g/mL, GP increased IL-1 production, but at a dose of 100 g/mL, GP dramatically increased NO, IL-6, and IL-12 production.

14.14.3 Timing of Administration

In addition to immunisation, immunomodulators can also be administered. Fish might be treated beforehand to get them ready for expected illness occurrences, including seasonal exposure to a virus. When the fish are being exposed to a particular antigen is the best moment to deliver immunostimulants and vaccines. Immunostimulants should frequently be given before immunisation to assist the fish become ready for exposure to the antigen. However, each substance's dose and time for administration should be chosen.

14.14.4 Drug Duration

In aquaculture, the effect of immunomodulators is considered to be of short duration (Galeotti, 1998). After 14 days of immunological activation, the impact in trout utilising glucans and chitosan was significantly diminished (Siwicki *et al.*, 1994). When administered for extended durations, the majority of immunostimulants did not enhance fish development in comparison to a single dosage. Tseng *et al.* noted that the protection induced in three spot gourami after 29 days of injection of 2 mg/100 g laminaran for 22 days with a single injection of 2 mg/100 g laminaran showed no appreciable improvement. The production of IL-1, IL-12, and NO were dramatically elevated within six hours, according to a time-dependent augmentation.

14.15 Limitations of Immunomodulators

Modern health and wellness industries heavily rely on immunomodulators in case of human; same applies on fishes, which is appropriate given that the immune system is the first line of defence against illness. In order to maintain homeostasis, immune systems in healthy organisms produce a variety of immunomodulators (El Enshasy and Hatti-Kaul., 2013). Despite their many benefits, they also have a number of disadvantages, including drug instability, a brief half-life, rapid drug clearance, and uncontrolled immunological reactions (Chauhan *et al.*, 2021). Since immunomodulator medications affect the immune system, they all have the potential to raise infection risk. This risk can be mild or severe, such as those caused by opportunistic agents. Every immunomodulator agent has different kinds of effects as they act on different levels of immune responses either by stimulating immunodeficiency or restrict the various diseases (Ben *et al.*, 2012).

14.16 Immunomodulators' Limitations in Aquaculture

Although the general side effects of immunoadjuvants are not as harmful in fish compared to mammals, there are some limitations to immunostimulants used in aquaculture. According to the most recent research, injection administration is less efficient than oral administration. Also, they are not effective against all diseases. In aquaculture, optimum dosage is very critical, as they induce immunosuppression when overdosed. Many immunoadjuvants that are broadly used in mammals have limited use in aquaculture due to many reasons, for instance. Immunoadjuvants like saponins are widely used in mammals, but due to their instability in the aqueous phase and high cytotoxicity, their importance in aquaculture is limited.

14.17 Nutrigenomics and Immune Function Modulation in Fish

A fish nutrition study is emerging as a key benefactor especially in aquaculture systems, where feeds are embodied to management of health and called as functional feeds. These functional feeds are heading to be of the opinion to reduce disease outbreaks, improve fish health status, and also enhance the post-infection recovery. When new transcriptomics (RNA-seq and microarray), proteomics, and other new omics technologies are used, the effect of nutrition on immune system becomes more comprehensible (Martin *et al.*, 2017).

Food provides both energy and nutrients, required for maintenance of immunity (Chandra., 1997). The host's immunity is unable to defend itself from viruses, germs, and parasites, etc; when adequate nutritional resources are missing in the diet (Marcos *et al.*, 2003). Exposure of fish to pathogens is considerably higher when compared with other non-aquatic vertebrates (Fuhrman., 1999), as the pathogenic exposure starts immediately after hatching and it further intensifies during the onset of exogenous feeding of the larvae specially at gut and mouth opening stages (Castro *et al.*, 2015). In aquaculture, bacteria (54.9%) is the dominating causative agent of infectious diseases, followed by viruses (22.6%), parasites (19.4%), and fungi (3.1%) (McLoughlin and Graham., 2007). The capability of fish to withstand pathogens depends mainly on their nutritional status (Figure 14.5).

Innate immunity and adaptive immunity are generally the two forms of immune responses in fish and other vertebrates. Considered the initial line of defence against pathogenic infections, the innate immune response. Along with physical barriers like skin, mucous, and scales, it also involves cellular and humoral reactions. The acquired immunity, also known as the adaptive immune response, is dependent on both cellular and humoral processes and is characterised by the detection of a specific pathogen-specific antigen that causes illness (Secombes, 2016). The cellular components comprise of macrophages, monocytes, mast cells, neutrophils, eosinophils, thrombocytes, natural killer cells, and T and B lymphocytes (Castro and Tafalla, 2015). In addition to this gill-associated lymphoid tissue (GIALT), skin-associated lymphoid tissue (SALT), gut-associated lymphoid tissue (GALT) are immune cells found in other mucosal tissues of skin, gut, and gills (Salinas and Miller, 2015). Because of nutritional deficiency, an imbalance can be created between adaptive and innate immune system responses, with an abnormally high impact seen in production of cytokine, phagocyte function, complement system, cell-mediated immunity, and antibody affinity (Marcos *et al.*, 2003).

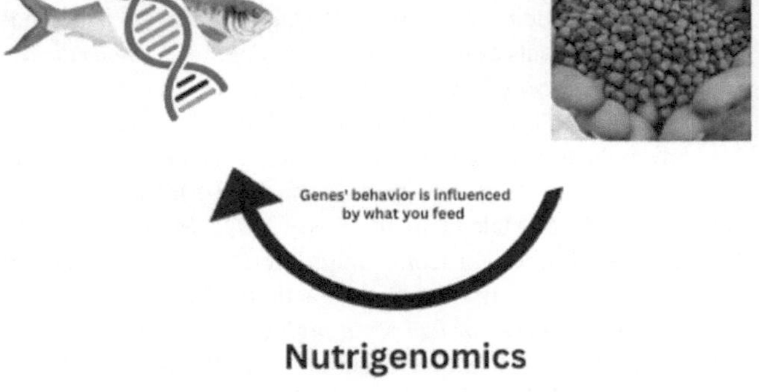

FIGURE 14.5 Interplay between nutrition and gene.

Using omics, a relationship between immune function and fasting can be well studied. Transcriptomic (microarray) studies showed that when Atlantic salmon (*Salmo salar*) fasted for about four weeks, it reduced utterance of genes associated with immunity and changed the responses of liver transcriptome to disease caused by *Aeromonas salmonicida* (Martin *et al.*, 2010). The reduced intakes of soyabean meal (SBM) and/or caloric restriction of fishmeal (40% of ad libitum food intake, FM-CR) have also shown to alter the expression in Atlantic salmon of both pro- and anti-inflammatory genes. This was revealed through transcriptomics (microarray) technology when tissue from liver and distal intestine was analyzed from the fish (Skugor *et al.*, 2011). Channel catfish (*Ictalurus punctatus*) also showed significant altered expression of immune factors of innate immunity. Omics technology including transcriptomics (microarray) using tissue from gills and skin were analyzed in this catfish. The seven days of fasting brought lower immune function as well as dysregulation in key genes involved in cell cycling proliferation and energy metabolism (Liu *et al.*, 2013). The above-mentioned studies indicate how clearly transcriptomics reveals that even a short-term fasting can affect immunity in fishes and hence fish nutrition needs careful assessment during industrial feed production and when feeding in commercial and/or experimental fish production are being performed.

Feed additives being beyond the standard requirement of normal health and growth of fish, are supplied in feeds, just to enhance the health and immunity in fishes. Such feeds are termed *functional feeds*. These feeds include various micro-minerals and components like zinc, selenium, vitamins, and components that act as immunostimulants, for example plant extracts, algal extracts, etc. (Martin *et al.*, 2017). These also include diets rich in phospholipids from krill or mineral oils, both of which can improve immune function (Martinez-Rubio *et al.*, 2013). As a result, these feeds are considered to be the main factor in aquaculture's maintenance of fish health (Tacchi *et al.*, 2011). Functional feeds used as part of dietary therapies improve both systemic and local (intestinal) immunity (Lazado and Caipang., 2014). To assess the effects of functional feeds on immunity using transcriptomics (RNA-seq), plant-derived additives were added to the meals of Atlantic salmon for roughly 21 days (1% immunostimulants and 3% anti-attachment substance). The tissue from head kidney and skin were analyzed. In fishes, feed additives were found to decrease sea lice infection *Caligusro gercresseyi*. Altered expressions of immune genes were seen including MHC-1 and MHC-2 transcript, which indicates an improvement in immunity (Núñez-Acuña *et al.*, 2015). Another study employing transcriptomics (microarray) and Atlantic salmon enriched with glucosinolates (GLs, 7.3 micromol/g and 26.4 micromol/g) was assessed. Fish that had been fed high doses of GLs for 17 to 18 days were sampled before infection, while fish that had been fed low doses of GLs and control diets were infected with parasites at the time of the sampling five weeks after infection. The tissue of the skull kidneys and liver was examined. The results of the study suggested that GLs reduce salmon lice (*Lepeophtheirus salmonis*) infection and enhance the gene expression related to heme and iron withdrawal responses, suggesting that depriving salmon lice of heme may be a useful anti-parasitic tactic (Skugor *et al.*, 2016).

Research indicates that Selenium (Se), a micronutrient can have a great influence on immune responses. The bioavailability of organic selenium like that derived from yeast (for example Sel-Plex® which is produced by company Alltec), is much higher than that of inorganic selenium. Transcriptomic profiling through microarray showed that fish fed with selenium have an enhanced immune stimulation when compared with normal standard diets (Pacitti *et al.*, 2016).

14.18 Negative Effects and Neutral Behaviour of Immunomodulators

Studies looking at the adverse effects of immunomodulators or immunostimulants are extremely rare. When administered at the recommended amount and frequency, it is thought that they have little to no inhibitory effects. The detrimental consequences of immunostimulants have seldom been studied. Immunostimulants may have adverse effects on fish immune systems, particularly immature immune systems of fish larvae, according to Bricknell and Dalmo's 2005 study. The immunostimulant and related PAMPs may develop a resistance, and continuous treatment may have unforeseen consequence on the larvae's capacity to develop an adaptive system. It is often suggested to employ pulse application rather than constantly adding the immunostimulant to the meal, even though this problem has not yet been empirically addressed (Magnadottir, B. 2010). Another study showed that immunosuppression can be caused by excessive immunostimulant dosages (Sakai, M. 1999). However, multiple immunostimulants given over a prolonged period of time do not seem to offer any significant benefits over a single dosage (Galeotti M., 1998). In a few incidents, immunostimulants may fail to provide improved protection or increased immunity (Sahoo P.K., 2007). (Dügenci, S. K. *et al.*, 2003) stated that some immunostimulants are ineffective when administered parenterally.

14.19 Conclusion

Disease outbreaks are a barrier to the growth of the aquaculture sector. Same as humans, aquatic animals (fish) rely on both specialised and non-specialized defence mechanisms to fend off invasive illnesses. The primary lines of non-specific defence and protection for fish are their skin and mucus, while other organs including the liver, spleen, gills, and kidney also play a critical part in infection defense. Adjuvants contribute significantly to the improvement of vaccines' efficacy to prevent infections in fish and shrimp by increasing the adaptive and innate immune responses to the deficient antigen. Numbers of adjuvants are available to enhance the efficacy and efficiency of vaccines for the fishes. The feed additives are utilized in aquaculture to boost the immunity of fish under culture and to promote the fish welfare. According to the scientific literature on this subject, the immunostimulatory effects of various feed additives differ depending on the kind of fish, the mode of administration, the dose, the duration, and the combination with other immunostimulants. However, limited information is available on negative side of immunomodulators in the scientific papers. Moreover, specificity, safety, effectiveness, side effects, availability, long-term efficacy, cost-effectiveness, and affordability are a few of the requirements that must be assessed for future application in order to treat illnesses properly and prevent disease, thereby reducing fish suffering brought on by immune-related disorders. To improve fish welfare, it is essential to look at the short and long-term adverse effects of immunomodulators. New gene-based technologies will highlight unanticipated functions for candidate genes and pathways as well as unique patterns of teleost mucosal gene expression. Utilizing these data will enhance methods for choosing disease-resistant broodstock and evaluating available preventative and therapeutic measures for teleostfishes. More research is needed for the evaluation of behaviour and underling mechanism of the action on the molecular.

REFERENCES

Abram, Q. H., Dixon, B., & Katzenback, B. A. (2017). Impacts of low temperature on the teleost immune system. *Biology*, 6(4), 39.

Agarwal, S. S., & V. K. Singh. (1999). Immunomodulators: A review of studies on Indian medicinal plants and synthetic peptides. Part-I: Medicinal plants. *Proceedings of the Indian National Science Academy-Part B: Biological Sciences*, 65(3–4), 179–204.

Agboola, J. O., Schiavone, M., Øverland, M., Morales-Lange, B., Lagos, L., Arntzen, M. Ø.,... ... & Hansen, J. Ø. (2021). Impact of down-stream processing on functional properties of yeasts and the implications on gut health of Atlantic salmon (Salmo salar). *Scientific reports*, 11(1), 1–14.

Alamgir, M., & Uddin, S. J. (2010). Recent advances on the ethnomedicinal plants as immunomodulatory agents. *Ethnomedicine: a source of complementary therapeutics*, 37(661), 2.

Allison, A. C., & Eugui, E. M. (2005). Mechanisms of action of mycophenolate mofetil in preventing acute and chronic allograft rejection. *Transplantation*, 80(2-S), S181–S190.

Almawi, W. Y., & Melemedjian, O. K. (2002). Molecular mechanisms of glucocorticoid antiproliferative effects: antagonism of transcription factor activity by glucocorticoid receptor. *Journal of leukocyte biology*, 71(1), 9–15.

Alvarez-Pellitero, P., Sitjà-Bobadilla, A., Bermúdez, R., & Quiroga, M. I. (2006). Levamisole activates several innate immune factors in Scophthalmus maximus (L.)(Teleostei).

International Journal of Immunopathology and Pharmacology, 19(4), 727–738.

Arya, Vikrant, & Vivek Kumar Gupta. (2011). A review on marine immunomodulators. *Int J Pharm Life Sci*, 2(5), 751–758.

Asherson, R. A., Gunter, K., Daya, D., & Shoenfeld, Y. (2008). Multiple autoimmune diseases in a young woman: tuberculosis and splenectomy as possible triggering factors? Another example of the "mosaic" of autoimmunity. *The Journal of rheumatology*, 35(6), 1224–1227.

Avorn, J. (2011). Learning about the safety of drugs—a half-century of evolution. *New England journal of medicine*, 365(23), 2151–2153.

Banstola, A., Jeong, J. H., & Yook, S. (2020). Immunoadjuvants for cancer immunotherapy: A review of recent developments. *Actabiomaterialia*, 114, 16–30.

Barman, D., Nen, P., Mandal, S. C., & Kumar, V. (2013). Immunostimulants for aquaculturehealth management. *J Marine Sci Res Dev*, 3(3), 1–11.

Barrett, R. (2022). Calcineurin inhibitors and related medicines: a cohort study examining England's primary care prescription changes during the COVID-19 pandemic (January 2019 to March 2021). *DARU Journal of Pharmaceutical Sciences*, 30(1), 59–66.

Bascones-Martinez, A., Mattila, R., Gomez-Font, R., & Meurman, J. H. (2014). Immunomodulatory drugs: Oral and systemic adverse effects. *Medicina oral, patologia oral y cirugiabucal*, 19(1), e24.

Ben Dhaou, B., Boussema, F., Aydi, Z., Baili, L., Tira, H., & Cherif, O. (2012). Corticosteroid-associated complications in elderly. *Tunis Med.*, 90, 774–777.

Benguira S., Leblond V. S., Weber J.-P., & Hontela A. (2002) Loss of capacity to elevate plasma cortisol in rainbow trout (Oncorhynchus mykiss) treated with a single injection of o, p'- dichlorodiphenyldichloroethane. *Environ Toxicol Chem*, 21, 1753–1756

Besse, Benjamin, et al. (2016). Dendritic cell-derived exosomes as maintenance immunotherapy after first line chemotherapy in NSCLC. *Oncoimmunology*, 5(4), 1071008.

Bode, Christian, et al. (2011). CpG DNA as a vaccine adjuvant. *Expert review of vaccines*, 10(4), 499–511.

Bricknell, I., Dalmo, R. A. (2005). The use of immunostimulants in fish larval aquaculture. *Fish& Shellfish immunology*, 19 (2005), 457–472.

Brinker, A. M., Ma, J., Lipsky, P. E., & Raskin, I. (2007). Medicinal chemistryand pharmacology of genus Tripterygium (Celastraceae). *Phytochemistry* 68,732–766.

Carmona-Ribeiro, A. M. (2017). Nanomaterials based on lipids for vaccine development. In *Micro and Nanotechnology in Vaccine Development* (pp. 241–257). William Andrew Publishing.

Carsons, S. E. (2008). Issues related to clinical trials of oral and biologic disease-modifying agents for Sjögren's syndrome. *Rheumatic Disease Clinics of North America*, 34(4), 1011–1023.

Castro, R., Tafalla, C. (2015). Overview of fish immunity (Book Chapter). *Mucosal Health in Aquaculture*. pp. 3–54.

Castro, R., Jouneau, L., Tacchi, L., Macqueen, D. J., Alzaid, A., Secombes, C. J., Martin, S. A., Boudinot, P. (2015). Disparate developmental patterns of immune responses to bacterial and viral infections in fish. *Sci. Rep*, 5, 15458.

Chandra, R. K. (1997). Nutrition and the immune system: an introduction. *Am. J. Clin. Nutr.*, 66, 460S–463S.

Chauhan, A., Khan, T., & Omri, A. (2021). Design and Encapsulation of Immunomodulators onto Gold Nanoparticles in Cancer Immunotherapy. *International Journal of Molecular Sciences*, 22(15), 8037.

Chen, Guo-Yun, et al. (2014). Broad and direct interaction between TLR and Siglec families of pattern recognition receptors and its regulation by Neu1. *Elife*, e04066.

Ching, Joo Jie, et al. (2021). Immunomodulatory activity of β-glucans in fish: Relationship between β-glucan administration parameters and immune response induced. *Aquaculture Research*, 52(5), 1824–1845.

Chu W. H. (2006) Adjuvant effect of propolis on immunisation by inactivated Aeromonas hydrophila in carp (Carassius auratus gibelio). *Fish Shellfish Immu*, 21, 113–117. doi:10.1016/j.fsi.2005.10.002

Chung, J., Choi, R., Seo, E.-K., Nam, J.-W., Dong, M.-S., Shin, E., et al. (2012). Anti-inflammatory effects of (Z)-ligustilide through suppressionof mitogen-activated protein kinases and nuclear factor-κBactivationpathways. *Arch. Pharm. Res.*, 35, 723–732.

Da Silva, C. A., Chalouni, C., Williams, A., Hartl, D., Lee, C. G., Elias, J. A. (2009). Chitin Is a Size-Dependent Regulator of Macrophage TNF and IL-10 Production. *J. Immunol*, 182, 3573–3582.

Dinesh, M. G., et al. (2011). Leaf and seed extracts of Bixa orellana L. exert anti-microbial activity against bacterial pathogens. *Journal of Applied Pharmaceutical Science* Issue, 116–120.

Djordjevic, B., Skugor, S., Jørgensen, S. M., Overland, M., Mydland, L. T., & Krasnov, A. (2009). Modulation of Splenic Immune Responses to Bacterial Lipopolysaccharide in Rainbow Trout (Oncorhynchus Mykiss) Fed Lentinan, a Beta-GlucanFrom Mushroom Lentinula Edodes. *Fish Shellfish Immunol*, 26(2), 201–209.

Dügenci, S. K., Arda, N., & Candan, A. (2003). Some medicinal plants as immunostimulant for fish. *Journal of Ethnopharmacology*, 88(1), 99–106. doi:10.1016/s0378-8741(03)00182-x

Dummer, R., Hauschild, A., Henseler, T., & Burg, G. (1998). Combined interferon-α and interleukin-2 as adjuvant treatment for melanoma. *The Lancet*, 352(9131), 908–909.

Duncan, P. L. and Klesius, P. H. (1996). Dietary immunostimulants enhance nonspecific immune responses in channel catfish but not resistance to Edwardsiellaictaluri. *Journal of Aquatic Animal Health*, 8(3), 241–248.

El Enshasy, H. A., & Hatti-Kaul, R. (2013). Mushroom immunomodulators: unique molecules with unlimited applications. *Trends in biotechnology*, 31(12), 668–677.

Emadi, A., Jones, R. J., & Brodsky, R. A. (2009). Cyclophosphamide and cancer: golden anniversary. *Nature reviews Clinical oncology*, 6(11), 638–647.

Fernandez-Tejada, Alberto, et al. (2014). Design, synthesis, and immunologic evaluation of vaccine adjuvant conjugates based on QS-21 and tucaresol. *Bioorganic & Medicinal Chemistry*, 22(21), 5917–5923.

Fireman, M., DiMartini, A. F., Armstrong, S. C., &Cozza, K. L. (2004). Immunosuppressants. *Psychosomatics*, 45(4), 354–360.

Fuhrman, J. A. (1999). Marine viruses and their biogeochemical and ecological effects. *Nature*, 399, 541–548.

Galeotti M. (1998) Some aspects of the application of immunostimulants and a critical review of methods for their evaluation. *J ApplIchthyo*, 114, 189–199.

Galindo-Villegas J., & Hosokawa H. (2004) Immunostimulants: towards temporary prevention of diseases in marine fish. In Cruz Suarez L. E., Ricque M. D., Nieto Lopez M. G., Villareal D., Scholz Y., Gonzalez M., Eds. *Advances en Nutricion. Acuicola VII Memorias del VII Simposium Internationale de Nutricion Acuícola*, (vols 16–19, pp 279–319).

Gjessing, M. C., Falk, K., Weli, S. C., Koppang, E. O., & Kvellestad, A. (2012). A sequential study of incomplete Freund's adjuvant-induced peritonitis in Atlantic cod. *Fish Shellfish Immunol*, 32, 141–150.

Gupta, A. K., Adamiak, A., & Chow, M. (2002). Tacrolimus: a review of its use for the management of dermatoses. *Journal of the European Academy of Dermatology and Venereology*, 16(2), 100–114.

Hamajima, K., Kojima, Y., Matsui, K., Toda, Y., Jounai, N., Ozaki, T., Xin, K. Q., Strong, P., Okuda, K. (2003). Chitin Micro-Particles: A Useful Adjuvant for Inducing Viral Specific Immunity when Delivered Intranasally with an HIV-DNA Vaccine. *Viral Immunol*, 16, 541–547.

Heo, Y., et al. (2004). Serum IgE elevation correlates with blood lead levels in battery manufacturing workers. *Human & Experimental Toxicology*, 23(5), 209–213.

Hoel, K., & Lillehaug, A. (1997). Adjuvant activity of polar glycopeptidolipids from Mycobacterium chelonae in experimental vaccines against Aeromonassalmonicidain salmonid fish. *Fish Shellfish Immunol*, 7, 365–376.

Hynes, Natasha A., et al. (2011). Immune response of Atlantic salmon to recombinant flagellin. *Vaccine*, 29(44), 7678–7687.

Ibrahim, M, Khan, M, Rinard, J, & Mustafa, A. (2015). Determination of effective dosage of Phyllanthus niruri to modulate stress in tilapia, Oreochromis niloticus. Bioeng Biosci 3, 68–71.

Jensen, I., Albuquerque, A., Sommer, A. I., & Robertsen, B. (2002). Effect of poly I:C on the expression of Mx proteins and resistance against infection by infectious salmon anaemia virus in Atlantic salmon. *Fish Shellfish Immunol*, 13, 311–326.

Jiao, X. D., Cheng, S., Hu, Y. H., & Sun, L. (2010). Comparative study of the effects of aluminum adjuvants and Freund's incomplete adjuvant on the immune response to an Edwardsiellatarda major antigen. *Vaccine*, 28, 1832–1837

Jimenez, N., Coll, J., Salguero, F. J., & Tafalla, C. (2006). Co-injection of interleukin 8 with the glycoprotein gene from viral haemorrhagic septicemia virus (VHSV) modulates the cytokine response in rainbow trout (Oncorhynchus mykiss). *Vaccine*, 24, 5615e26 .

Johnson, R. W., Kreis, H., Oberbauer, R., Brattstrom, C., Claesson, K., & Eris, J. (2001). Sirolimus allows early cyclosporine withdrawal in renal transplantation resulting in improved renal function and lower blood pressure1, 2, 10. *Transplantation*, 72(5), 777–786.

Juyal, P. D., & Singla, L. D. (2001). *Herbal immunomodulatory and therapeutic approaches to control parasitic infection in livestock* (pp. 1–8). Department of Veterinary Parasitology, College of Veterinary Science. Punjab Agricultural University, India

Kaufmann, E., Sanz, J., Dunn, J. L., Khan, N., Mendonça, L. E., Pacis, A.,… … & Divangahi, M. (2018). BCG educates hematopoietic stem cells to generate protective innate immunity against tuberculosis. *Cell*, 172(1-2), 176–190.

Kumar, R. A., Sridevi, K., Kumar, N. V., Nanduri, S., & Rajagopal, S. (2004). Anticancer and immunostimulatory compounds from Andrographispaniculata. *J. Ethnopharmacol*, 92, 291–295.

Kumar, P., Rai, S., Verma, S. K., Prakash, P. S., &Chitara, D. (2022). *Classification, Mode of Action and Uses of Various Immunomodulators. In Immunomodulators and Human Health* (pp. 3–38). Springer, Singapore.

Kyo E., Uda N., Suzuki A., Kakimoto M., Ushijima M., Kasuga & Itakura Y. (1998). Immunomodulation and antitumour activities of aged garlic extract. *Phytomedicine*, 5, 259–267. doi:10.1016/S0944-7113(98)80064-0

Lawrence, Gregor W., et al. (1997). Phase I trial in humans of an oil-based adjuvant SEPPIC MONTANIDE ISA 720. *Vaccine*, 15(2), 176–178.

Lazado, C. C., & Caipang, C. M. (2014). Mucosal immunity and probiotics in fish. *Fish Shellfish Immunol*, 39(1), 78–89.

Lee, S. J., Chinen, J., & Kavanaugh, A. (2010). Immunomodulator therapy: monoclonal antibodies, fusion proteins, cytokines, and immunoglobulins. *J Allergy ClinImmunol*, 125, S314–S323.

Lee, C. G., Da Silva, C. A., Lee, J. Y., Hartl, D., Elias, J. A. (2008). Chitin Regulation of Immune Responses: an Old Molecule with New Roles. *Curr. Opin. Immunol.*, 20, 684–689.

Liddicoat, A. M., & Lavelle, E. C. (2019). Modulation of innate immunity by cyclosporine A. *Biochemical pharmacology*, 163, 472–480.

Liu, X., Mei, Z., Qian, J., Zeng, Y., & Wang, M. (2013). Puerarinpartlycounteracts the inflammatory response after cerebral ischemia/reperfusion viaactivating the cholinergic anti-inflammatory pathway. *Neural Regen. Res.*, 8, 3203.

Magdelin S. M. (2005) Culture of ornamental fish, Black molly (*Poecilia sphenops*) using medicinal plants having immunostimulant characteristics. M.Phil Dissertation, Manonmaiam Sundaranar University, India

Magnadottir, B. (2010). Immunological Control of Fish Diseases. *Mar Biotechnol*, 12, 361–379. 10.1007/s10126-010-9279-x

Manicassamy S., & Pulendran B. (2009). Modulation of adaptive immunity with Toll-like receptors. *SeminImmunol*, 21, *185e93.*

Marcos, A., Nova, E., & Montero. A. (2003). Changes in the immune system are conditioned by nutrition. *Eur. J. Clin. Nutr.*, 57, S66–S69.

Martin, Samuel AM, & Elżbieta Król. (2017). Nutrigenomics and immune function in fish: new insights from omics technologies. *Developmental & Comparative Immunology*, 75, 86–98.

Martin, S. A. M., Douglas, A., Houlihan D. F., & Secombes, C. J. (2010). Starvation alters the liver transcriptome of the innate immune response in Atlantic salmon (Salmo salar). *BMC Genomics*, 11, 418.

Martinez-Rubio, Laura, et al. (2013). Effect of functional feeds on fatty acid and eicosanoid metabolism in liver and head kidney of Atlantic salmon (Salmo salar L.) with experimentally induced heart and skeletal muscle inflammation. *Fish & Shellfish Immunology*, 34(6), 1533–1545.

McLoughlin, M. F., & Graham, D. A. (2007). Alphavirus infections in salmonids—a review. *J Fish Dis*, 2007(30), 511–531.

Mehana, Elsayed E., Arshad H. Rahmani, & Salah M. Aly. (2015). Immunostimulants and fish culture: an overview. *Annual Research & Review in Biology*, 477–489.

Micol, Vicente, et al. (2005). The olive leaf extract exhibits antiviral activity against viral haemorrhagic septicaemia rhabdovirus (VHSV). *Antiviral Research*, 66 (2-3), 129–136.

Midtlyng, P., Reitan, L., & Speilberg, L. (1996). Experimental studies on the efficacy and side-effects of intraperitoneal vaccination of Atlantic salmon (Salmo salar L.) against furunculosis. *Fish Shellfish Immunol*, 6, 335–350.

Miller, M. T., Ventura, L., & Strömland, K. (2009). Thalidomide and misoprostol: ophthalmologic manifestations and associations both expected and unexpected. *Birth Defects Research Part A: Clinical and Molecular Teratology*, 85(8), 667–676.

Mitoma, H., Horiuchi, T., Tsukamoto, H., & Ueda, N. (2018). Molecular mechanisms of action of anti-TNF-α agents–Comparison among therapeutic TNF-α antagonists. *Cytokine*, 101, 56–63.

Mizel, S. B., & Bates, J. T. (2010). Flagellin as an adjuvant: cellular mechanisms and potential. *J Immunol*, 185, 5677–5682.

Nishimura, K., Nishimura, S., Nishi, N., Saiki, I., Tokura, S., Azuma, I. (1984). Immunological Activity of Chitin and its Derivatives. *Vaccine*, 2, 93–99.

Núñez-Acuña, G., Gonçalves, A. T., Valenzuela-Muñoz, V., Pino-Marambio, J., Wadsworth, S., Gallardo-Escárate, C. (2015). Transcriptome immunomodulation of in-feed additives in Atlantic salmon Salmo salar infested with sea lice Caligusrogercresseyi. *Fish Shellfish Immunol*, 47(1), 450–460.

Olivier, G., Evelyn, T. P., & Lallier, R. (1985). Immunity to Aeromonassalmonicida in coho salmon (Oncorhynchuskisutch) induced by modified Freund's complete adjuvant: its non-specific nature and the probable role of macrophages in the phenomenon. *Dev Comp Immunol*, 9, 419e32.

Opie, E. L., & Freund, J. (1937). An experimental study of protective inoculation with heat killed tubercle bacilli. *J Exp Med*, 66, 761–788.

Pacitti, D., Lawan, M. M., Feldmann, J., Sweetman, J., Wang, T., Martin, S. A., Secombes, C. J. (2016). Impact of selenium supplementation on fish antiviral responses: a whole transcriptomic analysis in rainbow trout (Oncorhynchus mykiss) fed supranutritional levels of Sel-Plex®. *BMC Genomics*., 17, 116.

Parra, David, Felipe E. Reyes-Lopez, & Lluis Tort. (2015). Mucosal immunity and B cells in teleosts: effect of vaccination and stress. *Frontiers in Immunology*, 6, 354.

Paulsen, Berit. (2001). Plant polysaccharides with immunostimulatory activities. *Current Organic Chemistry*, 5(9), 939–950.

Poppe, T. T., & Olav Breck. (1997). Pathology of Atlantic salmon Salmo salar intraperitoneally immunized with oil-adjuvanted vaccine. A case report. *Diseases of Aquatic Organisms*, 29(3), 219–226.

Porwal, O., Ozdemir, M., Kala, D., & Anwer, E. T. (2021). A review on medicinal plants as potential sources of natural immunomodulatory action. *Journal of Drug Delivery and Therapeutics*, 11(16), 324–331.

Praseetha (2005) Enrichment of brine shrimp *Artemia franciscana* with commercial probiotics and herbal extracts and their resistance against shrimp pathogen *Vibrio* sp. (*Vibrio parahaemolyticus* and *Vibrio damsela*), M.Phil Dissertation, Manonmaiam Sundaranar University, India

Preiss, J. C., & Zeitz, M. (2010). Use of methotrexate in patients with inflammatory bowel diseases. *Clin Exp Rheumatol*, 28, S151–S155.

Pujol, J., Vansteenkiste, J., Martino, T., Atanackovic, D., Reck, M., Thomeer, M. et al. (2015). Safety and immunogenicity of MAGE-A3 cancer immunotherapeutic with or without adjuvant chemotherapy in patients with resected stage IB to III MAGE-A3- positive non-small-cell lung cancer. *J Thoracic Oncol*, 10(10), 1458–1467.

Roy, Samaresh Pal, et al. (2022). Development and Screening of Anxiolytic and Antistress activity of Novel Health Promoting Tea Fortified with Herbs. *Journal of Biologically Active Products from Nature*, 12(2), 146–158.

Sageshima, J., Ciancio, G., Chen, L., & Burke III, G. W. (2009). Anti-interleukin-2 receptor antibodies—basiliximab and daclizumab—for the prevention of acute rejection in renal transplantation. *Biologics: Targets & Therapy*, 3, 319.

Sahoo, P. K. (2007). Role of Imunostimulants in disease resistance of fish. *CAB Review: Perspective in Aquaculture, Veterinary Science, Natural and National Resources*, 2(045), 1–3.

Sahoo, P. K., & Mukherjee S. C. (2002). The effect of dietary immunomodulation upon Edwardsiella tarda vaccination in healthy and immunocompromised Indian major carp (Labeo rohita). Fish & Shellfish Immunology, 12(1): 1–16.

Sakai, M. (1999). Current Status of Fish Immunostimulants. *Aquaculture*, 172, 63–92.

Salinas, I., & Miller, R. D. (2015). Comparative Phylogeny of the Mucosa-Associated Lymphoid Tissue (Book Chapter) Mucosal Immunology: Fourth Edition1-2, pp. 145–159.

Secombes, C. J. (2016). What is new in fish cytokine research? *Fish and Shellfish Immunology*, 53, 1–3.

Selvaraj, V., K. Sampath, & V. Sekar. (2005). Administration of yeast glucan enhances survival and some non-specific and specific immune parameters in carp (Cyprinus carpio) infected with Aeromonas hydrophila. *Fish & Shellfish Immunology*, 19(4), 293–306.

Seung-Cheol, J., Osamu, T., Gwan-Sik, J., Lee, S.-W., Ishimaru, K., Seoka, M., & Kenji, T. (2007). Dietary Medicinal Herbs Improve Growth and Some Non-Specific Immunity of Red Sea Bream Pagrus major. *Fisheries Science*, 73, 63–69.

Shahbaz Akhtar, M., Pal, A. K., Sahu, N. P., Alexander, C., Gupta, S. K., Choudhary, A. K.,… … & Rajan, M. G. (2010). Stress mitigating and immunomodulatory effect of dietary pyridoxine inLabeorohita (Hamilton) fingerlings. *Aquaculture Research*, 41(7), 991–1002.

Siqueiros-Cendón, T., Arévalo-Gallegos, S., Iglesias-Figueroa, B. F., García-Montoya, I. A., Salazar-Martínez, J., & Rascón-Cruz, Q. (2014). Immunomodulatory effects of lactoferrin. *Acta PharmacologicaSinica*, 35(5), 557–566.

Sitjà-Bobadilla, Ariadna, Oswaldo Palenzuela, & P. Alvarez-Pellitero. Immune response of turbot, Psetta maxima (L.) (Pisces: Teleostei), to formalin-killed scuticociliates (Ciliophora) and adjuvanted formulations. *Fish & Shellfish Immunology*, 24(1), 1–10.

Sivaram, V., et al. (2004). Growth and immune response of juvenile greasy groupers (Epinephelus tauvina) fed with herbal antibacterial active principle supplemented diets against Vibrio harveyi infections. *Aquaculture*, 237(1-4), 9–20.

Siwicki, A. K., Anderson, D. P., & Rumsey, G. L. (1994). Dietary intake of immunostimulants by rainbow trout affects non-specific immunity and protection against furunculosis.

Veterinary immunology and immunopathology, 41(1-2), 125–139.

Skugor, S., Grisdale-Helland, B., Refstie, S., Afanasyev, S., Vielma, J., & Krasnov, A. (2011). Gene expression responses to restricted feeding and extracted soybean meal in Atlantic salmon (Salmo salar L.). *Aquac. Nutr.*, 17, 505–517.

Skugor S., Jodaa Holm H., Bjelland A. K., Pino J., Evensen Ø, Krasnov A., Wadsworth S. (2016). Nutrigenomic effects of glucosinolates on liver, muscle and distal kidney in parasite-free and salmon louse infected Atlantic salmon. *Parasit Vectors*, 9(1), 639.

Song, Ying, et al. (2021). Evaluation of the optimum dietary iron level and its immunomodulatory effects on juvenile Chinese mitten crab, Eriocheir sinensis. *Aquaculture*, 544, 737122.

Sun, H. X., Xie, Y., & Ye, Y. P. (2009). Advances in saponin-based adjuvants. *Vaccine*, 27(12), 1787–1796.

Suzuki, S., Okawa, Y., Okura, Y., Hashimoto, K., Suzuki, M. (1982). Immunoadjuvant Effect of Chitin and Chitosan. In Hirano, S., & Tokura, S., Eds. *Chitin and Chitosan* (pp. 210–212). Japan Soc. Chitin Chitosan, Sapporo, Japan.

Tacchi, L., Bickerdike, R., Douglas, A., Secombes, C. J., & Martin, S. A. (2011). Transcriptomic responses to functional feeds in Atlantic salmon (Salmo salar). *Fish Shellfish Immunol*, 31(5), 704–715.

Tafalla, C., Bøgwald, J. & Dalmo, R. A. (2013). Adjuvants and immunostimulants in fish vaccines: current knowledge and future perspectives. *Fish & Shellfish Immunology*, 35(6), 1740–1750.

Timmermans, S., Souffriau, J., & Libert, C. (2019). A general introduction to glucocorticoid biology. *Frontiers in immunology*, 10, 1545.

Tokura, S., Tamura, H., Azuma, I. (1999). Immunological Aspects of Chitin and Chitin Derivatives Administered to Animals. In Jollès, P., & Muzzarelli, R. A. A., Eds. *Chitin and Chitinases*. (pp. 279–292). Birkhauser Verlag, Basel, Switzerland.

Tsan, M. F., & Gao, B. (2004). Cytokine function of heat shock proteins, *Am J Physiol Cell Physiol*, 286, C739–C744.

Vaibhav, K., Shrivastava, P., Javed, H., Khan, A., Ahmed, M. E., Tabassum, R.,et al. (2012). Piperine suppresses cerebral ischemia–reperfusion-inducedinflammation through the repression of COX-2, NOS-2, and NF-κB in middlecerebral artery occlusion rat model. *Mol. Cell. Biochem.*, 367, 73–84.

Vikrant A. & Gupta, V. K. (2011): A review on marine immunomodulators. *International journal of pharmacy and life sciences*, 2(5), May:2011, 751–758.

Wang, E., Liu, T., Wu, J., Wang, K., Chen, D., Geng, Y., Huang, X., Ouyang, P., Lai, W. & Ai, X. (2019). Molecular characterization, phylogenetic analysis and adjuvant effect of channel catfish interleukin-1βs against Streptococcus iniae. *Fish & Shellfish Immunology*, 87, 155–165.

Wikström, A. C. (2003). Glucocorticoid action and novel mechanisms of steroid resistance: role of glucocorticoid receptor-interacting proteins for glucocorticoid responsiveness. *The Journal of Endocrinology*, 178(3), 331–337.

Xu, Y.-K., Liao, S.-G., Na, Z., Hu, H.-B., Li, Y., & Luo, H.-R. (2012). Gelsemiumalkaloids, immunosuppressive agents from Gelsemium elegans. *Fitoterapia*, 83, 1120–1124.

Yin, Guojun, et al. (2009). Chinese herbs (Astragalus radix and Ganoderma lucidum) enhance immune response of carp, Cyprinus carpio, and protection against Aeromonas hydrophila. *Fish & Shellfish Immunology*, 26(1), 140–145.

Ytrestøyl, T., B. Finstad, & R. S. McKinley. (2001). Swimming performance and blood chemistry in Atlantic salmon spawners exposed to acid river water with elevated aluminium concentrations. *Journal of Fish Biology*, 58(4), 1025–1038.

Zia, Q., Fatima, N., Alam, M., Bisht, D., Yadav, P., Ahmad, I.,… … &Owais, M. (2010). Immunomodulators: potential in treatment of systemic fungal infections. In *Combating Fungal Infections* (pp. 397–421). Springer, Berlin, Heidelberg.

15

Current Status and Recent Advancements with Immunostimulants in Aquaculture

Parasuraman Aiya Subramani[1], S. Kalaivani Priyadarshini[2], Ramalakshmi Balasubramanian[3], M. Divya Gnaneswari[4], Devasree Ganesh Kumar[5], Priyatharsini Rajendran[3], Catherine Alexander[6], and R. Dinakaran Michael[1]

[1]*Centre for Fish Immunology, Vels Institute of Science, Technology, and Advanced Studies (VISTAS), Pallavaram, Chennai, Tamil Nadu*

[2]*Department of Biotechnology, Lady Doak College, Tallakulam, Tamil Nadu*

[3]*Department of Zoology, Lady Doak College, Tallakulam, Tamil Nadu*

[4]*Department of Zoology, Gargi College, New Delhi, India*

[5]*The Madura College, Vidya Nagar, Madurai, Tamil Nadu*

[6]*Jayaraj Annapackiam College for Women, Thamaraikulam, Periyakulam, Tamil Nadu*

15.1 Introduction

Intensification of aquaculture without good prophylactic practices may lead to diseases of microbial aetiology and colossal economic damage to the industry (Evensen 2016). Fish disease in aquaculture is not a mere association between pathogen and fish. Culture conditions, such as poor water quality or other stressors often contribute to the disease outbreak. These diseases are the main threat to sustainable aquaculture, the wakeup call for a profound investigation and finding appropriate solutions.

Traditionally, antibiotics and disinfectants were the preferred 'treatment of choice' against diseases in the aquaculture industry. However, antibiotics could lead to the development of antibiotic-resistant bacterial strains, its persistence in the environment, and food safety problems. The use of antibiotics negatively affects the health status of the fish (Panigrahi, Naveenkumar, and Das 2021). Hence, top aquaculture countries of the world switched to the immunoprophylactic measures to tackle the effects of high intensification in aquaculture. Immunoprophylactic control of fish diseases involves vaccines, pro- and prebiotic treatments, and the most effective immunostimulation. Immunoprophylactic agents prime the innate and/or the adaptive immune system and should be used ahead of infection (Tafalla, Bøgwald, and Dalmo 2013). Since an adaptive immune system in fish develops only after maturity, prophylaxis of infectious diseases in fish larvae and juveniles should focus on the stimulation of the innate immune system. Vaccines, especially the recombinant and inactivated vaccines, require repeated booster doses and adjuvants. Therefore, an alternative inexpensive and effective substitute is needed to enhance the efficacy of vaccines. Immunostimulant is a substance that confers protection against pathogens by stimulating the function of innate immune cells (Mehana, Rahmani, and Aly 2015). Immunostimulants are primarily found in natural products, boost immunity, lower disease susceptibility, and guard fish against stress and illness. This prevents the need for chemicals or pharmaceuticals and has a lesser negative effect on the environment. Additionally, it can increase consumer acceptance of aquaculture products (Mohan et al. 2019). Typically, an immunostimulant can be anything from a selenium atom, herbal plant extracts to a genetically engineered viral particle. These compounds/extracts have the potential to be applied for enhancing aquaculture production. As shown in Figure 15.1, immunostimulant is one of the hot research topics and the number of papers published per year is increasing annually. Cancer immunotherapy uses immunostimulants for a better result, but it is beyond scope of this chapter to review those compounds. The present review is to summarize the current state of immunostimulant research, their apparent actions shown in fish immune system.

As evident from Figure 15.2, almost all countries participate in immunostimulant research. Hence, this chapter is subdivided continent wise for providing a concise review of the status of immunostimulant research in aquaculture.

15.2 Immunostimulant Research in Asia

Asia dominates the aquaculture industry with a share of 87.9% in terms of production. The amount of Chinese aquaculture

DOI: 10.1201/9781003361183-19

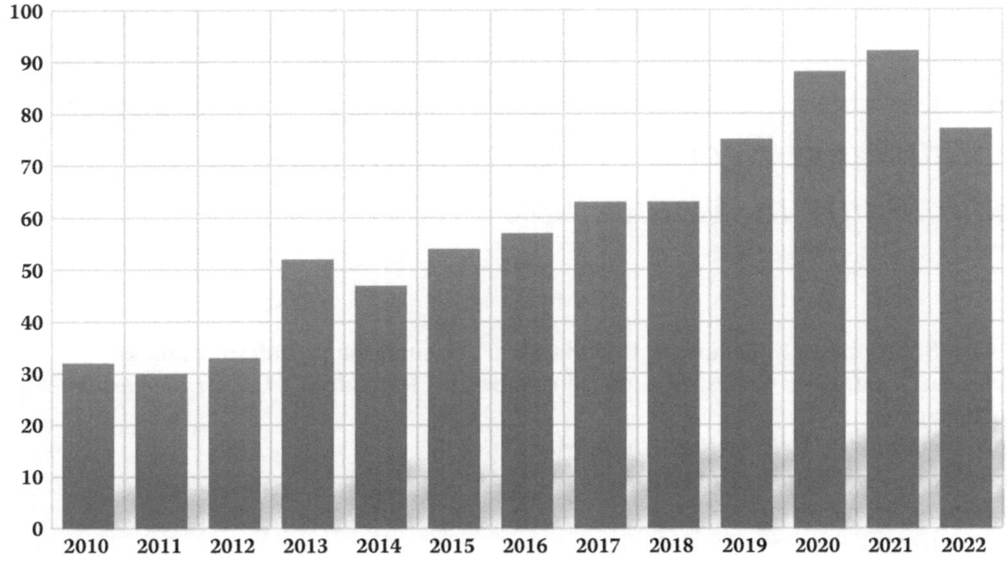

FIGURE 15.1 Year wise number of publications retrieved from Scopus with key term 'fish immunostimulant'.

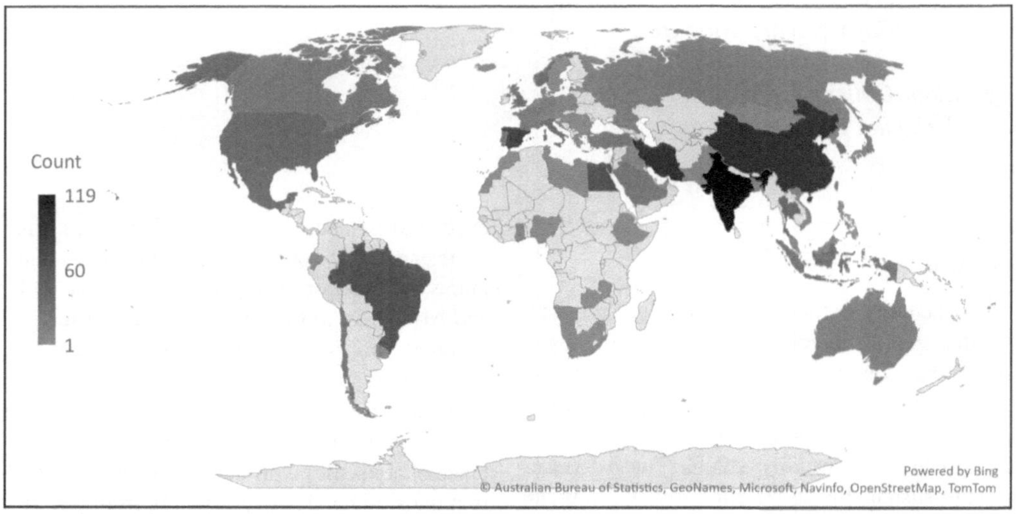

FIGURE 15.2 Country wise number of publications retrieved from Scopus with key term 'fish immunostimulant'.

production is greater than that of the rest of the world combined (Suzuki). Together, Asia produced 72.8 million tonnes of fishes in 2018 which could be of value more than US$200 billion (FAO 2020). At the same time, Asian aquaculture suffers the most from infectious diseases. Since the 1980s, the use of immunostimulants in aquaculture for the control of aquatic diseases has manifested as a critical area of research in Asian countries like China, India, Iran, Malaysia, South Korea, Thailand, and Turkey. Compounds like glucan, chitosan, selenium, levamisole, astaxanthin, sodium propionate, growth hormone, and vitamin C have been reported to be immunostimulatory in various culture fishes. Extracts from garlic, cherry, mushroom, marine algae, mistletoe, nettle, ginger, *Hybanthus*, *Spirulina*, *Astragalus*, *Andrographis paniculata*, *Origanum vulgare*, jujube, guava, and *Padina gymnospora* have been studied for their immunostimulatory properties in fish. Extracts/fractions from medicinal plants like *Achyranthes aspera, Solanum trilobatum, Ocimum sanctum, Phyllanthus niruri, Curcuma zedoaria, Zingiber zerumber, Ixora coccinea, Sophora flavescens, Sargassum whitti, Sophora flavescens, Ficus benghalensis, Euphorbia hirta, Curcuma longa, Tinospora cordifolia, Nyctanthes arbortristis,* and *Withania somnifera* have shown immunostimulatory properties in finfish aquaculture. Substances derived from plants like fruit peels such as apple peel-derived pectin, rambutan peel powder, pineapple peel powder, banana peel powder and fermented lemon peel supplemented diet have been shown to enhance the growth and immune response in fishes like common carp, striped catfish, Nile tilapia, Rohu, rainbow trout, Catla, and Asian seabass. The immunostimulants have been often found to activate the specific and nonspecific immune responses in terms of increased lysozyme, respiratory burst, and peroxidase activities, phagocytosis, and up-regulation of immune-related genes like interleukin (IL): IL-8, IL-10, IL-1β, interferon

(IFN): IFN-α, IFN-γ, MyD88, transforming growth factor (TGF)-β and tumour necrosis factor (TNF)-α expression, and better disease resistance in the treated fishes. The status of immunostimulant research in Asia is summarised in Table 15.1. Various novel immunostimulants are discussed below.

15.2.1 Plant and Plant Products

Oncorhynchus mykiss (rainbow trout) fed with 0.1% and 0.5%, cherry (Prunus spp) stem extract showed significantly enhanced superoxide dismutase (SOD) activity, catalase and haematocrit values of all the treated fish groups were significantly increased (Amoush et al. 2021). Likewise, aqueous methanol extract of *Prunus domestica* (thin-skinned plum) supplemented diet was found to increase the respiratory burst, bacterial killing activity, myeloperoxidase activity, and the expression of IL and cyclooxygenase-2 (COX2) genes and augmented survival against *Yersinia ruckeri* challenge in rainbow trout (Terzi et al. 2021). The use of fermented lemon (*Citrus limon*) peel as a supplement in aquafeed was reported to enhance the plasma lysozyme activity, growth, and intestinal morphology of Lates calcarifer (Asian seabass) (Zhuo et al. 2021). Inclusion of *Stachys lavandulifolia* Vahl (wood betony) extract in the diet of rainbow trout was found to significantly increase the red blood cell count, cell volume, and haemoglobin content, serum immunoglobulins, alternative complement activity, lysozyme activity (Sarvi Moghanlou et al. 2018). Dietary administration of *Excoecaria agallocha* (milky mangrove) leaf extract at 50 mg kg^{-1} concentration to *Oreochromis niloticus* (Nile tilapia) for 28 days enhanced the leucocyte count, phagocytosis, respiratory burst, serum lysozyme, and bactericidal activities and the relative percent survival against *Streptococcus agalactiae* (Laith et al. 2017). Likewise, the feed supplementation of a 6% *Achyranthes aspera* extract enriched diet for four weeks improved the phagocytic activity, and specific and nonspecific immune responses of treated rohu and reduced the mortality (20%) against *Pseudomonas fluorescens* (Hasan-Uj-Jaman et al. 2016). Intraperitoneal administration of the methanolic extract of *Solanum trilobatum* and *Ocimum sanctum* (1:1) has been reported to enhance the leucocyte count, phagocytic activity, and serum antiprotease activity in *Mystus keletius* (freshwater catfish) (Subeenabegum and Navaraj 2016b). Also, rainbow trout fed with a 2% *Lentinula edodes* (medicinal mushroom) extract supplemented diet for six weeks showed enhanced levels of serum total immunoglobulin and maximum disease resistance against *Lactococcus garvieae* infection (Baba, Uluköy, and Öntaş 2015).

India has a rich tradition of using herbal medicines in aquaculture practices. Kirubakaran, Subramani, and Michael (2016) have reported the immunostimulatory effect of the methanol extract of *Nyctanthes arbortristis* (coral jasmine) seeds (20 mg/kg) on the nonspecific, specific immunity and the disease resistance of *Oreochromis mossambicus* (Mozambique tilapia) against live virulent *A. hydrophila*. In a similar study, the intraperitoneal administration of the aqueous leaf extract of *Phyllanthus niruri* (stone breaker) was reported to enhance the neutrophil activation, the primary and secondary antibody response in Mozambique tilapia (Muthulakshmi, Subramani, and Michael 2016). Apart from terrestrial plants, marine macroalga has been widely studied for their immunostimulatory potential in finfish culture. The polysaccharide fraction extracted from *Padina gymnospora*, a marine macroalga was reported to stimulate the serum lysozyme, myeloperoxidase activities, and antibody response in Cyprinus carpio (common carp). The polysaccharide fraction also upregulated the expression of interleukin-1β (IL-1β) and lysozyme-C and elicited improved disease resistance against *A. hydrophila* and *Edwardsiella tarda* (Rajendran, Subramani, and Michael 2016). Similarly, the polysaccharides from *Caulerpa scalpelliformis* were reported to enhance the serum lysozyme, myeloperoxidase, antiprotease, and bactericidal activities and upregulate the expression of IL-1β, lysozyme, and TNF-α genes in the spleen and improved the relative percent survival of the treated Nile tilapia against *A. hydrophila* (Yengkhom et al. 2018).

Giri et al. (2016) and his team have studied the immunostimulatory effects of several indigenous plant products on the growth and immune functions of fish. In tropical countries like India, banana (*Musa* sp.) peel, a rich source of dietary fibre, essential amino acids, vitamins, potassium, and bioactive compounds, is discarded in large quantities as household and industrial food waste. A diet containing 5% of banana peel flour in *Labeo rohita* (rohu) fingerlings fed for 60 days showed a significant increase in the lysozyme, alternative complement system, leukocyte phagocytosis, superoxide dismutase, and catalase activities, and upregulated the expression of IL-1β, TNF-α, and heat shock protein (HSP) HSP-70 and increased the disease resistance against *A. hydrophila*. Similarly, rohu fingerlings fed with 3 g kg^{-1} *Hybanthus enneaspermus* (spade flower) aqueous extract supplemented diet exhibited enhanced serum lysozyme, phagocytic and respiratory burst activity, increased levels of RBC, WBC, neutrophils, total serum protein, and albumin and the highest survival rate against *A. hydrophila* infection (Giri et al. 2017). A diet containing 0.5% of *Psidium guajava* L. (guava) leaves was found to enhance the lysozyme, phagocyte, and alternate complement activity in rohu fed for 60 days. Among the immune response genes, the expression of IL-1β and TNF-α were up-regulated and IL-10, transforming growth factor-beta (TGF-β), inducible nitric oxide synthase (iNOS), cyclooxygenase-2 (COX-2), and transcription factor nuclear factor-κB (NF-κB) were down-regulated. Also, the post-challenge survival rate (66.66%) against *A. hydrophila* was higher compared to the control. Such indigenous plants and their parts are promising feed additives for carp in aquaculture (Giri et al. 2015).

Chinese herbal medicine has been used for thousands of years as a "green drug" and is replacing chemical therapeutics and antibiotics to control and prevent diseases in fish culture (Van Doan et al. 2021). These medicinal plants contain alkaloids, flavonoids, anthraquinone, and polysaccharides that can promote growth, immunity, disease resistance, and the quality of fish. Astragalus (milkvetch) is an herbal supplement that has been used for centuries in traditional Chinese medicine. Supplementation of Astragalus polysaccharides in the crucian carp diet has been proven to upregulate the expression of IL-8, IL-10, IL-1β, IFN-α, IFN-γ, MyD88, TGF-β, and TNF-α in the spleen, kidney, liver, and intestine, along with improved disease resistance against Spring Viraemia of Carp Virus (SVCV)

TABLE 15.1

Current Status of Immunostimulant Research in Asia

Immunostimulant	Name of the Fish	Immunomodulatory Effect on				Country	Reference
		Specific	Nonspecific	Functional Immunity	Gene Expression		
A. Microbial							
β-1, 3 glucan binding protein with Bacillus licheniformis β-glucan	*Oreochromis mossambicus, Pangasianodon hypophthalmus*	–	+	–	–	India	(Anjugam et al. 2022)
Microbial levan supplemented diet	Rohu *(Labeo rohita)* fingerlings.	+	+	–	+	India	(Gupta et al. 2020)
B. Plants							
Dietary rambutan (*Nephelium lappaceum* L.) peel powder	Striped catfish (*Pangasianodon hypophthalmus*)	–	+	–	+	Thailand	(Le Xuan et al. 2022)
Astragalus polysaccharides	Crucian carp	–	–	–	+	China	(Liu et al. 2022)
Aquatic fern (*Azolla caroliniana*)	Nile tilapia	–	+	+	–	Thailand	(Lumsangkul et al. 2022)
Dietary garlic (Allium sativum) powder	Narrow-clawed crayfish (*Postantacus leptodactylus*)	–	+	–	–	Iran	(Rezaei, Mehrgan, and Paknejad 2022)
Dietary Astragalus polysaccharides	Asian Seabass (*Lates calcarifer*)	–	+	+	–	China	(Yu et al. 2022)
Aegle marmelos	Common carp (*Cyprinus carpio*) and Nile tilapia	–	+	+	–	Thailand	(Wangkahart, Wachiraamonloed, et al. 2022)
Andrographis paniculata	Striped catfish	–	+	+	–	India	(Maiti et al. 2021)
Fermented lemon peel diet	Asian seabass	–	+	–	–	Taiwan	(Zhuo et al. 2021)
Pineapple peel powder	Nile tilapia	–	+	–	+	Thailand	(Van Doan et al. 2021)
Dietary Astaxanthin	Asian seabass	+	+	–	–	Malaysia	(Lim et al. 2021)
Apple peel-derived pectin	Common carp	+	+	–	–	Iran	(Hoseinifar et al. 2021)
Chinese herbal medicines	Crucian carp	–	+	+	–	China	(He et al. 2020)
Origanum vulgare	Rainbow trout	–	+	–	–	Iran	(Haghighi 2018)
Dietary *Stachys lavandulifolia* Vahl (wood betony) extract	Rainbow trout	+	+	–	–	Iran	(Sarvi Moghanlou et al. 2018)
Jujube (*Ziziphus jujube*) fruit extract	Common carp	+	+	–	–	Iran	(Hoseinifar, Khodadadian Zou, et al. 2018)
Polysaccharides from marine macroalgae, *Caulerpa scalpelliformis*	Nile tilapia	–	+	+	+	India	(Yengkhom et al. 2018)
Loquat (*Eriobotrya japonica*) leaf extract	Common carp	–	+	–	–	Iran	(Hoseinifar, Zou, et al. 2018)
Excoecaria agallocha	Nile tilapia	–	+	–	–	Malaysia	(Laith et al. 2017)
Hybanthus enneaspermus aqueous extract	Rohu	–	+	–	–	South Korea	(Giri et al. 2017)
Marine macroalga, *Padina gymnospora*	Common carp	+	+	+	+	India	(Rajendran, Subramani, and Michael 2016)
Achyranthes aspera	Rohu	–	–	+	–	India	(Hasan-Uj-Jaman et al. 2016)

Banana (*Musa acuminata*) peel flour	Rohu	–	+	–	–	South Korea	(Giri et al. 2016)
Plant polysaccharides from *Ficus carica*, *Radix isatidis*, and *Schisandra chinensis*	Crucian carp	–	+	–	–	China	(Wang et al. 2016)
Plantago asiatica, *Houttuynia* and *Mentha haplocalyx*	Cobia (*Rachycentron canadum*)	–	+	–	–	Taiwan	(Wu et al. 2016)
Solanum trilobatum and *Ocimum sanctum*	Mystus keletius	–	+	–	–	India	(Subeenabegum and Navaraj 2016a, 2016b)
Phyllanthus niruri	Mozambique tilapia (Oreochromis mossambicus)	+	+	–	–	India	(Muthulakshmi, Subramani, and Michael 2016)
Sargassum wightii	Striped catfish	–	+	–	–	India	(Prabu et al. 2016)
Nyctanthes arbortristis	Mozambique tilapia	–	+	+	–	India	(Kirubakaran, Subramani, and Michael 2016)
Plantago asiatica and *Houttuynia*	Cobia	–	+	–	–	Taiwan	(Wu et al. 2016)
Guava (*Psidium guajava* L.) leaves	Rohu	–	+	–	+	South Korea	(Giri et al. 2015)
Curcuma zedoaria and *Zingiber zerumber*	Epinephelus coioides.	–	+	–	–	Taiwan	(Nan et al. 2015)
C. Probiotics							
Probiotic (*Bacillus subtilis* KADR1)	Rohu	+	+	+	–	India	(Ramesh and Souissi 2018)
D. Others							
Organic and inorganic selenium	Nile tilapia (*Oreochromis niloticus*)	–	–	+	–	Thailand	(Wangkahart et al. 2022)
Dietary sodium propionate	Caspian white fish (Rutilus frisii kutum) fry	–	+	–	–	Iran	(Hoseinifar, Zoheiri, and Caipang 2016)
Dietary carotenoids obtained from shrimp exoskeleton	Common carp	–	+	–	–	India	(Sowmya and Sachindra 2015)

(Liu et al. 2022); and 0.10% of Astragalus polysaccharide supplemented diet has been found to significantly facilitate the antioxidant capacity, immune function and disease resistance against *Vibrio harveyi* in *Lates calcarifer* (Asian seabass) (Yu et al. 2022). Geniposide (iridoid glucoside), a common Chinese herbal medicine extracted from *Gardenia jasminoides* supplemented (100 mg kg-1) diet showed a significant increase in serum lysozyme, superoxide dismutase, complement C3, and catalase activity in *Carassius carassius* (crucian carp). The geniposide supplementation upregulated the expression of IL-1β, CRP, TNF-α, and TGF-β expression in the spleen and kidney. Also, the relative percent survival of the treated fish against *A. hydrophila* was 70.4% compared to the control (He et al. 2020). Wang et al. (2016) have compared the immunostimulatory activity of three plant polysaccharides (*Ficus carica* polysaccharides, *Radix isatidis* polysaccharides, and *Schisandra chinensis* polysaccharides) as feed supplements in crucian carp. In general, the plant polysaccharides significantly increased the leukocyte phagocytosis, serum bactericidal, lysozyme, complement, and superoxide dismutase activities, and total protein level. Among the plant polysaccharides tested by this group, *Ficus carica* polysaccharides showed the maximum RPS of 57.9% against *A. hydrophila* infection. In a similar comparative study, the dietary supplementation of the 10 g/kg or 20 g/kg of *Plantago asiatica* (Chinese plantain) and *Houttuynia cordata* (Heartleaf) were found to induce phagocytosis, reactive oxygen production, lysozyme activity, and superoxide dismutase production in *Rachycentron canadum* (cobia). However, the *Mentha haplocalyx* (Chinese mint) treatment didn't show any immunostimulatory activity (Wu et al. 2016). *Epinephelus coioides* (orange-spotted grouper) fed with *Curcuma zedoaria* (white turmeric) at 0.5 g/kg diet and *Zingiber zerumbet* (pinecone ginger) at 1 g/kg and 2.5 g/kg diets showed a significant increase in phagocytic rate, respiratory burst activity, superoxide dismutase and lysozyme activity (Nan et al. 2015).

Native plants and plant-derived immunostimulants are widely used to boost the health status of fish in Iranian aquaculture. Apples contain bioactive substrates, such as pectin and flavonoids which can be potent immunostimulants. Apple peel-derived pectin-supplemented diet was reported to significantly increase the total serum immunoglobulin and lysozyme activity of the skin mucus of treated common carp (Hoseinifar et al. 2021). *Ziziphus jujube* (Jujube) fruit extract supplemented diet showed a significant increase in immunoglobulin level, serum lysozyme, and alternate complement activity. In addition, the expression of TNF-α, IL-1β, and lysozyme was up-regulated, whereas, TGF-β and IL-10 were down-regulated in the treated common carp (Hoseinifar, Khodadadian Zou, et al. 2018). Similarly, in common carp fingerlings, the dietary supplementation of *Eriobotrya japonica* (loquat) leaf extract at 0.5% or 1% significantly increased serum immunoglobulin, lysozyme, alternate complement activity, up-regulation of TNF-alpha, IL1B, IL8, LYZ, and down-regulation of TGF-β (Hoseinifar, Zou, et al. 2018). In a similar recent study, the dietary supplement of *Allium sativum* (garlic) powder was reported to enhance the lysozyme activity, total plasma protein and phenoloxidase activity, superoxide dismutase, and nitric oxide synthase in treated *Postantacus leptodactylus* (narrow-clawed crayfish) (Rezaei, Mehrgan, and Paknejad 2022). Nile tilapia fed 10 g kg^{-1} *Ananas comosus* (pineapple) peel powder supplemented diet showed significantly increased lysozyme, peroxidase, alternative complement activity, phagocytosis, and respiratory burst activities and improved disease resistance against *Streptococcus agalactiae* (Van Doan et al. 2021).

Nephelium lappaceum L. (rambutan) is a common tropical fruit in Thailand with good medicinal value. Le Xuan et al. (2022) reported the immunostimulatory effect of dietary rambutan peel powder on the immune response in terms of increased serum lysozyme and peroxide activities, upregulation of IFN-2α, IFN-2β, and MHCII gene expression and improved growth performance of *Pangasianodon hypophthalmus* (striped catfish) cultured in biofloc system. Dietary supplementation of 100 g kg^{-1} of *Azolla caroliniana* (aquatic fern) promoted the growth rate and enhanced the lysozyme, peroxidase, alternative complement activity, phagocytosis index, and respiratory burst activity, and higher disease resistance against *Streptococcus agalactiae* in treated Nile tilapia fingerlings (Lumsangkul et al. 2022). Similarly, Nile tilapia supplemented with *Aegle marmelos* (Stone apple) fruit extract at a concentration of 20 g/kg showed significant increases in antioxidant and innate immune responses and a high survival rate against *S. agalactiae* (Wangkahart et al. 2022).

15.2.2 Chemical Compounds

Astaxanthin is a red pigmented carotene isolated from *Haematococcus pluvialis* with an immense role in the aquaculture industry as a functional additive and feed supplement. Dietary provisioning with astaxanthin in *Lates calcarifer* (Asian seabass) exhibited significant increase in the leucocyte count, erythrocyte count, haemoglobin level, hematocrit level, lysozyme, phagocytic activity, respiratory burst activity, and total serum immunoglobulin level and improved the post-challenge survival rate against *Vibrio alginolyticus* infection (Lim et al. 2021). Similarly, a carotenoid from shrimp exoskeleton supplemented diet containing 200 mg kg^{-1} of astaxanthin enhanced the leukocyte count, haemoglobin content, albumin content, respiratory burst activity, lysozyme activity, serum bactericidal activity, and leukocyte myeloperoxidase activity in common carp fingerlings, along with a 100% survival against *A. hydrophila* (Sowmya and Sachindra 2015).

Sodium propionate is a commonly used antifungal agent and food preservative. Dietary supplementation of diet with 0.25% and 0.5% of sodium propionate in the early stage of the *Rutilus frisii* (Caspian white fish or kutum) culture significantly improved the skin mucus immunity, lysozyme, and alternate complement activities as well as growth performance of the treated fish (Hoseinifar, Zoheiri, and Caipang 2016).

15.2.3 Prebiotics

The term 'prebiotics' include selectively fermented naturally occurring indigestible substances that encourage the growth and performance of beneficial microbes in the gastrointestinal tract (Bindels et al. 2015). The commonly used prebiotics in culture fish is fructose oligosaccharides, mannan oligosaccharides, and trans galactose oligosaccharides. Levan is a type of fructose oligosaccharide containing D-fructosyl residues and is found in plants and microbial products. Microbial levans extracted from *Acetobacter xylinum* supplemented diet (1.25%) significantly upregulated the mRNA levels of TLR22, β-2 M, and IFN-γ, and downregulated TGF-β in *A. hydrophila* infected rohu fingerlings (Gupta et al. 2020).

15.2.4 Probiotics

Probiotics are product or preparation containing live microorganisms that maintain and promotes the well-being of the host. The commonly used probiotics in aquaculture are *Bacillus* spp., *Lactobacillus* spp., *Lactococcus* spp., and *Saccharomyces cerevisiae* (Ramesh and Souissi 2018). In a recent study, Spirulina (*Arthrospira platensis*) meal in the diet supplement of European seabass (*Dicentrarchus labrax*) showed significantly elevated respiratory burst activity, lysozyme, myeloperoxidase activity, and disease resistance against *V. anguillarum*. Spirulina supplemented diet also upregulated the expression of IL-1β, IL-10, IL-6, IL-8, TNF-α, TGF-β, and COX-2 genes (Güroy et al. 2022). In another report, dietary administration of *Bacillus subtilis* KADR1 and its subcellular components at 10^8 $CFUg^{-1}$ in rohu has been found to effectively enhance the serum lysozyme, phagocytosis, serum total protein, respiratory burst, serum immunoglobulin levels, superoxide dismutase, and alternative complement pathway activity and increased the post-challenge survivability against *A. hydrophila* (Ramesh and Souissi 2018).

15.2.5 Others

Other functional additives that act as immunostimulants are β-glucans. Supplementation of β 1, 3-glucan from marine crab, *Portunus pelagicus* (Ppβ-GBP), and *Bacillus licheniformis* symbiotic enriched diet induced higher growth rate and enhanced lysozyme, protease, myeloperoxidase, and alkaline phosphatase activity and increased survival rate against *A. hydrophila* in *O. mossambicus* (Anjugam et al. 2022).

In general, the Asian aquaculture industries focus on the use of plant and plant product-based substances as immunostimulants. The traditional herbs and plants indigenous to Asia have been shown to enhance the growth and immune response in various fishes. The immunostimulants have been often found to activate the specific and nonspecific immune responses in terms of cellular, serological, and immune-related gene responses. This in turn leads to a better disease resistance in the treated fish.

15.3 Immunostimulant Research in Africa

African aquaculture production is trivial and contributes (~2.7%) very less towards world aquaculture (Halwart 2020). Between the 1940s and 1950s, colonial governments introduced aquaculture as a sustainable food production method throughout Africa (Brummett, Lazard, and Moehl 2008) with the goals of enhancing rural nutrition, generating additional income, diversifying to lower the risk of crop failure, and creating rural employment. African countries like Egypt, Nigeria, Uganda, and Ghana produce substantial quantities of fish (Halwart 2020). From 1995 to 2018, the region saw a 20-fold rise in production, from 110,200 to 2,196,000 tonnes, with a compound annual growth rate (CAGR) of 15.55% (FAO 2020). Due to the emergence and expansion of small and medium-sized businesses under the supervision of the private sector, aquaculture production has increased. The aquaculture industry in Africa is characterized by the culture of diverse species around forty-three species of finfish, shellfish, and aquatic plants were generated through aquaculture. The main species grown was Nile tilapia (Shaalan et al. 2017) followed by mullets (flathead grey mullet, *Mugil cephalus*, and thin lip grey mullet, *Chelon ramada*), common carp, and silver/big-head carp (*Hypophthalmichthys molitrix*/*Hypophthalmichthys nobilis*) and African catfish (*Clarias gariepinus*) (Soliman and Yacout 2016).

The African countries used natural and herbal components as resources and sources of immunostimulants rather than the chemicals in aquaculture activity, drugs, and hormones. Natural resources like plants, microbes, and polysaccharides were extensively studied to test the growth, nutrient utilization, and immunomodulatory effect of these substances on fish are tabulated in Table 15.2. Recent developments in Africa demonstrate that probiotics and immunostimulants can make a substantial contribution to the health management of farm fish through eco-friendly methods (Abdel-Tawwab 2016, Hassan et al. 2018).

15.3.1 Plant Extracts and Products

There is a great deal of interest in developing inexpensive feed additives to combat pathogenic diseases to sustain environmentally friendly aquaculture (Quagrainie, Cai, and Hishamunda 2017). In Nigeria, Adeshina, Adewale, and Tiamiyu (2017), evaluated the effect of dietary clove, *Eugenia caryophyllata*, buds extract (ECBE), supplementation on the growth performance, physiological, antioxidant, and immunity biomarkers of African catfish. A significant progressive increase in the values of RBC, haemoglobin, haematocrit, white blood cells, platelets, lymphocytes, and heterocytes was observed. Similarly, African catfish fed with *Curcuma longa* leaf-supplemented diet at the concentration of 2.5% had better growth and immunity against *Aeromonas hydrophila*. The oral administration of *Allium sativum* and *Ocimum sanctum* enriched diets enhanced the antioxidant and immunity capacity like superoxide dismutase activity, nitroblue tetrazolium (NBT), and immunoglobulin M (IgM). Additionally, after being exposed to a particularly virulent strain of *A. hydrophila*, fish fed these herbal immunostimulants experienced an increase in their relative percentage survival (RPS) (Nya, Udosen, and Otoh 2017).

Aquaculture research in Egypt provides a comprehensive account of the development of fish farming in the respective country with a focus on the social and economic dimensions. In this regard, the African catfish administered with *Ocimum gratissimum*, clove basil leaves extract (CBLE) at the concentration of 5, 10, or 15 g CBLE/kg diet for 12 weeks showed an enhanced antioxidant and immuno-activity and elevated the fish resistance against pathogenic bacteria, *Listeria monocytogenes* (Abdel-Tawwab et al. 2018). F.A. Emeish, Al-Amgad, and Ahmed (2018) explored the modulatory effects of

TABLE 15.2

Current Status of Immunostimulant Research in Africa

Immunostimulant	Name of the Fish	Immunomodulatory Effect on				Country	Reference
		Specific	Nonspecific	Functional Immunity	Gene Expression		
A. Microbial							
Spirulina platensis	Nile Tilapia	+	+	–	–	Egypt	(Amer 2016)
S. platensis and Betaine supplementation	Nile tilapia	–	+	+	+	Egypt	(Awad et al. 2022)
S. platensis	Common carp, Cyprinus carpio	+	+	+	+	Egypt	(Khalil, Reda, and Awad 2017)
β-carotene and phycocyanin extracted from *S. platensis*	Nile tilapia	–	+	–	+	Egypt	(Hassaan et al. 2021)
Cyanobacterium *S platensis* or the green alga *Chlorella vulgaris*	Nile tilapia	–	+	+	–	Egypt	(Fadl et al. 2017)
B. Plant							
Clove, *Eugenia caryophyllata*, buds extract (ECBE)	African catfish, *Clarias gariepinus*	–	+	+	–	Nigeria	(Adeshina et al. 2018)
clove basil, *Ocimum gratissimum*, leaf extract	African catfish	–	+	+	–	Egypt	(Abdel-Tawwab et al. 2018)
Turmeric *Curcuma longa* leaf	African catfish	+	+	–	–	Nigeria	(Adeshina, Adewale, and Tiamiyu 2017)
Allium sativum and *Ocimum sanctum*	African catfish	–	+	–	–	Nigeria	(Nya, Udosen, and Otoh 2017)
Tri-herbal combination Black cumin seeds (*Nigella sativa*), Peppermint (*Mentha piperita*) and Fennel (*Foeniculum vulgare*)	African catfish	–	+	+	–	Egypt	(Emeish, Al-Amgad, and Ahmed 2018)
Afsanteen (*Artemisia absinthium*) aqueous extract	Common carp	–	+	+	–	Iran	(Yousefi et al. 2021)
Laurel-leaf cistus (*Cistus laurifolius*) ethanolic extract	Common carp	–	+	–	+	Libya	(Bilen et al. 2021) – LIBYA
Rosemary leaf powder (RLP)	Nile tilapia	–	+	+	–	Egypt	(Naiel et al. 2020)
Aqueous extract of yucca, *Yucca schidigera* (YE)	Nile tilapia	–	+	–	–	Egypt	(Abdel-Tawwab, Mounes, et al. 2021)
Turmeric, Rosemary and Thyme (*Thymus vulgaris*)	Nile tilapia	–	+	–	–	Egypt	(Hassan et al. 2018)
C. Probiotics							
Yeast, *Saccharomyces cerevisiae*	Nile tilapia	–	+	+	+	Egypt	(Abdel-Tawwab, Adeshina, and Issa 2020)
Yucca schidigera extract and/or Yeast	Nile tilapia	+	+	–	–	Egypt	
Bacillus amyloliquefaciens spores	Nile tilapia	–	+	+	+	Egypt	(Selim & Reda 2015)

Probiotic Zado® (*Ruminococcus Flavefaciens*)	Nile tilapia	–	+	+	–	Egypt	(Gayed et al. 2021)
Garlic, Echinacea, and probiotics (Organic Green™ and Vet-Yeast™)	Nile tilapia	+	–	+	–	Egypt	(Aly et al. 2016)
Sanolife PRO-F®	Nile tilapia	–	+	+	–	Egypt	(Kord et al. 2021)
Pediococcus pentosaceus	Common carp	–	+	+	–	Egypt	(Ahmadifar et al. 2020)
Dietary probiotic *Lactobacillus plantarum* and whey protein concentrate (WPC)	Nile tilapia	–	+	+	–	Egypt	(Abou-El-Atta et al. 2019)
D. Others							
Ulva lactuca, *Jania rubens*, and *Pterocladia capillacea*	Striped catfish, *Pangasianodon hypophthalmus*	–	+	+	–	Egypt	(Abdelhamid et al. 2021)
Garlic crude polysaccharide (GPE)	African catfish	–	+	+	–	Namibia	(Gabriel et al. 2019
Aloe vera polysaccharides	African catfish fingerlings	–	+	+	–	Namibia	(Gabriel et al. 2019
Pistacia vera hulls derived polysaccharide	Nile tilapia	–	+	+	–	Egypt	(Mohammadi et al. 2020)
Gum Arabic (polysaccharide gum from Acacia Senegal)	Nile tilapia	–	+	+	+	Egypt	(Naiel et al. 2022)
Mannanoligosaccharide (MOS)	Nile tilapia	–	+	+	–	Egypt	(Dawood et al. 2020)

the incorporation of an herbal combination of black cumin seeds (*Nigella sativa*), peppermint (*Mentha piperita*), and fennel (*Foeniculum vulgare*), on African sharp tooth Catfish, *Clarias gariepinus*. A significant increase in the total antioxidant capacity, serum total proteins, albumins, and globulins were observed in the treated group. The herbal diet-fed group increased the body weight, lymphocytes, and neutrophil percentage as well. Catfish challenged with a *Vibrio vulnificus* isolate showed less mortality in the herbal diet than the control group. Common carp fed with Afsanteen (*Artemisia absinthium*) aqueous extract supplemented diets (at 0.5, 1, or 1.5% concentrations) increased growth performance and feed efficiency, improved the innate immunity (serum lysozyme, alternative complement, total immunoglobulin, superoxide dismutase, and total protein level)s, antioxidant response, and improved hepatic health showed a higher level of compared to the control diet (Yousefi, Zahedi, et al. 2021). Dietary rosemary leaf powder (RLP) supplementation at the dose of 10 g kg^{-1}, improved the growth, feed utilization, antioxidant activity, immune response, and disease resistance activity (against *A. hydrophila* challenge) in Nile tilapia (Naiel et al. 2020). Hassan et al. (2018) studied the effects of turmeric, rosemary, and thyme on growth performance, proximate composition, and some immune-haematological parameters on Nile tilapia. Groups treated with turmeric, rosemary and thyme showed a significant increase ($P<0.05$) in haematocrit, leukocrit levels, and nitroblue tetrazolium activity (NBT) in comparison with the control. Yet another study conducted by Abdel-Tawwab and colleagues in Egypt states that the dietary supplementation with garlic powder and chitosan significantly attenuated the toxic effects of feed-born zearalenone ZEN-induced oxidative stress, immunosuppression, and modulated transcriptomic responses of ZEN-exposed European seabass (Abdel-Tawwab, Khalil, et al. 2021).

In a recent study conducted in Libya, dietary administration of laurel-leaf cistus (*Cistus laurifolius*) ethanolic extract enhanced the growth performance, digestive enzyme activity, haematological profile, and nonspecific immune responses in common carp. The immune gene expression of IL-1β, anti-inflammatory cytokines, IL-10 and TGF-β were highly upregulated in the intestine and head kidney of the treated groups. The ethanol extract-fed group exhibited higher survival rate in all treatment groups than that of the control against experimental infection with *A. hydrophila* (Bilen et al. 2021).

15.3.2 Microorganisms and Its Products

Spirulina platensis is a photosynthetic, filamentous, blue-green microalgae and is generally regarded as a rich source of vitamins, essential amino acids, minerals, essential fatty acids (γ-linolenic acid), and antioxidant pigments such as carotenoids and phycocyanin (Jaime-Ceballos et al. 2006). Despite the underlying potential in the formulation of functional food/feed, extensive research and development efforts are still required before microalgae at large become a commercial reality in food and feed formulation. Several researchers evaluated the effect of *S. platensis* meal, as a feed additive on growth performance, the whole-body composition, lysozyme, IgG, IgM values, and catalase activity in serum and antioxidant status (Amer 2016, Awad et al. 2022, Khalil, Reda, and Awad 2017, Hassaan et al. 2021).

Levels of lysozyme, IgG, and catalase activity were higher in *S. platensis*-supplemented groups but do not have a significant effect on IgM levels. Thus, the *S. platensis* supplementation increased antioxidant protective capacities, and some innate and humoral immunity parameters thereby enhancing the growth performance of several culture fish in Africa. Awad et al. (2022) reported that the dietary inclusion of *Spirulina* and Betaine (0.5% and 0.3% concentrations) significantly improved growth performance, hepato-renal, and immune response (interleukins, TNF- alpha, and IL-10 levels). Significant protection was conferred against pathogenic *Aeromonas* infection was evidenced in all treated groups. Recently, Hassaan et al. (2021) investigated the roles of β-carotene and phycocyanin extracted from *S. platensis* on growth, serum biochemical, digestive enzymes, antioxidant defence, immune responses, and immune gene expression in Nile tilapia. Significant phagocytic, lysozyme, immunoglobulin M (IgM), superoxide dismutase, catalase, glutathione peroxidase (GPx) and total antioxidant capacity (T-AOC) activities were recorded in diet supplemented with phycocyanin. The transcripts of IFN-γ and IL-1β genes were upregulated in the liver of fish-fed diet supplemented with β-carotene and phycocyanin.

Regarding the gene expression studies, the mRNA levels of lysozyme, IL6, IL-1β, TGF, and TNF were higher in fish fed with *S. platensis* and probiotic bacterium *Bacillus licheniformis* as feed additives enhanced the disease resistance against *A. hydrophila* infection by stimulating the immune system in goldfish (Yousefi et al. 2022). *S. platensis* or the green alga *Chlorella vulgaris* administered in feed increased the immune status by the enhanced production of serum total contents of protein, albumin, and globulin level in challenged with *A. hydrophila.*

15.3.3 Probiotics

The term "probiotic" is used to describe bacteria that have favourable effects on the host. Probiotics were originally used in 1986 to see if they could speed up the growth of hydrobionts (organisms that live in water). Probiotics were later employed to reduce bacterial infections and enhance water quality (Martínez Cruz et al. 2012). Today, there is scientific proof that probiotics can raise stress tolerance, increase nutrient absorption, and promote reproduction (El-Saadony et al. 2021).

Dietary administration of *Bacillus amyloliquefaciens* spores at concentrations of 10^6 ($CFUg^{-1}$) enhanced the serum bactericidal activity, lysozyme activity, a nitric oxide assay ($mmol^{-1}$) and phagocytic activity in Nile Tilapia. It also caused significant increases in IL-1 and TNF α mRNA levels in the kidneys. Fish that were fed *B. amyloliquefaciens* exhibited better relative survival percentages than the controls when challenged by *Yersinia ruckeri* (Selim and Reda 2015). Similarly, yeast, *S. cerevisiae* (6 g kg^{-1} diet) supplemented through feed strongly improved growth, haematological, antioxidant, and immune responses (respiratory burst and lysozyme) of Nile tilapia and its resistance against pathogenic fungus, *A. flavus,* and infection (Abdel-Tawwab, Adeshina, and Issa 2020). The same group of researchers studied the administration of *Yucca schidigera* aqueous extract and *S. cerevisiae* yeast as an ideal inclusion as a water additive and significantly enhanced growth, haematological and immunological responses in Nile tilapia (Abdel-Tawwab, Mounes, et al. 2021).

Aly et al., investigated the role of probiotics Organic Green™ (contains 10^{11} bacterial cells each from *Lactobacillus acidophilus*, *Bacillus subtilis*, and *Aspergillus oryzae*) and Vet-Yeast™ (contains 10^{9} *S. cerevisiae* dried cells) in improving antibody titre response of overwintered tilapia to *A. hydrophila* vaccine. A significantly high antibody titre with subsequent reduction in the cumulative mortality was recognized in probiotic-treated tilapia against virulent *A. hydrophila* challenge during the winter (Salah et al. 2016).

Fish fed with Zado® (*Ruminococcus flavefaciens*, a cellulolytic probiotic bacteria) enriched diets at the dose of 2 g kg^{-1} for three weeks displayed considerably better haematological and leukocytic readings than fish fed with control diets. It also showed beneficial effects on immune parameters (phagocytosis, lysozyme, and nitric oxide), antioxidants and growth profiles (Gayed et al. 2021).

Sanolife® PRO-F is a highly concentrated microbial mixture containing selected *Bacillus* strains that grow and stimulate the immune system in the gut. Supplementation of Sanolife® PRO-F increased the levels of total serum proteins, globulin, immunoglobulin M, catalase, and super oxide dismutase in Nile tilapia (Kord et al. 2021).

Recently, Ahmadifar et al. (2020) investigated the potential impacts of feeding common carp with *Pediococcus pentosaceus* (10^{8} CFU g^{-1}) on growth performance, digestive enzyme activities, haematological indices, and innate immune responses. Fish fed with various doses of *P. pentosaceus* showed significantly higher serum total antibody levels, alternative complement, protease, and lysozyme activities, as well as antibacterial activity, as compared to the control group.

15.3.4 Others

Among the bioactive molecules of herbal or algal proteins and peptides, polysaccharides, polyphenols, polyunsaturated fatty acids, and pigments have potential advantages as immunostimulants (Øverland, Mydland, and Skrede 2018). However, very little attention has been paid to study the effects of complete and fractionated seaweed extracts on the fish immune system that may contain both these and other substances with immunostimulatory capacities. In Africa, there were very few reports on macroalgal extracts stimulating the nonspecific immune mechanisms such as phagocytic respiratory burst activity, serum lysozyme activity, etc. in different species of fishes. Brown seaweeds (Phaeophyceae) are a large group of multi-cellular macroalgae that are widely distributed in marine aquatic environments. They are abundant in several bioactive sulphated polysaccharides known as fucoidan (FCD) which helps in the growth, intestinal health, antioxidant capacity, and immune responses of several finfish and shellfish (Abdel-Latif et al. 2022). Dietary effects of seaweeds mixture (*Ulva lactuca, Jania rubens,* and *Pterocladia capillacea*) extract enhanced the growth performance, haemato-chemical parameters, antioxidant capacity, immune indices, and resistance of striped catfish fry against *A. hydrophila* infection (Abdelhamid et al. 2021). *Aloe vera* polysaccharides had improved haemato-biochemical indices and higher survival probability in African catfish fingerlings (Gabriel, Wilhelm, Habte-Tsion, Chimwamurombe, Omoregie, et al. 2019). Similarly, the same group of researchers investigated the potential effects of dietary garlic (*Allium sativum*) crude polysaccharide on growth, haematological parameters, whole body composition, and resistance against low pH in African catfish juveniles (Gabriel, Wilhelm, Habte-Tsion, Chimwamurombe, and Omoregie 2019). Dietary administration enhanced the survival rate, amylase activity, red blood cells, haemoglobin, serum total protein, and serum lysozyme activity in Nile tilapia (Mohammadi et al. 2020).

Gum Arabic (GA) is a mixture of polysaccharides and glycoproteins, which give it the properties of glue and binder (Mariod 2018). The carbohydrate fraction of GA is galactose, rhamnose, glucuronic acid, and Arabinose (Fadlelmoula 2018). Dietary supplementation with GA significantly upregulated IL-1β and IL10, whereas downregulated TNF-α mRNA expression in kidney and spleen tissues in Nile tilapia.

Dawood et al. (2020) evaluated the role of mannanoligosaccharide (MOS) in reducing the adverse effects of chlorpyrifos (CPF) toxicity in tilapia. Fish fed with MOS revealed higher growth, survival rates, and haematological (Hb, PCV, RBCs, and WBCs) parameters in Nile tilapia intoxication with CPF. Up-regulation of immune genes viz., CASP3, IFN-γ, and IL-8 genes were also observed in a MOS administered group.

To sum up, African nations primarily rely on studied natural resources including microorganisms, plants, and polysaccharides. For instance, dietary supplementation with plant materials such as *Yucca schidigera*, *Spirulina platensis*), clove, turmeric, rosemary, thyme, laurel-leaf cistus. Currently, this region uses polysaccharides made from GA (*Acacia senegal*), the hulls of *Pistacia vera*, and *Aloe vera*. Additionally, research was conducted on several commercially available immunostimulants, including live baker's yeast, *S. cerevisiae*, and fucoidan (sulphated polysaccharide). These studies revealed that diets fortified with plant and microbial based immunostimulants can improve growth performance, innate immune system, and disease resistance in several cultured fishes.

15.4 Immunostimulant Research in the Americas

Americas combined are the second largest aquaculture producer next to Asia both in terms of quantity and value. In 2019, Americas produced 4 million tonnes of fish through aquaculture which accounts for US$19 billion. In the Americas, countries like Brazil, Chile, Peru, Mexico, Canada, and the USA are currently using various types of immunostimulants for finfish aquaculture. The use of immunostimulants is preferred during periods of high stress, reproduction, transportation, and vaccination to enhance the innate defence mechanism of the fish (Bricknell and Dalmo 2005). As shown in Table 15.3, a wide range of products including plant extracts and products, chemical compounds, microbial substances, and nutritional factors were evaluated as immunostimulants in American aquaculture research laboratories.

TABLE 15.3

Current status of Immunostimulant Research in Americas

Immunostimulant	Name of the Fish	Immunomodulatory Effect on				Country	Reference
		Specific	Nonspecific	Functional Immunity	Gene Expression		
A. Microbial							
Yarrowia lipolytica N-6	*Lutjanus peru* (Pacific red snapper)	–	+	–	+	Brazil	Alamillo, Reyes-Becerril et al. 2017
Y. lipolytica biomass	Nile tilapia	–	+	–	–	Brazil	Neuls, Souza et al. 2021
Sterigmatomyces halophilus	*Sparus aurata L.* (gilthead seabream)	–	–	–	+	Mexico	Reyes-Becerril, Guluarte et al. 2017
Chlorella peruviana	Rainbow trout fingerlings	–	+	+	–	Peru	Arteaga Quico, Mariano Astocondor et al. 2021
Rubrivivax gelatinosus	*Piaractus mesopotamicus* (Pacu)	–	+	–	–	Brazil	Gallani, Valladão et al. 2017
B. Plant							
Origanum majorana	*Rhamdia quelen* (silver catfish)	–	+	+	–	Brazil	da Cunha et al. 2018
Thymus vulgaris (thyme), T. zygis (red thyme) and *Lippia sidoides* (pepper rosemary)	Nile tilapia	–	+	+	–	Brazil	Estaiano de Rezende, Soares et al. 2021
Cinnamomum	Nile tilapia	–	+	–	+	Brazil	M. dos Santos, S. de Brito et al. 2016
Futerpenol	Rainbow trout	–	–	+	–	Chile	Hernández, Romero et al. 2016
Chenopodium ambrosioides L	*Lutjanus peru* (Pacific red snapper)	–	+	–	–	Mexico	Maldonado-Garcia, Angulo et al. 2019, Reyes-Becerril, Angulo et al. 2019
Jatropha vernicosa	*Seriola rivoliana* (Longfin yellowtail)	–	+	+	+	Mexico	Silva-Jara, Angulo et al. 2020
Vitis vinifera	*Colossoma macropomum* (tambaqui)	–	+	–	–	Brazil	Morante, Copatti et al. 2021
Artemisia annua (wormwood)	Nile tilapia	–	+	+	–	Brazil	Soares, Cardoso et al. 2020
Aloe vera	Pacu	–	+	+	–	Brazil	Zanuzzo, Sabioni et al. 2017
C. Chemical compounds							
Levamisole	Pacu	+	+	+	–	Brazil	Pahor-Filho, Castillo, et al. 2017, Biller-Takahashi, Montassier et al. 2016
Levamisole	*Pseudoplatystoma reticulatum* (Striped Surubim)	–	+	–	–	Brazil	Zanon, Cerozi et al. 2014
D. Other substances							
Bee pollen	*Danio rerio* (Zebra fish)	–	+	+	–	Brazil	Di Chiacchio, Paiva et al. 2021
Propolis and *Aloe barbadensis*	Nile tilapia juveniles	–	+	–	–	Brazil	Dotta, de Andrade et al. 2014
β-1,3/1,6-glucan and Pdp11	*Sparus aurata* (gilthead seabream)	–	+	–	+	Brazil	Guzmán Villanueva, Tovar Ramírez et al. 2014

β-glucan	Nile tilapia	–	+	+	–	Brazil	Koch, de Oliveira et al. 2021
β-glucan from *Saccharomyces cerevisiae*	Pacu	–	+	+	–	Brazil	Lopes, Mello et al. 2022, Marinho de Mello, de Fátima Pereira de Faria et al. 2019
β-1,3/1,6-glucan	*Salmo salar* (*Atlantic salmon*)	–	–	–	+	Chile	Rodríguez, Valenzuela et al. 2016
Zymosan	Rainbow trout	–	–	–	+	Chile	Schmitt, Wacyk et al. 2015
Aquate Fish™®	*Arapaima gigas* (pirarucu)	–	+	+	–	Brazil	Reis Dias, Oba Yoshioka et al. 2019
Xanthophyllomyces dendrorhous (yeast) and *Hypericum perforatum* (Saint John's wort), *Melissa officinalis* (lemon balm), and *Salvia Rosmarinus* (rosemary)	Atlantic salmon	–	+	–	–	Chile	Reyes-Cerpa, Vallejos-Vidal et al. 2018

15.4.1 Plant Products

Medicinal plants and plant products are considered suitable supplements to improve the immune response of cultured fish (Maldonado-Garcia et al. 2019). In *Rhamdia quelen* (silver catfish) infected with *Aeromonas hydrophila*, 1 h immersion with essential oil extracted from *Origanum majorana* ($20\,\mu LL^{-1}$) and nanocapsules of the oil ($5\,\mu LL^{-1}$) for five successive days showed higher lymphocyte levels and improved survival rate in the infected fish (da Cunha et al. 2018).

A phytobiotic mixture of volatile oils of *Thymus vulgaris* (thyme), *T. zygis* (red thyme) and *Lippia sidoides* (pepper rosemary), supplemented diet for 20 days enhanced serum lysozyme concentration and the population of leukocytes, monocytes, and neutrophils and improved the resistance to *A. hydrophila* in juvenile Nile tilapia (Estaiano de Rezende et al. 2021).

Similarly, Nile tilapia fed with a 0.5% *Cinnamomum* sp. (cinnamon) powder-supplemented diet for 71 days improved the immune mechanism in terms of increased γ-globulin levels in the treated fish. Also, the dietary administration of the essential oil (0.15%) of cinnamon increased the levels of α1 and α2-globulins in the treated fish (M. dos Santos et al. 2016).

Dietary administration with a phyto-pharmaceutical formulation, Futerpenol® (contains fucoidans and labdane-type diterpenes) preceding challenge with *Piscirickettsia salmonis*, gave significant resistance against infection in the treated rainbow trout (Hernández et al. 2016).

The Mexican tea, *Chenopodium ambrosioides* L., a traditional medicine used for centuries is being explored for its immunostimulatory potential in farm fish. In *Lutjanus peru* (Pacific red snapper) dietary administration of *C. ambrosioides* L. supplemented diet increased the haemoglobin content, respiratory burst activity, and leukocyte blood count. The skin mucus of the treated fish exhibited enhanced anti-protease, lysozyme, and peroxidase activities and antimicrobial activity against *A. hydrophila* and *Vibrio parahaemolyticus* (Maldonado-Garcia et al. 2019). In another study, the splenocytes of Pacific red snapper treated with *C. ambrosioides* L showed increased production of superoxide anion and nitric oxide and enhanced catalase and superoxide dismutase activities (Reyes-Becerril et al. 2019). *Jatropha vernicosa* stem bark extract has been reported to increase respiratory burst activity, phagocytosis, nitric oxide production, upregulation of pro-inflammatory cytokine IL-1β, suppression of anti-inflammatory IL-10 gene, and disease resistance against opportunistic vibriosis in *Seriola rivoliana* (Longfin yellowtail) (Silva-Jara et al. 2020).

In Brazil, utilization of *Vitis vinifera* (grape) crude extract as dietary supplementation in *Colossoma macropomum* (tambaqui) juveniles showed increased levels of plasma triglycerides, and albumin and enhanced the lysozyme and protease activity of the treated fish conferring improved immunity. The utilization of crude grape extract, a by-product of juice or wine processing is considered to be a viable and sustainable alternative in aquaculture (Morante et al. 2021).

Bioactive compounds of *Artemisia annua* (wormwood) are globally recognized pharmacological supplements. In juvenile Nile tilapia, the dietary supplementation with alcoholic extract of 0.5% *A. annua* increased the lysozyme activity, respiratory burst of leukocyte, total plasma protein, blood thrombocytes, monocytes, and neutrophils and increased resistance to *A. hydrophila* (Soares et al. 2020).

Aloe vera is one of the most popular medicinal plants in the world. The *A. vera* extract-supplemented diet for 10 days, was reported to enhance innate immune parameters like complement system, serum lysozyme, and leukocyte respiratory burst activity and consequently support the fish to resist the pathogen amid the transport stress (Zanuzzo et al. 2017).

15.4.2 Chemical Compounds

Levamisole is an anthelmintic drug widely used as an immunostimulant. A study on the effect of dietary supplementation of levamisole hydrochloride for 15 days on the stress and innate immune responses of *Piaractus mesopotamicus* (pacu), showed an increased leukocyte respiratory burst, and haemolytic activity of the complement system. The levamisole administration regulated the level of cortisol secretion during stress, and enhanced the innate immune response and disease resistance against *A. hydrophila* infection in pacu (Pahor-Filho et al. 2017). In a similar study, levamisole was used as an adjuvant during immunization with *A. hydrophila* in pacu. Dietary supplementation of levamisole (125 or 250 mg/kg) for seven days increased the serum bactericidal activity, antibody titre, haematocrit value, and the population of leukocytes, thrombocytes, and red blood cells (Biller-Takahashi et al. 2016). Dietary supplementation of levamisole (247 mg per kg of feed) enhanced the lysozyme activity and improved the immunological response of *Pseudoplatystoma reticulatum* (striped surubim) (Zanon et al. 2014).

15.4.3 Probiotics (Microbial, Algal Products)

In teleost fish, the administration of probiotics of microbial and algal origin is commonly reported to stimulate immune responses, utilization of feed, growth performance, antioxidant potential, gene expression, gut morphology, disease resistance, anti-stress responses, and larval survival (Dimitroglou et al. 2011). In Mexico, the *Yarrowia lipolytica* N-6, an ascomycetous yeast isolated from a hypersaline environment in Guerrero Negro, Baja California Sur, has been tested for its immunostimulatory activity in the *Lutjanus peru* (Pacific red snapper) challenged with *Vibrio parahaemolyticus*. It was reported that the bacterial load was reduced in the spleen leucocytes of the treated fish. Other nonspecific parameters like phagocytosis, respiratory burst activity, production of nitric oxide and myeloperoxidase, and IL-1β transcription were also significantly increased. *Y. lipolytica* N-6 isolate (extreme yeast) improves the innate immune response and antioxidant response in the spleen and head-kidney leukocytes of the pacific red snapper (Alamillo et al. 2017). Similarly, dietary supplementation of *Yarrowia lipolytica* (yeast) fermented biomass was found to increase the lysozyme, myeloperoxidase activities, population of monocytes, neutrophils, production of nitrite/nitrate content in the blood of Nile tilapia (Neuls et al. 2021). Yeast isolated from hypersaline sediments, *Sterigmatomyces halophilus* supplemented diet upregulated the expression of trypsin and immune response genes (TNF-α, IL-1β, C3, IgM, and lysozyme) in the intestine and skin of *Sparus aurata L.* (gilthead seabream) (Reyes-Becerril et al. 2017).

Dietary supplementation with *Chlorella peruviana* (3.6 and 5.0 g kg^{-1}), freshwater algae improved the innate immune response, physiological parameters, growth performance, efficiency, and disease resistance to bacterial infection in rainbow trout fingerlings (3.72 ± 0.25 cm and 0.57 ± 0.13 g) (Arteaga Quico, Mariano Astocondor, and Aquino Ortega 2021).

Besides cleaning the effluent, *Rubrivivax gelatinosus* biomass is a potential immunostimulant feed in fish farming. *R. gelatinosus* biomass-supplemented feed (0.5 or 1.5 g kg^{-1}) improved the immune response of *Piaractus mesopotamicus* by increasing the number of thrombocytes and neutrophils in the blood and reducing the activation time of the complement system (Gallani et al. 2017).

15.4.4 Others

Bee pollen, a nutrient-rich resource has been found to influence the vertical immunity of offspring in the early stages, before complete development. The offspring of the *Danio rerio* (zebrafish) breeders fed with a bee pollen-supplemented diet showed a longer survival rate against viral infection and increased neutrophil migration to wound sites (Di Chiacchio et al. 2021). Hence, bee pollen is considered a potential immunostimulant. Similarly, the dietary supplement of Propolis and *Aloe barbadensis* mix (1:1) in Nile tilapia juveniles showed a higher number of thrombocytes (Dotta et al. 2014).

β-glucan is one of the widely used immunostimulants to enhance innate immunity in fish. Dietary supplementation of a mixture of β-1,3/1,6-glucan isolated from *Laminarina digitata* and Pdp 11 from *Shewanella putrefaciens*, increased the phagocytic activity and upregulated IL-1β and IFN-γ transcripts in the head kidney and stimulates the growth of gilthead seabream after two or four weeks of feeding (Guzmán-Villanueva et al. 2014). In Nile tilapia, β-glucan supplemented diet increased the respiratory burst activity and the lysozyme activity in the liver, plasma, and intestine, however, the haemolytic complement activity of the plasma and the myeloperoxidase activity was not modulated by the β-glucan supplemented diet. No mortality was recorded in the fish fed with β-glucan for 45 days and improved innate immune responses in tilapia (Koch, de Oliveira, and Zanuzzo 2021). Dietary β-glucan from *S. cerevisiae* enhanced the leukocyte respiratory activity and lysozyme activity in pacu juveniles inoculated with *A. hydrophila* (Lopes, Mello, and Urbinati 2022). Similarly, pacu fed with β-glucan (0.1%) supplemented diet for 15 days preceding transportation and inoculation with heat-killed *A. hydrophila*, resulted in increased complement activity at 24 h and reduced the lymphocytes and monocytes count in the peripheral blood at 3 and 24 h. Hence it was concluded that β-glucan regulates the bidirectional interface between the immune response and stress (Marinho de Mello et al. 2019). β-1,3/1,6-glucan diet was reported to activate the regulation of IL-12, IFN-γ, IFN-α, and Mx gene transcripts in *Salmo salar* (Atlantic salmon) (Rodríguez et al. 2016). In rainbow trout, the 0.3% of zymosan-supplemented diet increased the expression of host defence peptides, cathelicidin, and IL-1β in the epithelial RTgutGC cell line of the intestine (Schmitt et al. 2015).

Dietary supplementation with 12 g of Aquate Fish™®, an immunostimulant additive improved the immune response of *Arapaima gigas* (pirarucu) fingerlings, in terms of increased erythrocytes, thrombocytes, leukocytes, lymphocytes, and monocytes count, along with the enhanced respiratory burst of leukocytes, and disease resistance against *A. hydrophila* (Reis Dias et al. 2019).

In Atlantic salmon, dietary supplementation with *Xanthophyllomyces dendrorhous* (yeast) and *Hypericum perforatum* (Saint John's wort), *Melissa officinalis* (lemon balm), and *Salvia Rosmarinus* (rosemary) extract were reported to improve the antioxidant capacity and increase the Th2-like immune response thereby preventing cell damage during crowding stress (Reyes-Cerpa et al. 2018).

Substances like β-glucan, yeast, and levamisole are the most widely studied immunostimulatory compounds in America. Also, indigenous medicinal plants like *Origanum majorana*, thyme, red thyme, pepper rosemary, cinnamon, Mexican tea, *Aloe vera,* and grape are commonly studied for their immunostimulatory property in finfish aquaculture. Probiotics like yeast, *Chlorella,* and *Rubrivivax gelatinosus* biomass have been used as a nutrient as well as an immunostimulant in farm fish. In general, the administration of these immunostimulatory substances has been reported to enhance the nonspecific and specific immune response, disease resistance to bacterial and viral infections, and combat the adverse effects of environmental stress in fish.

15.5 Immunostimulant Research in Europe

Globally, Europe is the prime market for seafood wherein, marine mollusks and diadromous, marine, and freshwater fish constitute the bulk of aquaculture production. Aquaculture production reached 3.4 million tons with a US$16.6 billion in the year 2018. The top ten European countries which contribute to major aquaculture viz., Norway, Spain, Russian Federation, United Kingdom, France, Italy, Greece, Faroe Islands, and The Netherlands generated more than 50,000 tons annually during the period 2000–2018 (Hough 2022). European aquaculture focuses on fish species like Atlantic salmon, rainbow trout, European seabass, gilthead seabream, common carp, silver carp, and *Ctenopharyngodon idella* (grass carp). The aquaculture sector is often plagued by infectious disease epidemics, causing enormous loss and damage. In recent years, with the development of the aquaculture industry, various substances (as food additives) like plant extracts, prebiotics, and probiotics are extensively used to fight pathologies without negatively impacting the environment (Carbone and Faggio 2016). Table 15.4 shows the recently tested fish immunostimulants.

15.5.1 Plant Extracts and Products

The supplementation of aquafeeds with bioactive substances in medicinal plants acts as an alternative to antibiotics and guarantee a natural, safe, and sustainable aquaculture. In Portugal, the administration of *Portulaca oleracea* L. (purslane) aqueous and ethanolic extracts have been found to stimulate phagocytosis and bactericidal activity against *Vibrio harveyi*, *V. anguillarum*, and *Photobacterium damselae*. Oral administration of 2% purslane for 15 or 30 days increased the skin mucus IgM levels and the phagocytic capacity of head kidney leucocytes of gilthead seabream (Cámara-Ruiz et al. 2020). Similarly in the United Kingdom, dietary administration of 2% of *Pandanus tectorius* (screw pine) extract enhanced the total protein, myeloperoxidase, antiproteases, and bactericidal activities in rainbow trout, with a reduction in mortality against *Yersinia ruckeri* and upregulation of TNF, LYZ2, IL-8 and CD-4 genes in the head kidney of rainbow trout (Awad et al. 2019).

Aquaculture research in Italy focuses on the use of medicinal plant extracts and products. *Origanum vulgare* (Oregano) is an important culinary herb in Italian cuisine. Dietary administration of oregano extract supplemented feed was reported to improve the mucosal lysozyme, complement activity, immunoglobulin level, superoxide dismutase, and catalase activity of *Danio rerio* (zebrafish) (Rashidian et al. 2021). Elsewhere, oregano extract-supplemented diet (200 mg kg^{-1}) was reported to significantly increase the lysozyme, alkaline phosphatase, complement activities, and immunoglobulin levels and reduced the mortality against *Aeromonas hydrophila* in *Cyprinus carpio* (Yousefi, Ghafarifarsani, et al. 2021). Several medicinal plant extracts were tested for their immunostimulatory activity on the head kidney leukocytes of rainbow trout, wherein, the extracts of *Lavandula officinalis*, *Origanum vulgare*, and *Rheum officinale* were found to reduce the respiratory burst activity, and the ginseng extract stimulated the respiratory burst of leukocytes after 2 h exposure. Also, the *L. officinalis*, *R. officinale*, *Curcuma longa*, *Echinacea purpurea*, and *Panax ginseng* extracts stimulated the leukocyte proliferation after 72 h (Bulfon, Galeotti, and Volpatti 2018).

Several studies in Spain highlight the potential of plant extracts as immunostimulants in aquaculture. In a recent study, *Salvia officinalis* (sage) and *Lippia citriodora* (lemon verbena) leaf extract incorporated diet was reported to upregulate 285 genes in transcriptomic profiling, wherein the gut transcriptome revealed the activation and differentiation of T cells in *Gilthead Seabream* (Salomón et al. 2021). In a similar study, the oregano leaf powder (0.5% and 1%) supplemented feed was found to stimulate the IgM level, bactericidal, protease activity, and phagocytosis of head kidney leucocytes in gilthead seabream after 15 and 30 days of feeding (García Beltrán et al. 2020). However, *Rhodomyrtus tomentosa* (rose myrtle) leaf extract and its active compound, rhodomyrtone, decreased the cellular ROS levels in the head kidney macrophages of treated rainbow trout (Na-Phatthalung et al. 2018). Furthermore, two-week feeding of *Phoenix dactylifera* (date palm) extract-supplemented diet increased the complement activity of European seabass. Similarly, a mixture of date palm fruit extracts and probiotic (Pdp 11) showed an enhanced activity on the phagocytic ability of head kidney leucocytes of European seabass (Guardiola et al. 2016). A commercial diet supplemented with 10% of *Trigonella foenum graecum* (fenugreek) seeds for four weeks was reported to enhance the respiratory burst activity, leucocyte peroxidase content, complement activity, antiprotease, total protein, peroxidase, and IgM level and upregulation of MHC1, CSF-1R,

TABLE 15.4

Current Status of Immunostimulant Research in Europe

Immunostimulant	Name of the Fish	Immunomodulatory Effect on				Country	Reference
		Specific	Nonspecific	Functional Immunity	Gene Expression		
A. Microbial							
Escherichia coli HSP70 (DnaK)	Axenic European Sea bass larvae	+	–	–	+	Belgium	(Yaacob et al. 2020)
Flagellin	Rainbow trout	–	+	+	–	United Kingdom	(Wangkahart, Secombes, and Wang 2019)
Recombinant flagellin B (rFla) and ND1 domain (rND1) from *Vibrio anguillarum*	Gilthead seabream and Rainbow trout	–	+	+	–	Spain	(González-Stegmaier et al. 2015)
B. Plant							
Allium tuncelianum (Tunceli garlic oil)	Rainbow trout (*Oncorhynchus mykiss*)	+	+	–	–	Turkey	(Ispir et al. 2022)
Avena sativa (oat)	Common carp	+	+	–	–	Italy	(Baba et al. 2016)
Crude ethanol extracts of 5 plants *Phyllanthus amarus* (Bhumi amla); *Psidium guajava* (Guava,); *Mimosa pudica* (sensitive plant); *Azadirachta indica* (neem) and *Euphorbia hirta* (Athma plant)	Striped catfish (*Pangasianodon hypophthalmus*)	+	+	+	–	Belgium	(Nhu et al. 2020)
Date palm fruit extracts and probiotic enriched diet	European sea bass (*Dicentrarchus labrax*)	+	–	–	+	Spain	(Guardiola et al. 2016)
Dietary caper (*Capparis spinosa*) supplementation	Rainbow trout	+	+	+	+	Turkey	(Bilen et al. 2016)
Lemon balm (*Melissa officinalis*)	Rainbow trout	+	–	–	+	Turkey	(Bilen et al. 2019)
Oregano (Origanum vulgare) Extract	Zebrafish *(Danio rerio)*	+	+	–	–	Italy	(Rashidian et al. 2021)
Origanum vulgare	Gilthead seabream (*Sparus aurata* L.)	+	+	–	–	Spain	(Magouz et al. 2022; García Beltrán et al. 2020)
Pandanus tectorius (hala extract)	Rainbow trout	+	–	–	–	United Kingdom	(Awad et al. 2019)
Phyllanthus amarus, Psidium guajava, Mimosa pudica, Azadirachta indica Euphorbia hirta	Striped catfish	+	+	–	–	Belgium	(Nhu et al. 2019)
Portulaca oleracea (purslane)	Gilthead seabream	+	+	–	–	Spain	(Cámara-Ruiz et al. 2020)
Rhodomyrtus tomentosa extract	Rainbow trout	+	+	–	–	Spain	(Na-Phatthalung et al. 2018)
Salvia officinalis (sage) and *Lippia citriodora* (lemon verbena)	Gilthead seabream	+	+	–	–	Spain	(Salomón et al. 2021)
The geophyte plant, Tassel hyacinth (*Muscari comosum*)	Rainbow trout	+	+	–	–	Turkey	(Ulukoy et al. 2018)
Trigonella foenum graecum (fenugreek)	Gilthead seabream	+	+	+	+	United Kingdom	(Awad, Cerezuela, and Esteban 2015)

Gracilaria gracilis macroalgae	European sea bass	+	−	−	+	Portugal	(Passos et al. 2021)
Gracilaria vermiculophylla	Rainbow trout	+	−	−	−	Portugal	(Araújo et al. 2016)
C. Insect							
Insect meals (*Hermetia illucens* and *Tenebrio molitor*)	Tench (*Tinca tinca*)	−	+	+	−	Spain	(Hidalgo et al. 2022)
Yellow mealworm (*Tenebrio molitor*)	European sea bass	+	+	+	−	Greece	(Henry et al. 2018)
Yellow mealworm (*Tenebrio molitor*) larva	Blackspot sea bream (*Pagellus bogaraveo*)	−	+	−	−	Italy	(Iaconisi et al. 2017)
Black soldier fly (*Hermetia illucens*) larvae	Atlantic salmon	−	+	−	+	Norway	(Stenberg et al. 2019)
D. Others							
Alginate microparticles loaded with the model antigen fluorescein isothiocyanate BSA (FITC-BSA)	European sea bass	−	+	+	−	Belgium	(Yaacob et al. 2017)
17β-estradiol (E2)	Common carp	−	+	−	+	Poland	(Szwejser et al. 2017)
Mannan oligosaccharides (cMOS)	European sea bass juveniles	−	+	−	+	Spain	(Torrecillas et al. 2015)
Short-chain fructooligosaccharides (scFOS) with plant-based protein	Caspian white fish (*Rutilus frisii kutum*) fry	+	+	−	−	Portugal	(Azeredo et al. 2017)
Poly-β-hydroxybutyrate (PHB)	European sea bass	−	+	−	+	Germany	(Franke et al. 2017)
Two forms of β-Glucan namely, Gas1 mutant (beta-1,3-glucanosyltransferase activity) and Wild type-β-glucan	Rainbow Trout	+	+	+	+	France	(Cornet et al. 2021)
CpG	Rainbow trout	−	+	+	+	Norway	(Munang'andu et al. 2020)
CpG Oligodeoxynucleotides	Rainbow trout	+	−	−	+	Spain	(Simón et al. 2019)
CpGs	Atlantic salmon	−	−	+	+	Norway	(Iliev et al. 2019)

IL-8, and IgM genes in gilthead seabream (Awad, Cerezuela, and Esteban 2015).

Nhu *et al.*, Belgium, studied the effects of ethanolic extracts of *Phyllanthus amarus*, *Psidium guajava L.* (guava), *Mimosa pudica L.*, *Azadirachta indica* (neem) and *Euphorbia hirta* L. on the immune response and disease resistance of striped catfish. The diets enriched with extracts of *P. guajava* at 0.2 and 1.0%, or *M. pudica* at 2.0% for eight weeks stimulated the number of total red blood cells, white blood cells, lymphocytes, monocytes, and neutrophils, alternative complement (ACH_{50}), lysozyme activity, and the total immunoglobulin level of the treated fish with subsequent reduction in the cumulative mortality against *Edwardsiella ictaluri* infection (Nhu, Bich Hang, Bach, et al. 2019). In another study, the ethanolic extract of *Allium sativum* L., *Azadirachta indica*, *Euphorbia hirta* L., *Phyllanthus amarus*, and *Zingiber officinale* were reported to significantly increase the expression of pro-inflammatory (IL-1β), antiviral (IFN-γ 2a and b) and adaptive (MHC class II) cytokines in the peripheral blood mononuclear cells and head kidney leukocytes of striped catfish (Nhu, Bich Hang, Vinikas, et al. 2019). Later, they found the ethyl acetate and dichloromethane fractions of *P. guajava* to significantly enhance the lysozyme activity, respiratory burst, and nitric oxide synthase production in head kidney leukocytes of striped catfish. Also, the crude extracts and fractions of *P. guajava* were found to be stronger than those in *P. amarus* treatments (Nhu et al. 2020).

The potential of utilizing seaweed as an ecologically sustainable aquafeed supplement is being explored by researchers around the world. *Gracilaria sp.* has been widely studied in Portugal for its immunostimulatory potential. In European seabass, the dietary supplementation of 2.5% *Gracilaria gracilis* was found to enhance plasma lysozyme activity (Passos et al. 2021). Similarly, the dietary inclusion of *Gracilaria vermiculophylla* (5%) in rainbow trout for 91 days enhanced the peroxidase, ACH_{50}, and lysozyme activities (Araújo et al. 2016).

15.5.2 Prebiotics

The indigestible by-products generated by bacterial fermentation are known as 'prebiotics. β-glucans are widely used to enhance the immune response of fish. β-glucan from *Saccharomyces cerevisiae* (wild-type baker's yeast), especially the high dose, stimulated the activity of lysozyme and ACH_{50} activity and Ig level. Also, the null mutant Gas1-β-glucan increased the lysozyme activity, immunoglobulin level, and resistance against *Aeromonas salmonicida* achromogenes in juvenile rainbow trout (Cornet et al. 2021). Polysaccharides like mannanoligosaccharides (MOS), fructooligosaccharides (FOS), and inulin are commonly used immune supplements (Carbone and Faggio 2016). Dietary substitution of fish meal (FM) by plant protein (PP) was found to enhance the production of nitric oxide and lysozyme and plant protein with a 1% fructooligosaccharides supplemented diet increased the monocyte production. However, fish-fed xylooligosaccharides supplemented diet down-regulated the immune gene expression in juvenile European seabass (Azeredo et al. 2017). Similarly, poly-β-hydroxybutyrate (PHB), an energy storage compound produced by bacteria has been shown to enhance the survival, growth, and disease resistance of aquatic organisms. The expression of the insulin-like growth factor 1, antimicrobial peptides dicentracin (dic) and hepcidin (hep) as well as MHC-IIa and MHC-IIb was elevated in the PHB-fed European seabass larvae (Franke et al. 2017). Similarly, dietary mannan oligosaccharides (cMOS) supplemented diet was reported to upregulate the immunoglobulin (Ig), major histocompatibility complex class II (MHCII), T cell receptor β (TCR-β) and Caspase 3 (Casp-3) genes and down-regulate the TGF-β in European seabass (Torrecillas et al. 2015).

The extracellular extracts of *Bacillus subtilis* FI314, *B. vezelensis* FI436 and *B. pumilus* FI464 bacteria isolated from fish up-regulated the expression of IL-1β, IL-6, and COX-2 and improved the *in vitro* response of gilthead seabream immune cells to fight diseases. Specifically, in the *E. tarda* challenged cells, FI314 extracellular extract significantly increased the expression of IL-1β, IL-6, and COX-2, and FI436 and FI464 significantly increased the IL-6 expression (Santos et al. 2022).

15.5.3 Others

Several studies report on the potential of utilizing flagellin as an immunostimulant in aquaculture. Wild-type flagellin (F) from *Marinobacter algicola* and recombinant flagellin with a hypervariable region deletion (FredV2) has been found to upregulate the transcription of the immune-related genes, IL-1β1, IL-6, and IL-8 in the monocyte/macrophage-cell line (RTS-11) and an epithelial cell line from the intestine (RTgutGC) of rainbow trout. Also, the IFN-γ1, hepcidin, and TNF-α were upregulated and the duplication of viral haemorrhagic septicaemia virus was significantly reduced by the F and FredV2 flagellins in RTgutGC cells (Nuñez-Ortiz et al. 2022). Wangkahart *et al.*, reported that the recombinant flagellin from Yersinia ruckeri (YRF) strongly up-regulated the expression of IL-11, IL-23P19, IL-17C1, SAA, and cathelicidin-2 and enhanced the antibody response in rainbow trout (Wangkahart, Secombes, and Wang 2019). In a similar study, *Vibrio anguillarum* derived recombinant flagellin B and ND1 domain were found to up-regulate the expression of IL-1β, TNF-α, and IL-8 in gilthead seabream macrophages (González-Stegmaier et al. 2015).

Recently, insect-based meals are being used as an alternative protein source in aquaculture. The insect-based meals are also reported to confer an immunological boost to the farmed fish. Partial dietary replacement of fishmeal with 35% and 67% by *Tenebrio molitor* (yellow mealworm) showed an increase in the antioxidant, antibacterial activity, and trypsin inhibition in rainbow trout (Henry et al. 2018). However, in *Tinca tinca* (tench) inclusion of insect meal diet with 15 or 30% of *Tenebrio molitor* or *Hermetia illucens* showed the least influence on the immune parameters (Hidalgo et al. 2022).

Arena *et al.*, reported the immunostimulatory effect of honeybee pollen supercritical fluid extracted pollen (HBP_SFE) in gilthead seabream, wherein 1% inclusion of HBP_SFE was found to stimulate the upregulation of IL-1β, IL-6, and IL-8 gene expression (Arena et al. 2022).

In recent times, nanoliposomes are encapsulated with crude lipopolysaccharide (LPS) from *E. coli* and polyinosinic: polycytidylic acid (Poly I:C), a synthetic analogue of viral dsRNA and administrated by bath immersion in zebrafish larvae. The treated larvae showed upregulation of the immune-related genes (TNFα, IL1β, NOS2a, IRF1a, and PTGS2a) and survival against *A. hydrophila* lethal infection (Ji et al. 2019). Alginate microparticles loaded with the model antigen fluorescein isothiocyanate conjugated-bovine serum albumin administration to axenic European seabass larvae resulted in a significant upregulation of cxcr4, cas1, hep, and dic with significantly higher survival rates against *V. anguillarum* (Yaacob et al. 2020).

In conclusion, European aquaculture research employs an extensive range of immunostimulants like plant extracts, fractions, and products, microbes, and microbial products, flagellins, oligosaccharides, insect meal, macroalgae, nanoliposomes, and microparticles. Though a wide variety of medicinal plants like *M. officinalis*, *P. oleracea*, *Z. officinale*, *P. amarus*, *P. guajava*, *M. pudica*, *A. indica*, *E. hirta*, *O. vulgare*, *S. officinalis*, etc. are commonly used in aquaculture, the use of other bioactive substances, prebiotics, probiotics, chemical compounds, and their cocktails are explored for their immunostimulatory potential in European finfish aquaculture.

15.6 Immunostimulant Research in Australia

Domestic aquaculture in Australia has enormous potential to expand to meet the demand. From 2010–2019, the aquaculture production including freshwater and diadromous fish increased from 36573 to 61505 tonnes in Australia (FAO 2021). The increase in aquaculture in the year 2019–2020 was found to be by 11%, to 106,139 tonnes, accounting for 38% of total volume. According to Steven, Mobsby, and Curtotti (2021), the total fisheries and aquaculture production is 204,335 tonnes with the value of $1720 million USD. The total apparent consumption between 2019–2020 was 334,615 tonnes which comprises of 12.4 kg per person. In 2019, some of the main species produced through aquaculture includes, Atlantic salmon (56,989 tonnes), *Thunnus maccoyii* (8,252 tonnes), *Lates calcarifer* (3,365), *Osteichthyes* (2,951 tonnes). The various categories like seaweeds, algae, plant extract, animal extract, vitamins, prebiotics, β-glucan, and few other commercial products have been tested for its immunostimulant activity in fish (Table 15.5). Several combinations of different immunostimulant have also been reported with impressive results in enhancing innate immune mechanism of different fish species. The mode of administration for various immunostimulants reported from Australia includes mainly by feed, few reports on IP, *in vitro* studies, and bioencapsulation.

15.6.1 Microbial Immunostimulant

Thépot et al. (2021) tested several seaweeds such as *Caulerpa taxifolia*, *Halimeda* sp., *Ulva fasciata* (Chlorophyta), *Asparagopsis taxiformis*, *Hydropuntia perplexa*, *Kappaphycus alvarezii*, *Laurencia obtuse*, *Sarconema filiforme* (Rhodophyta), *Dictyota intermedia*, *Lobophora nigrescens*, and *Sargassum spinuligeru* (Phaeophyta). They tested alternate complement activity, serum lysozyme activity, respiratory and phagocytic activity in Mottled rabbitfish, *Siganus fuscescens* and compared it with standard immunostimulants (sodium alginate, hydrolysed yeast, *Spirulina* and *Haematococcus pluvialis*). They observed enhanced humoral, and cell mediated innate immune responses in the group fed with *Asparagopsis taxiformis* and *Dictyota intermedia.* The significant enhancement in tested immune parameters was also reported in the group fed with *Ulva fascioata*. Enhancement of nonspecific immune responses and modulation in immune stress related gene expression was observed in Atlantic salmon upon fed with either *Asparagopsis taxiformis* whole plant or methanolic extracts (Thépot et al. 2022). The maternal exposure to Atrazine results in immunotoxicity and the ability to overcome this after feeding it with Spirulina was measured in female brood stock and its offspring by Hedayatirad et al. (2020). They found that this seaweed diet enhanced the immune parameters in female and helped to restore their immune activity of the offspring after maternal exposure to toxin. Siddik et al. (2022a) studied the combined effect of yeast and lactic acid bacteria where they reported enhanced lysozyme activity, serum protein levels and enhanced expression of IL-10 and TNF-α in Juvenile barramundi, *Lates calcarifer.* Similar symbiotic feed supplementation of *Lactobacillus casei* and garlic, *Allium sativum* and enhancement in immune related gene expression in juvenile barramundi, *Lates calcarifer* was reported by Siddik et al. (2022b). When the striped catfish was fed with mixed probiotics of *Bacillus amyloliquefaciens* 54 A, and *B. pumilus* 47B, it showed enhanced phagocytic activity, respiratory burst and lysozyme activity and enhanced resistance to *E. ictaluri* infection (Truong Thy et al. 2017). The enhancement in several such immune parameters and stress tolerance in juvenile rainbow trout, after feeding with commercial prebiotic, Immunogen was reported by Yarahmadi et al. (2016).

15.6.2 Plant Immunostimulants

Several nonspecific immune parameters, immunoglobulin level, complement and survival rate after challenging with *A. hydrophila* was enhanced in common carp after feeding with different concentration of Figwort, *Scrophularia striata* (Hosseini et al. 2022). Similar results were obtained in Nile tilapia fingerlings upon dietary administration of miswak leaf powder (Naiel et al. 2021). Based on the reports of Doan et al. (2019) in Nile tilapia, survival and nonspecific immune responses have been enhanced in serum and mucus samples after feeding them with *Elephantopus scaber* extract. Ingelbrecht et al. (2020) highlighted the benefits of using garlic, rosemary, and commercially manufactured herbal products as anthelmintic in yellowtail kingfish, *Seriola lalandi*. Dietary supplementation of ginger, *Zingiber officinale,* enhancing nonspecific immune responses as well as lysozyme mRNA expression in zebrafish, *Danio rerio,* was reported by Ahmadifar et al. (2019).

15.6.3 Animal Extracts as Immunostimulants

There are only two reports under animal extract, which are tuna hydrolysate-poultry by-product meal in juvenile barramundi, *Lates calcarifer,* by Siddik et al. 2020, Siddik, Howieson, and Fotedar 2019. They observed upregulation of pro-inflammatory cytokines IL-1β and TNF-α and downregulation of anti-inflammatory cytokine IL-10, enhanced serum lysozyme level and survival against challenge with *Vibrio harveyi* in the tested group.

TABLE 15.5

Current Status of Immunostimulant Research in Australia

Immunostimulant	Name of the Fish	Immunomodulatory Effect on				Reference
		Specific	Nonspecific	Functional Immunity	Gene Expression	
A. Microbial						
Combined effects of probiotic yeast, *Saccharomyces cerevisiae*, and lactic acid bacteria, *Lactobacillus casei*	Juvenile barramundi, *Lates calcarifer*	–	+	–	+	Siddik et al. 2022a
L. casei and garlic	Juvenile barramundi	–	–	–	+	Siddik et al. 2022b
Red seaweed *Asparagopsis taxiformis*	Atlantic Salmon	–	+	–	+	Thépot et al. 2022
11 seaweeds, 4 chemicals	Mottled rabbitfish, *Siganus fuscescens*	+	+	–	–	Thépot et al. 2021
Nutraceutical, *Arthrospira platensis* (spirulina- SP)	Zebrafish, *Danio rerio*	–	+	+	–	Hedayatirad et al. 2020
Mixed probiotics of *Bacillus amyloliquefaciens* 54 A and *B. pumilus* 47B	Striped catfish	–	+	+	–	Truong Thy et al. 2017
Commercial prebiotic, Immunogen	Rainbow trout	–	+	–	–	Yarahmadi et al. 2016
B. Plant						
Figwort, *Scrophularia striata*	Common carp	+	+	+	–	Hosseini et al. 2022
Miswak leaf powder	Nile tilapia	–	+	+	–	Naiel et al. 2021
Garlic, rosemary, and commercially manufactured herbal products	Yellowtail kingfish, *Seriola lalandi*	–	–	+	–	Ingelbrecht et al. 2020
Elephantopus scaber extract	Nile tilapia fingerlings	–	+	+	–	Doan et al. 2019
Ginger, *Zingiber officinale*	Zebrafish	+	+	–	–	Ahmadifar et al. 2019
C. Animal						
Tuna hydrolysate-poultry by-product meal	Juvenile barramundi	–	–	–	+	Siddik et al. 2020
Tuna hydrolysate-poultry by-product meal	Juvenile barramundi	–	+	+	–	Siddik et al. 2019
D. Chemical						
Sodium Propionate	Goldfish, *Carassius auratus*	–	–	+	+	Sheikhzadeh et al. 2021
Glycyrrhizic acid (GA)	Silver carp, *Hypophthalmichthys molitrix*	–	+	–	+	Harikrishnan et al. 2021
Vitamin E	Caspian trout, *Salmo caspius*	–	+	–	–	Saheli et al. 2021
Chitosan and nano-chitosan (0.05, 0.5 and 5 g kg^{-1}) loaded in clinoptilolite (14.28 g kg^{-1})	Rainbow trout	–	+	–	+	Khani Oushani et al. 2020
Non-protein energy source	Barramundi	–	+	–	+	Fu et al. 2019

15.6.4 Chemical Immunostimulants

Praziquantel, an anti-helminth drug enhanced gene expression of TCR, IL-8, IL-1β, IFNα, γ, and MHC class I in Southern bluefin tuna, *Thunnus maccoyii,* and Atlantic salmon cell lines, *in vitro* (Polinski et al. 2014). The non-protein energy source such as fish oil and starch had been explored for its impact on the immune system, where it enhanced the nonspecific immune response in the fish Barramundi (Fu et al. 2019). Feed containing chitosan and nano-chitosan loaded in clinoptilolite enhanced serum lysozyme level, complement activity, Ig level in rainbow trout (Khani Oushani et al. 2020). The supplemented diet also reported to upregulate the gene expression of IgM, iNOS and IL-1β. Sodium propionate also upregulated the expression of gLys, iNOS, TNF-α2 and IL-1β gene in the skin of goldfish and also enhanced the disease resistance against *Ichthyophthirius multifiliis* infection (Sheikhzadeh et al. 2021). The enhanced phagocytic, lysozyme, respiratory burst activity and immunoglobulin production along with enhanced expression of IL-8, TNF- and IL-10 genes were reported in silver carp upon dietary administration of glycyrrhizic acid (Harikrishnan et al. 2021). Similar enhancements in nonspecific immune responses were reported by Saheli et al. (2021) after the dietary supplementation of vitamin E in Caspian trout, *Salmo caspius.*

15.7 Conclusion

Continent-wise data on the immunostimulant research studies are tabulated in this chapter. Apparently, it is easy to find that research on immunostimulants is dominated by Asian countries. This is because Asia is not only the largest aquaculture producer but also has the rich traditional system of medicine. Further, research interest of Western countries is currently focused in developing vaccines. In the long run, fish farmers of Europe and the Americas may start preferring affordable immunostimulants over expensive vaccines. For the same reason, pure, and expensive compounds are seldom used for aquaculture research. Of the different class of biological compounds, polysaccharides with their high complexity can activate special receptors called dectins and toll-like receptors. Certain polysaccharides can also activate B and T cells and enhance the specific immune response. Obviously, polysaccharides are the most widely researched immunostimulant compounds for human and veterinary use. In aquaculture, it is evident that microbial polysaccharides such as bacterial and fungal cell wall components, cell lysates, etc. are widely used as immunostimulants for research and development purposes. Also, most of the commercial immunostimulants available in the market belong to this category. Recent trend shows a shift from bacterial and fungal polysaccharides to algal and plant polysaccharides. Immunostimulant could become successful in aquaculture if we are able to apply the stimulant repeatedly to many small animals (larvae). Hence, research on formulating immunostimulant feed supplementation is given more importance. Certain studies reported the use of microbial biomass having immunostimulating activity as feed. However, immunology of fish gut is not well understood and new tools like gut cell lines are being developed. Such strategies could go a long way in developing marketable immunostimulant feed. Nanotechnology is another promising approach that could enhance the potential of immunostimulants. Further, nanotechnology-mediated targeted delivery will help in developing immunostimulant feed. Though the magnitude of research conducted on immunostimulants by scientists is huge, only a fraction of the research effort gets translated into products. Hence, future research studies should focus on converting the potential compounds into marketable products.

REFERENCES

Abdelhamid, Ahmed F., Hala F. Ayoub, Eman A. Abd El-Gawad, Mohamed F. Abdelghany, and Mohsen Abdel-Tawwab. 2021. "Potential effects of dietary seaweeds mixture on the growth performance, antioxidant status, immunity response, and resistance of striped catfish (Pangasianodon hypophthalmus) against Aeromonas hydrophila infection." *Fish & Shellfish Immunology* 119:76–83. doi: 10.1016/j.fsi.2021.09.043.

Abdel-Latif, Hany M. R., Mahmoud A. O. Dawood, Mahmoud Alagawany, Caterina Faggio, Joanna Nowosad, and Dariusz Kucharczyk. 2022. "Health benefits and potential applications of fucoidan (FCD) extracted from brown seaweeds in aquaculture: An updated review." *Fish & Shellfish Immunology* 122:115–130. doi: 10.1016/j.fsi.2022.01.039.

Abdel-Tawwab, Mohsen. 2016. *Feed supplementation to freshwater fish: Experimental approaches*: LAP Lambert Academic Publishing.

Abdel-Tawwab, Mohsen, Ibrahim Adeshina, and Zeenat A. Issa. 2020. "Antioxidants and immune responses, resistance to Aspergilus flavus infection, and growth performance of Nile tilapia, Oreochromis niloticus, fed diets supplemented with yeast, Saccharomyces serevisiae." *Animal Feed Science and Technology* 263. doi: 10.1016/j.anifeedsci.2020.114484.

Abdel-Tawwab, Mohsen, Hossam A. M. Mounes, Sherien H. H. Shady, and Kareem M. Ahmed. 2021. "Effects of yucca, Yucca schidigera, extract and/or yeast, Saccharomyces cerevisiae, as water additives on growth, biochemical, and antioxidants/oxidant biomarkers of Nile tilapia, Oreochromis niloticus." *Aquaculture* 533. doi: 10.1016/j.aquaculture.2020.736122.

Abdel-Tawwab, M., I. Adeshina, A. Jenyo-Oni, E. K. Ajani, and B. O. Emikpe. 2018. "Growth, physiological, antioxidants, and immune response of African catfish, Clarias gariepinus (B.), to dietary clove basil, Ocimum gratissimum, leaf extract and its susceptibility to Listeria monocytogenes infection." *Fish Shellfish Immunol* 78:346–354. doi: 10.1016/j.fsi.2018.04.057.

Abdel-Tawwab, Mohsen, Riad H. Khalil, Amany M. Diab, Mohamed A. Khallaf, Nashwa Abdel-Razek, Hany M. R. Abdel-Latif, and Eman Khalifa. 2021. "Dietary garlic and chitosan enhanced the antioxidant capacity, immunity, and modulated the transcription of HSP70 and Cytokine genes in Zearalenone-intoxicated European seabass." *Fish & Shellfish Immunology* 113:35–41. doi: 10.1016/j.fsi.2021.03.012.

Abou-El-Atta, Mohamed E., Mohsen Abdel-Tawwab, Nashwa Abdel-Razek, and Taghrid M. N. Abdelhakim. 2019. "Effects of dietary probiotic Lactobacillus plantarum and whey protein concentrate on the productive parameters, immunity response and susceptibility of Nile tilapia, Oreochromis niloticus (L.), to Aeromonas sobria infection." *Aquaculture Nutrition* 25 (6):1367–1377. doi: 10.1111/anu.12957.

Adeshina, I., Y. A. Adewale, and L. O. Tiamiyu. 2017. "Growth performance and innate immune response of Clarias gariepinus infected with Aeromonas hydrophila fed diets fortified with Curcuma longa leaf." *West African Journal of Applied Ecology* 25 (2):87–99.

Adeshina, Ibrahim, Adetola Jenyo-Oni, Benjamin O. Emikpe, Emmanuel K. Ajani, and Mohsen Abdel-Tawwab. 2018. "Stimulatory effect of dietary clove, Eugenia caryophyllata, bud extract on growth performance, nutrient utilization, antioxidant capacity, and tolerance of African catfish, Clarias gariepinus (B.), to Aeromonas hydrophila infection." *Journal of the World Aquaculture Society* 50 (2):390–405. doi: 10.1111/jwas.12565.

Ahmadifar, Ehsan, Najmeh Sheikhzadeh, Kambiz Roshanaei, Narges Dargahi, and Caterina Faggio. 2019. "Can dietary ginger (Zingiber officinale) alter biochemical and immunological parameters and gene expression related to growth, immunity and antioxidant system in zebrafish (Danio rerio)?" *Aquaculture* 507:341–348. doi: 10.1016/j.aquaculture.2019.04.049.

Ahmadifar, Ehsan, Toba Heydari Sadegh, Mahmoud A. O. Dawood, Maryam Dadar, and Najmeh Sheikhzadeh. 2020. "The effects of dietary Pediococcus pentosaceus on growth performance, hemato-immunological parameters and digestive enzyme activities of common carp (Cyprinus carpio)." *Aquaculture* 516. doi: 10.1016/j.aquaculture.2019.734656.

Alamillo, Erika, Martha Reyes-Becerril, Alberto Cuesta, and Carlos Angulo. 2017. "Marine yeast Yarrowia lipolytica improves the immune responses in Pacific red snapper (Lutjanus peru) leukocytes." *Fish & Shellfish Immunology* 70:48–56. doi: 10.1016/j.fsi.2017.08.036.

Aly, S. M., Al Zohairy, M. A., Rahmani, A. H., Fathi, M., and Atti, N. M. A. 2016. Trials to improve the response of Oreochromis niloticus to Aeromonas hydrophila vaccine using immunostimulants (garlic, Echinacea) and probiotics (Organic Green™ and Vet-Yeast™). *African Journal of Biotechnology* 15 (21): 989–994

Amer, Shimaa A. 2016. "Effect of Spirulina platensis as feed supplement on growth performance, immune response and antioxidant status of mono-sex Nile Tilapia (Oreochromis niloticus)." *Benha veterinary medical journal* 30 (1):1–10.

Amoush, Osama A. A., Soner Bilen, Adem Yavuz Sönmez, and Mahmut Elp. 2021. "Antioxidant and immunostimulant responses in rainbow trout (*Oncorhynchus mykiss*) fed with cherry stem extract." *Aquaculture Research* 53 (2):487–496. doi: 10.1111/are.15594.

Anjugam, Mahalingam, Arokiadhas Iswarya, Ashokkumar Sibiya, Chandrabose Selvaraj, Sanjeev Kumar Singh, Marimuthu Govindarajan, Naiyf S. Alharbi, Shine Kadaikunnan, Jamal M. Khaled, Jeyachandran Sivakamavalli, and Baskaralingam Vaseeharan. 2022. "Molecular interaction analysis of β-1, 3 glucan binding protein with Bacillus licheniformis and evaluation of its immunostimulant property in Oreochromis mossambicus." *Fish & Shellfish Immunology* 121:183–196. doi: 10.1016/j.fsi.2021.12.044.

Araújo, Mariana, Paulo Rema, Isabel Sousa-Pinto, Luís M. Cunha, Maria João Peixoto, Maria A. Pires, Fernanda Seixas, Vanda Brotas, Carolina Beltrán, and Luísa M. P. Valente. 2016. "Dietary inclusion of IMTA-cultivated Gracilaria vermiculophylla in rainbow trout (Oncorhynchus mykiss) diets: effects on growth, intestinal morphology, tissue pigmentation, and immunological response." *Journal of Applied Phycology* 28 (1):679–689. doi: 10.1007/s10811-015-0591-8.

Arena, R., A. C. L. de Medeiros, G. Secci, S. Mancini, S. Manuguerra, F. Bovera, A. Santulli, G. Parisi, C. M. Messina, and G. Piccolo. 2022. "Effects of Dietary Supplementation with Honeybee Pollen and Its Supercritical Fluid Extract on Immune Response and Fillet's Quality of Farmed Gilthead Seabream (Sparus aurata)." *Animals (Basel)* 12 (6). doi: 10.3390/ani12060675.

Arteaga Quico, Caren, Mauro Mariano Astocondor, and Ronald Aquino Ortega. 2021. "Dietary supplementation with Chlorella peruviana improve the growth and innate immune response of rainbow trout Oncorhynchus mykiss fingerlings." *Aquaculture* 533:736117. doi: 10.1016/j.aquaculture.2020.736117.

Awad, Elham, Rebeca Cerezuela, and M. Ángeles Esteban. 2015. "Effects of fenugreek (Trigonella foenum graecum) on gilthead seabream (Sparus aurata L.) immune status and growth performance." *Fish & Shellfish Immunology* 45 (2):454–464. doi: 10.1016/j.fsi.2015.04.035.

Awad, Elham, Dawn Austin, Alastair Lyndon, and Amani Awaad. 2019. "Possible effect of hala extract (Pandanus tectorius) on immune status, anti-tumour and resistance to Yersinia ruckeri infection in rainbow trout (Oncorhynchus mykiss)." *Fish & Shellfish Immunology* 87:620–626. doi: 10.1016/j.fsi.2019.02.012.

Awad, Laila Z., Heba S. El-Mahallawy, Noha S. Abdelnaeim, Manal M. A. Mahmoud, Amina A. Dessouki, and Noha I. ElBanna. 2022. "Role of dietary Spirulina platensis and betaine supplementation on growth, hematological, serum biochemical parameters, antioxidant status, immune responses, and disease resistance in Nile tilapia." *Fish & Shellfish Immunology* 126:122–130. doi: 10.1016/j.fsi.2022.05.040.

Azeredo, R., M. Machado, E. Kreuz, S. Wuertz, A. Oliva-Teles, P. Enes, and B. Costas. 2017. The European seabass (Dicentrarchus labrax) innate immunity and gut health are modulated by dietary plant-protein inclusion and prebiotic supplementation. *Fish & Shellfish Immunology* 60: 78–87.

Baba, Esin, Gülşen Uluköy, and Canan Öntaş. 2015. "Effects of feed supplemented with Lentinula edodes mushroom extract on the immune response of rainbow trout, Oncorhynchus mykiss, and disease resistance against Lactococcus garvieae." *Aquaculture* 448:476–482. doi: 10.1016/j.aquaculture.2015.04.031.

Baba, Esin, Ümit Acar, Canan Öntaş, Osman Sabri Kesbiç, and Sevdan Yilmaz. 2016. "The use ofAvena sativaextract againstAeromonas hydrophilaand its effect on growth performance, hematological and immunological parameters in common carp (Cyprinus carpio)." *Italian Journal of Animal Science* 15 (2):325–333. doi: 10.1080/1828051x.2016.1185977.

Bilen, Soner, Yasemin Celik Altunoglu, Ferhat Ulu, and Gouranga Biswas. 2016. "Innate immune and growth promoting responses to caper (Capparis spinosa) extract in rainbow trout (Oncorhynchus mykiss)." *Fish & Shellfish Immunology* 57:206–212. doi: 10.1016/j.fsi.2016.08.040.

Bilen, Soner, Gamaia Ali Mohamed Ali, Iman Daw Amhamed, and Ahmed Alhadi Almabrok. 2021. "Modulatory effects of laurel-leaf cistus (Cistus laurifolius) ethanolic extract on innate immune responses and disease resistance in common carp (Cyprinus carpio)." *Fish & Shellfish Immunology* 116:98–106. doi: 10.1016/j.fsi.2021.07.001.

Bilen, Soner, Tarek Abdalsalam Salem Altief, Keriman Yürüten Özdemir, Mohamed Omar Abdalla Salem, Ertugrul Terzi, and Kerim Güney. 2019. "Effect of lemon balm (Melissa officinalis) extract on growth performance, digestive and antioxidant enzyme activities, and immune responses in rainbow trout (Oncorhynchus mykiss)." *Fish Physiology and Biochemistry* 46 (1):471–481. doi: 10.1007/s10695-019-00737-z.

Biller-Takahashi, J. D., H. J. Montassier, L. S. Takahashi, and E. C. Urbinati. 2016. "Levamisole promotes an adjuvant effect on the immunity of pacu (Piaractus mesopotamicus) when immunized with Aeromonas hydrophila, even when provided in the diet." *Animal Feed Science and Technology* 211:164–173. doi: 10.1016/j.anifeedsci.2015.11.008.

Bindels, Laure B., Nathalie M. Delzenne, Patrice D. Cani, and Jens Walter. 2015. "Towards a more comprehensive concept for prebiotics." *Nature Reviews Gastroenterology & Hepatology* 12 (5):303–310. doi: 10.1038/nrgastro.2015.47.

Bricknell, I., and R. Dalmo. 2005. "The use of immunostimulants in fish larval aquaculture." *Fish & Shellfish Immunology* 19 (5):457–472. doi: 10.1016/j.fsi.2005.03.008.

Brummett, Randall E., Jérôme Lazard, and John Moehl. 2008. "African aquaculture: Realizing the potential." *Food Policy* 33 (5):371–385. doi: 10.1016/j.foodpol.2008.01.005.

Bulfon, C., M. Galeotti, and D. Volpatti. 2018. "Medicinal plant extracts modulate respiratory burst and proliferation activity of rainbow trout (Oncorhynchus mykiss) leukocytes." *Fish Physiol Biochem* 44 (1):109–117. doi: 10.1007/s10695-017-0417-5.

Cámara-Ruiz, María, José María García Beltrán, Francisco Antonio Guardiola, and María Ángeles Esteban. 2020. "In vitro and in vivo effects of purslane (Portulaca oleracea L.) on gilthead seabream (Sparus aurata L.)." *AIMS Agriculture and Food* 5 (4):799–824. doi: 10.3934/agrfood.2020.4.799.

Carbone, Donatella, and Caterina Faggio. 2016. "Importance of prebiotics in aquaculture as immunostimulants. Effects on immune system of Sparus aurata and Dicentrarchus labrax." *Fish & Shellfish Immunology* 54:172–178. doi: 10.1016/j.fsi.2016.04.011.

Cornet, V., T. D. Khuyen, S. N. M. Mandiki, S. Betoulle, P. Bossier, F. E. Reyes-López, L. Tort, and P. Kestemont. 2021. "GAS1: A New β-Glucan Immunostimulant Candidate to Increase Rainbow Trout (Oncorhynchus mykiss) Resistance to Bacterial Infections With Aeromonas salmonicida achromogenes." *Front Immunol* 12:693613. doi: 10.3389/fimmu.2021.693613.

da Cunha, Jessyka Arruda, Cecília de Ávila Scheeren, Viviane Pedroso Fausto, Larissa Daiane Willrich de Melo, Bruno Henneman, Clarissa Piccinin Frizzo, Rodrigo de Almeida Vaucher, Agueda Castagna de Vargas, and Bernardo Baldisserotto. 2018. "The antibacterial and physiological effects of pure and nanoencapsulated Origanum majorana essential oil on fish infected with Aeromonas hydrophila." *Microbial Pathogenesis* 124:116–121. doi: 10.1016/j.micpath.2018.08.040.

Dawood, Mahmoud A. O., Ibrahim S. El-Shamaa, Nagwa I. Abdel-Razik, Azza H. Elkomy, Mahmoud S. Gewaily, Safaa E. Abdo, Ali A. Soliman, Bilal Ahamad Paray, and Nevien Abdelkhalek. 2020. "The effect of mannanoligosaccharide on the growth performance, histopathology, and the expression of immune and antioxidative related genes in Nile tilapia reared under chlorpyrifos ambient toxicity." *Fish & Shellfish Immunology* 103:421–429. doi: 10.1016/j.fsi.2020.05.061.

Di Chiacchio, Isabela M., Isadora M. Paiva, Danilo J. M. de Abreu, Elisângela E. N. Carvalho, Pedro J. Martínez, Stephan M. Carvalho, Victoriano Mulero, and Luis David S. Murgas. 2021. "Bee pollen as a dietary supplement for fish: Effect on the reproductive performance of zebrafish and the immunological response of their offspring." *Fish & Shellfish Immunology* 119:300–307. doi: 10.1016/j.fsi.2021.10.012.

Dimitroglou, Arkadios, Daniel L. Merrifield , Oliana Carnevali, Simona Picchietti, Matteo Avella, Carly Daniels, Derya Güroy, Simon J. Davies. 2011. "Microbial manipulations to improve fish health and production – A Mediterranean perspective." *Fish & Shellfish Immunology* 30 (1): 1–16. doi: 10.1016/j.fsi.2010.08.009.

Doan, Hien Van, Seyed Hossein Hoseinifar, Korawan Sringarm, Sanchai Jaturasitha, Trisadee Khamlor, Mahmoud A. O. Dawood, Maria Ángeles Esteban, Mehdi Soltani, and Mohamed Saiyad Musthafa. 2019. "Effects of elephant's foot (Elephantopus scaber) extract on growth performance, immune response, and disease resistance of nile tilapia (Oreochromis niloticus) fingerlings." *Fish & Shellfish Immunology* 93:328–335. doi: 10.1016/j.fsi.2019.07.061.

Dotta, Geovana, Jaqueline Inês Alves de Andrade, Eduardo Luiz Tavares Gonçalves, Aline Brum, Jacó Joaquim Mattos, Marcelo Maraschin, and Maurício Laterça Martins. 2014. "Leukocyte phagocytosis and lysozyme activity in Nile tilapia fed supplemented diet with natural extracts of propolis and Aloe barbadensis." *Fish & Shellfish Immunology* 39 (2):280–284. doi: 10.1016/j.fsi.2014.05.020.

El-Saadony, Mohamed T., Mahmoud Alagawany, Amlan K. Patra, Indrajit Kar, Ruchi Tiwari, Mahmoud A. O. Dawood, Kuldeep Dhama, and Hany M. R. Abdel-Latif. 2021. "The functionality of probiotics in aquaculture: An overview." *Fish & Shellfish Immunology* 117:36–52. doi: 10.1016/j.fsi.2021.07.007.

Emeish, F.A. Walaa, Zeinab Al-Amgad, and Hassan Ahmed. 2018. "Antioxidant, immunostimulant and renal protective activities of tri-herbal combination in African Sharptooth Catfish, Clarias gariepinus." *Journal of Veterinary Medical Research* 25 (2):213–229. doi: 10.21608/jvmr.2017.43320.

Evensen, Øystein. 2016. "Development of Fish Vaccines: Focusing on Methods." In *Fish Vaccines*, edited by Alexandra Adams, 53–74. Basel: Springer Basel.

Fadl, Sabreen E., M. S. ElGohary, Abdelgawad Y. Elsadany, Doaa M. Gad, Farag F. Hanaa, and Nagwan M. El-Habashi. 2017. "Contribution of microalgae-enriched fodder for the Nile tilapia to growth and resistance to infection with Aeromonas hydrophila." *Algal Research* 27:82–88. doi: 10.1016/j.algal.2017.08.022.

Fadlelmoula, Abdelaziz A. 2018. "Dietary Gum Arabic as Animal Feed Additive." In *Gum Arabic*, 261–267.

FAO. 2020. *The State of World Fisheries and Aquaculture 2020.*

FAO. 2021. *FAO Yearbook: Fishery and Aquaculture Statistics 2019*: FAO.

Franke, Andrea, Catriona Clemmesen, Peter De Schryver, Linsey Garcia-Gonzalez, Joanna J. Miest, and Olivia Roth. 2017.

"Immunostimulatory effects of dietary poly-β-hydroxybutyrate in European sea bass postlarvae." *Aquaculture Research* 48 (12):5707–5717. doi: 10.1111/are.13393.

Fu, Zhengyi, Rui Yang, Xu Chen, Jian G. Qin, Zhifeng Gu, and Zhenhua Ma. 2019. "Dietary non-protein energy source regulates antioxidant status and immune response of barramundi (Lates calcarifer)." *Fish & Shellfish Immunology* 95:697–704. doi: 10.1016/j.fsi.2019.11.018.

Gabriel, Ndakalimwe Naftal, Margit R. Wilhelm, Habte-Michael Habte-Tsion, Percy Chimwamurombe, and Edosa Omoregie. 2019. "Dietary garlic (Allium sativum) crude polysaccharides supplementation on growth, haematological parameters, whole body composition and survival at low water pH challenge in African catfish (Clarias gariepinus) juveniles." *Scientific African* 5. doi: 10.1016/j.sciaf.2019.e00128.

Gabriel, Ndakalimwe Naftal, Margit R. Wilhelm, Habte-Michael Habte-Tsion, Percy Chimwamurombe, Edosa Omoregie, Linda N. Iipinge, and Kaspar Shimooshili. 2019. "Effect of dietary Aloe vera polysaccharides supplementation on growth performance, feed utilization, hemato-biochemical parameters, and survival at low pH in African catfish (Clarias gariepinus) fingerlings." *International Aquatic Research* 11 (1):57–72. doi: 10.1007/s40071-019-0219-8.

Gallani, Sílvia Umeda, Gustavo Moraes Ramos Valladão, Elisa Helena Giglio Ponsano, and Fabiana Pilarski. 2017. "Rubrivivax gelatinosus biomass as an immunostimulant for pacu Piaractus mesopotamicus." *Aquaculture Research* 48 (9):4836–4843. doi: 10.1111/are.13303.

García Beltrán, José María, Daniel González Silvera, Cristóbal Espinosa Ruiz, Vittorio Campo, Latifeh Chupani, Caterina Faggio, and María Ángeles Esteban. 2020. "Effects of dietary Origanum vulgare on gilthead seabream (Sparus aurata L.) immune and antioxidant status." *Fish & Shellfish Immunology* 99:452–461. doi: 10.1016/j.fsi.2020.02.040.

Gayed, Mohamed Abdel, Hiam Elabd, Mohamed Tageldin, and Amany Abbass. 2021. "Probiotic Zado® (Ruminococcus Flavefaciens) boosts hematology, immune, serum proteins, and growth profiles in Nile tilapia (Oreochromis niloticus)." *Fish and Shellfish Immunology Reports* 2. doi: 10.1016/j.fsirep.2021.100021.

Giri, Sib Sankar, Jin Woo Jun, Venkatachalam Sukumaran, and Se Chang Park. 2016. "Dietary Administration of Banana (Musa acuminata) Peel Flour Affects the Growth, Antioxidant Status, Cytokine Responses, and Disease Susceptibility of Rohu,Labeo rohita." *Journal of Immunology Research* 2016:1–11. doi: 10.1155/2016/4086591.

Giri, Sib Sankar, Jun Woo Jun, Venkatachalam Sukumaran, and Se Chang Park. 2017. "Evaluation of dietary Hybanthus enneaspermus (Linn F. Muell.) as a growth and haemato-immunological modulator in Labeo rohita." *Fish & Shellfish Immunology* 68:310–317. doi: 10.1016/j.fsi.2017.07.009.

Giri, Sib Sankar, Shib Sankar Sen, Cheng Chi, Hyoun Joong Kim, Saekil Yun, Se Chang Park, and V. Sukumaran. 2015. "Effect of guava leaves on the growth performance and cytokine gene expression of Labeo rohita and its susceptibility to Aeromonas hydrophila infection." *Fish & Shellfish Immunology* 46 (2):217–224. doi: 10.1016/j.fsi.2015.05.051.

González-Stegmaier, Roxana, Alex Romero, Amparo Estepa, Jana Montero, Victoriano Mulero, and Luis Mercado. 2015. "Effects of recombinant flagellin B and its ND1 domain from Vibrio anguillarum on macrophages from gilthead seabream (Sparus aurata L.) and rainbow trout (Oncorhynchus mykiss, W.)." *Fish & Shellfish Immunology* 42 (1):144–152. doi: 10.1016/j.fsi.2014.10.034.

Guardiola, F. A., C. Porcino, R. Cerezuela, A. Cuesta, C. Faggio, and M. A. Esteban. 2016. "Impact of date palm fruits extracts and probiotic enriched diet on antioxidant status, innate immune response and immune-related gene expression of European seabass (Dicentrarchus labrax)." *Fish & Shellfish Immunology* 52:298–308. doi: 10.1016/j.fsi.2016.03.152.

Gupta, Sanjay K., B. Sarkar, Manisha Priyam, Neeraj Kumar, S. Naskar, Md Javed Foysal, Shailesh Saurabh, and T. R. Sharma. 2020. "Inflammatory and stress biomarker response of Aeromonas hydrophila infected rohu, Labeo rohita fingerlings to dietary microbial levan." *Aquaculture* 521. doi: 10.1016/j.aquaculture.2020.735020.

Güroy, Betül, Derya Güroy, Soner Bilen, Osman Nezih Kenanoğlu, Izzet Şahin, Ertuğrul Terzi, Onur Karadal, and Serhan Mantoğlu. 2022. "Effect of dietary Spirulina (*Arthrospira platensis*) on the growth performance, immune-related gene expression and resistance to *Vibrio anguillarum* in European seabass (*Dicentrarchus labrax*)." *Aquaculture Research* 53 (6):2263–2274. doi: 10.1111/are.15745.

Guzmán-Villanueva, Laura T., Dariel Tovar-Ramírez, Enric Gisbert, Héctor Cordero, Francisco A. Guardiola, Alberto Cuesta, José Meseguer, Felipe Ascencio-Valle, and Maria A. Esteban. 2014. "Dietary administration of β-1,3/1,6-glucan and probiotic strain Shewanella putrefaciens, single or combined, on gilthead seabream growth, immune responses and gene expression." *Fish & Shellfish Immunology* 39 (1):34–41. doi: 10.1016/j.fsi.2014.04.024.

Haghighi, Masoud, Hamid Pourmoghim, and Mostafa Sharif Rohani. 2018. "Effect of Origanum vulgare extract on immune responses and heamatological parameters of rainbow trout (Oncorhynchus mykiss)." *Oceanography & Fisheries Open Access Journal* 6 (3): 71–76.

Halwart, Matthias. 2020. "Fish farming high on the global food system agenda in 2020." *FAO Aquaculture Newsletter* (61):II–III.

Harikrishnan, Ramasamy, Gunapathy Devi, Hien Van Doan, Sundaram Jawahar, Chellam Balasundaram, Kaliyaperumal Saravanan, Jesu Arockiaraj, Mehdi Soltani, and Sanchai Jaturasitha. 2021. "Study on antioxidant potential, immunological response, and inflammatory cytokines induction of glycyrrhizic acid (GA) in silver carp against vibriosis." *Fish & Shellfish Immunology* 119:193–208. doi: 10.1016/j.fsi.2021.09.040.

Hasan-Uj-Jaman, Md, Md Hossain, Shoumo Khondoker, Md Alam, Md Zaman, and Sanjoy Bappa. 2016. "Effects of Achyranthes aspera to the Immunity of Rohu (Labeo rohita) against Pseudomonas fluorescens." *British Journal of Pharmaceutical Research* 12 (5):1–9. doi: 10.9734/bjpr/2016/28110.

Hassaan, Mohamed S., Eman Y. Mohammady, Mohamed R. Soaudy, Soaad A. Sabae, Abeer M. A. Mahmoud, and Ehab R. El-Haroun. 2021. "Comparative study on the effect of dietary β-carotene and phycocyanin extracted from Spirulina platensis on immune-oxidative stress biomarkers, genes expression and intestinal enzymes, serum biochemical in Nile tilapia, Oreochromis niloticus." *Fish & Shellfish Immunology* 108:63–72. doi: 10.1016/j.fsi.2020.11.012.

Hassan, Ayman Abdel Mohsen, Mohamed Helmy Yacout, Mohamed Samir Khalel, Salma Hashim Abu Hafsa, Mostafa

Abdel Rahman Ibrahim, Dorina Nicoleta Mocuta, Adrian Turek Rahoveanu, and Lorena Dediu. 2018. "Effects of Some Herbal Plant Supplements on Growth Performance and the Immune Response in Nile Tilapia (Oreochromis Niloticus)." *"Agriculture for Life, Life for Agriculture" Conference Proceedings* 1 (1):134–141. doi: 10.2478/alife-2018-0020.

He, Maosheng, Gaoyang Liu, Yihang Liu, Kechen Yang, Xiaozhou Qi, Aiguo Huang, Tianqiang Liu, Gaoxue Wang, and Erlong Wang. 2020. "Effects of geniposide as immunostimulant on the innate immune response and disease resistance in crucian carp." *Aquaculture* 529. doi: 10.1016/j.aquaculture.2020.735713.

Hedayatirad, Maryam, Alireza Mirvaghefi, Mohammad Ali Nematollahi, Mohammad Navid Forsatkar, and Culum Brown. 2020. "Transgenerational disrupting impacts of atrazine in zebrafish: Beneficial effects of dietary spirulina." *Comparative Biochemistry and Physiology Part C: Toxicology & Pharmacology* 230. doi: 10.1016/j.cbpc.2019.108685.

Henry, M. A., F. Gai, P. Enes, A. Peréz-Jiménez, and L. Gasco. 2018. "Effect of partial dietary replacement of fishmeal by yellow mealworm (Tenebrio molitor) larvae meal on the innate immune response and intestinal antioxidant enzymes of rainbow trout (Oncorhynchus mykiss)." *Fish & Shellfish Immunology* 83:308–313. doi: 10.1016/j.fsi.2018.09.040.

Hernández, Adrián J., Alex Romero, Roxana Gonzalez-Stegmaier, and Patricio Dantagnan. 2016. "The effects of supplemented diets with a phytopharmaceutical preparation from herbal and macroalgal origin on disease resistance in rainbow trout against Piscirickettsia salmonis." *Aquaculture* 454:109–117. doi: 10.1016/j.aquaculture.2015.12.016.

Hidalgo, M. Carmen, Amalia E. Morales, Héctor J. Pula, Cristina Tomás-Almenar, M. José Sánchez-Muros, Federico Melenchón, Dmitri Fabrikov, and Gabriel Cardenete. 2022. "Oxidative metabolism of gut and innate immune status in skin and blood of tench (Tinca tinca) fed with different insect meals (Hermetia illucens and Tenebrio molitor)." *Aquaculture* 558:738384. doi: 10.1016/j.aquaculture.2022.738384.

Hoseinifar, Seyed Hossein, Fazel Zoheiri, and Christopher Marlowe Caipang. 2016. "Dietary sodium propionate improved performance, mucosal and humoral immune responses in Caspian white fish (Rutilus frisii kutum) fry." *Fish & Shellfish Immunology* 55:523–528. doi: 10.1016/j.fsi.2016.06.027.

Hoseinifar, Seyed Hossein, Hassan Khodadadian Zou, Hamed Paknejad, Ehsan Ahmadifar, and Hien Van Doan. 2018. "Non-specific immune responses and intestinal immunity of common carp (Cyprinus carpio) fed Jujube (Ziziphus jujube) fruit extract." *Aquaculture Research* 49 (9):2995–3003. doi: 10.1111/are.13759.

Hoseinifar, Seyed Hossein, Hassan Khodadadian Zou, Hien Van Doan, Hamed Kolangi Miandare, and Seyyed Morteza Hoseini. 2018. "Evaluation of some intestinal cytokines genes expression and serum innate immune parameters in common carp (Cyprinus carpio) fed dietary loquat (Eriobotrya japonica) leaf extract." *Aquaculture Research* 49 (1):120–127. doi: 10.1111/are.13440.

Hoseinifar, Seyed Hossein, Mohammad Amin Jahazi, Roghieh Mohseni, Morteza Yousefi, Mahsan Bayani, Mohammad Mazandarani, Hien Van Doan, and Ehab R. El-Haroun. 2021. "Dietary apple peel-derived pectin improved growth performance, antioxidant enzymes and immune response in common carp, Cyprinus carpio (Linnaeus, 1758)." *Aquaculture* 535. doi: 10.1016/j.aquaculture.2020.736311.

Hosseini, Hossein, Mehrdad Pooyanmehr, Azadeh Foroughi, Moha Esmaeili, Farzad Ghiasi, and Reza Lorestany. 2022. "Remarkable positive effects of figwort (Scrophularia striata) on improving growth performance, and immunohematological parameters of fish." *Fish & Shellfish Immunology* 120:111–121. doi: 10.1016/j.fsi.2021.11.020.

Hough, C. 2022. *Regional review on status and trends in aquaculture development in Europe – 2020, FAO Fisheries and Aquaculture Circular*. Rome, Italy: FAO.

Iaconisi, Valeria, Stefania Marono, Giuliana Parisi, Laura Gasco, Lucrezia Genovese, Giulia Maricchiolo, Fulvia Bovera, and Giovanni Piccolo. 2017. "Dietary inclusion of Tenebrio molitor larvae meal: Effects on growth performance and final quality treats of blackspot sea bream (Pagellus bogaraveo)." *Aquaculture* 476: 49–58. doi: 10.1016/j.aquaculture.2017.04.007.

Iliev, Dimitar B., Leidy Lagos, Hanna L. Thim, Sven M. Jørgensen, Aleksei Krasnov, and Jorunn B. Jørgensen. 2019. "CpGs Induce Differentiation of Atlantic Salmon Mononuclear Phagocytes Into Cells With Dendritic Morphology and a Proinflammatory Transcriptional Profile but an Exhausted Allostimulatory Activity." *Frontiers in Immunology* 10. doi: 10.3389/fimmu.2019.00378.

Ingelbrecht, Jack, Terrence L. Miller, Alan J. Lymbery, Masashi Maita, Shohei Torikai, and Gavin Partridge. 2020. "Anthelmintic herbal extracts as potential prophylactics or treatments for monogenean infections in cultured yellowtail kingfish (Seriola lalandi)." *Aquaculture* 520. doi: 10.1016/j.aquaculture.2019.734776.

Ispir, U., M. Ozcan, and E. Seker. 2022. "Immunomodulation function of Tunceli garlic (Allium tuncelianum) oil in Rainbow Trout (Oncorhynchus mykiss)." *International Journal of Agriculture, Environment and Food Sciences*: 7–12. doi: 10.31015/jaefs.2022.1.2.

Jaime-Ceballos, Barbarito J., Alfredo Hernández-Llamas, Tsai Garcia-Galano, and Humberto Villarreal. 2006. "Substitution of Chaetoceros muelleri by Spirulina platensis meal in diets for Litopenaeus schmitti larvae." *Aquaculture* 260 (1-4):215–220. doi: 10.1016/j.aquaculture.2006.06.002.

Ji, Jie, Susana Merino, Juan M. Tomás, and Nerea Roher. 2019. "Nanoliposomes encapsulating immunostimulants modulate the innate immune system and elicit protection in zebrafish larvae." *Fish & Shellfish Immunology* 92:421–429. doi: 10.1016/j.fsi.2019.06.016.

Khalil, Samah R., Rasha M. Reda, and Ashraf Awad. 2017. "Efficacy of Spirulina platensis diet supplements on disease resistance and immune-related gene expression in Cyprinus carpio L. exposed to herbicide atrazine." *Fish & Shellfish Immunology* 67:119–128. doi: 10.1016/j.fsi.2017.05.065.

Khani Oushani, Ali, Mehdi Soltani, Najmeh Sheikhzadeh, Mehdi Shamsaie Mehrgan, and Houman Rajabi Islami. 2020. "Effects of dietary chitosan and nano-chitosan loaded clinoptilolite on growth and immune responses of rainbow trout (Oncorhynchus mykiss)." *Fish & Shellfish Immunology* 98:210–217. doi: 10.1016/j.fsi.2020.01.016.

Kirubakaran, C. John Wesly, Parasuraman Aiya Subramani, and R. Dinakaran Michael. 2016. "Methanol extract of Nyctanthes arbortristis seeds enhances non-specific immune responses and protects Oreochromis mossambicus (Peters) against

Aeromonas hydrophila infection." *Research in Veterinary Science* 105:243–248. doi: 10.1016/j.rvsc.2016.02.013.

Koch, João Fernando Albers, Carlos Alberto Ferreira de Oliveira, and Fábio Sabbadin Zanuzzo. 2021. "Dietary β-glucan (MacroGard®) improves innate immune responses and disease resistance in Nile tilapia regardless of the administration period." *Fish & Shellfish Immunology* 112:56–63. doi: 10.1016/j.fsi.2021.02.014.

Kord, M. I., T. M. Srour, E. A. Omar, A. A. Farag, A. A. M. Nour, and H. S. Khalil. 2021. "The Immunostimulatory Effects of Commercial Feed Additives on Growth Performance, Non-specific Immune Response, Antioxidants Assay, and Intestinal Morphometry of Nile tilapia, Oreochromis niloticus." *Front Physiol* 12:627499. doi: 10.3389/fphys.2021.627499.

Laith, A. A., A. G. Mazlan, A. W. Effendy, M. A. Ambak, W. W. I. Nurhafizah, A. S. Alia, A. Jabar, and M. Najiah. 2017. "Effect of Excoecaria agallocha on non-specific immune responses and disease resistance of Oreochromis niloticus against Streptococcus agalactiae." *Research in Veterinary Science* 112:192–200. doi: 10.1016/j.rvsc.2017.04.020.

Le Xuan, Chinh, Supreya Wannavijit, Piyatida Outama, Napatsorn Montha, Chompunut Lumsangkul, Sudaporn Tongsiri, Chanagun Chitmanat, Seyed Hossein Hoseinifar, and Hien Van Doan 2022. "Effects of dietary rambutan (Nephelium lappaceum L.) peel powder on growth performance, immune response and immune-related gene expressions of striped catfish (Pangasianodon hypophthalmus) raised in biofloc system." *Fish & Shellfish Immunology* 124:134–141. doi: 10.1016/j.fsi.2022.03.039.

Lim, Keng Chin, Fatimah Md Yusoff, Mohamed Shariff, and Mohd Salleh Kamarudin. 2021. "Dietary astaxanthin augments disease resistance of Asian seabass, Lates calcarifer (Bloch, 1790), against Vibrio alginolyticus infection." *Fish & Shellfish Immunology* 114:90–101. doi: 10.1016/j.fsi.2021.03.025.

Liu, Jia, Peijun Zhang, Bo Wang, Yuting Lu, Liang Li, Yuehong Li, and Shaojun Liu. 2022. "Evaluation of the effects of Astragalus polysaccharides as immunostimulants on the immune response of crucian carp and against SVCV in vitro and in vivo." *Comparative Biochemistry and Physiology Part C: Toxicology & Pharmacology* 253. doi: 10.1016/j.cbpc.2021.109249.

Lopes, Larissa Maria Frazão, Mariana Maluli Marinho de Mello, and Elisabeth Criscuolo Urbinati. 2022. "β-Glucan reduces cortisol plasma levels, enhances innate immune system after A. hydrophila inoculation, and has lipolytic effects on the pacu (Piaractus mesopotamicus)." *Aquaculture* 546:737411. doi: 10.1016/j.aquaculture.2021.737411.

Lumsangkul, Chompunut, Nguyen Vu Linh, Fapailin Chaiwan, Mohsen Abdel-Tawwab, Mahmoud A. O. Dawood, Caterina Faggio, Sanchai Jaturasitha, and Hien Van Doan. 2022. "Dietary treatment of Nile tilapia (Oreochromis niloticus) with aquatic fern (Azolla caroliniana) improves growth performance, immunological response, and disease resistance against Streptococcus agalactiae cultured in bio-floc system." *Aquaculture Reports* 24. doi: 10.1016/j.aqrep.2022.101114.

M. dos Santos, Welliene, Túlio S. de Brito, Samuel de A. Prado, Camila G. de Oliveira, Andréia C. De Paula, Daniela C. de Melo, and Paula A. P. Ribeiro. 2016. "Cinnamon (Cinnamomum sp.) inclusion in diets for Nile tilapia submitted to acute hypoxic stress." *Fish & Shellfish Immunology* 54:551–555. doi: 10.1016/j.fsi.2016.04.135.

Magouz Fawzy I., Asem A. Amer, Alaa Faisal, Hani Sewilam, Salama M. Aboelenin, Mahmoud A.O. Dawood. 2022. The effects of dietary oregano essential oil on the growth performance, intestinal health, immune, and antioxidative responses of Nile tilapia under acute heat stress, *Aquaculture* 548 (1): 737632. doi: 10.1016/j.aquaculture.2021.737632.

Maldonado-Garcia, Minerva, Carlos Angulo, Juan Vazquez-Martinez, Veronica Sanchez, Mercedes G. Lopez, and Martha Reyes-Becerril. 2019. "Antioxidant and immunostimulant potentials of Chenopodium ambrosioides L. in Pacific red snapper (Lutjanus peru)." *Aquaculture* 513:734414. doi: 10.1016/j.aquaculture.2019.734414.

Marinho de Mello, Mariana Maluli, Camila de Fátima Pereira de Faria, Fábio Sabbadin Zanuzzo, and Elisabeth Criscuolo Urbinati. 2019. "β-glucan modulates cortisol levels in stressed pacu (Piaractus mesopotamicus) inoculated with heat-killed Aeromonas hydrophila." *Fish & Shellfish Immunology* 93:1076–1083. doi: 10.1016/j.fsi.2019.07.068.

Mariod, Abdalbasit A. 2018. "Enhancement of Color Stability in Foods by Gum Arabic." In *Gum Arabic*, 143–150.

Maiti, Sourav, Suvendu Saha, Prasanta Jana, Arka Chowdhury, Sanjib Khatua, and Tapas Kumar Ghosh. 2021. "Effect of dietary *Andrographis paniculata* leaf extract on growth, immunity, and disease resistance against *Aeromonas hydrophila* in *Pangasianodon hypopthalmus*." *Journal of Applied Aquaculture* 1–25. doi: 10.1080/10454438.2021.1959861.

Martínez Cruz, Patricia, Ana L. Ibáñez, Oscar A. Monroy Hermosillo, and Hugo C. Ramírez Saad. 2012. "Use of Probiotics in Aquaculture." *ISRN Microbiology* 2012:1–13. doi: 10.5402/2012/916845.

Mehana, Elsayed, Arshad Rahmani, and Salah Aly. 2015. "Immunostimulants and Fish Culture: An Overview." *Annual Research & Review in Biology* 5 (6):477–489. doi: 10.9734/arrb/2015/9558.

Mohammadi, Ghasem, Gholamreza Rafiee, Mohammed F. El Basuini, Hany M. R. Abdel-Latif, and Mahmoud A. O. Dawood. 2020. "The growth performance, antioxidant capacity, immunological responses, and the resistance against Aeromonas hydrophila in Nile tilapia (Oreochromis niloticus) fed Pistacia vera hulls derived polysaccharide." *Fish & Shellfish Immunology* 106:36–43. doi: 10.1016/j.fsi.2020.07.064.

Mohan, Kannan, Samuthirapandian Ravichandran, Thirunavukkarasu Muralisankar, Venkatachalam Uthayakumar, Ramachandran Chandirasekar, Palaniappan Seedevi, Ramu Ganesan Abirami, and Durairaj Karthick Rajan. 2019. "Application of marine-derived polysaccharides as immunostimulants in aquaculture: A review of current knowledge and further perspectives." *Fish & Shellfish Immunology* 86:1177–1193. doi: 10.1016/j.fsi.2018.12.072.

Morante, Vitor Hugo Penariol, Carlos Eduardo Copatti, Antônio Ramires Lyra Souza, Mateus Matiuzzi da Costa, Luís Gustavo Tavares Braga, Anderson Miranda Souza, Fúlvio Viegas Santos Teixeira de Melo, Antonio Cleber da Silva Camargo, and José Fernando Bibiano Melo. 2021. "Assessment the crude grape extract as feed additive for tambaqui (Colossoma macropomum), an omnivorous fish." *Aquaculture* 544. doi: 10.1016/j.aquaculture.2021.737068.

Munang'andu, Hetron M., Irene Salinas, Carolina Tafalla, and Roy Ambli Dalmo. 2020. "Editorial: Vaccines and Immunostimulants for Finfish." *Frontiers in Immunology* 11. doi: 10.3389/fimmu.2020.573771.

Muthulakshmi, M., P. A. Subramani, and R. D. Michael. 2016. "Immunostimulatory effect of the aqueous leaf extract of Phyllanthus niruri on the specific and nonspecific immune responses of Oreochromis mossambicus Peters." *Iran J Vet Res* 17 (3):200–202.

Naiel, Mohammed A. E., Samah A. A. Abd El-hameed, Ahmed H. Arisha, and Samar S. Negm. 2022. "Gum Arabic-enriched diet modulates growth, antioxidant defenses, innate immune response, intestinal microbiota and immune related genes expression in tilapia fish." *Aquaculture* 556. doi: 10.1016/j.aquaculture.2022.738249.

Naiel, Mohammed A. E., Nahla E. M. Ismael, Samar S. Negm, Mohamed S. Ayyat, and Adham A. Al-Sagheer. 2020. "Rosemary leaf powder–supplemented diet enhances performance, antioxidant properties, immune status, and resistance against bacterial diseases in Nile Tilapia (Oreochromis niloticus)." *Aquaculture* 526. doi: 10.1016/j.aquaculture.2020.735370.

Naiel, Mohammed A. E., Mohamed K. Khames, Nashwa Abdel-Razek, Amany A. Gharib, and Khaled A. El-Tarabily. 2021. "The dietary administration of miswak leaf powder promotes performance, antioxidant, immune activity, and resistance against infectious diseases on Nile tilapia (Oreochromis niloticus)." *Aquaculture Reports* 20. doi: 10.1016/j.aqrep.2021.100707.

Nan, Fan-Hua, A. S. Agus Putra, Bargir Margie, and M. C. Lee. 2015. "The effects of Curcuma zedoaria and Zingiber zerumbet on non-specific immune responses of grouper Epinephelus coioides." *Iranian Journal of Fisheries Sciences.*

Na-Phatthalung, Pinanong, Mariana Teles, Supayang Piyawan Voravuthikunchai, Lluís Tort, and Camino Fierro-Castro. 2018. "Immunomodulatory effects of Rhodomyrtus tomentosa leaf extract and its derivative compound, rhodomyrtone, on head kidney macrophages of rainbow trout (Oncorhynchus mykiss)." *Fish Physiology and Biochemistry* 44 (2):543–555. doi: 10.1007/s10695-017-0452-2.

Neuls, Luciane, Valmir José de Souza, Silvia Romão, Thiago Bergler Bitencourt, Carlos José Raupp Ramos, Jorge Erick Garcia Parra, and Luisa Helena Cazarolli. 2021. "Immunomodulatory effects of Yarrowia lipolytica as a food additive in the diet of Nile tilapia." *Fish & Shellfish Immunology* 119:272–279. doi: 10.1016/j.fsi.2021.10.011.

Nhu, Truong Quynh, Bui Thi Bich Hang, Le Thi Bach, Bui Thi Buu Hue, Joëlle Quetin-Leclercq, Marie-Louise Scippo, Nguyen Thanh Phuong, and Patrick Kestemont. 2019. "Plant extract-based diets differently modulate immune responses and resistance to bacterial infection in striped catfish (Pangasianodon hypophthalmus)." *Fish & Shellfish Immunology* 92:913–924. doi: 10.1016/j.fsi.2019.07.025.

Nhu, Truong Quynh, Bui Thi Bich Hang, Anais Vinikas, Le Thi Bach, Bui Thi Buu Hue, Do Thi Thanh Huong, Joëlle Quetin-Leclercq, Marie-Louise Scippo, Nguyen Thanh Phuong, and Patrick Kestemont. 2019. "Screening of immuno-modulatory potential of different herbal plant extracts using striped catfish (Pangasianodon hypophthalmus) leukocyte-based in vitro tests." *Fish & Shellfish Immunology* 93:296–307. doi: 10.1016/j.fsi.2019.07.064.

Nhu, Truong Quynh, Nguyen Phuc Dam, Bui Thi Bich Hang, Le Thi Bach, Do Thi Thanh Huong, Bui Thi Buu Hue, Marie-Louise Scippo, Nguyen Thanh Phuong, Joëlle Quetin-Leclercq, and Patrick Kestemont. 2020. "Immunomodulatory potential of extracts, fractions and pure compounds from Phyllanthus amarus and Psidium guajava on striped catfish (Pangasianodon hypophthalmus) head kidney leukocytes." *Fish & Shellfish Immunology* 104:289–303. doi: 10.1016/j.fsi.2020.05.051.

Nuñez-Ortiz, Noelia, Patricia Díaz-Rosales, Jose A. García, Claudia R. Serra, Paula Enes, Carolina Tafalla, and Eduardo Gomez-Casado. 2022. "Immunostimulant properties of full-length and truncated Marinobacter algicola flagellins, and their effects against viral hemorrhagic septicemia virus (VHSV) in trout." *Fish & Shellfish Immunology* 128:695–702. 10.1016/j.fsi.2022.08.018.

Nya, Elijah, Inyang Udosen, and Aniekan Otoh. 2017. "Effect of Herbal Based Immunostimulant Diets for Disease Control in African Catfish Clarias gariepinus against Aeromonas hydrophila Infections." *Journal of Biology, Agriculture and Healthcare* 7 (16):49–54.

Øverland, Margareth, Liv T. Mydland, and Anders Skrede. 2018. "Marine macroalgae as sources of protein and bioactive compounds in feed for monogastric animals." *Journal of the Science of Food and Agriculture* 99 (1):13–24. doi: 10.1002/jsfa.9143.

Pahor-Filho, Eduardo, Adriana Soliris Corredor Castillo, Nycolas Levy Pereira, Fabiana Pilarski, and Elisabeth Criscuolo Urbinati. 2017. "Levamisole enhances the innate immune response and prevents increased cortisol levels in stressed pacu (Piaractus mesopotamicus)." *Fish & Shellfish Immunology* 65:96–102. doi: 10.1016/j.fsi.2017.04.003.

Panigrahi, Akshaya, R. Naveenkumar, and R. R. Das. 2021. "Immunoprophylactic Measures in Aquaculture." In *Advances in Fisheries Biotechnology*, 263–288.

Passos, Ricardo, Ana Patrícia Correia, Damiana Pires, Pedro Pires, Inês Ferreira, Marco Simões, Beatriz do Carmo, Paulo Santos, Ana Pombo, Clélia Afonso, and Teresa Baptista. 2021. "Potential use of macroalgae Gracilaria gracilis in diets for European seabass (Dicentrarchus labrax): Health benefits from a sustainable source." *Fish & Shellfish Immunology* 119:105–113. doi: 10.1016/j.fsi.2021.09.033.

Polinski, Mark, Andrew Bridle, Lukas Neumann, and Barbara Nowak. 2014. "Preliminary evidence of transcriptional immunomodulation by praziquantel in bluefin tuna and Atlantic salmon in vitro cultures." *Fish & Shellfish Immunology* 38 (1):42–46. doi: 10.1016/j.fsi.2014.02.023.

Prabu, Dhanasekaran Linga, Narottam P. Sahu, Asim K. Pal, Subrata Dasgupta, and Ashalaxmi Narendra. 2016. "Immunomodulation and interferon gamma gene expression in sutchi cat fish,Pangasianodon hypophthalmus: Effect of dietary fucoidan rich seaweed extract (FRSE) on pre and post challenge period." *Aquaculture Research* 47 (1): 199–218. doi: 10.1111/are.12482.

Quagrainie, K., J. Cai, and Nathanael Hishamunda. 2017. *Social and economic performance of tilapia farming in Africa*: Food and Agriculture Organization of the United Nations.

Rajendran, Priyatharsini, Parasuraman Aiya Subramani, and Dinakaran Michael. 2016. "Polysaccharides from marine macroalga, Padina gymnospora improve the nonspecific and specific immune responses of Cyprinus carpio and protect it from different pathogens." *Fish & Shellfish Immunology* 58:220–228. doi: 10.1016/j.fsi.2016.09.016.

Ramesh, Dharmaraj, and Sami Souissi. 2018. "Effects of potential probiotic Bacillus subtilis KADR1 and its subcellular components on immune responses and disease resistance in Labeo rohita." *Aquaculture Research* 49 (1):367–377. doi: 10.1111/are.13467.

Rashidian, Ghasem, Javad Tahmasebi Boldaji, Simona Rainis, Marko D. Prokić, and Caterina Faggio. 2021. "Oregano (Origanum vulgare) Extract Enhances Zebrafish (Danio rerio) Growth Performance, Serum and Mucus Innate Immune Responses and Resistance against Aeromonas hydrophila Challenge." *Animals* 11 (2):299.

Reis Dias, Marcia Kelly, Eliane Tie Oba Yoshioka, Anselmo Fortunato Ruiz Rodriguez, Ricardo Amaral Ribeiro, Fernando Sérgio Escócio Drummond Viana Faria, Rodrigo Otávio Almeida Ozório, and Marcos Tavares-Dias. 2019. "Growth, physiological and immune responses of Arapaima gigas (Arapaimidae) to Aeromonas hydrophila challenge and handling stress following feeding with immunostimulant supplemented diets." *Fish & Shellfish Immunology* 84:843–847. doi: 10.1016/j.fsi.2018.10.045.

Reyes-Becerril, Martha, Crystal Guluarte, Diana Ceballos-Francisco, Carlos Angulo, and M. Ángeles Esteban. 2017. "Dietary yeast Sterigmatomyces halophilus enhances mucosal immunity of gilthead seabream (Sparus aurata L.)." *Fish & Shellfish Immunology* 64:165–175. doi: 10.1016/j.fsi.2017.03.027.

Reyes-Becerril, Martha, Carlos Angulo, Veronica Sanchez, Juan Vázquez-Martínez, and Mercedes G. López. 2019. "Antioxidant, intestinal immune status and anti-inflammatory potential of Chenopodium ambrosioides L. in fish: In vitro and in vivo studies." *Fish & Shellfish Immunology* 86:420–428. doi: 10.1016/j.fsi.2018.11.059.

Reyes-Cerpa, Sebastián, Eva Vallejos-Vidal, María José Gonzalez-Bown, Jonathan Morales-Reyes, Diego Pérez-Stuardo, Deborah Vargas, Mónica Imarai, Víctor Cifuentes, Eugenio Spencer, Ana María Sandino, and Felipe E. Reyes-López. 2018. "Effect of yeast (Xanthophyllomyces dendrorhous) and plant (Saint John's wort, lemon balm, and rosemary) extract based functional diets on antioxidant and immune status of Atlantic salmon (Salmo salar) subjected to crowding stress." *Fish & Shellfish Immunology* 74:250–259. doi: 10.1016/j.fsi.2017.12.061.

Rezaei, Omid, Mehdi Shamsaie Mehrgan, and Hamed Paknejad. 2022. "Dietary garlic (Allium sativum) powder improved zootechnical performance, digestive enzymes activity, and innate immunity in narrow-clawed crayfish (Postantacus leptodactylus)." *Aquaculture Reports* 24. doi: 10.1016/j.aqrep.2022.101129.

Estaiano de Rezende, Renata Antunes, Michelly Pereira Soares, Fernanda Garcia Sampaio, Israel Luz Cardoso, Márcia Mayumi Ishikawa, Bruno Stéfano Lima Dallago, Francisco Tadeu Rantin, and Marta Cristina Teixeira Duarte. 2021. "Phytobiotics blend as a dietary supplement for Nile tilapia health improvement." *Fish & Shellfish Immunology* 114:293–300. doi: 10.1016/j.fsi.2021.05.010.

Rodríguez, Felipe E., Beatriz Valenzuela, Ana Farías, Ana María Sandino, and Mónica Imarai. 2016. "β-1,3/1,6-Glucan-supplemented diets antagonize immune inhibitory effects of hypoxia and enhance the immune response to a model vaccine." *Fish & Shellfish Immunology* 59:36–45. doi: 10.1016/j.fsi.2016.10.020.

Saheli, Morteza, Houman Rajabi Islami, Mahmoud Mohseni, and Mehdi Soltani. 2021. "Effects of dietary vitamin E on growth performance, body composition, antioxidant capacity, and some immune responses in Caspian trout (Salmo caspius)." *Aquaculture Reports* 21. doi: 10.1016/j.aqrep.2021.100857.

Salah, Mesalhy Aly, A. Al Zohairy Mohamed, H. Rahmani Arshad, Fathi Mohamed, and M. Abdel Atti Nashwa. 2016. "Trials to improve the response of Orechromis niloticus to Aeromonas hydrophila vaccine using immunostimulants (garlic, Echinacea) and probiotics (Organic GreenTM and Vet-YeastTM)." *African Journal of Biotechnology* 15 (21):989–994. doi: 10.5897/ajb2015.15155.

Salomón, Ricardo, Felipe E. Reyes-López, Lluis Tort, Joana P. Firmino, Carmen Sarasquete, Juan B. Ortiz-Delgado, José C. Quintela, José M. Pinilla-Rosas, Eva Vallejos-Vidal, and Enric Gisbert. 2021. "Medicinal Plant Leaf Extract From Sage and Lemon Verbena Promotes Intestinal Immunity and Barrier Function in Gilthead Seabream (Sparus aurata)." *Frontiers in Immunology* 12. doi: 10.3389/fimmu.2021.670279.

Santos, Rafaela A., Nuno Mariz-Ponte, Nicole Martins, Rui Magalhães, Russell Jerusik, Maria J. Saavedra, Helena Peres, Aires Oliva-Teles, and Cláudia R. Serra. 2022. "In vitro modulation of gilthead seabream (Sparus aurata L.) leukocytes by Bacillus spp. extracellular molecules upon bacterial challenge." *Fish & Shellfish Immunology* 121:285–294. doi: 10.1016/j.fsi.2022.01.002.

Sarvi Moghanlou, Kourosh, Elyas Nasr Isfahani, Salar Dorafshan, Amir Tukmechi, and Mohammad Sadegh Aramli. 2018. "Effects of dietary supplementation with Stachys lavandulifolia Vahl extract on growth performance, hemato-biochemical and innate immunity parameters of rainbow trout (Oncorhynchus mykiss)." *Animal Feed Science and Technology* 237:98–105. doi: 10.1016/j.anifeedsci.2018.01.016.

Satia, Benedict P. Regional review on status and trends in aquaculture development in Sub-Saharan Africa -- 2010 =: Revue régionale sur la situation et les tendances dans l'aquaculture en Afrique subsaharienne.

Schmitt, Paulina, Jurij Wacyk, Byron Morales-Lange, Verónica Rojas, Fanny Guzmán, Brian Dixon, and Luis Mercado. 2015. "Immunomodulatory effect of cathelicidins in response to a β-glucan in intestinal epithelial cells from rainbow trout." *Developmental & Comparative Immunology* 51 (1):160–169. doi: 10.1016/j.dci.2015.03.007.

Selim, Khaled M., and Rasha M. Reda. 2015. "Improvement of immunity and disease resistance in the Nile tilapia, Oreochromis niloticus, by dietary supplementation with Bacillus amyloliquefaciens." *Fish & Shellfish Immunology* 44 (2):496–503. doi: 10.1016/j.fsi.2015.03.004.

Shaalan, Mohamed, Magdy El-Mahdy, Mona Saleh, and Mansour El-Matbouli. 2017. "Aquaculture in Egypt: Insights on the Current Trends and Future Perspectives for Sustainable Development." *Reviews in Fisheries Science & Aquaculture* 26 (1):99–110. doi: 10.1080/23308249.2017.1358696.

Sheikhzadeh, Najmeh, Ehsan Ahmadifar, Mahmoud A. O. Dawood, and Mehdi Soltani. 2021. "Dietary sodium propionate enhanced the growth performance, immune-related genes expression, and resistance against Ichthyophthirius multifiliis in goldfish (Carassius auratus)." *Aquaculture* 540. doi: 10.1016/j.aquaculture.2021.736720.

Siddik, Muhammad A. B., Janet Howieson, and Ravi Fotedar. 2019. "Beneficial effects of tuna hydrolysate in poultry by-product meal diets on growth, immune response, intestinal health and disease resistance to Vibrio harveyi in juvenile barramundi, Lates calcarifer." *Fish & Shellfish Immunology* 89:61–70. doi: 10.1016/j.fsi.2019.03.042.

Siddik, Muhammad A. B., Md Javed Foysal, Ravi Fotedar, David S. Francis, and Sanjay K. Gupta. 2022a. "Probiotic yeast Saccharomyces cerevisiae coupled with Lactobacillus casei modulates physiological performance and promotes gut microbiota in juvenile barramundi, Lates calcarifer." *Aquaculture* 546. doi: 10.1016/j.aquaculture.2021.737346.

Siddik, Muhammad A. B., Janet Howieson, S. M. Majharul Islam, and Ravi Fotedar. 2022b. "Synbiotic feed supplementation improves antioxidant response and innate immunity of juvenile barramundi, Lates calcarifer subjected to bacterial infection." *Aquaculture* 552. doi: 10.1016/j.aquaculture.2022.737965.

Siddik, Muhammad A. B., Md Reaz Chaklader, Md Javed Foysal, Janet Howieson, Ravi Fotedar, and Sanjay K. Gupta. 2020. "Influence of fish protein hydrolysate produced from industrial residues on antioxidant activity, cytokine expression and gut microbial communities in juvenile barramundi Lates calcarifer." *Fish & Shellfish Immunology* 97:465–473. doi: 10.1016/j.fsi.2019.12.057.

Silva-Jara, Jorge, Carlos Angulo, María Esther Macias, Carlos Velazquez, Crystal Guluarte, and Martha Reyes-Becerril. 2020. "First screening report of immune and protective effect of non-toxic Jatropha vernicosa stem bark against Vibrio parahaemolyticus in Longfin yellowtail Seriola rivoliana leukocytes." *Fish & Shellfish Immunology* 101:106–114. doi: 10.1016/j.fsi.2020.03.048.

Simón, Rocío, Patricia Díaz-Rosales, Esther Morel, Diana Martín, Aitor G. Granja, and Carolina Tafalla. 2019. "CpG Oligodeoxynucleotides Modulate Innate and Adaptive Functions of IgM+ B Cells in Rainbow Trout." *Frontiers in Immunology* 10. doi: 10.3389/fimmu.2019.00584.

Soares, Michelly Pereira, Israel Luz Cardoso, Márcia Mayumi Ishikawa, Adriana da Silva Santos de Oliveira, Adilson Sartoratto, Claudio Martin Jonsson, Sonia Claudia do Nascimento de Queiroz, Marta Cristina Teixeira Duarte, Francisco Tadeu Rantin, and Fernanda Garcia Sampaio. 2020. "Effects of Artemisia annua alcohol extract on physiological and innate immunity of Nile tilapia (Oreochromis niloticus) to improve health status." *Fish & Shellfish Immunology* 105:369–377. doi: 10.1016/j.fsi.2020.07.035.

Soliman, Naglaa F., and Dalia M. M. Yacout. 2016. "Aquaculture in Egypt: status, constraints and potentials." *Aquaculture International* 24 (5):1201–1227. doi: 10.1007/s10499-016-9989-9.

Sowmya, Rama, and Nakkarike Manjabhat Sachindra. 2015. "Enhancement of non-specific immune responses in common carp,Cyprinus carpio, by dietary carotenoids obtained from shrimp exoskeleton." *Aquaculture Research* 46 (7):1562–1572. doi: 10.1111/are.12310.

Stenberg, Oda Kvalsvik, Elisabeth Holen, Luisa Piemontese, Nina S. Liland, Erik-Jan Lock, Marit Espe, and Ikram Belghit. 2019. "Effect of dietary replacement of fish meal with insect meal on in vitro bacterial and viral induced gene response in Atlantic salmon (Salmo salar) head kidney leukocytes." *Fish & Shellfish Immunology* 91: 223–232. doi: 10.1016/j.fsi.2019.05.042.

Steven, A. H., D. Mobsby, and R. Curtotti. 2021. Australian fisheries and aquaculture statistics 2020. Canberra: Fisheries Research and Development Corporation project.

Subeenabegum, S., and P. S. Navaraj. 2016. "Studies on the immunostimulatory effect of extract of Solanum trilobatum and Ocimum sanctum in Mystus keletius." *International Journal of Fisheries and Aquatic Studies* 4 (2):376–381.

Subeenabegum, S, and P Navaraj. 2016a. "Dietary supplement of mixture of medicinal plant leaf extracts on immune response of fresh water fish Mystus keletius." *International Journal of Applied Research* 2 (2): 361–364.

Subeenabegum, S, and PS Navaraj. 2016b. "Studies on the immunostimulatory effect of extract of Solanum trilobatum and Ocimum sanctum in Mystus keletius." *International Journal of Fisheries and Aquatic Studies* 4 (2): 376–381.

Suzuki, Aya. "Rising Importance of Aquaculture in Asia: Current Status, Issues, and Recommendations."

Szwejser, Ewa, Magdalena Maciuszek, Ayako Casanova-Nakayama, Helmut Segner, B. M. Lidy Verburg-van Kemenade, and Magdalena Chadzinska. 2017. "A role for multiple estrogen receptors in immune regulation of common carp." *Developmental & Comparative Immunology* 66: 61–72. doi: 10.1016/j.dci.2016.04.003.

Tafalla, Carolina, Jarl Bøgwald, and Roy A. Dalmo. 2013. "Adjuvants and immunostimulants in fish vaccines: Current knowledge and future perspectives." *Fish & Shellfish Immunology* 35 (6):1740–1750. doi: 10.1016/j.fsi.2013.02.029.

Terzi, E., B. Kucukkosker, S. Bilen, O. N. Kenanoglu, O. Corum, M. Ozbek, and S. S. Parug. 2021. "A novel herbal immunostimulant for rainbow trout (Oncorhynchus mykiss) against Yersinia ruckeri." *Fish Shellfish Immunol* 110:55–66. doi: 10.1016/j.fsi.2020.12.019.

Thépot, Valentin, Alexandra H. Campbell, Nicholas A. Paul, and Michael A. Rimmer. 2021. "Seaweed dietary supplements enhance the innate immune response of the mottled rabbitfish, Siganus fuscescens." *Fish & Shellfish Immunology* 113:176–184. doi: 10.1016/j.fsi.2021.03.018.

Thépot, Valentin, Alexandra H. Campbell, Michael A. Rimmer, Martina Jelocnik, Colin Johnston, Brad Evans, and Nicholas A. Paul. 2022. "Dietary inclusion of the red seaweed Asparagopsis taxiformis boosts production, stimulates immune response and modulates gut microbiota in Atlantic salmon, Salmo salar." *Aquaculture* 546. doi: 10.1016/j.aquaculture.2021.737286.

Torrecillas, Silvia, Daniel Montero, Maria José Caballero, Lidia Robaina, Maria Jesús Zamorano, John Sweetman, and Marisol Izquierdo. 2015. "Effects of dietary concentrated mannan oligosaccharides supplementation on growth, gut mucosal immune system and liver lipid metabolism of European sea bass (Dicentrarchus labrax) juveniles." *Fish & Shellfish Immunology* 42 (2):508–516. doi: 10.1016/j.fsi.2014.11.033.

Truong Thy, Ho Thi, Nguyen Nhu Tri, Ong Moc Quy, Ravi Fotedar, Korntip Kannika, Sasimanas Unajak, and Nontawith Areechon. 2017. "Effects of the dietary supplementation of mixed probiotic spores of Bacillus amyloliquefaciens 54A, and Bacillus pumilus 47B on growth, innate immunity and stress responses of striped catfish (Pangasianodon hypophthalmus)." *Fish & Shellfish Immunology* 60:391–399. doi: 10.1016/j.fsi.2016.11.016.

Ulukoy, G., Kubilay, A., Didinen, B. I., Metin, S., Altun, S., Diler, O., & Dullac, A. 2018. Immunostimulant effects of geophyte plant extract on non-specific defence mechanisms of rainbow trout (Oncorhynchus mykiss). *Journal of Limnology and Freshwater Fisheries Research*, 4 (1): 36–41.

Van Doan, H., S. H. Hoseinifar, R. Harikrishnan, T. Khamlor, M. Punyatong, W. Tapingkae, M. Yousefi, J. Palma, and E. El-Haroun. 2021. "Impacts of pineapple peel powder on growth performance, innate immunity, disease resistance, and relative immune gene expression of Nile tilapia, Oreochromis niloticus." *Fish Shellfish Immunol* 114:311–319. doi: 10.1016/j.fsi.2021.04.002.

Wang, E., X. Chen, K. Wang, J. Wang, D. Chen, Y. Geng, W. Lai, and X. Wei. 2016. "Plant polysaccharides used as immunostimulants enhance innate immune response and disease resistance against Aeromonas hydrophila infection in fish." *Fish Shellfish Immunol* 59:196–202. doi: 10.1016/j.fsi.2016.10.039.

Wangkahart, Eakapol, Christopher J. Secombes, and Tiehui Wang. 2019. "Studies on the Use of Flagellin as an Immunostimulant and Vaccine Adjuvant in Fish Aquaculture." *Frontiers in Immunology* 9. doi: 10.3389/fimmu.2018.03054.

Wangkahart, Eakapol, Brecht Bruneel, Anut Chantiratikul, Matthijs de Jong, Noppakun Pakdeenarong, and Parasuraman Aiya Subramani. 2022. "Optimum dietary sources and levels of selenium improve growth, antioxidant status, and disease resistance: re-evaluation in a farmed fish species, Nile tilapia (Oreochromis niloticus)." *Fish & Shellfish Immunology* 121:172–182. doi: 10.1016/j.fsi.2021.12.003.

Wangkahart, Eakapol, Suriyet Wachiraamonloed, Po-Tsang Lee, Parasuraman Aiya Subramani, Zhitao Qi, and Bei Wang. 2022. "Impacts of Aegle marmelos fruit extract as a medicinal herb on growth performance, antioxidant and immune responses, digestive enzymes, and disease resistance against Streptococcus agalactiae in Nile tilapia (Oreochromis niloticus)." *Fish & Shellfish Immunology* 120:402–410. doi: 10.1016/j.fsi.2021.11.015.

Wu, Yu-Sheng, Yin-Yu Chen, Pien-Sheng Ueng, and Fan-Hua Nan. 2016. "Effects of medicinal herbs " Plantago asiatica", " Houttuynia cordata" and " Mentha haplocalyx" on non-specific immune responses of cobia (Rachycentron canadum)." *Fish & Shellfish Immunology* 58:406–414. doi: 10.1016/j.fsi.2016.09.043.

Yaacob E. N., J. Goethals, A. Bajek, K. Dierckens, P. Bossier, B. G. De Geest, D. Vanrompay. 2017. Preparation and Characterization of Alginate Microparticles Containing a Model Protein for Oral Administration in Gnotobiotic European Sea Bass (Dicentrarchus labrax) Larvae. *Mar Biotechnol (NY)* 19 (4): 391–400. doi: 10.1007/s10126-017-9758-4. Epub 2017 Jun 22. PMID: 28643227.

Yaacob, Eamy Nursaliza, Parisa Norouzitallab, Bruno G. De Geest, Aline Bajek, Kristof Dierckens, Peter Bossier, and Daisy Vanrompay. 2020. "Recombinant DnaK Orally Administered Protects Axenic European Sea Bass Against Vibriosis." *Frontiers in Immunology* 10. doi: 10.3389/fimmu.2019.03162.

Yarahmadi, Peyman, Hamed Ghafari Farsani, Amin Khazaei, Mohammad Khodadadi, Ghasem Rashidiyan, and M. Ali Jalali. 2016. "Protective effects of the prebiotic on the immunological indicators of rainbow trout (Oncorhynchus mykiss) infected with Aeromonas hydrophila." *Fish & Shellfish Immunology* 54:589–597. doi: 10.1016/j.fsi.2016.05.010.

Yengkhom, O., K. S. Shalini, P. A. Subramani, and R. D. Michael. 2018. "Non-specific immunity and disease resistance are enhanced by the polysaccharide fraction of a marine chlorophycean macroalga inOreochromis niloticus(Linnaeus, 1758)." *Journal of Applied Ichthyology* 34 (3):556–567. doi: 10.1111/jai.13606.

Yousefi, M., H. Ghafarifarsani, S. H. Hoseinifar, G. Rashidian, and H. Van Doan. 2021. "Effects of dietary marjoram, Origanum majorana extract on growth performance, hematological, antioxidant, humoral and mucosal immune responses, and resistance of common carp, Cyprinus carpio against Aeromonas hydrophila." *Fish Shellfish Immunol* 108:127–133. doi: 10.1016/j.fsi.2020.11.019.

Yousefi, Morteza, Saeed Zahedi, Miriam Reverter, Hossein Adineh, Seyyed Morteza Hoseini, Hien Van Doan, Ehab R. El-Haroun, and Seyed Hossein Hoseinifar. 2021. "Enhanced growth performance, oxidative capacity and immune responses of common carp, Cyprinus carpio fed with Artemisia absinthium extract-supplemented diet." *Aquaculture* 545. doi: 10.1016/j.aquaculture.2021.737167.

Yousefi, Morteza, Mehdi Ahmadifar, Sedigheh Mohammadzadeh, Naser Kalhor, Delaram Eslimi Esfahani, Azadeh Bagheri, Nika Mashhadizadeh, Mohsen Shahriari Moghadam, and Ehsan Ahmadifar. 2022. "Individual and combined effects of the dietary Spirulina platensis and Bacillus licheniformis supplementation on growth performance, antioxidant capacity, innate immunity, relative gene expression and resistance of goldfish, Carassius auratus to Aeromonas hydrophila." *Fish & Shellfish Immunology* 127:1070–1078. doi: 10.1016/j.fsi.2022.07.015.

Yu, Wei, Yukai Yang, Qicun Zhou, Xiaolin Huang, Zhong Huang, Tao Li, Qiaer Wu, Chuanpeng Zhou, Zhenhua Ma, and Heizhao Lin. 2022. "Effects of dietary Astragalus polysaccharides on growth, health and resistance to Vibrio harveyi of Lates calcarifer." *International Journal of Biological Macromolecules* 207:850–858. doi: 10.1016/j.ijbiomac.2022.03.176.

Zanon, Ricardo, Brunno Cerozi, J. Eurico Cyrino, and Tarcila Silva. 2014. "Dietary Levamisole as Immunostimulant for Striped Surubim, Pseudoplatystoma reticulatum." *Journal of the World Aquaculture Society* 45:672–680. doi: 10.1111/jwas.12156.

Zanuzzo, Fábio S., Rafael E. Sabioni, Luz Natalia F. Montoya, Gisele Favero, and Elisabeth C. Urbinati. 2017. "Aloe vera enhances the innate immune response of pacu (Piaractus mesopotamicus) after transport stress and combined heat killed Aeromonas hydrophila infection." *Fish & Shellfish Immunology* 65:198–205. doi: 10.1016/j.fsi.2017.04.013.

Zhuo, Li-Chao, Dayang Nur Jazlyn binti Abang Zamhari, Annita Seok Kian Yong, Rossita Shapawi, and Yu-Hung Lin. 2021. "Effects of fermented lemon peel supplementation in diet on growth, immune responses, and intestinal morphology of Asian sea bass, Lates calcarifer." *Aquaculture Reports* 21. doi: 10.1016/j.aqrep.2021.100801.

Index

Page numbers in **bold** indicate tables.

A. salmonicida, 8, 18, 35, 36, 79, 96, 97, 126, 140, 141, 190, 191
A. veronii, 143, 146
Acinetobacter, 147, 155
Acquired immunity, 18, 59, 80, 186, 194, 218, 222, 227
Adaptive immune system, 3, 15, 17, 18, 30, 44, 60, 62, 104, 111, 120, 132, 133, 136, 224, 233
Adaptive immunity, 14, 39, 59, 65, 80, 87, 88, 173, 175, 187, 193, 225, 227, 230
Adjuvants, 9, 25, 26, 38, 69, 74, 80, 101, 106, 111, 121, 205, 213, 214, 219, 220, 223, 224, 226, 228, 230, 232, 233
Administration, 8, 11, 12, 15, 19, 22, 25, 27, 35, 36, 37, 38, 51, 53, 54, 63, 67, 68, 80, 81, 82, 83, 85, 87, 88, 95, 96, 98, 99, 100, 106, 111, 115, 123, 124, 131, 137, 139, 140, 142, 143, 144, 145, 146, 147, 149, 150, 151, 153, 154, 155, 156, 158, 159, 160, 161, 162, 165, 175, 178, 180, 181, 183, 184, 191, 192, 193, 200, 202, 204, 205, 222, 225, 226, 228, 235, 239, 242, 243, 245, 246, 247, 251, 256, 259
Aerococcus, 143
Aeromonas, 13, 15, 17, 18, 22, 25, 26, 27, 35, 37, 41, 42, 51, 54, 55, 65, 74, 82, 85, 86, 87, 88, 90, 95, 96, 98, 99, 100, 101, 102, 106, 107, 112, 113, 114, 120, 121, 122, 123, 124, 125, 126, 127, 130, 131, 133, 134, 135, 137, 138, 139, 140, 143, 144, 146, 147, 150, 152, 153, 154, 155, 157, 158, 159, 160, 161, 162, 164, 165, 167, 180, 181, 182, 184, 188, 190, 192, 194, 198, 202, 207, 221, 224, 227, 229, 231, 232, 239, 242, 245, 247, 250, 253, 254, 255, 256, 258, 259, 260, 262
Aeromonas hydrophila, 18, 22, 35, 42, 51, 82, 85, 87, 88, 95, 98, 107, 113, 120, 121, 122, 124, 130, 134, 139, 150, 152, 188, 190, 192, 194, 221, 224, 239, 245, 247
Aeromonas media, 147, 157, 159
Aeromonas salmonicida, 13, 35, 37, 82, 88, 106, 146, 227, 250
Aeromonas sobria, 96, 144, 146, 254
Aflatoxin, 6, 14, 106
Agarivorans albus, 147
Agricultural pesticide, 23
Agrochemicals, 5
Algae, 30, 39, 59, 66, 67, 69, 70, 72, 74, 77, 93, 98, 102, 133, 158, 234, 246, 251
Alginate, 69, 73, 98, 112, 148, 154, 158, 251
Aliivibrio, 146, 159, 204, 209
Alkaloids, 67, 126, 209, 216
Alkylating agents, 214
Allium sativum, 19, 51, 53, 54, 120, 130, 219, 236, 238, 239, 250, 251, 256, 260
Aloe vera, 98, 120
Alteromonas, 146
Alum Magnesium, **218**
Aluminum hydroxide, 218
Andrographis paniculata, 27, 234, 236, 258
Annually, 233, 247
Anthelmintic drug, 22, 246
Antibacterial, 19, 22, 27, 33, 64, 65, 66, 67, 68, 71, 73, 75, 78, 98, 119, 120, 121, 122, 123, 124, 129, 130, 134, 135, 147, 157, 172, 175, 186, 191, 202, 204, 215, 243, 250, 255
Antibacterial immunity, **33**
Antibiotic drugs, 20
Antibiotic resistant bacteria, 105
Antibiotic(s), 6, 11, 13, 15, 16, 17, 18, 19, 20, 22, 25, 26, 29, 38, 41, 42, 45, 51, 60, 62, 65, 67, 68, 71, 77, 78, 79, 80, 81, 89, 90, 93, 98, 103, 105, 111, 119, 129, 136, 137, 138, 152, 155, 156, 169, 200, 204, 213, 233, 235, 247
Antibody, 4, 7, 8, 9, 20, 21, 22, 29, 30, 33, 36, 37, 40, 51, 61, 62, 78, 90, 91, 95, 97, 98, 99, 104, 105, 121, 124, 139, 148, 149, 152, 170, 175, 178, 181, 188, 190, 194, 196, 198, 199, 200, 206, 213, 214, 215, 217, 219, 222, 224, 227, 235, 243, 246, 250
Antibody responses, 7, 22, 105, 206
Anticancer, 40, 65, 68, 70, 72, 73, 102, 133
Anti-diabetic, 98, 215
Antifungal, 98, 119, 120, 121, 122, 124, 129, 131, 204, 238
Antigen, 4, 8, 10, 20, 21, 26, 30, 32, 33, 37, 60, 61, 62, 70, 81, 90, 91, 94, 102, 104, 107, 124, 133, 136, 149, 159, 162, 186, 188, 190, 193, 196, 205, 206, 213, 217, 218, 219, 220, 221, 224, 225, 226, 227, 228, 230, 249, 251
Antigen processing, 21, 30, 90, 107
Antigen-presenting cells, 61
Antimetabolites, 119
Antimicrobial, 3, 4, 11, 27, 33, 45, 60, 62, 64, 65, 68, 71, 72, 73, 74, 75, 77, 78, 88, 89, 91, 93, 95, 98, 102, 103, 107, 108, 109, 110, 113, 114, 120, 122, 129, 131, 132, 136, 141, 142, 145, 149, 153, 160, 173, 175, 191, 196, 245
Antimicrobial capability, 120
Antimicrobial peptide, 33, 107, 191, 196
Antimicrobial peptides, 62, 71, 77, 136, 175
Antimicrobial resistance, 45, 89
Antioxidant, 16, 23, 36, 37, 41, 42, 43, 44, 51, 52, 54, 55, 65, 67, 80, 97, 100, 104, 106, 108, 109, 110, 112, 113, 114, 115, 120, 121, 122, 123, 124, 125, 128, 129, 130, 131, 132, 133, 134, 135, 139, 140, 141, 142, 143, 144, 146, 153, 154, 156, 157, 165, 166, 170, 171, 172, 173, 175, 176, 177, 179, 180, 181, 182, 183, 184, 185, 193, 200, 201, 202, 203, 204, 205, 219, 238, 239, 242, 243, 246, 247, 250, 253, 254, 255, 256, 257, 258, 259, 260, 261, 262
Antiparasitic, 65, 119, 120, 121, 124
Antiproteases, 61, 247
Antistress, 110, 119, 120
Antiviral, 9, 33, 35, 65, 66, 68, 69, 98, 119, 120, 121, 129, 131, 145, 162, 174, 183, 191, 194, 197, 198, 199, 220, 224, 231, 250
Antiviraltherapies, 213
Apoptosis, 6, 16, 17, 33, 34, 35, 62, 69, 73, 74, 107, 134, 172, 173, 174, 176, 178, 179, 183, 185, 192, 193, 204
Aquaculture, 7, 8, 11, 12, 13, 14, 15, 16, 17, 18, 19, 22, 23, 24, 25, 26, 27, 29, 37, 38, 39, 42, 44, 45, 46, 47, 49, 51, 52, 53, 54, 57, 76, 77, 78, 80, 81, 82, 83, 84, 86, 87, 88, 89, 90, 91, 92, 93, 95, 96, 97, 98, 99, 100, 101, 103, 104, 105, 106, 107, 109, 110, 111, 112, 117, 119, 121, 122, 123, 124, 125, 129, 130, 131, 132, 133, 136, 137, 138, 139, 141, 143, 144, 145, 146, 148, 150, 152, 153, 154, 155, 156, 157, 158, 159, 160, 161, 162, 163, 164, 165, 166, 167, 169, 177, 178, 179, 182, 183, 186, 192, 193, 195, 198, 200, 201, 202, 203, 205, 206, 207, 208, 209, 221, 222, 223, 225, 226, 227, 228, 229, 233, 234, 235, 239, 243, 245, 247, 250, 251, 253, 254, 255, 256, 257, 258, 259, 260, 261, 262
Aquaculture industry, 7, 27, 29, 38, 45, 77, 80, 89, 99, 111, 129, 132, 136, 137, 143, 150, 153, 183, 192, 193, 200, 206, 233, 239, 247
Aquafeed, 6, 108, 121, 125, 171, 172, 235, 250
Aquaneem, 120
Arabinoxylan, 136, 157, 160
Aremonas hydrophila, 219
Aromonas media, 147
Artemia, 142, 143, 150, 231
Arthrobacter, 146, 147, 166
Asian sea bass, 141, 144, 153, 154, 262
Aspergillus niger, 151

Aspergillus oryzae, 148, 158, 170, 182, 243
Astragalus membranaceus, 51, 202
Atlantic cod, 39, 52, 53, 147, 230
Atlantic salmon, 8, 9, 12, 15, 16, 22, 27, 34, 35, 36, 38, 53, 65, 66, 80, 85, 86, 87, 88, 91, 101, 106, 108, 138, 140, 141, 143, 146, 150, 151, 157, 159, 166, 170, 171, 175, 179, 181, 184, 191, 193, 194, 195, 198, 199, 222, 224, 225, 227, 228, 230, 231, 232, 246, 247, 249, 251, 253, 259, 260, 261
Autoantibodies, 20
Autoimmune, 3, 5, 6, 20, 24, 60, 62, 69, 94, 104, 213, 214, 221, 229
Autoimmune diseases, 3, 5, 60, 213, 214, 229
Auto-immune disorders, 104
Azadirachta indica, 19, 120, 129, 215, 248, 250
Azathioprine, 214

B. aerius, 138, 153
B. amyloliquefaciens, 138, 140, 141, 153, 242
B. aryabhattai, 141, 142
B. atrophaeus, 138
B-cell(s), 23, 30, 33, 104, 189, 214, 217
B cell receptor, 30, 60
B. clausii, 140
B. firmus, 141
B. flexus, 138, 141, 142
B. fusiformimis, 141
B. infantis, 141
B. polymyxa, 141
B. velezensis, 138, 140, 141, 153
Bacillus altitudinis, 138
Bacillus cereus, 114, 138, 155
Bacillus circulans, 138, 166
Bacillus coagulans, 138
Bacillus licheniformis, 95, 138, 155, 157, 163, 236, 239, 242, 254, 262
Bacillus megaterium, 138
Bacillus pumilus, 138, 162, 167, 261
Bacillus subtilis, 96, 138, 139, 151, 153, 154, 155, 156, 157, 158, 159, 160, 161, 162, 163, 165, 166, 167, 168, 239, 243, 250
Bacillus tequilensis, 138
*Bacillus th*uringiensis, 138, 154, 164
Bacterial derivatives, 145
Bactericidal, 5, 7, 9, 19, 22, 36, 45, 87, 90, 95, 105, 106, 107, 108, 114, 124, 128, 140, 141, 142, 144, 147, 152, 153, 175, 204, 205, 222, 235, 238, 242, 246, 247
Bactericidal activities, 7, 36, 95, 105, 106, 107, 108, 114, 124, 128, 140, 144, 204, 235, 247
Barramundi, 219, 251, 252, 261
Bay scallop, 147
Bifidobacteria, 152
Bifidobacterium bifidium, 143
Bighead Catfish, 155
Bioactive compounds, 245
Biochemical, 4, 7, 14, 16, 36, 41, 53, 112, 113, 115, 122, 124, 131, 133, 134, 151, 156, 157, 159, 164, 170, 176, 177, 182, 184, 200, 202, 204, 242, 243, 253, 254, 256, 260
Bioencapsulation, 226, 251
Biological response modifiers, 29
Biomarkers, 3, 14, 15, 54, 171, 192, 202, 203, 204, 253, 256
Black tiger shrimp, 120, 121, 141, 142, 144, 145, 149, 162, 163, 165, 166, 172, 222
Blood, 5, 6, 8, 14, 15, 21, 28, 34, 35, 41, 47, 48, 50, 53, 54, 55, 60, 61, 62, 65, 67, 70, 71, 74, 81, 90, 91, 94, 95, 96, 98, 102, 104, 108, 112, 119, 120, 123, 124, 128, 130, 135, 148, 159, 161, 173, 176, 178, 179, 181, 184, 188, 190, 194, 195, 200, 202, 204, 219, 222, 224, 225, 230, 232, 235, 239, 243, 245, 246, 250, 257
Blue mussel, 23, 147
Brevibacillus agri, 138
Brochothrix thermosphacta, 144

Cancer, 24, 26, 39, 59, 63, 64, 69, 70, 72, 73, 74, 75, 100, 132, 213, 214, 215, 217, 218, 219, 220, 229, 231
Carcinogenic, 24
Carnobacterium, 143
Carnobacterium inhibens, 143
Carnobacterium maltaromaticum, 143
Carotenoids, 69, 128, 177, 179, 182, 242, 261
Carrageenan, 66, 70, 73, 75, 98, 112
Caspase-3, 6, 24, 203
Catalase, 37, 108, 124, 139, 140, 142, 143, 145, 150, 176, 235, 238, 242, 243, 245, 247
Catfish, 5, 10, 12, 14, 34, 35, 42, 51, 80, 85, 86, 88, 96, 97, 100, 102, 106, 107, 112, 113, 114, 115, 125, 134, 135, 138, 140, 147, 150, 156, 161, 162, 171, 172, 175, 176, 177, 180, 182, 184, 185, 192, 194, 198, 199, 204, 205, 227, 234, 235, 236, 237, 238, 239, 240, 241, 243, 244, 245, 248, 250, 251, 252, 253, 254, 256, 258, 259, 261
Catla, 90, 109, 114, 124, 133, 140, 156, 224, 234
CCL22 gene, **21**
CD26 expression, **20**
CD8+, 71, 190, 216, 218, 219
CD8a, 224
Cell differentiation, 33, 71, 190, 198, 217, 218
Cell-mediated defence, 4
Cell-mediated reactions, 18
Cellular immunity, 27, 78
Cellulose, 138, 152
Cell-wall skeletons, 218
Challenge, 11, 17, 34, 35, 54, 86, 87, 99, 103, 106, 107, 108, 109, 125, 141, 142, 144, 145, 146, 147, 148, 149, 152, 154, 158, 159, 162, 163, 164, 165, 171, 172, 174, 175, 176, 179, 180, 181, 214, 224, 225, 235, 238, 239, 242, 243, 245, 251, 259, 260
Channel catfish, 10, 22, 27, 31, 36, 150, 172, 180, 192, 195, 197, 198, 219, 225, 229, 232
Chemical agents, 7
Chemicals, 19, 23, 25, 43, 44, 51, 60, 64, 65, 68, 69, 70, 71, 77, 89, 90, 97, 105, 111, 119, 121, 169, 213, 218, 222, 233, 239, 252
Chemokines, 30, 33, 68, 71, 94, 124, 188, 193, 194, 195, 196
Chemokinesis, 12, 99, 222
Chemotherapeutic agents, 22, 29
Chemotherapy, 17, 26, 39, 44, 54, 68, 70, 73, 77, 83, 84, 132, 136, 152, 213, 217, 218, 222, 229, 231
Chinese shrimp, 142
Chinese white shrimp, 149
Chitin, 8, 10, 11, 25, 33, 51, 60, 66, 80, 82, 86, 87, 97, 103, 105, 106, 152, 221, 222
Chitinase, 138
Chitooligosaccharide, 152
Chitosan, 8, 25, 33, 66, 71, 75, 81, 82, 85, 86, 87, 97, 101, 103, 112, 113, 148, 152, 158, 160, 202, 204, 222, 226, 234, 242, 252, 253, 257
Chromium, 28, 37, 42, 92, 124, 134, 179
Chronic peritonitis, 219
Cirrhinus, 37, 42, 90, 148, 165, 176, 183
Clonorchis sinensisis, 149
Common carp, 5, 6, 14, 15, 16, 17, 27, 34, 35, 53, 80, 85, 95, 96, 101, 106, 113, 124, 125, 129, 133, 134, 135, 139, 140, 143, 147, 150, 151, 152, 153, 158, 159, 161, 162, 164, 174, 179, 180, 183, 184, 189, 190, 192, 193, 199, 204, 222, 234, 235, 238, 239, 242, 243, 247, 251, 254, 255, 257, 261, 262
Complement system, 4, 60, 61, 62, 67, 78, 99, 105, 124, 136, 193, 235, 246
Consumer, 196
Consumption, 196
Corticosteroids, 5, 35, 46, 103, 129
Cortisol, 8, 35, 36, 37, 41, 42, 43, 46, 47, 48, 49, 50, 51, 52, 53, 54, 95, 102, 104, 109, 112, 119, 127, 128, 178, 180, 183, 184, 202, 229, 246, 258, 259
CpG–Toll-like receptor 9 agonist, 220
C-reactive protein, 3, 12, 61, 82, 88, 90, 183, 188
Cyanobacteria, 6, 16, 68
Cyanogenic glycosides, 6
Cyclophosphamide, 59, 70, 73, 214
Cyclosporine, 59, 214, 230
Cyprinus carpio, 5, 11, 14, 15, 16, 17, 27, 35, 53, 54, 74, 79, 85, 87, 88, 91, 101, 107, 112, 113, 126, 129, 133, 134, 135, 139, 150, 151, 152, 153, 154, 156, 158, 159, 160, 161, 162, 163, 164, 167, 172, 173, 180, 181, 182, 183, 184, 185, 192, 193, 194, 195, 197, 208, 231, 232, 235, 236, 240, 247, 254, 255, 257, 259, 261, 262
Cytochrome P450, 71
Cytokine, 33, 100, 188, 190, 195, 197, 198, 199, 231, 232, 253, 256
Cytokine inhibitors, 214
Cytotoxic agents, 214
Cytotoxic cells, 23, 61, 62, 90, 104, 194, 198, 201
Cytotoxic T cells, 20, 30, 59, 61, 188, 217, 218, 219

Damage-associated molecular pattern, 218
Danger-associated molecular patterns, 32
Dectins, 253
Defense, 14, 18, 24, 29, 30, 37, 38, 41, 42, 59, 60, 61, 65, 66, 70, 71, 73, 86, 91, 106, 112, 119, 124, 135, 137, 166, 171, 172, 174, 177, 190, 198, 201, 204, 205, 228
Defensin, 65, 71, 74, 139, 140, 175, 191
Dendritic cells, 5, 20, 27, 34, 59, 61, 63, 65, 68, 75, 94, 149, 160, 188, 191, 214, 217, 225
Diadromous, 247, 251
Dichlorodiphenyldichloroethane, 222, 229
Dietary, 5, 6, 9, 10, 13, 14, 15, 25, 26, 28, 29, 30, 36, 37, 38, 41, 42, 51, 52, 53, 54, 69, 80, 82, 86, 87, 88, 89, 90, 95, 96, 98, 100, 101, 102, 106, 109, 110, 111, 112, 113, 114, 115, 120, 121, 122, 123, 125, 129, 130, 131, 132, 133, 134, 135, 140, 149, 150, 151, 153, 154, 155, 156, 157, 158, 159, 160, 161, 162, 163, 164, 165, 166, 167, 169, 172, 173, 175, 176, 177, 179, 180, 181, 182, 183, 184, 185, 202, 204, 222, 227, 231, 232, 235, 238, 239, 242, 243, 245, 246, 247, 250, 251, 253, 254, 255, 256, 257, 258, 259, 260, 261, 262
Dietary supplementation, 36, 37, 41, 51, 52, 95, 96, 101, 110, 112, 120, 122, 123, 130, 133, 150, 151, 153, 154, 156, 157, 158, 159, 161, 164, 165, 238, 242, 243, 245, 247, 250, 253, 260
Diets, 19, 22, 23, 27, 30, 35, 36, 37, 38, 52, 78, 81, 83, 92, 96, 99, 107, 110, 113, 114, 115, 122, 130, 133, 134, 145, 148, 155, 157, 158, 160, 161, 162, 164, 166, 167, 171, 172, 176, 177, 178, 179, 180, 182, 183, 200, 205, 221, 225, 227, 228, 231, 238, 239, 242, 243, 250, 253, 254, 257, 258, 259, 260, 261
Diferuloylmethane, 120
Disease-modifying anti-rheumatic drugs, 214
Diseases, 6, 7, 12, 13, 18, 20, 23, 25, 29, 30, 37, 38, 40, 45, 51, 54, 59, 60, 69, 70, 71, 72, 73, 76, 77, 78, 79, 80, 82, 83, 84, 85, 87, 89, 91, 92, 93, 94, 96, 99, 103, 104, 105, 106, 109, 111, 114, 119, 120, 121, 124, 130, 131, 136, 140, 143, 144, 145, 148, 153, 155, 169, 175, 179, 182, 186, 200, 201, 202, 205, 206, 213, 214, 219, 221, 222, 223, 226, 227, 230, 231, 233, 234, 235, 239, 250, 259
DNA damage, 6, 14, 43, 51
Dosage, 23, 81, 84, 99, 105, 111, 123, 128, 132, 136, 140, 143, 144, 145, 147, 148, 149, 226, 228, 230
Drug, 3, 12, 14, 42, 65, 68, 70, 72, 73, 74, 99, 105, 133, 186, 193, 200, 202, 205, 206, 214, 225, 226, 235, 253

E. coli, 66, 71, 108, 120, 188, 250
E. ictaluri, 18, 36, 88, 192, 251
E26 transformation-specific protein, 34
Eastern oyster, 147, 159, 164
Eclipta allea, 120
Edwardsiella, 10, 18, 22, 35, 38, 42, 79, 85, 88, 90, 96, 107, 123, 133, 138, 140, 175, 190, 192, 193, 197, 199, 225, 231, 235, 250
Edwardsiella ictalurid, 35
Efalizumb, 214
Efficacy, 4, 12, 22, 36, 38, 42, 44, 69, 78, 82, 83, 84, 86, 87, 93, 98, 99, 111, 122, 125, 131, 134, 137, 140, 141, 142, 144, 146, 148, 150, 153, 164, 165, 173, 201, 202, 204, 205, 220, 221, 223, 224, 226, 228, 231, 233
Eicosanoids, 91
Endotoxins, 8, 23
Ent. faecalis, 143, 144, 145
Ent. faecium, 143
Enterobacter, 146, 147, 157
Enterobacter cloacae, 147
Enterococcus, 138, 139, 143, 163, 167
Enterococcus faecium, 139, 143
Enterovibrio coralii, 146
Environment, 3, 5, 13, 22, 24, 25, 29, 38, 44, 46, 48, 49, 50, 66, 68, 76, 78, 84, 92, 96, 101, 119, 125, 129, 132, 137, 142, 152, 153, 169, 200, 205, 206, 218, 224, 233, 246, 247
Epizootic ulcerative syndrome, 120
Essential oils, 9, 19, 52, 98, 108, 134, 214
European sea bass, 6, 16, 49, 52, 146, 150, 161, 165, 193, 196, 248, 249, 256, 261
Extracellular microbes, 31, 32, 33

FAO, 45, 76, 87, 88, 93, 105, 186, 196, 234, 239, 251, 255, 256, 257
FCA, 19, 23, 79, 88, 96, 106, 218, 220, 222, 223, 224
Feed additives, 7, 14, 16, 44, 45, 51, 52, 78, 101, 119, 125, 131, 133, 152, 228, 235, 239, 242
Feed conversion ratio, 124, 149, 221
Feed efficiency, 29, 35, 38, 81, 155, 163, 242
Finfish, 15, 16, 25, 52, 101, 111, 132, 133, 137, 138, 139, 140, 141, 143, 146, 147, 148, 155, 163, 165, 223, 234, 235, 239, 243, 247, 251
Fish cytokines, 17, 32, 40, 41, 186, 198, 199
Fish farming, 51, 60, 77, 81, 85, 90, 99, 239, 246
Fish Health, 4, 6, 8, 25, 30, 36, 44, 46, 60, 62, 64, 68, 86, 87, 90, 94, 98, 100, 104, 106, 120, 122, 124, 152, 170, 186, 187, 191, 202, 204, 214, 234
Fish Immune System, 41, 60, 104, 187
Fishery, 24, 90
FK-565, 23, 33, 106, 222
Flagellin, 31, 138, 220, 221, 224, 230, 248, 250, 256
Flavobacteria, 120
Flavobacterium psychrophilum, 147
Flavonoids, 19, 22, 23, 26, 67, 98, 108, 119, 122, 124, 126, 127, 128, 201, 202, 214, 235, 238
Flavouring agents, 120
Flexibacter columnaris, 35
Folin-Ciocalteu method, 122
Forage species, 24
Freund's Complete Adjuvant, 23, 79, 96, 220
Freund's Incomplete Adjuvant, **220**
Fructooligosaccharide, 113, 153, 154, 157, 162
Fucoidan, 14, 66, 69, 74, 114, 243, 253, 259
Fungal, 10, 12, 16, 35, 72, 89, 95, 97, 101, 103, 120, 122, 129, 144, 147, 178, 200, 205, 232, 253
Furunculosis, 96, 146, 158, 226, 231

Galactooligosaccharide, 144, 157
Gamma, 34, 35, 39, 40, 108, 139, 191, 195, 197, 199, 214, 217, 259
GATA-binding protein-3, 33
Geniposide, 238
Genotoxicity, 6
Giant freshwater prawn, 26, 142, 145, 147
GIFT tilapia, 120, 129, 134, 185
Gill-associated lymphoid tissue, 224, 227
Gilthead sea bream, 15, 54, 140, 151, 155, 172, 178, 180, 183, 184
Ginger, 19, 121, 252
Ginseng, 39, 120, 214, 215, 247
Glucan, 10, 11, 12, 14, 16, 19, 22, 23, 25, 27, 35, 41, 42, 51, 52, 54, 63, 67, 68, 70, 74, 79, 80, 81, 82, 85, 86, 87, 88, 95, 100, 102, 103, 105, 106, 107, 108, 111, 112, 113, 136, 142, 150, 153, 156, 158, 164, 166, 167, 173, 174, 175, 180, 181, 182, 183, 184, 221, 222, 224, 225, 229, 231, 234, 236, 239, 244, 246, 247, 249, 250, 251, 254, 256, 258, 260
Glucocorticoid receptor, 228, 232
Glucocorticoids, 216
Glucomannan, 150
Glutathione, 37, 107, 108, 120, 124, 128, 139, 150, 171, 173, 176, 177, 202, 242
Glutathione peroxidase, 124, 139, 150, 171, 173, 176, 177, 202, 242
Glycoprotein, 4, 61, 62, 71, 95, 97, 102, 148, 190, 222, 230
Gobal warming, 23, 154
Granulocytes, 4, 60, 61, 72, 90, 91, 94, 124, 136, 180, 187, 188
Granulomas, 219
Grass carp, 34, 36, 39, 40, 41, 42, 86, 94, 109, 114, 139, 143, 146, 149, 157, 159, 165, 166, 172, 173, 175, 176, 180, 181, 183, 184, 185, 192, 193, 198, 247
Great sturgeon, 125, 144, 163
Growth hormone, 7, 8, 16, 17, 35, 97, 115, 179, 182, 185, 234
Gut- associated lymphoid tissue, 153, 224

Haematocrit, 108, 112, 128, 148, 235, 239, 242, 246
Haematological, 5, 102, 112, 113, 114, 120, 124, 130, 134, 148, 151, 154, 156, 167, 242, 243, 256
Haematology, 28, 148, 165
Hard clam, 147
Head kidney, 5, 8, 9, 15, 16, 35, 36, 40, 65, 82, 87, 91, 108, 149, 171, 172, 173, 174, 175, 176, 178, 179, 180, 181, 185, 194, 199, 224, 228, 230, 242, 246, 247, 250, 259, 261

Health management, 14, 25, 27, 42, 77, 80, 84, 93, 106, 111, 112, 117, 144, 152, 222, 239
Heat shock protein, 100, 108, 113, 114, 138, 139, 140, 142, 143, 144, 150, 177, 202, 218, 219, 232, 235
Helper T, 30, 197, 199, 217, 224
Helper T1 cells, 214
Helper T2 cells, 214
Hemicelluloses, 138
Hemocyanin, 64, 72, 172, 204, 222
Herbal, 22, 27, 38, 42, 119, 120, 121, 123, 124, 125, 126, 128, 129, 130, 131, 132, 133, 201, 230, 257, 259
Herbal compounds, 120
Heterocytes, 239
Histocompatibility, 3, 4, 39, 60, 61, 62, 69, 95, 102, 128, 140, 196
HLA-DR, **21**
Homeostasis, 24, 46, 48, 68, 70, 72, 95, 97, 133, 169, 177, 178, 226
Honey, 51, 152, 154, 163, 166
Hormones, 6, 8, 9, 19, 46, 97, 106, 169, 171, 173, 178, 222
Host response, 29, 34, 92, 184
HSPs, 46, 47, 109, 177, 219
Humoral defence, 3, 171, 172
Humoral immunity, 27, 59, 77, 78, 79, 148, 172, 216, 218, 220, 242
Humpback grouper, 144
Hybrid surubim, 151, 161, 162, 166
Hyperthermia, 109, 112
Hypothalamic-pituitary-interrenal, 109

Ichthyophthirius multifiliis, 144, 162, 260
IFN, 5, 6, 9, 20, 21, 30, 31, 32, 33, 34, 35, 39, 40, 41, 59, 62, 65, 66, 67, 68, 91, 94, 95, 107, 108, 136, 143, 148, 171, 172, 178, 188, 190, 193, 194, 199, 217, 224, 235, 238, 239, 242, 243, 246, 250
lgG, 21
lgM, 21
IgM titer, 36, 149
IL-1b, 109, 143, 144, 146, 224
IL-6, 5, 20, 21, 33, 40, 59, 65, 66, 67, 68, 69, 70, 71, 95, 143, 144, 146, 171, 172, 173, 174, 177, 188, 189, 190, 191, 195, 196, 197, 198, 215, 216, 224, 226, 239, 250
IL-8, 21, 40, 65, 70, 95, 127, 128, 139, 143, 146, 149, 171, 172, 188, 190, 205, 224, 235, 239, 243, 247, 250, 253
IL-I-ß, 20
Illness, 6, 22, 43, 61, 69, 77, 78, 79, 80, 92, 213, 218, 226, 227, 233
Immersion, 8, 11, 13, 22, 25, 82, 83, 86, 88, 99, 100, 140, 141, 142, 143, 144, 148, 154, 223, 225, 226, 245, 250
Immune adjuvants, 4
Immune response, 3, 4, 5, 6, 9, 10, 11, 12, 13, 14, 15, 16, 17, 18, 19, 22, 23, 24, 25, 27, 28, 29, 30, 33, 36, 37, 38, 40, 41, 42, 43, 44, 51, 52, 53, 55, 59, 60, 61, 62, 64, 69, 72, 73, 77, 79, 81, 82, 85, 86, 87, 88, 90, 91, 94, 95, 96, 99, 100, 101, 102, 103, 104, 105, 106, 109, 111, 113, 114, 115, 121, 123, 124, 125, 130, 131, 134, 135, 136, 140, 141, 147, 149, 152, 153, 154, 155, 156, 158, 159, 160, 161, 162, 163, 165, 166, 167, 168, 170, 172, 173, 174, 175, 176, 177, 179, 180, 181, 182, 183, 184, 185, 186, 187, 188, 189, 190, 192, 193, 194, 195, 196, 198, 199, 200, 201, 202, 203, 204, 205, 206, 213, 214, 217, 218, 219, 221, 223, 224, 227, 229, 230, 231, 232, 234, 235, 238, 239, 242, 245, 246, 247, 250, 253, 254, 255, 256, 257, 258, 259, 260, 261, 262
Immune system, 3, 4, 5, 6, 7, 8, 9, 10, 11, 12, 13, 14, 15, 16, 17, 18, 19, 20, 22, 23, 24, 25, 26, 27, 29, 30, 33, 34, 35, 37, 38, 39, 40, 41, 43, 44, 46, 51, 55, 59, 60, 61, 62, 64, 65, 66, 67, 68, 69, 70, 72, 74, 77, 78, 79, 80, 81, 82, 83, 85, 86, 87, 90, 91, 92, 93, 94, 95, 96, 98, 100, 101, 102, 103, 104, 105, 106, 108, 109, 111, 119, 124, 133, 136, 137, 138, 145, 147, 150, 151, 152, 157, 159, 169, 171, 172, 175, 177, 178, 179, 180, 181, 182, 184, 185, 186, 187, 190, 192, 193, 194, 195, 196, 199, 200, 205, 206, 213, 214, 217, 218, 219, 221, 222, 226, 227, 228, 229, 230, 233, 242, 243, 253, 255, 257, 258, 261
Immune-boosting herbs, 20
Immune-suppression, 119
Immunisation, 13, 81, 83, 91, 226, 229
Immunity, 3, 6, 7, 8, 9, 11, 12, 13, 14, 15, 16, 17, 18, 19, 22, 23, 24, 26, 27, 28, 30, 33, 34, 36, 37, 38, 39, 40, 41, 42, 51, 54, 59, 61, 62, 64, 65, 67, 68, 70, 72, 73, 78, 79, 80, 81, 82, 85, 86, 87, 88, 89, 90, 91, 92, 93, 95, 96, 97, 98, 99, 100, 101, 102, 104, 106, 107, 109, 110, 111, 112, 113, 114, 121, 123, 124, 125, 129, 130, 131, 133, 134, 135, 136, 137, 138, 139, 140, 141, 142, 143, 144, 145, 146, 147, 148, 149, 150, 153, 155, 156, 157, 158, 159, 160, 161, 162, 163, 164, 165, 166, 167, 169, 170, 171, 172, 173, 175, 176, 178, 179, 180, 183, 184, 185, 186, 187, 188, 195, 196, 197, 198, 201, 202, 203, 204, 205, 215, 217, 218, 220, 221, 222, 227, 228, 229, 230, 231, 233, 235, 238, 239, 242, 245, 246, 253, 254, 255, 257, 258, 260, 261, 262
Immunization, 21, 30, 62, 148, 149, 158, 224, 225, 246
Immunoadjuvants, 59, 214, 218, 219, 220, 226, 229
Immunocompetence, 28, 51, 106, 112, 181, 184
Immunodeficiencies, 214
Immunoglobulin E, 214
Immunoglobulin G, 214
immunoglobulin M (IgM), 5, 202, 239, 242
Immunological markers, 19, 79, 81
Immunological memory, 30, 61, 90
Immunomodulation, 23, 24, 25, 38, 42, 77, 79, 81, 82, 84, 108, 111, 145, 146, 151, 153, 169, 170, 171, 173, 200, 202, 213, 231
Immunomodulators, 3, 4, 8, 19, 23, 25, 29, 30, 33, 34, 35, 37, 38, 42, 43, 44, 59, 66, 77, 78, 79, 80, 103, 105, 107, 111, 131, 173, 213, 214, 221, 222, 224, 225, 226, 228, 229
Immunostimulant, 4, 8, 9, 11, 12, 13, 18, 22, 23, 25, 37, 38, 45, 51, 66, 67, 69, 70, 78, 80, 81, 82, 95, 98, 99, 100, 101, 105, 106, 107, 108, 112, 124, 130, 167, 174, 183, 193, 197, 202, 219, 226, 228, 229, 230, 233, 234, 246, 247, 250, 251, 253, 254, 255, 256
Immunostimulation, 54, 82, 87, 99, 164
Immunostimulatory, 9, 10, 16, 19, 22, 38, 63, 64, 66, 80, 81, 82, 84, 98, 111, 112, 122, 138, 139, 140, 141, 142, 143, 145, 146, 149, 153, 157, 169, 218, 219, 221, 228, 230, 234, 235, 238, 243, 247, 250, 251, 261
Immuno-stimulatory complexes, 219
Immunosuppression, 4, 12, 15, 18, 37, 38, 45, 66, 72, 78, 99, 104, 105, 226, 228, 242
Immunosuppressive, 4, 5, 12, 34, 70, 77, 81, 105, 106, 178, 216, 218, 219, 222, 232
Immunotherapy, 59, 68, 72, 73, 100, 218, 229, 233
Infection, 4, 5, 6, 8, 9, 14, 15, 16, 17, 18, 19, 20, 22, 30, 33, 36, 37, 38, 39, 40, 41, 42, 43, 44, 45, 51, 54, 55, 62, 67, 77, 79, 80, 85, 86, 87, 89, 90, 91, 92, 93, 94, 95, 96, 97, 98, 100, 101, 102, 106, 107, 110, 112, 113, 114, 120, 121, 123, 124, 125, 130, 131, 133, 134, 135, 139, 140, 141, 142, 143, 144, 145, 146, 147, 148, 149, 153, 155, 156, 157, 158, 159, 160, 161, 162, 164, 165, 167, 168, 171, 172, 174, 175, 176, 177, 180, 181, 182, 186, 187, 188, 190, 191, 192, 193, 194, 196, 197, 198, 199, 202, 217, 219, 220, 221, 222, 226, 227, 228, 230, 233, 235, 238, 242, 243, 245, 246, 250, 251, 253, 254, 255, 256, 258, 259, 261, 262
Infectious, 7, 14, 18, 27, 34, 37, 51, 76, 77, 78, 79, 80, 84, 85, 89, 91, 92, 93, 94, 99, 100, 103, 104, 105, 109, 136, 141, 142, 144, 148, 150, 157, 161, 167, 169, 170, 175, 186, 191, 192, 194, 195, 198, 205, 206, 213, 218, 221, 222, 223, 230, 233, 234, 247, 259
Infectious hematopoietic necrosis, 198
Infectious pancreatic necrosis virus, 34, 148, 206
Inflammation, 26, 73, 208, 209
Inflammatory, 23, 101, 182, 209, 256
Inflammatory cytokines, 5, 6, 20, 64, 65, 66, 67, 68, 95, 143, 144, 146, 171, 172, 174, 175, 178, 179, 190, 191, 216, 219, 242, 251, 256
Injection, 8, 11, 12, 13, 14, 19, 22, 25, 26, 34, 37, 45, 63, 70, 79, 81, 82, 83, 99,

100, 123, 148, 172, 178, 183, 194, 206, 219, 220, 225, 226, 229, 230
Innate Immunity, 14, 157, 183
insulin-like growth factor-I, 8
Interferon, 4, 5, 7, 23, 29, 39, 41, 59, 60, 91, 94, 108, 121, 139, 149, 171, 172, 191, 193, 194, 195, 197, 198, 214, 224, 229, 259
Interferon alpha, 128
Interferon gamma, 39, 197, 259
Interleukin, 4, 5, 17, 30, 34, 37, 39, 40, 59, 66, 67, 74, 94, 102, 108, 124, 128, 171, 172, 173, 175, 177, 179, 188, 190, 191, 192, 194, 195, 196, 197, 198, 199, 202, 204, 206, 222, 224, 226, 229, 230, 231, 232, 235
Interleukin-1 Family, 188
Interleukin-1β, 67, 179, 188, 199, 232, 235
Interleukin-2 Family, 188
Interleukin-6, 17, 67, 74, 196, 197, 198, 199
Interleukin, 10, 173, 175, 199
Interleukin-11, 190
Interleukin-12, 32, 191
Interleukin-15, 188, 191
Interleukin-17 Family, 191
Interleukin-22 and -26, 191
Inulin, 136, 150, 151, 153, 155, 157, 158, 160, 161, 163, 167, 250
Isomaltooligosaccharide, 152
Isoprinosine, 214, 217

Japanese abalone, 145, 147
Japanese flounder, 19, 30, 31, 34, 36, 39, 79, 87, 95, 96, 97, 112, 115, 151, 167, 175, 181, 190, 192, 194, 197, 219, 220
Juvenile, 5, 16, 17, 25, 28, 35, 41, 45, 50, 51, 52, 53, 112, 114, 130, 133, 134, 152, 153, 154, 155, 157, 158, 161, 164, 165, 166, 168, 171, 172, 173, 174, 175, 176, 178, 180, 181, 182, 183, 184, 185, 231, 232, 245, 250, 251, 261

Kaempferol, 37, 85
Killing, 4, 34, 80, 106, 110, 204, 235
Koi carp, 139, 148, 158, 193
Koi herpesvirus, 148, 156
Kuruma shrimp, 141, 144, 145, 156, 160, 165

L. fermentum, 145
Labeo rohita, 8, 22, 27, 36, 37, 41, 42, 53, 54, 85, 87, 95, 97, 98, 101, 102, 106, 107, 108, 109, 112, 113, 114, 121, 122, 123, 125, 126, 128, 129, 130, 131, 133, 135, 139, 146, 152, 157, 158, 161, 163, 173, 183, 184, 190, 219, 231, 235, 236, 256, 260
Lactobacillus, 8, 52, 70, 138, 139, 143, 144, 145, 148, 149, 153, 154, 156, 157, 159, 161, 162, 163, 164, 165, 166, 167, 209, 239, 241, 243, 251, 252, 253, 261
Lactobacillus fructivorans, 138
Lactobacillus plantarum, 52, 138, 154, 156, 157, 159, 161, 162, 164, 165, 166, 241, 253
Lactobacillus rhamnosus, 139
Lactococcus, 96, 112, 113, 127, 138, 143, 144, 156, 159, 161, 162, 163, 166, 235, 239, 254
Lactococcus garvieae, 96, 144, 235
Lactoferrin, 7, 36, 38, 41, 80, 101, 105, 226, 231
Latent, 18, 45, 51, 93, 105
Lb. acidophilus, 143, 144, 145, 148
Lb. bulgaricus, 144, 145
Lb. casei, 143, 148
Lb. curvatus, 144
Lb. delbrueckii, 143, 144
Lb. pentosus, 145
Lb. sakei, 143, 144, 145
Lb. sporogenes, 148
Leaf extract, 14, 27, 87, 100, 108, 114, 120, 121, 124, 127, 131, 134, 135, 231, 235, 236, 238, 240, 247, 257, 258, 259
Lectin, 42, 61, 62, 72, 73, 87, 88, 150, 173, 188, 192, 195
Leopard grouper, 151
Leu. mesenteroides, 143
Leuconostoc, 143, 144
Lecytes, 15, 17, 19, 22, 28, 34, 35, 36, 37, 40, 41, 45, 47, 55, 59, 61, 94, 97, 98, 104, 121, 140, 143, 161, 164, 173, 175, 178, 182, 183, 194, 195, 222, 224, 225, 245, 246, 247, 250, 254, 255, 259, 260, 261
Leukotriene, 21
Levamisole, 7, 15, 22, 27, 33, 37, 38, 42, 54, 59, 80, 81, 82, 85, 86, 87, 88, 99, 100, 102, 105, 106, 109, 112, 114, 164, 217, 226, 234, 246, 247
Levan, 95, 101, 152, 158, 162, 236, 256
Lipopolysaccharide, 15, 16, 26, 39, 42, 43, 53, 55, 60, 66, 85, 86, 94, 96, 108, 111, 136, 142, 174, 184, 190, 250
Litopenaeus stylirostris, 164
Lobster, 142, 145, 156, 161, 162
LPS, 8, 16, 17, 19, 21, 26, 30, 31, 33, 34, 37, 39, 51, 60, 61, 66, 68, 69, 70, 73, 85, 86, 87, 96, 106, 126, 127, 128, 136, 160, 174, 190, 191, 216, 222, 225, 250
Lumpfish, 146
Lymphocyte gene expression, 214
Lymphocytes, 3, 8, 9, 10, 21, 23, 30, 34, 35, 36, 59, 60, 61, 62, 63, 64, 65, 67, 68, 70, 78, 82, 90, 91, 94, 95, 96, 97, 104, 105, 107, 108, 136, 140, 141, 143, 169, 177, 179, 181, 187, 188, 194, 201, 216, 217, 218, 219, 225, 227, 239, 242, 246, 247
Lymphoid organs, 4, 20, 29, 61, 91, 178, 187, 195, 196, 197, 198, 224
Lysozyme, 3, 5, 6, 7, 8, 9, 10, 12, 16, 19, 22, 23, 35, 36, 38, 41, 61, 67, 78, 79, 80, 81, 85, 91, 97, 98, 99, 105, 106, 107, 108, 110, 120, 121, 124, 125, 126, 127, 128, 134, 136, 139, 140, 141, 142, 143, 144, 145, 146, 147, 148, 149, 150, 151, 152, 153, 170, 171, 172, 173, 174, 175, 176, 177, 178, 179, 190, 201, 202, 204, 222, 224, 234, 235, 238, 239, 242, 243, 245, 246, 247, 250, 251, 253, 255

Macroalgae, 69, 73, 236, 243, 249, 251, 259
Macrophage activating factors, 4
Macrophage-like phagocytic cells, 3
Macrophages, 4, 6, 8, 12, 14, 16, 18, 20, 26, 27, 30, 33, 34, 35, 37, 51, 59, 60, 61, 62, 64, 66, 67, 68, 69, 70, 71, 72, 75, 80, 81, 82, 85, 86, 87, 90, 91, 94, 95, 96, 97, 106, 124, 128, 136, 173, 174, 175, 181, 183, 187, 188, 190, 193, 194, 201, 204, 214, 216, 217, 218, 219, 221, 222, 224, 225, 227, 247, 250, 259
Magnetite, 202, 203, 208
Major carp/rohu, 146
Major histocompatibility complex, 3, 4, 60, 61, 69, 95, 128
Malignancies, 214
Mannanoligosaccharide, 241
Mast cell degranulation, 21
Mealworm, 171, 181, 249, 250, 257
Medicinal, 26, 38, 98, 113, 130, 131, 132, 133, 207, 213, 228, 229, 231, 245, 255, 260
Medicinal herbs, 51, 131, 133, 164, 262
Melanin-concentrating hormone, 8
Melanoma, 30, 218, 229
Melanomacrophage, 30, 225
Meliantroil, 120
Memory cells, 30, 62
Memory T-cell, 5
Mercury, 6, 23, 27, 105, 203
Metabolites, 24, 43, 70, 72, 73, 74, 80, 106, 112, 122, 134, 144, 150
Metals, 5, 6, 10, 14, 15, 16, 23, 24, 28, 82, 177, 183, 202, 203
MHC-Ia, 224
MHCII, 127, 238, 250
MHC-IIa, 224, 250
Microalgae, 63, 69, 72, 74, 75
Microbes, 19, 23, 31, 62, 63, 67, 77, 82, 93, 98, 104, 120, 122, 169, 172, 173, 186, 187, 221, 238, 239, 251
Microbial, 16, 30, 34, 59, 61, 65, 67, 71, 73, 76, 78, 87, 101, 103, 106, 109, 110, 111, 120, 121, 123, 136, 137, 138, 145, 150, 156, 157, 160, 162, 174, 177, 182, 193, 198, 204, 205, 219, 229, 233, 239, 243, 246, 251, 253, 256, 261
Microbial derivatives, 78, 103
Microbiota, 11, 45, 60, 63, 67, 68, 70, 73, 96, 101, 107, 111, 113, 135, 136, 137, 138, 139, 144, 145, 150, 153, 154, 155, 156, 157, 158, 159, 160, 161, 165, 166, 173, 259, 261
Microcystin, 6
Microparticles, 221, 249, 251
Mitogenesis, 20, 35, 36
Modulating, 34, 65, 108, 153, 200, 201, 219
Monoclonal, 213, 214, 217, 230
Monocyte(s), 4, 5, 14, 20, 21, 27, 35, 61, 64, 65, 72, 75, 80, 90, 91, 94, 95, 99, 107, 124, 126, 136, 190, 194, 206, 217, 222, 224, 227, 245, 246, 247, 250
Monogeneans, 6, 100
Montanide, 220, 224
Moritella viscosa, 150
MPL, 218

mRNA, 5, 6, 9, 16, 50, 71, 107, 114, 127, 135, 149, 173, 175, 176, 178, 191, 222, 239, 242, 243, 251
Mucosal-associated lymphoid tissues, 224
Mucus, 3, 4, 15, 17, 36, 61, 68, 71, 73, 98, 108, 110, 113, 124, 131, 135, 136, 137, 147, 149, 150, 165, 186, 224, 228, 238, 245, 247, 251
Mud crab, 112, 142, 145, 166, 167, 185
Multidrug-resistant, 45, 67
Muramyl dipeptide, 34, 95, 99, 222
Musculoaponeurotic fibrosarcoma, 223
Mx, 6, 35, 88, 146, 180, 183, 194, 224, 230, 246
Mycobacterial, 218, 223
Mycobacterium bovis BCG, 87, 218
Mycolic acids, 218
Myeloperoxidase, 78, 98, 107, 124, 128, 139, 140, 144, 147, 151, 177, 203, 204, 235, 238, 239, 246, 247
Mytilus edulis, 23, 28, 63, 147

N-acetylmuramic acid, 36, 96
NALP3, 218, 219
Nano-chitosan, 202
Nano-materials, 201
Nano-molecules, 201
Nanoparticle, 207, 208, 209
Nanotechnology, 130, 200, 201, 203, 205, 206, 207, 208, 209, 210, 229, 253
Natural killer, 3, 4, 7, 16, 20, 26, 34, 60, 62, 67, 78, 80, 90, 94, 97, 99, 105, 136, 169, 191, 196, 218, 219, 227
NBT-positive cells, 23
Necrosis, 5, 18, 30, 32, 34, 37, 40, 59, 66, 67, 79, 91, 94, 105, 108, 124, 128, 139, 142, 144, 148, 161, 166, 167, 171, 172, 173, 175, 188, 189, 192, 193, 195, 196, 197, 198, 204, 206, 221, 224, 235
Neoantigens, 24
Neurodegenerative diseases, 54
Neuroendocrine, 8, 48, 104, 180
Neutracutical, 201
Neutrophil, 5, 33, 37, 51, 70, 71, 106, 107, 110, 139, 153, 173, 174, 175, 180, 235, 242, 246
NF-kB signaling, 30
Nile tilapia, 147, 203
Nimbin, 120
Nitrosomonas sp., 149
NK cells, 8, 14, 20, 32, 33, 34, 61, 94, 96, 97, 191, 214, 217, 219, 222
NOD-like receptor, 107
Nodularin, 17
Non-specific defence, 3, 7, 10, 25, 45, 106, 184
NPs, 177, 200, 202, 203, 204, 205, 206
Nucleotides, 9, 10, 42, 52, 80, 106, 110, 115
NUPR1, 47
Nutraceuticals, 100, 103, 119, 120, 121
Nutrigenomics, 13, 17, 227, 230

O-antigens, 8
Ocimum sanctum, 19, 22, 27, 111, 120, 124, 130, 214, 215, 234, 235, 237, 239, 261
Oestrogen, 6, 110
Oligosaccharide, 8, 96, 109, 113, 136, 140, 150, 152, 153, 154, 158, 160, 163, 167, 239
Olive flounder, 15, 98, 101, 139, 140, 144, 155, 159, 189
Orange spotted grouper, 140, 152, 189
Oxidative damage, 5, 36, 44, 176
Oxidative stress, 6, 14, 17, 23, 24, 37, 43, 44, 47, 51, 52, 53, 54, 64, 97, 101, 106, 113, 115, 119, 124, 134, 140, 160, 181, 184, 202, 204, 242, 256

P. acidilactici, 143, 144
Pacific oyster, 147
Pacu, 99, 244
Paenibacillus polymyxa, 140, 141, 154
Pam3CSK4, **218**
PAMPs, 30, 31, 32, 136, 155, 183, 228
Pangasianodon, 172
Parasites, 181, 193, 199
Parasitic worms, 31, 32, 33
Pathogen, 6, 13, 14, 18, 23, 24, 30, 34, 36, 37, 47, 51, 59, 60, 61, 62, 64, 65, 77, 78, 81, 82, 85, 86, 87, 92, 94, 99, 102, 104, 133, 134, 136, 138, 139, 141, 150, 153, 158, 161, 164, 167, 175, 178, 188, 192, 195, 197, 201, 204, 205, 218, 220, 227, 231, 233
Pathogen Associated Molecular Patterns, 60
Pathway, 11, 14, 16, 17, 26, 30, 31, 34, 37, 40, 61, 62, 66, 68, 69, 70, 73, 74, 75, 87, 124, 142, 146, 147, 149, 151, 152, 160, 173, 177, 178, 181, 189, 192, 193, 196, 218, 219, 230, 239
PCR, 48
Pediococcus acidilactici, 110, 115, 139, 144, 154, 158
Penaus indicus, 145
Peptides, 8, 33, 40, 59, 62, 64, 65, 66, 70, 71, 72, 73, 75, 191, 214, 228, 243, 246, 250
Peptidoglucan, 19, 106
Peptidoglycans, 30, 33, 36, 51, 93
Persian sturgeon, 140, 144, 148, 154, 156, 158
Pesticides, 5, 24, 105
Peudoalteromonas sp., 147
Phaeobacter, 146, 147, 156
Phagocytic, 7, 8, 9, 10, 12, 19, 20, 22, 23, 28, 30, 36, 37, 38, 45, 51, 66, 69, 71, 78, 80, 85, 90, 95, 96, 97, 99, 102, 105, 106, 107, 108, 110, 113, 121, 124, 126, 139, 140, 141, 142, 143, 146, 147, 150, 151, 171, 173, 175, 176, 178, 183, 188, 190, 194, 202, 221, 222, 225, 226, 235, 238, 242, 243, 246, 247, 251
Phagocytic cells, 7, 10, 12, 22, 30, 37, 38, 78, 99, 183, 188
Phagocytizing, 18
Phagocytosis, 9, 10, 11, 12, 17, 22, 23, 30, 35, 36, 59, 61, 70, 71, 80, 82, 90, 91, 96, 97, 98, 99, 105, 106, 110, 121, 124, 126, 127, 139, 140, 141, 142, 144, 146, 147, 153, 173, 176, 179, 187, 201, 204, 216, 217, 218, 219, 222, 234, 235, 238, 239, 245, 246, 247, 255
Phenolic, 10, 23, 26, 44, 67, 97, 98, 108, 112, 120, 121, 122, 126, 128, 171, 214
Photobacterium, 140, 151, 155, 175, 247
Phototherapy, 218
Phytase, 138
Phytobiotics, 120, 121, 122, 123, 125, 129, 260
Phytochemicals, 42, 202, 216
Phytohemagglutinin, 20, 94
Phytotoxins, 6
Picrorhiza kurroa, 22, 120, 129
Plant-derived, 54, 218
Poly IC – Toll-like receptor 3 agonist, 220
Polychlorinated Biphenyls, 21
Polyclonal antibodies, 214, 217
Polycyclic aromatic hydrocarbons, 23, 24, 43
Polypeptides, 35, 93, 94, 98, 191
Polyphenolic compound, 120
Polyphenols, 23, 24, 28, 67, 98, 102, 103, 108, 114, 124, 126, 132, 135, 201, 243
Polysaccharides, 7, 9, 10, 15, 16, 18, 22, 30, 33, 62, 63, 66, 67, 68, 69, 70, 71, 72, 73, 74, 77, 88, 95, 99, 101, 102, 109, 114, 124, 134, 145, 150, 151, 173, 202, 219, 231, 235, 236, 237, 238, 239, 241, 243, 253, 256, 258, 262
POPs, 23, 28
Prebiotics, 8, 11, 14, 101, 136, 137, 140, 149, 150, 152, 153, 154, 155, 160, 161, 165, 166, 167, 238, 247, 250, 251, 255
Probiotics, 7, 8, 9, 10, 11, 16, 17, 63, 68, 70, 86, 101, 110, 111, 123, 131, 136, 137, 138, 139, 140, 141, 142, 143, 144, 145, 146, 147, 148, 152, 153, 154, 155, 156, 157, 158, 159, 160, 161, 162, 163, 164, 165, 166, 167, 169, 200, 205, 230, 231, 239, 241, 242, 243, 246, 247, 251, 252, 254, 255, 260, 261
Prolactin, 8, 60, 97, 102, 105, 110, 115, 178, 183
Prophylactic agent, 105
Prophylaxis, 89, 91, 99
Protease inhibitors, 62, 136
Proteases, 136, 148
Proteome, 38, 113, 133
Proteomics, 14, 30, 39, 111, 179, 181, 227
Pseudoalteromonas, 75, 146, 160, 164, 166
Pseudomonas, 82, 108, 120, 137, 138, 143, 146, 147, 154, 156, 157, 162, 165, 220, 235, 256
Pseudomonas aeruginosa, 108, 143, 146, 147, 165
Pseudomonas aestumarina, 147
Pseudomonas angulliseptica, 146
Pseudomonas fluorescenes, 120, 146, 147, 235
Pseudomonas monteilii, 146
Pseudomonas synxantha, 147
Psychrobacter maritimus, 147
Psychrobacter namhaensis, 147
Publications, 79, 234

Quaternary ammonium compounds, 23, 98

Rainbow trout, 11, 17, 19, 22, 23, 25, 26, 27, 28, 34, 35, 36, 37, 38, 40, 41, 42, 47, 49, 52, 53, 54, 80, 82, 86, 87, 88, 90, 91, 95, 96, 97, 98, 99, 101, 102, 106, 112, 113, 114, 120, 124, 125, 130, 132, 133, 134, 135, 138, 139, 140,

141, 143, 144, 146, 147, 148, 149, 150, 151, 152, 153, 154, 155, 156, 157, 158, 159, 161, 162, 163, 164, 167, 170, 172, 173, 175, 176, 177, 178, 179, 180, 181, 182, 183, 184, 185, 188, 189, 190, 191, 193, 194, 195, 196, 197, 198, 199, 202, 204, 205, 219, 220, 221, 222, 225, 229, 230, 231, 234, 235, 245, 246, 247, 250, 251, 253, 254, 255, 256, 257, 259, 260, 261, 262
Rambutan, 135, 234, 236, 238, 258
Razor clam, 147
Receptors, 3, 4, 6, 30, 31, 33, 34, 39, 47, 60, 61, 62, 67, 68, 78, 79, 88, 91, 94, 101, 104, 137, 138, 141, 149, 150, 151, 166, 173, 178, 183, 186, 187, 188, 192, 194, 196, 197, 199, 201, 205, 216, 217, 218, 229, 230, 253, 261
Recombinant, 7, 34, 35, 40, 96, 148, 149, 159, 161, 190, 191, 194, 195, 197, 214, 230, 233, 250, 256
Recombinant cytokines, 7
Red abalone, 147
Red drum, 130, 150, 151, 155, 167
Red spot disease, 120
Relative percent survival, 223
Relative percentage survival, 107, 222, 239
Reproduction, 35, 75, 90, 92, 97, 110, 115, 119, 169, 178, 184, 242, 243
Resistance, 7, 8, 9, 10, 11, 12, 13, 14, 15, 16, 17, 18, 19, 22, 23, 25, 27, 29, 34, 35, 36, 37, 38, 41, 42, 45, 46, 50, 51, 65, 67, 73, 77, 78, 79, 80, 82, 85, 87, 88, 92, 93, 96, 97, 98, 99, 101, 102, 103, 104, 105, 106, 107, 108, 109, 111, 112, 113, 114, 119, 120, 121, 123, 124, 125, 129, 130, 131, 133, 134, 135, 136, 137, 138, 139, 140, 141, 142, 143, 144, 145, 146, 147, 148, 149, 150, 152, 153, 154, 155, 156, 157, 158, 159, 160, 161, 162, 163, 164, 165, 166, 167, 168, 169, 170, 171, 172, 173, 174, 175, 176, 177, 178, 180, 181, 182, 183, 184, 185, 191, 192, 194, 200, 201, 204, 205, 220, 221, 222, 226, 228, 229, 230, 231, 232, 235, 238, 239, 242, 243, 245, 246, 247, 250, 251, 253, 254, 255, 256, 257, 258, 259, 260, 261, 262
Resistance genes, 137, 192
Resistant, 19, 45, 67, 68, 70, 78, 80, 82, 85, 92, 93, 103, 105, 123, 133, 152, 181, 186, 198, 220, 228, 233
Respiratory burst activity, 5, 9, 11, 12, 35, 36, 81, 85, 97, 98, 107, 108, 120, 124, 125, 126, 144, 147, 151, 173, 175, 177, 194, 221, 222, 235, 238, 239, 243, 245, 246, 247, 253
Rhodococcus, 146, 151
Rhodopseudomonas, 146, 147, 160
Rhodopseudomonas palustris, 147
Rohu, 36, 98, 133, 161, 173, 175, 176, 190, 219, 234, 236, 237, 256
Roseobacter, 146, 147, 163
Roseobacter gallaeciensis, 147
Saccharomyces cerevisiae, 11, 27, 37, 87, 95, 101, 145, 148, 150, 156, 158, 160, 173, 174, 239, 240, 250, 252, 261
Salannin, 120
Salmon alpha virus, 35
Saponins, 19, 23, 119, 121, 126, 127, 214, 226
Saprolegina paracitica, 147
Saprolegnia parasitica, 97, 108, 114, 144, 162
Sea bass, 36, 39, 86, 110, 115, 140, 146, 163, 165, 192, 194, 219
Seaweeds, 9, 14, 66, 67, 69, 72, 74, 85, 87, 243, 251, 252, 253
Secondary metabolites, 106
Selenium, 37, 42, 51, 92, 134, 138, 170, 175, 176, 177, 180, 182, 183, 184, 200, 202, 203, 204, 227, 228, 231, 233, 234, 262
Serological, 239
Sh. olleyana, 147
Sh. xiamenensis, 146
Shewanella, 146, 155, 158, 246, 256
Shewanella putrefaciens, 146, 246
Shrimp, 13, 16, 18, 19, 22, 25, 26, 27, 44, 66, 71, 74, 76, 80, 83, 87, 88, 99, 101, 113, 115, 120, 124, 131, 134, 136, 138, 141, 142, 143, 144, 145, 147, 149, 150, 152, 154, 155, 156, 157, 158, 159, 160, 161, 162, 163, 165, 166, 167, 168, 172, 173, 180, 185, 221, 222, 228, 231, 238, 261
Siberian sturgeon, 36, 41, 135, 144, 151, 157, 158, 163
Signal transduction, 71, 80, 106, 173
Signaling inhibitors, 214
Silica, 201, 204
Silver catfish, 156
Silver perch, 147
Sirolimus, 216, 230
Skin-associated lymphoid tissue, 224, 227
Snakehead, 122, 150, 165, 176, 185
Sobaity seabream, 152, 161
Specific growth rate, 120, 128, 147, 221
Spirulina platensis, 19, 25, 240, 242, 243, 254, 256, 257, 262
Spleen, 4, 5, 15, 20, 30, 36, 61, 65, 66, 70, 90, 91, 97, 136, 143, 148, 149, 172, 173, 176, 178, 179, 180, 181, 185, 187, 190, 191, 193, 194, 197, 198, 219, 224, 225, 228, 235, 238, 243, 246
Steroids, 6, 19, 109, 119, 126, 181, 214
Stimulant, 78, 108, 123, 253
Strepcoccus, 146
Streptococcosis, 17, 143, 144, 146, 149, 155, 164, 226
Streptococcus, 18, 66, 79, 87, 98, 99, 102, 112, 127, 135, 138, 139, 140, 143, 145, 146, 155, 156, 158, 159, 160, 164, 167, 192, 208, 221, 232, 235, 238, 258, 262
Streptococcus agalactiae, 99, 139, 192, 235, 238
Streptococcus cremoris, 145
Streptococcus iniae, 87, 98, 139, 146, 155, 158, 159, 232
Stressors, 5, 18, 23, 36, 43, 46, 47, 48, 49, 50, 51, 52, 53, 77, 92, 93, 99, 104, 105, 106, 109, 119, 122, 136, 138, 140, 148, 179, 202, 204, 233
Sub-lethal, 14, 45, 93
Subtilis, 8, 138, 139, 140, 141, 142, 145, 148, 149, 153, 156, 158, 166, 167, 168
Sulfamethoxazole, 6, 15
Sulfated, 63, 66, 67, 72, 74
Superoxide dismutase, 9, 37, 107, 108, 124, 128, 139, 140, 142, 144, 145, 146, 147, 150, 151, 152, 153, 170, 171, 183, 202, 216, 235, 238, 239, 242, 245, 247
Surubim hybrid, 149
Sustainable, 15, 17, 19, 38, 57, 72, 84, 101, 106, 112, 125, 132, 133, 152, 183, 201, 233, 239, 245, 247, 259
Synthetic chemical, 106

T and B cells, 4, 178
T cell receptor, 30, 60, 196, 250
Tambaqui, 38, 151, 156, 162, 244, 258
Tannase, 138, 165
Tannin, 120, 122
Taurine, 64
T-cell receptors, 3
T-cells, 5, 32, 104, 188, 189, 191, 205, 214, 216
Terpenes, 70, 120, 245
Terpenoids, 19, 98, 108, 119, 120, 121, 126, 214
Tetramisole, 8, 22
TGF, 20, 30, 33, 34, 40, 91, 94, 95, 100, 139, 143, 146, 149, 171, 172, 173, 177, 188, 190, 205, 224, 235, 238, 239, 242, 250
Th17 cells, 24, 33, 34, 94, 191
Thymic stromal lymphopoietein, 33
Thymus, 4, 21, 30, 35, 38, 61, 90, 91, 104, 105, 136, 148, 175, 178, 187, 194, 196, 224, 225
Tilapia, 5, 6, 14, 15, 17, 19, 23, 25, 28, 34, 37, 42, 51, 73, 80, 95, 97, 98, 100, 101, 102, 106, 108, 109, 110, 112, 113, 114, 115, 120, 121, 123, 125, 130, 131, 132, 133, 134, 135, 138, 139, 140, 141, 143, 145, 146, 147, 148, 149, 150, 151, 152, 153, 154, 157, 158, 160, 163, 164, 165, 166, 167, 172, 175, 176, 177, 178, 179, 180, 182, 184, 189, 192, 194, 197, 199, 202, 204, 205, 230, 234, 235, 236, 237, 238, 239, 240, 241, 242, 243, 244, 245, 246, 251, 252, 253, 254, 255, 256, 258, 259, 260, 261, 262
Tilapia Lake Virus, 119
Tinospora, 98, 214, 215, 234
TLRs, 30, 31, 39, 94, 173, 177, 187, 193, 201, 218, 224
TNF-a, 20, 21, 143, 144, 146, 171, 172, 217, 224
Toll receptors, 30
Toll-like receptor, 8, 181, 220, 224
Total ammonia nitrogen, 120
Transcriptome, 38, 53, 87, 133, 142, 177, 179, 192, 193, 195, 197, 198, 199, 230, 247
Transcriptomics, 14, 30, 39, 111, 227, 228
Transferring factor, 3
Triiodothyronine, **19**
Trout, 8, 9, 13, 19, 22, 25, 26, 27, 31, 35, 36, 41, 42, 49, 53, 81, 86, 91, 97, 110, 112,

113, 115, 124, 133, 135, 139, 148, 149, 159, 161, 164, 167, 176, 181, 183, 185, 188, 189, 190, 191, 194, 199, 219, 226, 236, 244, 248, 249, 252, 253, 259, 260, 262
Tumor necrosis factor, 34, 128, 204
Tumor necrosis factor alpha, 34
Tumour, 5, 32, 66, 91, 94, 97, 138, 139, 149, 188, 218, 219, 224, 235, 254
Type I interferon, 194
Type II interferon, 101

Ulva, 63, 67, 69, 241, 243, 251
United Nations, 45, 105, 162, 196, 259
Upregulation, 32, 108, 142, 144, 145, 146, 150, 245, 247, 250, 251

V. alginolyticus, 145, 147
V. campbellii, 143
V. harveyi, 79, 80, 141, 142, 144, 148, 154
V. parahaemolyticus, 140, 142, 144, 145, 147, 148
Vaccination, 7, 13, 19, 38, 41, 44, 45, 77, 81, 82, 84, 88, 99, 103, 111, 148, 149, 156, 162, 165, 167, 193, 205, 206, 218, 219, 220, 223, 224, 231, 243
Vaccines, 13, 22, 26, 27, 29, 38, 42, 44, 69, 77, 78, 84, 98, 99, 103, 105, 111, 131, 148, 153, 164, 192, 200, 205, 206, 213, 219, 220, 221, 222, 223, 224, 226, 228, 229, 230, 232, 233, 253, 261
Vagococcus, 143
Vepol, 120
Vertebrates, 3, 14, 29, 30, 34, 35, 36, 41, 55, 90, 104, 136, 151, 178, 188, 227
Vibrio, 14, 18, 19, 25, 26, 35, 36, 37, 41, 71, 82, 87, 88, 90, 96, 98, 102, 106, 107, 108, 112, 120, 121, 124, 129, 130, 131, 137, 138, 140, 141, 142, 144, 145, 146, 147, 151, 152, 154, 155, 156, 157, 158, 159, 160, 162, 163, 164, 165, 166, 168, 171, 173, 179, 181, 185, 191, 192, 193, 209, 219, 222, 231, 238, 242, 245, 246, 247, 248, 250, 251, 256, 258, 261, 262
Vibrio anguillarum, 18, 19, 35, 36, 88, 90, 96, 106, 120, 140, 146, 147, 191, 248, 250
Vibrio diabolicus, 147
Vibrio ordalii, 146
Vibrio owensii, 142, 145
Vibrio parahaemolyticus, 231
Vibrio salmonicida, 14, 35, 88
Vibrio splendidus, 145
Vibrio tubiashii, 147
Vibriosis, 96, 119, 140, 144, 145, 146, 147, 156, 164, 165, 204, 226, 245, 256
Viraemia, 235
Viral haemorrhagic septicemia, 119, 230
Viruses, 18, 19, 46, 79, 90, 92, 104, 153, 188, 194, 201, 206, 227, 229
Vitamin C, 10, 19, 36, 53, 96, 100, 106, 107, 110, 175, 176, 182, 183, 207, 222
Vitamin E, 9, 10, 12, 16, 36, 41, 81, 97, 110, 115, 134, 175, 180, 181, 183, 184, 185, 222, 253
Vitamins, 10, 36, 96
VLP, 205, 206

WBC, 98, 107, 108, 120, 124, 125, 126, 127, 128, 151, 215, 235
Weissella, 143, 144, 145, 149, 157, 163
Weissella cibaria, 144, 149
Weissella confuse, 144, 145
Western king prawn, 147
White spot viral disease, 119, 142, 145, 149
WHO, 45
Withania somnifera, 26, 108, 113, 120, 123, 131, 133, 222, 234

Xenobiotics, 6, 17, 18, 23, 24, 28
Xylooligosaccharide, 152

Y. ruckeri, 22, 36, 88, 97, 108, 109, 139, 141
Yeast, 10, 11, 12, 15, 16, 37, 74, 80, 81, 86, 87, 95, 101, 106, 112, 136, 137, 139, 140, 141, 144, 145, 150, 152, 156, 160, 166, 167, 173, 174, 175, 177, 181, 222, 224, 225, 228, 231, 242, 243, 246, 247, 250, 251, 252, 253, 254, 260, 261
Yellow catfish, 192
Yellow croaker, 139, 140, 151, 153, 191, 192, 193, 194, 196, 197
Yellow head disease, 221
Yellowfin sea bream, 36, 41
Yersinia ruckeri, 13, 18, 27, 35, 79, 88, 91, 96, 106, 108, 134, 140, 146, 149, 153, 162, 164, 167, 190, 193, 202, 235, 242, 247, 250, 254, 261
Yucca, 134, 240, 243, 253

Zebrafish, 6, 15, 16, 17, 30, 31, 39, 40, 51, 53, 65, 72, 88, 91, 110, 115, 140, 144, 146, 157, 160, 175, 183, 188, 189, 190, 191, 192, 193, 194, 195, 196, 197, 199, 202, 246, 247, 251, 254, 255, 257
Zinc, 10, 28, 37, 42, 92, 105, 108, 170, 175, 177, 179, 181, 182, 183, 201, 204, 218, 227
Zingiber officinale, 19, 25, 37, 98, 120, 122, 130, 131, 221, 250, 251, 252, 254
Zingiberaceae, 120, 130
ZnONPs, 204
Zymosan, 34, 174, 183, 246